Algebra

Notations	Definition	Examples	Explanation
$\equiv$	$A \equiv B$, indicates that A is defined by B	$f(x) \equiv x^2$	$f(x)$ is defined by x^2.
$:=$	$A := B$, indicates that A is defined by B	$f(x) := x^3$	$f(x)$ is defined by x^3.
∞	infinity	$\lim_{n \to \infty} n^2 = \infty$	When n approaches infinity, n^2 also approaches infinity.
$\lfloor x \rfloor$	$\lfloor x \rfloor$ is the greatest integer that is less than or equal to x	$\lfloor 2.3 \rfloor = 2$, $\lfloor -2.3 \rfloor = -3$	2 is the greatest integer that is less than 2.3. -3 is the greatest integer that is less than -2.3.
$[x]$	$[x]$ is the greatest integer that is less than or equal to x	$[2.3] = 2$, $[-2.3] = -3$	2 is the greatest integer that is less than 2.3. -3 is the greatest integer that is less than -2.3.
Σ	Summation(Sigma)	$\sum_{n=1}^{10} n = 55$	$1 + 2 + \cdots + 10 = 55$
Π	continued multiplication (Pi)	$\prod_{n=1}^{5} n = 120$	$1 \times 2 \ldots \times 5 = 120$
e	$e := \sum_{n=0}^{\infty} \frac{1}{n!}$, (Euler's number)	$e^2 \cdot e^3 = e^5$	Based on Exponential Law, $e^2 \cdot e^3 = e^5$.
π	$\pi = 3.1415926 \ldots$ (Pi, Ratio of the circumference of a circle to its diameter)	$\sin \frac{\pi}{2} = 1$	When the angle of θ equals $\frac{\pi}{2}$, $\sin \theta$ equals 1

Permutation

Notations	Definition	Examples	Explanation
!	$n! = 1 \times 2 \times \ldots \times n$	$3!$	$3! = 1 \times 2 \times 3 = 6$
P_k^n	$P_k^n = \dfrac{n!}{(n-k)!}$	$P_2^5 = \dfrac{5!}{3!}$	$P_2^5 = \dfrac{5!}{3!} = 5 \times 4 = 20$
C_k^n	$C_k^n = \dfrac{n!}{k!(n-k)!}$	$C_2^5 = \dfrac{5!}{3!\,2!}$	$C_2^5 = \dfrac{5!}{3!\,2!} = 10$

Set

Notations	Definition	Examples	Explanation
{ }	set	{1,2,3,5}	The set consist of these elements 1,2,3,5.
∪	union	{1,3} ∪ {2,5} = {1,2,3,5}	Union of {1,3} and {2,5} is equal to {1,2,3,5}.
∩	intersect	{1,3,5} ∩ {1,2,5,7} = {1,5}	Intersection of {1,3,5} and {1,2,5,7} is equal to {1,5}.
⊂	subset	{1,5} ⊂ {1,3,5}	{1,5} is the subset of {1,3,5}.
⊄	not a subset	{1,3,5} ⊄ {1,5}	Because 3 ∉ {1,5}, {1,3,5} is not the subset of {1,5}.
⊆	subset or equal	{1,5} ⊆ {1,3,5}, {1,5} ⊆ {1,5}, {1,3,5} ⊈ {1,5}	{1,5} is the subset of {1,3,5}. Because {1,5} = {1,5}, {1,5} ⊆ {1,5}. {1,3,5} is not the subset of {1,5}.
A^c	A's complement set	$\sqrt{3} \in \mathbb{Q}^c$	$\sqrt{3}$ can't be written in a fraction. Hence, it's an irrational number.
A\B	A set, excluding A ∩ B elements	{1,3,5}\{5,7} = {1,3}	The intersection for the two sets is {5}, so taking away {5} from {1,3,5} equals {1,3}.
∈	belong to	$a \in \{a, b, c\}$	a belongs to the set $\{a, b, c\}$.

$\notin$	does not belong to	$d \notin \{a,b,c\}$	d doesn't belong to the set $\{a,b,c\}$.
N	positive integers	N = $\{1,2,3,...\}$, $3 \in$ N	By definition, the set of positive integers is equal to $\{1,2,3,...\}$. 3 belong to the set of positive integers.
Z	integers	Z = $\{...,-2,-1,0,1,2,...\}$, $-2 \in$ Z	By definition, we have Z = $\{...,-2,-1,0,1,2,...\}$. -2 belongs to the set of integers.
Q	rational numbers	$Q = \{\frac{m}{n} : m \in Z, n \in Z\setminus\{0\}\}$, $-\frac{4}{3} \in Q$	Rational numbers set includes all numbers with a fraction format. $-\frac{4}{3}$ is a rational number.
Q^c	irrational numbers	$\sqrt{3} \in Q^c$	$\sqrt{3}$ can't be written in a fraction. Hence, it's an irrational number.
R	real numbers	R = $Q \cup Q^c$, $-\frac{4}{3} \in R, \sqrt{3} \in R$	Real numbers are the union of irrational numbers and rational numbers. $-\frac{4}{3}$ is a real number, $\sqrt{3}$ is a real number

Logic

Notations	Definition	Examples	Explanation
$\vee$	or	$x \in A \vee x \in B \Rightarrow x \in A \cup B$	If x belongs to A or B then x belongs to $A \cup B$.
$\wedge$	and	$x < 1 \wedge x > 0 \Rightarrow 0 < x < 1$	If $x < 1$ and $x > 0$ then $0 < x < 1$.
$\Rightarrow$	implies	$x < 1 \Rightarrow x < 2$	$x < 1$ implies $x < 2$.

	if and only if	$\|x\| < 1 \Leftrightarrow -1 < x < 1$	If $\|x\| < 1$ then $-1 < x < 1$. Conversely, if $-1 < x < 1$ then $\|x\| < 1$.
$\Leftrightarrow$			
$\because$	because	$\because x < 1 \;\; \therefore x < 2$	Because $x < 1, x < 2$.
$\therefore$	so	$\because x < 1 \;\; \therefore x < 2$	$x < 1$, so $x < 2$.
$\forall$	for each	$\forall x \in Q, \exists m \in Z, n \in Z\setminus\{0\}$ s.t. $x = \dfrac{m}{n}$	By definition of ration numbers, for each rational number x, there extis $m \in Z, n \in Z\setminus\{0\}$ such that $x = \dfrac{m}{n}$.
$\exists$	There exists	$\forall x \in Q, \exists m \in Z, n \in Z\setminus\{0\}$ s.t. $x = \dfrac{m}{n}$	By definition of ration numbers, for each rational number x, there extis $m \in Z, n \in Z\setminus\{0\}$ such that $x = \dfrac{m}{n}$.

Calculus

Notations	Definition	Examples	Explanation
$\lim_{x \to c} f(x) = L$	$\forall \varepsilon > 0, \exists \delta > 0$ such that $0 < \|x - c\| < \delta \Rightarrow \|f(x) - L\| < \varepsilon$	$\lim_{x \to 2} x^2 = 4$	When x approaches 2, the limit of x^2 is 4.
ε	epsilon	$\forall \varepsilon > 0$	for each $\varepsilon > 0$
δ	delta	$\forall \varepsilon > 0,$ $\exists \delta > 0$ s.t. ...	for each $\varepsilon > 0$, there exist $\delta > 0$ such that ....
e	$e := \sum_{n=0}^{\infty} \dfrac{1}{n!},$ (Euler's number)	$\lim_{x \to \infty} \left(1 + \dfrac{1}{x}\right)^x = e$	It can be proven with L'Hôpital's rule.
$f'(x)$	derivative	$(x^2)' = 2x$	x^2 has derivative $(x^2)' = 2x$.

$\dfrac{\partial f(x,y)}{\partial x}$	partial derivative	$\dfrac{\partial}{\partial x}(x^2 y) = 2xy$	$x^2 y$ has the partial derivative $\dfrac{\partial}{\partial x}(x^2 y) = 2xy$.
$\displaystyle\int$	integral	$\displaystyle\int 2x\, dx = x^2$	$2x$ has indefinite integral x^2.
$\displaystyle\iint$	double integral	Find $\displaystyle\iint_R y e^{xy}\, dA =?$	Find the double integral of $y e^{xy}$ in the range R.
$\displaystyle\iiint$	triple integral	Find $\displaystyle\iiint_V x^2\, dV =?$	Find the triple integral of x^2 in the range V
$\displaystyle\oint$	closed contour / line integral	Find $\displaystyle\oint y^3\, dx + x^3\, dy =?$, $C: x^2 + y^2 = r^2$	Find the closed line integral $\displaystyle\oint y^3\, dx + x^3\, dy =?$, where $C: x^2 + y^2 = r^2$.
$\displaystyle\oiint$	closed surface integral	Find $\displaystyle\oiint_S \vec{F} \cdot \vec{n}\, dA =?$, where $\vec{F} = (x, y, z)$	Find the closed surface integral $\displaystyle\oiint_S \vec{F} \cdot \vec{n}\, dA =?$, $\vec{F} = (x, y, z)$.

CHAPTER 7 DIFFERENTIATION AND APPLICATIONS OF MULTIVARIABLE FUNCTIONS 1

7.1 Find the Limits of Multivariable Functions 3
7.2 Determine the Continuity of Multivariable Functions 18
7.3 Calculate Partial Derivatives of Multivariable Functions 32
7.4 Determine the Differentiability of Multivariable Functions and Chain Rule 45
7.5 Find Directional Derivatives 73
7.6 Find Normal and Tangent Equations of Surfaces 80
7.7 Find the Extrema of Multivariable Functions 85
7.7.1 Use the Arithmetic-Geometric Mean Inequality to Find Extrema 91
7.7.2 Find Extrema Using Cauchy's Inequality 93
7.7.3 Find Extrema without Constraints 94
7.7.4 Find Extrema with Equality Constraints 98
7.7.5 Find Extrema with Inequality Constraints 112
7.7.6 Find Extrema on Compact Domains 121

CHAPTER 8 MULTIPLE INTEGRALS 128
8.1 Iterated Integrals 130
8.2 Multiple Integrals and Fubini's Theorem 137
8.3 Methods for Solving Double Integrals 143
8.3.1 Convert Double Integrals to Iterated Integrals 143
8.3.2 Change the Order of Integration 149
8.3.3 Use Polar Coordinates to Compute Double Integrals 168
8.3.4 Use Generalized Coordinates to Compute Double Integrals 186
8.4 Methods for Solving Triple Integrals 213
8.4.1 Directly Compute the Iterated Integrals without Changing the Order 214
8.4.2 Use Polar Coordinates to Compute Triple Integrals 229
8.4.3 Use Spherical Coordinates to Compute Triple Integrals 265

CHAPTER 9 VECTOR CALCULUS 302
9.1 Parametric Equations 304
9.2 Given the Vector Representation of a Curve, Find Its Derivative 307
9.3 Line Integrals 312
9.3.1 For Scalar Functions, Find the Line Integral along Curve C 312
9.3.2 For Vector Functions, Find the Line Integral along Curve C 326
9.4 Green's Theorem 353

9.4.1	The Line Integral Is Path-Independent	359
9.4.2	Find the Area of the Closed Region R	365
9.4.3	Convert Closed Line Integrals into Surface Integrals	369
9.4.4	Use Coordinate Transformation to Compute Surface Integrals	383
9.4.5	Convert Surface Integrals into Closed Line Integrals	399
9.4.6	Create a Closed Region without Singularities	412
9.5	Surface integrals	427
9.5.1	Representation of Surfaces S in Space and Smooth Surfaces	427
9.5.2	Find the Surface Area of a Closed Region on Surface S	436
9.5.3	For Scalar Functions, Find the Mass of Surface S	482
9.5.4	For Vector Functions, Find the Flux across Surface S	504
9.6	Stokes' Theorem	536
9.6.1	When $\nabla \times \mathbf{F} = 0$, the Line Integral Is Path-Independent	541
9.6.2	Convert Closed Line Integrals into Surface Integrals	551
9.6.3	Convert Surface Integrals into Closed Line Integrals	569
9.7	Gauss's Theorem	589
9.7.1	Convert Closed Surface Integrals into Volume Integrals($\nabla \cdot \mathbf{F} = c$)	595
9.7.2	Convert Closed Surface Integrals into Volume Integrals($\nabla \cdot \mathbf{F} \neq c$)	600
9.7.3	Convert Volume Integrals into Closed Surface Integrals	625

CHAPTER 7 DIFFERENTIATION AND APPLICATIONS OF MULTIVARIABLE FUNCTIONS

First, we will introduce the definition of the limit of multivariable functions. When determining whether the limit of a function at the origin exists, let $y = mx$, and then observe whether $\lim_{(x,y)\to(0,0)} f(x, mx)$ depends on the value of m. If it does, find exactly two distinct m values to demonstrate that the limit does not exist. If $\lim_{(x,y)\to(0,0)} f(x, mx)$ does not depend on m, it is possible to approach $(0,0)$ via a curve, which might indicate that the limit does not exist. If the limit value obtained by approaching $(0,0)$ via a curve still does not depend on m, then there is a high possibility that the limit of $f(x,y)$ exists at $(0,0)$. In this case, use the ε, δ definition to prove the existence of the limit.

Next, the definition of continuity for multivariable functions will be introduced. The process of determining whether a multivariable function is continuous is similar to determining whether its limit exists. If $\lim_{(x,y)\to(0,0)} f(x, mx)$ depends on the value of m, find exactly two distinct m values to demonstrate that the limit does not exist, which implies that $f(x,y)$ is not continuous at $(0,0)$. If $\lim_{(x,y)\to(0,0)} f(x, mx)$ does not depend on m, it is possible to approach $(0,0)$ via a curve, which might indicate that the limit does not exist and thereby show that $f(x,y)$ is not continuous at $(0,0)$. If the limit value obtained by approaching $(0,0)$ via a curve still does not depend on m and equals $f(0,0)$, then there is a high possibility that $f(x,y)$ is continuous at $(0,0)$. In this case, use the ε, δ definition to prove the continuity of $f(x,y)$ at $(0,0)$.

Next, we discuss partial derivatives of multivariable functions, differentiability of multivariable functions, and the Chain Rule for multivariable functions. The process of determining whether a multivariable function is differentiable requires calculating the partial derivatives of the function. Therefore, efficiently and accurately computing partial derivatives of multivariable functions is important. From the definition of differentiability for a function of two variables, we observe that if $f(x,y)$ is differentiable at $(0,0)$, then the limit of approaching $(0,0)$ from any direction (h,k) must be zero. Thus, if we let $h = mk$, when $f(x,y)$ is differentiable at $(0,0)$, then

$$\lim_{h\to 0, k\to 0} \left| \frac{f(mk,k) - f(0,0) - \frac{\partial f}{\partial x}(0,0)mk - \frac{\partial f}{\partial y}(0,0)k}{\sqrt{(mk)^2 + k^2}} \right| = 0$$

7.1 Find the Limits of Multivariable Functions 2

and this limit is independent of the value of m. Additionally, the existence of partial derivatives is a necessary condition for the differentiability of a function of two variables. If it depends on m, then try to find $m \in R$ such that the limit is not zero to demonstrate that $f(x, y)$ is not differentiable at $(0,0)$. If it does not depend on m, then let $\varepsilon > 0$, and choose a specific δ such that $h^2 + k^2 < \delta$ implies

$$\left| \frac{f(h,k) - f(0,0) - \frac{\partial f}{\partial x}(0,0)h - \frac{\partial f}{\partial y}(0,0)k}{\sqrt{h^2 + k^2}} \right| < \varepsilon.$$

Furthermore, the Chain Rule for multivariable functions can be viewed as the differentiation of multivariable composite functions.

Next, we discuss the problem of finding extrema for multivariable functions. From a broad perspective, the two most important theorems are the second-order derivative test for finding extrema of functions of two variables and the method of Lagrange multipliers. For unconstrained optimization of a function of two variables, we use the second-order derivative test, which involves Taylor's Formula for multivariable functions. Additionally, the proof of Taylor's Formula for multivariable functions relies on Taylor's Formula for single-variable functions. Thus, the second-order derivative test for finding extrema of functions of two variables under unconstrained conditions ultimately uses Taylor's Formula for single-variable functions. For constrained optimization, the method of Lagrange multipliers is used, and its proof involves the Implicit Function Theorem for multivariable functions. The Implicit Function Theorem's proof, in turn, references the Inverse Function Theorem for multivariable functions, which is proved using the Inverse Function Theorem for single-variable functions. In other words, the proof of the method of Lagrange multipliers ultimately relies on the Inverse Function Theorem for single-variable functions.

Finally, we explain how to use these two theorems to find extrema of multivariable functions. Such problems can be divided into unconstrained and constrained cases. For unconstrained optimization of a function of two variables, the second-order derivative test is typically used. Constrained cases usually involve equality or inequality constraints. For equality constraints, the method of Lagrange multipliers is used directly. For inequality constraints (such as in a closed region), the problem is divided into two parts: finding extrema within the closed region and finding extrema on the boundary of the closed region (with equality constraints). For finding extrema within the closed region, the second-order derivative test for functions of two variables is used to identify critical points, which are then checked to see if they lie within the closed region. On the other hand, for finding extrema on the boundary of the closed region, the method of Lagrange multipliers is employed. Finally, compare the critical points found within the closed region with those on the boundary. If both sets of points are relative extrema within their respective constraints, choose the larger as the absolute maximum and similarly determine the absolute minimum. Additionally, if given several points, the extrema within the region enclosed by these points need to be found, which requires solving the equations of the closed region.

7.1 Find the Limits of Multivariable Functions

【Definition】The Definition of the Limit of a Function of Two Variables

The limit of $f(x, y)$ as (x, y) approaches (x_0, y_0) exists and is equal to L

$\Leftrightarrow \lim\limits_{(x,y) \to (x_0,y_0)} f(x, y) = L$

$\Leftrightarrow \forall \varepsilon > 0, \exists \delta > 0,$ such that $\sqrt{(x - x_0)^2 + (y - y_0)^2} < \delta \Rightarrow |f(x, y) - L| < \varepsilon$

【Definition】The Definition of the Limit for Multivariable Functions

The limit of $f(x_1, \ldots, x_n)$ as $(x_1, \ldots, x_n)$ approaches $(c_1, \ldots, c_n)$ exists and is equal to L

$\Leftrightarrow \lim\limits_{(x_1,\ldots,x_n) \to (c_1,\ldots,c_n)} f(x_1, \ldots, x_n) = L$

$\Leftrightarrow \forall \varepsilon > 0, \exists \delta > 0,$ such that $\sqrt{(x_1 - c_1)^2 + \cdots + (x_n - c_n)^2} < \delta \Rightarrow |f(x_1, \ldots, x_n) - L| < \varepsilon$

The process for determining whether the limit of a function $f(x, y)$ exists at (x_0, y_0) is as follows: From the definition of the limit for a function of two variables, it can be observed that if the limit of $f(x, y)$ as (x, y) approaches (x_0, y_0) exists, then the limit must be the same when approaching (x_0, y_0) from any direction. A common approach is to let

$$y - y_0 = m(x - x_0)$$

represent a line passing through (x_0, y_0). When the limit of $f(x, y)$ at (x_0, y_0) exists, then

$$\lim\limits_{(x,y) \to (x_0,y_0)} f(x, y) = \lim\limits_{(x,y) \to (x_0,y_0)} f(x, m(x - x_0) + y_0)$$

and

$$\lim\limits_{(x,y) \to (x_0,y_0)} f(x, m(x - x_0) + y_0)$$

is independent of the value of m. It is important to note that while the independence of m in

$$\lim\limits_{(x,y) \to (x_0,y_0)} f(x, m(x - x_0) + y_0)$$

is a necessary condition for the existence of the limit of $f(x, y)$ at (x_0, y_0), it is not a sufficient condition. Below is an example where the limit of $f(x,y)$ at (0,0) is independent of m but the limit does not exist.

Examples:

Determine whether $f(x,y) = \dfrac{xy^2}{x^2+y^4}$ has a limit as $(x,y) \to (0,0)$.

Let $y = mx$ then $f(x,y) = \dfrac{xy^2}{x^2+y^4} = \dfrac{m^2x^3}{x^2+m^4x^4}$

$\therefore \lim\limits_{(x,y)\to(0,0)} f(x,mx)$ is independent of the value of m and $\lim\limits_{(x,y)\to(0,0)} f(x,mx) = 0$

Let $x = y^2$ then $f(x,y) = \dfrac{y^4}{y^4+y^4} = \dfrac{1}{2}$ and $\lim\limits_{(x,y)\to(0,0)} f(y^2,y) = \dfrac{1}{2}$

∵ if the limit of $f(x,y)$ as (x,y) approaches (0,0) exists
∴ the limit must be the same when approaching (0,0) from any direction
∴ the limit of $f(x,y)$ at (0,0) does not exist.

A common exam type involves limits at $(x_0, y_0) = (0,0)$. Suppose $(x_0, y_0) = (0,0)$ and the limit of $f(x,y)$ as (x,y) approaches (0,0) exists. Let $y = mx$, then
$$\lim_{(x,y)\to(0,0)} f(x,y) = \lim_{(x,y)\to(0,0)} f(x,mx)$$
and
$$\lim_{(x,y)\to(0,0)} f(x,mx)$$
is independent of the value of m. Therefore, the limit
$$\lim_{(x,y)\to(0,0)} f(x,mx)$$
being independent of m is a necessary condition for the existence of the limit of $f(x,y)$ at (0,0). Moreover, if there exist $\exists\, m_1 \neq m_2 \in R$ such that
$$\lim_{(x,y)\to(0,0)} f(x,m_1x) \neq \lim_{(x,y)\to(0,0)} f(x,m_2x)$$
the the limit $f(x,y)$ at (0,0) does not exist. In other words, if
$$\lim_{(x,y)\to(0,0)} f(x,mx)$$
depends on m, then finding $m_1 \neq m_2 \in R$ such that
$$\lim_{(x,y)\to(0,0)} f(x,m_1x) \neq \lim_{(x,y)\to(0,0)} f(x,m_2x)$$
can be used to prove that the limit of $f(x,y)$ at (0,0) does not exist. Among exam types involving the limits of multivariable functions, problems involving limits of functions of two variables are the most common. If readers become familiar with problems involving functions of two variables, they will be able to handle problems involving functions of three variables as well.

7.1 Find the Limits of Multivariable Functions

To summarize the above points, the following introduces the exam types and solution methods for determining whether the limit of a function of two variables exists at the origin (0,0).

Question Types:

Type 1.
Finding the limit of a function $f(x,y)$ as (x,y) approaches $(0,0)$, i. e.,
$$\lim_{(x,y)\to(0,0)} f(x,y) = ?$$
where the function $f(x,y)$ satisfies:
$$\lim_{(x,y)\to(0,0)} f(x,mx) = L \text{ is independent of } m.$$

Problem-Solving Process:
Let $\varepsilon > 0$, choose δ or find the specific value of δ such that:
$$x^2 + y^2 < \delta \Rightarrow |f(x,y) - L| < \varepsilon$$

Remark:

The proof process often uses the inequality $x^2 + y^2 \geq 2|xy|$, to derive

$$|f(x,y) - L| < g(x^2 + y^2) < g(\delta) < \varepsilon,$$

and identifying the explicit form of g can help determine δ. It is important to note that when

$$\lim_{(x,y)\to(0,0)} f(x,mx) = L \text{ is independent of } m,\text{ it only indicates that the limit of the function at}$$

$(0,0)$ may exist. However, it is also possible for the limit to not exist despite approaching $(0,0)$ along different paths. If the limit along curves approaching $(0,0)$ still yields L, then there is a high probability that the limit exists. In this case, follow the above process: Let $\varepsilon > 0$, explicitly choose δ such that

$$x^2 + y^2 < \delta \Rightarrow |f(x,y) - L| < \varepsilon$$

Examples:

(I) Find $\lim\limits_{(x,y)\to(0,0)} \dfrac{3x^2 y}{x^2 + y^2} = ?$

Let $y = mx$ then $\lim\limits_{(x,y)\to(0,0)} \dfrac{3x^2(mx)}{x^2 + (mx)^2} = \lim\limits_{(x,y)\to(0,0)} \dfrac{3mx}{1+m^2} = 0$ is independent of m

$\because x^2 + y^2 \geq 2|xy| \quad \therefore \dfrac{1}{x^2+y^2} \leq \dfrac{1}{2|xy|} \Rightarrow \left|\dfrac{3x^2 y}{x^2+y^2}\right| \leq \left|\dfrac{3x^2 y}{2xy}\right| = \left|\dfrac{3x}{2}\right|$

7.1 Find the Limits of Multivariable Functions

Claim: $\lim\limits_{(x,y)\to(0,0)} \dfrac{3x^2y}{x^2+y^2} = 0$

Let $\varepsilon > 0$ choose $\delta < (\dfrac{2\varepsilon}{3})^2$ then

$$x^2+y^2 < \delta \Rightarrow \left|\dfrac{3x^2y}{x^2+y^2}\right| \le \left|\dfrac{3x^2y}{2xy}\right| = \left|\dfrac{3x}{2}\right| < \left|\dfrac{3\sqrt{x^2+y^2}}{2}\right| < \dfrac{3\sqrt{\delta}}{2} < \dfrac{3}{2}\cdot\dfrac{2\varepsilon}{3} = \varepsilon$$

$\therefore \lim\limits_{(x,y)\to(0,0)} \dfrac{3x^2y}{x^2+y^2} = 0$

(II) Find $\lim\limits_{(x,y)\to(0,0)} \dfrac{x^4+y^4}{x^2+y^2} = ?$

Let $y = mx$ then $\lim\limits_{(x,y)\to(0,0)} \dfrac{x^4+y^4}{x^2+y^2} = \lim\limits_{(x,y)\to(0,0)} \dfrac{x^2+m^4x^2}{1+m^2} = 0$ is independent of m

$\because x^4+y^4 = (x^2+y^2)^2 - 2x^2y^2$

$\therefore \lim\limits_{(x,y)\to(0,0)} \dfrac{x^4+y^4}{x^2+y^2} = \lim\limits_{(x,y)\to(0,0)} \dfrac{(x^2+y^2)^2 - 2x^2y^2}{x^2+y^2}$

$= \lim\limits_{(x,y)\to(0,0)} x^2+y^2 - \lim\limits_{(x,y)\to(0,0)} \dfrac{2x^2y^2}{x^2+y^2} = -\lim\limits_{(x,y)\to(0,0)} \dfrac{2x^2y^2}{x^2+y^2}$

Claim: $\lim\limits_{(x,y)\to(0,0)} \dfrac{2x^2y^2}{x^2+y^2} = 0$

$\because x^2+y^2 \ge 2|xy| \quad \therefore \dfrac{1}{x^2+y^2} \le \dfrac{1}{2|xy|} \Rightarrow \left|\dfrac{2x^2y^2}{x^2+y^2}\right| \le \left|\dfrac{2x^2y^2}{2xy}\right| = |xy|$

Let $\varepsilon > 0$ choose $\dfrac{\delta}{2} < \varepsilon$ then

$$x^2+y^2 < \delta \Rightarrow \left|\dfrac{2x^2y^2}{x^2+y^2}\right| \le \left|\dfrac{2x^2y^2}{2xy}\right| = |xy| < \left|\dfrac{x^2+y^2}{2}\right| < \dfrac{\delta}{2} < \varepsilon$$

$\therefore \lim\limits_{(x,y)\to(0,0)} \dfrac{2x^2y^2}{x^2+y^2} = 0 \Rightarrow \lim\limits_{(x,y)\to(0,0)} \dfrac{x^4+y^4}{x^2+y^2} = 0$

Type 2.
Finding the limit of a function $f(x,y)$ as (x,y) approaches $(0,0)$, i. e.,
$$\lim\limits_{(x,y)\to(0,0)} f(x,y) = ?$$
where the function $f(x,y)$ satisfies:
$$\text{If } \lim\limits_{(x,y)\to(0,0)} f(x,mx) = L \text{ depends on } m.$$

Problem-Solving Process:
Choose $m_1 \neq m_2$ such that $\lim_{(x,y)\to(0,0)} f(x, m_1 x) \neq \lim_{(x,y)\to(0,0)} f(x, m_2 x)$

$\therefore \lim_{(x,y)\to(0,0)} f(x,y)$ does not exist

Examples:

(I) Find $\lim_{(x,y)\to(0,0)} \dfrac{x^2 - 2y^2}{x^2 + y^2} = ?$

Let $y = mx$ then $\lim_{(x,y)\to(0,0)} \dfrac{x^2 - 2y^2}{x^2 + y^2} = \lim_{(x,y)\to(0,0)} \dfrac{1 - 2m^2}{1 + m^2}$ depends on m

Choose $m = 1$ then $\lim_{(x,y)\to(0,0)} \dfrac{x^2 - 2y^2}{x^2 + y^2} = \lim_{(x,y)\to(0,0)} \dfrac{1 - 2m^2}{1 + m^2} = -\dfrac{1}{2}$

Choose $m = \sqrt{2}$ then $\lim_{(x,y)\to(0,0)} \dfrac{x^2 - 2y^2}{x^2 + y^2} = \lim_{(x,y)\to(0,0)} \dfrac{1 - 2m^2}{1 + m^2} = -1$

$\therefore \lim_{(x,y)\to(0,0)} \dfrac{x^2 - 2y^2}{x^2 + y^2}$ does not exist

(II) Find $\lim_{(x,y)\to(0,0)} \dfrac{x^3 y}{x^6 + 2y^2} = ?$

Let $y = mx$ then $\lim_{(x,y)\to(0,0)} \dfrac{x^3 y}{x^6 + 2y^2} = \lim_{(x,y)\to(0,0)} \dfrac{m}{1 + 2m^2}$ depends on m

Choose $m = -1$ then $\lim_{(x,y)\to(0,0)} \dfrac{x^3 y}{x^6 + 2y^2} = \lim_{(x,y)\to(0,0)} \dfrac{m}{1 + 2m^2} = -\dfrac{1}{3}$

Choose $m = 1$ then $\lim_{(x,y)\to(0,0)} \dfrac{x^3 y}{x^6 + 2y^2} = \lim_{(x,y)\to(0,0)} \dfrac{m}{1 + 2m^2} = \dfrac{1}{3}$

$\therefore \lim_{(x,y)\to(0,0)} \dfrac{x^3 y}{x^6 + 2y^2}$ does not exist

Type 3.
Finding the limit of a function $f(x, y, z)$ as (x, y, z) approaches $(0,0,0)$, i. e.,
$$\lim_{(x,y,z)\to(0,0,0)} f(x, y, z) = ?$$
where the function $f(x, y, z)$ satisfies:

If $\lim_{(x,y,z)\to(0,0,0)} f(x, mx, mx) = L$ is independent of m.

Problem-Solving Process:
Let $\varepsilon > 0$, choose δ or find the specific value of δ such that
$$x^2 + y^2 + z^2 < \delta \Rightarrow |f(x, y, z) - L| < \varepsilon$$

7.1 Find the Limits of Multivariable Functions

Examples:

(I) Find $\lim\limits_{(x,y,z)\to(0,0,0)} \dfrac{x^2 y}{x^2 + y^4 + z^6} = ?$

Let $y = z = mx$ then $\dfrac{x^2 y}{x^2 + y^4 + z^6} = \dfrac{mx^3}{x^2 + m^4 x^4 + m^6 x^6} = \dfrac{mx}{1 + m^4 x^2 + m^6 x^4}$

$\therefore \lim\limits_{(x,y,z)\to(0,0,0)} f(x, mx, mx) = 0$ is independent of m

$\because x^2 + y^4 + z^6 \geq x^2 \quad \therefore \left|\dfrac{x^2 y}{x^2 + y^4 + z^6}\right| \leq \left|\dfrac{x^2 y}{x^2}\right| = |y|$

Let $\varepsilon > 0$ choose $\delta^{\frac{1}{2}} < \varepsilon$ then

$x^2 + y^2 + z^2 < \delta \Rightarrow \left|\dfrac{x^2 y}{x^2 + y^4 + z^6}\right| < |y| < (x^2 + y^2 + z^2)^{\frac{1}{2}} < \delta^{\frac{1}{2}} < \varepsilon$

$\therefore \lim\limits_{(x,y,z)\to(0,0,0)} \dfrac{x^2 y}{x^2 + y^4 + z^6} = 0$

It is important to note that $\lim\limits_{(x,y,z)\to(0,0,0)} f(x, mx, mx) = 0$ being independent of m is a necessary condition for the existence of the limit of $f(x, y, z)$ at $(0,0,0)$, but it is not a sufficient condition. In other words, if during the reasoning process one encounters the statement: "Since $\lim\limits_{(x,y,z)\to(0,0,0)} f(x, mx, mx) = 0$ is independent of m, the limit of $f(x, y, z)$ at $(0,0,0)$ exists," this is an incorrect argument. Below is an example of a multivariable function to illustrate this.

Examples:

Determine whether $f(x, y, z) = \dfrac{xyz}{x^2 + y^4 + z^4}$ has a limit as $(x, y, z) \to (0,0,0)$

Let $y = z = mx$ then $f(x, y, z) = \dfrac{xyz}{x^2 + y^4 + z^4} = \dfrac{m^3 x^3}{x^2 + m^4 x^4 + m^4 x^4} = \dfrac{m^3 x}{1 + m^4 x^2 + m^4 x^2}$

$\therefore \lim\limits_{(x,y,z)\to(0,0,0)} f(x, mx, mx) = 0$ is independent of m

and

$$\lim\limits_{(x,y,z)\to(0,0,0)} f(x, mx, mx) = 0.$$

However, let $x = mt^2$, $y = z = t$ then $f(x, y, z) = \dfrac{mt^4}{(m^2 + 2)t^4} = \dfrac{m}{m^2 + 2}$

Choose $m = -1$ then $\lim\limits_{(x,y,z)\to(0,0,0)} \dfrac{xyz}{x^2 + y^4 + z^4} = \lim\limits_{(x,y,z)\to(0,0,0)} \dfrac{m}{m^2 + 2} = -\dfrac{1}{3}$

Choose $m = 1$ then $\lim_{(x,y,z)\to(0,0,0)} \dfrac{xyz}{x^2 + y^4 + z^4} = \lim_{(x,y,z)\to(0,0,0)} \dfrac{m}{m^2 + 2} = \dfrac{1}{3}$

$\therefore \lim_{(x,y,z)\to(0,0,0)} \dfrac{xyz}{x^2 + y^4 + z^4}$ does not exist

Type 4.

Finding the limit of a function $f(x, y, z)$ as (x, y, z) approaches $(0,0,0)$, i.e.,
$$\lim_{(x,y,z)\to(0,0,0)} f(x, y, z) =?$$
where the function $f(x, y, z)$ satisfies:

If $\lim_{(x,y,z)\to(0,0,0)} f(x, mx, mx) = L$ depends on m.

Problem-Solving Process:

Choose $m_1 \neq m_2$ s.t. $\lim_{(x,y,z)\to(0,0,0)} f(x, m_1x, m_1x) \neq \lim_{(x,y,z)\to(0,0,0)} f(x, m_2x, m_2x)$

$\therefore \lim_{(x,y,z)\to(0,0,0)} f(x, y, z)$ does not exist

Examples:

(I) Find $\lim_{(x,y,z)\to(0,0,0)} \dfrac{3xy}{x^2 + y^2 + z^2} =?$

Let $y = z = mx$ then $\dfrac{3xy}{x^2 + y^2 + z^2} = \dfrac{3mx^2}{x^2 + m^2x^2 + m^2x^2} = \dfrac{3m}{2m^2 + 1}$

Choose $m = -1$ then $\lim_{(x,y,z)\to(0,0,0)} \dfrac{3xy}{x^2 + y^2 + z^2} = \lim_{(x,y,z)\to(0,0,0)} \dfrac{3m}{2m^2 + 1} = -1$

Choose $m = 1$ then $\lim_{(x,y,z)\to(0,0,0)} \dfrac{3xy}{x^2 + y^2 + z^2} = \lim_{(x,y,z)\to(0,0,0)} \dfrac{3m}{2m^2 + 1} = 1$

$\therefore \lim_{(x,y,z)\to(0,0,0)} \dfrac{3xy}{x^2 + y^2 + z^2}$ does not exist

Example 1.

Find $\lim_{(x,y)\to(0,0)} \dfrac{x^2 - 2y^2}{x^2 + y^2} =?$

【Solution】

Let $y = mx$ then $\dfrac{x^2 - 2y^2}{x^2 + y^2} = \dfrac{x^2 - 2m^2x^2}{x^2 + m^2x^2} = \dfrac{x^2(1 - 2m^2)}{x^2(1 + m^2)} = \dfrac{1 - 2m^2}{1 + m^2}$

$$\therefore \lim_{(x,y)\to(0,0)} \frac{x^2-2y^2}{x^2+y^2} = \lim_{(x,y)\to(0,0)} \frac{1-2m^2}{1+m^2}$$

Choose $m=1$ then $\lim_{(x,y)\to(0,0)} \frac{x^2-2y^2}{x^2+y^2} = \lim_{(x,y)\to(0,0)} \frac{1-2m^2}{1+m^2} = -\frac{1}{2}$

Choose $m=\sqrt{2}$ then $\lim_{(x,y)\to(0,0)} \frac{x^2-2y^2}{x^2+y^2} = \lim_{(x,y)\to(0,0)} \frac{1-2m^2}{1+m^2} = -1$

$\therefore \lim_{(x,y)\to(0,0)} \frac{x^2-2y^2}{x^2+y^2}$ does not exist

Example 2.

Find $\lim_{(x,y)\to(0,0)} \left(\frac{x^2-y^2}{x^2+2y^2}\right)^2 = ?$

【Solution】

Let $y=mx$ then $\left(\frac{x^2-y^2}{x^2+2y^2}\right)^2 = \left(\frac{x^2-m^2x^2}{x^2+2m^2x^2}\right)^2 = \left(\frac{x^2(1-m^2)}{x^2(1+2m^2)}\right)^2 = \frac{(1-m^2)^2}{(1+2m^2)^2}$

$\therefore \lim_{(x,y)\to(0,0)} \left(\frac{x^2-y^2}{x^2+2y^2}\right)^2 = \lim_{(x,y)\to(0,0)} \frac{(1-m^2)^2}{(1+2m^2)^2}$

Choose $m=1$ then $\lim_{(x,y)\to(0,0)} \left(\frac{x^2-y^2}{x^2+2y^2}\right)^2 = \lim_{(x,y)\to(0,0)} \frac{(1-m^2)^2}{(1+2m^2)^2} = 0$

Choose $m=\sqrt{2}$ then $\lim_{(x,y)\to(0,0)} \left(\frac{x^2-y^2}{x^2+2y^2}\right)^2 = \lim_{(x,y)\to(0,0)} \frac{(1-m^2)^2}{(1+2m^2)^2} = \frac{1}{25}$

$\therefore \lim_{(x,y)\to(0,0)} \left(\frac{x^2-y^2}{x^2+2y^2}\right)^2$ does not exist

Example 3.

Find $\lim_{(x,y)\to(0,0)} \frac{x^2+y^2}{|x|+|y|} = ?$

【Solution】

$\because (|x|+|y|)^2 = |x|^2+|y|^2+2|xy| \geq x^2+y^2 \quad \therefore |x|+|y| \geq \sqrt{x^2+y^2}$

$\therefore \frac{x^2+y^2}{|x|+|y|} \leq \frac{x^2+y^2}{\sqrt{x^2+y^2}} = \sqrt{x^2+y^2}$

Claim: $\lim_{(x,y)\to(0,0)} \frac{x^2+y^2}{|x|+|y|} = 0$

Let $\varepsilon > 0$ choose $\delta^{\frac{1}{2}} < \varepsilon$ then $x^2 + y^2 < \delta \Rightarrow \dfrac{x^2 + y^2}{|x| + |y|} \leq \dfrac{x^2 + y^2}{\sqrt{x^2 + y^2}} = \sqrt{x^2 + y^2} < \delta^{\frac{1}{2}} < \varepsilon$

$\therefore \lim\limits_{(x,y) \to (0,0)} \dfrac{x^2 + y^2}{|x| + |y|} = 0$

Example 4.

Find $\lim\limits_{(x,y) \to (0,0)} \dfrac{xy \cos(x^2 + y^2)}{\sqrt{x^2 + y^2}} = ?$

【Solution】

$\because x^2 + y^2 \geq 2|xy| \quad \therefore \dfrac{1}{\sqrt{x^2 + y^2}} \leq \dfrac{1}{\sqrt{2|xy|}} \Rightarrow \left| \dfrac{xy \cos(x^2 + y^2)}{\sqrt{x^2 + y^2}} \right| \leq \left| \dfrac{xy}{\sqrt{2|xy|}} \right| < |xy|^{\frac{1}{2}}$

Claim: $\lim\limits_{(x,y) \to (0,0)} \dfrac{xy \cos x^2 + y^2}{\sqrt{x^2 + y^2}} = 0$

Let $\varepsilon > 0$ choose $\delta^{\frac{1}{2}} < \varepsilon$ then

$x^2 + y^2 < \delta \Rightarrow \left| \dfrac{xy \cos(x^2 + y^2)}{\sqrt{x^2 + y^2}} \right| \leq \left| \dfrac{xy}{\sqrt{2|xy|}} \right| < |xy|^{\frac{1}{2}} < (x^2 + y^2)^{\frac{1}{2}} < \delta^{\frac{1}{2}} < \varepsilon$

$\therefore \lim\limits_{(x,y) \to (0,0)} \dfrac{xy \cos x^2 + y^2}{\sqrt{x^2 + y^2}} = 0$

Example 5.

Find $\lim\limits_{(x,y,z) \to (0,0,0)} \dfrac{\sin(x^2 + y^2 + z^2)}{\sqrt{x^2 + y^2 + z^2}} = ?$

【Solution】

Let $t = x^2 + y^2 + z^2$ then $\dfrac{\sin(x^2 + y^2 + z^2)}{\sqrt{x^2 + y^2 + z^2}} = \dfrac{\sin t}{\sqrt{t}} = \dfrac{\sqrt{t} \sin t}{t}$

$\lim\limits_{(x,y,z) \to (0,0,0)} \dfrac{\sin(x^2 + y^2 + z^2)}{\sqrt{x^2 + y^2 + z^2}} = \lim\limits_{t \to 0} \dfrac{\sqrt{t} \sin t}{t}$

By L'Hôpital's Rule, $\lim\limits_{t \to 0} \dfrac{\sin t}{t} = \lim\limits_{t \to 0} \dfrac{\cos t}{1} = 1 \Rightarrow \lim\limits_{t \to 0} \dfrac{\sqrt{t} \sin t}{t} = 0$

$\therefore \lim\limits_{(x,y,z) \to (0,0,0)} \dfrac{\sin x^2 + y^2 + z^2}{\sqrt{x^2 + y^2 + z^2}} = 0$

7.1 Find the Limits of Multivariable Functions

Example 6.

Find $\lim_{(x,y)\to(0,0)} \dfrac{3x^2y}{x^2+y^2} = ?$

【Solution】

$\because x^2+y^2 \geq 2|xy| \quad \therefore \dfrac{1}{x^2+y^2} \leq \dfrac{1}{2|xy|} \Rightarrow \left|\dfrac{3x^2y}{x^2+y^2}\right| \leq \left|\dfrac{3x^2y}{2xy}\right| = \left|\dfrac{3x}{2}\right|$

Claim: $\lim_{(x,y)\to(0,0)} \dfrac{3x^2y}{x^2+y^2} = 0$

Let $\varepsilon > 0$ choose $\delta < (\dfrac{2\varepsilon}{3})^2$ then

$x^2+y^2 < \delta \Rightarrow \left|\dfrac{3x^2y}{x^2+y^2}\right| \leq \left|\dfrac{3x^2y}{2xy}\right| = \left|\dfrac{3x}{2}\right| < \left|\dfrac{3\sqrt{x^2+y^2}}{2}\right| < \dfrac{3\sqrt{\delta}}{2} < \dfrac{3}{2} \cdot \dfrac{2\varepsilon}{3} = \varepsilon$

$\therefore \lim_{(x,y)\to(0,0)} \dfrac{3x^2y}{x^2+y^2} = 0$

Example 7.

Find $\lim_{(x,y)\to(0,0)} \dfrac{2xy^2}{x^2+y^2} = ?$

【Solution】

$\because x^2+y^2 \geq 2|xy| \quad \therefore \dfrac{1}{x^2+y^2} \leq \dfrac{1}{2|xy|} \Rightarrow \left|\dfrac{2xy^2}{x^2+y^2}\right| \leq \left|\dfrac{2xy^2}{2xy}\right| = |y|$

Claim: $\lim_{(x,y)\to(0,0)} \dfrac{2xy^2}{x^2+y^2} = 0$

Let $\varepsilon > 0$ choose $\delta < \varepsilon$ then

$x^2+y^2 < \delta \Rightarrow \left|\dfrac{2xy^2}{x^2+y^2}\right| \leq \left|\dfrac{2xy^2}{2xy}\right| = |y| < \left|\sqrt{x^2+y^2}\right| < \delta < \varepsilon$

$\therefore \lim_{(x,y)\to(0,0)} \dfrac{2xy^2}{x^2+y^2} = 0$

Example 8.

Find $\lim_{(x,y,z)\to(0,0,0)} \dfrac{xyz}{x^2+y^4+z^4} = ?$

【Solution】

Let $x = mt^2$, $y = z = t$ then $\dfrac{mt^4}{(m^2+2)t^4} = \dfrac{m}{m^2+2}$

Choose $m = -1$ then $\lim\limits_{(x,y,z) \to (0,0,0)} \dfrac{xyz}{x^2+y^4+z^4} = \lim\limits_{(x,y,z) \to (0,0,0)} \dfrac{m}{m^2+2} = -\dfrac{1}{3}$

Choose $m = 1$ then $\lim\limits_{(x,y,z) \to (0,0,0)} \dfrac{xyz}{x^2+y^4+z^4} = \lim\limits_{(x,y,z) \to (0,0,0)} \dfrac{m}{m^2+2} = \dfrac{1}{3}$

$\therefore \lim\limits_{(x,y,z) \to (0,0,0)} \dfrac{xyz}{x^2+y^4+z^4}$ does not exist

Example 9.

Find $\lim\limits_{(x,y,z) \to (0,0,0)} \dfrac{x^2 y}{x^2+y^4+z^6} = ?$

【Solution】

$\because x^2 + y^4 + z^6 \geq x^2 \quad \therefore \left| \dfrac{x^2 y}{x^2+y^4+z^6} \right| \leq \left| \dfrac{x^2 y}{x^2} \right| = |y|$

Let $\varepsilon > 0$ choose $\delta^{\frac{1}{2}} < \varepsilon$ then

$x^2 + y^2 + z^2 < \delta \Rightarrow \left| \dfrac{x^2 y}{x^2+y^4+z^6} \right| < |y| < (x^2+y^2+z^2)^{\frac{1}{2}} < \delta^{\frac{1}{2}} < \varepsilon$

$\therefore \lim\limits_{(x,y,z) \to (0,0,0)} \dfrac{x^2 y}{x^2+y^4+z^6} = 0$

Example 10.

Find $\lim\limits_{(x,y) \to (0,0)} \dfrac{x^4+y^4}{x^2+y^2} = ?$

【Solution】

$\because x^4 + y^4 = (x^2+y^2)^2 - 2x^2 y^2$

$\therefore \lim\limits_{(x,y) \to (0,0)} \dfrac{x^4+y^4}{x^2+y^2} = \lim\limits_{(x,y) \to (0,0)} \dfrac{(x^2+y^2)^2 - 2x^2 y^2}{x^2+y^2}$

$= \lim\limits_{(x,y) \to (0,0)} x^2+y^2 - \lim\limits_{(x,y) \to (0,0)} \dfrac{2x^2 y^2}{x^2+y^2} = -\lim\limits_{(x,y) \to (0,0)} \dfrac{2x^2 y^2}{x^2+y^2}$

Claim: $\lim\limits_{(x,y) \to (0,0)} \dfrac{2x^2 y^2}{x^2+y^2} = 0$

$\because x^2 + y^2 \geq 2|xy| \quad \therefore \dfrac{1}{x^2+y^2} \leq \dfrac{1}{2|xy|} \Rightarrow \left| \dfrac{2x^2 y^2}{x^2+y^2} \right| \leq \left| \dfrac{2x^2 y^2}{2xy} \right| = |xy|$

7.1 Find the Limits of Multivariable Functions

Let $\varepsilon > 0$ choose $\dfrac{\delta}{2} < \varepsilon$ then

$$x^2 + y^2 < \delta \Rightarrow \left|\dfrac{2x^2y^2}{x^2+y^2}\right| \le \left|\dfrac{2x^2y^2}{2xy}\right| = |xy| < \left|\dfrac{x^2+y^2}{2}\right| < \dfrac{\delta}{2} < \varepsilon$$

$$\therefore \lim_{(x,y)\to(0,0)} \dfrac{2x^2y^2}{x^2+y^2} = 0 \Rightarrow \lim_{(x,y)\to(0,0)} \dfrac{x^4+y^4}{x^2+y^2} = 0$$

Example 11.

Find $\lim\limits_{(x,y)\to(0,0)} \dfrac{x^3y}{x^6+2y^2} = ?$

【Solution】

Let $y = mx^3$ then $\dfrac{x^3y}{x^6+2y^2} = \dfrac{x^3(mx^3)}{x^6+2m^2x^6} = \dfrac{m}{1+2m^2}$

Choose $m = -1$ then $\lim\limits_{(x,y)\to(0,0)} \dfrac{x^3y}{x^6+2y^2} = \lim\limits_{(x,y)\to(0,0)} \dfrac{m}{1+2m^2} = -\dfrac{1}{3}$

Choose $m = 1$ then $\lim\limits_{(x,y)\to(0,0)} \dfrac{x^3y}{x^6+2y^2} = \lim\limits_{(x,y)\to(0,0)} \dfrac{m}{1+2m^2} = \dfrac{1}{3}$

$\therefore \lim\limits_{(x,y)\to(0,0)} \dfrac{x^3y}{x^6+2y^2}$ does not exist

Example 12.

Find $\lim\limits_{(x,y)\to(0,0)} \dfrac{x-y^2}{x+2y^2} = ?$

【Solution】

Let $y = m\sqrt{x}$ then $\dfrac{x-y^2}{x+2y^2} = \dfrac{x-m^2x}{x+2m^2x} = \dfrac{1-m^2}{1+2m^2}$

Choose $m = \sqrt{2}$ then $\lim\limits_{(x,y)\to(0,0)} \dfrac{x-y^2}{x+2y^2} = \lim\limits_{(x,y)\to(0,0)} \dfrac{1-m^2}{1+2m^2} = \dfrac{-1}{5}$

Choose $m = 1$ then $\lim\limits_{(x,y)\to(0,0)} \dfrac{x-y^2}{x+2y^2} = \lim\limits_{(x,y)\to(0,0)} \dfrac{1-m^2}{1+2m^2} = 0$

$\therefore \lim\limits_{(x,y)\to(0,0)} \dfrac{x-y^2}{x+2y^2}$ does not exist

Example 13.

Find $\lim\limits_{(x,y)\to(0,0)} \dfrac{xy^5}{x^2+y^{10}} = ?$

【Solution】

Let $y = (mx)^{\frac{1}{5}}$ then $\dfrac{xy^5}{x^2+y^{10}} = \dfrac{mx^2}{(1+m^2)x^2} = \dfrac{m}{1+m^2}$

Choose $m = -1$ then $\lim\limits_{(x,y)\to(0,0)} \dfrac{xy^5}{x^2+y^{10}} = \lim\limits_{(x,y)\to(0,0)} \dfrac{m}{1+m^2} = -\dfrac{1}{2}$

Choose $m = 1$ then $\lim\limits_{(x,y)\to(0,0)} \dfrac{xy^5}{x^2+y^{10}} = \lim\limits_{(x,y)\to(0,0)} \dfrac{m}{1+m^2} = \dfrac{1}{2}$

∴ $\lim\limits_{(x,y)\to(0,0)} \dfrac{xy^5}{x^2+y^{10}}$ does not exist

Example 14.

Find $\lim\limits_{(x,y)\to(0,0)} \dfrac{x^2y^2}{\sqrt{x^2+y^2}} = ?$

【Solution】

$\because x^2+y^2 \geq 2|xy|$ ∴ $\dfrac{1}{\sqrt{x^2+y^2}} \leq \dfrac{1}{\sqrt{2|xy|}} \Rightarrow \left|\dfrac{x^2y^2}{\sqrt{x^2+y^2}}\right| \leq \left|\dfrac{x^2y^2}{\sqrt{2|xy|}}\right| = |xy|^{\frac{3}{2}}$

Claim: $\lim\limits_{(x,y)\to(0,0)} \dfrac{x^2y^2}{\sqrt{x^2+y^2}} = 0$

Let $\varepsilon > 0$ choose $\delta^{\frac{3}{2}} < \varepsilon$ then $x^2+y^2 < \delta$

$\Rightarrow \left|\dfrac{x^2y^2}{\sqrt{x^2+y^2}}\right| \leq \left|\dfrac{x^2y^2}{\sqrt{2|xy|}}\right| = |xy|^{\frac{3}{2}} < (x^2+y^2)^{\frac{3}{2}} < \delta^{\frac{3}{2}} < \varepsilon$

∴ $\lim\limits_{(x,y)\to(0,0)} \dfrac{x^2y^2}{\sqrt{x^2+y^2}} = 0$

Example 15.

Find $\lim\limits_{(x,y)\to(0,0)} \dfrac{(xy)^p}{\sqrt{x^2+y^2}} = ?,\ \forall p > \dfrac{1}{2}$

【Solution】

Let $p > \dfrac{1}{2}$

7.1 Find the Limits of Multivariable Functions

$\because x^2 + y^2 \geq 2|xy| \quad \therefore \dfrac{1}{\sqrt{x^2+y^2}} \leq \dfrac{1}{\sqrt{2|xy|}} \Rightarrow \left|\dfrac{(xy)^p}{\sqrt{x^2+y^2}}\right| \leq \left|\dfrac{(xy)^p}{\sqrt{2|xy|}}\right| < |xy|^{p-\frac{1}{2}}$

Claim: $\lim\limits_{(x,y)\to(0,0)} \dfrac{(xy)^p}{\sqrt{x^2+y^2}} = 0$

Let $\varepsilon > 0$ choose $\delta^{p-\frac{1}{2}} < \varepsilon$ then

$x^2 + y^2 < \delta \Rightarrow \left|\dfrac{(xy)^p}{\sqrt{x^2+y^2}}\right| \leq \left|\dfrac{(xy)^p}{\sqrt{2|xy|}}\right| < |xy|^{p-\frac{1}{2}} < (x^2+y^2)^{p-\frac{1}{2}} < \delta^{p-\frac{1}{2}} < \varepsilon$

$\therefore \lim\limits_{(x,y)\to(0,0)} \dfrac{(xy)^p}{\sqrt{x^2+y^2}} = 0$

Example 16.

Let $f(x,y) = (x+y)\sin\dfrac{1}{x}\sin\dfrac{1}{y}$. Find $\lim\limits_{x\to 0}\left(\lim\limits_{y\to 0} f(x,y)\right) = ?$

【Solution】

Claim: $\lim\limits_{y\to 0} f(x,y)$ does not exist

Let $y_{1,n} = \dfrac{1}{2n\pi + \dfrac{\pi}{2}}$ and $y_{2,n} = \dfrac{1}{2n\pi + \dfrac{3\pi}{2}}$ then $\lim\limits_{n\to\infty} y_{1,n} = 0$ and $\lim\limits_{n\to\infty} y_{2,n} = 0$

then $\lim\limits_{n\to\infty} f(x, y_{1,n}) = \lim\limits_{n\to\infty}\left(x + \dfrac{1}{2n\pi + \dfrac{\pi}{2}}\right)\sin\dfrac{1}{x}\sin\left(2n\pi + \dfrac{\pi}{2}\right) = \sin\dfrac{1}{x}$

$\lim\limits_{n\to\infty} f(x, y_{2,n}) = \lim\limits_{n\to\infty}\left(x + \dfrac{1}{2n\pi + \dfrac{3\pi}{2}}\right) + \sin\dfrac{1}{x}\sin\left(2n\pi + \dfrac{3\pi}{2}\right) = -\sin\dfrac{1}{x}$

Let $x_m = \dfrac{1}{2m\pi + \dfrac{\pi}{2}}$ then $\lim\limits_{m\to\infty} x_m = 0$

$\lim\limits_{m\to\infty}\left(\lim\limits_{n\to\infty} f(x_m, y_{1,n})\right) = \lim\limits_{m\to\infty} \sin\left(2m\pi + \dfrac{\pi}{2}\right) = 1$

$\lim\limits_{m\to\infty}\left(\lim\limits_{n\to\infty} f(x_m, y_{2,n})\right) = -\lim\limits_{m\to\infty} \sin\left(2m\pi + \dfrac{\pi}{2}\right) = -1$

$\therefore \lim\limits_{x\to 0}\left(\lim\limits_{y\to 0} f(x,y)\right)$ does not exist

Example 17.

Find $\lim\limits_{y\to 0}\left(\lim\limits_{x\to 0}\dfrac{\cos x \tan y}{x-y}\right)=?$

【Solution】

$\because \lim\limits_{x\to 0}\dfrac{\cos x \tan y}{x-y}=\dfrac{\cos 0 \tan y}{-y}=\dfrac{\tan y}{-y}$

$\therefore \lim\limits_{y\to 0}\left(\lim\limits_{x\to 0}\dfrac{\cos x \tan y}{x-y}\right)=\lim\limits_{y\to 0}\left(\dfrac{\tan y}{-y}\right)=-\lim\limits_{y\to 0}\left(\dfrac{\tan y}{y}\right)=-1$

Example 18.

Find $\lim\limits_{(x,y,z)\to(0,0,0)}\dfrac{4xy+xz+z^2}{x^2+2y^2}=?$

【Solution】

Let $z=y=mx$ then $\dfrac{4mx^2+mx^2+m^2x^2}{(1+2m^2)x^2}=\dfrac{5m+m^2}{1+2m^2}$

Choose $m=-1$ then $\lim\limits_{(x,y,z)\to(0,0,0)}\dfrac{4xy+xz+z^2}{x^2+2y^2}=\lim\limits_{(x,y,z)\to(0,0,0)}\dfrac{5m+m^2}{1+2m^2}=-\dfrac{4}{3}$

Choose $m=1$ then $\lim\limits_{(x,y,z)\to(0,0,0)}\dfrac{4xy+xz+z^2}{x^2+2y^2}=\lim\limits_{(x,y,z)\to(0,0,0)}\dfrac{5m+m^2}{1+2m^2}=\dfrac{6}{3}=2$

$\therefore \lim\limits_{(x,y,z)\to(0,0,0)}\dfrac{4xy+xz+z^2}{x^2+2y^2}$ does not exist

Example 19.

Find $\lim\limits_{(x,y,z)\to(0,0,0)}\dfrac{3xy}{x^2+y^2+z^2}=?$

【Solution】

Let $y=mx$ and $z=mx$ then $\dfrac{3xy}{x^2+y^2+z^2}=\dfrac{3mx^2}{x^2+m^2x^2+m^2x^2}=\dfrac{3m}{2m^2+1}$

Choose $m=-1$ then $\lim\limits_{(x,y,z)\to(0,0,0)}\dfrac{3xy}{x^2+y^2+z^2}=\lim\limits_{(x,y,z)\to(0,0,0)}\dfrac{3m}{2m^2+1}=-1$

Choose $m=1$ then $\lim\limits_{(x,y,z)\to(0,0,0)}\dfrac{3xy}{x^2+y^2+z^2}=\lim\limits_{(x,y,z)\to(0,0,0)}\dfrac{3m}{2m^2+1}=1$

$\therefore \lim\limits_{(x,y,z)\to(0,0,0)}\dfrac{3xy}{x^2+y^2+z^2}$ does not exist

Example 20.

Find $\lim\limits_{(x,y,z)\to(0,0,0)} \dfrac{x^4+y^4}{x^2+y^2+z^4} = ?$

【Solution】

$\because x^2+y^2+z^2 \geq x^2+y^2 \quad \therefore \left|\dfrac{x^4+y^4}{x^2+y^2+z^4}\right| \leq \left|\dfrac{x^4+y^4}{x^2+y^2}\right|$

Claim: $\lim\limits_{(x,y)\to(0,0)} \dfrac{x^4+y^4}{x^2+y^2} = 0$

$\because x^4+y^4 = (x^2+y^2)^2 - 2x^2y^2$

$\therefore \lim\limits_{(x,y)\to(0,0)} \dfrac{x^4+y^4}{x^2+y^2} = \lim\limits_{(x,y)\to(0,0)} \dfrac{(x^2+y^2)^2 - 2x^2y^2}{x^2+y^2}$

$= \lim\limits_{(x,y)\to(0,0)} x^2+y^2 - \lim\limits_{(x,y)\to(0,0)} \dfrac{2x^2y^2}{x^2+y^2} = -\lim\limits_{(x,y)\to(0,0)} \dfrac{2x^2y^2}{x^2+y^2}$

Claim: $\lim\limits_{(x,y)\to(0,0)} \dfrac{2x^2y^2}{x^2+y^2} = 0$

$\because x^2+y^2 \geq 2|xy| \quad \therefore \dfrac{1}{x^2+y^2} \leq \dfrac{1}{2|xy|} \Rightarrow \left|\dfrac{2x^2y^2}{x^2+y^2}\right| \leq \left|\dfrac{2x^2y^2}{2xy}\right| = |xy|$

Let $\varepsilon > 0$ choose $\dfrac{\delta}{2} < \varepsilon$ then

$x^2+y^2 < \delta \Rightarrow \left|\dfrac{2x^2y^2}{x^2+y^2}\right| \leq \left|\dfrac{2x^2y^2}{2xy}\right| = |xy| < \left|\dfrac{x^2+y^2}{2}\right| < \dfrac{\delta}{2} < \varepsilon$

$\therefore \lim\limits_{(x,y)\to(0,0)} \dfrac{2x^2y^2}{x^2+y^2} = 0 \Rightarrow \lim\limits_{(x,y)\to(0,0)} \dfrac{x^4+y^4}{x^2+y^2} = 0$

$\Rightarrow \lim\limits_{(x,y,z)\to(0,0,0)} \dfrac{x^4+y^4}{x^2+y^2+z^4} = 0$

7.2 Determine the Continuity of Multivariable Functions

【Definition】The Definition of Continuity for Bivariate Functions

$f(x,y)$ is continuous at $(x_0, y_0) \Leftrightarrow \lim\limits_{(x,y)\to(x_0,y_0)} f(x,y) = f(x_0, y_0)$

$\Leftrightarrow \forall \varepsilon > 0,\ \exists \delta > 0,$ such that $\sqrt{(x-x_0)^2 + (y-y_0)^2} < \delta \Rightarrow |f(x,y) - f(x_0, y_0)| < \varepsilon$

【Definition】The Definition of Continuity for Multivariable Functions

7.2 Determine the Continuity of Multivariable Functions

$f(x_1, \ldots, x_n)$ is continuous at $(c_1, \ldots, c_n)$ $\Leftrightarrow \lim_{(x_1,\ldots,x_n) \to (c_1,\ldots,c_n)} f(x_1, \ldots, x_n) = f(c_1, \ldots, c_n)$

$\Leftrightarrow \forall \varepsilon > 0, \exists \delta > 0,$ such that

$\sqrt{(x_1 - c_1)^2 + \cdots + (x_n - c_n)^2} < \delta \Rightarrow |f(x_1, \ldots, x_n) - f(c_1, \ldots, c_n)| < \varepsilon$

From the definition of continuity for a function of two variables, we can observe that if $f(x, y)$ is continuous at (x_0, y_0), then the limit of $f(x, y)$ as (x, y) approaches (x_0, y_0) from any direction must equal $f(x_0, y_0)$. Therefore, if we let $y - y_0 = m(x - x_0)$ represent a line passing through (x_0, y_0), and if $f(x, y)$ is continuous at (x_0, y_0), then

$$\lim_{(x,y) \to (x_0,y_0)} f(x, m(x - x_0) + y_0) = f(x_0, y_0)$$

and

$$\lim_{(x,y) \to (x_0,y_0)} f(x, m(x - x_0) + y_0)$$

is independent of the value of m.

A common exam problem involves $(x_0, y_0) = (0,0)$. Suppose $f(x, y)$ is continuous at $(0,0)$. Let $y = mx$. Then

$$\lim_{(x,y) \to (0,0)} f(x, mx) = f(0,0)$$

and

$$\lim_{(x,y) \to (0,0)} f(x, mx)$$

is independent of m. Therefore, the independence of m in

$$\lim_{(x,y) \to (0,0)} f(x, mx)$$

is a necessary condition for the continuity of $f(x, y)$ at $(0,0)$, but it is not a sufficient condition. An example of a function $f(x, y)$ that illustrates this will be provided below.

Examples:

Let $f(x, y) = \begin{cases} \dfrac{xy^2}{x^2 + y^4}, & (x, y) \neq 0 \\ 0, & (x, y) = 0 \end{cases}$. Determine whether $f(x, y)$ is continuous at $(0,0)$.

Let $y = mx$ then $f(x, y) = \dfrac{xy^2}{x^2 + y^4} = \dfrac{m^2 x^3}{x^2 + m^4 x^4}$

$\therefore \lim_{(x,y) \to (0,0)} f(x, mx)$ is independent of m and $\lim_{(x,y) \to (0,0)} f(x, mx) = 0$

Let $x = y^2$ then $f(x, y) = \dfrac{y^4}{y^4 + y^4} = \dfrac{1}{2}$ and $\lim_{(x,y) \to (0,0)} f(y^2, y) = \dfrac{1}{2} \neq f(0,0) = 0$

$\therefore f(x, y)$ is not continuous at $(0,0)$.

Moreover, if $\exists\ m \in R$ such that
$$\lim_{(x,y)\to(0,0)} f(x,mx) \neq f(0,0)$$
then $f(x,y)$ is discontinuous at $(0,0)$. Therefore, if $\lim_{(x,y)\to(0,0)} f(x,mx)$ depends on m, then choose $m \in R$ such that:
$$\lim_{(x,y)\to(0,0)} f(x,mx) \neq f(0,0)$$
to show that $f(x,y)$ is discontinuous at $(0,0)$. In exam questions about the continuity of multivariable functions, problems involving functions of two variables are the most common. If you are familiar with these two-variable problems, you will be able to handle other types as well. To summarize the above points, the following explains the exam types and problem-solving process for determining whether a function $f(x,y)$ is continuous at the origin $(0,0)$.

Question Types:

Type 1.

Given a function $f(x,y)$, determine whether $f(x,y)$ is continuous at $(0,0)$, where $f(x,y)$ satisfies
$$\lim_{(x,y)\to(0,0)} f(x,mx) = f(0,0) \text{ is independent of } m.$$

Problem-Solving Process:

Let $\varepsilon > 0$, explicitly choose δ such that $x^2 + y^2 < \delta \Rightarrow |f(x,y) - f(0,0)| < \varepsilon$

Remark:

It is common to use the inequality $x^2 + y^2 \geq 2|xy|$ to derive
$$|f(x,y) - f(0,0)| < g(x^2 + y^2) < g(\delta) < \varepsilon$$
where g is a function with a specific form. The final inequality $g(\delta) < \varepsilon$ helps to precisely define δ.

Examples:

(I) Let $f(x,y) = \begin{cases} \dfrac{3x^2 y}{x^2 + y^2}, & (x,y) \neq 0 \\ 0, & (x,y) = 0 \end{cases}$.

Determine whether $f(x,y)$ is continuous at $(0,0)$.

Let $y = mx$ then $\lim_{(x,y)\to(0,0)} \dfrac{3x^2(mx)}{x^2 + (mx)^2} = \lim_{(x,y)\to(0,0)} \dfrac{3mx}{1 + m^2} = 0$ is independent of m

Claim: $\left|\dfrac{3x^2 y}{x^2 + y^2}\right| \leq \left|\dfrac{3\sqrt{x^2 + y^2}}{2}\right|$

7.2 Determine the Continuity of Multivariable Functions

$\because x^2 + y^2 \geq 2|xy| \quad \therefore \dfrac{1}{x^2+y^2} \leq \dfrac{1}{2|xy|} \Rightarrow \left|\dfrac{3x^2y}{x^2+y^2}\right| \leq \left|\dfrac{3x^2y}{2xy}\right| = \left|\dfrac{3x}{2}\right| \leq \left|\dfrac{3\sqrt{x^2+y^2}}{2}\right|$

Let $\varepsilon > 0$ choose $\delta < \left(\dfrac{2\varepsilon}{3}\right)^2$ then

$x^2 + y^2 < \delta \Rightarrow \left|\dfrac{3x^2y}{x^2+y^2}\right| \leq \left|\dfrac{3\sqrt{x^2+y^2}}{2}\right| < \dfrac{3\sqrt{\delta}}{2} < \dfrac{3}{2} \cdot \dfrac{2\varepsilon}{3} = \varepsilon$

$\therefore f(x,y)$ is continuous at $(0,0)$

(II) Let $f(x,y) = \begin{cases} \dfrac{x^3 - y^3}{x^2 + y^2}, & (x,y) \neq 0 \\ 0, & (x,y) = 0 \end{cases}$.

Determine whether $f(x,y)$ is continuous at $(0,0)$.

Let $y = mx$ then $\lim\limits_{(x,y)\to(0,0)} \dfrac{x^3 - y^3}{x^2 + y^2} = \lim\limits_{(x,y)\to(0,0)} \dfrac{x - m^3 x}{1 + m^2} = 0$ is independent of m

$\because x^3 - y^3 = (x-y)(x^2 + xy + y^2)$ and $x^2 + y^2 \geq 2|xy|$

$\therefore |x^3 - y^3| \leq |x-y||x^2 + |xy| + y^2| \leq |x - y|\left|x^2 + \dfrac{x^2+y^2}{2} + y^2\right|$

$\therefore |x^3 - y^3| \leq \dfrac{3}{2}|x-y||x^2 + y^2| \Rightarrow \left|\dfrac{x^3 - y^3}{x^2 + y^2}\right| \leq \dfrac{3}{2}|x - y|$

$\because |x - y| = \sqrt{|x-y|^2} \leq \sqrt{x^2 + y^2 + |2xy|} \leq \sqrt{2(x^2+y^2)}$

Let $\varepsilon > 0$ choose $3\sqrt{\dfrac{\delta}{2}} < \varepsilon$ then

$x^2 + y^2 < \delta \Rightarrow \left|\dfrac{x^3 - y^3}{x^2 + y^2}\right| \leq \dfrac{3}{2}|x-y| \leq \dfrac{3}{2}\sqrt{2(x^2+y^2)} < 3\sqrt{\dfrac{\delta}{2}} < \varepsilon$

$\therefore f(x,y)$ is continuous at $(0,0)$

Type 2.
Given a function $f(x,y)$, determine whether $f(x,y)$ is continuous at $(0,0)$, where $f(x,y)$ satisfies:

$$\lim_{(x,y)\to(0,0)} f(x, mx) = f(0,0) \text{ depends on } m.$$

Problem-Solving Process:
Choose m such that $\lim\limits_{(x,y)\to(0,0)} f(x,mx) \neq f(0,0) \quad \therefore f(x,y)$ is discontinuous at $(0,0)$

Examples:

7.2 Determine the Continuity of Multivariable Functions

(I) Let $f(x,y) = \begin{cases} \dfrac{x^2 - y^2}{x^2 + y^2}, & (x,y) \neq 0 \\ 0, & (x,y) = 0 \end{cases}$.

Determine whether $f(x,y)$ is continuous at $(0,0)$.

Let $y = mx$ then $\lim\limits_{(x,y) \to (0,0)} \dfrac{x^2 - y^2}{x^2 + y^2} = \lim\limits_{(x,y) \to (0,0)} \dfrac{1 - m^2}{1 + m^2}$ depends on m

Choose $m = \sqrt{2}$ then $\lim\limits_{(x,y) \to (0,0)} f(x,y) = \lim\limits_{(x,y) \to (0,0)} \dfrac{1 - m^2}{1 + m^2} = -\dfrac{1}{3} \neq f(0,0) = 0$

∴ $f(x,y)$ is discontinuous at $(0,0)$

(II) Let $f(x,y) = \begin{cases} \dfrac{x^2 y}{x^3 + y^3}, & (x,y) \neq 0 \\ 0, & (x,y) = 0 \end{cases}$.

Determine whether $f(x,y)$ is continuous at $(0,0)$.

Let $y = mx$ then $\lim\limits_{(x,y) \to (0,0)} \dfrac{x^2 y}{x^3 + y^3} = \lim\limits_{(x,y) \to (0,0)} \dfrac{m}{1 + m^3}$ depends on m

Choose $m = 1$ then $\lim\limits_{(x,y) \to (0,0)} f(x,y) = \lim\limits_{(x,y) \to (0,0)} \dfrac{m}{1 + m^3} = \dfrac{1}{2} \neq f(0,0)$

∴ $f(x,y)$ is discontinuous at $(0,0)$

Type 3.

Given a function $f(x,y,z)$, to determine whether $f(x,y,z)$ is continuous at $(0,0,0)$, where $f(x,y,z)$ satisfies:

$$\lim\limits_{(x,y,z) \to (0,0,0)} f(x, mx, mx) = L \text{ is independent of } m.$$

Problem-Solving Process:

Let $\varepsilon > 0$, explicitly choose δ such that $x^2 + y^2 + z^2 < \delta \Rightarrow |f(x,y,z) - f(0,0,0)| < \varepsilon$

Examples:

(I) Let $f(x,y,z) = \begin{cases} \dfrac{x^2 y}{x^2 + y^4 + z^6}, & (x,y,z) \neq 0 \\ 0, & (x,y,z) = 0 \end{cases}$.

Determine whether $f(x,y,z)$ is continuous at $(0,0,0)$.

Let $y = z = mx$ then $\dfrac{x^2 y}{x^2 + y^4 + z^6} = \dfrac{mx^3}{x^2 + m^4 x^4 + m^6 x^6} = \dfrac{mx}{1 + m^4 x^2 + m^6 x^4}$

∴ $\lim\limits_{(x,y,z) \to (0,0,0)} f(x, mx, mx) = 0$ is independent of m

∵ $x^2 + y^4 + z^6 \geq x^2$ ∴ $\left|\dfrac{x^2 y}{x^2 + y^4 + z^6}\right| \leq \left|\dfrac{x^2 y}{x^2}\right| = |y|$

Let $\varepsilon > 0$ choose $\delta^{\frac{1}{2}} < \varepsilon$ then

$$x^2 + y^2 + z^2 < \delta \Rightarrow \left|\frac{x^2 y}{x^2 + y^4 + z^6}\right| < |y| < (x^2 + y^2 + z^2)^{\frac{1}{2}} < \delta^{\frac{1}{2}} < \varepsilon$$

$\therefore$ $f(x, y, z)$ is continuous at $(0,0,0)$

It is important to note that $\lim_{(x,y,z)\to(0,0,0)} f(x, mx, mx) = 0$ being independent of m is a necessary condition for the continuity of $f(x, y, z)$ at $(0,0,0)$, but it is not a sufficient condition. In other words, if during the reasoning process one encounters the statement:
"Since $\lim_{(x,y,z)\to(0,0,0)} f(x, mx, mx) = 0$ is independent of m, $f(x, y, z)$ is continuous at $(0,0,0)$," this is an incorrect argument. An example of a multivariable function will be provided below to illustrate this.

Examples:

Let $f(x, y, z) = \begin{cases} \dfrac{xyz}{x^2 + y^4 + z^4}, & (x, y, z) \neq 0 \\ 0, & (x, y, z) = 0 \end{cases}$.

Determine whether $f(x, y, z)$ is continuous at $(0,0,0)$.

Let $y = z = mx$ then $f(x, y, z) = \dfrac{xyz}{x^2 + y^4 + z^4} = \dfrac{m^3 x^3}{x^2 + m^4 x^4 + m^4 x^4} = \dfrac{m^3 x}{1 + m^4 x^2 + m^4 x^2}$

$\therefore$ $\lim_{(x,y,z)\to(0,0,0)} f(x, mx, mx) = 0$ is independent of m

and

$\lim_{(x,y,z)\to(0,0,0)} f(x, mx, mx) = 0$. However,

Let $x = mt^2$, $y = z = t$ then $f(x, y, z) = \dfrac{mt^4}{(m^2 + 2)t^4} = \dfrac{m}{m^2 + 2}$

Choose $m = -1$ then $\lim_{(x,y,z)\to(0,0,0)} \dfrac{xyz}{x^2 + y^4 + z^4} = \lim_{(x,y,z)\to(0,0,0)} \dfrac{m}{m^2 + 2} = -\dfrac{1}{3}$

Choose $m = 1$ then $\lim_{(x,y,z)\to(0,0,0)} \dfrac{xyz}{x^2 + y^4 + z^4} = \lim_{(x,y,z)\to(0,0,0)} \dfrac{m}{m^2 + 2} = \dfrac{1}{3}$

$\therefore$ $\lim_{(x,y,z)\to(0,0,0)} \dfrac{xyz}{x^2 + y^4 + z^4}$ does not exist $\therefore$ $f(x, y, z)$ is discontinuous at $(0,0,0)$

Type 4.
Given a function $f(x, y, z)$, to determine whether $f(x, y, z)$ is continuous at $(0,0,0)$, where $f(x, y, z)$ satisfies

7.2 Determine the Continuity of Multivariable Functions

$\lim_{(x,y,z)\to(0,0,0)} f(x, mx, mx) = L$ depends on m.

Problem-Solving Process:

Choose $m_1 \neq m_2$ such that $\lim_{(x,y,z)\to(0,0,0)} f(x, m_1x, m_1x) \neq \lim_{(x,y,z)\to(0,0,0)} f(x, m_2x, m_2x)$

$\therefore f(x, y, z)$ is discontinuous at $(0,0,0)$

Examples:

(I) Let $f(x, y, z) = \begin{cases} \dfrac{3xy}{x^2 + y^2 + z^2}, & (x, y, z) \neq 0 \\ 0, & (x, y, z) = 0 \end{cases}$.

Determine whether $f(x, y, z)$ is continuous at $(0,0,0)$.

Let $y = z = mx$ then $\dfrac{3xy}{x^2 + y^2 + z^2} = \dfrac{3mx^2}{x^2 + m^2x^2 + m^2x^2} = \dfrac{3m}{2m^2 + 1}$

Choose $m = -1$ then $\lim_{(x,y,z)\to(0,0,0)} \dfrac{3xy}{x^2 + y^2 + z^2} = \lim_{(x,y,z)\to(0,0,0)} \dfrac{3m}{2m^2 + 1} = -1$

Choose $m = 1$ then $\lim_{(x,y,z)\to(0,0,0)} \dfrac{3xy}{x^2 + y^2 + z^2} = \lim_{(x,y,z)\to(0,0,0)} \dfrac{3m}{2m^2 + 1} = 1$

$\therefore \lim_{(x,y,z)\to(0,0,0)} \dfrac{3xy}{x^2 + y^2 + z^2}$ does not exist $\quad \therefore f(x, y, z)$ is discontinuous at $(0,0,0)$

Example 1.

Let $f(x, y) = \begin{cases} (x + y) \sin\dfrac{1}{x} \sin\dfrac{1}{y}, & (x, y) \neq (0,0) \\ 0, & (x, y) = (0,0) \end{cases}$.

Determine whether $f(x, y)$ is continuous at $(0,0)$.

【Solution】

$\because (x + y)^2 = x^2 + y^2 + 2xy \leq (|x| + |y|)^2 = x^2 + y^2 + 2|xy|$ and $x^2 + y^2 \geq 2|xy|$

$\therefore (x + y)^2 \leq 2(x^2 + y^2)$

Let $\varepsilon > 0$ choose $\sqrt{2\delta} < \varepsilon$ then

$x^2 + y^2 < \delta \Rightarrow \left|(x + y)\sin\dfrac{1}{x}\sin\dfrac{1}{y}\right| \leq |x + y| < \sqrt{2(x^2 + y^2)} < \sqrt{2\delta} < \varepsilon$

$\therefore \lim_{(x,y)\to(0,0)} f(x, y) = 0 = f(0,0), \quad \therefore f(x, y)$ is continuous at $(0,0)$

Example 2.

Let $f(x,y) = \begin{cases} \dfrac{x^2 - y^2}{x^2 + y^2}, & (x,y) \neq 0 \\ 0, & (x,y) = 0 \end{cases}$.

Determine whether $f(x,y)$ is continuous at $(0,0)$.

【Solution】

Let $y = mx$ then $\dfrac{x^2 - y^2}{x^2 + y^2} = \dfrac{x^2 - m^2 x^2}{x^2 + m^2 x^2} = \dfrac{x^2(1 - m^2)}{x^2(1 + m^2)} = \dfrac{1 - m^2}{1 + m^2}$

$\therefore \lim\limits_{(x,y) \to (0,0)} \dfrac{x^2 - y^2}{x^2 + y^2} = \lim\limits_{(x,y) \to (0,0)} \dfrac{1 - m^2}{1 + m^2}$

Choose $m = \sqrt{2}$ then

$\lim\limits_{(x,y) \to (0,0)} f(x,y) = \lim\limits_{(x,y) \to (0,0)} \dfrac{x^2 - y^2}{x^2 + y^2} = \lim\limits_{(x,y) \to (0,0)} \dfrac{1 - m^2}{1 + m^2} = -\dfrac{1}{3} \neq f(0,0) = 0$

$\therefore f(x,y)$ is discontinuous at $(0,0)$

Example 3.

Let $f(x,y) = \begin{cases} \dfrac{x^2 y}{x^4 + y^2}, & (x,y) \neq 0 \\ 0, & (x,y) = 0 \end{cases}$.

Determine whether $f(x,y)$ is continuous at $(0,0)$.

【Solution】

Let $y = mx^2$ then $\dfrac{x^2 y}{x^4 + y^2} = \dfrac{x^2(m^2 x^2)}{x^4 + m^4 x^4} = \dfrac{m^2}{1 + m^4}$

$\therefore \lim\limits_{(x,y) \to (0,0)} \dfrac{x^2 y}{x^4 + y^2} = \lim\limits_{(x,y) \to (0,0)} \dfrac{m^2}{1 + m^4}$

Choose $m = 1$ then $\lim\limits_{(x,y) \to (0,0)} f(x,y) = \lim\limits_{(x,y) \to (0,0)} \dfrac{x^2 y}{x^4 + y^2} = \lim\limits_{(x,y) \to (0,0)} \dfrac{m^2}{1 + m^4} = \dfrac{1}{2} \neq f(0,0)$

$\therefore f(x,y)$ is discontinuous at $(0,0)$

Example 4.

Let $f(x,y) = \begin{cases} \dfrac{x^2 y}{x^3 + y^3}, & (x,y) \neq 0 \\ 0, & (x,y) = 0 \end{cases}$.

Determine whether $f(x,y)$ is continuous at $(0,0)$.

【Solution】

Let $y = mx$ then $\dfrac{x^2y}{x^3+y^3} = \dfrac{mx^3}{x^3+m^3x^3} = \dfrac{m}{1+m^3}$

$\therefore \lim\limits_{(x,y)\to(0,0)} \dfrac{x^2y}{x^3+y^3} = \lim\limits_{(x,y)\to(0,0)} \dfrac{m}{1+m^3}$

Choose $m = 1$ then $\lim\limits_{(x,y)\to(0,0)} f(x,y) = \lim\limits_{(x,y)\to(0,0)} \dfrac{x^2y}{x^3+y^3} = \lim\limits_{(x,y)\to(0,0)} \dfrac{m}{1+m^3} = \dfrac{1}{2} \neq f(0,0)$

$\therefore f(x,y)$ is discontinuous at $(0,0)$

Example 5.

Let $f(x,y) = \begin{cases} \dfrac{x^3 - y^3}{x^2 + y^2}, & (x,y) \neq 0 \\ 0, & (x,y) = 0 \end{cases}$.

Determine whether $f(x,y)$ is continuous at $(0,0)$.

【Solution】

$\because x^3 - y^3 = (x-y)(x^2 + xy + y^2)$ and $x^2 + y^2 \geq 2|xy|$

$\therefore |x^3 - y^3| \leq |x-y||x^2 + |xy| + y^2| \leq |x-y|\left|x^2 + \dfrac{x^2+y^2}{2} + y^2\right|$

$\therefore |x^3 - y^3| \leq \dfrac{3}{2}|x-y||x^2 + y^2| \Rightarrow \left|\dfrac{x^3-y^3}{x^2+y^2}\right| \leq \dfrac{3}{2}|x-y|$

$\because |x-y| = \sqrt{|x-y|^2} \leq \sqrt{x^2+y^2+|2xy|} \leq \sqrt{2(x^2+y^2)}$

Let $\varepsilon > 0$ choose $3\sqrt{\dfrac{\delta}{2}} < \varepsilon$ then

$x^2 + y^2 < \delta \Rightarrow \left|\dfrac{x^3-y^3}{x^2+y^2}\right| \leq \dfrac{3}{2}|x-y| \leq \dfrac{3}{2}\sqrt{2(x^2+y^2)} < 3\sqrt{\dfrac{\delta}{2}} < \varepsilon$

$\therefore f(x,y)$ is continuous at $(0,0)$.

Example 6.

Let $f(x,y) = \begin{cases} \dfrac{3x^2y}{x^2 + y^2}, & (x,y) \neq 0 \\ 0, & (x,y) = 0 \end{cases}$.

Determine whether $f(x,y)$ is continuous at $(0,0)$.

【Solution】

Claim: $\left|\dfrac{3x^2y}{x^2+y^2}\right| \le \left|\dfrac{3\sqrt{x^2+y^2}}{2}\right|$

$\because x^2+y^2 \ge 2|xy| \quad \therefore \dfrac{1}{x^2+y^2} \le \dfrac{1}{2|xy|} \Rightarrow \left|\dfrac{3x^2y}{x^2+y^2}\right| \le \left|\dfrac{3x^2y}{2xy}\right| = \left|\dfrac{3x}{2}\right| \le \left|\dfrac{3\sqrt{x^2+y^2}}{2}\right|$

Let $\varepsilon > 0$ choose $\delta < \left(\dfrac{2\varepsilon}{3}\right)^2$ then

$x^2+y^2 < \delta \Rightarrow \left|\dfrac{3x^2y}{x^2+y^2}\right| \le \left|\dfrac{3\sqrt{x^2+y^2}}{2}\right| < \dfrac{3\sqrt{\delta}}{2} < \dfrac{3}{2} \cdot \dfrac{2\varepsilon}{3} = \varepsilon$

$\therefore f(x,y)$ is continuous at $(0,0)$

Example 7.

Let $f(x,y) = \begin{cases} \dfrac{2xy^2}{x^2+y^2}, & (x,y) \ne 0 \\ 0, & (x,y) = 0 \end{cases}$.

Determine whether $f(x,y)$ is continuous at $(0,0)$.

【Solution】

Claim: $\left|\dfrac{2xy^2}{x^2+y^2}\right| \le \left|\sqrt{x^2+y^2}\right|$

$\because x^2+y^2 \ge 2|xy| \quad \therefore \dfrac{1}{x^2+y^2} \le \dfrac{1}{2|xy|} \Rightarrow \left|\dfrac{2xy^2}{x^2+y^2}\right| \le \left|\dfrac{2xy^2}{2xy}\right| = |y| < \left|\sqrt{x^2+y^2}\right|$

Let $\varepsilon > 0$ choose $\sqrt{\delta} < \varepsilon$ then

$x^2+y^2 < \delta \Rightarrow \left|\dfrac{2xy^2}{x^2+y^2}\right| \le \left|\sqrt{x^2+y^2}\right| < \sqrt{\delta} < \varepsilon \Rightarrow f(x,y)$ is continuous at $(0,0)$

Example 8.

Let $f(x,y) = \begin{cases} \dfrac{x^4+y^4}{x^2+y^2}, & (x,y) \ne 0 \\ 0, & (x,y) = 0 \end{cases}$.

Determine whether $f(x,y)$ is continuous at $(0,0)$.

【Solution】

Claim: $\left|\dfrac{x^4+y^4}{x^2+y^2}\right| \le \dfrac{3}{2}|x^2+y^2|$

7.2 Determine the Continuity of Multivariable Functions 28

$\because x^4 + y^4 = (x^2+y^2)^2 - 2x^2y^2 \quad \therefore \dfrac{x^4+y^4}{x^2+y^2} = \dfrac{(x^2+y^2)^2 - 2x^2y^2}{x^2+y^2} = x^2+y^2 - \dfrac{2x^2y^2}{x^2+y^2}$

$\because x^2+y^2 \ge 2|xy| \quad \therefore \dfrac{1}{x^2+y^2} \le \dfrac{1}{2|xy|} \Rightarrow \left|\dfrac{2x^2y^2}{x^2+y^2}\right| \le \left|\dfrac{2x^2y^2}{2xy}\right| = |xy| \le \left|\dfrac{x^2+y^2}{2}\right|$

$\therefore \left|\dfrac{x^4+y^4}{x^2+y^2}\right| = \left|x^2+y^2 - \dfrac{2x^2y^2}{x^2+y^2}\right| \le |x^2+y^2| + \left|\dfrac{2x^2y^2}{x^2+y^2}\right| \le \dfrac{3}{2}|x^2+y^2|$

Let $\varepsilon > 0$ choose $\dfrac{3\delta}{2} < \varepsilon$ then

$x^2+y^2 < \delta \Rightarrow \left|\dfrac{x^4+y^4}{x^2+y^2}\right| \le \dfrac{3}{2}|x^2+y^2| < \dfrac{3\delta}{2} < \varepsilon \Rightarrow f(x,y)$ is continuous at $(0,0)$

Example 9.

Let $f(x,y) = \begin{cases} \dfrac{xy^4}{x^2+y^8}, & (x,y) \neq 0 \\ 0, & (x,y) = 0 \end{cases}$.

Determine whether $f(x,y)$ is continuous at $(0,0)$.

【Solution】

Let $y = (mx)^{\frac{1}{4}}$ then $\dfrac{xy^4}{x^2+y^8} = \dfrac{mx^2}{(1+m^2)x^2} = \dfrac{m}{1+m^2}$

Choose $m = -1$ then $\lim\limits_{(x,y)\to(0,0)} \dfrac{xy^4}{x^2+y^8} = \lim\limits_{(x,y)\to(0,0)} \dfrac{m}{1+m^2} = -\dfrac{1}{2} \neq 0 = f(0,0)$

$\therefore f(x,y)$ is discontinuous at $(0,0)$

Example 10.

Let $f(x,y) = \begin{cases} \dfrac{x^2y^2}{\sqrt{x^2+y^2}}, & (x,y) \neq 0 \\ 0, & (x,y) = 0 \end{cases}$.

Determine whether $f(x,y)$ is continuous at $(0,0)$.

【Solution】

$\because x^2+y^2 \ge 2|xy| \quad \therefore \dfrac{1}{\sqrt{x^2+y^2}} \le \dfrac{1}{\sqrt{2|xy|}} \Rightarrow \left|\dfrac{x^2y^2}{\sqrt{x^2+y^2}}\right| \le \left|\dfrac{x^2y^2}{\sqrt{2|xy|}}\right| < |xy|^{\frac{3}{2}}$

Claim: $\lim\limits_{(x,y)\to(0,0)} \dfrac{x^2y^2}{\sqrt{x^2+y^2}} = 0$

Let $\varepsilon > 0$ choose $\delta^{\frac{3}{2}} < \varepsilon$ then

$$x^2+y^2<\delta \Rightarrow \left|\frac{x^2y^2}{\sqrt{x^2+y^2}}\right| \leq \left|\frac{x^2y^2}{\sqrt{2|xy|}}\right| < |xy|^{\frac{3}{2}} < (x^2+y^2)^{\frac{3}{2}} < \delta^{\frac{3}{2}} < \varepsilon$$

$$\therefore \lim_{(x,y)\to(0,0)} \frac{x^2y^2}{\sqrt{x^2+y^2}} = 0 \Rightarrow f(x,y) \text{ is continuous at } (0,0)$$

Example 11.

Let $f(x,y) = \begin{cases} \dfrac{(xy)^p}{\sqrt{x^2+y^2}}, & (x,y) \neq 0 \\ 0, & (x,y) = 0 \end{cases}$.

Determine whether $f(x,y)$ is continuous at $(0,0)$, $\forall p > \dfrac{1}{2}$.

【Solution】

Let $p > \dfrac{1}{2}$

$$\because x^2+y^2 \geq 2|xy| \quad \therefore \frac{1}{\sqrt{x^2+y^2}} \leq \frac{1}{\sqrt{2|xy|}} \Rightarrow \left|\frac{(xy)^p}{\sqrt{x^2+y^2}}\right| \leq \left|\frac{(xy)^p}{\sqrt{2|xy|}}\right| < |xy|^{p-\frac{1}{2}}$$

Claim: $\lim_{(x,y)\to(0,0)} \dfrac{(xy)^p}{\sqrt{x^2+y^2}} = 0$

Let $\varepsilon > 0$ choose $\delta^{p-\frac{1}{2}} < \varepsilon$ then

$$x^2+y^2 < \delta \Rightarrow \left|\frac{(xy)^p}{\sqrt{x^2+y^2}}\right| \leq \left|\frac{(xy)^p}{\sqrt{2|xy|}}\right| < |xy|^{p-\frac{1}{2}} < (x^2+y^2)^{p-\frac{1}{2}} < \delta^{p-\frac{1}{2}} < \varepsilon$$

$$\therefore \lim_{(x,y)\to(0,0)} \frac{(xy)^p}{\sqrt{x^2+y^2}} = 0 \Rightarrow f(x,y) \text{ is continuous at } (0,0)$$

Example 12.

Let $f(x,y) = \begin{cases} \dfrac{x^2+y^2}{|x|+|y|}, & (x,y) \neq 0 \\ 0, & (x,y) = 0 \end{cases}$.

Determine whether $f(x,y)$ is continuous at $(0,0)$.

【Solution】

$\because (|x|+|y|)^2 = |x|^2+|y|^2+2|xy| \geq x^2+y^2 \quad \therefore |x|+|y| \geq \sqrt{x^2+y^2}$

$\therefore \dfrac{x^2+y^2}{|x|+|y|} \leq \dfrac{x^2+y^2}{\sqrt{x^2+y^2}} = \sqrt{x^2+y^2}$

7.2 Determine the Continuity of Multivariable Functions

Claim: $\lim\limits_{(x,y)\to(0,0)} \dfrac{x^2+y^2}{|x|+|y|} = 0$

Let $\varepsilon > 0$ choose $\delta^{\frac{1}{2}} < \varepsilon$ then $x^2+y^2 < \delta \Rightarrow \dfrac{x^2+y^2}{|x|+|y|} \leq \dfrac{x^2+y^2}{\sqrt{x^2+y^2}} = \sqrt{x^2+y^2} < \delta^{\frac{1}{2}} < \varepsilon$

$\therefore \lim\limits_{(x,y)\to(0,0)} \dfrac{x^2+y^2}{|x|+|y|} = 0 \Rightarrow f(x,y)$ is continuous at $(0,0)$

Example 13.

Let $f(x,y) = \begin{cases} \dfrac{xy\cos(x^2+y^2)}{\sqrt{x^2+y^2}}, & (x,y) \neq 0 \\ 0, & (x,y) = 0 \end{cases}$.

Determine whether $f(x,y)$ is continuous at $(0,0)$.

【Solution】

$\because x^2+y^2 \geq 2|xy| \quad \therefore \dfrac{1}{\sqrt{x^2+y^2}} \leq \dfrac{1}{\sqrt{2|xy|}} \Rightarrow \left|\dfrac{xy\cos x^2+y^2}{\sqrt{x^2+y^2}}\right| \leq \left|\dfrac{xy}{\sqrt{2|xy|}}\right| < |xy|^{\frac{1}{2}}$

Claim: $\lim\limits_{(x,y)\to(0,0)} \dfrac{xy\cos(x^2+y^2)}{\sqrt{x^2+y^2}} = 0$

Let $\varepsilon > 0$ choose $\delta^{\frac{1}{2}} < \varepsilon$ then

$x^2+y^2 < \delta \Rightarrow \left|\dfrac{xy\cos(x^2+y^2)}{\sqrt{x^2+y^2}}\right| \leq \left|\dfrac{xy}{\sqrt{2|xy|}}\right| < |xy|^{\frac{1}{2}} < (x^2+y^2)^{\frac{1}{2}} < \delta^{\frac{1}{2}} < \varepsilon$

$\therefore \lim\limits_{(x,y)\to(0,0)} \dfrac{xy\cos(x^2+y^2)}{\sqrt{x^2+y^2}} = 0 \Rightarrow f(x,y)$ is continuous at $(0,0)$

Example 14.

Let $f(x,y,z) = \begin{cases} \dfrac{\sin(x^2+y^2+z^2)}{\sqrt{x^2+y^2+z^2}}, & (x,y,z) \neq 0 \\ 0, & (x,y,z) = 0 \end{cases}$.

Determine whether $f(x,y,z)$ is continuous at $(0,0,0)$.

【Solution】

Let $t = x^2+y^2+z^2$ then $\dfrac{\sin(x^2+y^2+z^2)}{\sqrt{x^2+y^2+z^2}} = \dfrac{\sin t}{\sqrt{t}} = \dfrac{\sqrt{t}\sin t}{t}$

$$\lim_{(x,y,z)\to(0,0,0)} \frac{\sin(x^2+y^2+z^2)}{\sqrt{x^2+y^2+z^2}} = \lim_{t\to 0} \frac{\sqrt{t}\sin t}{t}$$

By L'Hôpital's Rule, $\lim_{t\to 0}\dfrac{\sin t}{t} = \lim_{t\to 0}\dfrac{\cos t}{1} = 1 \Rightarrow \lim_{t\to 0}\dfrac{\sqrt{t}\sin t}{t} = 0$

$\therefore \lim_{(x,y,z)\to(0,0,0)} \dfrac{\sin(x^2+y^2+z^2)}{\sqrt{x^2+y^2+z^2}} = 0 \Rightarrow f(x,y,z)$ is continuous at $(0,0,0)$

Example 15.

Let $f(x,y,z) = \begin{cases} \dfrac{xyz}{x^2+y^4+z^4} & (x,y,z) \neq 0 \\ 0, & (x,y,z) = 0 \end{cases}$.

Determine whether $f(x,y,z)$ is continuous at $(0,0,0)$.

【Solution】

Let $x = mt^2$, $y = z = t$ then $\dfrac{mt^4}{(m^2+2)t^4} = \dfrac{m}{m^2+2}$

Choose $m = 1$ then $\lim_{(x,y,z)\to(0,0,0)} \dfrac{xyz}{x^2+y^4+z^4} = \lim_{(x,y,z)\to(0,0,0)} \dfrac{m}{m^2+2} = \dfrac{1}{3} \neq 0 = f(0,0,0)$

$\therefore f(x,y,z)$ is discontinuous at $(0,0,0)$

Example 16.

Let $f(x,y,z) = \begin{cases} \dfrac{x^2y}{x^2+y^4+z^4} & (x,y,z) \neq 0 \\ 0, & (x,y,z) = 0 \end{cases}$.

Determine whether $f(x,y,z)$ is continuous at $(0,0,0)$.

【Solution】

$\because x^2+y^4+z^4 \geq x^2 \quad \therefore \left|\dfrac{x^2y}{x^2+y^4+z^4}\right| \leq \left|\dfrac{x^2y}{x^2}\right| = |y|$

Let $\varepsilon > 0$ choose $\delta^{\frac{1}{4}} < \varepsilon$ then

$x^2+y^4+z^4 < \delta \Rightarrow \left|\dfrac{x^2y}{x^2+y^4+z^4}\right| < |y| < (x^2+y^4+z^4)^{\frac{1}{4}} < \delta^{\frac{1}{4}} < \varepsilon$

$\therefore \lim_{(x,y,z)\to(0,0,0)} \dfrac{x^2y}{x^2+y^4+z^4} = 0 \Rightarrow f(x,y,z)$ is continuous at $(0,0,0)$

Example 17.

Let $f(x, y, z) = \begin{cases} \dfrac{3xy + xz + z^2}{x^2 + 2y^2}, & (x, y, z) \neq 0 \\ 0, & (x, y, z) = 0 \end{cases}$.

Determine whether $f(x, y, z)$ is continuous at $(0,0,0)$.

【Solution】

Let $z = y = mx$ then $\dfrac{3mx^2 + mx^2 + m^2x^2}{(1 + 2m^2)x^2} = \dfrac{4m + m^2}{1 + 2m^2}$

choose $m = 1$ then $\lim\limits_{(x,y,z) \to (0,0,0)} \dfrac{3xy + xz + z^2}{x^2 + 2y^2} = \lim\limits_{(x,y,z) \to (0,0,0)} \dfrac{4m + m^2}{1 + 2m^2} = \dfrac{5}{3} \neq f(0,0,0) = 0$

∴ $f(x, y, z)$ is discontinuous at $(0,0,0)$

7.3 Calculate Partial Derivatives of Multivariable Functions

This section introduces how to find the partial derivatives of multivariable functions. It is important to note that in the next section, where we determine whether a multivariable function is differentiable, the values of the partial derivatives will be needed. Therefore, readers should practice partial derivative examples extensively to become more efficient in determining differentiability.

【Definition】 Definition of the partial derivative of a function of two variables

(i) A function $f(x, y)$ is partially differentiable with respect to x at (x_0, y_0)

$\Leftrightarrow \lim\limits_{h \to 0} \dfrac{f(h + x_0, y_0) - f(x_0, y_0)}{h}$ exists.

If this limit exists, it is denoted by $\dfrac{\partial f}{\partial x}(x_0, y_0)$ or $f_x(x_0, y_0)$.

(ii) A function $f(x, y)$ is partially differentiable with respect to y at (x_0, y_0)

$\Leftrightarrow \lim\limits_{h \to 0} \dfrac{f(x_0, y_0 + h) - f(x_0, y_0)}{h}$ exists.

If this limit exists, it is denoted by $\dfrac{\partial f}{\partial y}(x_0, y_0)$ or $f_y(x_0, y_0)$.

【Definition】 Definition of partial derivatives of multivariable functions

$f(x_1, \dots, x_n)$ is partially differentiable with respect to x_j at $(c_1, \dots, c_n)$

$\Leftrightarrow \lim\limits_{h \to 0} \dfrac{f(c_1, c_2, \dots, h + c_j, \dots, c_n) - f(c_1, c_2, \dots, c_j, \dots, c_n)}{h}$ exists, $1 \leq j \leq n$.

If this limit exists, it is denoted by $\dfrac{\partial f}{\partial x_j}(c_1, \ldots, c_n)$ or $f_{x_j}(c_1, \ldots, c_n)$

In the types of exam questions involving partial derivatives of multivariable functions, problems involving functions of two variables are the most common. If readers are familiar with these two-variable problems, they will be able to handle other types as well.

Question Types:

Type 1.

Given a function $f(x, y)$, find $\dfrac{\partial f}{\partial x}(x_0, y_0) = ?$ and $\dfrac{\partial f}{\partial y}(x_0, y_0) = ?$

Problem-Solving Process:

Step 1.

Use the definition to find

$$\lim_{h \to 0} \dfrac{f(h + x_0, y_0) - f(x_0, y_0)}{h} = ?$$

and

$$\lim_{h \to 0} \dfrac{f(x_0, y_0 + h) - f(x_0, y_0)}{h} = ?$$

Step 2.

If $\lim\limits_{h \to 0} \dfrac{f(h + x_0, y_0) - f(x_0, y_0)}{h}$ exists, then $\dfrac{\partial f}{\partial x}(x_0, y_0) = \lim\limits_{h \to 0} \dfrac{f(h + x_0, y_0) - f(x_0, y_0)}{h}$

If $\lim\limits_{h \to 0} \dfrac{f(x_0, y_0 + h) - f(x_0, y_0)}{h}$ exists, then $\dfrac{\partial f}{\partial y}(x_0, y_0) = \lim\limits_{h \to 0} \dfrac{f(x_0, y_0 + h) - f(x_0, y_0)}{h}$

If $\lim\limits_{h \to 0} \dfrac{f(h + x_0, y_0) - f(x_0, y_0)}{h}$ does not exist, then

$\dfrac{\partial f}{\partial x}(x_0, y_0)$ does not exist, i. e., the partial derivative does not exist.

If $\lim\limits_{h \to 0} \dfrac{f(x_0, y_0 + h) - f(x_0, y_0)}{h}$ does not exist, then

$\dfrac{\partial f}{\partial y}(x_0, y_0)$ does not exist, i. e., the partial derivative does not exist.

Type 2.

Given a function $f(x, y)$, find $f_{xx}(x_0, y_0)$, $f_{xy}(x_0, y_0)$, $f_{yx}(x_0, y_0)$, and $f_{yy}(x_0, y_0)$.
Below is the method for finding $f_{yx}(x_0, y_0)$ and the methods for the other three are similar.

Problem-Solving Process:

Step 1.

7.3 Calculate Partial Derivatives of Multivariable Functions

$$f_{yx}(0,0) := \lim_{h \to 0} \frac{f_y(h,0) - f_y(0,0)}{h}$$

Step2.

Find $f_y(h,0) =?$ and $f_y(0,0) =?$

where $f_y(h,0) = \lim_{t \to 0} \frac{f(h, 0+t) - f(h,0)}{t}$ and $f_y(0,0) = \lim_{t \to 0} \frac{f(0, 0+t) - f(0,0)}{t}$

Step3.

Substitute $f_y(h,0)$ and $f_y(0,0)$ into Step1 to find $\lim_{h \to 0} \frac{f_y(h,0) - f_y(0,0)}{h} =?$

Example 1.

(1) Let $f(x,y) = 3x - x^2y^2 + 2x^3y + y$. Find $\frac{\partial f}{\partial x}(x,y)$, $\frac{\partial f}{\partial y}(x,y) =?$

(2) Let $f(x,y) = (x^2y + xy + x)^2$. Find $\frac{\partial f}{\partial x}(x,y) =?$

【Solution】

(1)

$$\frac{\partial f}{\partial x} = 3 - 2xy^2 + 6x^2y, \quad \frac{\partial f}{\partial y} = -2x^2y + 2x^3 + 1$$

(2)

$$\frac{\partial f}{\partial x} = 2(x^2y + xy + x)(2xy + y + 1)$$

Example 2.

Let $f(x,y) = \tan^{-1}\frac{y}{x}$. Find $\frac{\partial f}{\partial x}(3,1) =?$, $\frac{\partial f}{\partial y}(2,2) =?$

【Solution】

$$\frac{\partial f}{\partial x}(3,1) = \lim_{h \to 0} \frac{f(h+3, 1) - f(3,1)}{h} = \lim_{h \to 0} \frac{\tan^{-1}\frac{1}{3+h} - \tan^{-1}\frac{1}{3}}{h} = \lim_{h \to 0} \frac{-(3+h)^{-2}}{1 + \left(\frac{1}{3+h}\right)^2} = \frac{-1}{10}$$

$$\frac{\partial f}{\partial y}(2,2) = \lim_{h \to 0} \frac{f(2, 2+h) - f(2,2)}{h} = \lim_{h \to 0} \frac{\tan^{-1}\frac{2+h}{2} - \tan^{-1} 1}{h}$$

$$= \lim_{h \to 0} \frac{\frac{1}{2}}{1 + \left(\frac{2+h}{2}\right)^2} = \frac{1}{4}$$

Example 3.

Let $f(x,y) = \int_x^{y^3} \cos t \, dt$. Find $\dfrac{\partial f}{\partial x}(x,y) - \dfrac{\partial f}{\partial y}(x,y) = ?$

【Solution】

$\dfrac{\partial f}{\partial x} - \dfrac{\partial f}{\partial y} = -\cos x - 3y^2 \cos y^3$

Example 4.

Let $f(x,y) = \sin(x^4 + x^2 y^2) + \cos(xy + y^3)$. Find $f_x = ?$

【Solution】

$f_x = \cos(x^3 + x^2 y)(4x^3 + 2xy^2) - y\sin(xy + y^3)$

Example 5.

Let $f(x,y,z) = x^{y^z}$ $(x > 0, y > 0, z > 0)$. Find $f_x = ?, \ f_y = ?, \ f_z = ?$

【Solution】

$\because f(x,y,z) = x^{y^z} = e^{y^z \ln x}$ $\ \therefore f_x(x,y,z) = x^{y^z - 1} \cdot y^z$

$\therefore f_y(x,y,z) = e^{y^z \ln x}(z\, y^{z-1} \ln x) = x^{y^z}(z\, y^{z-1} \ln x)$

$\therefore f_z(x,y,z) = e^{y^z \ln x}(y^z \ln x \ln y)$

Example 6.

Let $f(x,y) = x^3 y + e^{xy^2}$. Find $f_x = ?, \ f_y = ?, \ f_{xy} = ?, \ f_{yx} = ?$

【Solution】

$f_x(x,y) = 3x^2 y + y^2 e^{xy^2}, \ f_y(x,y) = x^3 + 2xy e^{xy^2}$

$f_{xy}(x,y) = 3x^2 + 2y e^{xy^2} + 2xy^3 e^{xy^2}, \ f_{yx}(x,y) = 3x^2 + 2y e^{xy^2} + 2xy^3 e^{xy^2}$

Example 7.

Let $f(x,y) = x^2 \tan^{-1} \dfrac{y}{x}$. Find $\dfrac{\partial^2 f}{\partial x \partial y}(1,0) = ?$

【Solution】

7.3 Calculate Partial Derivatives of Multivariable Functions

$$\therefore \frac{\partial f}{\partial y}(x,y) = \frac{x}{1+\left(\frac{y}{x}\right)^2} = \frac{x^3}{x^2+y^2}, \quad \frac{\partial^2 f}{\partial x \partial y}(x,y) = \frac{3x^2(x^2+y^2) - x^3(2x)}{(x^2+y^2)^2} = \frac{x^4 + 3x^2y^2}{(x^2+y^2)^2}$$

$$\therefore \frac{\partial^2 f}{\partial x \partial y}(1,0) = 1$$

Example 8.

Let $f(x,y) = x^2 - 2x + y^2 + 2y + 3$. Find $\frac{\partial f}{\partial x}(x,y) = ?$, $\frac{\partial f}{\partial y}(x,y) = ?$

【Solution】

$$\frac{\partial f}{\partial x}(x,y) = 2x - 2, \quad \frac{\partial f}{\partial y}(x,y) = 2y + 2$$

Example 9.

Let $f(x,y) = x^2 e^{2y} - \cos\frac{x}{y} + x^2 y^3$. Find $f_{xy}(x,y) = ?$, $f_{yy}(x,y) = ?$

【Solution】

$$\frac{\partial f}{\partial x}(x,y) = 2xe^{2y} + \frac{1}{y}\sin\frac{x}{y} + 2xy^3, \quad \frac{\partial f}{\partial y}(x,y) = 2x^2 e^{2y} + \left(\frac{-x}{y^2}\right)\sin\frac{x}{y} + 3x^2 y^2$$

$$f_{xy}(x,y) = \frac{\partial}{\partial y}\left(\frac{\partial f(x,y)}{\partial x}\right) = \frac{\partial}{\partial y}\left(2xe^{2y} + \frac{1}{y}\sin\frac{x}{y} + 2xy^3\right)$$

$$= 4xe^{2y} - \frac{1}{y^2}\sin\frac{x}{y} - \frac{x}{y^3}\cos\frac{x}{y} + 6xy^2$$

$$f_{yy}(x,y) = \frac{\partial}{\partial y}\left(\frac{\partial f(x,y)}{\partial y}\right) = \frac{\partial}{\partial y}\left(2x^2 e^{2y} + \left(\frac{-x}{y^2}\right)\sin\frac{x}{y} + 3x^2 y^2\right)$$

$$= 4x^2 e^{2y} + (2xy^3)\sin\frac{x}{y} + \left(\frac{x^2}{y^4}\right)\cos\frac{x}{y} + 6x^2 y$$

Example 10.

Let $f(x,y) = \sin x^2 + y^2$. Find $f_{xx}(x,y) = ?$, $f_{yx}(x,y) = ?$, $f_{yy}(x,y) = ?$

【Solution】

$f_x(x,y) = 2x\cos(x^2 + y^2), \quad f_y(x,y) = 2y\cos(x^2 + y^2)$

$$f_{xx}(x,y) = \frac{\partial}{\partial x}\left(\frac{\partial f(x,y)}{\partial x}\right) = \frac{\partial}{\partial x}(2x\cos(x^2+y^2)) = 2\cos(x^2+y^2) - 4x^2\sin(x^2+y^2)$$

$$f_{yx}(x,y) = \frac{\partial}{\partial x}\left(\frac{\partial f(x,y)}{\partial y}\right) = \frac{\partial}{\partial x}(2y\cos(x^2+y^2)) = -4xy\sin(x^2+y^2)$$

$$f_{yy}(x,y) = \frac{\partial}{\partial y}\left(\frac{\partial f(x,y)}{\partial y}\right) = \frac{\partial}{\partial y}(2y\cos(x^2+y^2)) = 2\cos(x^2+y^2) - 4y^2\sin(x^2+y^2)$$

Example 11.

Let $f(x,y) = e^{ny}\sin nx$. Find $f_{xx} + f_{yy}$ and prove $f_{xy} = f_{yx}$.

【Solution】

Claim: $f_{xx} + f_{yy} = 0$

∵ $f_x = ne^{ny}\cos nx$ and $f_y = ne^{ny}\sin nx$

∴ $f_{xx} = \dfrac{\partial f_x}{\partial x} = -n^2 e^{ny}\sin nx$ and $f_{yy} = \dfrac{\partial f_y}{\partial y} = n^2 e^{ny}\sin nx \Rightarrow f_{xx} + f_{yy} = 0$

Claim: $f_{xy} = f_{yx}$

∵ $f_{xy} = \dfrac{\partial f_x}{\partial y} = n^2 e^{ny}\cos nx$ and $f_{yx} = \dfrac{\partial f_y}{\partial x} = n^2 e^{ny}\cos nx$ ∴ $f_{xy} = f_{yx}$

Example 12.

Let $f(x,y) = \dfrac{x}{x^2+y^2}$. Find $\dfrac{\partial^2 f}{\partial x^2}(x,y) + \dfrac{\partial^2 f}{\partial y^2}(x,y) = ?$

【Solution】

$$\frac{\partial f}{\partial x}(x,y) = \frac{\partial}{\partial x}(x(x^2+y^2)^{-1}) = (x^2+y^2)^{-1} - 2x^2(x^2+y^2)^{-2}$$

$$\frac{\partial f}{\partial y}(x,y) = \frac{\partial}{\partial y}(x(x^2+y^2)^{-1}) = -2xy(x^2+y^2)^{-2}$$

$$\frac{\partial^2 f}{\partial x^2}(x,y) = \frac{\partial}{\partial x}\left(\frac{\partial f}{\partial x}(x,y)\right) = -2x(x^2+y^2)^{-2} - 4x(x^2+y^2)^{-2} + 8x^3(x^2+y^2)^{-3}$$

$$\frac{\partial^2 f}{\partial y^2}(x,y) = \frac{\partial}{\partial y}\left(\frac{\partial f}{\partial y}(x,y)\right) = -2x(x^2+y^2)^{-2} + 8xy^2(x^2+y^2)^{-3}$$

∵ $8x^3(x^2+y^2)^{-3} + 8xy^2(x^2+y^2)^{-3} = 8x(x^2+y^2)^{-2}$ ∴ $\dfrac{\partial^2 f(x,y)}{\partial x^2} + \dfrac{\partial^2 f(x,y)}{\partial y^2} = 0$

Example 13.

7.3 Calculate Partial Derivatives of Multivariable Functions

Let $f(x,y) = \dfrac{e^{-\sqrt{x^2+y^2}}}{\sqrt{x^2+y^2}}$. Find $\dfrac{\partial f}{\partial x}(x,y) = ?$

【Solution】

Let $r = \sqrt{x^2+y^2}$ then $f(r) = \dfrac{e^{-r}}{r}$

$\dfrac{\partial f}{\partial x}(x,y) = \dfrac{\partial f(r)}{\partial r} \cdot \dfrac{\partial r}{\partial x}, \quad \dfrac{\partial f(r)}{\partial r} = -\dfrac{e^{-r}}{r} - \dfrac{e^{-r}}{r^2}, \quad \dfrac{\partial r}{\partial x} = x(x^2+y^2)^{-\frac{1}{2}} = \dfrac{x}{r}$

$\dfrac{\partial f}{\partial x}(x,y) = \dfrac{\partial f(r)}{\partial r} \cdot \dfrac{\partial r}{\partial x} = -\dfrac{xe^{-\sqrt{x^2+y^2}}}{x^2+y^2} - \dfrac{xe^{-\sqrt{x^2+y^2}}}{(x^2+y^2)^{\frac{3}{2}}}$

Example 14.

Let $f(x,y,z) = (x^2+y^2+z^2)^{-\frac{1}{2}}$. Prove $\dfrac{\partial^2 f}{\partial x^2} + \dfrac{\partial^2 f}{\partial y^2} + \dfrac{\partial^2 f}{\partial z^2} = 0$.

【Solution】

$\because \dfrac{\partial f}{\partial x} = -\dfrac{2x}{2}(x^2+y^2+z^2)^{-\frac{3}{2}} = -x(x^2+y^2+z^2)^{-\frac{3}{2}}$

$\dfrac{\partial f}{\partial y} = -y(x^2+y^2+z^2)^{-\frac{3}{2}}$ and $\dfrac{\partial f}{\partial z} = -z(x^2+y^2+z^2)^{-\frac{3}{2}}$

$\therefore \dfrac{\partial^2 f}{\partial x^2} = -(x^2+y^2+z^2)^{-\frac{3}{2}} + \dfrac{3x}{2}(x^2+y^2+z^2)^{-\frac{5}{2}}(2x)$

$= -(x^2+y^2+z^2)^{-\frac{3}{2}} + 3x^2(x^2+y^2+z^2)^{-\frac{5}{2}}$

$\dfrac{\partial^2 f}{\partial y^2} = -(x^2+y^2+z^2)^{-\frac{3}{2}} + \dfrac{3y}{2}(x^2+y^2+z^2)^{-\frac{5}{2}}(2y)$

$= -(x^2+y^2+z^2)^{-\frac{3}{2}} + 3y^2(x^2+y^2+z^2)^{-\frac{5}{2}}$

$\dfrac{\partial^2 f}{\partial z^2} = -(x^2+y^2+z^2)^{-\frac{3}{2}} + \dfrac{3z}{2}(x^2+y^2+z^2)^{-\frac{5}{2}}(2z)$

$= -(x^2+y^2+z^2)^{-\frac{3}{2}} + 3z^2(x^2+y^2+z^2)^{-\frac{5}{2}}$

$\therefore \dfrac{\partial^2 f}{\partial x^2} + \dfrac{\partial^2 f}{\partial y^2} + \dfrac{\partial^2 f}{\partial z^2}$

$= -3(x^2+y^2+z^2)^{-\frac{3}{2}} + 3x^2(x^2+y^2+z^2)^{-\frac{5}{2}} + 3y^2(x^2+y^2+z^2)^{-\frac{5}{2}}$

$+ 3z^2(x^2+y^2+z^2)^{-\frac{5}{2}}$

$= -3(x^2+y^2+z^2)^{-\frac{3}{2}} + 3(x^2+y^2+z^2)^{-\frac{5}{2}}(x^2+y^2+z^2)$

$$= -3(x^2 + y^2 + z^2)^{-\frac{3}{2}} + 3(x^2 + y^2 + z^2)^{-\frac{3}{2}} = 0$$

Example 15.

(1) Let $f(x,y) = (x^4 + y^4)^{\frac{1}{4}}$. Find $\dfrac{\partial f}{\partial x}(0,0)$, $\dfrac{\partial f}{\partial y}(0,0) =?$

(2) Let $f(x,y) = \begin{cases} \dfrac{\tan(2x+y)}{2x+y}, & (x,y) \neq (0,0) \\ 0, & (x,y) = (0,0) \end{cases}$. Find $\dfrac{\partial f}{\partial x}(0,0) =?, \dfrac{\partial f}{\partial y}(0,0) =?$

【Solution】

(1)

$$\because \frac{\partial f}{\partial x}(0,0) = \lim_{h \to 0} \frac{f(h+0,0) - f(0,0)}{h} = \lim_{h \to 0} \frac{(h^4 + 0^4)^{\frac{1}{4}} - 0}{h} = \lim_{h \to 0} \frac{|h|}{h} = \pm 1$$

$$\frac{\partial f}{\partial y}(0,0) = \lim_{h \to 0} \frac{f(0, h+0) - f(0,0)}{h} = \lim_{h \to 0} \frac{(h^4 + 0^4)^{\frac{1}{4}} - 0}{h} = \lim_{h \to 0} \frac{|h|}{h} = \pm 1$$

$$\therefore \frac{\partial f}{\partial x}(0,0) \text{ and } \frac{\partial f}{\partial y}(0,0) \text{ do not exist}$$

(2)

$$\because \frac{\partial f}{\partial x}(0,0) = \lim_{h \to 0} \frac{f(h+0,0) - f(0,0)}{h} = \lim_{h \to 0} \frac{\frac{\tan 2h}{2h} - 0}{h} = \lim_{h \to 0} \frac{1 - 0}{h} = \pm\infty$$

$$\frac{\partial f}{\partial y}(0,0) = \lim_{h \to 0} \frac{f(0, h+0) - f(0,0)}{h} = \lim_{h \to 0} \frac{\frac{\tan h}{h} - 0}{h} = \lim_{h \to 0} \frac{1 - 0}{h} = \pm\infty$$

$$\therefore \frac{\partial f}{\partial x}(0,0) \text{ and } \frac{\partial f}{\partial y}(0,0) \text{ do not exist}$$

Example 16.

Let $\dfrac{\partial f}{\partial x} = x^2 + y^2$, $\dfrac{\partial f}{\partial y} = 2xy$ and $f(0,1) = 5$. Find $f(x,y) =?$

【Solution】

$\because \dfrac{\partial f}{\partial x} = x^2 + y^2 \quad \therefore f(x,y) = \dfrac{x^3}{3} + xy^2 + c$

$\because \dfrac{\partial f}{\partial y} = 2xy \quad \therefore f(x,y) = xy^2 + c \quad \therefore f(x,y) = \dfrac{x^3}{3} + xy^2 + c$

$\because f(0,1) = 5 \quad \therefore c = 5 \quad \therefore f(x,y) = \dfrac{x^3}{3} + xy^2 + 5$

Example 17.

Let $f(x,y) = \begin{cases} \sqrt{xy}, & (x,y) \neq 0 \\ 0, & (x,y) = 0 \end{cases}$. Find $\dfrac{\partial f}{\partial x}(0,0) =?$, $\dfrac{\partial f}{\partial y}(0,0) =?$

【Solution】

$f_x(0,0) = \lim\limits_{h \to 0} \dfrac{f(h+0,0) - f(0,0)}{h} = \lim\limits_{h \to 0} \dfrac{\sqrt{h \cdot 0} - 0}{h} = 0$

$f_y(0,0) = \lim\limits_{h \to 0} \dfrac{f(0,h+0) - f(0,0)}{h} = \lim\limits_{h \to 0} \dfrac{\sqrt{h \cdot 0} - 0}{h} = 0$

Example 18.

Let $f(x,y) = \begin{cases} \dfrac{x^2 - y^2}{\sqrt{x^2 + y^2}}, & (x,y) \neq 0 \\ 0, & (x,y) = 0 \end{cases}$. Find $\dfrac{\partial f}{\partial x}(0,0) =?$, $\dfrac{\partial f}{\partial y}(0,0) =?$

【Solution】

$f_x(0,0) = \lim\limits_{h \to 0} \dfrac{f(h+0,0) - f(0,0)}{h} = \lim\limits_{h \to 0} \dfrac{\frac{h^2}{h} - 0}{h} = 1$

$f_y(0,0) = \lim\limits_{h \to 0} \dfrac{f(0,h+0) - f(0,0)}{h} = \lim\limits_{h \to 0} \dfrac{\frac{-h^2}{h} - 0}{h} = -1$

Example 19.

Let $f(x,y) = \begin{cases} \dfrac{x^3 - y^3}{x^2 + y^2}, & (x,y) \neq 0 \\ 0, & (x,y) = 0 \end{cases}$. Find $\dfrac{\partial f}{\partial x}(0,0) =?$, $\dfrac{\partial f}{\partial y}(0,0) =?$

【Solution】

$f_x(0,0) = \lim\limits_{h \to 0} \dfrac{f(h+0,0) - f(0,0)}{h} = \lim\limits_{h \to 0} \dfrac{\frac{h^3}{h^2} - 0}{h} = 1$

$f_y(0,0) = \lim\limits_{h \to 0} \dfrac{f(0,h+0) - f(0,0)}{h} = \lim\limits_{h \to 0} \dfrac{\frac{-h^3}{h^2} - 0}{h} = -1$

Example 20.

(1) Let $f(x,y) = \begin{cases} \dfrac{2x^2y}{x^5+y^2}, & (x,y) \neq 0 \\ 0, & (x,y) = 0 \end{cases}$. Find $\dfrac{\partial f}{\partial x}(0,0) =?$, $\dfrac{\partial f}{\partial y}(0,0) =?$

(2) Let $f(x,y) = \begin{cases} \dfrac{3x^2y}{x^4+y^3}, & (x,y) \neq 0 \\ 0, & (x,y) = 0 \end{cases}$. Find $\dfrac{\partial f}{\partial x}(0,0) =?$, $\dfrac{\partial f}{\partial y}(0,0) =?$

【Solution】

(1)

$\because \dfrac{\partial f}{\partial x}(0,0) = \lim\limits_{h \to 0} \dfrac{f(h+0,0) - f(0,0)}{h} = \lim\limits_{h \to 0} \dfrac{\dfrac{2h^2 \cdot 0}{h^5 + 0^2} - 0}{h} = \lim\limits_{h \to 0} \dfrac{0}{h^6} = 0$

$\dfrac{\partial f}{\partial y}(0,0) = \lim\limits_{h \to 0} \dfrac{f(0,h+0) - f(0,0)}{h} = \lim\limits_{h \to 0} \dfrac{\dfrac{2 \cdot 0^2 \cdot h}{0^5 + h^2} - 0}{h} = \lim\limits_{h \to 0} \dfrac{0}{h^3} = 0$

$\therefore \dfrac{\partial f}{\partial x}(0,0) = 0, \ \dfrac{\partial f}{\partial y}(0,0) = 0$

(2)

$\because \dfrac{\partial f}{\partial x}(0,0) = \lim\limits_{h \to 0} \dfrac{f(h+0,0) - f(0,0)}{h} = \lim\limits_{h \to 0} \dfrac{\dfrac{3 \cdot h^2 \cdot 0}{h^4 + 0^3} - 0}{h} = \lim\limits_{h \to 0} \dfrac{0}{h^3} = 0$

$\dfrac{\partial f}{\partial y}(0,0) = \lim\limits_{h \to 0} \dfrac{f(0,h+0) - f(0,0)}{h} = \lim\limits_{h \to 0} \dfrac{\dfrac{3 \cdot 0^2 \cdot h}{0^4 + h^3} - 0}{h} = \lim\limits_{h \to 0} \dfrac{0}{h^4} = 0$

$\therefore \dfrac{\partial f}{\partial x}(0,0) = 0, \ \dfrac{\partial f}{\partial y}(0,0) = 0$

Example 21.

(1) Let $f(x,y) = \begin{cases} \dfrac{x^5 - y^5}{x^5 + y^5}, & (x,y) \neq 0 \\ 0, & (x,y) = 0 \end{cases}$. Find $\dfrac{\partial f}{\partial x}(0,0) =?$, $\dfrac{\partial f}{\partial y}(0,0) =?$

(2) Let $f(x,y) = \begin{cases} \dfrac{5x^4 - 2y^3}{x^2 + 3y^2}, & (x,y) \neq 0 \\ 0, & (x,y) = 0 \end{cases}$. Find $\dfrac{\partial f}{\partial x}(0,0) =?$, $\dfrac{\partial f}{\partial y}(0,0) =?$

【Solution】

(1)

$\because \dfrac{\partial f}{\partial x}(0,0) = \lim\limits_{h \to 0} \dfrac{f(h+0,0) - f(0,0)}{h} = \lim\limits_{h \to 0} \dfrac{\dfrac{h^5 - 0^5}{h^5 + 0^5} - 0}{h} = \lim\limits_{h \to 0} \dfrac{1}{h} = \pm\infty$

$$\frac{\partial f}{\partial y}(0,0) = \lim_{h \to 0} \frac{f(0, h+0) - f(0,0)}{h} = \lim_{h \to 0} \frac{\frac{0^5 - h^5}{0^5 + h^5} - 0 - 0}{h} = \lim_{h \to 0} \frac{-1}{h} = \pm\infty$$

$$\therefore \frac{\partial f}{\partial x}(0,0) \text{ and } \frac{\partial f}{\partial y}(0,0) \text{ do not exist}$$

(2)

$$\frac{\partial f}{\partial x}(0,0) = \lim_{h \to 0} \frac{f(h+0,0) - f(0,0)}{h} = \lim_{h \to 0} \frac{\frac{5h^4 - 2 \cdot 0^3}{h^2 + 3 \cdot 0^2} - 0}{h} = \lim_{h \to 0} \frac{5h^2}{h} = 0$$

$$\frac{\partial f}{\partial y}(0,0) = \lim_{h \to 0} \frac{f(0, h+0) - f(0,0)}{h} = \lim_{h \to 0} \frac{\frac{5 \cdot 0^4 - 2 \cdot h^3}{0^2 + 3 \cdot h^2} - 0}{h} = \lim_{h \to 0} \frac{-2h}{3h} = \frac{-2}{3}$$

$$\therefore \frac{\partial f}{\partial x}(0,0) = 0, \quad \frac{\partial f}{\partial y}(0,0) = \frac{-2}{3}$$

Example 22.

Let $f(x,y) = \begin{cases} xy\left(\dfrac{x^2 - y^2}{x^2 + y^2}\right), & (x,y) \neq 0 \\ 0, & (x,y) = 0 \end{cases}$. Find

(1) $f_x(0,0)$ (2) $f_y(0,0)$ (3) $f_{xx}(0,0)$ (4) $f_{yx}(0,0)$ (5) $f_{xy}(0,0)$ (6) $f_{yy}(0,0)$

【Solution】

(1) $f_x(0,0) = \lim_{h \to 0} \dfrac{f(h+0,0) - f(0,0)}{h} = \lim_{h \to 0} \dfrac{0 - 0}{h} = 0$

(2) $f_y(0,0) = \lim_{h \to 0} \dfrac{f(0,h) - f(0,0)}{h} = \lim_{h \to 0} \dfrac{0 - 0}{h} = 0$

(3) $f_{xx}(0,0) = \lim_{h \to 0} \dfrac{f_x(h,0) - f_x(0,0)}{h}$

$\because f_x(h,0) = \lim_{k \to 0} \dfrac{f(h+k,0) - f(h,0)}{k} = \lim_{h \to 0} \dfrac{0 - 0}{h} = 0$

$\therefore f_{xx}(0,0) = \lim_{h \to 0} \dfrac{f_x(h,0) - f_x(0,0)}{h} = \lim_{h \to 0} \dfrac{0 - 0}{h} = 0$

(4) $f_{yx}(0,0) = \lim_{h \to 0} \dfrac{f_y(h,0) - f_y(0,0)}{h}$

$\because f_y(h,0) = \lim_{k \to 0} \dfrac{f(h,k) - f(h,0)}{k} = \lim_{k \to 0} \dfrac{hk\left(\dfrac{h^2 - k^2}{h^2 + k^2}\right) - 0}{k} = h$

$\therefore f_{yx}(0,0) = \lim_{h \to 0} \dfrac{f_y(h,0) - f_y(0,0)}{h} = \lim_{h \to 0} \dfrac{h - 0}{h} = 1$

7.3 Calculate Partial Derivatives of Multivariable Functions

(5) $f_{xy}(0,0) = \lim\limits_{h \to 0} \dfrac{f_x(0,h) - f_x(0,0)}{h}$

$\because f_x(0,h) = \lim\limits_{k \to 0} \dfrac{f(k,h) - f(0,h)}{k} = \lim\limits_{k \to 0} \dfrac{hk\left(\dfrac{k^2 - h^2}{h^2 + k^2}\right)}{k} = -h$

$\therefore f_{xy}(0,0) = \lim\limits_{h \to 0} \dfrac{f_x(0,h) - f_x(0,0)}{h} = \lim\limits_{h \to 0} \dfrac{-h - 0}{h} = -1$

(6) $f_{yy}(0,0) = \lim\limits_{h \to 0} \dfrac{f_y(0,h) - f_y(0,0)}{h}$

$\because f_y(0,h) = \lim\limits_{k \to 0} \dfrac{f(0,h+k) - f(0,h)}{k} = \lim\limits_{h \to 0} \dfrac{0 - 0}{h} = 0$

$\therefore f_{yy}(0,0) = \lim\limits_{h \to 0} \dfrac{f_y(0,h) - f_y(0,0)}{h} = \lim\limits_{h \to 0} \dfrac{0 - 0}{h} = 0$

Example 23.

Let $f(x,y) = \begin{cases} (x^2 + y^2)\sin\left(\dfrac{1}{x^2+y^2}\right), & (x,y) \neq 0 \\ 0, & (x,y) = 0 \end{cases}$.

Find $\dfrac{\partial f}{\partial x}(0,0) = ?$, $\dfrac{\partial f}{\partial y}(0,0) = ?$

【Solution】

$f_x(0,0) = \lim\limits_{h \to 0} \dfrac{f(h+0,0) - f(0,0)}{h} = \lim\limits_{h \to 0} \dfrac{h^2 \sin(\dfrac{1}{h^2}) - 0}{h} = 0$

$f_y(0,0) = \lim\limits_{h \to 0} \dfrac{f(0,0+h) - f(0,0)}{h} = \lim\limits_{h \to 0} \dfrac{h^2 \sin(\dfrac{1}{h^2}) - 0}{h} = 0$

Example 24.

Let $f(x,y) = \begin{cases} (x^2 + y^2)^p \sin\left(\dfrac{1}{x^2+y^2}\right), & (x,y) \neq 0 \\ 0, & (x,y) = 0 \end{cases}$, $\forall p > \dfrac{1}{2}$.

Find $\dfrac{\partial f}{\partial x}(0,0) = ?$, $\dfrac{\partial f}{\partial y}(0,0) = ?$

【Solution】

Let $p > \dfrac{1}{2}$, $f_x(0,0) = \lim\limits_{h \to 0} \dfrac{f(h+0,0) - f(0,0)}{h} = \lim\limits_{h \to 0} \dfrac{h^{2p} \sin(\dfrac{1}{h^2}) - 0}{h} = \lim\limits_{h \to 0} h^{2p-1} \sin(\dfrac{1}{h^2})$

7.3 Calculate Partial Derivatives of Multivariable Functions

$\because p > \dfrac{1}{2}$ $\therefore \lim\limits_{h \to 0} h^{2p-1} \sin(\dfrac{1}{h^2}) = 0 \Rightarrow f_x(0,0) = 0$

$f_y(0,0) = \lim\limits_{h \to 0} \dfrac{f(0,0+h) - f(0,0)}{h} = \lim\limits_{h \to 0} \dfrac{h^{2p} \sin(\dfrac{1}{h^2}) - 0}{h} = \lim\limits_{h \to 0} h^{2p-1} \sin(\dfrac{1}{h^2})$

$\because p > \dfrac{1}{2}$ $\therefore \lim\limits_{h \to 0} h^{2p-1} \sin(\dfrac{1}{h^2}) = 0 \Rightarrow \dfrac{\partial f}{\partial y}(0,0) = 0$

Example 25.

Let $f(x,y) = \begin{cases} x|y|, & (x,y) \neq 0 \\ 0, & (x,y) = 0 \end{cases}$. Find $\dfrac{\partial f}{\partial x}(0,0) = ?$, $\dfrac{\partial f}{\partial y}(0,0) = ?$

【Solution】

$f_x(0,0) = \lim\limits_{h \to 0} \dfrac{f(h+0,0) - f(0,0)}{h} = \lim\limits_{h \to 0} \dfrac{0-0}{h} = 0$

$f_y(0,0) = \lim\limits_{h \to 0} \dfrac{f(0,0+h) - f(0,0)}{h} = \lim\limits_{h \to 0} \dfrac{0-0}{h} = 0$

Example 26.

Let $f(x,y) = \begin{cases} x^p|y|, & (x,y) \neq 0 \\ 0, & (x,y) = 0 \end{cases}$, $\forall p > 0$. Find $\dfrac{\partial f}{\partial x}(0,0) = ?$, $\dfrac{\partial f}{\partial y}(0,0) = ?$

【Solution】

$f_x(0,0) = \lim\limits_{h \to 0} \dfrac{f(h+0,0) - f(0,0)}{h} = \lim\limits_{h \to 0} \dfrac{0-0}{h} = 0$

$f_y(0,0) = \lim\limits_{h \to 0} \dfrac{f(0,0+h) - f(0,0)}{h} = \lim\limits_{h \to 0} \dfrac{0-0}{h} = 0$

Example 27.

Let $f(x,y) = \begin{cases} (xy)^p \dfrac{x^2 - y^2}{x^2 + y^2}, & (x,y) \neq 0 \\ 0, & (x,y) = 0 \end{cases}$, $\forall p > 0$. Find $\dfrac{\partial f}{\partial x}(0,0) = ?$, $\dfrac{\partial f}{\partial y}(0,0) = ?$

【Solution】

$\dfrac{\partial f}{\partial x}(0,0) = \lim\limits_{h \to 0} \dfrac{f(h+0,0) - f(0,0)}{h} = \lim\limits_{h \to 0} \dfrac{0-0}{h} = 0$

$\dfrac{\partial f}{\partial y}(0,0) = \lim\limits_{h \to 0} \dfrac{f(0,h+0) - f(0,0)}{h} = \lim\limits_{h \to 0} \dfrac{0-0}{h} = 0$

Example 28.

Let $f(x, y) = \begin{cases} \dfrac{x^2 y}{x^2 + y^2}, & (x, y) \neq 0 \\ 0, & (x, y) = 0 \end{cases}$. Find $\dfrac{\partial f}{\partial x}(0,0) = ?, \dfrac{\partial f}{\partial y}(0,0) = ?$

【Solution】

$\dfrac{\partial f}{\partial x}(0,0) = \lim_{h \to 0} \dfrac{f(h+0,0) - f(0,0)}{h} = \lim_{h \to 0} \dfrac{0-0}{h} = 0$

$\dfrac{\partial f}{\partial y}(0,0) = \lim_{h \to 0} \dfrac{f(0,h+0) - f(0,0)}{h} = \lim_{h \to 0} \dfrac{0-0}{h} = 0$

Example 29.

Let $f(x, y) = \sin(x - ay)$. Prove $\dfrac{\partial^2 f}{\partial y^2}(x,y) - a^2 \dfrac{\partial^2 f}{\partial x^2}(x,y) = 0$.

【Solution】

$\dfrac{\partial f}{\partial x}(x, y) = \cos(x - ay), \quad \dfrac{\partial f}{\partial y}(x, y) = -a \cos(x - ay)$

$\dfrac{\partial^2 f}{\partial x^2}(x, y) = \dfrac{\partial}{\partial x}\left(\dfrac{\partial f}{\partial x}(x, y)\right) = -\sin(x - ay)$

$\dfrac{\partial^2 f}{\partial y^2}(x, y) = \dfrac{\partial}{\partial y}\left(\dfrac{\partial f}{\partial y}(x, y)\right) = -a^2 \sin(x - ay)$

$\therefore \dfrac{\partial^2 f}{\partial y^2}(x, y) - a^2 \dfrac{\partial^2 f}{\partial x^2}(x, y) = 0$

7.4 Determine the Differentiability of Multivariable Functions and Chain Rule

【Definition】Differentiability of a Function of Two Variables

A function $f(x, y)$ is differentiable at $(x_0, y_0) \Leftrightarrow$

(i) both $\dfrac{\partial f}{\partial x}(x_0, y_0)$ and $\dfrac{\partial f}{\partial y}(x_0, y_0)$ exist, and

(ii) $\lim_{h \to 0, k \to 0} \left| \dfrac{f(h + x_0, k + y_0) - f(x_0, y_0) - \dfrac{\partial f}{\partial x}(x_0, y_0)h - \dfrac{\partial f}{\partial y}(x_0, y_0)k}{\sqrt{h^2 + k^2}} \right| = 0$

【Definition】Differentiability of a Function of Multiple Variables

A function $f(x_1, \ldots, x_n)$ is differentiable at $(c_1, \ldots, c_n) \Leftrightarrow$

(i) all partial derivatives $\dfrac{\partial f}{\partial x_j}(c_1, \ldots, c_n)$ exist, $\forall 1 \leq j \leq n$

(ii) $\displaystyle\lim_{(h_1,\ldots,h_n)\to(0,\ldots,0)} \left| \dfrac{f(h_1+c_1, \ldots, h_n+c_n) - f(c_1, \ldots, c_n) - \sum_{j=1}^{n} \dfrac{\partial f}{\partial x_j}(c_1, \ldots, c_n)h_j}{\sqrt{h_1^2 + \cdots + h_n^2}} \right| = 0$

From the definition of differentiability for functions of two variables, we can observe that if $f(x,y)$ is differentiable at (x_0, y_0), then the limit of (h,k) approaching (x_0, y_0) from any direction must be zero. Therefore, let $h = mk$, then if $f(x,y)$ is differentiable at (x_0, y_0)

$$\lim_{h\to 0, k\to 0} \left| \dfrac{f(mk+x_0, k+y_0) - f(x_0, y_0) - \dfrac{\partial f}{\partial x}(x_0, y_0)mk - \dfrac{\partial f}{\partial y}(x_0, y_0)k}{\sqrt{(mk)^2 + k^2}} \right| = 0$$

and this limit is independent of m. Additionally, the existence of partial derivatives is a necessary condition for differentiability of a function of two variables, but it is not a sufficient condition. An example of a function $f(x,y)$ is provided below to illustrate this point.

Examples:
(I) Suppose $f(x,y)$ is defined by

$$f(x,y) = \begin{cases} \dfrac{x^2 y}{x^2+y^2}, & (x,y) \neq (0,0) \\ 0, & (x,y) = (0,0) \end{cases}.$$

Determine whether $f(x,y)$ is differentiable at $(0,0)$.
To find the partial derivatives at $(0,0)$:

$\dfrac{\partial f}{\partial x}(0,0) = \displaystyle\lim_{h\to 0} \dfrac{f(h+0,0) - f(0,0)}{h} = \lim_{h\to 0} \dfrac{0-0}{h} = 0$

$\dfrac{\partial f}{\partial y}(0,0) = \displaystyle\lim_{h\to 0} \dfrac{f(0,h+0) - f(0,0)}{h} = \lim_{h\to 0} \dfrac{0-0}{h} = 0$

Let $h = mk$ then $\displaystyle\lim_{h\to 0, k\to 0} \left| \dfrac{f(mk,k) - f(0,0) - \dfrac{\partial f}{\partial x}(0,0)mk - \dfrac{\partial f}{\partial y}(0,0)k}{\sqrt{(mk)^2 + k^2}} \right|$

$$= \lim_{h \to 0, k \to 0} \left| \frac{\frac{(mk)^2 k}{(mk)^2 + k^2}}{\sqrt{(mk)^2 + k^2}} \right| = \lim_{h \to 0, k \to 0} \left| \frac{\frac{m^2}{m^2+1}}{\sqrt{m^2+1}} \right| \text{ depends on } m.$$

Choose $h = k$, i.e., $m = 1$, then

$$\lim_{h \to 0, k \to 0} \left| \frac{f(h+0, k+0) - f(0,0) - \frac{\partial f}{\partial x}(0,0)h - \frac{\partial f}{\partial y}(0,0)k}{\sqrt{h^2+k^2}} \right| = \lim_{h \to 0, k \to 0} \left| \frac{\frac{m^2}{m^2+1}}{\sqrt{m^2+1}} \right| = \frac{1}{2\sqrt{2}} \neq 0$$

Therefore, while the partial derivatives of $f(x,y)$ exist at $(0,0)$, $f(x,y)$ is not differentiable at $(0,0)$.

Assume that $f(x,y)$ is differentiable at $(0,0)$. Let $y = mx$, then

$$\lim_{h \to 0, k \to 0} \left| \frac{f(mk, k) - f(0,0) - \frac{\partial f}{\partial x}(0,0)mk - \frac{\partial f}{\partial y}(0,0)k}{\sqrt{(mk)^2 + k^2}} \right| = 0$$

and this limit is independent of m. Therefore, the independence of this limit from m is a necessary condition for $f(x,y)$ to be differentiable at $(0,0)$. Additionally, if there exists $m \in R$ such that

$$\lim_{h \to 0, k \to 0} \left| \frac{f(mk, k) - f(0,0) - \frac{\partial f}{\partial x}(0,0)mk - \frac{\partial f}{\partial y}(0,0)k}{\sqrt{(mk)^2 + k^2}} \right| \neq 0$$

then $f(x,y)$ is not differentiable at $(0,0)$. In other words, if

$$\lim_{h \to 0, k \to 0} \left| \frac{f(mk, k) - f(0,0) - \frac{\partial f}{\partial x}(0,0)mk - \frac{\partial f}{\partial y}(0,0)k}{\sqrt{(mk)^2 + k^2}} \right|$$

depends on m, then try to find $m \in R$ such that this limit is not equal to 0 to show that $f(x,y)$ is not differentiable at $(0,0)$.

Next, we introduce the Chain Rule, which can be viewed as the differentiation of composite functions in multiple variables.

【Theorem】Chain Rule

(i) Suppose that the function f is continuously differentiable in an open set A.
(ii) Let $\vec{r} = \vec{r}(t)$ be a differentiable curve contained within the open set A

$\Rightarrow f(\vec{r}(t))$ is differentiable and $\dfrac{d}{dt} f(\vec{r}(t)) = \nabla f(\vec{r}(t)) \cdot \vec{r}'(t).$

7.4 Determine the Differentiability of Multivariable Functions and Chain Rule

Proof:

Claim: $\lim\limits_{h\to 0} \dfrac{f(\vec{r}(t+h)) - f(\vec{r}(t))}{h} = \nabla f(\vec{r}(t)) \cdot \vec{r}'(t)$

∵ A is an open set and $\vec{r}(t)$ is continuous

Choose $h > 0$ s.t. the line segment joining $\vec{r}(t)$ to $\vec{r}(t+h)$ lies entirely in A

∵ By mean-value theorem, ∃ a point $\vec{c}(h)$ between $\vec{r}(t)$ and $\vec{r}(t+h)$ such that

$f(\vec{r}(t+h)) - f(\vec{r}(t)) = \nabla f(\vec{c}(h)) \cdot (\vec{r}(t+h) - \vec{r}(t))$

$\Rightarrow \dfrac{f(\vec{r}(t+h)) - f(\vec{r}(t))}{h} = \nabla f(\vec{c}(h)) \cdot \left(\dfrac{\vec{r}(t+h) - \vec{r}(t)}{h} \right)$

Let $h \to 0$, $\vec{c}(h) \to \vec{r}(t)$, ∵ ∇f is continuous ∴ $\nabla f(\vec{c}(h)) \to \nabla f(\vec{r}(t))$

∵ $\dfrac{\vec{r}(t+h) - \vec{r}(t)}{h} \to \vec{r}'(t)$ ∴ $\lim\limits_{h\to 0} \dfrac{f(\vec{r}(t+h)) - f(\vec{r}(t))}{h} = \nabla f(\vec{r}(t)) \cdot \vec{r}'(t)$

Question Types:

Type 1.

Given a function $f(x,y)$, determine whether $f(x,y)$ is differentiable at $(0,0)$, where $f(x,y)$ satisfies the following condition: Let $h = mk$. Then

$$\lim_{h\to 0, k\to 0} \left| \dfrac{f(mk, k) - f(0,0) - \dfrac{\partial f}{\partial x}(0,0)mk - \dfrac{\partial f}{\partial y}(0,0)k}{\sqrt{(mk)^2 + k^2}} \right| \text{ does not depend on } m$$

Problem-Solving Process:

Let $\varepsilon > 0$, choose a specific value of δ such that $h^2 + k^2 < \delta$ implies:

$$\left| \dfrac{f(h,k) - f(0,0) - \dfrac{\partial f}{\partial x}(0,0)h - \dfrac{\partial f}{\partial y}(0,0)k}{\sqrt{h^2 + k^2}} \right| < \varepsilon$$

Remark:

In the process, it is common to derive the inequality $x^2 + y^2 \geq 2|xy|$, which leads to:

$$\left| \dfrac{f(h,k) - f(0,0) - \dfrac{\partial f}{\partial x}(0,0)h - \dfrac{\partial f}{\partial y}(0,0)k}{\sqrt{h^2 + k^2}} \right| < g(h^2 + k^2) < g(\delta) < \varepsilon$$

and to identify the specific form of the function g. The final inequality $g(\delta) < \varepsilon$ helps to clearly define δ.

Examples:

(I) Suppose $f(x,y) = \begin{cases} xy\left(\dfrac{x^2-y^2}{x^2+y^2}\right), & (x,y) \neq (0,0) \\ 0, & (x,y) = (0,0) \end{cases}$.

Determine whether $f(x,y)$ is differentiable at $(0,0)$.

$\dfrac{\partial f}{\partial x}(0,0) = \lim\limits_{h \to 0} \dfrac{f(h+0,0) - f(0,0)}{h} = \lim\limits_{h \to 0} \dfrac{0-0}{h} = 0$

$\dfrac{\partial f}{\partial y}(0,0) = \lim\limits_{h \to 0} \dfrac{f(0,h+0) - f(0,0)}{h} = \lim\limits_{h \to 0} \dfrac{0-0}{h} = 0$

Let $h = mk$ then $\lim\limits_{h \to 0, k \to 0} \left| \dfrac{f(mk,k) - f(0,0) - \dfrac{\partial f}{\partial x}(0,0)mk - \dfrac{\partial f}{\partial y}(0,0)k}{\sqrt{(mk)^2 + k^2}} \right|$

$= \lim\limits_{h \to 0, k \to 0} \left| \dfrac{mk^2\left(\dfrac{(mk)^2 - k^2}{(mk)^2 + k^2}\right)}{\sqrt{(mk)^2 + k^2}} \right| = 0$ is independent of m

Claim: $\lim\limits_{h \to 0, k \to 0} \left| \dfrac{f(h+0, k+0) - f(0,0) - \dfrac{\partial f}{\partial x}(0,0)h - \dfrac{\partial f}{\partial y}(0,0)k}{\sqrt{h^2 + k^2}} \right| = 0$

$\dfrac{\partial f}{\partial x}(0,0) = \lim\limits_{h \to 0} \dfrac{f(h+0,0) - f(0,0)}{h} = \lim\limits_{h \to 0} \dfrac{0-0}{h} = 0$

$\dfrac{\partial f}{\partial y}(0,0) = \lim\limits_{h \to 0} \dfrac{f(0,h+0) - f(0,0)}{h} = \lim\limits_{h \to 0} \dfrac{0-0}{h} = 0$

$\because \left| \dfrac{f(h+0, k+0) - f(0,0) - \dfrac{\partial f}{\partial x}(0,0)h - \dfrac{\partial f}{\partial y}(0,0)k}{\sqrt{h^2 + k^2}} \right| = \left| \dfrac{hk\left(\dfrac{h^2 - k^2}{h^2 + k^2}\right)}{\sqrt{h^2 + k^2}} \right|$

$\leq \left| \dfrac{hk\left(\dfrac{h^2 + k^2}{h^2 + k^2}\right)}{\sqrt{h^2 + k^2}} \right| = \left| \dfrac{hk}{\sqrt{h^2 + k^2}} \right| \leq \left| \dfrac{hk}{\sqrt{2hk}} \right| \leq \left| \dfrac{\sqrt{hk}}{\sqrt{2}} \right| \leq \sqrt{\dfrac{h^2 + k^2}{4}}$

and $\lim\limits_{h \to 0, k \to 0} \sqrt{\dfrac{h^2 + k^2}{4}} = 0$

$\therefore \lim\limits_{h \to 0, k \to 0} \left| \dfrac{f(h+0, k+0) - f(0,0) - \dfrac{\partial f}{\partial x}(0,0)h - \dfrac{\partial f}{\partial y}(0,0)k}{\sqrt{h^2 + k^2}} \right| = 0$

∴ $f(x,y)$ is differentiable at $(0,0)$

Type 2.
Given a function $f(x,y)$, determine whether $f(x,y)$ is differentiable at $(0,0)$, where $f(x,y)$ satisfies the following condition: Let $h = mk$. Then

$$\lim_{h\to 0, k\to 0} \left| \frac{f(mk,k) - f(0,0) - \frac{\partial f}{\partial x}(0,0)mk - \frac{\partial f}{\partial y}(0,0)k}{\sqrt{(mk)^2 + k^2}} \right| \text{ depends on } m.$$

Problem-Solving Process:
Step1.

Choose m such that $\lim_{h\to 0, k\to 0} \left| \dfrac{f(mk,k) - f(0,0) - \frac{\partial f}{\partial x}(0,0)mk - \frac{\partial f}{\partial y}(0,0)k}{\sqrt{(mk)^2 + k^2}} \right| \neq 0$

to prove $f(x,y)$ is not differentiable at $(0,0)$

Examples:

(I) Let $f(x,y) = \begin{cases} \dfrac{x^3 - y^3}{x^2 + y^2}, & (x,y) \neq (0,0) \\ 0, & (x,y) = (0,0) \end{cases}$.

Determine whether $f(x,y)$ is differentiable at $(0,0)$.

$f_x(0,0) = \lim\limits_{h\to 0} \dfrac{f(h+0,0) - f(0,0)}{h} = \lim\limits_{h\to 0} \dfrac{\frac{h^3}{h^2} - 0}{h} = 1$

$f_y(0,0) = \lim\limits_{h\to 0} \dfrac{f(0,h+0) - f(0,0)}{h} = \lim\limits_{h\to 0} \dfrac{\frac{-h^3}{h^2} - 0}{h} = -1$

Let $h = mk$ then $\lim\limits_{h\to 0, k\to 0} \left| \dfrac{f(mk,k) - f(0,0) - \frac{\partial f}{\partial x}(0,0)mk - \frac{\partial f}{\partial y}(0,0)k}{\sqrt{(mk)^2 + k^2}} \right|$

$= \lim\limits_{h\to 0, k\to 0} \left| \dfrac{\frac{(mk)^3 - k^3}{(mk)^2 + k^2} - mk + k}{\sqrt{(mk)^2 + k^2}} \right| = \lim\limits_{h\to 0, k\to 0} \left| \dfrac{\frac{m^3 - 1}{m^2 + 1} - m + 1}{\sqrt{m^2 + 1}} \right|$ depends on m

choose $h = 2k$, i.e., $m = 2$,

$$\lim_{h\to 0, k\to 0} \left| \frac{f(h+0, k+0) - f(0,0) - \frac{\partial f}{\partial x}(0,0)h - \frac{\partial f}{\partial y}(0,0)k}{\sqrt{h^2 + k^2}} \right| = \left| \frac{\frac{7}{5} - 1}{\sqrt{5}} \right| \neq 0$$

$\therefore f(x,y)$ is not differentiable at $(0,0)$

(II) Let $f(x,y) = \begin{cases} \frac{x^2 y}{x^2 + y^2}, & (x,y) \neq (0,0) \\ 0, & (x,y) = (0,0) \end{cases}$.

Determine whether $f(x,y)$ is differentiable at $(0,0)$.

$$\frac{\partial f}{\partial x}(0,0) = \lim_{h\to 0} \frac{f(h+0,0) - f(0,0)}{h} = \lim_{h\to 0} \frac{0-0}{h} = 0$$

$$\frac{\partial f}{\partial y}(0,0) = \lim_{h\to 0} \frac{f(0, h+0) - f(0,0)}{h} = \lim_{h\to 0} \frac{0-0}{h} = 0$$

Let $h = mk$ then $\lim_{h\to 0, k\to 0} \left| \frac{f(mk, k) - f(0,0) - \frac{\partial f}{\partial x}(0,0)mk - \frac{\partial f}{\partial y}(0,0)k}{\sqrt{(mk)^2 + k^2}} \right|$

$$= \lim_{h\to 0, k\to 0} \left| \frac{\frac{(mk)^2 k}{(mk)^2 + k^2}}{\sqrt{(mk)^2 + k^2}} \right| = \lim_{h\to 0, k\to 0} \left| \frac{\frac{m^2}{m^2 + 1}}{\sqrt{m^2 + 1}} \right| \text{ depends on } m$$

choose $h = k$, i.e., $m = 1$,

$$\lim_{h\to 0, k\to 0} \left| \frac{f(h+0, k+0) - f(0,0) - \frac{\partial f}{\partial x}(0,0)h - \frac{\partial f}{\partial y}(0,0)k}{\sqrt{h^2 + k^2}} \right| = \lim_{h\to 0, k\to 0} \left| \frac{\frac{m^2}{m^2 + 1}}{\sqrt{m^2 + 1}} \right|$$

$$= \frac{1}{2\sqrt{2}} \neq 0$$

$\therefore f(x,y)$ is not differentiable at $(0,0)$

Type 3.
Given a multivariable function $f(x,y,z)$ and a vector function $\vec{r}(t) = (x(t), y(t), z(t))$, find $\frac{df}{dt}(\vec{r}(t)) = ?$

Problem-Solving Process:

Step1.
$$\nabla f = \left(\frac{\partial f}{\partial x}, \frac{\partial f}{\partial y}, \frac{\partial f}{\partial z} \right) \text{ and } \vec{r}'(t) = \left(\frac{dx}{dt}, \frac{dy}{dt}, \frac{dz}{dt} \right)$$

Step2.

7.4 Determine the Differentiability of Multivariable Functions and Chain Rule

$$\therefore \frac{df}{dt}(\vec{r}(t)) = \nabla f(\vec{r}(t)) \cdot \vec{r}'(t) = \frac{\partial f}{\partial x}\left(\frac{dx}{dt}\right) + \frac{\partial f}{\partial y}\left(\frac{dy}{dt}\right) + \frac{\partial f}{\partial z}\left(\frac{dz}{dt}\right)$$

Examples:

(I) Let $f(x,y) = \dfrac{x^4 + y^4}{4}$, $r(t) = (a\cos t, b\sin t)$. Find $\dfrac{df}{dt}(\vec{r}(t)) = ?$

$\because \nabla f = (x^3, y^3) \quad \therefore \nabla f(\vec{r}(t)) = (a^3 \cos^3 t, b^3 \sin^3 t)$

$\because \vec{r}'(t) = (-a\sin t, b\cos t)$

$\therefore \dfrac{df}{dt}(\vec{r}(t)) = \nabla f(\vec{r}(t)) \cdot \vec{r}'(t) = (a^3 \cos^3 t, b^3 \sin^3 t) \cdot (-a\sin t, b\cos t)$

$= \sin t \cos t \, (b^4 \sin^2 t - a^4 \cos^2 t)$

(II) $f(x,y,z) = x^2 y + z\cos x$, $r(t) = (t, t^2, t^3)$, Find $\dfrac{df}{dt}(\vec{r}(t)) = ?$

$\because \nabla f = (2xy - z\sin x, x^2, \cos x) \quad \therefore \nabla f(\vec{r}(t)) = (2t^3 - t^3 \sin t, t^2, \cos t)$

$\because \vec{r}'(t) = (1, 2t, 3t^2)$

$\therefore \dfrac{df}{dt}(\vec{r}(t)) = \nabla f(\vec{r}(t)) \cdot \vec{r}'(t) = (2t^3 - t^3 \sin t, t^2, \cos t) \cdot (1, 2t, 3t^2)$

$= 4t^3 - t^3 \sin t + 3t^2 \cos t$

Type 4.
Given a multivariable function $f(x,y,z)$ where $x = x(t,s), y = y(t,s)$, and $z = z(t,s)$, find $\dfrac{\partial f}{\partial t} = ?$ and $\dfrac{\partial f}{\partial s} = ?$

Problem-Solving Process:
Step1.

$\because \nabla f = \left(\dfrac{\partial f}{\partial x}, \dfrac{\partial f}{\partial y}, \dfrac{\partial f}{\partial z}\right), \quad \dfrac{\partial \vec{r}}{\partial t} = \left(\dfrac{\partial x}{\partial t}, \dfrac{\partial y}{\partial t}, \dfrac{\partial z}{\partial t}\right)$ and $\dfrac{\partial \vec{r}}{\partial s} = \left(\dfrac{\partial x}{\partial s}, \dfrac{\partial y}{\partial s}, \dfrac{\partial z}{\partial s}\right)$

Step2.

$\therefore \dfrac{\partial f}{\partial t} = \dfrac{\partial f}{\partial x}\left(\dfrac{\partial x}{\partial t}\right) + \dfrac{\partial f}{\partial y}\left(\dfrac{\partial y}{\partial t}\right) + \dfrac{\partial f}{\partial z}\left(\dfrac{\partial z}{\partial t}\right)$

$\therefore \dfrac{\partial f}{\partial s} = \dfrac{\partial f}{\partial x}\left(\dfrac{\partial x}{\partial s}\right) + \dfrac{\partial f}{\partial y}\left(\dfrac{\partial y}{\partial s}\right) + \dfrac{\partial f}{\partial z}\left(\dfrac{\partial z}{\partial s}\right)$

Examples:

(I) Let $f(x,y) = x^2 - 2xy + y^3$, $x = s^2 \ln t$, $y = 2st^3$. Find $\dfrac{\partial f}{\partial t} = ?$ and $\dfrac{\partial f}{\partial s} = ?$

$\because \dfrac{\partial f}{\partial x} = 2x - 2y, \quad \dfrac{\partial f}{\partial y} = -2x + 6y^2, \quad \dfrac{\partial x}{\partial t} = \dfrac{s^2}{t}, \quad \dfrac{\partial x}{\partial s} = 2s\ln t, \quad \dfrac{\partial y}{\partial t} = 6st^2, \quad \dfrac{\partial y}{\partial s} = 2t^3$

$$\therefore \frac{\partial f}{\partial t} = \frac{\partial f}{\partial x}\left(\frac{\partial x}{\partial t}\right) + \frac{\partial f}{\partial y}\left(\frac{\partial y}{\partial t}\right) = (2x-2y)\frac{s^2}{t} + (-2x+6y^2)6st^2$$

$$\therefore \frac{\partial f}{\partial s} = \frac{\partial f}{\partial x}\left(\frac{\partial x}{\partial s}\right) + \frac{\partial f}{\partial y}\left(\frac{\partial y}{\partial s}\right) = (2x-2y)2s\ln t + (-2x+6y^2)2t^3$$

(II) Let $f(x,y,z) = x^2 y^3 e^{xz}$, $x = t^2 + s^2$, $y = 2st$, $z = s\ln t$. Find $\dfrac{\partial f}{\partial s} = ?$

$$\because \frac{\partial f}{\partial s} = \frac{\partial f}{\partial x}\left(\frac{\partial x}{\partial s}\right) + \frac{\partial f}{\partial y}\left(\frac{\partial y}{\partial s}\right) + \frac{\partial f}{\partial z}\left(\frac{\partial z}{\partial s}\right)$$

$$\because \frac{\partial f}{\partial x} = 2xy^3 e^{xz} + zx^2 y^3 e^{xz}, \quad \frac{\partial f}{\partial y} = 3x^2 y^2 e^{xz}, \quad \frac{\partial f}{\partial z} = x^3 y^3 e^{xz},$$

$$\because \frac{\partial x}{\partial s} = 2s, \quad \frac{\partial y}{\partial s} = 2t, \quad \frac{\partial z}{\partial s} = \ln t$$

$$\therefore \frac{\partial f}{\partial s} = \frac{\partial f}{\partial x}\left(\frac{\partial x}{\partial s}\right) + \frac{\partial f}{\partial y}\left(\frac{\partial y}{\partial s}\right) + \frac{\partial f}{\partial z}\left(\frac{\partial z}{\partial s}\right)$$
$$= (2xy^3 e^{xz} + zx^2 y^3 e^{xz})2s + (3x^2 y^2 e^{xz})2t + (x^3 y^3 e^{xz})\ln t$$

Example 1.

(1) Let $f(x,y) = \sqrt[3]{xy}$.

Prove that both $f_x(0,0)$ and $f_y(0,0)$ exist, but $f(x,y)$ is not differentiable at $(0,0)$.

(2) Let $f(x,y) = \begin{cases} \dfrac{x^4 - y^4}{x^3 + y^3}, & (x,y) \neq (0,0) \\ 0, & (x,y) = (0,0) \end{cases}$.

Prove that $f(x,y)$ is not differentiable at $(0,0)$.

【Solution】

(1)
Claim: both $f_x(0,0)$ and $f_y(0,0)$ exist

$$f_x(0,0) = \lim_{h \to 0} \frac{f(h+0,0) - f(0,0)}{h} = \lim_{h \to 0} \frac{\sqrt[3]{h \cdot 0} - 0}{h} = 0$$

$$f_y(0,0) = \lim_{h \to 0} \frac{f(0, h+0) - f(0,0)}{h} = \lim_{h \to 0} \frac{\sqrt[3]{0 \cdot h} - 0}{h} = 0$$

Claim: $\lim_{h \to 0, k \to 0} \left| \dfrac{f(h+0, k+0) - f(0,0) - \dfrac{\partial f}{\partial x}(0,0)h - \dfrac{\partial f}{\partial y}(0,0)k}{\sqrt{h^2 + k^2}} \right| \neq 0$

7.4 Determine the Differentiability of Multivariable Functions and Chain Rule

$$\therefore \left| \frac{f(h+0,k+0) - f(0,0) - \frac{\partial f}{\partial x}(0,0)h - \frac{\partial f}{\partial y}(0,0)k}{\sqrt{h^2+k^2}} \right| = \left| \frac{\sqrt[3]{hk}}{\sqrt{h^2+k^2}} \right|$$

Choose $h = k^2$ then $\lim\limits_{h \to 0, k \to 0} \left| \frac{f(h+0,k+0) - f(0,0) - \frac{\partial f}{\partial x}(0,0)h - \frac{\partial f}{\partial y}(0,0)k}{\sqrt{h^2+k^2}} \right|$

$$= \lim\limits_{h \to 0, k \to 0} \left| \frac{\sqrt[3]{hk}}{\sqrt{h^2+k^2}} \right| = \lim\limits_{h \to 0, k \to 0} \left| \frac{\sqrt[3]{k^3}}{\sqrt{k^4+k^2}} \right| = \lim\limits_{k \to 0} \left| \frac{1}{\sqrt{k^2+1}} \right| = 1 \neq 0$$

$\therefore f(x, y)$ is not differentiable at $(0,0)$

(2)
Claim: both $f_x(0,0)$ and $f_y(0,0)$ exist

$$f_x(0,0) = \lim\limits_{h \to 0} \frac{f(h+0,0) - f(0,0)}{h} = \lim\limits_{h \to 0} \frac{\frac{h^4}{h^3} - 0}{h} = 1$$

$$f_y(0,0) = \lim\limits_{h \to 0} \frac{f(0, h+0) - f(0,0)}{h} = \lim\limits_{h \to 0} \frac{\frac{-h^4}{h^3} - 0}{h} = -1$$

Claim: $\lim\limits_{h \to 0, k \to 0} \left| \frac{f(h+0,k+0) - f(0,0) - \frac{\partial f}{\partial x}(0,0)h - \frac{\partial f}{\partial y}(0,0)k}{\sqrt{h^2+k^2}} \right| \neq 0$

$$\therefore \left| \frac{f(h+0,k+0) - f(0,0) - \frac{\partial f}{\partial x}(0,0)h - \frac{\partial f}{\partial y}(0,0)k}{\sqrt{h^2+k^2}} \right| = \left| \frac{\frac{h^4-k^4}{h^3+k^3} - h + k}{\sqrt{h^2+k^2}} \right|$$

Choose $h = 2k$ then

$$\lim\limits_{h \to 0, k \to 0} \left| \frac{f(h+0,k+0) - f(0,0) - \frac{\partial f}{\partial x}(0,0)h - \frac{\partial f}{\partial y}(0,0)k}{\sqrt{h^2+k^2}} \right| = \lim\limits_{h \to 0, k \to 0} \left| \frac{\frac{2}{3}k}{\sqrt{5k}} \right| = \frac{2}{3\sqrt{5}} \neq 0$$

$\therefore f(x, y)$ is not differentiable at $(0,0)$

Example 2.

Let $f(x, y) = \begin{cases} (x^2 + y^2)\sin\left(\frac{1}{x^2+y^2}\right), & (x, y) \neq (0,0) \\ 0, & (x, y) = (0,0) \end{cases}$.

(1) Prove that both $f_x(x, y)$ and $f_y(x, y)$ exist but are discontinuous at $(0,0)$.
(2) Prove that $f(x, y)$ is differentiable at $(0,0)$.

【Solution】
(1)
$$f_x(x,y) = 2x \sin\left(\frac{1}{x^2+y^2}\right) + (x^2+y^2)\cos\left(\frac{1}{x^2+y^2}\right)\frac{-2x}{(x^2+y^2)^{-2}}$$
$$= 2x \sin\left(\frac{1}{x^2+y^2}\right) + \cos\left(\frac{1}{x^2+y^2}\right)\frac{-2x}{x^2+y^2}$$
$$f_y(x,y) = 2y \sin\left(\frac{1}{x^2+y^2}\right) + (x^2+y^2)\cos\left(\frac{1}{x^2+y^2}\right)\frac{-2y}{(x^2+y^2)^{-2}}$$
$$= 2y \sin\left(\frac{1}{x^2+y^2}\right) + \cos\left(\frac{1}{x^2+y^2}\right)\frac{-2y}{x^2+y^2}$$

Claim: $\lim_{x\to 0, y\to 0} f_x(x,y) \neq f_x(0,0)$ and $\lim_{x\to 0, y\to 0} f_y(x,y) \neq f_y(0,0)$

$$f_x(0,0) = \lim_{h\to 0}\frac{f(h+0,0)-f(0,0)}{h} = \lim_{h\to 0}\frac{h^2 \sin(\frac{1}{h^2}) - 0}{h} = 0$$

$$f_y(0,0) = \lim_{h\to 0}\frac{f(0,0+h)-f(0,0)}{h} = \lim_{h\to 0}\frac{h^2 \sin(\frac{1}{h^2}) - 0}{h} = 0$$

Choose $x_n = y_n = \frac{1}{\sqrt{n\pi}}$

$$\lim_{n\to\infty} f_x(x_n, y_n) = \lim_{n\to\infty} \cos\left(\frac{1}{x_n^2+y_n^2}\right)\frac{-2x_n}{(x_n^2+y_n^2)} = \lim_{n\to\infty} \cos(2n\pi)\frac{-4n\pi}{\sqrt{n\pi}} = -\infty$$

$$\lim_{n\to\infty} f_y(x_n, y_n) = \lim_{n\to\infty} \cos\left(\frac{1}{x_n^2+y_n^2}\right)\frac{-2y_n}{(x_n^2+y_n^2)} = \lim_{n\to\infty} \cos(2n\pi)\frac{-4n\pi}{\sqrt{n\pi}} = -\infty$$

∴ $\lim_{x\to 0, y\to 0} f_x(x,y) \neq f_x(0,0)$ and $\lim_{x\to 0, y\to 0} f_y(x,y) \neq f_y(0,0)$

∴ $f_x(x,y)$ and $f_y(x,y)$ are discontinuous at $(0,0)$

(2)

Claim: $\lim_{h\to 0, k\to 0} \left|\frac{f(h+0,k+0)-f(0,0)-\frac{\partial f}{\partial x}(0,0)h-\frac{\partial f}{\partial y}(0,0)k}{\sqrt{h^2+k^2}}\right| = 0$

∵ $\left|\frac{f(h+0,k+0)-f(0,0)-\frac{\partial f}{\partial x}(0,0)h-\frac{\partial f}{\partial y}(0,0)k}{\sqrt{h^2+k^2}}\right| = \left|\frac{(h^2+k^2)\sin\left(\frac{1}{h^2+k^2}\right)}{\sqrt{h^2+k^2}}\right|$

$= \left|(\sqrt{h^2+k^2})\sin\left(\frac{1}{h^2+k^2}\right)\right| \leq \sqrt{h^2+k^2}$

$$\because \lim_{h\to 0, k\to 0} |\sqrt{h^2+k^2}\,| = 0 \therefore \lim_{h\to 0, k\to 0} \left| \frac{f(h+0, k+0) - f(0,0) - \frac{\partial f}{\partial x}(0,0)h - \frac{\partial f}{\partial y}(0,0)k}{\sqrt{h^2+k^2}} \right| = 0$$

$\therefore f(x,y)$ is differentiable at $(0,0)$

Example 3.

Let $f(x,y) = |x|y$. Prove the following: (1) both $f_x(x,y)$ and $f_y(x,y)$ are continuous. (2) $f(x,y)$ is differentiable at $(0,0)$.

【Solution】

(1)
As $x > 0$, $f_x(x,y) = y$ and $f_y(x,y) = x$
As $x < 0$, $f_x(x,y) = -y$ and $f_y(x,y) = -x$
Claim: $\lim_{x\to 0, y\to 0} f_x(x,y) = f_x(0,0)$ and $\lim_{x\to 0, y\to 0} f_y(x,y) = f_y(0,0)$

$\because f_x(0,0) = \lim_{h\to 0} \frac{f(h+0,0) - f(0,0)}{h} = \lim_{h\to 0} \frac{0-0}{h} = 0$

$f_y(0,0) = \lim_{h\to 0} \frac{f(0,0+h) - f(0,0)}{h} = \lim_{h\to 0} \frac{0-0}{h} = 0$

$\lim_{x\to 0, y\to 0} f_x(x,y) = \lim_{x\to 0, y\to 0} y = \lim_{x\to 0, y\to 0} -y = 0$

$\lim_{x\to 0, y\to 0} f_y(x,y) = \lim_{x\to 0, y\to 0} x = \lim_{x\to 0, y\to 0} -x = 0$

$\therefore \lim_{x\to 0, y\to 0} f_x(x,y) = f_x(0,0)$ and $\lim_{x\to 0, y\to 0} f_y(x,y) = f_y(0,0)$

$\therefore f_x(x,y)$ and $f_y(x,y)$ are continuous

(2)

Claim: $\lim_{h\to 0, k\to 0} \left| \frac{f(h+0, k+0) - f(0,0) - \frac{\partial f}{\partial x}(0,0)h - \frac{\partial f}{\partial y}(0,0)k}{\sqrt{h^2+k^2}} \right| = 0$

$\because \left| \frac{f(h+0, k+0) - f(0,0) - \frac{\partial f}{\partial x}(0,0)h - \frac{\partial f}{\partial y}(0,0)k}{\sqrt{h^2+k^2}} \right|$

$= \left| \frac{|h|k - f(0,0) - \frac{\partial f}{\partial x}(0,0)h - \frac{\partial f}{\partial y}(0,0)k}{\sqrt{h^2+k^2}} \right| = \left| \frac{|h|k}{\sqrt{h^2+k^2}} \right| \leq \left| \frac{hk}{\sqrt{2hk}} \right| = \left| \frac{\sqrt{hk}}{\sqrt{2}} \right|$

$$\because \lim_{h\to 0, k\to 0} \left|\frac{\sqrt{hk}}{\sqrt{2}}\right| = 0$$

$$\therefore \lim_{h\to 0, k\to 0} \left|\frac{f(h+0, k+0) - f(0,0) - \frac{\partial f}{\partial x}(0,0)h - \frac{\partial f}{\partial y}(0,0)k}{\sqrt{h^2+k^2}}\right| = 0$$

$\therefore f(x,y)$ is differentiable at $(0,0)$

Example 4.

$$\text{Let } f(x,y) = \begin{cases} (xy)^p \left(\frac{x^2-y^2}{x^2+y^2}\right), & (x,y) \neq (0,0) \\ 0, & (x,y) = (0,0) \end{cases}$$

Prove that $f(x,y)$ is differentiable at $(0,0)$, $\forall p > \frac{1}{2}$.

【Solution】

Claim: $\lim_{h\to 0, k\to 0} \left|\frac{f(h+0, k+0) - f(0,0) - \frac{\partial f}{\partial x}(0,0)h - \frac{\partial f}{\partial y}(0,0)k}{\sqrt{h^2+k^2}}\right| = 0$

$$\frac{\partial f}{\partial x}(0,0) = \lim_{h\to 0} \frac{f(h+0, 0) - f(0,0)}{h} = \lim_{h\to 0} \frac{0-0}{h} = 0$$

$$\frac{\partial f}{\partial y}(0,0) = \lim_{h\to 0} \frac{f(0, h+0) - f(0,0)}{h} = \lim_{h\to 0} \frac{0-0}{h} = 0$$

Let $p > \frac{1}{2}$

$$\because \left|\frac{f(h+0, k+0) - f(0,0) - \frac{\partial f}{\partial x}(0,0)h - \frac{\partial f}{\partial y}(0,0)k}{\sqrt{h^2+k^2}}\right| = \left|\frac{(hk)^p \left(\frac{h^2-k^2}{h^2+k^2}\right)}{\sqrt{h^2+k^2}}\right|$$

$$\leq \left|\frac{(hk)^p \left(\frac{h^2+k^2}{h^2+k^2}\right)}{\sqrt{h^2+k^2}}\right| = \left|\frac{(hk)^p}{\sqrt{h^2+k^2}}\right| = \left|\frac{(hk)^p}{\sqrt{2hk}}\right| \leq \left|\frac{(hk)^{p-\frac{1}{2}}}{\sqrt{2}}\right|$$

and $\lim_{h\to 0, k\to 0} (hk)^{p-\frac{1}{2}} = 0$, $\forall p > \frac{1}{2}$

$$\therefore \lim_{h\to 0, k\to 0} \left|\frac{f(h+0, k+0) - f(0,0) - \frac{\partial f}{\partial x}(0,0)h - \frac{\partial f}{\partial y}(0,0)k}{\sqrt{h^2+k^2}}\right| = 0$$

∴ $f(x,y)$ is differentiable at $(0,0)$, $\forall p > \dfrac{1}{2}$

Example 5.

Let $f(x,y) = \begin{cases} \dfrac{xy}{\sqrt{x^2+y^2}}, & (x,y) \neq (0,0) \\ 0, & (x,y) = (0,0) \end{cases}$.

Prove that $f(x,y)$ is not differentiable at $(0,0)$.

【Solution】

Claim: $\displaystyle\lim_{h\to 0, k\to 0} \left| \dfrac{f(h+0, k+0) - f(0,0) - \dfrac{\partial f}{\partial x}(0,0)h - \dfrac{\partial f}{\partial y}(0,0)k}{\sqrt{h^2+k^2}} \right| \neq 0$

$\dfrac{\partial f}{\partial x}(0,0) = \displaystyle\lim_{h\to 0} \dfrac{f(h+0,0) - f(0,0)}{h} = \lim_{h\to 0} \dfrac{0-0}{h} = 0$

$\dfrac{\partial f}{\partial y}(0,0) = \displaystyle\lim_{h\to 0} \dfrac{f(0,h+0) - f(0,0)}{h} = \lim_{h\to 0} \dfrac{0-0}{h} = 0$

∵ $\left| \dfrac{f(h+0, k+0) - f(0,0) - \dfrac{\partial f}{\partial x}(0,0)h - \dfrac{\partial f}{\partial y}(0,0)k}{\sqrt{h^2+k^2}} \right| = \left| \dfrac{\dfrac{hk}{\sqrt{h^2+k^2}}}{\sqrt{h^2+k^2}} \right|$

Choose $h = k$

$\displaystyle\lim_{h\to 0, k\to 0} \left| \dfrac{f(h+0, k+0) - f(0,0) - \dfrac{\partial f}{\partial x}(0,0)h - \dfrac{\partial f}{\partial y}(0,0)k}{\sqrt{h^2+k^2}} \right|$

$= \displaystyle\lim_{h\to 0, k\to 0} \left| \dfrac{\dfrac{hk}{\sqrt{h^2+k^2}}}{\sqrt{h^2+k^2}} \right| = \lim_{h\to 0, k\to 0} \left| \dfrac{h^2}{2h^2} \right| = \dfrac{1}{2} \neq 0$

∴ $f(x,y)$ is not differentiable at $(0,0)$

Example 6.

Let $f(x,y) = \begin{cases} \dfrac{2x\sqrt{y}}{\sqrt{x^4+y^2}}, & (x,y) \neq (0,0) \\ 0, & (x,y) = (0,0) \end{cases}$.

Prove $f(x,y)$ is not differentiable at $(0,0)$.

【Solution】

Claim: $\lim\limits_{h\to 0, k\to 0} \left| \dfrac{f(h+0, k+0) - f(0,0) - \dfrac{\partial f}{\partial x}(0,0)h - \dfrac{\partial f}{\partial y}(0,0)k}{\sqrt{h^2+k^2}} \right| \neq 0$

$\dfrac{\partial f}{\partial x}(0,0) = \lim\limits_{h\to 0} \dfrac{f(h+0,0) - f(0,0)}{h} = \lim\limits_{h\to 0} \dfrac{\dfrac{2h\cdot 0}{\sqrt{h^4+0^2}} - 0}{h} = \lim\limits_{h\to 0} \dfrac{0}{h^2} = 0$

$\dfrac{\partial f}{\partial y}(0,0) = \lim\limits_{h\to 0} \dfrac{f(0,h+0) - f(0,0)}{h} = \lim\limits_{h\to 0} \dfrac{\dfrac{2\cdot 0\cdot \sqrt{h}}{\sqrt{0^4+h^2}} - 0}{h} = \lim\limits_{h\to 0} \dfrac{0}{h^{\frac{3}{2}}} = 0$

$\therefore \dfrac{\partial f}{\partial x}(0,0) = 0, \ \dfrac{\partial f}{\partial y}(0,0) = 0$

$\therefore \left| \dfrac{f(h+0, k+0) - f(0,0) - \dfrac{\partial f}{\partial x}(0,0)h - \dfrac{\partial f}{\partial y}(0,0)k}{\sqrt{h^2+k^2}} \right| = \left| \dfrac{\dfrac{2h\sqrt{k}}{\sqrt{h^4+k^2}}}{\sqrt{h^2+k^2}} \right|$

Choose $k = h^2$

$\lim\limits_{h\to 0, k\to 0} \left| \dfrac{f(h+0, k+0) - f(0,0) - \dfrac{\partial f}{\partial x}(0,0)h - \dfrac{\partial f}{\partial y}(0,0)k}{\sqrt{h^2+k^2}} \right|$

$= \lim\limits_{h\to 0, k\to 0} \left| \dfrac{\dfrac{2h\sqrt{k}}{\sqrt{h^4+k^2}}}{\sqrt{h^2+k^2}} \right| = \lim\limits_{h\to 0, k\to 0} \left| \dfrac{\dfrac{2h^2}{\sqrt{2h^2}}}{\sqrt{h^2+h^4}} \right| = \infty \neq 0$

$\therefore f(x,y)$ is not differentiable at $(0,0)$

Example 7.

Let $f(x,y) = \begin{cases} xy\left(\dfrac{x^2-y^2}{x^2+y^2}\right), & (x,y) \neq (0,0) \\ 0, & (x,y) = (0,0) \end{cases}$.

Prove that (1) $f(x,y)$ is continuous at $(x,y) = (0,0)$ (2) $f(x,y)$ is differentiable at $(0,0)$

【Solution】

(1)

$\because x^2 + y^2 \geq 2|xy| \quad \therefore \dfrac{1}{x^2+y^2} \leq \dfrac{1}{2|xy|}$

7.4 Determine the Differentiability of Multivariable Functions and Chain Rule

$$\therefore \left|xy\left(\frac{x^2-y^2}{x^2+y^2}\right)\right| \le \left|xy\left(\frac{x^2-y^2}{2xy}\right)\right| \le \frac{x^2+y^2}{2}$$

Claim: $\lim_{x\to 0, y\to 0} f(x,y) = f(0,0)$

$\therefore \lim_{x\to 0, y\to 0} \frac{x^2+y^2}{2} = 0 \quad \therefore \lim_{x\to 0, y\to 0} \left|xy\left(\frac{x^2-y^2}{x^2+y^2}\right)\right| = 0$

$\therefore f(x,y)$ is continuous at $(x,y) = (0,0)$

(2)

Claim: $\lim_{h\to 0, k\to 0} \left| \dfrac{f(h+0, k+0) - f(0,0) - \frac{\partial f}{\partial x}(0,0)h - \frac{\partial f}{\partial y}(0,0)k}{\sqrt{h^2+k^2}} \right| = 0$

$\dfrac{\partial f}{\partial x}(0,0) = \lim_{h\to 0} \dfrac{f(h+0,0) - f(0,0)}{h} = \lim_{h\to 0} \dfrac{0-0}{h} = 0$

$\dfrac{\partial f}{\partial y}(0,0) = \lim_{h\to 0} \dfrac{f(0,h+0) - f(0,0)}{h} = \lim_{h\to 0} \dfrac{0-0}{h} = 0$

$\therefore \left| \dfrac{f(h+0, k+0) - f(0,0) - \frac{\partial f}{\partial x}(0,0)h - \frac{\partial f}{\partial y}(0,0)k}{\sqrt{h^2+k^2}} \right| = \left| \dfrac{hk\left(\frac{h^2-k^2}{h^2+k^2}\right)}{\sqrt{h^2+k^2}} \right|$

$\le \left| \dfrac{hk\left(\frac{h^2+k^2}{h^2+k^2}\right)}{\sqrt{h^2+k^2}} \right| = \left| \dfrac{hk}{\sqrt{h^2+k^2}} \right| \le \left| \dfrac{hk}{\sqrt{2hk}} \right| \le \left| \dfrac{\sqrt{hk}}{\sqrt{2}} \right| \le \sqrt{\dfrac{h^2+k^2}{4}}$

and $\lim_{h\to 0, k\to 0} \sqrt{\dfrac{h^2+k^2}{4}} = 0$

$\therefore \lim_{h\to 0, k\to 0} \left| \dfrac{f(h+0, k+0) - f(0,0) - \frac{\partial f}{\partial x}(0,0)h - \frac{\partial f}{\partial y}(0,0)k}{\sqrt{h^2+k^2}} \right| = 0$

$\therefore f(x,y)$ is differentiable at $(0,0)$

Example 8.

Let $f(x,y) = \begin{cases} \dfrac{x^2 y}{x^2+y^2}, & (x,y) \neq (0,0) \\ 0, & (x,y) = (0,0) \end{cases}$.

Prove (1) $f(x,y)$ is continuous at $(x,y) = (0,0)$ (2) $f(x,y)$ is not differentiable at $(0,0)$.

【Solution】

(1)

$\because x^2 + y^2 \geq 2|xy| \quad \therefore \dfrac{1}{x^2+y^2} \leq \dfrac{1}{2|xy|} \Rightarrow \left|\dfrac{x^2y}{x^2+y^2}\right| \leq \left|\dfrac{x^2y}{2xy}\right| \leq \left|\dfrac{x}{2}\right|$

Claim: $\lim\limits_{x\to 0, y\to 0} f(x,y) = f(0,0)$

$\because \lim\limits_{x\to 0, y\to 0}\left|\dfrac{x}{2}\right| = 0 \quad \therefore \lim\limits_{x\to 0, y\to 0}\left|\dfrac{x^2y}{x^2+y^2}\right| = 0 \quad \therefore f(x,y)$ is continuous at $(x,y) = (0,0)$

(2)

Claim: $\lim\limits_{h\to 0, k\to 0}\left|\dfrac{f(h+0, k+0) - f(0,0) - \dfrac{\partial f}{\partial x}(0,0)h - \dfrac{\partial f}{\partial y}(0,0)k}{\sqrt{h^2+k^2}}\right| \neq 0$

$\dfrac{\partial f}{\partial x}(0,0) = \lim\limits_{h\to 0}\dfrac{f(h+0,0)-f(0,0)}{h} = \lim\limits_{h\to 0}\dfrac{0-0}{h} = 0$

$\dfrac{\partial f}{\partial y}(0,0) = \lim\limits_{h\to 0}\dfrac{f(0,h+0)-f(0,0)}{h} = \lim\limits_{h\to 0}\dfrac{0-0}{h} = 0$

$\therefore \left|\dfrac{f(h+0, k+0) - f(0,0) - \dfrac{\partial f}{\partial x}(0,0)h - \dfrac{\partial f}{\partial y}(0,0)k}{\sqrt{h^2+k^2}}\right| = \left|\dfrac{\dfrac{h^2k}{h^2+k^2}}{\sqrt{h^2+k^2}}\right|$

Choose $h = k$

$\lim\limits_{h\to 0, k\to 0}\left|\dfrac{f(h+0, k+0) - f(0,0) - \dfrac{\partial f}{\partial x}(0,0)h - \dfrac{\partial f}{\partial y}(0,0)k}{\sqrt{h^2+k^2}}\right|$

$= \lim\limits_{h\to 0, k\to 0}\left|\dfrac{\dfrac{h^2k}{h^2+k^2}}{\sqrt{h^2+k^2}}\right| = \lim\limits_{h\to 0, k\to 0}\left|\dfrac{\dfrac{h^3}{2h^2}}{\sqrt{2h^2}}\right| = \dfrac{1}{2\sqrt{2}} \neq 0$

$\therefore f(x,y)$ is not differentiable at $(0,0)$

Example 9.

Let $f(x,y) = \begin{cases} \dfrac{2x^2y}{x^4+y^2}, & (x,y) \neq (0,0) \\ 0, & (x,y) = (0,0) \end{cases}$.

Prove that $f(x,y)$ is not differentiable at $(0,0)$.

【Solution】

7.4 Determine the Differentiability of Multivariable Functions and Chain Rule

Claim: $\lim\limits_{h\to 0, k\to 0} \left| \dfrac{f(h+0, k+0) - f(0,0) - \dfrac{\partial f}{\partial x}(0,0)h - \dfrac{\partial f}{\partial y}(0,0)k}{\sqrt{h^2 + k^2}} \right| \neq 0$

$\dfrac{\partial f}{\partial x}(0,0) = \lim\limits_{h\to 0} \dfrac{f(h+0,0) - f(0,0)}{h} = \lim\limits_{h\to 0} \dfrac{\dfrac{2h^2 \cdot 0}{h^4 + 0^2} - 0}{h} = \lim\limits_{h\to 0} \dfrac{0}{h^5} = 0$

$\dfrac{\partial f}{\partial y}(0,0) = \lim\limits_{h\to 0} \dfrac{f(0, h+0) - f(0,0)}{h} = \lim\limits_{h\to 0} \dfrac{\dfrac{2 \cdot 0^2 \cdot h}{0^4 + h^2} - 0}{h} = \lim\limits_{h\to 0} \dfrac{0}{h^3} = 0$

$\therefore \dfrac{\partial f}{\partial x}(0,0) = 0, \ \dfrac{\partial f}{\partial y}(0,0) = 0$

$\because \left| \dfrac{f(h+0, k+0) - f(0,0) - \dfrac{\partial f}{\partial x}(0,0)h - \dfrac{\partial f}{\partial y}(0,0)k}{\sqrt{h^2 + k^2}} \right| = \left| \dfrac{\dfrac{2h^2 k}{h^4 + k^2}}{\sqrt{h^2 + k^2}} \right|$

Choose $k = h^2$

$\lim\limits_{h\to 0, k\to 0} \left| \dfrac{f(h+0, k+0) - f(0,0) - \dfrac{\partial f}{\partial x}(0,0)h - \dfrac{\partial f}{\partial y}(0,0)k}{\sqrt{h^2 + k^2}} \right|$

$= \lim\limits_{h\to 0, k\to 0} \left| \dfrac{\dfrac{2h^2 k}{h^4 + k^2}}{\sqrt{h^2 + k^2}} \right| = \lim\limits_{h\to 0, k\to 0} \left| \dfrac{\dfrac{2h^4}{2h^4}}{\sqrt{h^2 + h^4}} \right| = \infty \neq 0$

$\therefore f(x, y)$ is not differentiable at $(0,0)$

Example 10.

Let $f(x, y) = \begin{cases} \dfrac{3x^2 y}{x^3 + y^3}, & (x, y) \neq (0,0) \\ 0, & (x, y) = (0,0) \end{cases}$.

Prove that $f(x, y)$ is not differentiable at $(0,0)$.

【Solution】

Claim: $\lim\limits_{h\to 0, k\to 0} \left| \dfrac{f(h+0, k+0) - f(0,0) - \dfrac{\partial f}{\partial x}(0,0)h - \dfrac{\partial f}{\partial y}(0,0)k}{\sqrt{h^2 + k^2}} \right| \neq 0$

$\dfrac{\partial f}{\partial x}(0,0) = \lim\limits_{h\to 0} \dfrac{f(h+0,0) - f(0,0)}{h} = \lim\limits_{h\to 0} \dfrac{\dfrac{3 \cdot h^2 \cdot 0}{h^3 + 0^3} - 0}{h} = \lim\limits_{h\to 0} \dfrac{0}{h^4} = 0$

$$\frac{\partial f}{\partial y}(0,0) = \lim_{h\to 0}\frac{f(0,h+0)-f(0,0)}{h} = \lim_{h\to 0}\frac{\frac{3\cdot 0^2\cdot h}{0^3+h^3}-0}{h} = \lim_{h\to 0}\frac{0}{h^4} = 0$$

$$\therefore \frac{\partial f}{\partial x}(0,0) = 0, \quad \frac{\partial f}{\partial y}(0,0) = 0$$

$$\because \left|\frac{f(h+0,k+0)-f(0,0)-\frac{\partial f}{\partial x}(0,0)h-\frac{\partial f}{\partial y}(0,0)k}{\sqrt{h^2+k^2}}\right| = \left|\frac{\frac{3h^2k}{h^3+k^3}}{\sqrt{h^2+k^2}}\right|$$

Choose $k = h$

$$\lim_{h\to 0, k\to 0}\left|\frac{f(h+0,k+0)-f(0,0)-\frac{\partial f}{\partial x}(0,0)h-\frac{\partial f}{\partial y}(0,0)k}{\sqrt{h^2+k^2}}\right|$$

$$= \lim_{h\to 0, k\to 0}\left|\frac{\frac{3h^2k}{h^3+k^3}}{\sqrt{h^2+k^2}}\right| = \lim_{h\to 0, k\to 0}\left|\frac{\frac{3h^3}{2h^3}}{\sqrt{h^2+h^2}}\right| = \infty \neq 0$$

$\therefore f(x,y)$ is not differentiable at $(0,0)$

Example 11.

(1) Let $f(x,y) = \dfrac{x^2}{x^2+y^2}$, $x = 3\cos t$, and $y = 5\sin t$. Find $\dfrac{df}{dt} = ?$

(2) Let $f(x,y) = \sin^{-1}(5x+2y)$, $x = r^2 e^s$, and $y = \sin(rs)$. Find $\dfrac{\partial f}{\partial r}, \dfrac{\partial f}{\partial s} = ?$

【Solution】

(1)

$$\because \frac{df}{dt} = \frac{\partial f}{\partial x}\frac{\partial x}{\partial t} + \frac{\partial f}{\partial y}\frac{\partial y}{\partial t}$$

$$\frac{\partial f}{\partial x} = \frac{2x(x^2+y^2)-x^2(2x)}{(x^2+y^2)^2} = \frac{2xy^2}{(x^2+y^2)^2}, \quad \frac{\partial f}{\partial y} = \frac{-x^2(2y)}{(x^2+y^2)^2} = \frac{-2yx^2}{(x^2+y^2)^2}$$

$$\frac{\partial x}{\partial t} = -3\sin t, \quad \frac{\partial y}{\partial t} = 5\cos t$$

$$\therefore \frac{df}{dt} = \frac{\partial f}{\partial x}\frac{\partial x}{\partial t} + \frac{\partial f}{\partial y}\frac{\partial y}{\partial t} = \frac{2xy^2}{(x^2+y^2)^2}(-3\sin t) + \frac{-2yx^2}{(x^2+y^2)^2}(5\cos t)$$

(2)

$$\because \frac{\partial f}{\partial r} = \frac{\partial f}{\partial x}\frac{\partial x}{\partial r} + \frac{\partial f}{\partial y}\frac{\partial y}{\partial r} \quad \text{and} \quad \frac{\partial f}{\partial s} = \frac{\partial f}{\partial x}\frac{\partial x}{\partial s} + \frac{\partial f}{\partial y}\frac{\partial y}{\partial s}$$

$$\frac{\partial f}{\partial x} = \frac{5}{\sqrt{1-(5x+2y)^2}}, \quad \frac{\partial f}{\partial y} = \frac{2}{\sqrt{1-(5x+2y)^2}}$$

$$\frac{\partial x}{\partial r} = 2re^s, \quad \frac{\partial y}{\partial r} = s\cos(rs), \quad \frac{\partial x}{\partial s} = r^2 e^s, \quad \frac{\partial y}{\partial s} = r\cos(rs)$$

$$\therefore \frac{\partial f}{\partial r} = \frac{\partial f}{\partial x}\frac{\partial x}{\partial r} + \frac{\partial f}{\partial y}\frac{\partial y}{\partial r} = \frac{10re^s + 2s\cos(rs)}{\sqrt{1-(5x+2y)^2}}$$

and $\dfrac{\partial f}{\partial s} = \dfrac{\partial f}{\partial x}\dfrac{\partial x}{\partial s} + \dfrac{\partial f}{\partial y}\dfrac{\partial y}{\partial s} = \dfrac{5r^2 e^s + 2r\cos(rs)}{\sqrt{1-(5x+2y)^2}}$

Example 12.

Let $F(x,y) = x^2 y^2 + x + y + 2$, $x(t) = t^4 - 1$, $y(t) = t^2 - t^3$, and $G(t) = F(x(t), y(t))$. Find $G'(t) = ?$.

【Solution】

$$\because G'(t) = \frac{\partial G}{\partial x}\frac{\partial x}{\partial t} + \frac{\partial G}{\partial y}\frac{\partial y}{\partial t}, \frac{\partial G}{\partial x} = 2xy^2 + 1, \frac{\partial G}{\partial y} = 2yx^2 + 1, \frac{\partial x}{\partial t} = 4t^3, \text{and } \frac{\partial y}{\partial t} = 2t - 3t^2$$

$$\therefore G'(t) = 4t^3(2xy^2 + 1) + (2t - 3t^2)(2yx^2 + 1)$$

Example 13.

Let $z = g(x^2 - y^2)$. Prove that $y\dfrac{\partial z}{\partial x} + x\dfrac{\partial z}{\partial y} = 0$.

【Solution】

$$\because \frac{\partial z}{\partial x} = g'(x^2 - y^2)(2x), \quad \frac{\partial z}{\partial y} = g'(x^2 - y^2)(-2y)$$

$$\therefore y\frac{\partial z}{\partial x} + x\frac{\partial z}{\partial y} = yg'(x^2 - y^2)(2x) + xg'(x^2 - y^2)(-2y) = 0$$

Example 14.

Let $z = x^2 + g(xy)$. Prove that $x\dfrac{\partial z}{\partial x} - y\dfrac{\partial z}{\partial y} = 2x^2$.

【Solution】

$$\because \frac{\partial z}{\partial x} = 2x + g'(xy)y, \quad \frac{\partial z}{\partial y} = g'(xy)x$$

$$\therefore x\frac{\partial z}{\partial x} - y\frac{\partial z}{\partial y} = x(2x + g'(xy)y) - y(g'(xy)x) = 2x^2$$

Example 15.

Let $f(x,y,z) = \dfrac{1}{\sqrt{x^2+y^2+z^2}}$. Prove that $f_{xx}+f_{yy}+f_{zz}=0$.

【Solution】

$\because f_x = -\dfrac{1}{2}(x^2+y^2+z^2)^{-\frac{3}{2}}(2x) = -x(x^2+y^2+z^2)^{-\frac{3}{2}}$

$f_y = -\dfrac{1}{2}(x^2+y^2+z^2)^{-\frac{3}{2}}(2y) = -y(x^2+y^2+z^2)^{-\frac{3}{2}}$

$f_z = -\dfrac{1}{2}(x^2+y^2+z^2)^{-\frac{3}{2}}(2z) = -z(x^2+y^2+z^2)^{-\frac{3}{2}}$

$f_{xx} = -(x^2+y^2+z^2)^{-\frac{3}{2}} + 3x^2(x^2+y^2+z^2)^{-\frac{5}{2}}$

$= (x^2+y^2+z^2)^{-\frac{5}{2}}(3x^2 - (x^2+y^2+z^2)) = (x^2+y^2+z^2)^{-\frac{5}{2}}(2x^2-y^2-z^2)$

$f_{yy} = (x^2+y^2+z^2)^{-\frac{5}{2}}(2y^2-x^2-z^2)$

$f_{zz} = (x^2+y^2+z^2)^{-\frac{5}{2}}(2z^2-x^2-y^2)$

$\therefore f_{xx}+f_{yy}+f_{zz}=0$

Example 16.

Let $f(x,y,z) = \left(\dfrac{x-y+z}{x+y-z}\right)^n$. Prove that $xf_x+yf_y+zf_z=0$.

【Solution】

$f_x(x,y,z) = n\left(\dfrac{x-y+z}{x+y-z}\right)^{n-1} \dfrac{(x+y-z)-(x-y+z)}{(x+y-z)^2} = 2n\left(\dfrac{x-y+z}{x+y-z}\right)^{n-1} \dfrac{y-z}{(x+y-z)^2}$

$f_y(x,y,z) = -2n\left(\dfrac{x-y+z}{x+y-z}\right)^{n-1} \dfrac{x-z}{(x+y-z)^2}$

$f_z(x,y,z) = 2n\left(\dfrac{x-y+z}{x+y-z}\right)^{n-1} \dfrac{x-y}{(x+y-z)^2}$

$\therefore xf_x+yf_y+zf_z = 2n\left(\dfrac{x-y+z}{x+y-z}\right)^{n-1}(x(y-z)-y(x-z)+z(x-y)) = 0$

Example 17.

Let $g(x,y) = f(xy^2)$. Prove that $4x^2 g_{xx} - y^2 g_{yy} + yg_y = 0$.

【Solution】

$\because g_x = f'(xy^2)y^2, \ g_y = f'(xy^2)2xy$

$\therefore g_{xx} = f''(xy^2)y^4, \ g_{yy} = f''(xy^2)4x^2y^2 + 2xf'(xy^2)$

$\therefore 4x^2 g_{xx} - y^2 g_{yy} + yg_y$

$= 4x^2 f''(xy^2)y^4 - y^2(f''(xy^2)4x^2y^2 + 2xf'(xy^2)) + yf'(xy^2)2xy = 0$

Example 18.

Let $w = f(u), u = \dfrac{xy}{x^2+y^2}$. Prove that $x\dfrac{\partial w}{\partial x} + y\dfrac{\partial w}{\partial y} = 0$.

【Solution】

$\dfrac{\partial w}{\partial x} = f'(u)\dfrac{y(x^2+y^2) - xy(2x)}{(x^2+y^2)^2} = f'(u)\dfrac{y^3 - x^2y}{(x^2+y^2)^2}$

$\dfrac{\partial w}{\partial y} = f'(u)\dfrac{x^3 - y^2x}{(x^2+y^2)^2}$

$x\dfrac{\partial w}{\partial x} + y\dfrac{\partial w}{\partial y} = x\left(f'(u)\dfrac{y^3 - x^2y}{(x^2+y^2)^2}\right) + y\left(f'(u)\dfrac{x^3 - y^2x}{(x^2+y^2)^2}\right) = 0$

Example 19.

Let $w = f(x-y, y-x)$. Prove that $\dfrac{\partial w}{\partial x} + \dfrac{\partial w}{\partial y} = 0$.

【Solution】

Let $u = x - y, \ v = y - x$

$\because \dfrac{\partial w}{\partial x} = f_u(u,v)\dfrac{\partial u}{\partial x} + f_v(u,v)\dfrac{\partial v}{\partial x} = f_u(u,v) - f_v(u,v)$

$\dfrac{\partial w}{\partial y} = f_u(u,v)\dfrac{\partial u}{\partial y} + f_v(u,v)\dfrac{\partial v}{\partial y} = -f_u(u,v) + f_v(u,v)$

$\therefore \dfrac{\partial w}{\partial x} + \dfrac{\partial w}{\partial y} = 0$

Example 20.

Let $u = f(x,y), x = r\cos\theta$, and $y = r\sin\theta$. Prove that

$\dfrac{\partial^2 u}{\partial x^2} + \dfrac{\partial^2 u}{\partial y^2} = \dfrac{\partial^2 u}{\partial r^2} + \dfrac{1}{r^2}\dfrac{\partial^2 u}{\partial \theta^2} + \dfrac{1}{r}\dfrac{\partial u}{\partial r}.$

【Solution】

$$\because \frac{\partial u}{\partial r} = \frac{\partial u}{\partial x} \cdot \frac{\partial x}{\partial r} + \frac{\partial u}{\partial y} \cdot \frac{\partial y}{\partial r} = \frac{\partial u}{\partial x} \cdot \cos\theta + \frac{\partial u}{\partial y} \cdot \sin\theta$$

$$\frac{\partial u}{\partial \theta} = \frac{\partial u}{\partial x} \cdot \frac{\partial x}{\partial \theta} + \frac{\partial u}{\partial y} \cdot \frac{\partial y}{\partial \theta} = -\frac{\partial u}{\partial x} \cdot r\sin\theta + \frac{\partial u}{\partial y} \cdot r\cos\theta$$

$$\frac{\partial^2 u}{\partial r^2} = \frac{\partial}{\partial r}\left(\frac{\partial u}{\partial r}\right) = \frac{\partial}{\partial r}\left(\frac{\partial u}{\partial x} \cdot \cos\theta + \frac{\partial u}{\partial y} \cdot \sin\theta\right) = \frac{\partial^2 u}{\partial x^2} \cdot \cos^2\theta + \frac{\partial^2 u}{\partial y^2} \cdot \sin^2\theta$$

$$\frac{\partial^2 u}{\partial \theta^2} = \frac{\partial}{\partial \theta}\left(\frac{\partial u}{\partial \theta}\right) = \frac{\partial}{\partial \theta}\left(-\frac{\partial u}{\partial x} \cdot r\sin\theta + \frac{\partial u}{\partial y} \cdot r\cos\theta\right)$$

$$= \frac{\partial^2 u}{\partial x^2} r^2 \sin^2\theta + \frac{\partial^2 u}{\partial y^2} \cdot r^2 \cos^2\theta - \frac{\partial u}{\partial x} \cdot r\cos\theta - \frac{\partial u}{\partial y} \cdot r\sin\theta$$

$$\therefore \frac{\partial^2 u}{\partial r^2} + \frac{1}{r^2}\frac{\partial^2 u}{\partial \theta^2} + \frac{1}{r}\frac{\partial u}{\partial r} = \frac{\partial^2 u}{\partial x^2} + \frac{\partial^2 u}{\partial y^2}$$

Example 21.

Let $f(x, y, z) = g(x - y, y - z, z - x)$. Prove that $\dfrac{\partial f}{\partial x} + \dfrac{\partial f}{\partial y} + \dfrac{\partial f}{\partial z} = 0$.

【Solution】

Let $t_1 = x - y$, $t_2 = y - z$, $t_3 = z - x$

$$\because \frac{\partial f}{\partial x} = \frac{\partial g}{\partial t_1}\frac{\partial t_1}{\partial x} + \frac{\partial g}{\partial t_2}\frac{\partial t_2}{\partial x} + \frac{\partial g}{\partial t_3}\frac{\partial t_3}{\partial x}, \quad \frac{\partial f}{\partial y} = \frac{\partial g}{\partial t_1}\frac{\partial t_1}{\partial y} + \frac{\partial g}{\partial t_2}\frac{\partial t_2}{\partial y} + \frac{\partial g}{\partial t_3}\frac{\partial t_3}{\partial y}$$

$$\frac{\partial f}{\partial z} = \frac{\partial g}{\partial t_1}\frac{\partial t_1}{\partial z} + \frac{\partial g}{\partial t_2}\frac{\partial t_2}{\partial z} + \frac{\partial g}{\partial t_3}\frac{\partial t_3}{\partial z} \quad \text{and} \quad \left(\frac{\partial t_1}{\partial x}, \frac{\partial t_2}{\partial x}, \frac{\partial t_3}{\partial x}\right) = (1, 0, -1),$$

$$\left(\frac{\partial t_1}{\partial y}, \frac{\partial t_2}{\partial y}, \frac{\partial t_3}{\partial y}\right) = (-1, 1, 0), \quad \left(\frac{\partial t_1}{\partial z}, \frac{\partial t_2}{\partial z}, \frac{\partial t_3}{\partial z}\right) = (0, -1, 1)$$

$$\therefore \frac{\partial f}{\partial x} + \frac{\partial f}{\partial y} + \frac{\partial f}{\partial z} = \frac{\partial g}{\partial t_1} - \frac{\partial g}{\partial t_3} - \frac{\partial g}{\partial t_1} + \frac{\partial g}{\partial t_2} - \frac{\partial g}{\partial t_2} + \frac{\partial g}{\partial t_3} = 0$$

Example 22.

Let $f(x, y) = g(x^2 - y^2, y^2 - x^2)$. Prove that $y\dfrac{\partial f}{\partial x} + x\dfrac{\partial f}{\partial y} = 0$.

【Solution】

Let $t_1 = x^2 - y^2$, $t_2 = y^2 - x^2$

$$\because \frac{\partial f}{\partial x} = \frac{\partial g}{\partial t_1}\frac{\partial t_1}{\partial x} + \frac{\partial g}{\partial t_2}\frac{\partial t_2}{\partial x}, \quad \frac{\partial f}{\partial y} = \frac{\partial g}{\partial t_1}\frac{\partial t_1}{\partial y} + \frac{\partial g}{\partial t_2}\frac{\partial t_2}{\partial y}$$

7.4 Determine the Differentiability of Multivariable Functions and Chain Rule

and $\left(\dfrac{\partial t_1}{\partial x}, \dfrac{\partial t_2}{\partial x}\right) = (2x, -2x)$, $\left(\dfrac{\partial t_1}{\partial y}, \dfrac{\partial t_2}{\partial y}\right) = (-2y, 2y)$

$\therefore y\dfrac{\partial f}{\partial x} + x\dfrac{\partial f}{\partial y} = y\left(\dfrac{\partial g}{\partial t_1}(2x) + \dfrac{\partial g}{\partial t_2}(-2x)\right) + x\left(\dfrac{\partial g}{\partial t_1}(-2y) + \dfrac{\partial g}{\partial t_2}(2y)\right) = 0$

Example 23.

Let $f(u, v) = g(x(u, v), y(u, v))$ and $g(x, y) = \dfrac{1}{(x^2 y^2 - 4x + 3y + 1)^p}$ $(p > 0)$,

where $\begin{cases} x(u, v) = u^2 - 3uv + 2v^2 \\ y(u, v) = u^4 + 3uv^3 - 4v^4 \end{cases}$. Find $\dfrac{\partial f}{\partial u}(1,1) = ?$

【Solution】

Let $p > 0$

$\therefore \dfrac{\partial f}{\partial u} = \dfrac{\partial g}{\partial x}\dfrac{\partial x}{\partial u} + \dfrac{\partial g}{\partial y}\dfrac{\partial y}{\partial u}$, $\dfrac{\partial g}{\partial x} = -\dfrac{p(2xy^2 - 4)}{(x^2 y^2 - 4x + 3y + 1)^{2p}}$

$\dfrac{\partial g}{\partial y} = -\dfrac{p(2yx^2 + 3)}{(x^2 y^2 - 4x + 3y + 1)^{2p}}$, $\dfrac{\partial x}{\partial u} = 2u - 3v$, $\dfrac{\partial y}{\partial u} = 4u^3 + 3v^3$

$\because (u, v) = (1,1) \quad \therefore (x, y) = (0,0)$

$\therefore \dfrac{\partial f}{\partial u}(1,1) = \dfrac{\partial g}{\partial x}(0,0)\dfrac{\partial x}{\partial u}(1,1) + \dfrac{\partial g}{\partial y}(0,0)\dfrac{\partial y}{\partial u}(1,1) = 4p(-1) + (-3p)7 = -25p$

Example 24.

Let $f(x, y) = \dfrac{x^2 y}{1 + x^2 + y^2}$, $x, y \in R$ and $g(u, v) = f(uv, u^2 + 2v^3)$, $u, v \in R$.

Find $\dfrac{\partial g}{\partial v}(\sqrt{2}, -1) = ?$

【Solution】

Let $x = uv$, $y = u^2 + 2v^3$ then $\dfrac{\partial g}{\partial v} = \dfrac{\partial f}{\partial x}\dfrac{\partial x}{\partial v} + \dfrac{\partial f}{\partial y}\dfrac{\partial y}{\partial v}$

$\therefore \dfrac{\partial f}{\partial x} = \dfrac{2xy(1 + x^2 + y^2) - x^2 y(2x)}{(1 + x^2 + y^2)^2} = \dfrac{2xy(1 + y^2)}{(1 + x^2 + y^2)^2}$

$\dfrac{\partial f}{\partial y} = \dfrac{x^2(1 + x^2 + y^2) - x^2 y(2y)}{(1 + x^2 + y^2)^2} = \dfrac{x^2 + x^4 - x^2 y^2}{(1 + x^2 + y^2)^2}$,

and $\dfrac{\partial x}{\partial v} = u$, $\dfrac{\partial y}{\partial v} = 6v^2$

Let $(u, v) = (\sqrt{2}, -1)$ then $(x, y) = (-\sqrt{2}, 0)$

$$\therefore \frac{\partial g}{\partial v}(-1,-1) = \frac{\partial f}{\partial x}(-\sqrt{2},0)\frac{\partial x}{\partial v}(\sqrt{2},-1) + \frac{\partial f}{\partial y}(-\sqrt{2},0)\frac{\partial y}{\partial v}(\sqrt{2},-1) = (0)(\sqrt{2}) + \frac{6}{9}(6)$$
$$= 4$$

Example 25.

Let $f(x,y) = \begin{cases} \dfrac{x^2 - y^2}{\sqrt{x^2+y^2}}, & (x,y) \neq (0,0) \\ 0, & (x,y) = (0,0) \end{cases}$.

Prove that $f(x,y)$ is not differentiable at $(0,0)$.

【Solution】

$$f_x(0,0) = \lim_{h \to 0}\frac{f(h+0,0) - f(0,0)}{h} = \lim_{h \to 0}\frac{\frac{h^2}{h} - 0}{h} = 1$$

$$f_y(0,0) = \lim_{h \to 0}\frac{f(0,h+0) - f(0,0)}{h} = \lim_{h \to 0}\frac{\frac{-h^2}{h} - 0}{h} = -1$$

Claim: $\lim\limits_{h \to 0, k \to 0} \left| \dfrac{f(h+0, k+0) - f(0,0) - \frac{\partial f}{\partial x}(0,0)h - \frac{\partial f}{\partial y}(0,0)k}{\sqrt{h^2+k^2}} \right| \neq 0$

$$\because \left| \frac{f(h+0, k+0) - f(0,0) - \frac{\partial f}{\partial x}(0,0)h - \frac{\partial f}{\partial y}(0,0)k}{\sqrt{h^2+k^2}} \right| = \left| \frac{\frac{h^2-k^2}{\sqrt{h^2+k^2}} - h + k}{\sqrt{h^2+k^2}} \right|$$

$$= \left| \frac{\frac{(h^2-k^2) - (h-k)(\sqrt{h^2+k^2})}{\sqrt{h^2+k^2}}}{\sqrt{h^2+k^2}} \right|$$

Choose $h = 2k$ then $\left| \dfrac{\frac{(h^2-k^2) - (h-k)(\sqrt{h^2+k^2})}{\sqrt{h^2+k^2}}}{\sqrt{h^2+k^2}} \right| = \left| \dfrac{3k^2 - \sqrt{5}k^2}{5k^2} \right|$

$$\therefore \lim_{h \to 0, k \to 0} \left| \frac{f(h+0, k+0) - f(0,0) - \frac{\partial f}{\partial x}(0,0)h - \frac{\partial f}{\partial y}(0,0)k}{\sqrt{h^2+k^2}} \right|$$

$$= \lim_{h \to 0, k \to 0} \left| \frac{3k^2 - \sqrt{5}k^2}{5k^2} \right| = \frac{3 - \sqrt{5}}{5} \neq 0$$

∴ $f(x,y)$ is not differentiable at $(0,0)$

Example 26.

Let $f(x,y) = \begin{cases} (x^2+y^2)^p \sin\left(\dfrac{1}{x^2+y^2}\right), & (x,y) \neq (0,0) \\ 0, & (x,y) = (0,0) \end{cases}$, $p > \dfrac{1}{2}$.

Prove that $f(x,y)$ is differentiable at $(0,0)$.

【Solution】

Let $p > \dfrac{1}{2}$

$f_x(0,0) = \lim\limits_{h \to 0} \dfrac{f(h+0,0) - f(0,0)}{h} = \lim\limits_{h \to 0} \dfrac{h^{2p} \sin(\frac{1}{h^2}) - 0}{h} = \lim\limits_{h \to 0} h^{2p-1} \sin(\dfrac{1}{h^2})$

$\because p > \dfrac{1}{2}$ ∴ $\lim\limits_{h \to 0} h^{2p-1} \sin(\dfrac{1}{h^2}) = 0 \Rightarrow f_x(0,0) = 0$

$f_y(0,0) = \lim\limits_{h \to 0} \dfrac{f(0,0+h) - f(0,0)}{h} = \lim\limits_{h \to 0} \dfrac{h^{2p} \sin(\frac{1}{h^2}) - 0}{h} = \lim\limits_{h \to 0} h^{2p-1} \sin(\dfrac{1}{h^2})$

$\because p > \dfrac{1}{2}$ ∴ $\lim\limits_{h \to 0} h^{2p-1} \sin(\dfrac{1}{h^2}) = 0 \Rightarrow f_y(0,0) = 0$

Claim: $\lim\limits_{h \to 0, k \to 0} \left| \dfrac{f(h+0, k+0) - f(0,0) - \frac{\partial f}{\partial x}(0,0)h - \frac{\partial f}{\partial y}(0,0)k}{\sqrt{h^2+k^2}} \right| = 0$

$\because \left| \dfrac{f(h+0, k+0) - f(0,0) - \frac{\partial f}{\partial x}(0,0)h - \frac{\partial f}{\partial y}(0,0)k}{\sqrt{h^2+k^2}} \right| = \left| \dfrac{(h^2+k^2)^p \sin\frac{1}{h^2+k^2}}{\sqrt{h^2+k^2}} \right|$

$= \left| (h^2+k^2)^{p-\frac{1}{2}} \sin\dfrac{1}{h^2+k^2} \right| \leq (h^2+k^2)^{p-\frac{1}{2}}$

∴ $\lim\limits_{h \to 0, k \to 0} (h^2+k^2)^{p-\frac{1}{2}} = 0$

∴ $\lim\limits_{h \to 0, k \to 0} \left| \dfrac{f(h+0, k+0) - f(0,0) - \frac{\partial f}{\partial x}(0,0)h - \frac{\partial f}{\partial y}(0,0)k}{\sqrt{h^2+k^2}} \right| = 0$

∴ $f(x,y)$ is differentiable at $(0,0)$

Example 27.

Let $f(x,y) = (x|y|)^p$, $p > \frac{1}{2}$. Prove that $f(x,y)$ is differentiable at $(0,0)$.

【Solution】

Let $p > \frac{1}{2}$

$$f_x(0,0) = \lim_{h \to 0} \frac{f(h+0,0) - f(0,0)}{h} = \lim_{h \to 0} \frac{0-0}{h} = 0$$

$$f_y(0,0) = \lim_{h \to 0} \frac{f(0,0+h) - f(0,0)}{h} = \lim_{h \to 0} \frac{0-0}{h} = 0$$

Claim: $\lim_{h \to 0, k \to 0} \left| \frac{f(h+0, k+0) - f(0,0) - \frac{\partial f}{\partial x}(0,0)h - \frac{\partial f}{\partial y}(0,0)k}{\sqrt{h^2 + k^2}} \right| = 0$

$\because \left| \frac{f(h+0, k+0) - f(0,0) - \frac{\partial f}{\partial x}(0,0)h - \frac{\partial f}{\partial y}(0,0)k}{\sqrt{h^2 + k^2}} \right|$

$= \left| \frac{h|k| - f(0,0) - \frac{\partial f}{\partial x}(0,0)h - \frac{\partial f}{\partial y}(0,0)k}{\sqrt{h^2 + k^2}} \right| = \left| \frac{(h|k|)^p}{\sqrt{h^2 + k^2}} \right| \leq \left| \frac{(h|k|)^p}{\sqrt{2|hk|}} \right| < |hk|^{p-\frac{1}{2}}$

$\because \lim_{h \to 0, k \to 0} |hk|^{p-\frac{1}{2}} = 0$

$\therefore \lim_{h \to 0, k \to 0} \left| \frac{f(h+0, k+0) - f(0,0) - \frac{\partial f}{\partial x}(0,0)h - \frac{\partial f}{\partial y}(0,0)k}{\sqrt{h^2 + k^2}} \right| = 0$

$\therefore f(x,y)$ is differentiable at $(0,0)$

Example 28.

Let $f(x,y) = \begin{cases} \dfrac{3x\sqrt{y}}{\sqrt{x^3 + y^3}}, & (x,y) \neq (0,0) \\ 0, & (x,y) = (0,0) \end{cases}$.

Prove that $f(x,y)$ is not differentiable at $(0,0)$.

【Solution】

Claim: $\lim_{h \to 0, k \to 0} \left| \frac{f(h+0, k+0) - f(0,0) - \frac{\partial f}{\partial x}(0,0)h - \frac{\partial f}{\partial y}(0,0)k}{\sqrt{h^2 + k^2}} \right| \neq 0$

$$\frac{\partial f}{\partial x}(0,0) = \lim_{h \to 0} \frac{f(h+0,0) - f(0,0)}{h} = \lim_{h \to 0} \frac{\frac{3 \cdot h^2 \cdot 0}{\sqrt{h^3 + 0^3}} - 0}{h} = \lim_{h \to 0} \frac{0}{h^{\frac{1}{2}}} = 0$$

$$\frac{\partial f}{\partial y}(0,0) = \lim_{h \to 0} \frac{f(0,h+0) - f(0,0)}{h} = \lim_{h \to 0} \frac{\frac{3 \cdot 0^2 \cdot \sqrt{h}}{\sqrt{0^3 + h^3}} - 0}{h} = \lim_{h \to 0} \frac{0}{h^2} = 0$$

$$\therefore \frac{\partial f}{\partial x}(0,0) = 0, \quad \frac{\partial f}{\partial y}(0,0) = 0$$

$$\because \left| \frac{f(h+0,k+0) - f(0,0) - \frac{\partial f}{\partial x}(0,0)h - \frac{\partial f}{\partial y}(0,0)k}{\sqrt{h^2 + k^2}} \right| = \left| \frac{3h\sqrt{k}}{\sqrt{h^3 + k^3}\sqrt{h^2 + k^2}} \right|$$

Choose $k = h$

$$\lim_{h \to 0, k \to 0} \left| \frac{f(h+0,k+0) - f(0,0) - \frac{\partial f}{\partial x}(0,0)h - \frac{\partial f}{\partial y}(0,0)k}{\sqrt{h^2 + k^2}} \right|$$

$$= \lim_{h \to 0, k \to 0} \left| \frac{3h\sqrt{k}}{\sqrt{h^3 + k^3}\sqrt{h^2 + k^2}} \right| = \lim_{h \to 0, k \to 0} \left| \frac{3h^{\frac{3}{2}}}{\sqrt{2h^{\frac{3}{2}}}\sqrt{h^2 + h^2}} \right| = \infty \neq 0$$

$\therefore f(x,y)$ is not differentiable at $(0,0)$

Example 29.

Let $f: R \to R$ be a differentiable function, such that $f(0) = 0$, $f'(0) = 2$, and $F(x) = \int_0^x t^2 f(x^3 - t^3) \, dt$. Find $\lim_{x \to 0} \frac{F(x)}{x^6} = ?$

【Solution】

Using L'Hôpital's Rule, $\lim_{x \to 0} \frac{F(x)}{x^6} = \lim_{x \to 0} \frac{F'(x)}{6x^5}$

By Leibniz's Rule for differentiation, $F'(x) = 3x^2 \int_0^x t^2 f'(x^3 - t^3) \, dt$

$$\therefore \lim_{x \to 0} \frac{F'(x)}{6x^5} = \lim_{x \to 0} \frac{3x^2 \int_0^x t^2 f'(x^3 - t^3) \, dt}{6x^5} = \lim_{x \to 0} \frac{\int_0^x t^2 f'(x^3 - t^3) \, dt}{2x^3}$$

$$= \lim_{x \to 0} \frac{3x^2 \int_0^x t^2 f''(x^3 - t^3) \, dt + x^2 f'(x^3 - x^3)}{6x^2} = \lim_{x \to 0} \frac{x^2 f'(x^3 - x^3)}{6x^2} = \frac{f'(0)}{6} = \frac{1}{3}$$

7.5 Find Directional Derivatives

【Definition】 Directional Derivative for a Function of Two Variables

(i) Given a function $f(x, y)$, and a unit vector $\vec{u} = (u_1, u_2)$, $(x_0, y_0) \in R^2$

(ii) If the limit $\lim\limits_{h \to 0} \dfrac{f(x_0 + hu_1, y_0 + hu_2) - f(x_0, y_0)}{h}$ exists, then

the directional derivative of $f(x, y)$ at (x_0, y_0) in the direction of $\vec{u} = (u_1, u_2)$ exists and is denoted by $D_{\vec{u}} f(x_0, y_0)$.

【Definition】 Directional Derivative for a Multivariable Function

(i) Given a multivariable function $f(x_1, \ldots, x_n)$, and a unit vector $\vec{u} = (u_1, \ldots, u_n)$, where $(c_1, \ldots, c_n) \in R^n$

(ii) If the limit $\lim\limits_{h \to 0} \dfrac{f(c_1 + hu_1, \ldots, c_n + hu_n) - f(c_1, \ldots, c_n)}{h}$ exists, then the directional

derivative of $f(x_1, \ldots, x_n)$ at $(c_1, \ldots, c_n)$ in the direction of $\vec{u} = (u_1, \ldots, u_n)$ exists and is denoted by $D_{\vec{u}} f(c_1, \ldots, c_n)$.

Question Types:

Type 1.

Given a function $f(x, y)$ of two variables, find the directional derivative of f at (x_0, y_0) in the direction of the vector $\vec{u} = (u_1, u_2)$, $D_{\vec{u}} f(x_0, y_0) = ?$

Problem-Solving Process:

Use the definition of the directional derivative. If

$$\lim_{h \to 0} \frac{f(x_0 + hu_1, y_0 + hu_2) - f(x_0, y_0)}{h} \text{ exists,}$$

then the directional derivative of the function $f(x, y)$ at (x_0, y_0) in the direction of (u_1, u_2) is

$$\lim_{h \to 0} \frac{f(x_0 + hu_1, y_0 + hu_2) - f(x_0, y_0)}{h},$$

which can be expressed as

$$D_{\vec{u}} f(x_0, y_0) = \lim_{h \to 0} \frac{f(x_0 + hu_1, y_0 + hu_2) - f(x_0, y_0)}{h}.$$

Additionally, a common problem type is to find the directional derivative at $(0,0)$ in the direction of (u_1, u_2), which involves finding

7.5 Find Directional Derivatives

$$\lim_{h \to 0} \frac{f(hu_1, hu_2) - f(0,0)}{h}.$$

Type 2.
Given a function $f(x, y, z)$ of three variables, find the directional derivative of f at (x_0, y_0, z_0) in the direction of the vector $\vec{u} = (u_1, u_2, u_3)$.

Problem-Solving Process:
Use the definition to find the directional derivative. If

$$\lim_{h \to 0} \frac{f(x_0 + hu_1, y_0 + hu_2, z_0 + hu_3) - f(x_0, y_0, z_0)}{h} \text{ exists,}$$

then the directional derivative of the function $f(x, y, z)$ at (x_0, y_0, z_0) in the direction of (u_1, u_2, u_3) is

$$\lim_{h \to 0} \frac{f(x_0 + hu_1, y_0 + hu_2, z_0 + hu_3) - f(x_0, y_0, z_0)}{h}$$

which can be written as

$$D_{\vec{u}} f(x_0, y_0, z_0) = \lim_{h \to 0} \frac{f(x_0 + hu_1, y_0 + hu_2, z_0 + hu_3) - f(x_0, y_0, z_0)}{h}.$$

Additionally, a common problem type is to find the directional derivative at (0,0,0) in the direction of $\vec{u} = (u_1, u_2, u_3)$, which involves finding

$$\lim_{h \to 0} \frac{f(hu_1, hu_2, hu_3) - f(0,0,0)}{h}.$$

Example 1.

Let $f(x, y) = \begin{cases} \dfrac{x^2 y}{x^2 + y^2}, & (x, y) \neq 0 \\ 0, & (x, y) = 0 \end{cases}$.

Find the directional derivative of $f(x, y)$ at $(0,0)$ in the direction of the unit vector $\vec{u} = (u_1, u_2)$.

【Solution】

$$\because D_{\vec{u}} f(x, y) = \lim_{h \to 0} \frac{f(x + hu_1, y + hu_2) - f(x, y)}{h}, \quad \vec{u} = (u_1, u_2)$$

$$\therefore D_{\vec{u}} f(0,0) = \lim_{h \to 0} \frac{f(hu_1, hu_2) - f(0,0)}{h} = \lim_{h \to 0} \frac{\frac{(hu_1)^2 hu_2}{(hu_1)^2 + (hu_2)^2}}{h} = \frac{u_1^2 u_2}{u_1^2 + u_2^2}$$

Example 2.

Let $f(x,y) = \begin{cases} \dfrac{xy^3}{x^3+y^6}, & (x,y) \neq 0 \\ 0, & (x,y) = 0 \end{cases}$.

Find the directional derivative of $f(x,y)$ at $(0,0)$ in the direction of the unit vector $\vec{u} = (s,t)$.

【Solution】

$\because D_{\vec{u}} f(x,y) = \lim\limits_{h \to 0} \dfrac{f(x+hs, y+ht) - f(x,y)}{h}$, $\vec{u} = (s,t)$

$\therefore D_{\vec{u}} f(0,0) = \lim\limits_{h \to 0} \dfrac{f(hs, ht) - f(0,0)}{h} = \lim\limits_{h \to 0} \dfrac{\frac{hs(ht)^3}{(hs)^3 + (ht)^6}}{h} = \lim\limits_{h \to 0} \dfrac{h^4 st^3}{h^4 s^3 + h^7 t^6}$

$= \lim\limits_{h \to 0} \dfrac{st^3}{s^3 + h^3 t^6} = \dfrac{t^3}{s^2}$

Example 3.

Let $f(x,y,z) = x^2 + 3y^2 + 4z^2$. Find the directional derivative of $f(x,y,z)$ at $(2,0,1)$ in the direction of the vector $(2,-1,0)$.

【Solution】

$\because D_{\vec{u}} f(x,y,z) = \lim\limits_{h \to 0} \dfrac{f(x+hu_1, y+hu_2, z+hu_3) - f(x,y,z)}{h}$,

$\vec{u} = (u_1, u_2, u_3) = \left(\dfrac{2}{\sqrt{5}}, \dfrac{-1}{\sqrt{5}}, \dfrac{0}{\sqrt{5}}\right)$, by L'Hôpital's Rule,

$D_{\vec{u}} f(2,0,1) = \lim\limits_{h \to 0} \dfrac{f(2+hu_1, hu_2, 1+hu_3) - f(2,0,1)}{h}$

$= \lim\limits_{h \to 0} \dfrac{(2+hu_1)^2 + 3(hu_2)^2 + 4(1+hu_3)^2 - 8}{h}$

$= \lim\limits_{h \to 0} 2u_1(2+hu_1) + 8u_3(1+hu_3) = 4u_1 + 8u_3 = \dfrac{8}{\sqrt{5}}$

Example 4.

Let $f(x,y) = \begin{cases} \sqrt{xy}\left(\dfrac{x^2-y^2}{x^2+y^2}\right), & (x,y) \neq 0 \\ 0, & (x,y) = 0 \end{cases}$.

Find the directional derivative of $f(x,y)$ at $(0,0)$ in the direction of the unit vector $\vec{u} = (s,t)$.

【Solution】

7.5 Find Directional Derivatives

$$\because D_{\vec{u}}f(x,y) = \lim_{h\to 0}\frac{f(x+hs, y+ht) - f(x,y)}{h}, \quad \vec{u} = (s,t)$$

$$\therefore D_{\vec{u}}f(0,0) = \lim_{h\to 0}\frac{f(hs, ht) - f(0,0)}{h} = \lim_{h\to 0}\frac{\sqrt{h^2 st}\left(\frac{(hs)^2 - (ht)^2}{(hs)^2 + (ht)^2}\right)}{h} = \lim_{h\to 0}\frac{\sqrt{st}(s^2 - t^2)}{s^2 + t^2}$$

$$= \frac{\sqrt{st}(s^2 - t^2)}{s^2 + t^2}$$

Example 5.

Let $f(x,y) = \begin{cases} \dfrac{x^3 - y^3}{x^2 + y^3}, & (x,y) \neq 0 \\ 0, & (x,y) = 0 \end{cases}$.

Find the directional derivative of $f(x,y)$ at $(0,0)$ in the direction of the unit vector $\vec{u} = (s,t)$.

【Solution】

$$\because D_{\vec{u}}f(x,y) = \lim_{h\to 0}\frac{f(x+hs, y+ht) - f(x,y)}{h}, \quad \vec{u} = (s,t)$$

$$\therefore D_{\vec{u}}f(0,0) = \lim_{h\to 0}\frac{f(hs, ht) - f(0,0)}{h} = \lim_{h\to 0}\frac{\frac{(hs)^3 - (ht)^3}{(hs)^2 + (ht)^3}}{h} = \lim_{h\to 0}\frac{s^3 - t^3}{s^2} = \frac{s^3 - t^3}{s^2}$$

Example 6.

Let $f(x,y) = \begin{cases} \dfrac{x^2 - y^2}{\sqrt{x^2 + y^3}}, & (x,y) \neq 0 \\ 0, & (x,y) = 0 \end{cases}$.

Find the directional derivative of $f(x,y)$ at $(0,0)$ in the direction of the unit vector $\vec{u} = (s,t)$.

【Solution】

$$\because D_{\vec{u}}f(x,y) = \lim_{h\to 0}\frac{f(x+hs, y+ht) - f(x,y)}{h}, \quad \vec{u} = (s,t)$$

$$\therefore D_{\vec{u}}f(0,0) = \lim_{h\to 0}\frac{f(hs, ht) - f(0,0)}{h} = \lim_{h\to 0}\frac{\frac{(hs)^2 - (ht)^2}{\sqrt{(hs)^2 + (ht)^3}}}{h} = \lim_{h\to 0}\frac{s^2 - t^2}{s} = \frac{s^2 - t^2}{s}$$

Example 7.

Let $f(x, y) = \begin{cases} \dfrac{\sin(x^2 - y^2)}{x + y}, & (x, y) \neq 0 \\ 0, & (x, y) = 0 \end{cases}$.

Find the directional derivative of $f(x, y)$ at $(0,0)$ in the direction of the unit vector $\vec{u} = (s, t)$.

【Solution】

$\because D_{\vec{u}}f(x, y) = \lim\limits_{h \to 0} \dfrac{f(x + hs, y + ht) - f(x, y)}{h}$, $\vec{u} = (s, t)$

$\therefore D_{\vec{u}}f(0,0) = \lim\limits_{h \to 0} \dfrac{f(hs, ht) - f(0,0)}{h} = \lim\limits_{h \to 0} \dfrac{\frac{\sin((hs)^2 - (ht)^2)}{hs + ht}}{h}$

$= \lim\limits_{h \to 0} \dfrac{(hs - ht) \cdot \frac{\sin((hs)^2 - (ht)^2)}{(hs)^2 - (ht)^2}}{h}$

$\because \lim\limits_{h \to 0} \dfrac{\sin((hs)^2 - (ht)^2)}{(hs)^2 - (ht)^2} = 1 \quad \therefore D_{\vec{u}}f(0,0) = \lim\limits_{h \to 0} \dfrac{(hs - ht)\left(\frac{\sin(hs)^2 - (ht)^2}{(hs)^2 - (ht)^2}\right)}{h} = s - t$

Example 8.

Let $f(x, y) = xe^y - ye^x$. Find the directional derivative of $f(x, y)$ at $(1, 0)$ in the direction of the unit vector $\vec{u} = \left(\dfrac{3}{5}, \dfrac{4}{5}\right)$.

【Solution】

$\because D_{\vec{u}}f(x, y) = \lim\limits_{h \to 0} \dfrac{f(x + hu_1, y + hu_2) - f(x, y)}{h}$, $\vec{u} = \left(\dfrac{3}{5}, \dfrac{4}{5}\right) = (u_1, u_2)$

By L'Hôpital's Rule,

$D_{\vec{u}}f(0,0) = \lim\limits_{h \to 0} \dfrac{f(1 + u_1h, u_2h) - f(1,0)}{h} = \lim\limits_{h \to 0} \dfrac{(1 + u_1h)e^{u_2h} - u_2he^{1+u_1h} - 1}{h}$

$= \lim\limits_{h \to 0} u_1 e^{u_2h} + u_2(1 + u_1h)e^{u_2h} - u_2 e^{1+u_1h} - u_1 u_2 h e^{1+u_1h} = \dfrac{7}{5} - \dfrac{4e}{5}$

Example 9.

Let $f(x, y, z) = xy^2 + yz^3$. Find the directional derivative of $f(x, y, z)$ at $(2, -1, 1)$ in the direction of the unit vector (u_1, u_2, u_3).

【Solution】

$\because D_{\vec{u}}f(x, y, z) = \lim\limits_{h \to 0} \dfrac{f(x + hu_1, y + hu_2, z + hu_3) - f(x, y, z)}{h}$, where $\vec{u} = (u_1, u_2, u_3)$,

By L'Hôpital's Rule,
$$D_{\vec{u}}f(2,-1,1) = \lim_{h \to 0} \frac{f(2+hu_1, -1+hu_2, 1+hu_3) - f(2,-1,1)}{h}$$
$$= \lim_{h \to 0} \frac{(2+hu_1)(-1+hu_2)^2 + (-1+hu_2)(1+hu_3)^3 - 1}{h}$$
$$= \lim_{h \to 0} u_1(-1+hu_2)^2 + (2+hu_1)2u_2(-1+hu_2) + u_2(1+hu_3)^3$$
$$+(-1+hu_2)3u_3(1+hu_3)^2$$
$$= u_1 - 4u_2 + u_2 - 3u_3 = u_1 - 3u_2 - 3u_3$$

Example 10.

Let $f(x,y) = \begin{cases} \sqrt{xy}\sin\dfrac{x^2-y^2}{x^2+y^2}, & (x,y) \neq 0 \\ 0, & (x,y) = 0 \end{cases}$.

Find the directional derivative of $f(x,y)$ at $(0,0)$ in the direction of the unit vector $\vec{u} = (s,t)$.

【Solution】

$\because D_{\vec{u}}f(x,y) = \lim_{h \to 0} \dfrac{f(x+hs, y+ht) - f(x,y)}{h}$, $\vec{u} = (s,t)$

$\therefore D_{\vec{u}}f(0,0) = \lim_{h \to 0} \dfrac{f(hs, ht) - f(0,0)}{h} = \lim_{h \to 0} \dfrac{\sqrt{h^2 st}\dfrac{\sin((hs)^2 - (ht)^2)}{(hs)^2 + (ht)^2}}{h}$

$= \lim_{h \to 0} \dfrac{\sqrt{h^2 st}((hs)^2 - (ht)^2)\dfrac{\sin((hs)^2 - (ht)^2)}{((hs)^2 + (ht)^2)((hs)^2 - (ht)^2)}}{h}$

$\because \lim_{h \to 0} \dfrac{\sin((hs)^2 - (ht)^2)}{(hs)^2 - (ht)^2} = 1$

$\therefore D_{\vec{u}}f(0,0) = \lim_{h \to 0} \dfrac{\sqrt{h^2 st}((hs)^2 - (ht)^2)\dfrac{\sin(hs)^2 - (ht)^2}{((hs)^2 + (ht)^2)((hs)^2 - (ht)^2)}}{h}$

$= \lim_{h \to 0} \dfrac{\dfrac{\sqrt{h^2 st}((hs)^2 - (ht)^2)}{((hs)^2 + (ht)^2)}}{h} = \dfrac{\sqrt{st}(s^2 - t^2)}{s^2 + t^2}$

Example 11.

Let $f(x,y,z) = xyz$.

(1) Find the directional derivative of $f(x,y,z)$ at $(1,3,2)$ in the direction of the unit vector (u_1, u_2, u_3).

(2) Find the maximum directional derivative at $(x,y,z) = (1,3,2)$.

【Solution】

(1)
$$\because D_{\vec{u}}f(x,y,z) = \lim_{h \to 0} \frac{f(x+hu_1, y+hu_2, z+hu_3) - f(x,y,z)}{h}, \quad \vec{u} = (u_1, u_2, u_3),$$

By L'Hôpital's Rule,
$$D_{\vec{u}}f(1,3,2) = \lim_{h \to 0} \frac{f(1+hu_1, 3+hu_2, 2+hu_3) - f(1,3,2)}{h}$$
$$= \lim_{h \to 0} \frac{(1+hu_1)(3+hu_2)(2+hu_3) - 6}{h}$$
$$= \lim_{h \to 0} u_1(3+hu_2)(2+hu_3) + u_2(1+hu_1)(2+hu_3) + u_3(1+hu_1)(3+hu_2)$$
$$= 6u_1 + 2u_2 + 3u_3$$

(2)
$$\because D_{\vec{u}}f(1,3,2) = 6u_1 + 2u_2 + 3u_3$$

By Cauchy inequality, $6u_1 + 2u_2 + 3u_3 \leq \sqrt{(6^2 + 2^2 + 3^2)(u_1^2 + u_2^2 + u_3^2)}$

$\because u_1^2 + u_2^2 + u_3^2 = 1$ and equality holds when $\dfrac{u_1}{6} = \dfrac{u_2}{2} = \dfrac{u_3}{3}$

$\therefore$ the maximum directional derivative occurs along the vector $\left(\dfrac{6}{7}, \dfrac{2}{7}, \dfrac{3}{7}\right)$ with a value of 7

Example 12.

Let $f(x,y,z) = xy + yz + xz$.

(1) Find the directional derivative of $f(x,y,z)$ at $(1,1,1)$ in the direction of the unit vector (u_1, u_2, u_3).

(2) Find the maximum directional derivative at $(x,y,z) = (1,1,1)$.

【Solution】

(1)
$$\because D_{\vec{u}}f(x,y,z) = \lim_{h \to 0} \frac{f(x+hu_1, y+hu_2, z+hu_3) - f(x,y,z)}{h}, \vec{u} = (u_1, u_2, u_3),$$

By L'Hôpital's Rule,
$$D_{\vec{u}}f(1,1,1) = \lim_{h \to 0} \frac{f(1+hu_1, 1+hu_2, 1+hu_3) - f(1,1,1)}{h}$$
$$= \lim_{h \to 0} \frac{(1+hu_1)(1+hu_2) + (1+hu_2)(1+hu_3) + (1+hu_1)(1+hu_3) - 3}{h}$$
$$= \lim_{h \to 0}(u_1 + u_2 + 2hu_1u_2) + (u_2 + u_3 + 2hu_3u_2) + (u_1 + u_3 + 2hu_1u_3) = 2(u_1 + u_2 + u_3)$$

(2)

∵ $D_{\vec{u}}f(1,1,1) = 2(u_1 + u_2 + u_3)$

By Cauchy inequality, $2u_1 + 2u_2 + 2u_3 \leq \sqrt{(2^2 + 2^2 + 2^2)(u_1^2 + u_2^2 + u_3^2)}$

∵ $u_1^2 + u_2^2 + u_3^2 = 1$ and equality holds when $\dfrac{u_1}{2} = \dfrac{u_2}{2} = \dfrac{u_3}{2}$

∴ the maximum directional derivative occurs along the vector $\left(\dfrac{1}{\sqrt{3}}, \dfrac{1}{\sqrt{3}}, \dfrac{1}{\sqrt{3}}\right)$ with a value of $\sqrt{12}$

Example 13.

Let $f(x,y) = \begin{cases} \dfrac{\sin(x^3 - y^3)}{x^2 + xy + y^2}, & (x,y) \neq 0 \\ 0, & (x,y) = 0 \end{cases}$.

Find the directional derivative of $f(x,y)$ at $(0,0)$ in the direction of the unit vector $\vec{u} = (s,t)$.

【Solution】

∵ $D_{\vec{u}}f(x,y) = \lim\limits_{h \to 0} \dfrac{f(x+hs, y+ht) - f(x,y)}{h}$, $\vec{u} = (s,t)$

∴ $D_{\vec{u}}f(0,0) = \lim\limits_{h \to 0} \dfrac{f(hs, ht) - f(0,0)}{h} = \lim\limits_{h \to 0} \dfrac{\dfrac{\sin((hs)^3 - (ht)^3)}{(hs)^2 + h^2st + (ht)^2}}{h}$

$= \lim\limits_{h \to 0} \dfrac{(hs - ht)\dfrac{\sin((hs)^3 - (ht)^3)}{(hs)^3 - (ht)^3}}{h}$

∵ $\lim\limits_{h \to 0} \dfrac{\sin((hs)^3 - (ht)^3)}{(hs)^3 - (ht)^3} = 1$ ∴ $D_{\vec{u}}f(0,0) = \lim\limits_{h \to 0} \dfrac{(hs - ht)\dfrac{\sin((hs)^3 - (ht)^3)}{(hs)^3 - (ht)^3}}{h} = s - t$

7.6 Find Normal and Tangent Equations of Surfaces

Question Types:

Type 1.

Given a surface defined by $f(x,y,z) = 0$, find the unit normal vector, the tangent plane, and the normal line equation at the point (x_0, y_0, z_0).

Problem-Solving Process:

7.6 Find Normal and Tangent Equations of Surfaces

Step1.
Find $\nabla f(x, y, z)$. Assume $\nabla f(x, y, z) = \left(f_x(x, y, z), f_y(x, y, z), f_z(x, y, z)\right)$.

Step2.
Thus, the normal vector at (x_0, y_0, z_0) is
$$\nabla f(x_0, y_0, z_0) = (f_x(x_0, y_0, z_0), f_y(x_0, y_0, z_0), f_z(x_0, y_0, z_0))$$

Step3.
Therefore, the unit normal vector at (x_0, y_0, z_0) is
$$\left(\frac{f_x(x_0, y_0, z_0)}{\sqrt{f_x^2 + f_y^2 + f_z^2}}, \frac{f_y(x_0, y_0, z_0)}{\sqrt{f_x^2 + f_y^2 + f_z^2}}, \frac{f_z(x_0, y_0, z_0)}{\sqrt{f_x^2 + f_y^2 + f_z^2}}\right)$$

Step4.
The tangent plane at (x_0, y_0, z_0) is given by
$$f_x(x_0, y_0, z_0)(x - x_0) + f_y(x_0, y_0, z_0)(y - y_0) + f_z(x_0, y_0, z_0)(z - z_0) = 0$$

Step5.
The normal line equation at (x_0, y_0, z_0) is
$$\frac{x - x_0}{f_x(x_0, y_0, z_0)} = \frac{y - y_0}{f_y(x_0, y_0, z_0)} = \frac{z - z_0}{f_z(x_0, y_0, z_0)}.$$

Type 2.
Given two surfaces $f_1(x, y, z) = 0$ and $f_2(x, y, z) = 0$, find the unit tangent vector to the curve of intersection of the two surfaces at the point (x_0, y_0, z_0).

Problem-Solving Process:

Step1.
Find $\nabla f_1(x, y, z)$ and $\nabla f_2(x, y, z)$.
Assume $\nabla f_1(x_0, y_0, z_0) = (a, b, c)$ and $\nabla f_2(x_0, y_0, z_0) = (d, e, f)$

Step2.
∵ the tangent vector to the intersection curve is perpendicular to the unit normal vectors of both surfaces
∴ the tangent vector is given by

$$(a, b, c) \times (d, e, f) = \begin{vmatrix} \vec{i} & \vec{j} & \vec{k} \\ a & b & c \\ d & e & f \end{vmatrix}$$

Step3.
The unit tangent vector is obtained by dividing the above tangent vector by its magnitude.

7.6 Find Normal and Tangent Equations of Surfaces

Example 1.
Find the unit normal vector of the surface $2x^2 + 4yz - 5z^2 = -1$ at point $(0,1,1)$.

【Solution】

Let $f(x,y,z) = 2x^2 + 4yz - 5z^2 + 1$ then $\nabla f(x,y,z) = (4x, 4z, 4y - 10z)$
∴ the normal vector at $((0,1,1) = \nabla f(0,1,1) = (0,4,-6)$
∴ the unit normal vector through $(0,1,1) = \left(0, \dfrac{2}{\sqrt{13}}, \dfrac{-3}{\sqrt{13}}\right)$

Example 2.
Find the unit normal vector of the surface $z = x^3y^3 + x + 3$ at point $(1,1,5)$.

【Solution】

Let $f(x,y,z) = x^3y^3 + x + 3 - z$ then $\nabla f(x,y,z) = (3x^2y^3 + 1, 3y^2x^3, -1)$
∴ the normal vector at $(1,1,5)$ is $\nabla f(1,1,5) = (4,3,-1)$
∴ the unit normal vector through $(1,1,5) = \left(\dfrac{4}{\sqrt{26}}, \dfrac{3}{\sqrt{26}}, \dfrac{-1}{\sqrt{26}}\right)$

Example 3.
Find the equation of the tangent plane and the normal line to the surface
$2xz^2 - 3yx - 4x = 5$ at point $(-1,1,1)$.

【Solution】

Let $f(x,y,z) = 2xz^2 - 3yx - 4x$ then $\nabla f(x,y,z) = (2z^2 - 3y - 4, -3x, 4xz)$
∴ the normal vector at $(-1,1,1)$ is $\nabla f(-1,1,1) = (-5,3,-4)$
∴ the equation of the tangent plane at $(-1,1,1)$ is $-5(x+1) + 3(y-1) - 4(z-1) = 0$
∴ the equation of the normal line at $(-1,1,1)$ is $\dfrac{x+1}{-5} = \dfrac{y-1}{3} = \dfrac{z-1}{-4}$

Example 4.
Find the equation of the tangent plane and the normal line to the surface
$x^2 + 2y^2 + z^2 = 4$ at point $(1,1,1)$.

【Solution】

Let $f(x,y,z) = x^2 + 2y^2 + z^2$ then $\nabla f(x,y,z) = (2x, 4y, 2z)$
∴ the normal vector at $(1,1,1)$ is $\nabla f(1,1,1) = (2,4,2)$
∴ the equation of the tangent plane at $(1,1,1)$ is $1(x-1) + 2(y-1) + 1(z-1) = 0$

∴ the equation of the normal line at $(1,1,1)$ is $\dfrac{x-1}{1} = \dfrac{y-1}{2} = \dfrac{z-1}{1}$

Example 5.

Find the unit tangent vector at point $(1,1,0)$ on the curve formed by the intersection of the surfaces $x^3 + 2y^3 + 3z^3 - 2xyz = 3$ and $x^2 + y^2 + z^2 = 2$.

【Solution】

Let $f_1(x,y,z) = x^3 + 2y^3 + 3z^3 - 2xyz$ then
$\nabla f_1(x,y,z) = (3x^2 - 2yz, 6y^2 - 2xz, 9z^2 - 2xy)$
Let $f_2(x,y,z) = x^2 + y^2 + z^2$ then $\nabla f_2(x,y,z) = (2x, 2y, 2z)$
∴ $\nabla f_1(1,1,0) = (3,6,-2)$, $\nabla f_2(1,1,0) = (2,2,0)$
∵ the unit tangent vector of the curve is perpendicular to the unit normal vectors of the two surfaces and $(3,6,-2) \times (2,2,0) = \begin{vmatrix} i & j & k \\ 3 & 6 & -2 \\ 2 & 2 & 0 \end{vmatrix} = (4,-4,-6)$

∴ the unit tangent vector is $= \left(\dfrac{2}{\sqrt{17}}, \dfrac{-2}{\sqrt{17}}, \dfrac{-3}{\sqrt{17}}\right)$

Example 6.

Find the unit tangent vector at point $\left(2, \dfrac{1}{3}, \dfrac{4}{3}\right)$ on the curve formed by the intersection of the surfaces $\dfrac{x^2}{9} + y^2 + \dfrac{z^2}{4} = 1$ and $x^3 = 9yz + 4$.

【Solution】

Let $f_1(x,y,z) = \dfrac{x^2}{9} + y^2 + \dfrac{z^2}{4}$ then $\nabla f_1(x,y,z) = \left(\dfrac{2x}{9}, 2y, \dfrac{z}{2}\right)$
Let $f_2(x,y,z) = x^3 - 9yz - 4$ then $\nabla f_2(x,y,z) = (3x^2, -9z, -9y)$
∴ $\nabla f_1\left(2, \dfrac{1}{3}, \dfrac{4}{3}\right) = \left(\dfrac{4}{9}, \dfrac{2}{3}, \dfrac{2}{3}\right)$, $\nabla f_2\left(2, \dfrac{1}{3}, \dfrac{4}{3}\right) = (12, -12, -3)$
∵ the unit tangent vector of the curve is perpendicular to the unit normal vectors of the two surfaces

and $(2,3,3) \times (4,-4,-1) = \begin{vmatrix} i & j & k \\ 2 & 3 & 3 \\ 4 & -4 & -1 \end{vmatrix} = (9,14,-20)$

∴ the unit tangent vector is $= \left(\dfrac{9}{\sqrt{677}}, \dfrac{14}{\sqrt{677}}, \dfrac{-20}{\sqrt{677}}\right)$

7.6 Find Normal and Tangent Equations of Surfaces

Example 7.
Given the surface $z^2 + xy - 2x - y^2 = 3$, find the points on the surface where the tangent plane is parallel to $z = 3$.

【Solution】

Let $f(x, y, z) = z^2 + xy - 2x - y^2 - 3$ then $\nabla f(x, y, z) = (y - 2, x - 2y, 2z)$
Let the tangent plane pass through (x_0, y_0, z_0), then its normal vector is
$(y_0 - 2, x_0 - 2y_0, 2z_0)$
∵ the tangent plane is parallel to $z = 3$ ∴ $(y_0 - 2, x_0 - 2y_0) = (0,0) \Rightarrow y_0 = 2, x_0 = 4$
∵ the surface $z^2 + xy - 2x - y^2 = 3$ passes through the point $(4, 2, z_0)$
∴ $z_0^2 + 8 - 8 - 4 = 3 \Rightarrow z_0 = \pm\sqrt{7}$
∴ the tangent plane at $(4, 2, \pm\sqrt{7})$ is parallel to $z = 3$

Example 8.
Given the surface $x^2 - 4xy - 2y^2 + 12x - 12y - z - 3 = 0$, find the points on the surface where the tangent plane is parallel to $z = 2$.

【Solution】

Let $f(x, y, z) = x^2 - 4xy - 2y^2 + 12x - 12y - z - 3$
then $\nabla f(x, y, z) = (2x - 4y + 12, -4x - 4y - 12, -1)$
Let the tangent plane pass through (x_0, y_0, z_0), then its normal vector is
$(2x_0 - 4y_0 + 12, -4x_0 - 4y_0 - 12, -1)$
∵ the tangent plane is parallel to $z = 2$
∴ $(2x_0 - 4y_0 + 12, -4x_0 - 4y_0 - 12) = (0,0) \Rightarrow y_0 = 1, x_0 = -4$,
∵ the surface $x^2 - 4xy - 2y^2 + 12x - 12y - z - 3 = 0$ passes through the point $(-4, 1, z_0)$
∴ $4^2 - 4(-4) - 2 + 12(-4) - 12 - z_0 - 3 = 0 \Rightarrow z_0 = -33$
∴ the tangent plane at $(-4, 1, -33)$ is parallel to $z = 2$

Example 9.
Find the equation of the tangent plane and the normal line to the surface
$x^2yz + 3y^2 = 2xz^2 - 8z$ at point $(1, 2, -1)$.

【Solution】

Let $f(x, y, z) = x^2yz + 3y^2 - 2xz^2 + 8z$
then $\nabla f(x, y, z) = (2xyz - 2z^2, x^2z + 6y, x^2y - 4xz + 8)$
∴ the normal vector at $(1, 2, -1)$ is $\nabla f(1, 2, -1) = (-6, 11, 14)$
∴ the equation of the tangent plane at $(1, 2, -1)$ is $-6(x - 1) + 11(y - 2) + 14(z + 1) = 0$

∴ the equation of the normal line at $(1, 2, -1)$ is $\dfrac{x-1}{-6} = \dfrac{y-2}{11} = \dfrac{z+1}{14}$

7.7 Find the Extrema of Multivariable Functions

The types of problems involving finding extrema of multivariable functions include:
1. **Find Extrema Using the Arithmetic Mean-Geometric Mean Inequality (AM-GM Inequality).**
2. **Find Extrema Using Cauchy's Inequality.**
3. **Find Extrema of Unconstrained Multivariable Functions.**
4. **Find Extrema of Multivariable Functions with Equality Constraints (using Lagrange multipliers).**
5. **Find Extrema of Multivariable Functions with Inequality Constraints.**
6. **Find Extrema of Multivariable Functions over a Bounded and Closed Domain.**

For unconstrained multivariable functions, the problem mainly involves the second derivative test for multivariable functions. For multivariable functions with equality constraints, the main method used is Lagrange multipliers. When dealing with inequality constraints, it involves simultaneously handling both the unconstrained and equality-constrained problems, thus employing both methods. For finding extrema over a bounded and closed domain, it generally requires explicitly stating the inequality constraints before finding the extrema.

For convenience, the following notations are introduced. If a real-valued function $f: R^n \to R$ has all first-order partial derivatives at $c \in R^n$, then

$$f^{(1)}(c; s) := \sum_{i=1}^{n} D_i f(c) s_i, \quad \forall s \in R^n.$$

If a real-valued function $f: R^n \to R$ has all second-order partial derivatives at $c \in R^n$ then

$$f^{(2)}(c; s) := \sum_{i=1}^{n} \sum_{j=1}^{n} D_{i,j} f(c) s_i s_j, \quad \forall s \in R^n.$$

If a real-valued function $f: R^n \to R$ has all third-order partial derivatives at $c \in R^n$ then

$$f^{(3)}(c; s) := \sum_{i=1}^{n} \sum_{j=1}^{n} \sum_{k=1}^{n} D_{i,j,k} f(c) s_i s_j s_k, \quad \forall s \in R^n.$$

The notation continues similarly for higher-order derivatives.

To illustrate the second derivative test for finding extrema of multivariable functions, we first introduce the Taylor's Formula for multivariable functions. The derivation of Taylor's Formula for multivariable functions relies on Taylor's Formula for single-variable functions.

【Theorem】 Taylor's Formula for Multivariable Functions

(i) Given a real-valued function $f: R^n \to R$, suppose that all partial derivatives up to order $m - 1$ are differentiable in an open set B.

(ii) If $x, y \in B$ and $L(x, y) \subseteq B$ then

$$\exists z \in L(x, y) \text{ s.t. } f(y) = f(x) + \sum_{k=1}^{m-1} \frac{f^{(k)}(x: y - x)}{k!} + \frac{f^{(m)}(z: y - x)}{m!}.$$

Proof:

Let $x, y \in B$. $\because B$ is open $\therefore \exists \delta > 0$ s.t. $x + s(y - x) \in B$, $\forall s \in (-\delta, 1 + \delta)$

Let $h(t) = f(x + s(y - x))$, $\forall s \in (-\delta, 1 + \delta)$, then $f(y) - f(x) = h(1) - h(0)$

By one-dimension Taylor formula, we have

$$h(1) - h(0) = \sum_{k=1}^{m-1} \frac{h^{(k)}(0)}{k!} + \frac{h^{(m)}(s')}{m!}, \text{ where } 0 < s' < 1.$$

Let $q(s) = x + s(y - x)$, then $h(s) = f(q(s))$ and $q'_k(s) = y_k - x_k$.

By using Chain Rule, we have $h'(s) = \sum_{j=1}^{n} D_j f(q(s))(y_j - x_j) = f'(q(s): y - x)$

Applying Chain Rule to $h'(s)$, we have

$$h''(s) = \sum_{i=1}^{n} \sum_{j=1}^{n} D_{i,j} f(q(s))(y_j - x_j)(y_i - x_i) = f''(q(s): y - x)$$

$\therefore h^{(m)}(s) = f^{(m)}(q(s): y - x)$

Let $z = x + s'(y - x) \in L(x, y)$ then

$$f(y) - f(x) = h(1) - h(0) = \sum_{k=0}^{m-1} \frac{h^{(k)}(0)}{k!} + \frac{h^{(m)}(s')}{m!} = \sum_{k=0}^{m-1} \frac{f^{(k)}(x: y - x)}{k!} + \frac{f^{(m)}(z: y - x)}{m!}$$

【Definition】 Stationary Point and Saddle Point

(i) If the function f is differentiable at c and $\nabla f(c) = 0$ then c is called a stationary point of f.

(ii) If, for any open ball $B_\delta(c)$ centered at the stationary point c, there exist points $x_0, x_1 \in B_\delta(c)$ such that $f(x_0) > f(c)$ and $f(x_1) < f(c)$, then c is called a saddle point.

It is worth noting that if we consider the case where $m=2$ and c is a stationary point, then
$$f(c+s)-f(c)=\nabla f(c)\cdot s+\frac{1}{2}f''(z:s) \text{ where } z\in L(c,c+s)$$
$\because c$ is a stationary point $\quad\therefore \nabla f(c)=0,$ the expression can be rewritten as
$$f(c+s)-f(c)=\frac{1}{2}f''(z:s)=\frac{1}{2}f''(c:s)+\left(\frac{1}{2}f''(z:s)-\frac{1}{2}f''(c:s)\right)$$
$$\because \frac{1}{2}(f''(z:s)-f''(c:s))\le \frac{1}{2}\sum_{i=1}^{n}\sum_{j=1}^{n}|D_{i,j}f(z)-D_{i,j}f(c)|\,\|s\|^2$$
If all second-order partial derivatives of f are continuous at c, then
$$\lim_{s\to 0}\frac{1}{2}(f''(z:s)-f''(c:s))=0.$$
This observation leads to the derivation of methods for finding extrema of multivariable functions.

【Theorem】Second Derivative Test for Multivariable Functions

Given a real-valued function $f:R^n\to R$, suppose there exists $\delta>0$ such that all second-order partial derivatives of f exist within $B_\delta(c)$ and are continuous at the stationary point c. Define
$$P(s):=\frac{1}{2}f''(c:s)=\frac{1}{2}\sum_{i=1}^{n}\sum_{j=1}^{n}D_{i,j}f(c)s_is_j.$$
Then,
(i) If $P(s)<0,\ \forall s\ne 0,$ then f has a relative maximum at c.
(ii) if $P(s)>0,\ \forall s\ne 0,$ then f has a relative minimum at c.
(iii) If there exist $s_1,s_2\ne 0$ s.t. $P(s_1)<0$ and $P(s_2)>0,$ then c is a saddle point of f
Proof:
(i) Assume $P(s)<0,\ \forall s\ne \vec{0}.$ Let $B=\{s:\|s\|=1\},$ then $P(s)<0,\ \forall s\in B$
$\because P(s)$ is continuous on B and B is a compact set.
$\therefore P(s)$ has a maximum M on B and $M<0.$
$\because P(\alpha s)=\alpha^2 P(s),\ \forall \alpha\in R$ and choose $\alpha=\frac{1}{\|s\|},$ then $\alpha s\in B$
$\therefore \alpha^2 P(s)=P(\alpha s)\le M\Rightarrow P(s)\le M\|s\|^2$
$\because f(c+s)-f(c)=P(s)+R(s)\le M\|s\|^2+|R(s)|$

7.7 Find the Extrema of Multivariable Functions

where $R(s) = \frac{1}{2}f''(z:s) - \frac{1}{2}f''(c:s)$

$\because |R(s)| \leq \frac{1}{2}\sum_{i=1}^{n}\sum_{j=1}^{n}|D_{i,j}f(z) - D_{i,j}f(c)|\,\|s\|^2$ and f is continuous at c

Choose $\delta_1 > 0$ s.t. $|R(s)| < \dfrac{-M\|s\|^2}{2}$, $\forall 0 < \|s\| < \delta_1$

then $f(c+s) - f(c) \leq M\|s\|^2 + |R(s)| \leq \dfrac{M\|s\|^2}{2} < 0$, $\forall 0 < \|s\| < \delta_1$

$\Rightarrow f$ has a relative maximum at c

(ii) To prove (ii), we apply part (i) to $-f$

(iii) $\forall \alpha > 0$, we have $f(c + \alpha s) - f(c) = \alpha^2 P(s) + R(\alpha s)$

$\because \lim_{t \to 0} R(t) = 0$, choose $\delta_2 > 0$ s.t. $0 < |R(\alpha s)| < \dfrac{\alpha^2 |P(s)|}{2}$, $\forall 0 < \alpha < \delta_2$

$\therefore f(c + \alpha s) - f(c)$ and $P(s)$ has the same sign

$\therefore$ if $\exists s_1, s_2 \neq 0$ s.t. $P(s_1) < 0$ and $P(s_2) > 0$ then f has a saddle point at c.

Among the various types of exam questions for finding extrema of multivariable functions, finding extrema of functions of two variables is the most common.

【Theorem】Second Derivative Test for Functions of Two Variables

Given a real-valued function $f(x, y): R^2 \to R$, suppose there exists $\delta > 0$ such that all second-order partial derivatives of f exist and are continuous within $B_\delta(c)$ at the stationary point c. Define

$$\Delta(c) := \det\begin{bmatrix} f_{xx}(c) & f_{xy}(c) \\ f_{xy}(c) & f_{yy}(c) \end{bmatrix} = f_{xx}(c)f_{yy}(c) - \bigl(f_{xy}(c)\bigr)^2.$$

Then,
(i) If $\Delta(c) > 0$ and $f_{xx}(c) > 0$, then f has a relative minimum at c.
(ii) If $\Delta(c) > 0$ and $f_{xx}(c) < 0$, then f has a relative maximum at c.
(iii) If $\Delta(c) < 0$, then c is a saddle point of f.
Proof:

$\because n = 2 \quad \therefore P(x, y) = \dfrac{\alpha x^2 + 2\beta xy + \gamma y^2}{2}$, where $\alpha = f_{xx}(c)$, $\beta = f_{xy}(c)$, $\gamma = f_{yy}(c)$

Let $\alpha \neq 0$, then

$P(x, y) = \dfrac{\alpha x^2 + 2\beta xy + \gamma y^2}{2} = \dfrac{\alpha^2 x^2 + 2\beta \alpha xy + \alpha \gamma y^2}{2\alpha}$

$= \dfrac{(\alpha x + \beta y)^2 + (2\beta \alpha xy + \alpha \gamma y^2 - 2\alpha x \beta y - \beta^2 y^2)}{2\alpha} = \dfrac{(\alpha x + \beta y)^2 + \Delta(c) \cdot y^2}{2\alpha}$

∴ If $\Delta(c) > 0$ and $\alpha > 0$, then $P(x,y) > 0, \forall (x,y) \neq (0,0)$
$\Rightarrow f$ has a relative minimum at c.
∴ If $\Delta(c) > 0$ and $\alpha < 0$, then $P(x,y) < 0, \forall (x,y) \neq (0,0)$
$\Rightarrow f$ has a relative maximum at c.
Assume $\Delta(c) < 0$, then $\exists$ two line s.t. $P(x,y) = 0$
∵ $P(x,y)$ is continuous at c
∴ $\forall r > 0, \exists (x_0, y_0), (x_1, y_1) \in B_r(c)$ s.t. $P(x_0, y_0) < 0$ and $P(x_1, y_1) > 0$
∴ f has a saddle point at c

The methods described above are primarily used for finding extrema of multivariable functions without constraints. The following Lagrange multiplier method is mainly used for problems with constraints.

【Theorem】Lagrange multiplier

Assume $f: R^n \to R$, x_0 is a local extreme point of f subject to $g_1(x) = \cdots = g_m(x) = 0$ and

$$\begin{vmatrix} \frac{\partial g_1(x_0)}{\partial x_{r_1}} & \frac{\partial g_1(x_0)}{\partial x_{r_2}} & \cdots & \frac{\partial g_1(x_0)}{\partial x_{r_m}} \\ \frac{\partial g_2(x_0)}{\partial x_{r_1}} & \frac{\partial g_2(x_0)}{\partial x_{r_2}} & \cdots & \frac{\partial g_2(x_0)}{\partial x_{r_m}} \\ \vdots & \vdots & \ddots & \vdots \\ \frac{\partial g_m(x_0)}{\partial x_{r_1}} & \frac{\partial g_m(x_0)}{\partial x_{r_2}} & \cdots & \frac{\partial g_m(x_0)}{\partial x_{r_m}} \end{vmatrix} \neq 0, \text{ (Eq1)}$$

for at leats one choice of $r_1 < r_2 < \cdots < r_m$ in $\{1,2,\ldots,n\}$ where $m < n$. Then there exist constants $\lambda_1, \lambda_2, \ldots, \lambda_m$ such that

$$\frac{\partial f(x_0)}{\partial x_i} - \sum_{j=1}^{m} \lambda_j \frac{\partial g_j(x_0)}{\partial x_i} = 0, \ \forall 1 \leq i \leq n.$$

Proof:

Let $r_p = p, 1 \leq p \leq m$ and $A = \begin{bmatrix} \frac{\partial g_1(x_0)}{\partial x_1} & \frac{\partial g_1(x_0)}{\partial x_2} & \cdots & \frac{\partial g_1(x_0)}{\partial x_m} \\ \frac{\partial g_2(x_0)}{\partial x_1} & \frac{\partial g_2(x_0)}{\partial x_2} & \cdots & \frac{\partial g_2(x_0)}{\partial x_m} \\ \vdots & \vdots & \ddots & \vdots \\ \frac{\partial g_m(x_0)}{\partial x_1} & \frac{\partial g_m(x_0)}{\partial x_2} & \cdots & \frac{\partial g_m(x_0)}{\partial x_m} \end{bmatrix}$

then (Eq1) becomes

$$\det(A) \neq 0 \ (\text{Eq2}).$$

7.7 Find the Extrema of Multivariable Functions

Let $x_0 = (x_0'; t_0)$, $x_0' = (x_{0,1}, x_{0,2}, \ldots, x_{0,m})$ and $t_0 = (x_{0,m+1}, x_{0,m+2}, \ldots, x_{0,n})$. Apply the Implicit Function Theorem to (Eq2) then there exist continuously differentiable functions $h_p = h_p(t)$, $1 \le p \le m$, defined on a neighborhood $B_\delta(t_0)$ s.t.
$(h_1(t), h_2(t), \ldots, h_m(t), t) \in D$, $\forall t \in B_\delta(t_0)$, $(h_1(t_0), h_2(t_0), \ldots, h_m(t_0), t_0) = x_0$ (Eq3)
and
$$g_p(h_1(t), h_2(t), \ldots, h_m(t), t) = 0, \quad \forall t \in B_\delta(t_0), \quad 1 \le p \le m \quad (Eq4).$$

Let $A^T \begin{bmatrix} \lambda_1 \\ \lambda_2 \\ \vdots \\ \lambda_m \end{bmatrix} = \begin{bmatrix} f_{x_1}(x_0) \\ f_{x_2}(x_0) \\ \vdots \\ f_{x_m}(x_0) \end{bmatrix}$ with (Eq1), then $\dfrac{\partial f(x_0)}{\partial x_i} - \sum_{j=1}^{m} \lambda_j \dfrac{\partial g_j(x_0)}{\partial x_i} = 0$, $\forall 1 \le i \le m$.

For $m + 1 \le i \le n$, differentiating (Eq4) with respect to x_i and using (Eq3) yields
$$\dfrac{\partial g_p(x_0)}{\partial x_i} + \sum_{j=1}^{m} \dfrac{\partial g_p(x_0)}{\partial x_j} \dfrac{\partial h_j(x_0)}{\partial x_j} = 0, \quad \forall 1 \le p \le m.$$

∵ x_0 is a local extreme point of f subject to $g_1(x) = g_1(x) = \cdots = g_m(x) = 0$
∴ t_0 is an unconstrained local extreme point of $f(h_1(t), h_2(t), \ldots, h_m(t), t)$

∴ $\dfrac{\partial f(x_0)}{\partial x_i} + \sum_{j=1}^{m} \dfrac{\partial f(x_0)}{\partial x_j} \dfrac{\partial h_j(x_0)}{\partial x_i} = 0$

∴ ∃ nontrivial Solution $\left(1, \dfrac{\partial h_1(x_0)}{\partial x_1}, \dfrac{\partial h_2(x_0)}{\partial x_2}, \ldots, \dfrac{\partial h_m(x_0)}{\partial x_m}\right)$ such that

Let $B = \begin{bmatrix} \dfrac{\partial f(x_0)}{\partial x_i} & \dfrac{\partial g_1(x_0)}{\partial x_i} & \dfrac{\partial g_2(x_0)}{\partial x_i} & \cdots & \dfrac{\partial g_m(x_0)}{\partial x_i} \\ \dfrac{\partial f(x_0)}{\partial x_1} & \dfrac{\partial g_1(x_0)}{\partial x_1} & \dfrac{\partial g_2(x_0)}{\partial x_1} & \cdots & \dfrac{\partial g_m(x_0)}{\partial x_1} \\ \dfrac{\partial f(x_0)}{\partial x_2} & \dfrac{\partial g_1(x_0)}{\partial x_2} & \dfrac{\partial g_2(x_0)}{\partial x_2} & \cdots & \dfrac{\partial g_m(x_0)}{\partial x_2} \\ \vdots & \vdots & \vdots & \ddots & \vdots \\ \dfrac{\partial f(x_0)}{\partial x_m} & \dfrac{\partial g_1(x_0)}{\partial x_m} & \dfrac{\partial g_2(x_0)}{\partial x_m} & \cdots & \dfrac{\partial g_m(x_0)}{\partial x_m} \end{bmatrix}$ then $B^T \begin{bmatrix} 1 \\ \dfrac{\partial h_1(x_0)}{\partial x_1} \\ \dfrac{\partial h_2(x_0)}{\partial x_2} \\ \vdots \\ \dfrac{\partial h_m(x_0)}{\partial x_m} \end{bmatrix} = 0$

∴ $\det(B^T) = 0 \Rightarrow \det(B) = 0$
∴ ∃ not all zero constants $c = (c_0, c_1, c_2, \ldots, c_m)$ such that $Bc = 0$
If $c_0 = 0$ then using (Eq2) implies $c_1 = c_2 = \cdots = c_m = 0$, a contradition. ∴ $c_0 \ne 0$

$$\therefore B \begin{bmatrix} c_0 \\ c_1 \\ c_2 \\ \vdots \\ c_m \end{bmatrix} = 0 \Rightarrow B \begin{bmatrix} 1 \\ c'_1 \\ c'_2 \\ \vdots \\ c'_m \end{bmatrix} = 0 \Rightarrow A^T \begin{bmatrix} -c'_1 \\ -c'_2 \\ \vdots \\ -c'_m \end{bmatrix} = \begin{bmatrix} \frac{\partial f(x_0)}{\partial x_1} \\ \frac{\partial f(x_0)}{\partial x_2} \\ \vdots \\ \frac{\partial f(x_0)}{\partial x_m} \end{bmatrix}, \text{ where } c'_j = \frac{c_j}{c_0}$$

By (Eq2) and $\dfrac{\partial f(x_0)}{\partial x_i} - \sum_{j=1}^{m} \lambda_j \dfrac{\partial g_j(x_0)}{\partial x_i} = 0, \ \forall 1 \le i \le m$, we have $c'_j = -\lambda_j, \ \forall 1 \le j \le m$

$$\Rightarrow B \begin{bmatrix} 1 \\ -\lambda_1 \\ -\lambda_2 \\ \vdots \\ -\lambda_m \end{bmatrix} = 0. \text{ Computing the top row of vector on the left, we have}$$

$$\frac{\partial f(x_0)}{\partial x_i} - \sum_{j=1}^{m} \lambda_j \frac{\partial g_j(x_0)}{\partial x_i} = 0, \ \forall m+1 \le i \le n.$$

7.7.1 Use the Arithmetic-Geometric Mean Inequality to Find Extrema

Question Types:

Type 1.

Given $x + y + z = r, \ x > 0, y > 0,$ and $z > 0,$ find the maximum value of $x^a y^b z^c$, $\forall a \cdot b \cdot c \in N$

Problem-Solving Process:

Step1.

$$\because x + y + z = \sum_{i=1}^{a} \frac{x}{a} + \sum_{i=1}^{b} \frac{y}{b} + \sum_{i=1}^{c} \frac{z}{c} \text{ and } \frac{\sum_{i=1}^{n} x_i}{n} \ge \sqrt[n]{\prod_{i=1}^{n} x_i}$$

Step2.

$$\therefore \frac{\sum_{i=1}^{a} \frac{x}{a} + \sum_{i=1}^{b} \frac{y}{b} + \sum_{i=1}^{c} \frac{z}{c}}{a+b+c} \ge \sqrt[a+b+c]{\left(\frac{x}{a}\right)^a \left(\frac{y}{b}\right)^b \left(\frac{z}{c}\right)^c}$$

Step3.

$$\therefore \left(\frac{r}{a+b+c}\right)^{a+b+c} \ge \left(\frac{x}{a}\right)^a \left(\frac{y}{b}\right)^b \left(\frac{z}{c}\right)^c \Rightarrow \left(\frac{r}{a+b+c}\right)^{a+b+c} a^a b^b c^c \ge x^a y^b z^c$$

7.7.1 Use the Arithmetic-Geometric Inequality to Find Extrema 92

Step4.

When $\dfrac{x}{a} = \dfrac{y}{b} = \dfrac{z}{c}$, equality holds in the inequality above. The maximum value of $x^a y^b z^c$ is $\left(\dfrac{r}{a+b+c}\right)^{a+b+c} a^a b^b c^c$

Example 1.

Given $x + y + z = 9$, $x > 0, y > 0$, and $z > 0$, find the maximum value of xyz.

【Solution】

$$\because \dfrac{\sum_{i=1}^n x_i}{n} \geq \sqrt[n]{\prod_{i=1}^n x_i}, \quad \forall n \in N \quad \therefore \dfrac{x+y+z}{3} \geq \sqrt[3]{xyz} \Rightarrow \dfrac{9}{3} \geq \sqrt[3]{xyz} \Rightarrow 3^3 \geq xyz$$

∵ When $x = y = z$, equality holds in the inequality above and $x + y + z = 9$
∴ $x = y = z = 3$ ∴ The maximum value of $xyz = 27$

Example 2.

Given $x + y + z = 1$, $x > 0, y > 0$, and $z > 0$, find the maximum value of $xy^2 z^3$.

【Solution】

$$\because \dfrac{\sum_{i=1}^n x_i}{n} \geq \sqrt[n]{\prod_{i=1}^n x_i}, \quad \forall n \in N$$

$$\therefore x + y + z = \dfrac{6x + 3y + 3y + 2z + 2z + 2z}{6} \geq \sqrt[6]{6x(3y)^2 (2z)^3}$$

∵ When $6x = 3y = 2z$, equality holds in the inequality above and $x + y + z = 1$

∴ $x = \dfrac{1}{6}, y = \dfrac{1}{3}, z = \dfrac{1}{2}$

∴ The maximum value of $xy^2 z^3 = \dfrac{1}{432}$

Example 3.

Given $x + y + z = 6$, $xx > 0, y > 0$, and $z > 0$, find the maximum value of $xy^2 z^3$.

【Solution】

$$\therefore \frac{\sum_{i=1}^{n} x_i}{n} \geq \sqrt[n]{\prod_{i=1}^{n} x_i}, \quad \forall n \in N$$

$$\therefore x + y + z = \frac{6x + 3y + 3y + 2z + 2z + 2z}{6} \geq \sqrt[6]{6x(3y)^2(2z)^3}$$

∵ When $6x = 3y = 2z$ equality holds in the inequality above and $x + y + z = 6$

$\therefore x = 1, y = 2, z = 3$

∴ The maximum value of $xy^2z^3 = 108$

7.7.2 Find Extrema Using Cauchy's Inequality

Question Types:

Type 1.

Given $f(x, y, z) = ax + by + cz$ and $x^2 + y^2 + z^2 = \alpha^2$, $\alpha > 0$, find the extrema of f.

Problem-Solving Process:

Step1.

Using the Cauchy Inequality,

$(ax + by + cz)^2 \leq (a^2 + b^2 + c^2)(x^2 + y^2 + z^2) = \alpha^2(a^2 + b^2 + c^2)$

Step2.

The maximum value of $f(x, y, z)$ is $\alpha\sqrt{(a^2 + b^2 + c^2)}$,

The minimum value of $f(x, y, z)$ is $-\alpha\sqrt{(a^2 + b^2 + c^2)}$

Example 1.

　　Given $f(x, y, z) = x + 2y + 3z$ and $x^2 + y^2 + z^2 = 1$, find the extrema of f.

【Solution】

By Cauchy inequality,

$(x + 2y + 3z)^2 \leq (1^2 + 2^2 + 3^2)(x^2 + y^2 + z^2) \Rightarrow |x + 2y + 3z| \leq \sqrt{14}$

∴ The maximum value of $f(x, y) = \sqrt{14}$, the minimum value of $f(x, y) = -\sqrt{14}$

Example 2.

　　Given $f(x, y, z) = 3x + 2y + 1z$ and $x^2 + y^2 + z^2 = 2$, find the extrema of f.

【Solution】

By Cauchy inequality,

$(3x + 2y + 1z)^2 \leq (3^2 + 2^2 + 1^2)(x^2 + y^2 + z^2) \Rightarrow |3x + 2y + z| \leq \sqrt{28}$

∴ The maximum value of $f(x,y) = \sqrt{28}$, the minimum value of $f(x,y) = -\sqrt{28}$

Example 3.
Given $f(x,y,z) = x + 3y + 2z$ and $x^2 + y^2 + z^2 = 3$, find the extrema of f.

【Solution】

By Cauchy inequality,
$(1x + 3y + 2z)^2 \leq (1^2 + 3^2 + 2^2)(x^2 + y^2 + z^2) \Rightarrow |x + 3y + 2z| \leq \sqrt{42}$
∴ The maximum value of $f(x,y) = \sqrt{42}$, the minimum value of $f(x,y) = -\sqrt{42}$

7.7.3 Find Extrema without Constraints

First, let's review the theorem for finding extrema of a function of two variables without constraints:

【Theorem】 Second Derivative Test for Functions of Two Variables

Given a real-valued function $f(x,y): R^2 \to R$, suppose there exist $\delta > 0$ such that all second-order partial derivatives of f exist and are continuous within $B_\delta(c)$ at the stationary point c. Define

$$\Delta(c) := \det \begin{bmatrix} f_{xx}(c) & f_{xy}(c) \\ f_{xy}(c) & f_{yy}(c) \end{bmatrix} = f_{xx}(c)f_{yy}(c) - \left(f_{xy}(c)\right)^2.$$

Then
(i) If $\Delta(c) > 0$ and $f_{xx}(c) > 0$, then f has a relative minimum at c.
(ii) if $\Delta(c) > 0$ and $f_{xx}(c) < 0$, then f has a relative maximum at c.
(iii) If $\Delta(c) < 0$, then c is a saddle point of f.

Question Types:
Type 1.
Given a function $f(x,y)$, find the minimum and maximum values of $f(x,y)$.
Problem-Solving Process:
Step1.
Let $\nabla f(x_0, y_0) = (0,0)$. Find (x_0, y_0) s.t. $\nabla f(x_0, y_0) = (0,0)$
Step2.
Calculate $\Delta(x_0, y_0) = ?$ and $f_{xx}(x_0, y_0) = ?$
Step3.
If $\Delta(x_0, y_0) > 0$ and $f_{xx}(x_0, y_0) > 0$ then f has a relative minimum at (x_0, y_0)

If $\Delta(x_0, y_0) > 0$ and $f_{xx}(x_0, y_0) < 0$ then f has a relative maximum at (x_0, y_0)

Example 1.
Given $f(x, y) = 3x^2 - 2xy + y^2 - 8y + 4$, find the relative extrema of f.
【Solution】
$\because \nabla f(x, y) = \left(\dfrac{\partial f}{\partial x}, \dfrac{\partial f}{\partial y}\right) = (6x - 2y, -2x + 2y - 8)$

Let $\nabla f(x, y) = (0,0)$ then $(x, y) = (2,6)$

$\because \Delta(x, y) = \begin{vmatrix} f_{xx} & f_{xy} \\ f_{yx} & f_{yy} \end{vmatrix} = \begin{vmatrix} 6 & -2 \\ -2 & 2 \end{vmatrix} = 8 > 0$ and $f_{xx}(2,6) = 6 > 0$

$\therefore f(x, y)$ has a relative minimum at $(2,6)$ with $f(2,6) = -20$

Example 2.
Given $f(x, y) = 4xy - x^4 - y^4 + 1$, find the relative extrema and saddle points of f.
【Solution】
$\because \nabla f(x, y) = \left(\dfrac{\partial f}{\partial x}, \dfrac{\partial f}{\partial y}\right) = (4y - 4x^3, 4x - 4y^3)$

Let $\nabla f(x, y) = (0,0)$ then $(x, y) = (0,0), (1,1), (-1,-1)$

$\because \Delta(x, y) = \begin{vmatrix} f_{xx} & f_{xy} \\ f_{yx} & f_{yy} \end{vmatrix} = \begin{vmatrix} -12x^2 & 4 \\ 4 & -12y^2 \end{vmatrix} = 144x^2 y^2 - 16$

$\therefore \Delta(0,0) = -16$, $\Delta(1,1) = \Delta(-1,-1) = 128 > 0$ and $f_{xx}(1,1) = f_{xx}(-1,-1) = -12 < 0$
$\therefore f(x, y)$ has relative maxima at $(1,1)(-1,-1)$ with $f(1,1) = f(-1,-1) = 3$

Example 3.
Given $f(x, y) = x^2 + y^2 + xy - 3x - 3y + 2$, find the relative extrema and saddle points of f.
【Solution】
$\because \nabla f(x, y) = \left(\dfrac{\partial f}{\partial x}, \dfrac{\partial f}{\partial y}\right) = (2x + y - 3, 2y + x - 3)$

Let $\nabla f(x, y) = (0,0)$ then $(x, y) = (1,1)$

$\because \Delta(x, y) = \begin{vmatrix} f_{xx} & f_{xy} \\ f_{yx} & f_{yy} \end{vmatrix} = \begin{vmatrix} 2 & 1 \\ 1 & 2 \end{vmatrix} = 3 > 0 \quad \therefore \Delta(1,1) = 3 > 0$ and $f_{xx}(1,1) = 2 > 0$

$\therefore f(x, y)$ has a relative minimum at $(1,1)$ with $f(1,1) = -1$

7.7.3 Find Extrema without Constraints

Example 4.

Given $f(x,y) = x^3 + y^3 - 3x - 12y + 22$, find the relative extrema and saddle points of f.

【Solution】

$\because \nabla f(x,y) = \left(\dfrac{\partial f}{\partial x}, \dfrac{\partial f}{\partial y}\right) = (3x^2 - 3, 3y^2 - 12)$

Let $\nabla f(x,y) = (0,0)$ then $(x,y) = (1,2), (1,-2), (-1,2), (-1,-2)$

$\because \Delta(x,y) = \begin{vmatrix} f_{xx} & f_{xy} \\ f_{yx} & f_{yy} \end{vmatrix} = \begin{vmatrix} 6x & 0 \\ 0 & 6y \end{vmatrix} = 36xy$

$\therefore \Delta(1,2) = \Delta(-1,-2) = 72,\ \Delta(1,-2) = \Delta(-1,2) = -72$

and $f_{xx}(-1,-2) = -6 < 0,\ f_{xx}(1,2) = 6 > 0$

$\therefore f(x,y)$ has a relative maximum at $(-1,-2)$ with $f(-1,-2) = 40$,

$f(x,y)$ has a relative minimum at $(1,2)$ with $f(1,2) = 4$

Example 5.

Given $f(x,y) = x^3 + y^2 + 2xy - 4x - 3y + 6$, find the relative extrema and saddle points of f.

【Solution】

$\because \nabla f(x,y) = \left(\dfrac{\partial f}{\partial x}, \dfrac{\partial f}{\partial y}\right) = (3x^2 + 2y - 4, 2y + 2x - 3)$

Let $\nabla f(x,y) = (0,0)$ then $(x,y) = \left(-\dfrac{1}{3}, \dfrac{11}{6}\right), \left(1, \dfrac{1}{2}\right)$

$\because \Delta(x,y) = \begin{vmatrix} f_{xx} & f_{xy} \\ f_{yx} & f_{yy} \end{vmatrix} = \begin{vmatrix} 6x & 2 \\ 2 & 2 \end{vmatrix} = 12x - 4 > 0$

$\therefore \Delta\left(1, \dfrac{1}{2}\right) = 8 > 0,\ \Delta\left(-\dfrac{1}{3}, \dfrac{11}{6}\right) = -8 < 0$ and $f_{xx}\left(1, \dfrac{1}{2}\right) = 6 > 0$

$\therefore f(x,y)$ has a relative minimum at $\left(1, \dfrac{1}{2}\right)$ with $f\left(1, \dfrac{1}{2}\right) = \dfrac{11}{4}$

Example 6.

Given $f(x,y) = x^3 + y^3 - 3xy - 1$, find the relative extrema and saddle points of f.

【Solution】

$\because \nabla f(x,y) = \left(\dfrac{\partial f}{\partial x}, \dfrac{\partial f}{\partial y}\right) = (3x^2 - 3y, 3y^2 - 3x)$

Let $\nabla f(x,y) = (0,0)$ then $(x,y) = (1,1), (0,0)$

7.7.3 Find Extrema without Constraints 97

$$\therefore \Delta(x,y) = \begin{vmatrix} f_{xx} & f_{xy} \\ f_{yx} & f_{yy} \end{vmatrix} = \begin{vmatrix} 6x & -3 \\ -3 & 6y \end{vmatrix} = 36xy - 9$$

$\therefore \Delta(1,1) = 27 > 0$, $\Delta(0,0) = -9 < 0$ and $f_{xx}(1,1) = 6 > 0$

$\therefore f(x,y)$ has a relative minimum at $(1,1)$ with $f(1,1) = -2$

Example 7.
  Given $f(x,y) = e^{-(x^2+y^2+2x)}$, find the relative extrema and saddle points of f.

【Solution】

$\therefore \nabla f(x,y) = \left(\dfrac{\partial f}{\partial x}, \dfrac{\partial f}{\partial y}\right) = (e^{-(x^2+y^2+2x)}(-2x-2), e^{-(x^2+y^2+2x)}(-2y))$

Let $\nabla f(x,y) = (0,0)$ then $(x,y) = (-1,0)$

$$\therefore \Delta(x,y) = \begin{vmatrix} f_{xx} & f_{xy} \\ f_{yx} & f_{yy} \end{vmatrix}$$

$$= \begin{vmatrix} e^{-(x^2+y^2+2x)}((-2x-2)^2 - 2) & (e^{-(x^2+y^2+2x)}(-2x-2)(-2y) \\ e^{-(x^2+y^2+2x)}(-2y)(-2x-2) & e^{-(x^2+y^2+2x)}((-2y)^2 - 2) \end{vmatrix}$$

$\therefore \Delta(-1,0) = 4e > 0$ and $f_{xx}(-1,0) = -2e < 0$

$\therefore f(x,y)$ has a relative maximum at $(-1,0)$ with $f(-1,0) = e$

Example 8.
  Given $f(x,y) = \sin x + \sin y + e$, $0 < x, y < \pi$, find the relative extrema of f.

【Solution】

$\therefore \nabla f(x,y) = \left(\dfrac{\partial f}{\partial x}, \dfrac{\partial f}{\partial y}\right) = (\cos x, \cos y)$

Let $\nabla f(x,y) = (0,0)$ then $(x,y) = \left(\dfrac{\pi}{2}, \dfrac{\pi}{2}\right)$

$$\therefore \Delta(x,y) = \begin{vmatrix} f_{xx} & f_{xy} \\ f_{yx} & f_{yy} \end{vmatrix} = \begin{vmatrix} -\sin x & 0 \\ 0 & -\sin y \end{vmatrix} = \sin x \sin y$$

$\therefore \Delta\left(\dfrac{\pi}{2}, \dfrac{\pi}{2}\right) = 1 > 0$ and $f_{xx}\left(\dfrac{\pi}{2}, \dfrac{\pi}{2}\right) = -1 < 0$

$\therefore f(x,y)$ has a relative minimum at $\left(\dfrac{\pi}{2}, \dfrac{\pi}{2}\right)$ with $f\left(\dfrac{\pi}{2}, \dfrac{\pi}{2}\right) = 2 + e$

Example 9.
  Given $f(x,y) = \sin x + \sin y + \sin(x+y) + 1, 0 < x, y < \pi$, find the relative extrema of f.

【Solution】
$$\because \nabla f(x,y) = \left(\frac{\partial f}{\partial x}, \frac{\partial f}{\partial y}\right) = (\cos x + \cos(x+y), \cos y + \cos(x+y))$$
$$= \left(2\cos\left(x+\frac{y}{2}\right)\cos\frac{y}{2}, 2\cos\left(y+\frac{x}{2}\right)\cos\frac{x}{2}\right)$$

Let $\nabla f(x,y) = (0,0)$ then $(x,y) = \left(\frac{\pi}{3}, \frac{\pi}{3}\right)$

$$\because \Delta(x,y) = \begin{vmatrix} -\sin x - \sin(x+y) & -\sin(x+y) \\ -\sin(x+y) & -\sin y - \sin(x+y) \end{vmatrix}$$
$$= \sin(x+y)(\sin x + \sin y) + \sin x \sin y$$

$$\therefore \Delta\left(\frac{\pi}{3}, \frac{\pi}{3}\right) = \frac{\sqrt{3}}{2} \cdot \sqrt{3} + \frac{3}{4} = \frac{9}{4} > 0 \text{ and } f_{xx}\left(\frac{\pi}{3}, \frac{\pi}{3}\right) = -\sqrt{3} < 0$$

$\therefore f(x,y)$ has a relative maximum at $\left(\frac{\pi}{3}, \frac{\pi}{3}\right)$ with $f\left(\frac{\pi}{3}, \frac{\pi}{3}\right) = \frac{3\sqrt{3}}{2} + 1$

7.7.4 Find Extrema with Equality Constraints

Review the previously introduced method for finding extrema of multivariable functions with equality constraints.

【Theorem】Lagrange multiplier

Assume $f: R^n \to R$, x_0 is a local extreme point of f subject to $g_1(x) = \cdots = g_m(x) = 0$ and

$$\begin{vmatrix} \dfrac{\partial g_1(x_0)}{\partial x_{r_1}} & \dfrac{\partial g_1(x_0)}{\partial x_{r_2}} & \cdots & \dfrac{\partial g_1(x_0)}{\partial x_{r_m}} \\ \dfrac{\partial g_2(x_0)}{\partial x_{r_1}} & \dfrac{\partial g_2(x_0)}{\partial x_{r_2}} & \cdots & \dfrac{\partial g_2(x_0)}{\partial x_{r_m}} \\ \vdots & \vdots & \ddots & \vdots \\ \dfrac{\partial g_m(x_0)}{\partial x_{r_1}} & \dfrac{\partial g_m(x_0)}{\partial x_{r_2}} & \cdots & \dfrac{\partial g_m(x_0)}{\partial x_{r_m}} \end{vmatrix} \neq 0, \quad (\text{Eq1})$$

for at leats one choice of $r_1 < r_2 < \cdots < r_m$ in $\{1, 2, \ldots, n\}$ where $m < n$. Then there exist constants $\lambda_1, \lambda_2, \ldots, \lambda_m$ such that

$$\frac{\partial f(x_0)}{\partial x_i} - \sum_{j=1}^{m} \lambda_j \frac{\partial g_j(x_0)}{\partial x_i} = 0, \quad \forall 1 \leq i \leq n.$$

Question Types:
Type 1.
Given a function $f(x,y)$ with a constraint $C_1(x,y) = 0$, use the Lagrange multiplier

method to find the extrema of f subject to this constraint.

Problem-Solving Process:

Step1.

Let $L(x, y, \lambda) = f(x, y) + \lambda C_1(x, y)$. If $L(x, y, \lambda)$ has extrema, then solve: $\begin{cases} \frac{\partial L}{\partial x}(x, y) = 0 \\ \frac{\partial L}{\partial y}(x, y) = 0 \\ C_1(x, y) = 0 \end{cases}$

Step2.

Find (x_0, y_0) that satisfies: $\begin{cases} \frac{\partial L}{\partial x}(x, y) = 0 \\ \frac{\partial L}{\partial y}(x, y) = 0 \\ C_1(x, y) = 0 \end{cases}$

Step3.

There may be multiple solutions (x_0, y_0). Compare each $f(x_0, y_0)$ to determine the relative maxima and minima of $f(x, y)$

Type 2.

Given the curve $C_1(x, y) = 0$, find the coordinates of the points on the curve that are closest to and farthest from the origin.

Problem-Solving Process:

Step1.

The distance from a point on the curve to the origin is $d = \sqrt{x^2 + y^2}$. Let $L(x, y, \lambda) = x^2 + y^2 + \lambda C_1(x, y)$. If $L(x, y, \lambda)$ has extrema, then solve:

$\begin{cases} \frac{\partial L}{\partial x}(x, y) = 0 \\ \frac{\partial L}{\partial y}(x, y) = 0 \\ C_1(x, y) = 0 \end{cases}$

Step2.

Find (x_0, y_0) that satisfies: $\begin{cases} \frac{\partial L}{\partial x}(x, y) = 0 \\ \frac{\partial L}{\partial y}(x, y) = 0 \\ C_1(x, y) = 0 \end{cases}$

Step3.

7.7.4 Find Extrema with Equality Constraints

There may be multiple solutions (x_0, y_0). Use $\sqrt{x_0^2 + y_0^2}$ to compare the distances and determine the closest and farthest points from the origin.

Type 3.
Given the surface $C_1(x, y, z) = 0$, find the coordinates on the surface that are closest to the plane $ax + by + cz = 0$.

Problem-Solving Process:

Step1.

The distance from a point on the surface to the plane is given by $\dfrac{|ax + by + cz|}{\sqrt{a^2 + b^2 + c^2}}$.

Let $L(x, y, z, \lambda) = ax + by + cz + \lambda C_1(x, y, z)$
If $L(x, y, z, \lambda)$ has extrema, then solve:

$$\begin{cases} \dfrac{\partial L}{\partial x}(x, y, z) = 0 \\ \dfrac{\partial L}{\partial y}(x, y, z) = 0 \\ \dfrac{\partial L}{\partial z}(x, y, z) = 0 \\ C_1(x, y, z) = 0 \end{cases}$$

Step2.

Find (x_0, y_0, z_0) that satisfies:

$$\begin{cases} \dfrac{\partial L}{\partial x}(x, y, z) = 0 \\ \dfrac{\partial L}{\partial y}(x, y, z) = 0 \\ \dfrac{\partial L}{\partial z}(x, y, z) = 0 \\ C_1(x, y, z) = 0 \end{cases}$$

Step3.
There may be multiple solutions (x_0, y_0, z_0). use $|ax_0 + by_0 + cz_0|$ to compare the distances and determine the minimum distance.

Type 4.
Given two surfaces $C_1(x, y, z) = 0$ and $C_2(x, y, z) = 0$, find the points on the intersection curve of the to surfaces where the height is maximized and minimized.

Problem-Solving Process:

Step1.
Let $L(x, y, z, \lambda_1, \lambda_2) = z + \lambda_1 C_1(x, y, z) + \lambda_2 C_2(x, y, z)$
If $L(x, y, z, \lambda_1, \lambda_2)$ has extrema, then solve:
$$\begin{cases} \dfrac{\partial L}{\partial x}(x, y, z, \lambda_1, \lambda_2) = 0 \\ \dfrac{\partial L}{\partial y}(x, y, z, \lambda_1, \lambda_2) = 0 \\ \dfrac{\partial L}{\partial z}(x, y, z, \lambda_1, \lambda_2) = 0 \\ C_1(x, y, z) = 0 \\ C_2(x, y, z) = 0 \end{cases}$$

Step2.
Find (x_0, y_0, z_0) that satisfies:
$$\begin{cases} \dfrac{\partial L}{\partial x}(x, y, z, \lambda_1, \lambda_2) = 0 \\ \dfrac{\partial L}{\partial y}(x, y, z, \lambda_1, \lambda_2) = 0 \\ \dfrac{\partial L}{\partial z}(x, y, z, \lambda_1, \lambda_2) = 0 \\ C_1(x, y, z) = 0 \\ C_2(x, y, z) = 0 \end{cases}$$

Step3.
There may be multiple solutions (x_0, y_0, z_0). Use z_0 to compare the heights and determine the maximum and minimum values.

Type 5.
Given two surfaces $C_1(x, y, z) = 0$ and $C_2(x, y, z) = 0$, find the shortest distance from the origin (0,0,0) to the line of intersection of these two surfaces.
Problem-Solving Process:
Step1.
The distance from the origin to the line is $d = \sqrt{x^2 + y^2 + z^2}$.
Let $L(x, y, z, \lambda_1, \lambda_2) = x^2 + y^2 + z^2 + \lambda_1 C_1(x, y, z) + \lambda_2 C_2(x, y, z)$
If $L(x, y, z, \lambda_1, \lambda_2)$ has an extremum, then solve:

7.7.4 Find Extrema with Equality Constraints

$$\begin{cases} \dfrac{\partial L}{\partial x}(x,y,z,\lambda_1,\lambda_2) = 0 \\ \dfrac{\partial L}{\partial y}(x,y,z,\lambda_1,\lambda_2) = 0 \\ \dfrac{\partial L}{\partial z}(x,y,z,\lambda_1,\lambda_2) = 0 \\ C_1(x,y,z) = 0 \\ C_2(x,y,z) = 0 \end{cases}$$

Step2.

Find (x_0, y_0, z_0) that satisfies:

$$\begin{cases} \dfrac{\partial L}{\partial x}(x,y,z,\lambda_1,\lambda_2) = 0 \\ \dfrac{\partial L}{\partial y}(x,y,z,\lambda_1,\lambda_2) = 0 \\ \dfrac{\partial L}{\partial z}(x,y,z,\lambda_1,\lambda_2) = 0 \\ C_1(x,y,z) = 0 \\ C_2(x,y,z) = 0 \end{cases}$$

Step3.

There may be more than one (x_0, y_0, z_0). Compare $\sqrt{x_0^2 + y_0^2 + z_0^2}$ to find the shortest distance.

Example 1.

Given $x^2 + y^2 = 1$, find the extrema of xy.

【Solution】

Let $L(x, y, \lambda) = xy + \lambda(x^2 + y^2 - 1)$

The extrema of $L(x, y, \lambda)$ occur

$$\Rightarrow \begin{cases} \dfrac{\partial L}{\partial x} = y + 2\lambda x = 0 \\ \dfrac{\partial L}{\partial y} = x + 2\lambda y = 0 \\ x^2 + y^2 = 1 \end{cases} \Rightarrow y^2 = x^2 \quad \therefore x^2 + y^2 = 1 \text{ and } y^2 = x^2$$

$\therefore (x, y) = \left(\dfrac{1}{\sqrt{2}}, \dfrac{1}{\sqrt{2}}\right), \left(\dfrac{-1}{\sqrt{2}}, \dfrac{1}{\sqrt{2}}\right), \left(\dfrac{1}{\sqrt{2}}, \dfrac{-1}{\sqrt{2}}\right), \left(\dfrac{-1}{\sqrt{2}}, \dfrac{-1}{\sqrt{2}}\right)$

$\Rightarrow xy$ has a maximum value of $\dfrac{1}{2}$ and a minimum value of $\dfrac{-1}{2}$

Example 2.

Let $f(x, y, z) = xyz$, where $2xz + 2yz + xy = 12$. Find the maximum value of f.

【Solution】

Let $L(x, y, \lambda) = xyz + \lambda(2xz + 2yz + xy - 12)$

The extrema of $L(x, y, z, \lambda)$ occur

$$\Rightarrow \begin{cases} \dfrac{\partial L}{\partial x} = yz + \lambda(2z + y) = 0 & (1) \\ \dfrac{\partial L}{\partial y} = xz + \lambda(2z + x) = 0 & (2) \\ \dfrac{\partial L}{\partial z} = xy + \lambda(2x + 2y) = 0 & (3) \\ 2xz + 2yz + xy = 12 & (4) \end{cases}$$

From (1)(2), we have $z(x - y) + \lambda(x - y) = 0$
From (2)(3), we have $x(y - 2z) + \lambda(2y - 4z) = 0$
From (1)(3), we have $y(x - 2z) + \lambda(2x - 4z) = 0$

$\therefore \lambda = -\dfrac{x}{2} = -\dfrac{y}{2} = -z \Rightarrow x = y = -2\lambda, z = -\lambda$

$\therefore 2xz + 2yz + xy = 12 \Rightarrow 4\lambda^2 + 4\lambda^2 + 4\lambda^2 = 12 \Rightarrow \lambda = \pm 1$

$\therefore (x, y, z) = (-2, -2, -1), (2, 2, 1) \quad \therefore f$ has a maximum value of 4

Example 3.

Let $f(x, y) = xy^3$, where $x^2 + 3y^2 = 16$. Find the maximum value of f.

【Solution】

Let $L(x, y, \lambda) = xy^3 + \lambda(x^2 + 3y^2 - 16)$

The extrema of $L(x, y, \lambda)$ occur $\Rightarrow \begin{cases} \dfrac{\partial L}{\partial x} = y^3 + 2x\lambda = 0 & (1) \\ \dfrac{\partial L}{\partial y} = 3xy^2 + 6y\lambda = 0 & (2) \\ x^2 + 3y^2 - 16 = 0 & (3) \end{cases}$

By (1)(2), we have $\lambda = -\dfrac{y^3}{2x} = -\dfrac{xy}{2} \Rightarrow x^2 = y^2$

$\because x^2 + 3y^2 = 16 \quad \therefore 4x^2 = 16 \Rightarrow (x, y) = (2,2), (-2,2), (2,-2), (-2,-2)$

$\therefore f$ has a maximum value of 16

Example 4.

7.7.4 Find Extrema with Equality Constraints 104

Let $f(x,y) = 4xy$, where $\dfrac{x^2}{9} + \dfrac{y^2}{25} = 1$. Find the extrema of f.

【Solution】

Let $L(x,y,\lambda) = 4xy + \lambda\left(\dfrac{x^2}{9} + \dfrac{y^2}{25} - 1\right)$

The extrema of $L(x,y,\lambda)$ occur $\Rightarrow \begin{cases} \dfrac{\partial L}{\partial x} = 4y + \dfrac{2x\lambda}{9} = 0 & (1) \\ \dfrac{\partial L}{\partial y} = 4x + \dfrac{2y\lambda}{25} = 0 & (2) \\ \dfrac{x^2}{9} + \dfrac{y^2}{25} - 1 = 0 & (3) \end{cases}$

By (1)(2), we have $\lambda = \dfrac{-18y}{x} = \dfrac{-50x}{y} \Rightarrow 9y^2 = 25x^2$

$\because \dfrac{x^2}{9} + \dfrac{y^2}{25} - 1 = 0 \quad \therefore \dfrac{25x^2 + 9y^2}{225} = 1$

$\Rightarrow (x,y) = \left(\dfrac{3}{\sqrt{2}}, \dfrac{5}{\sqrt{2}}\right), \left(\dfrac{3}{\sqrt{2}}, -\dfrac{5}{\sqrt{2}}\right), \left(-\dfrac{3}{\sqrt{2}}, \dfrac{5}{\sqrt{2}}\right), \left(-\dfrac{3}{\sqrt{2}}, -\dfrac{5}{\sqrt{2}}\right)$

$\therefore f$ has a maximum value of 30 and a minimum value of -30

Example 5.
  Find the shortest distance from the point (1,2,0) to the surface $z^2 = x^2 + y^2$.

【Solution】

The distance from the surface to the point $= d = \sqrt{(x-1)^2 + (y-2)^2 + z^2}$

Let $L(x,y,\lambda) = (x-1)^2 + (y-2)^2 + z^2 + \lambda(x^2 + y^2 - z^2)$

The extrema of $L(x,y,z,\lambda)$ exist $\Rightarrow \begin{cases} \dfrac{\partial L}{\partial x} = 2(x-1) + 2\lambda x = 0 & (1) \\ \dfrac{\partial L}{\partial y} = 2(y-2) + 2\lambda y = 0 & (2) \\ \dfrac{\partial L}{\partial z} = 2z - 2\lambda z = 0 & (3) \\ z^2 - x^2 + y^2 = 0 & (4) \end{cases}$

From (3), we have $\lambda = 1$ or $z = 0$

As $\lambda = 1$, $2(x-1) + 2\lambda x = 0$, $2(y-2) + 2\lambda y = 0$ and $z^2 = x^2 + y^2$

$\therefore 4x - 2 = 0$ and $4y - 4 = 0 \Rightarrow (x,y,z) = \left(\dfrac{1}{2}, 1, \pm\dfrac{\sqrt{5}}{2}\right)$

As $z = 0$, $2(x-1) + 2\lambda x = 0$, $2(y-2) + 2\lambda y = 0$ and $0 = x^2 + y^2 \Rightarrow (x,y,z) = (0,0,0)$

As $(x, y, z) = (0,0,0)$, $d = \sqrt{(x-1)^2 + (y-2)^2 + z^2} = \sqrt{5}$

As $(x, y, z) = \left(\dfrac{1}{2}, 1, \pm\dfrac{\sqrt{5}}{2}\right)$, the shortest distance $d = \sqrt{(x-1)^2 + (y-2)^2 + z^2} = \dfrac{\sqrt{10}}{2}$

Example 6.

Find the shortest distance from the origin $(0,0,0)$ to the surface $z^2 + 2 = (x - y)^2$.

【Solution】

The distance from the surface to the origin $(0,0,0) = d = \sqrt{x^2 + y^2 + z^2}$

Let $L(x, y, z, \lambda) = x^2 + y^2 + z^2 + \lambda((x - y)^2 - z^2 - 2)$

The extrema of $L(x, y, z, \lambda)$ exist $\Rightarrow$
$\begin{cases} \dfrac{\partial L}{\partial x} = 2x + 2\lambda(x - y) = 0 & (1) \\ \dfrac{\partial L}{\partial y} = 2y - 2\lambda(x - y) = 0 & (2) \\ \dfrac{\partial L}{\partial z} = 2z - 2\lambda z = 0 & (3) \\ z^2 + 2 - (x - y)^2 = 0 & (4) \end{cases}$

From (3), we have $\lambda = 1$ or $z = 0$

As $\lambda = 1$, $4x - 2y = 0$, $4y - 2x = 0$ and $z^2 + 2 = (x - y)^2$ $\Rightarrow$ (x, y, z) does not exist

As $z = 0$, $x + y = 0$ and $2 = (x - y)^2$

$\Rightarrow (x, y, z) = \left(\dfrac{1}{\sqrt{2}}, \dfrac{-1}{\sqrt{2}}, 0\right), \left(\dfrac{-1}{\sqrt{2}}, \dfrac{1}{\sqrt{2}}, 0\right)$

$\therefore$ the shortest distance $d = \sqrt{x^2 + y^2 + z^2} = \sqrt{\dfrac{1}{2} + \dfrac{1}{2} + z^2} = 1$

Example 7.

(i) Suppose $f(x, y, z) = xyz$, where $x^2 + \dfrac{y^2}{4} + \dfrac{z^2}{9} = r^2, (r > 0)$. Find the extrema of f.

(ii) Suppose $f(x, y, z) = xyz$, where $x^3 + y^3 + z^3 = 24$. Find the extrema of f.

【Solution】

(i)

Let $r > 0$, $L(x, y, z, \lambda) = xyz + \lambda\left(x^2 + \dfrac{y^2}{4} + \dfrac{z^2}{9} - r^2\right)$

7.7.4 Find Extrema with Equality Constraints

The extrema of $L(x,y,z,\lambda)$ exist $\Rightarrow$
$$\begin{cases} \dfrac{\partial L}{\partial x} = yz + 2\lambda x = 0 & (1) \\ \dfrac{\partial L}{\partial y} = xz + \lambda \dfrac{y}{2} = 0 & (2) \\ \dfrac{\partial L}{\partial z} = xy + \lambda \dfrac{2z}{9} = 0 & (3) \\ x^2 + \dfrac{y^2}{4} + \dfrac{z^2}{9} - r^2 = 0 & (4) \end{cases}$$

By (1)(2)(3), we have $\lambda = -\dfrac{yz}{2x} = -\dfrac{2xz}{y} = -\dfrac{9xy}{2z}$

Substitute into (4) gives $x^2 + \dfrac{4x^2}{4} + \dfrac{9x^2}{9} = r^2$ $\quad \therefore (x,y,z) = \left(\pm \dfrac{r}{\sqrt{3}}, \pm \dfrac{2r}{\sqrt{3}}, \pm \dfrac{3r}{\sqrt{3}}\right)$

$\therefore f\left(\dfrac{r}{\sqrt{3}}, \dfrac{2r}{\sqrt{3}}, \dfrac{3r}{\sqrt{3}}\right) = \dfrac{2r^3}{\sqrt{3}}$

$f\left(-\dfrac{r}{\sqrt{3}}, \dfrac{2r}{\sqrt{3}}, \dfrac{3r}{\sqrt{3}}\right) = f\left(\dfrac{r}{\sqrt{3}}, -\dfrac{2r}{\sqrt{3}}, \dfrac{3r}{\sqrt{3}}\right) = f\left(\dfrac{r}{\sqrt{3}}, \dfrac{2r}{\sqrt{3}}, -\dfrac{3r}{\sqrt{3}}\right) = -\dfrac{2r^3}{\sqrt{3}}$

$f\left(-\dfrac{r}{\sqrt{3}}, -\dfrac{2r}{\sqrt{3}}, \dfrac{3r}{\sqrt{3}}\right) = f\left(\dfrac{r}{\sqrt{3}}, -\dfrac{2r}{\sqrt{3}}, -\dfrac{3r}{\sqrt{3}}\right) = f\left(-\dfrac{r}{\sqrt{3}}, \dfrac{2r}{\sqrt{3}}, -\dfrac{3r}{\sqrt{3}}\right) = \dfrac{2r^3}{\sqrt{3}}$

$f\left(-\dfrac{r}{\sqrt{3}}, -\dfrac{2r}{\sqrt{3}}, -\dfrac{3r}{\sqrt{3}}\right) = -\dfrac{2r^3}{\sqrt{3}}$

$\therefore f$ has a maximum value of $\dfrac{2r^3}{\sqrt{3}}$, and a minimum value of $-\dfrac{2r^3}{\sqrt{3}}$

(ii)
Let $L(x,y,z,\lambda) = xyz + \lambda(x^3 + y^3 + z^3 - 24)$

The extrema of $L(x,y,z,\lambda)$ exist $\Rightarrow$
$$\begin{cases} \dfrac{\partial L}{\partial x} = yz + 3\lambda x^2 = 0 & (1) \\ \dfrac{\partial L}{\partial y} = xz + 3\lambda y^2 = 0 & (2) \\ \dfrac{\partial L}{\partial z} = xy + 3\lambda z^2 = 0 & (3) \\ x^3 + y^3 + z^3 = 24 & (4) \end{cases}$$

By (1)(2)(3), we have $\lambda = -\dfrac{yz}{3x^2} = -\dfrac{xz}{3y^2} = -\dfrac{xy}{3z^2}$

Substitute into (4) gives $x^3 + x^3 + x^3 = 24$

$\therefore (x,y,z) = (2,2,2)$, $\quad \therefore f$ has a maximum value of 8

Example 8.

(i) Find the coordinates of the points on the curve $17x^2 + 12xy + 8y^2 = 400$ that are nearest to and farthest from the origin.

(ii) Find the coordinates of the points on the curve $7x^2 - 6\sqrt{3}xy + 13y^2 = 64$ that are nearest to and farthest from the origin.

【Solution】

(i)

The distance from the curve to the origin $= d = \sqrt{x^2 + y^2}$

Let $L(x, y, \lambda) = x^2 + y^2 + \lambda(17x^2 + 12xy + 8y^2 - 400)$

The extrema of $L(x, y, \lambda)$ exist

$$\Rightarrow \begin{cases} \dfrac{\partial L}{\partial x} = 2x + \lambda(34x + 12y) = 0 \quad (1) \\ \dfrac{\partial L}{\partial y} = 2y + \lambda(12x + 16y) = 0 \quad (2) \\ 17x^2 + 12xy + 8y^2 - 400 = 0 \quad (3) \end{cases}$$

By (1)(2), we have $\lambda = \dfrac{x}{17x + 6y} = \dfrac{y}{6x + 8y} \Rightarrow x(6x + 8y) = y(17x + 6y)$

$\therefore (x - 2y)(2x + y) = 0$

(a) As $x - 2y = 0$

Substitute into (3) gives $(x, y) = (4, 2), (-4, -2)$ and $d = \sqrt{4^2 + 2^2} = 2\sqrt{5}$

(b) As $2x + y = 0$

Substitute into (3) gives $(x, y) = (4, -8), (-4, 8)$ and $d = \sqrt{4^2 + 8^2} = \sqrt{80} = 4\sqrt{5}$

∴ The points closest to the origin are at $(4, 2)$ and $(-4, -2)$ and the points farthest from the origin are at $(4, -8)$ and $(-4, 8)$

(ii)

The distance from the curve to the origin $= d = \sqrt{x^2 + y^2}$

Let $L(x, y, \lambda) = x^2 + y^2 + \lambda(7x^2 - 6\sqrt{3}xy + 13y^2 - 64)$

The extrema of $L(x, y, \lambda)$ exist $\Rightarrow \begin{cases} \dfrac{\partial L}{\partial x} = 2x + \lambda(14x - 6\sqrt{3}y) = 0 \quad (1) \\ \dfrac{\partial L}{\partial y} = 2y + \lambda(-6\sqrt{3}x + 26y) = 0 \quad (2) \\ 7x^2 - 6\sqrt{3}xy + 13y^2 - 64 = 0 \quad (3) \end{cases}$

By (1)(2), we have $\lambda = \dfrac{-x}{7x - 3\sqrt{3}y} = \dfrac{-y}{-3\sqrt{3}x + 13y}$

$\Rightarrow x(-3\sqrt{3}x + 13y) = y(7x - 3\sqrt{3}y) \quad \therefore (7x - 3\sqrt{3}y)(-3\sqrt{3}x + 13y) = 0$

(a) As $7x - 3\sqrt{3}y = 0$

7.7.4 Find Extrema with Equality Constraints

$\because 7x^2 - 6\sqrt{3}xy + 13y^2 - 64 = 0$

$\therefore 7\left(\dfrac{3\sqrt{3}}{7}y\right)^2 - 6\sqrt{3}\left(\dfrac{3\sqrt{3}}{7}y\right)y + 13y^2 - 64 = 0 \quad \therefore y = \pm\sqrt{7}$

$\therefore (x,y) = \left(\dfrac{3\sqrt{21}}{7}, \sqrt{7}\right), \left(-\dfrac{3\sqrt{21}}{7}, -\sqrt{7}\right)$ and $d = \sqrt{\left(\dfrac{3\sqrt{21}}{7}\right)^2 + (\sqrt{7})^2} = 2\sqrt{19}$

(b) As $-3\sqrt{3}x + 13y = 0$

$\because 7x^2 - 6\sqrt{3}xy + 13y^2 - 64 = 0$

$\therefore 7\left(\dfrac{13}{3\sqrt{3}}y\right)^2 - 6\sqrt{3}\left(\dfrac{13}{3\sqrt{3}}y\right)y + 13y^2 - 64 = 0 \quad \therefore y = \pm\sqrt{\dfrac{27}{13}}$

$\therefore (x,y) = \left(\sqrt{13}, \sqrt{\dfrac{27}{13}}\right), \left(-\sqrt{13}, -\sqrt{\dfrac{27}{13}}\right)$ and $d = \sqrt{(\sqrt{13})^2 + \left(\sqrt{\dfrac{27}{13}}\right)^2} = 2\sqrt{6}$

∴ The points closest to the origin are at $\left(\sqrt{13}, \sqrt{\dfrac{27}{13}}\right)$ and $\left(-\sqrt{13}, -\sqrt{\dfrac{27}{13}}\right)$, and

the points farthest to the origin are at $\left(\dfrac{3\sqrt{21}}{7}, \sqrt{7}\right)$ and $\left(-\dfrac{3\sqrt{21}}{7}, -\sqrt{7}\right)$

Example 9.

Find the highest point on the intersection of the cone $x^2 + y^2 = z^2$ and the plane $x + 14z = 10$.

【Solution】

Let $L(x, y, z, \lambda_1, \lambda_2) = z + \lambda_1(x^2 + y^2 - z^2) + \lambda_2(x + 14z - 10)$

The extrema of $L(x, y, z, \lambda_1, \lambda_2)$ exist $\Rightarrow$
$\begin{cases} \dfrac{\partial L}{\partial x} = 2x\lambda_1 + \lambda_2 = 0 & (1) \\ \dfrac{\partial L}{\partial y} = 2y\lambda_1 = 0 & (2) \\ \dfrac{\partial L}{\partial z} = 1 - 2z\lambda_1 + 14\lambda_2 = 0 & (3) \\ x^2 + y^2 - z^2 = 0 & (4) \\ x + 14z - 10 = 0 & (5) \end{cases}$

By (1)(2), we have $\lambda_1 \neq 0, y = 0$

By (4) $x = \pm z$, $\because x + 14z - 10 = 0$, $(x, z) = \left(\dfrac{2}{3}, \dfrac{2}{3}\right), \left(-\dfrac{10}{13}, \dfrac{10}{13}\right)$

$$\Rightarrow (x, y, z) = \left(\frac{2}{3}, 0, \frac{2}{3}\right), \left(\frac{-10}{13}, 0, \frac{10}{13}\right) \quad \therefore \text{ the highest point} = \frac{10}{13}$$

Example 10.

Find the point on the surface $x^2 + y^2 + 5 = z$ that is closest to the plane $x + 2y - z = 0$.

【Solution】

The distance from the surface to the plane $= d = \dfrac{|x + 2y - z|}{\sqrt{6}}$

Let $L(x, y, \lambda) = x + 2y - z + \lambda(x^2 + y^2 + 5 - z)$

The extremum $of\ L(x, y, z, \lambda)$ exist $\Rightarrow \begin{cases} \dfrac{\partial L}{\partial x} = 1 + 2x\lambda = 0 & (1) \\ \dfrac{\partial L}{\partial y} = 2 + 2y\lambda = 0 & (2) \\ \dfrac{\partial L}{\partial z} = -1 - \lambda = 0 & (3) \\ x^2 + y^2 + 5 - z = 0 & (4) \end{cases}$

From (1)(2)(3), we get $\lambda = -1, x = \dfrac{1}{2}$, and $y = 1$. Substituting these into (4) gives $z = \dfrac{25}{4}$

$\Rightarrow$ the closest point on the surface to the plane is $\left(\dfrac{1}{2}, 1, \dfrac{25}{4}\right)$

Example 11.

Find the point on the surface $x^2 + y^2 + z - 5 = 0$ that is closest to the plane $x + 2y + 3z = 0$.

【Solution】

The distance from the surface to the plane $= d = \dfrac{|x + 2y + 3z|}{\sqrt{14}}$

Let $L(x, y, \lambda) = x + 2y + 3z + \lambda(x^2 + y^2 + z - 5)$

The extremum of $L(x, y, z, \lambda)$ exist $\Rightarrow \begin{cases} \dfrac{\partial L}{\partial x} = 1 + 2x\lambda = 0 & (1) \\ \dfrac{\partial L}{\partial y} = 2 + 2y\lambda = 0 & (2) \\ \dfrac{\partial L}{\partial z} = 3 + \lambda = 0 & (3) \\ x^2 + y^2 + z - 5 = 0 & (4) \end{cases}$

7.7.4 Find Extrema with Equality Constraints

From (1)(2)(3), we get $\lambda = -3, x = \frac{1}{6}$, and $y = \frac{1}{3}$. Substituting these into (4) gives $z = \frac{175}{36}$

$\Rightarrow$ the closest point on the surface to the plane is $\left(\frac{1}{6}, \frac{1}{3}, \frac{175}{36}\right)$

Example 12.

Find the shortest distance from the origin (0,0,0) to the line of intersection between the planes $x + y + z = 2$ and $x - y - z = 2$.

【Solution】

The distance from the origin to the line of intersection $= d = \sqrt{x^2 + y^2 + z^2}$

Let $L(x, y, z, \lambda_1, \lambda_2) = x^2 + y^2 + z^2 + \lambda_1(x + y + z - 2) + \lambda_2(x - y - z - 2)$

The extremum of $L(x, y, z, \lambda_1, \lambda_2)$ exist $\Rightarrow$
$\begin{cases} \frac{\partial L}{\partial x} = 2x + \lambda_1 + \lambda_2 = 0 \quad (1) \\ \frac{\partial L}{\partial y} = 2y + \lambda_1 - \lambda_2 = 0 \quad (2) \\ \frac{\partial L}{\partial z} = 2z + \lambda_1 - \lambda_2 = 0 \quad (3) \\ x + y + z - 2 = 0 \quad (4) \\ x - y - z - 2 = 0 \quad (5) \end{cases}$

From (2)(3), we have $y = z$. Substituting these into (4)(5) gives $(x, y, z) = (2, 0, 0)$.

∴ The shortest distance $= d = \sqrt{2^2 + 0^2 + 0^2} = 2$

Example 13.

(i) On the intersection line of the cone $x^2 + y^2 = z^2$ and the plane $x + y + 2z = 2$, what are the highest and lowest points in terms of height?

(ii) On the intersection line of the cone $x^2 + y^2 = z^2$ and the plane $4x - 3y - z = 5$, what are the highest and lowest points in terms of height?

【Solution】

(i)

Let $L(x, y, z, \lambda_1, \lambda_2) = z + \lambda_1(x^2 + y^2 - z^2) + \lambda_2(x + y + 2z - 2)$

The extreme values of $L(x,y,z,\lambda_1,\lambda_2)$ exist $\Rightarrow$
$$\begin{cases} \dfrac{\partial L}{\partial x} = 2x\lambda_1 + \lambda_2 = 0 & (1) \\ \dfrac{\partial L}{\partial y} = 2y\lambda_1 + \lambda_2 = 0 & (2) \\ \dfrac{\partial L}{\partial z} = 1 - 2z\lambda_1 + 2\lambda_2 = 0 & (3) \\ x^2 + y^2 - z^2 = 0 & (4) \\ x + y + 2z - 2 = 0 & (5) \end{cases}$$

From (1)(2), we get $(\lambda_1, \lambda_2) = (0,0)$ or $x = y$

Since $(\lambda_1, \lambda_2) = (0,0)$ contradicts equation (3), no solution exists in this case.

For $x = y$, substituting into (4) and (5) gives

$(x, y, z) = (-1 - \sqrt{2}, -1 - \sqrt{2}, 2 + \sqrt{2}), (\sqrt{2} - 1, \sqrt{2} - 1, 2 - \sqrt{2})$

∴ the highest point is at $(-1 - \sqrt{2}, -1 - \sqrt{2}, 2 + \sqrt{2})$ and the lowest point is at $(\sqrt{2} - 1, \sqrt{2} - 1, 2 - \sqrt{2})$

(ii)

Let $L(x,y,z,\lambda_1,\lambda_2) = z + \lambda_1(x^2 + y^2 - z^2) + \lambda_2(4x - 3y - z - 5)$

The extreme values of $L(x,y,z,\lambda_1,\lambda_2)$ exist $\Rightarrow$
$$\begin{cases} \dfrac{\partial L}{\partial x} = 2x\lambda_1 + 4\lambda_2 = 0 & (1) \\ \dfrac{\partial L}{\partial y} = 2y\lambda_1 - 3\lambda_2 = 0 & (2) \\ \dfrac{\partial L}{\partial z} = 1 - 2z\lambda_1 - \lambda_2 = 0 & (3) \\ x^2 + y^2 - z^2 = 0 & (4) \\ 4x - 3y - z - 5 = 0 & (5) \end{cases}$$

By (1)(2), we get $(\lambda_1, \lambda_2) = (0,0)$ or $y = \dfrac{-3x}{4}$

Since $(\lambda_1, \lambda_2) = (0,0)$ contradicts equation (3), no solution exists in this case.

For $y = \dfrac{-3x}{4}$, substituting into (4) and (5) gives $(x, y, z) = \left(1, -\dfrac{3}{4}, \dfrac{5}{4}\right), \left(\dfrac{2}{3}, -\dfrac{1}{2}, -\dfrac{5}{6}\right)$

∴ the highest point is at $\left(1, -\dfrac{3}{4}, \dfrac{5}{4}\right)$ and the lowest point is at $\left(\dfrac{2}{3}, -\dfrac{1}{2}, -\dfrac{5}{6}\right)$

Example 14.

Given $f(x, y) = x^2 + xy + 2y^2$, where $x^2 + y^2 = 1$, find the maximum value of f.

【Solution】

Let $L(x, y, \lambda) = x^2 + xy + 2y^2 + \lambda(x^2 + y^2 - 1)$

7.7.4 Find Extrema with Equality Constraints

The extreme values of $L(x, y, \lambda)$ exist $\Rightarrow$ $\begin{cases} \dfrac{\partial L}{\partial x} = 2x + y + 2x\lambda = 0 & (1) \\ \dfrac{\partial L}{\partial y} = x + 4y + 2y\lambda = 0 & (2) \\ x^2 + y^2 - 1 = 0 & (3) \end{cases}$

By (1)(2), we have $\lambda = -\dfrac{2x+y}{2x} = -\dfrac{x+4y}{2y} \Rightarrow 1 + \dfrac{y}{2x} = \dfrac{x}{2y} + 2$

Let $t = \dfrac{y}{x} \Rightarrow t^2 - 2t - 1 = 0 \Rightarrow t = 1 \pm \sqrt{2}$

(i) As $\dfrac{y}{x} = 1 + \sqrt{2}$,

$\because \dfrac{y}{x} = 1 + \sqrt{2}$ $\therefore y = (1+\sqrt{2})x \Rightarrow xy = (1+\sqrt{2})x^2$ and $y^2 = (1+\sqrt{2})xy$

$\because x^2 + y^2 - 1 = 0$ $\therefore x^2 + (3 + 2\sqrt{2})x^2 = 1 \Rightarrow x^2 = \dfrac{2-\sqrt{2}}{4}, y^2 = \dfrac{2+\sqrt{2}}{4}$

$\because y^2 = (1+\sqrt{2})xy$ $\therefore xy = \dfrac{\frac{2+\sqrt{2}}{4}}{1+\sqrt{2}} = \dfrac{\sqrt{2}}{4}$

$\therefore f(x,y) = x^2 + xy + 2y^2 = 1 + y^2 + xy = \dfrac{3+\sqrt{2}}{2}$

(ii) As $\dfrac{y}{x} = 1 - \sqrt{2}$,

$\because \dfrac{y}{x} = 1 - \sqrt{2}$ $\therefore y = (1-\sqrt{2})x \Rightarrow xy = (1-\sqrt{2})x^2$ and $y^2 = (1-\sqrt{2})xy$

$\because x^2 + y^2 - 1 = 0$ $\therefore x^2 + (3 - 2\sqrt{2})x^2 = 1 \Rightarrow x^2 = \dfrac{2+\sqrt{2}}{4}, y^2 = \dfrac{2-\sqrt{2}}{4}$

$\because y^2 = (1-\sqrt{2})xy$ $\therefore xy = \dfrac{\frac{2-\sqrt{2}}{4}}{1-\sqrt{2}} = -\dfrac{\sqrt{2}}{4}$

$\therefore f(x,y) = x^2 + xy + 2y^2 = 1 + y^2 + xy = 1 + \dfrac{2-\sqrt{2}}{4} - \dfrac{\sqrt{2}}{4} = \dfrac{3-\sqrt{2}}{2}$

By (1)(2), f has a maximum value of $\dfrac{3+\sqrt{2}}{2}$

7.7.5 Find Extrema with Inequality Constraints

If the constraints for finding extrema are inequalities (i.e., $C_1(x,y) \leq 0$), then decompose it into $C_1(x,y) < 0$ and the equality constraint $C_1(x,y) = 0$ to find extrema. Use the second derivative test for functions of two variables to find extrema under $C_1(x,y) < 0$

and apply Lagrange multipliers to find the extrema of the function under the equality $C_1(x, y) = 0$.

Question Types:
Type 1.
Given a function $f(x, y)$ with the constraint $C_1(x, y) \leq 0$, find the extrema of $f(x, y)$ subject to this constraint.
Problem-Solving Process:
Step1.
Decompose the problem into $C_1(x, y) < 0$ and the equality constraint $C_1(x, y) = 0$, and find extrema under both constraints.
Step2.
Finding extrema for $C_1(x, y) < 0$ is similar to finding extrema without constraints.
After determining the extremum point (x_0, y_0), verify if it lies within $C_1(x, y) < 0$.
Let $\nabla f(x_0, y_0) = (0,0)$ and find (x_0, y_0). Calculate $\Delta(x_0, y_0)$ and $f_{xx}(x_0, y_0)$.
If $\Delta(x_0, y_0) > 0$ and $f_{xx}(x_0, y_0) > 0$, then f has a relative minimum at (x_0, y_0).
if $\Delta(x_0, y_0) > 0$ and $f_{xx}(x_0, y_0) < 0$, then f has a relative maximum at (x_0, y_0).
Confirm if $C_1(x_0, y_0) < 0$.
Step3.
Use Lagrange multipliers to find extrema under the equality constraint $C_1(x, y) = 0$.
Let $L(x, y, \lambda) = f(x, y) + \lambda C_1(x, y)$. If $L(x, y, \lambda)$ has an extremum, solve:
$$\begin{cases} \dfrac{\partial L}{\partial x}(x, y) = 0 \\ \dfrac{\partial L}{\partial y}(x, y) = 0 \\ C_1(x, y) = 0 \end{cases}$$
Find (x_0, y_0) that satisfies the above conditions. The (x_0, y_0) values may be more than one; use $f(x_0, y_0)$ to compare and determine the relative maxima and minima.
Step4.
Finally, compare the extremum points from Step 2 and Step 3. If both are relative maxima under their respective constraints, choose the larger one as the absolute maximum within $C_1(x, y) \leq 0$. Apply a similar approach for finding the absolute minimum.

Type 2.
Given a function $f(x, y)$ with the constraints $C_1(x, y) \leq 0$ and $C_2(x, y) \leq 0$, find the extrema of $f(x, y)$ under these constraints.

7.7.5 Find Extrema with Inequality Constraints

Problem-Solving Process:

Step1.

Decompose the problem into finding extrema under $C_1(x,y) < 0 \cdot C_2(x,y) < 0$ and the equality constraints $C_1(x,y) = 0 \cdot C_2(x,y) = 0$

Step2.

Finding extrema under $C_1(x,y) < 0$ and $C_2(x,y) < 0$ is similar to finding extrema without constraints. However, after determining the extremum point (x_0, y_0), check if it lies within the region defined by $C_1(x,y) < 0$ and $C_2(x,y) < 0$:

Let $\nabla f(x_0, y_0) = (0,0)$ and find (x_0, y_0). Compuate $\Delta(x_0, y_0)$ and $f_{xx}(x_0, y_0)$.
If $\Delta(x_0, y_0) > 0$ and $f_{xx}(x_0, y_0) > 0$, then f has a relative minimum at (x_0, y_0).
If $\Delta(x_0, y_0) > 0$ and $f_{xx}(x_0, y_0) < 0$, then f has a relative maximum at (x_0, y_0).
Verify that $C_1(x_0, y_0) < 0$ and $C_2(x_0, y_0) < 0$.

Step3.

Use Lagrang Multiplier to find extrema under the equality constraints $C_1(x,y) = 0$ and $C_2(x,y) = 0$:

Let $L(x, y, \lambda_1, \lambda_2) = f(x,y) + \lambda C_1(x,y) + \lambda C_2(x,y)$
If $L(x, y, \lambda_1, \lambda_2)$ has an extremum, solve:

$$\begin{cases} \dfrac{\partial L}{\partial x}(x, y, \lambda_1, \lambda_2) = 0 \\ \dfrac{\partial L}{\partial y}(x, y, \lambda_1, \lambda_2) = 0 \\ C_1(x,y) = 0 \\ C_2(x,y) = 0 \end{cases}$$

Find (x_0, y_0) that satisfies the above conditions. The values of (x_0, y_0) may not be unique; use $f(x_0, y_0)$ to compare and determine the relative maxima and minima.

Step4.

Finally, compare the extrema found in Step 2 and Step 3. If both are relative maxima under their respective constraints, choose the larger one as the absolute maximum within the region defined by $C_1(x,y) \le 0$ and $C_2(x,y) \le 0$. Apply a similar approach to determine the absolute minimum.

Example 1.

Suppose $f(x,y) = x^2 + y^2 - 3xy + 1$ and $x^2 + y^2 - 6 \le 0$. Find the extreme values of f.

【Solution】

(1)
As $x^2 + y^2 - 6 < 0$,

let $f_x = f_y = 0$ then $\begin{cases} f_x = 2x - 3y = 0 \\ f_y = 2y - 3x = 0 \end{cases}$ $\Rightarrow (x, y) = (0,0)$ is a critical point, and $f(0,0) = 1$

$\because \Delta(0,0) = \begin{vmatrix} f_{xx}(0,0) & f_{xy}(0,0) \\ f_{yx}(0,0) & f_{yy}(0,0) \end{vmatrix} = \begin{vmatrix} 2 & -3 \\ -3 & 2 \end{vmatrix} < 0$ $\therefore (0,0)$ is not an extreme point

(2)
As $x^2 + y^2 - 6 = 0$,
let $L(x, y, \lambda) = x^2 + y^2 - 3xy + 1 + \lambda(x^2 + y^2 - 6)$

The extreme values of $L(x, y, \lambda)$ exist $\Rightarrow \begin{cases} \dfrac{\partial L}{\partial x} = 2x - 3y + 2\lambda x = 0 \\ \dfrac{\partial L}{\partial y} = 2y - 3x + 2\lambda y = 0 \\ x^2 + y^2 - 6 = 0 \end{cases}$ $\Rightarrow x^2 - y^2 = 0$

$\because x^2 - y^2 = 0$ and $x^2 + y^2 - 6 = 0$
$\therefore (x, y) = (\sqrt{3}, \sqrt{3}), (-\sqrt{3}, \sqrt{3}), (\sqrt{3}, -\sqrt{3}), (-\sqrt{3}, -\sqrt{3})$
$\Rightarrow f(\sqrt{3}, \sqrt{3}) = -2, f(-\sqrt{3}, -\sqrt{3}) = -2, f(-\sqrt{3}, \sqrt{3}) = 16, f(\sqrt{3}, -\sqrt{3}) = 16$

$\therefore f$ has a minimum value of -2 and a maximum value of 16

Example 2.
Given $f(x, y) = x^2 + 3y^2 + y + 1$, $x^2 + y^2 \leq 1$, find the extreme values of f.

【Solution】
(1)
As $x^2 + y^2 - 1 < 0$

Let $f_x = f_y = 0$ then $\begin{cases} f_x = 2x = 0 \\ f_y = 6y + 1 = 0 \end{cases}$ $\Rightarrow (x, y) = \left(0, -\dfrac{1}{6}\right)$ is a critical point

and $f\left(0, -\dfrac{1}{6}\right) = \dfrac{11}{12}$

$\because \Delta\left(0, -\dfrac{1}{6}\right) = \begin{vmatrix} f_{xx}\left(0, -\dfrac{1}{6}\right) & f_{xy}\left(0, -\dfrac{1}{6}\right) \\ f_{yx}\left(0, -\dfrac{1}{6}\right) & f_{yy}\left(0, -\dfrac{1}{6}\right) \end{vmatrix} = \begin{vmatrix} 2 & 0 \\ 0 & 6 \end{vmatrix} = 12 > 0$ and $f_{xx} = 2 > 0$

$\therefore f$ has a relative minimum of $\dfrac{11}{12}$ at $\left(0, -\dfrac{1}{6}\right)$

(2)

As $x^2 + y^2 - 1 = 0$, let $L(x, y, \lambda) = x^2 + 3y^2 + y + 1 + \lambda(x^2 + y^2 - 1)$

The extreme values of $L(x, y, \lambda)$ exist $\Rightarrow$
$\begin{cases} \dfrac{\partial L}{\partial x} = 2x + 2\lambda x = 0 \\ \dfrac{\partial L}{\partial y} = 6y + 1 + 2\lambda y = 0 \\ x^2 + y^2 - 1 = 0 \end{cases}$ $\Rightarrow 4xy + x = 0$

$\because 4xy + x = 0$ and $x^2 + y^2 - 1 = 0$ $\quad \therefore (x, y) = \left(\pm \dfrac{\sqrt{15}}{4}, -\dfrac{1}{4}\right), (0,1), (0,-1)$

$\Rightarrow f\left(\pm \dfrac{\sqrt{15}}{4}, -\dfrac{1}{4}\right) = \dfrac{15}{8}, \quad f(0,1) = 5, \quad f(0,-1) = 3$

$\therefore f$ has a relative maximum of 5 at $(0,1)$

Example 3.

Suppose $f(x, y) = x^2 - y^2 + 3$, $x^2 + \dfrac{y^2}{4} \leq 1$. Find the extreme values of f.

【Solution】

(1)

As $x^2 + \dfrac{y^2}{4} - 1 < 0$,

let $f_x = f_y = 0$ then $\begin{cases} f_x = 2x = 0 \\ f_y = -2y = 0 \end{cases}$ $\Rightarrow (x, y) = (0,0)$ is a critical point and $f(0,0) = 3$

$\because \Delta(0,0) = \begin{vmatrix} f_{xx}(0,0) & f_{xy}(0,0) \\ f_{yx}(0,0) & f_{yy}(0,0) \end{vmatrix} = \begin{vmatrix} 2 & 0 \\ 0 & -2 \end{vmatrix} = -4 < 0$ $\quad \therefore (0,0)$ is not an extreme point

(2)

As $x^2 + \dfrac{y^2}{4} - 1 = 0$, let $L(x, y, \lambda) = x^2 - y^2 + 3 + \lambda\left(x^2 + \dfrac{y^2}{4} - 1\right)$

The extreme values of $L(x, y, \lambda)$ exist $\Rightarrow$ $\begin{cases} \dfrac{\partial L}{\partial x} = 2x + 2\lambda x = 0 \\ \dfrac{\partial L}{\partial y} = -2y + \dfrac{\lambda y}{2} = 0 \Rightarrow xy = 0 \\ x^2 + \dfrac{y^2}{4} - 1 = 0 \end{cases}$

$\because xy = 0$ and $x^2 + \dfrac{y^2}{4} - 1 = 0$ $\quad \therefore (x, y) = (0, \pm 2), (\pm 1, 0)$

$\Rightarrow f(0, \pm 2) = -1, f(\pm 1, 0) = 4$

∴ f has a relative maximum of 4 at $(\pm 1, 0)$, f has a relative minimum of -1 at $(0, \pm 2)$

Example 4.

Given $f(x, y) = x^2 + 2y^2 - x + 1$ and $x^2 + y^2 - 4 \leq 0$. Find the extreme values of f.

【Solution】

(1)

As $x^2 + y^2 - 4 < 0$,

let $f_x = f_y = 0$ then $\begin{cases} f_x = 2x - 1 = 0 \\ f_y = 4y = 0 \end{cases} \Rightarrow (x, y) = \left(\frac{1}{2}, 0\right)$ is a critical point and $f\left(\frac{1}{2}, 0\right) = \frac{3}{4}$

∵ $\Delta\left(\frac{1}{2}, 0\right) = \begin{vmatrix} f_{xx}\left(\frac{1}{2}, 0\right) & f_{xy}\left(\frac{1}{2}, 0\right) \\ f_{yx}\left(\frac{1}{2}, 0\right) & f_{yy}\left(\frac{1}{2}, 0\right) \end{vmatrix} = \begin{vmatrix} 2 & 0 \\ 0 & 4 \end{vmatrix} > 0$ and $f_{xx}\left(\frac{1}{2}, 0\right) = 2 > 0$

∴ As $x^2 + y^2 - 4 < 0$, f has a relative minimum of $\frac{3}{4}$ at $\left(\frac{1}{2}, 0\right)$

(2)

As $x^2 + y^2 - 4 = 0$, let $L(x, y, \lambda) = x^2 + 2y^2 - x + 1 + \lambda(x^2 + y^2 - 4)$

The extreme values of $L(x, y, \lambda)$ exist $\Rightarrow \begin{cases} \dfrac{\partial L}{\partial x} = 2x - 1 + 2\lambda x = 0 \\ \dfrac{\partial L}{\partial y} = 4y + 2\lambda y = 0 \\ x^2 + y^2 - 4 = 0 \end{cases} \Rightarrow 2xy + y = 0$

∵ $2xy + y = 0$ and $x^2 + y^2 - 4 = 0$ ∴ $(x, y) = \left(-\dfrac{1}{2}, \pm\dfrac{\sqrt{15}}{2}\right), (\pm 2, 0)$

$\Rightarrow f\left(-\dfrac{1}{2}, \pm\dfrac{\sqrt{15}}{2}\right) = \dfrac{37}{4}$, $f(2, 0) = 3$, $f(-2, 0) = 7$

∴ f has a relative minimum of $\dfrac{3}{4}$ at $\left(\dfrac{1}{2}, 0\right)$ and f has a relative maximum of $\dfrac{37}{4}$ at $\left(-\dfrac{1}{2}, \pm\dfrac{\sqrt{15}}{2}\right)$

Example 5.

Suppose $f(x, y) = 4x^2 + 4x - 2y^2 + 6$ and $x^2 + y^2 - 3 \leq 0$. Find the extreme values of f.

【Solution】

7.7.5 Find Extrema with Inequality Constraints

(1)

As $x^2 + y^2 - 3 < 0$, let $f_x = f_y = 0$ then $\begin{cases} f_x = 8x + 4 = 0 \\ f_y = -4y = 0 \end{cases}$

$\Rightarrow (x, y) = \left(-\dfrac{1}{2}, 0\right)$ is a critical point and $f\left(-\dfrac{1}{2}, 0\right) = 5$

$\because \Delta\left(-\dfrac{1}{2}, 0\right) = \begin{vmatrix} f_{xx}\left(-\dfrac{1}{2},0\right) & f_{xy}\left(-\dfrac{1}{2},0\right) \\ f_{yx}\left(-\dfrac{1}{2},0\right) & f_{yy}\left(-\dfrac{1}{2},0\right) \end{vmatrix} = \begin{vmatrix} 8 & 0 \\ 0 & -4 \end{vmatrix} < 0$

$\therefore \left(-\dfrac{1}{2}, 0\right)$ is not an extreme point

(2)

As $x^2 + y^2 - 3 = 0$, let $L(x, y, \lambda) = 4x^2 + 4x - 2y^2 + 6 + \lambda(x^2 + y^2 - 3)$

The extreme values of $L(x, y, \lambda)$ exist $\Rightarrow \begin{cases} \dfrac{\partial L}{\partial x} = 8x + 4 + 2\lambda x = 0 \\ \dfrac{\partial L}{\partial y} = -4y + 2\lambda y = 0 \\ x^2 + y^2 - 3 = 0 \end{cases} \Rightarrow 3xy + y = 0$

$\because 3xy + y = 0$ and $x^2 + y^2 - 3 = 0$ $\therefore (x, y) = \left(-\dfrac{1}{3}, \pm\dfrac{\sqrt{26}}{3}\right), (\pm\sqrt{3}, 0)$

$\Rightarrow f\left(-\dfrac{1}{3}, \pm\dfrac{\sqrt{26}}{3}\right) = -\dfrac{2}{3}$, $f(\sqrt{3}, 0) = 18 + 4\sqrt{3}$, $f(-\sqrt{3}, 0) = 18 - 4\sqrt{3}$

$\therefore f$ has a relative minimum of $-\dfrac{2}{3}$ and f has a relative maximum of $18 + 4\sqrt{3}$

Example 6.

Given $f(x, y) = x^2 + y^2 - 2x - 2y + 2$, and $2x + y - 4 \geq 0$, $x + 2y - 4 \geq 0$, find the minimum of f.

【Solution】

(1)

As $2x + y - 4 > 0$ and $x + 2y - 4 > 0$,

let $f_x = f_y = 0$ then $\begin{cases} f_x = 2x - 2 = 0 \\ f_y = 2y - 2 = 0 \end{cases}$ $\Rightarrow (x, y) = (1,1)$ is a critical point and $f(1,1) = 0$

$\therefore (1,1)$ does not satisfy $2x + y - 4 > 0$ and $x + 2y - 4 > 0$

(2)

As $2x + y - 4 = 0$ and $x + 2y - 4 = 0$,
let $L(x, y, \lambda_1, \lambda_2) = x^2 + y^2 - 2x - 2y + 2 + \lambda_1(2x + y - 4) + \lambda_2(x + 2y - 4)$

The extreme values of $L(x, y, \lambda_1, \lambda_2)$ exist $\Rightarrow$
$\begin{cases} \dfrac{\partial L}{\partial x} = 2x - 2 + 2\lambda_1 + \lambda_2 = 0 \\ \dfrac{\partial L}{\partial y} = 2y - 2 + \lambda_1 + 2\lambda_2 = 0 \\ 2x + y - 4 = 0 \\ x + 2y - 4 = 0 \end{cases}$

$\because 2x + y - 4 = 0$ and $x + 2y - 4 = 0$ $\therefore (x, y) = \left(\dfrac{4}{3}, \dfrac{4}{3}\right)$

$\Rightarrow 2\lambda_1 + \lambda_2 = \dfrac{2}{3}$, $\lambda_1 + 2\lambda_2 = \dfrac{2}{3}$ $\therefore (\lambda_1, \lambda_2) = \left(\dfrac{2}{9}, \dfrac{2}{9}\right)$

$\therefore f$ has a relative minimum value of $\dfrac{2}{9}$ at $\left(\dfrac{4}{3}, \dfrac{4}{3}\right)$

Example 7.
Let $f(x, y) = 2x^2 + y^2 - 2y + 3$ and $x^2 + y^2 \leq 4$. Find the extreme values of f.

【Solution】
(1)
As $x^2 + y^2 - 4 < 0$,

let $f_x = f_y = 0$ then $\begin{cases} f_x = 4x = 0 \\ f_y = 2y - 2 = 0 \end{cases}$ $\Rightarrow (x, y) = (0, 1)$ is a critical point and $f(0, 1) = 0$

$\because \Delta(0,1) = \begin{vmatrix} f_{xx}(0,1) & f_{xy}(0,1) \\ f_{yx}(0,1) & f_{yy}(0,1) \end{vmatrix} = \begin{vmatrix} 4 & 0 \\ 0 & 2 \end{vmatrix} = 8 > 0$ and $f_{xx}(0,1) = 4 > 0$

$\therefore f$ has a minimum value of 2 at $(0, 1)$

(2)
As $x^2 + y^2 - 4 = 0$, let $L(x, y, \lambda) = 2x^2 + y^2 - 2y + 3 + \lambda(x^2 + y^2 - 4)$

The extreme values of $L(x, y, \lambda)$ exist $\Rightarrow$ $\begin{cases} \dfrac{\partial L}{\partial x} = 4x + 2\lambda x = 0 \\ \dfrac{\partial L}{\partial y} = 2y - 2 + 2\lambda y = 0 \\ x^2 + y^2 - 4 = 0 \end{cases}$ $\Rightarrow 2xy + 2x = 0$

$\because 2xy + 2x = 0$ and $x^2 + y^2 - 4 = 0$ $\therefore (x, y) = (\pm\sqrt{3}, -1) \Rightarrow f(\pm\sqrt{3}, -1) = 12$

$\therefore f$ has a relative maximum of 12 at $(\pm\sqrt{3}, -1)$

Example 8.

7.7.5 Find Extrema with Inequality Constraints 120

Suppose $f(x,y) = e^{x^2 y}$, where $x^2 + y^2 \leq 1$. Find the extreme values of f.

【Solution】

(1)

As $x^2 + y^2 - 1 < 0$,

let $f_x = f_y = 0$ then $\begin{cases} f_x = e^{x^2 y} 2xy = 0 \\ f_y = e^{x^2 y} x^2 = 0 \end{cases}$ $\Rightarrow (x,y) = (0,y)$ is a critical point and $f(0,y) = 1$

$\therefore \Delta(x,y) = \begin{vmatrix} f_{xx} & f_{xy} \\ f_{yx} & f_{yy} \end{vmatrix} = e^{2x^2 y} \begin{vmatrix} (2xy)^2 + 2y & 2x^3 y + 2x \\ 2x^3 y + 2x & x^4 \end{vmatrix} = 12 > 0$

$\therefore \Delta(0,y) = 0 \quad \therefore (0,y)$ is not an extreme point

(2)

As $x^2 + y^2 - 1 = 0$, let $L(x,y,\lambda) = e^{x^2 y} + \lambda(x^2 + y^2 - 1)$

The extreme values of $L(x,y,\lambda)$ exist $\Rightarrow \begin{cases} \dfrac{\partial L}{\partial x} = e^{x^2 y} 2xy + 2\lambda x = 0 \\ \dfrac{\partial L}{\partial y} = e^{x^2 y} x^2 + 2\lambda y = 0 \\ x^2 + y^2 - 1 = 0 \end{cases} \Rightarrow x(2y^2 - x^2) = 0$

$\therefore x(2y^2 - x^2)$ and $x^2 + y^2 - 1 = 0 \quad \therefore (x,y) = \left(\pm\sqrt{\dfrac{2}{3}}, \pm\sqrt{\dfrac{1}{3}}\right), (0, \pm 1)$

$\Rightarrow f\left(\pm\sqrt{\dfrac{2}{3}}, \sqrt{\dfrac{1}{3}}\right) = e^{\frac{2}{3\sqrt{3}}}, \quad f\left(\pm\sqrt{\dfrac{2}{3}}, -\sqrt{\dfrac{1}{3}}\right) = e^{\frac{-2}{3\sqrt{3}}}, \quad f(0, \pm 1) = e^0 = 1$

$\therefore f$ has a relative maximum of $e^{\frac{2}{3\sqrt{3}}}$ at $\left(\pm\sqrt{\dfrac{2}{3}}, \sqrt{\dfrac{1}{3}}\right)$ and f has a relative mainimum of

$e^{\frac{-2}{3\sqrt{3}}}$ at $\left(\pm\sqrt{\dfrac{2}{3}}, -\sqrt{\dfrac{1}{3}}\right)$

Example 9.

Suppose $f(x,y) = 6x^2 - 8x + 2y^2 - 4$, $x^2 + y^2 \leq 1$. Find the extreme values of f.

【Solution】

(1)

As $x^2 + y^2 - 1 < 0$, let $f_x = f_y = 0$ then $\begin{cases} f_x = 12x - 8 = 0 \\ f_y = 4y = 0 \end{cases}$

$\Rightarrow (x, y) = \left(\dfrac{2}{3}, 0\right)$ is a critical point and $f\left(\dfrac{2}{3}, 0\right) = -\dfrac{20}{3}$

$\therefore \Delta\left(\dfrac{2}{3}, 0\right) = \begin{vmatrix} f_{xx}\left(\dfrac{2}{3}, 0\right) & f_{xy}\left(\dfrac{2}{3}, 0\right) \\ f_{yx}\left(\dfrac{2}{3}, 0\right) & f_{yy}\left(\dfrac{2}{3}, 0\right) \end{vmatrix} = \begin{vmatrix} 12 & 0 \\ 0 & 4 \end{vmatrix} = 48 > 0$ and $f_{xx}\left(\dfrac{2}{3}, 0\right) = 12$

$\therefore f$ has a relative minimum value of $-\dfrac{20}{3}$ at $\left(\dfrac{2}{3}, 0\right)$

(2)
As $x^2 + y^2 - 1 = 0$, let $L(x, y, \lambda) = 6x^2 - 8x + 2y^2 - 4 + \lambda(x^2 + y^2 - 1)$

The extreme values of $L(x, y, \lambda)$ exist $\Rightarrow \begin{cases} \dfrac{\partial L}{\partial x} = 12x - 8 + 2\lambda x = 0 \\ \dfrac{\partial L}{\partial y} = 4y + 2\lambda y = 0 \end{cases} \Rightarrow 8xy - 8y = 0$

$\because xy - y = 0$ and $x^2 + y^2 - 1 = 0$ $\therefore (x, y) = (\pm 1, 0)$
$\Rightarrow f(-1, 0) = 10, f(1, 0) = -6$,
$\therefore f$ has a relative maximum of 10 at $(-1, 0)$

7.7.6 Find Extrema on Compact Domains

When the constraints for finding extrema are given as a closed region, first determine the explicit inequality constraints (i.e., $C_1(x, y) \le 0$). Then decompose this into $C_1(x, y) < 0$ and the equality constraint $C_1(x, y) = 0$ to find the extrema.

Question Types:
Type 1.
Given a function $f(x, y)$,
(1) find the extrema of $f(x, y)$ in the closed region bounded by the vertices $(0,0), (a, 0)$ and $(0, b)$
(2) find the extrema of $f(x, y)$ in the closed region bounded by the vertices $\{(x, y): 0 \le x \le a, 0 \le y \le b\}$
(3) find the extrema of $f(x, y)$ in the closed region bounded by the vertices $\{(x, y): 0 \le x \le a, 0 \le y \le b, y + mx \le c\}$.
Problem-Solving Process:

7.7.6 Find Extrema on Compact Domains

Identify $C_1(x, y)$ so that the original problem can be transformed into finding extrema of a multivariable function with inequality constraints $C_1(x, y) \leq 0$.

Example 1.

Suppose $f(x, y) = 3xy - 6x - 3y + 8$. Find the extreme values of f within the closed region formed by the three vertices $(0,0)$, $(3,0)$, and $(0,5)$.

【Solution】

The closed region formed by the three vertices $= \{(x, y): x \geq 0, y \geq 0, y \leq -\frac{5}{3}(x - 3)\}$

(1)

As $x > 0, y > 0$ and $y < -\frac{5}{3}(x - 3)$,

let $f_x = f_y = 0$ then $\begin{cases} f_x = 3y - 6 = 0 \\ f_y = 3x - 3 = 0 \end{cases}$ $\Rightarrow (x, y) = (1, 2)$ is the critical point

and $(1, 2) \in \{(x, y): x > 0, y > 0, y < -\frac{5}{3}(x - 3)\}$, $f(1, 2) = 2$

$\because \Delta(1, 2) = \begin{vmatrix} f_{xx}(1,2) & f_{xy}(1,2) \\ f_{yx}(1,2) & f_{yy}(1,2) \end{vmatrix} = \begin{vmatrix} 0 & 3 \\ 3 & 0 \end{vmatrix} < 0 \Rightarrow (1, 2)$ is not an extreme point

(2)

As $(x, y) \in \{(x, 0): 0 \leq x \leq 3\} \cup \{(0, y): 0 \leq y \leq 5\}$

$\cup \{(x, y): y = -\frac{5(x - 3)}{3}, 0 \leq y \leq 5, 0 \leq x \leq 3\}$

(a)

If $(x, y) \in \{(x, 0): 0 \leq x \leq 3\}$ then $f(x, y) = -6x + 8$, $f(0, 0) = 8$, $f(3, 0) = -10$

(b)

If $(x, y) \in \{(0, y): 0 \leq y \leq 5\}$ then $f(x, y) = -3y + 8$, $f(0, 0) = 8$, $f(0, 5) = -7$

(c)

If $(x, y) \in \{(x, y): y = -\frac{5}{3}(x - 3), 0 \leq y \leq 5, 0 \leq x \leq 3\}$ then

$f(x, y) = 3x\left(-\frac{5}{3}(x - 3)\right) - 6x - 3\left(-\frac{5}{3}(x - 3)\right) + 8 = -5x^2 + 14x - 7$

$= -5\left(x - \frac{7}{5}\right)^2 + \frac{14}{5}$

$\therefore f(x, y)$ has a relative maximum value of $\frac{14}{5}$ at $\left(\frac{7}{5}, \frac{8}{3}\right)$

$\Rightarrow f(x, y)$ has a maximum value of 8 at $(0,0)$ and a minimum value of -10 at $(3,0)$

Example 2.

Suppose $f(x,y) = x^2 - 2xy + 2y$. Find the extreme values of f within the region $\{(x,y): 0 \le x \le 3, 0 \le y \le 2\}$.

【Solution】

(1)

As $0 < x < 3,\ 0 < y < 2$,

let $f_x = f_y = 0$ then $\begin{cases} f_x = 2x - 2y = 0 \\ f_y = -2x + 2 = 0 \end{cases} \Rightarrow (x,y) = (1,1)$

$\because \Delta(1,1) = \begin{vmatrix} f_{xx}(1,1) & f_{xy}(1,1) \\ f_{yx}(1,1) & f_{yy}(1,1) \end{vmatrix} = \begin{vmatrix} 2 & -2 \\ -2 & 0 \end{vmatrix} = -4 < 0 \quad \therefore (1,1)$ is a saddle point

(2)

As $(x,y) \in \{(x,y): 0 \le x \le 3, y = 0 \vee 2\} \cup \{(x,y): x = 0 \vee 3, 0 \le y \le 2\}$

(a)

If $(x,y) \in \{(x,0): 0 \le x \le 3\}$ then $f(x,0) = x^2$

$\therefore f$ has a relative minimum value of 0 at (0,0) and a relative maximum value of 9 at (3,0)

(b)

If $(x,y) \in \{(x,2): 0 \le x \le 3\}$ then $f(x,2) = x^2 - 4x + 4$

Let $\dfrac{\partial f(x,2)}{\partial x} = 0$ then $2x - 4 = 0 \Rightarrow x = 2$

$\therefore f$ has a relative minimum value of 0 at (2,2) and a relative maximum value of 4 at (0,2)

(c)

If $(x,y) \in \{(0,y): , 0 \le y \le 2\}$ then $f(0,y) = 2y$

$\therefore f$ has a relative minimum value of 0 at (0,0) and a relative maximum value of 4 at (0,2)

(d)

If $(x,y) \in \{(3,y): , 0 \le y \le 2\}$ then $f(3,y) = 9 - 4y$

$\therefore f$ has a relative minimum value of 1 at (3,2) and a relative maximum value of 9 at (3,0)

$\therefore f$ has the maximum value of 9 at (3,0)

Example 3.

Suppose $f(x,y) = x + y^2 - xy$, find the extreme values of f within the region $\{(x,y): 0 \le x \le 1, 0 \le y \le 2, 2x + y \le 2\}$.

【Solution】

(1)

7.7.6 Find Extrema on Compact Domains

As $(x,y) \in \{(x,y): 0 < x < 1, 0 < y < 2, 2x + y < 2\}$,

let $f_x = f_y = 0$ then $\begin{cases} f_x = 1 - y = 0 \\ f_y = 2y - x = 0 \end{cases} \Rightarrow (x,y) = (2,1)$

$\because \Delta(2,1) = \begin{vmatrix} f_{xx}(2,1) & f_{xy}(2,1) \\ f_{yx}(2,1) & f_{yy}(2,1) \end{vmatrix} = \begin{vmatrix} 0 & -1 \\ -1 & 2 \end{vmatrix} = -1 < 0 \Rightarrow (2,1)$ is a saddle point

(2)
As $(x,y) \in \{(x,0): 0 \le x \le 1\} \cup \{(0,y): 0 \le y \le 2\} \cup \{(x,y): 2x + y = 2\}$

(a)
If $(x,y) \in \{(x,0): 0 \le x \le 1\}$ then $f(x,0) = x$
$\therefore f$ has a relative minimum value of 0 at (0,0) and a relative maximum value of 1 at (1,0)

(b)
If $(x,y) \in \{(0,y): 0 \le y \le 2\}$ then $f(0,y) = y^2$
$\therefore f$ has a relative minimum value of 0 at (0,0) and a relative maximum value of 4 at (0,2)

(c)
If $(x,y) \in \{(x,y): 2x + y = 2, 0 \le x \le 1, 0 \le y \le 2\}$
then $f(x, 2 - 2x) = x + (2 - 2x)^2 - x(2 - 2x)$

Let $\dfrac{\partial f(x,y)}{\partial x} = 0$ then $1 - 2(4 - 4x) - 2 + 4x = 0 \Rightarrow x = \dfrac{3}{4}, y = \dfrac{1}{2}$

$\therefore f$ has a relative minimum value of $\dfrac{5}{8}$ at $\left(\dfrac{3}{4}, \dfrac{1}{2}\right)$ and a relative maximum value of 4 at (0,2)

$\therefore f$ has the minimum value of 0 at (0,0) and the maximum value of 4 at (0,2)

Example 4.
  Given $f(x,y) = 4xy^2 - x^2y^2 - xy^3$. Find the extreme values of f within the closed region formed by the three vertices (0,0), (0,6), (6,0).

【Solution】
The closed region formed by the three vertices $= \{(x,y): x \ge 0, y \ge 0, y \le -(x - 6)\}$
(1)
As $x > 0, y > 0$ and $y < -(x - 6)$,

let $f_x = f_y = 0$ then $\begin{cases} f_x = 4y^2 - 2xy^2 - y^3 = 0 \\ f_y = 8xy - 2x^2y - 3xy^2 = 0 \end{cases}$

$\Rightarrow \begin{cases} f_x = y^2(4 - 2x - y) = 0 \\ f_y = xy(8 - 2x - 3y) = 0 \end{cases} \Rightarrow \begin{cases} f_x = 4 - 2x - y = 0 \\ f_y = 8 - 2x - 3y = 0 \end{cases} \Rightarrow (x,y) = (1,2)$

$$\because \Delta(x,y) = \begin{vmatrix} f_{xx} & f_{xy} \\ f_{yx} & f_{yy} \end{vmatrix} = \begin{vmatrix} -2y^2 & 8y - 4xy - 3y^2 \\ 8x - 4xy - 3y^2 & 8x - 2x^2 - 6xy \end{vmatrix}$$

and $\Delta(1,2) > 0$ and $f_{xx}(1,2) < 0$

∴ f has a relative maximum value of 4 at $(1,2)$

(2)

As $(x,y) \in \{(x,0): 0 \leq x \leq 6\} \cup \{(0,y): 0 \leq y \leq 6\}$
$\cup \{(x,y): y = -(x-6), 0 \leq y \leq 6, 0 \leq x \leq 6\}$

(a)
If $(x,y) \in \{(x,0): 0 \leq x \leq 6\}$ then $f(x,y) = 0$

(b)
If $(x,y) \in \{(0,y): 0 \leq y \leq 6\}$ then $f(x,y) = 0$

(c)
If $(x,y) \in \{(x,y): y = -(x-6), 0 \leq y \leq 6, 0 \leq x \leq 6\}$
then $f(x,y) = 4x(6-x)^2 - x^2(6-x)^2 - x(6-x)^3 = -2x^3 + 24x^2 - 72x$

Let $\dfrac{\partial f(x,y)}{\partial x} = 0$ then $-6x^2 + 48x - 72 = 0 \Rightarrow x = 2$ or $6 \Rightarrow (x,y) = (2,4)(6,0)$

∵ $f(2,4) = -64$ and $f(6,0) = 0$

∴ f has the maximum value of 4 at $(1,2)$ and the minimum value of -64 at $(2,4)$

Example 5.
Suppose $f(x,y) = x^4 + y^4 - 4xy + 3$, find the extreme values of f within the region $\{(x,y): 0 \leq x \leq 3, 0 \leq y \leq 2\}$.

【Solution】

(1)

As $0 < x < 3, 0 < y < 2$, let $f_x = f_y = 0$ then $\begin{cases} f_x = 4x^3 - 4y = 0 \\ f_y = 4y^3 - 4x = 0 \end{cases}$

$\Rightarrow (x,y) = (1,1), (-1,-1), (0,0)$ and $(-1,-1) \notin \{(x,y): 0 < x < 3, 0 < y < 2\}$

$\because \Delta(x,y) = \begin{vmatrix} f_{xx} & f_{xy} \\ f_{yx} & f_{yy} \end{vmatrix} = \begin{vmatrix} 12x^2 & -4 \\ -4 & 12y^2 \end{vmatrix} = 144x^2y^2 - 16$

and $\Delta(1,1) = 128 > 0$, $f_{xx}(1,1) = 12 > 0$

∴ f has a relative minimum value of 1 at $(1,1)$

(2)

As $(x,y) \in \{(x,y): 0 \leq x \leq 3, y = 0 \vee 2\} \cup \{(x,y): x = 0 \vee 3, 0 \leq y \leq 2\}$

(a)

If $(x, y) \in \{(x, 0): 0 \le x \le 3\}$ then $f(x, 0) = x^4 + 3$
$\Rightarrow f$ has a relative minimum value of 3 at (0,0), and a relative maximum value of 84 at (3,0)

(b)
If $(x, y) \in \{(x, 2): 0 \le x \le 3\}$ then $f(x, 2) = x^4 - 8x + 19$

Let $\dfrac{\partial f(x, 2)}{\partial x} = 0$ then $4x^3 - 8 = 0 \Rightarrow x = 2^{\frac{1}{3}}$

$\therefore f$ has a relative minimum value of $2^{\frac{4}{3}} + 19 - 8 \cdot 2^{\frac{1}{3}}$ at $\left(2^{\frac{1}{3}}, 2\right)$, and

a relative maximum value of 76 at (3,2)

(c)
If $(x, y) \in \{(0, y): 0 \le y \le 2\}$ then $f(0, y) = y^4 + 3$
$\therefore f$ has a relative minimum value of 3 at (0,0), and a relative maximum value of 19 at (0,2)

(d)
If $(x, y) \in \{(3, y): 0 \le y \le 2\}$ then $f(3, y) = y^4 - 12y + 84$

Let $\dfrac{\partial f(x, y)}{\partial y} = 0$ then $4y^3 - 12 = 0 \Rightarrow y = 3^{\frac{1}{3}}$

$\therefore f$ has a relative minimum value of $84 + 3^{\frac{4}{3}} - 12 \cdot 3^{\frac{1}{3}}$ at $\left(3, 3^{\frac{1}{3}}\right)$, and

a relative maximum value of 76 at (3,2)
$\therefore f$ has the minimum value of 1 at (1,1), and the maximum value of 84 at (3,0)

Example 6.
Given $f(x, y) = 4xy - y^2 - x^2 - 6x + 1$, find the extreme values of f within the closed region formed by the three vertices (0,0), (2,0), (2,6).

【Solution】

The closed region formed by the three vertices $= \{(x, y): 0 \le x \le 2, 0 \le y \le 6, y \ge 3x\}$
(1) $(x, y) \in \{(x, y): 0 < x < 2, 0 < y < 6, y > 3x\}$

Let $f_x = f_y = 0$ then $\begin{cases} f_x = 4y - 2x - 6 = 0 \\ f_y = 4x - 2y = 0 \end{cases} \Rightarrow (x, y) = (1, 2)$

$\because \Delta(1,2) = \begin{vmatrix} f_{xx}(1,2) & f_{xy}(1,2) \\ f_{yx}(1,2) & f_{yy}(1,2) \end{vmatrix} = \begin{vmatrix} -2 & 4 \\ 4 & -2 \end{vmatrix} = -20 < 0 \Rightarrow (1,2)$ is a saddle point

(2) $(x, y) \in \{(x, 0): 0 \le x \le 2\} \cup \{(2, y): 0 \le y \le 6\} \cup \{(x, y): y = 3x, 0 \le y \le 6, 0 \le x \le 2\}$
(a)

If $(x, y) \in \{(x, 0): 0 \leq x \leq 2\}$ then $f(x, 0) = -x^2 - 6x + 1$

Let $\dfrac{\partial f(x, 0)}{\partial x} = 0$ then $-2x - 6 = 0 \quad \therefore x = -3 \notin (0,2)$

$\therefore f$ has a relative maximum value of 1 at (0,0), and a relative minimum value of -15 at (2,0)

(b)
If $(x, y) \in \{(2, y): 0 \leq y \leq 6\}$ then $f(2, y) = 8y - y^2 - 15$

Let $\dfrac{\partial f(2, y)}{\partial y} = 0$ then $8 - 2y = 0 \quad \therefore y = 4$

$\therefore f$ has a relative minimum value of -15 at (2,0), and a relative maximum value of 1 at (2,4)

(c)
If $(x, y) \in \{(x, y): y = 3x, 0 \leq y \leq 6, 0 \leq x \leq 2\}$
then $f(x, y) = 4x \cdot 3x - 9x^2 - x^2 - 6x + 1 = 2x^2 - 6x + 1$

Let $\dfrac{\partial f(x, y)}{\partial x} = 0$ then $4x - 6 = 0 \Rightarrow x = \dfrac{3}{2}, y = \dfrac{9}{2}$

$\therefore f$ has a relative minimum value of $-\dfrac{7}{2}$ at $\left(\dfrac{3}{2}, \dfrac{9}{2}\right)$, and a relative maximum value of 1 at (0,0)

$\therefore f$ has the minimum value of -15 at (2,0), and the maximum value of 1 at (2,4) and (0,0)

CHAPTER 8 MULTIPLE INTEGRALS

This chapter sequentially introduces iterated integrals, the definition of multiple integrals, Fubini's Theorem, and types of exam problems related to multiple integrals. Iterated integrals can be viewed as the computation of multiple single-variable definite integrals. In the section on multiple integrals, the definition of a Riemann integrable function in multiple variables is first explained, which is similar to the definition for single-variable functions. Next, the necessary and sufficient conditions for Riemann integrability are described: a function $f(x)$ is Riemann integrable over an interval $I \Leftrightarrow f(x)$ satisfies the Riemann condition on $I \Leftrightarrow$ the upper Riemann integral of $f(x)$ equals the lower Riemann integral of $f(x)$ on I. However, these conditions are not always easy to use. In practice, any continuous function defined on a compact set is Riemann integrable. Therefore, the criteria for determining the continuity of a function, as introduced in Chapter 7, become quite important. Specifically, in Section 7.2, any multivariable continuous function defined on a compact set is Riemann integrable. Furthermore, when calculating multiple integrals of multivariable functions, Fubini's Theorem is primarily used to transform the problem into computing multiple single-variable definite integrals. It is worth noting that a prerequisite for using Fubini's Theorem is that the multivariable function must be Riemann integrable. Common types of integrals tested include double and triple integrals. Both double and triple integrals are based on Fubini's Theorem, which converts the problem of calculating multiple integrals into computing several single-variable definite integrals. Therefore, during the solution process, methods for evaluating definite integrals, such as substitution and integration by parts, may be used.

The types of exam problems involving double integrals can be categorized into: (1) problems that can be evaluated without changing the order of integration, (2) problems that require changing the order of integration for evaluation, (3) problems that need conversion to polar coordinates for evaluation, and (4) problems that require conversion to generalized coordinates for evaluation. It is crucial to be familiar with these various problem types. When estimating integrals where direct evaluation is not feasible, ensure you can determine the new limits of integration after changing the order. Additionally, after converting to polar or generalized coordinates, be able to quickly and accurately find the new limits of integration. Mastery of these techniques will enable you to handle various double integral problems with ease. Triple integrals can be categorized into: (1) problems that can be directly converted into double integrals and then iterated, (2) problems that involve expressions like $x^2 + y^2$ after converting to double integrals and require polar coordinates, and (3) problems that require spherical coordinates for evaluation.

Next, let's discuss the relationship between multiple integrals and vector calculus. Key chapters in vector calculus include line integrals, surface integrals, Green's Theorem, Stokes' Theorem, and the Divergence Theorem. Line integrals can be divided into those involving scalar functions along a curve C and those involving vector functions along C. Both types of line integrals ultimately reduce to single-variable definite integrals. Green's Theorem is introduced next. When the first partial derivatives of the integrand exist and are continuous, Green's Theorem is an important tool for computing integrals along a closed curve in 2D and double integrals. This theorem links the line integral along a simple closed plane curve C with the double integral over the plane region R it encloses, facilitating the conversion between line integrals and double integrals. If calculating a line integral is difficult, the theorem can convert it to a double integral, and conversely, if a double integral is challenging, it can be converted to a line integral for easier computation.

Surface integrals can be divided into three main types: finding the surface area of a closed region in space, determining the mass of a surface S given a scalar function, and calculating the flux through a surface given a vector function. Ultimately, all three types of surface integrals are reduced to double integrals for computation, making familiarity with double integrals quite important. Stokes' Theorem is not only a generalization of the Fundamental Theorem of Calculus to higher dimensions but also a higher-dimensional extension of Green's Theorem. While Green's Theorem converts a double integral over a planar region to a line integral along its boundary, Stokes' Theorem converts a line integral around a closed curve in three-dimensional space to a surface integral over a non-closed surface S. When applying Stokes' Theorem to transform a line integral in three-dimensional space into a surface integral over a non-closed surface, it often requires projection techniques to convert the problem into a double integral for calculation, further emphasizing the importance of proficiency in double integrals. Additionally, the Divergence Theorem allows the conversion of the flux of a vector field $\vec{F}$ through a closed surface S into the triple integral of the divergence of $\vec{F}$ over the enclosed volume V, highlighting the importance of skills in evaluating triple integrals.

Overall, the ability to use these theorems to transform line integrals and surface integrals into double and triple integrals, respectively, underscores the need to develop strong skills in evaluating multiple integrals.

8.1 Iterated Integrals

Question Types:

Type 1.

Given $f(x,y)$, $g_1(x)$ and $g_2(x)$, find $\int_a^b \int_{g_1(x)}^{g_2(x)} f(x,y) dy dx$.

Problem-Solving Process:

Step1.

Treat x as a constant in $f(x,y)$ and compute $\int_{g_1(x)}^{g_2(x)} f(x,y) dy$.

Step2.

Compute $\int_a^b h(x) dx$, where $h(x) = \int_{g_1(x)}^{g_2(x)} f(x,y) dy$.

Type 2.

Given $f(x,y)$, $g_1(y)$ and $g_2(y)$, find $\int_a^b \int_{g_1(y)}^{g_2(y)} f(x,y) dx dy$.

Problem-Solving Process:

Step1.

Treat y as a constant in $f(x,y)$ and compute $\int_{g_1(y)}^{g_2(y)} f(x,y) dx = ?$

Step2.

Compute $\int_a^b h(y) dy = ?$, where $h(y) = \int_{g_1(y)}^{g_2(y)} f(x,y) dx$.

Example 1.

Find $\int_0^1 \int_0^x e^{x^2} dy dx = ?$

【Solution】

$\int_0^1 \int_0^x e^{x^2} dy dx = \int_0^1 x e^{x^2} dx = \frac{1}{2} e^{x^2} \Big|_0^1 = \frac{e-1}{2}$

Example 2.

(1) Find $\int_0^{\frac{\pi}{2}} \int_0^x \frac{\sin x}{x} dy dx =?$ (2) Find $\int_0^{\frac{\pi}{4}} \int_0^x \frac{\sec^2 x}{x} dy dx =?$

【Solution】

(1)
$$\int_0^{\frac{\pi}{2}} \int_0^x \frac{\sin x}{x} dy dx = \int_0^{\frac{\pi}{2}} \left(\frac{\sin x}{x}\right) y\big|_{y=0}^{y=x} dx = \int_0^{\frac{\pi}{2}} \left(\frac{\sin x}{x}\right) x dx = \int_0^{\frac{\pi}{2}} \sin x\, dx = -\cos x\big|_{x=0}^{x=\frac{\pi}{2}} = 1$$

(2)
$$\int_0^{\frac{\pi}{4}} \int_0^x \frac{\sec^2 x}{x} dy dx = \int_0^{\frac{\pi}{4}} \left(\frac{\sec^2 x}{x}\right) y\big|_{y=0}^{y=x} dx = \int_0^{\frac{\pi}{4}} \left(\frac{\sec^2 x}{x}\right) x dx = \int_0^{\frac{\pi}{4}} \sec^2 x\, dx = \tan x\big|_{x=0}^{x=\frac{\pi}{4}}$$
$$= 1$$

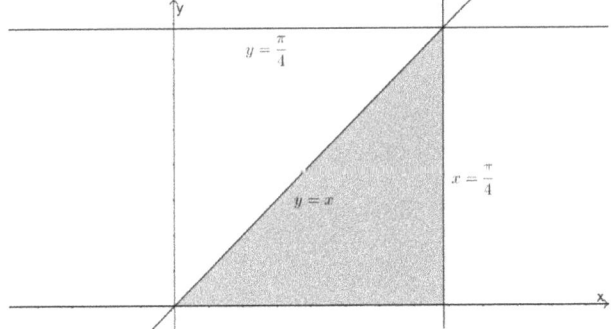

Example 3.

Find $\int_0^1 \int_0^x \frac{1}{1+x^4} dy dx =?$

【Solution】

8.1 Iterated Integrals

$$\int_0^1 \int_0^x \frac{1}{1+x^4} dy dx = \int_0^1 \frac{x}{1+x^4} dx = \left.\frac{\tan^{-1} x^2}{2}\right|_0^1 = \frac{\pi}{8}$$

Example 4.

Find $\int_0^2 \int_0^{2y} e^{y^2} dx dy = ?$

【Solution】

$$\int_0^2 \int_0^{2y} e^{y^2} dx dy = \int_0^2 2y e^{y^2} dy = \left.e^{y^2}\right|_0^2 = e^4 - 1$$

Example 5.

Find $\int_0^1 \int_0^x e^{\frac{y}{x}} dy dx = ?$

【Solution】

$$\int_0^1 \int_0^x e^{\frac{y}{x}} dy dx = \int_0^1 \left.xe^{\frac{y}{x}}\right|_{y=0}^{y=x} dx = (e-1)\int_0^1 x\, dx = \frac{e-1}{2}$$

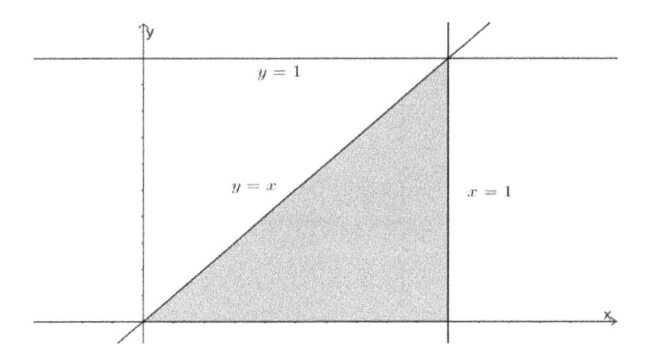

Example 6.

Find $\displaystyle\int_0^2 \int_0^x x\sqrt{x^3+1}\,dy\,dx =?$

【Solution】

$$\int_0^2 \int_0^x x\sqrt{x^3+1}\,dy\,dx = \int_0^2 x^2\sqrt{x^3+1}\,dx = \left.\frac{2(x^3+1)^{\frac{3}{2}}}{9}\right|_{x=0}^{x=2} = \frac{2(9)^{\frac{3}{2}}}{9} - \frac{2}{9} = \frac{52}{9}$$

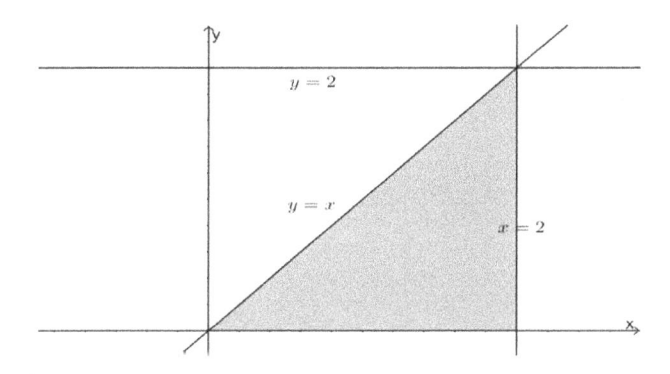

Example 7.

Find $\displaystyle\int_0^\pi \int_0^y \frac{\sin y}{y} \cos\frac{x}{y}\,dx\,dy =?$

【Solution】

$$\int_0^\pi \int_0^y \frac{\sin y}{y} \cos\frac{x}{y}\,dx\,dy = \int_0^\pi \left.\left(\sin y \sin\frac{x}{y}\right)\right|_{x=0}^{x=y} dy = \int_0^\pi \sin y\,(\sin 1)\,dy = \sin 1\,(-\cos y)|_0^\pi$$
$$= 2\sin 1$$

8.1 Iterated Integrals

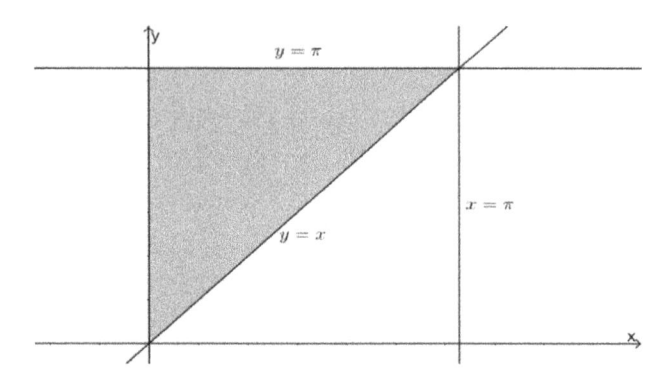

Example 8.

Find $\displaystyle\int_0^{\frac{\pi}{2}}\int_0^y \frac{\cos y}{y}\sin\frac{x}{y}\,dxdy =?$

【Solution】

$$\int_0^{\frac{\pi}{2}}\int_0^y \frac{\cos y}{y}\sin\frac{x}{y}\,dxdy = \int_0^{\frac{\pi}{2}}\left(-\cos y\cos\frac{x}{y}\right)\bigg|_{x=0}^{x=y}dy = \int_0^{\frac{\pi}{2}}\cos y\,(1-\cos 1)\,dy$$

$$= (1-\cos 1)\sin y\big|_0^{\frac{\pi}{2}} = 1-\cos 1$$

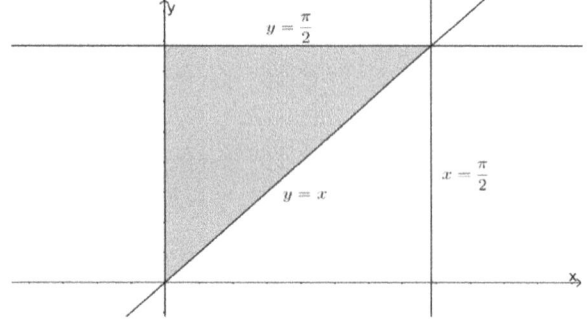

Example 9.

Find $\displaystyle\int_0^2\int_0^y \sin y^2\,dxdy =?$

【Solution】

$$\int_0^2\int_0^y \sin y^2\,dxdy = \int_0^2 y\sin y^2\,dy = -\frac{\cos y^2}{2}\bigg|_0^2 = \frac{1}{2}(1-\cos 4)$$

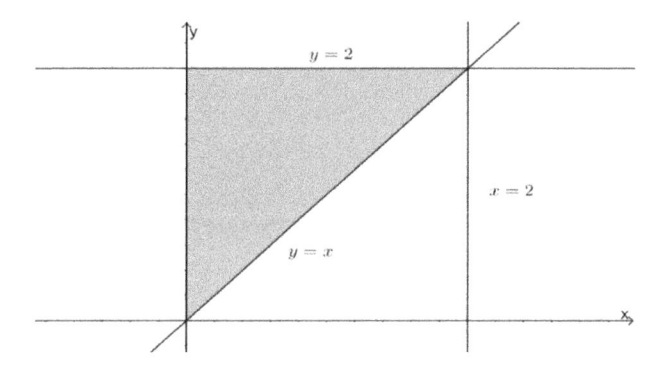

Example 10.

Find $\displaystyle\int_0^{\sqrt{3}}\int_0^x \cos\frac{\pi x^2}{2} dy dx = ?$

【Solution】

$$\int_0^{\sqrt{3}}\int_0^x \cos\frac{\pi x^2}{2} dy dx = \int_0^{\sqrt{3}} x\cos\frac{\pi x^2}{2} dx = \frac{1}{\pi}\sin\frac{\pi x^2}{2}\bigg|_0^{\sqrt{3}} = -\frac{1}{\pi}$$

Example 11.

Find $\displaystyle\int_0^{\sqrt{3}}\int_0^x \sin\frac{\pi x^2}{2} dy dx = ?$

【Solution】

$$\int_0^{\sqrt{3}}\int_0^x \sin\frac{\pi x^2}{2} dy dx = \int_0^{\sqrt{3}} x\sin\frac{\pi x^2}{2} dx = \frac{-1}{\pi}\cos\frac{\pi x^2}{2}\bigg|_0^{\sqrt{3}} = \frac{1}{\pi}$$

8.1 Iterated Integrals

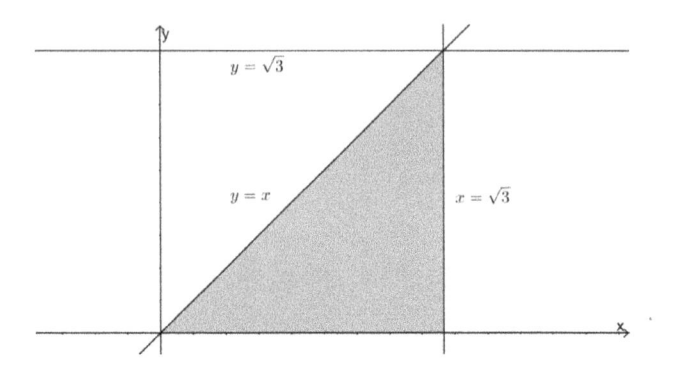

Example 12.

Find $\int_0^2 \int_0^y \cos y^2 \, dx dy = ?$

【Solution】

$$\int_0^2 \int_0^y \cos y^2 \, dx dy = \int_0^2 y \cos y^2 \, dy = \left.\frac{\sin y^2}{2}\right|_0^2 = \frac{\sin 4}{2}$$

Example 13.

Find $\int_0^1 \int_0^x e^{-x^2} \, dy dx = ?$

【Solution】

$$\int_0^1 \int_0^x e^{-x^2} \, dy dx = \int_0^1 x e^{-x^2} \, dx = \left.-\frac{e^{-x^2}}{2}\right|_0^1 = \frac{1}{2}(1 - e^{-1})$$

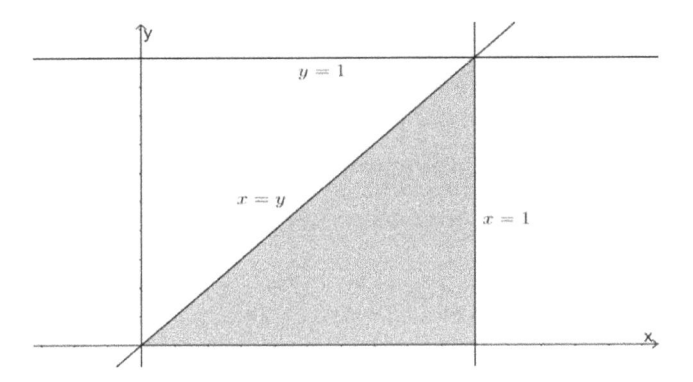

8.2 Multiple Integrals and Fubini's Theorem

【Definition】 n-dimensional interval

Let I_j be an interval in one-dimensional space, $\forall j = 1,..,n$. Then
$$I = I_1 \times \cdots \times I_n = \{(x_1, \ldots, x_n): x_j \in I_j, \forall j = 1,..,n\}$$
is called an n-dimensional interval. Here, I_j is not restricted to being open, closed, or bounded.

【Definition】 n-dimensional measure

Given $I = I_1 \times \cdots \times I_n$ where I_j is bounded, $\forall j = 1,..,n$,
$$\mu(I) := \mu(I_1) \cdot \mu(I_2) \cdots \mu(I_n)$$
is called the n-dimensional measure of the interval I. Here, $\mu(I_j)$ is the length of the one-dimensional interval I_j, $\forall j = 1,..,n$.

【Definition】 Partition of a Compact Set in R^n

(i) Suppose $I = I_1 \times \cdots \times I_n$ is a compact set in R^n. If P_k is a partition of I_k, then cartesian product $P = P_1 \times \cdots \times P_n$ is a partition of I. If P_k divides I_k into n_k one-dimensional subintervals, then P divides I into $n_1 \cdots n_k$ n-dimensional intervals.
(ii) If $P \subseteq P'$ then the partition P' of I is said to be finer than the partition P.

【Definition】 Riemann sum

8.2 Multiple Integrals and Fubini's Theorem

Let f be defined on a compact interval $I \subseteq \mathbb{R}^n$. If the partition P divides I into k subintervals $I_1, \ldots, I_k$, then

$$R(P, f) = \sum_{j=1}^{k} f(s_j) \mu(I_j) \text{ is called the Riemann sum.}$$

Here, $s_j \in I_j$.

【Definition】 Riemann Integrable

(i) A function $f(x)$ is Riemann integrable on $I \subseteq \mathbb{R}^n$
$\Leftrightarrow \exists A \in \mathbb{R}$ such that $\forall \varepsilon > 0, \exists$ Partition P_ε of I s.t. $\forall$ Partition P finer than P_ε, we have $|R(P, f) - A| < \varepsilon$, for all Riemann sums $R(P, f)$.

(ii) When A exists, the Riemann integral is denoted by $\int_I f(x) dx$, and we have

$$\int_I f(x) dx = A.$$

【Definition】 Upper Riemann Sum

For a partition P of the set $I \subseteq \mathbb{R}^n$, the upper Riemann sum of a function $f(x)$ is defined as

$$U(P, f) = \sum_{i=1}^{n} \mu(I_k) \sup_{x \in I_k} f(x).$$

【Definition】 Lower Riemann Sum

For a partition P of the set $I \subseteq \mathbb{R}^n$, the lower Riemann sum of a function $f(x)$ is defined as

$$L(P, f) = \sum_{i=1}^{n} \mu(I_k) \inf_{x \in I_k} f(x).$$

To illustrate the necessary and sufficient condition for Riemann integrability, we first introduce the Riemann condition, upper Riemann integral, and lower Riemann integral.

【Definition】 Riemann Condition

A function $f(x)$ satisfies the Riemann condition on I
$\Leftrightarrow \forall \varepsilon > 0, \exists$ partition P_ε of I such that if P is finer than P_ε then $0 \leq U(P, f) - L(P, f) < \varepsilon$

【Definition】Upper Riemann Integral

$$(U)\int_I f(x)dx = \inf\{U(P,f): \forall \text{ partition } P \in P(I)\}$$

【Definition】Lower Riemann Integral

$$(L)\int_I f(x)dx = \sup\{L(P,f): \forall \text{ partition } P \in P(I)\}$$

The proof of the necessary and sufficient condition for a function to be Riemann integrable in R^n is analogous to that in R^1. Readers are encouraged to practice this proof independently.

【Definition】Necessary and Sufficient Condition for Riemann Integrability

Let $f: I \subseteq R^n \to R$. Then
$f(x)$ is Riemann integrable on $I \Leftrightarrow$ the function $f(x)$ on I satisfies the Riemann condition

$$\Leftrightarrow (U)\int_I f(x)dx = (L)\int_I f(x)dx$$

Up to this point, we have mainly discussed the conditions under which a function $f: R^n \to R$ is Riemann integrable. Next, we will focus on how to compute double integrals in two-dimensional space.

【Theorem】Fubini's Theorem

Let f be a Riemann integrable function defined on $K = [a,b] \times [c,d]$.

If $\int_a^b f(x,y)dx$ exists, $\forall y \in [c,d]$ and $\int_c^d f(x,y)dy$ exists, $\forall x \in [a,b]$ then

(i) both $\int_a^b \int_c^d f(x,y)dydx$ and $\int_c^d \int_a^b f(x,y)dxdy$ exist.

(ii) $\iint_K f(x,y)dxdy = \int_a^b \int_c^d f(x,y)dydx = \int_c^d \int_a^b f(x,y)dxdy$

Proof:

Let $\varphi(x) = \int_c^d f(x,y)dy$,

Claim: $\varphi(x)$ is integrable on $[a,b]$ and $\iint_K f(x,y)dxdy = \int_a^b \int_c^d f(x,y)dydx$

8.2 Multiple Integrals and Fubini's Theorem

Let P_x, P_y be the two partition of $[a,b]$ and $[c,d]$ where $P_x = \{a = x_0 < x_1 < \cdots x_n = b\}$, $P_y = \{c = y_0 < y_1 < \cdots y_n = d\}$. Then $P = P_x \times P_y$ is a partition of K.

Let $m_{ij} = \inf_{K_{ij}} f(x,y)$ and $M_{ij} = \sup_{K_{ij}} f(x,y)$.

Let $m_j(x) = \inf_{y_{j-1} < y < y_j} f(x,y)$ and $M_j(x) = \sup_{y_{j-1} < y < y_j} f(x,y)$, $\forall x \in [x_{i-1}, x_i]$.

then $m_{ij} \leq m_j(x) \leq M_j(x) \leq M_{ij}$.

$$\therefore \sum_j^n m_{ij} \Delta y_j \leq \sum_j^n m_j(x) \Delta y_j \leq \sum_j^n M_j(x) \Delta y_j \leq \sum_j^n M_{ij} \Delta y_j$$

$$\therefore \sum_j^n m_j(x) \Delta y_j \leq \varphi(x) = \int_c^d f(x,y) dy \leq \sum_j^n M_j(x) \Delta y_j$$

$$\therefore \sum_j^n m_{ij}(x) \Delta y_j \leq \varphi(x) \leq \sum_j^n M_{ij}(x) \Delta y_j, \quad \forall x \in [a,b]$$

$$\therefore \sum_j^n m_{ij}(x) \Delta y_j \leq \inf_{x \in [x_{i-1}, x_i]} \varphi(x) \leq \sup_{x \in [x_{i-1}, x_i]} \varphi(x) \leq \sum_j^n M_{ij}(x) \Delta y_j, \forall x \in [x_{i-1}, x_i],$$

$i = 1, 2, \ldots, n$

$$\therefore \sum_i^n \left(\sum_j^n m_{ij}(x) \Delta y_j \right) \Delta x_i \leq \sum_i^n \left(\inf_{x \in [x_{i-1}, x_i]} \varphi(x) \right) \Delta x_i \leq \sum_i^n \left(\sup_{x \in [x_{i-1}, x_i]} \varphi(x) \right) \Delta x_i$$

$$\leq \sum_i^n \left(\sum_j^n M_{ij}(x) \Delta y_j \right) \Delta x_i$$

$\therefore L(P, f) \leq L(P_x, f) \leq U(P_x, f) \leq U(P, f)$

$\therefore L(P_x, f) \leq (L) \int_a^b \varphi(x) dx \leq (U) \int_a^b \varphi(x) dx \leq U(P_x, f)$

$\therefore L(P, f) \leq (L) \int_a^b \varphi(x) dx \leq (U) \int_a^b \varphi(x) dx \leq U(P, f)$, $\forall$ partition P

$\therefore (L) \iint_K f(x,y) dx dy \leq (L) \int_a^b \varphi(x) dx \leq (U) \int_a^b \varphi(x) dx \leq (U) \iint_K f(x,y) dx dy$

$\because f$ is Riemann integrable on $K = [a,b] \times [c,d]$

$\therefore (L) \iint_K f(x,y) dx dy = (U) \iint_K f(x,y) dx dy = \iint_K f(x,y) dx dy$

$\therefore (L) \int_a^b \varphi(x) dx = (U) \int_a^b \varphi(x) dx = \iint_K f(x,y) dx dy \Rightarrow \int_a^b \varphi(x) dx = \iint_K f(x,y) dx dy$

By the same way, $\iint_K f(x,y)dxdy = \int_c^d \int_a^b f(x,y)dxdy$.

For Fubini's Theorem, it is crucial that f is a Riemann integrable function on $K = [a,b] \times [c,d]$. Next, we will demonstrate an example where a function $f(x,y)$ meets the conditions:

$$\int_a^b f(x,y)dx \text{ exists, } \forall y \in [c,d] \text{ and } \int_c^d f(x,y)dy \text{ exists, } \forall x \in [a,b].$$

However,

$$\iint_K f(x,y)dxdy \neq \int_a^b \int_c^d f(x,y)dydx \neq \int_c^d \int_a^b f(x,y)dxdy.$$

Examples:

Let $f(x,y) = \begin{cases} -\dfrac{1}{x^2}, & 0 < y \leq x \leq 1 \\ \dfrac{1}{y^2}, & 0 < x \leq y \leq 1 \end{cases}$. Then f is not Riemann integrable on $K = [0,1] \times [0,1]$.

We find that

$$\int_0^1 f(x,y)dx = \int_0^y \frac{1}{y^2}dx + \int_y^1 -\frac{1}{x^2}dx = 1$$

and

$$\int_0^1 f(x,y)dy = \int_0^x -\frac{1}{x^2}dy + \int_x^1 \frac{1}{y^2}dy = -1.$$

If f were Riemann integrable on $K = [0,1] \times [0,1]$ then

$$\iint_K f(x,y)dxdy = \int_0^1 \int_0^1 f(x,y)dydx = \int_0^1 \int_0^1 f(x,y)dxdy, \text{ a contradiction.}$$

$\therefore f$ is not Riemann integrable on $K = [0,1] \times [0,1]$.

So far, Fubini's Theorem helps to transform the problem of computing double integrals into an iterated integral problem. The crucial condition is that f is Riemann integrable on $K = [a,b] \times [c,d]$. Next, we will prove that if $f(x,y)$ is a continuous function on $K = [a,b] \times [c,d]$, then f is Riemann integrable on $K = [a,b] \times [c,d]$ and

$$\iint_K f(x,y)dA = \int_a^b \int_c^d f(x,y)dydx = \int_c^d \int_a^b f(x,y)dxdy.$$

8.2 Multiple Integrals and Fubini's Theorem

【Theorem】

Let $f(x,y)$ be a continuous function on $R = [a,b] \times [c,d]$. Then
$$\iint_R f(x,y)dA = \int_a^b \int_c^d f(x,y)dydx = \int_c^d \int_a^b f(x,y)dxdy.$$

Proof:

Let P be the partition of $[a,b] \times [c,d]$

Let $R_{ij} = [x_{i-1}, x_i] \times [y_{j-1}, y_j]$, $\forall \, 0 \leq i \leq m, 0 \leq j \leq n$

then $[a,b] \times [c,d] = \bigcup_{i=0}^{m} \bigcup_{j=0}^{n} [x_{i-1}, x_i] \times [y_{j-1}, y_j] = \bigcup_{i=0}^{m} \bigcup_{j=0}^{n} R_{ij}$

Let $m_{ij} = \inf_{R_{ij}} f(x,y)$ and $M_{ij} = \sup_{R_{ij}} f(x,y)$

$\because m_{ij} \leq f(x,y) \leq M_{ij}, \forall (x,y) \in R_{ij}, \forall \, 0 \leq i \leq m, 0 \leq j \leq n$

$\therefore m_{ij}\Delta y_j \leq \int_{y_{j-1}}^{y_j} f(x,y)dy \leq M_{ij}\Delta y_j, \forall x \in [x_{i-1}, x_i], \forall \, 0 \leq i \leq m, 0 \leq j \leq n$

$\therefore \sum_{j=0}^{n} m_{ij}\Delta y_j \leq \int_c^d f(x,y)dy \leq \sum_{j=0}^{n} M_{ij}\Delta y_j, \forall x \in [x_{i-1}, x_i], \forall \, 0 \leq i \leq m, 0 \leq j \leq n$

$\therefore \sum_{j=0}^{n} m_{ij}\Delta x_i \, \Delta y_j \leq \int_{x_{i-1}}^{x_i} \int_c^d f(x,y)dydx \leq \sum_{j=0}^{n} M_{ij}\Delta x_i \, \Delta y_j$

$\therefore \sum_{i=0}^{m} \sum_{j=0}^{n} m_{ij}\Delta x_i \, \Delta y_j \leq \int_a^b \int_c^d f(x,y)dydx \leq \sum_{i=0}^{m} \sum_{j=0}^{n} M_{ij}\Delta x_i \, \Delta y_j$

$\Rightarrow L(f,P) \leq \int_a^b \int_c^d f(x,y)dydx \leq U(f,P), \forall \text{ partition } P$

$\because L(f,P) \leq \iint_R f(x,y)dA \leq U(f,P), \forall \text{ partition } P$

$\therefore \iint_R f(x,y)dA = \int_a^b \int_c^d f(x,y)dydx = \int_c^d \int_a^b f(x,y)dxdy$

【Theorem】Properties of Multiple Integrals

Let f, g be functions defined on $I \subseteq R^n$ and both are Riemann integrable.

(i) $\int_I \alpha f(x) \pm \beta f(x) dx = \alpha \int_I f(x)\,dx \pm \beta \int_I f(x)\,dx, \forall \, \alpha, \beta \in R$

(ii) $\int_I f(x)\,dx \geq 0$ if $f(x) \geq 0, \forall x \in R^n$

(iii) $\int_I f(x)\,dx \geq \int_I g(x)\,dx$ if $f(x) \geq g(x), \forall x \in R^n$

8.3 Methods for Solving Double Integrals

Types of Exam Questions on Double Integrals:

1. **Convert Double Integrals to Iterated Integrals:**
   This involves transforming a double integral into iterated integrals and then calculating the result.
2. **Change the Order of Integration:**
   This involves switching the order of integration when it is not feasible to directly evaluate the integral.
3. **Use Polar Coordinates to Evaluate Double Integrals:**
   This involves converting the double integral into polar coordinates to simplify the calculation.
4. **Use General Coordinate Transformations to Evaluate Double Integrals:**
   This involves applying a general coordinate transformation to compute the value of the double integral.

8.3.1 Convert Double Integrals to Iterated Integrals

Question Types:

Type 1.

When applying Fubini's Theorem, double integrals can be converted into iterated integrals for direct evaluation. For example, consider

$$\iint_R f(x,y)\,dA \quad \text{where } R = \{(x,y): a \leq x \leq b, c \leq y \leq d\}.$$

Problem-Solving Process:

Step 1.

Using Fubinis Theorem, we convert the double integral:

$$\iint_R f(x,y)\,dA = \int_a^b \int_c^d f(x,y)\,dy\,dx = \int_c^d \int_a^b f(x,y)\,dx\,dy$$

Step 2.

8.3.1 Convert Double Integrals to Iterated Integrals

Compute either $\int_a^b \int_c^d f(x,y)dydx =?$ or $\int_c^d \int_a^b f(x,y)dxdy =?$

Type 2.
After converting the double integral to iterated integrals using Fubini's Theorem, integration by parts can also be employed. Consider
$$R = \{(x,y): a \le x \le b, g_1(x) \le y \le g_2(x)\}.$$
Problem-Solving Process:

Step1.

Apply Fubinis Theorem, $\iint_R f(x,y)dA = \int_a^b \int_{g_1(x)}^{g_2(x)} f(x,y)dydx$

Step2.

Assume $\int_a^b \int_{g_1(x)}^{g_2(x)} f(x,y)dydx = \int_a^b u(x) v'(x)dx.$ Then, use integration by parts,

$\int_a^b u(x)v'(x)dx = u(x)v(x)|_{x=a}^{x=b} - \int_a^b u'(x)v(x)dx.$

Step3.

Find $u(x)v(x)|_{x=a}^{x=b} - \int_a^b u'(x)v(x)dx$

Examples:

(I) Find $\int_0^2 \int_0^{x^2} e^{\frac{y}{x}} dydx =?$

$\int_0^2 \int_0^{x^2} e^{\frac{y}{x}} dydx = \int_0^2 xe^{\frac{y}{x}}\big|_{y=0}^{y=x^2} dx = \int_0^2 x(e^x - 1)dx = \int_0^2 xe^x dx - \int_0^2 xdx$

Let $u = x$, $dv = e^x dx$ then $du = dx$, $v = e^x$, by integration by parts,

$\int_0^2 xe^x dx = e^2 + 1$ and $\int_0^2 xdx = 2$

(II) Find $\int_0^1 \int_x^1 \frac{1}{1+y^2} dydx =?$

$\int_0^1 \int_x^1 \frac{1}{1+y^2} dydx = \int_0^1 \tan^{-1} y\big|_x^1 dx = \int_0^1 \frac{\pi}{4} - \tan^{-1} x \; dx$

Let $u = \tan^{-1} x$, $dv = dx$ then $du = \frac{dx}{1+x^2}$, $v = x$, by integration by parts,

$$\int_0^1 \tan^{-1} x \, dx = \frac{\pi}{4} - \frac{\ln 2}{2}$$

Example 1.

Find $\iint_R 4 - x - y \, dA = ?$, $R = \{(x,y): 0 \le x \le 2, 0 \le y \le 2\}$.

【Solution】

∵ $f(x,y) = 4 - x - y$ is continuous on R, by Fubini's Theorem,

$$\iint_R 4 - x - y \, dA = \int_0^2 \int_0^2 4 - x - y \, dy \, dx = \int_0^2 4y - xy - \frac{y^2}{2} \Big|_{y=0}^{y=2} dx = \int_0^2 8 - 2x - 2 \, dx$$
$$= (6x - x^2)\Big|_{x=0}^{x=2} = 12 - 4 = 8$$

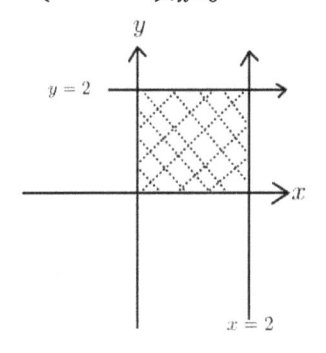

Example 2.

Find $\iint_R (1 + 8xy) \, dA = ?$, $R = \{(x,y): 1 \le x \le 2, 0 \le y \le 2\}$.

【Solution】

∵ $f(x,y) = 1 + 8xy$ is continuous on R, by Fubini's Theorem,

$$\iint_R (1 + 8xy) \, dA = \int_0^2 \int_1^2 (1 + 8xy) \, dx \, dy = \int_0^2 (x + 4x^2 y)\Big|_{x=1}^{x=2} dy = \int_0^2 1 + 12y \, dy$$
$$= (y + 6y^2)\Big|_{y=0}^{y=2} = 2 + 6 \cdot 4 = 26$$

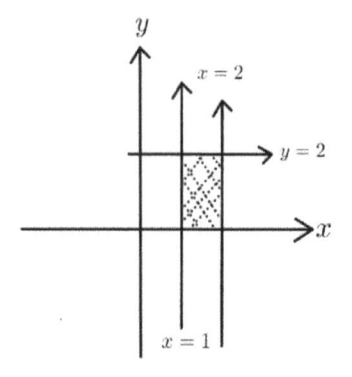

8.3.1 Convert Double Integrals to Iterated Integrals

Example 3.

Find $\iint_R \dfrac{1}{(x+y)^2} dxdy = ?$, $R = \{(x,y): 1 \le x \le 3, 0 \le y \le 2\}$.

【Solution】

$\because f(x,y) = \dfrac{1}{(x+y)^2}$ is continuous on R, by Fubini's Theorem,

$\iint_R \dfrac{1}{(x+y)^2} dxdy = \int_1^3 \int_0^2 \dfrac{1}{(x+y)^2} dydx = \int_1^3 \dfrac{-1}{(x+y)}\Big|_{y=0}^{y=2} dx = -\int_1^3 \dfrac{1}{x+2} - \dfrac{1}{x} dx$

$= -(\ln(x+2) - \ln x)|_{x=1}^{x=3} = -(\ln 5 - \ln 3 - \ln 3) = 2\ln 3 - \ln 5$

Example 4.

Find $\iint_R xy dA = ?$, $R = \{(x,y): 0 \le x \le 2, 2x - 4 \le y \le 0\}$.

【Solution】

$\because f(x,y) = xy$ is continuous on R, by Fubini's Theorem,

$\iint_R xy dA = \int_0^2 \int_{2x-4}^0 xy \, dydx = \int_0^2 \dfrac{xy^2}{2}\Big|_{y=2x-4}^{y=0} dx = -\int_0^2 \dfrac{x(2x-4)^2}{2} dx$

$= -2\int_0^2 x(x-2)^2 dx = -2\int_0^2 x^3 - 4x^2 + 4x \, dx = -2\left(\dfrac{x^4}{4} - \dfrac{4x^3}{3} + 2x^2\right)\Big|_{x=0}^{x=2}$

$= -2\left(4 - \dfrac{32}{3} + 8\right) = \dfrac{-8}{3}$

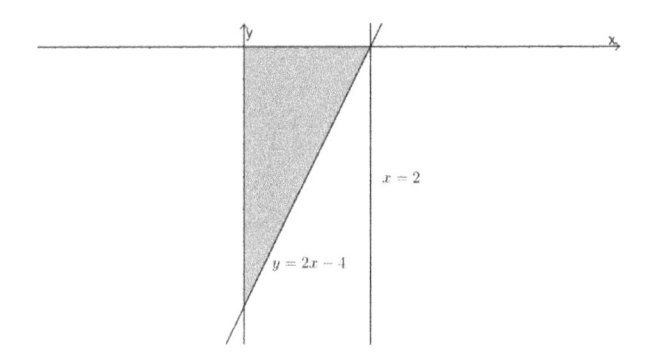

Example 5.

Find $\iint_R e^{\frac{y}{x}} dA = ?$, $R = \{(x, y): 0 \leq x \leq 2, 0 \leq y \leq x^2\}$.

【Solution】

$\because f(x, y) = e^{\frac{y}{x}}$ is continuous on R, by Fubini's Theorem,

$$\iint_R e^{\frac{y}{x}} dA = ? = \int_0^2 \int_0^{x^2} e^{\frac{y}{x}} dy dx = \int_0^2 x e^{\frac{y}{x}} \Big|_{y=0}^{y=x^2} dx = \int_0^2 x(e^x - 1) dx = \int_0^2 x e^x dx - \int_0^2 x dx$$

Let $u = x$, $dv = e^x dx$ then $du = dx$, $v = e^x$, by integration by parts,

then $\int_0^2 x e^x dx = x e^x \big|_{x=0}^{x=2} - \int_0^2 e^x dx = 2e^2 - (e^2 - 1) = e^2 + 1$

$\because \int_0^2 x dx = 2$ $\therefore \int_0^2 \int_0^{x^2} e^{\frac{y}{x}} dy dx = \int_0^2 x e^x dx - \int_0^2 x dx = e^2 + 1 - 2 = e^2 - 1$

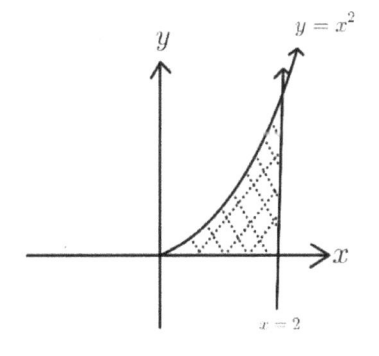

Example 6.

Find $\iint_R e^{y^2} dA = ?$, $R = \{(x, y): x - y \leq 0, y \leq 3, x + 3y \geq 0\}$.

【Solution】

8.3.1 Convert Double Integrals to Iterated Integrals

∵ $f(x,y) = e^{y^2}$ is continuous on R, by Fubini's Theorem,

$$\iint_R e^{y^2} dA = \int_0^3 \int_{-3y}^{y} e^{y^2} dxdy = \int_0^3 4ye^{y^2} dy = 2e^{y^2}\Big|_{y=0}^{y=3} = 2(e^9 - 1)$$

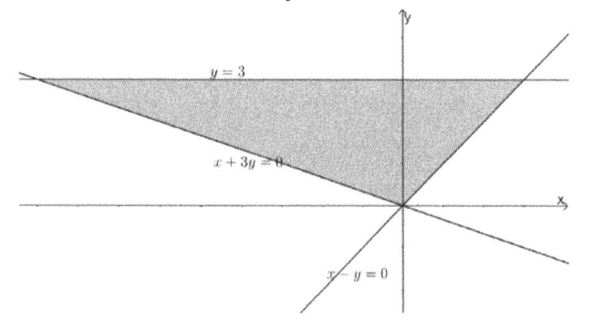

Example 7.

Find $\iint_R ye^{xy} dA =?$, $R = \{(x,y): 0 \le x \le 1, 0 \le y \le 3\}$.

【Solution】

∵ $f(x,y) = ye^{xy}$ is continuous on R, by Fubini's Theorem,

then $\iint_R ye^{xy} dA = \int_0^3 \int_0^1 ye^{xy} dxdy = \int_0^3 e^{xy}\Big|_{x=0}^{x=1} dy = \int_0^3 e^y - 1 dy = (e^3 - 1) - 3$
$= e^3 - 4$

Example 8.

Find $\iint_R \frac{1}{1+y^2} dA =?$, $R = \{(x,y): 0 \le x \le 1, x \le y \le 1\}$.

【Solution】

∵ $f(x,y) = \frac{1}{1+y^2}$ is continuous on R, by Fubini's Theorem,

$$\iint_R \frac{1}{1+y^2} dA = \int_0^1 \int_x^1 \frac{1}{1+y^2} dydx = \int_0^1 \tan^{-1} y\Big|_x^1 dx = \int_0^1 \frac{\pi}{4} - \tan^{-1} x \ dx$$

Let $u = \tan^{-1} x$, $dv = dx$ then $du = \frac{dx}{1+x^2}$, $v = x$, by integration by parts,

$$\int_0^1 \tan^{-1} x \ dx = x\tan^{-1} x\Big|_0^1 - \int_0^1 \frac{xdx}{1+x^2} = \frac{\pi}{4} - \frac{1}{2}\ln(1+x^2)\Big|_0^1 = \frac{\pi}{4} - \frac{\ln 2}{2}$$

$$\therefore \int_0^1 \frac{\pi}{4} - \tan^{-1} x \, dx = \frac{\pi}{4} - \left(\frac{\pi}{4} - \frac{\ln 2}{2}\right) = \frac{\ln 2}{2}$$

Example 9.

Find $\iint_R 2x - y^2 dA = ?$, $R = \{(x,y): x - y + 1 \leq 0, y \leq 2, x + y - 1 \geq 0\}$

【Solution】

$\because f(x,y) = 2x - y^2$ is continuous on R, by Fubini's Theorem,

$$\iint_R 2x - y^2 dA = \int_1^2 \int_{1-y}^{y-1} 2x - y^2 dx dy = \int_1^2 x^2 - y^2 x \Big|_{x=1-y}^{x=y-1} dy$$

$$= \int_1^2 (y-1)^2 - y^2(y-1) - ((1-y)^2 - y^2(1-y)) \, dy = \int_1^2 -2y^3 + 2y^2 \, dy$$

$$= \frac{-y^4}{2}\Big|_{y=1}^{y=2} + \frac{2y^3}{3}\Big|_{y=1}^{y=2} = \frac{-17}{6}$$

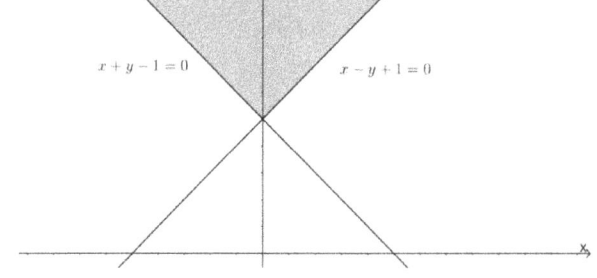

8.3.2 Change the Order of Integration

Question Types:

Type 1.

8.3.2 Change the Order of Integration

If $R = \{(x,y): a \leq x \leq b, f_1(x) \leq y \leq f_2(x)\}$, and you cannot directly calculate:
$$\int_a^b \int_{f_1(x)}^{f_2(x)} f(x,y)dydx = ?$$

Problem-Solving Process:

Step1.
Find c, d, f_3, f_4 such that $R = \{(x,y): f_3(y) \leq x \leq f_4(y), c \leq y \leq d\}$

Step2.
Thus, $\int_a^b \int_{f_1(x)}^{f_2(x)} f(x,y)dydx = \int_c^d \int_{f_3(y)}^{f_4(y)} f(x,y)dxdy$

Step3.
Calculate the integral: $\int_c^d \int_{f_3(y)}^{f_4(y)} f(x,y)dxdy$

Step4.
If you cannot calculate:
$$\int_c^d \int_{f_3(y)}^{f_4(y)} f(x,y)dxdy$$
then typically use variable substitution or integration by parts to find the value.

Type 2.
If $R = \{(x,y): f_1(y) \leq x \leq f_2(y), c \leq y \leq d\}$, and you cannot directly calculate:
$$\int_c^d \int_{f_1(y)}^{f_2(y)} f(x,y)dxdy$$

Problem-Solving Process:

Step1.
Find $a, b, f_3,$ and f_4 such that $R = \{(x,y): a \leq x \leq b, f_3(x) \leq y \leq f_4(x)\}$

Thus, $\int_c^d \int_{f_1(y)}^{f_2(y)} f(x,y)dxdy = \int_a^b \int_{f_3(x)}^{f_4(x)} f(x,y)dydx$

Step2.
Calculate the integral, $\int_a^b \int_{f_3(x)}^{f_4(x)} f(x,y)dydx$

Step3.
If you cannot calculate:
$$\int_a^b \int_{f_3(x)}^{f_4(x)} f(x,y)dydx$$

then typically use variable substitution or integration by parts to find the value.

Example 1.

$$\text{Find } \int_0^1 \int_y^1 e^{x^2} \, dx\, dy = ?$$

【Solution】

∵ $\{(x,y): y \leq x \leq 1, 0 \leq y \leq 1\} = \{(x,y): 0 \leq x \leq 1, 0 \leq y \leq x\}$

∴ $\int_0^1 \int_y^1 e^{x^2}\, dx\, dy = \int_0^1 \int_0^x e^{x^2}\, dy\, dx = \int_0^1 x e^{x^2}\, dx = \frac{1}{2} e^{x^2} \Big|_0^1 = \frac{e-1}{2}$

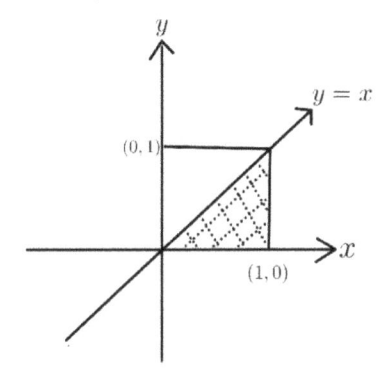

Example 2.

$$\text{Find } \int_0^2 \int_{\sqrt{x}}^2 e^{y^3}\, dy\, dx = ?$$

【Solution】

∵ $\{(x,y): 0 \leq x \leq 2, \sqrt{x} \leq y \leq 2\} = \{(x,y): 0 \leq x \leq y^2, 0 \leq y \leq 2\}$

∴ $\int_0^2 \int_{\sqrt{x}}^2 e^{y^3}\, dy\, dx = \int_0^2 \int_0^{y^2} e^{y^3}\, dx\, dy = \int_0^2 y^2 e^{y^3}\, dy = \frac{e^{y^3}}{3}\Big|_{y=0}^{y=2} = \frac{1}{3}(e^8 - 1)$

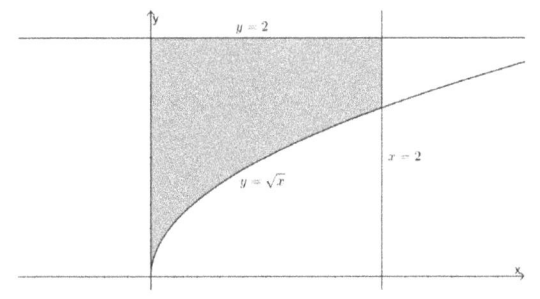

8.3.2 Change the Order of Integration

Example 3.

Find $\int_0^4 \int_{\sqrt{y}}^2 \sqrt{x^3 + 1}\, dx\, dy = ?$

【Solution】

$\because \{(x,y): \sqrt{y} \leq x \leq 2, 0 \leq y \leq 4\} = \{(x,y): 0 \leq x \leq 2, 0 \leq y \leq x^2\}$

$\therefore \int_0^4 \int_{\sqrt{y}}^2 \sqrt{x^3 + 1}\, dx\, dy = \int_0^2 \int_0^{x^2} \sqrt{x^3 + 1}\, dy\, dx = \int_0^2 x^2 \sqrt{x^3 + 1}\, dx = \left. \frac{2(x^3 + 1)^{\frac{3}{2}}}{9} \right|_0^2 = \frac{54 - 2}{9}$

$= \frac{52}{9}$

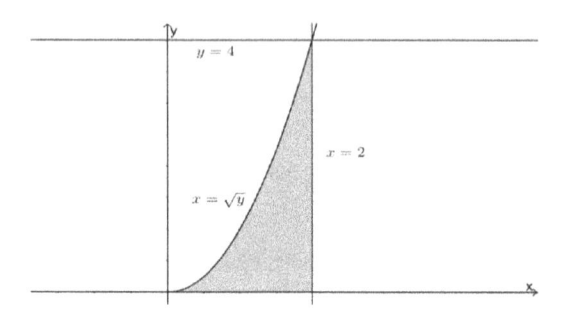

Example 4.

Find $\int_0^4 \int_{\sqrt{y}}^2 \frac{y\sqrt{x^2 + 1}}{x^3}\, dx\, dy = ?$

【Solution】

$\because \{(x,y): \sqrt{y} \leq x \leq 2, 0 \leq y \leq 4\} = \{(x,y): 0 \leq x \leq 2, 0 \leq y \leq x^2\}$

$\therefore \int_0^4 \int_{\sqrt{y}}^2 \frac{y\sqrt{x^2 + 1}}{x^3}\, dx\, dy = \int_0^2 \int_0^{x^2} \frac{y\sqrt{x^2 + 1}}{x^3}\, dy\, dx = \int_0^2 \frac{x^4 \sqrt{x^2 + 1}}{2x^3}\, dx = \int_0^2 \frac{x\sqrt{x^2 + 1}}{2}\, dx$

$= \left. \frac{(x^2 + 1)^{\frac{3}{2}}}{6} \right|_0^2 = \frac{5\sqrt{5} - 1}{6}$

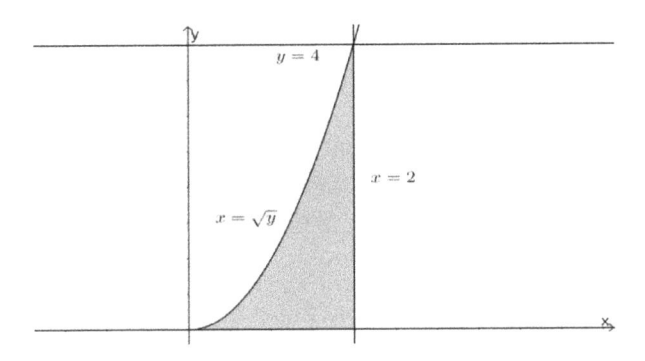

Example 5.

(1) Find $\int_0^{\frac{\pi}{2}} \int_y^{\frac{\pi}{2}} \frac{\sin x}{x} dx dy = ?$ (2) Find $\int_0^{\frac{\pi}{4}} \int_y^{\frac{\pi}{4}} \frac{\sec^2 x}{x} dx dy = ?$

【Solution】

(1)

$\because \{(x,y): y \le x \le \frac{\pi}{2}, 0 \le y \le \frac{\pi}{2}\} = \{(x,y): 0 \le x \le \frac{\pi}{2}, 0 \le y \le x\}$

$\therefore \int_0^{\frac{\pi}{2}} \int_y^{\frac{\pi}{2}} \frac{\sin x}{x} dx dy = \int_0^{\frac{\pi}{2}} \int_0^x \frac{\sin x}{x} dy dx = \int_0^{\frac{\pi}{2}} \left(\frac{\sin x}{x}\right) y\Big|_{y=0}^{y=x} dx = \int_0^{\frac{\pi}{2}} \left(\frac{\sin x}{x}\right) x \, dx$

$= \int_0^{\frac{\pi}{2}} \sin x \, dx = -\cos x \Big|_{x=0}^{x=\frac{\pi}{2}} = 1$

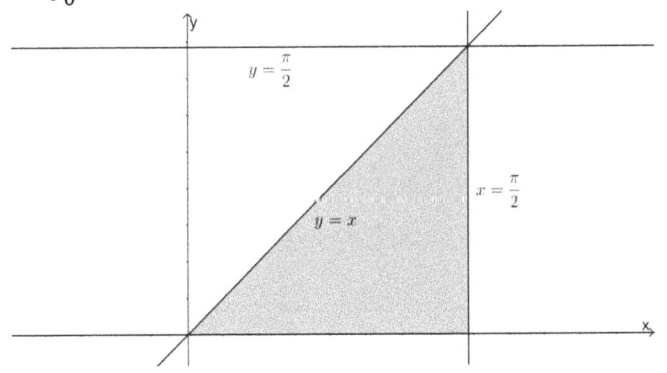

(2)

$\because \{(x,y): y \le x \le \frac{\pi}{4}, 0 \le y \le \frac{\pi}{4}\} = \{(x,y): 0 \le x \le \frac{\pi}{4}, 0 \le y \le x\}$

$\therefore \int_0^{\frac{\pi}{4}} \int_y^{\frac{\pi}{4}} \frac{\sec^2 x}{x} dx dy = \int_0^{\frac{\pi}{4}} \int_0^x \frac{\sec^2 x}{x} dy dx = \int_0^{\frac{\pi}{4}} \left(\frac{\sec^2 x}{x}\right) y\Big|_{y=0}^{y=x} dx = \int_0^{\frac{\pi}{4}} \left(\frac{\sec^2 x}{x}\right) x \, dx$

8.3.2 Change the Order of Integration

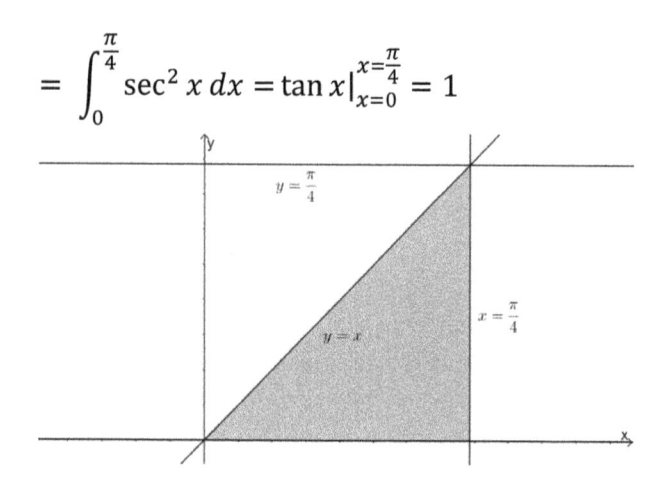

$$= \int_0^{\frac{\pi}{4}} \sec^2 x \, dx = \tan x \Big|_{x=0}^{x=\frac{\pi}{4}} = 1$$

Example 6.

Find $\int_0^9 \int_{\sqrt{y}}^3 y \cos x^5 \, dxdy = ?$

【Solution】

$\because \{(x,y): \sqrt{y} \leq x \leq 3, 0 \leq y \leq 9\} = \{(x,y): 0 \leq x \leq 3, 0 \leq y \leq x^2\}$

$\therefore \int_0^9 \int_{\sqrt{y}}^3 y \cos x^5 \, dxdy = \int_0^3 \int_0^{x^2} y \cos x^5 \, dydx = \int_0^3 (\cos x^5) \frac{y^2}{2} \Big|_0^{x^2} dx = \frac{1}{2} \int_0^3 (\cos x^5) x^4 \, dx$

$= \frac{1}{10} \sin x^5 \Big|_0^3 = \frac{\sin 243}{10}$

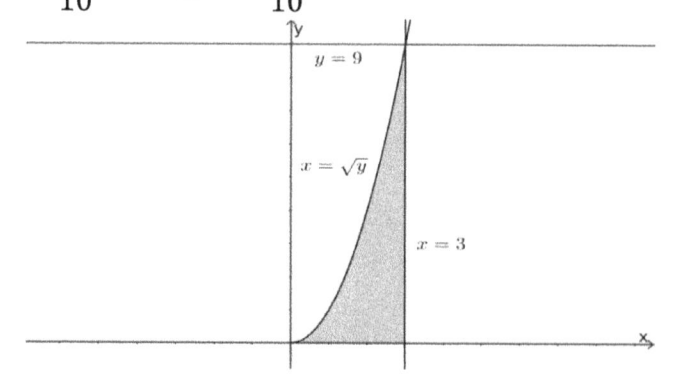

Example 7.

Find $\int_0^1 \int_y^1 \frac{1}{1+x^4} \, dxdy = ?$

【Solution】

$\because \{(x,y): y \leq x \leq 1, 0 \leq y \leq 1\} = \{(x,y): 0 \leq x \leq 1, 0 \leq y \leq x\}$

$\therefore \int_0^1 \int_y^1 \frac{1}{1+x^4} dx dy = \int_0^1 \int_0^x \frac{1}{1+x^4} dy dx = \int_0^1 \frac{x}{1+x^4} dx = \frac{\tan^{-1} x^2}{2}\Big|_0^1 = \frac{\pi}{8}$

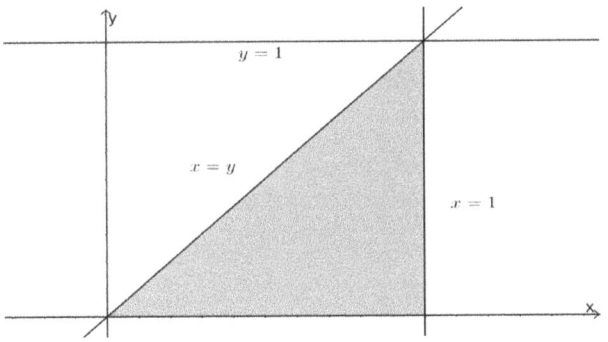

Example 8.

Find $\int_0^4 \int_{\frac{x}{2}}^2 e^{y^2} dy dx = ?$

【Solution】

$\because \{(x,y): 0 \leq x \leq 4, \frac{x}{2} \leq y \leq 2\} = \{(x,y): 0 \leq x \leq 2y, 0 \leq y \leq 2\}$

$\therefore \int_0^4 \int_{\frac{x}{2}}^2 e^{y^2} dy dx = \int_0^2 \int_0^{2y} e^{y^2} dx dy = \int_0^2 2y e^{y^2} dy = e^{y^2}\Big|_0^2 = e^4 - 1$

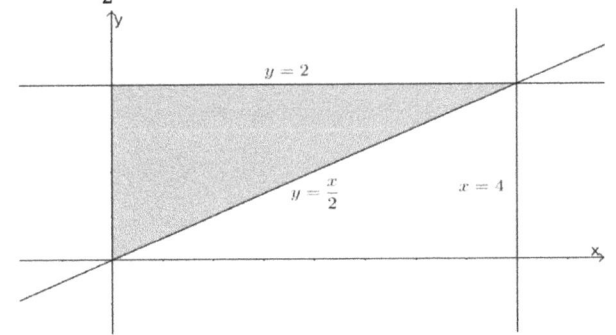

Example 9.

Find $\int_0^1 \int_y^1 e^{\frac{y}{x}} dx dy = ?$

8.3.2 Change the Order of Integration

【Solution】

$\because \{(x,y): y \le x \le 1, 0 \le y \le 1\} = \{(x,y): 0 \le x \le 1, 0 \le y \le x\}$

$\therefore \int_0^1 \int_y^1 e^{\frac{y}{x}} dx dy = \int_0^1 \int_0^x e^{\frac{y}{x}} dy dx = \int_0^1 xe^{\frac{y}{x}}\Big|_{y=0}^{y=x} dx = (e-1)\int_0^1 x\,dx = \frac{e-1}{2}$

Example 10.

Find $\int_0^{\ln 10} \int_{e^x}^{10} \frac{1}{\ln y} dy dx = ?$

【Solution】

$\because \{(x,y): 0 \le x \le \ln 10, e^x \le y \le 10\} = \{(x,y): 0 \le x \le \ln y, 1 \le y \le 10\}$

$\therefore \int_0^{\ln 10} \int_{e^x}^{10} \frac{1}{\ln y} dy dx = \int_1^{10} \int_0^{\ln y} \frac{1}{\ln y} dx dy = \int_1^{10} dy = 9$

Example 11.

Find $\int_0^2 \int_y^2 x\sqrt{x^3+1}\,dx dy = ?$

【Solution】

∵ $\{(x,y): y \leq x \leq 2, 0 \leq y \leq 2\} = \{(x,y): 0 \leq x \leq 2, 0 \leq y \leq x\}$

∴ $\int_0^2 \int_y^2 x\sqrt{x^3+1}\,dxdy = \int_0^2 \int_0^x x\sqrt{x^3+1}\,dydx = \int_0^2 x^2\sqrt{x^3+1}\,dx = \left.\dfrac{2(x^3+1)^{\frac{3}{2}}}{9}\right|_{x=0}^{x=2}$

$= \dfrac{2(9)^{\frac{3}{2}}}{9} - \dfrac{2}{9} = \dfrac{52}{9}$

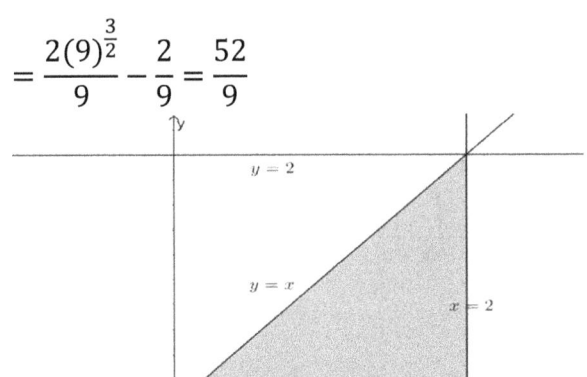

Example 12.

(1) Find $\displaystyle\int_0^2 \int_y^2 e^{\max(x^2,y^2)}\,dxdy = ?$ (2) Find $\displaystyle\int_0^2 \int_0^2 e^{\max(x^2,y^2)}\,dxdy = ?$

【Solution】

(1)

∵ $\{(x,y): y \leq x \leq 2, 0 \leq y \leq 2\} = \{(x,y): 0 \leq x \leq 2, 0 \leq y \leq x\}$

∴ $\displaystyle\int_0^2 \int_y^2 e^{\max(x^2,y^2)}\,dxdy = \int_0^2 \int_0^x e^{\max(x^2,y^2)}\,dydx = \int_0^2 \int_0^x e^{x^2}\,dydx = \int_0^2 e^{x^2} x\,dx$

$= \left.\dfrac{e^{x^2}}{2}\right|_{x=0}^{x=2} = \dfrac{e^4-1}{2}$

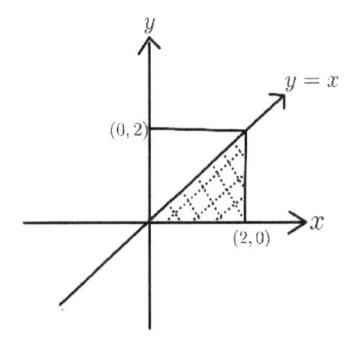

(2)

8.3.2 Change the Order of Integration 158

$$\because \int_0^2 e^{\max(x^2,y^2)}dx = \int_0^y e^{\max(x^2,y^2)}dx + \int_y^2 e^{\max(x^2,y^2)}dx, \forall 0 < y < 2$$

$$\therefore \int_0^2\int_0^2 e^{\max(x^2,y^2)}dxdy = \int_0^2\left(\int_0^y e^{\max(x^2,y^2)}dx + \int_y^2 e^{\max(x^2,y^2)}dx\right)dy$$

$$\because \int_0^2\int_0^y e^{\max(x^2,y^2)}dx\,dy = \int_0^2\left(\int_0^y e^{y^2}dx\right)dy = \int_0^2 ye^{y^2}dy = \frac{e^{y^2}}{2}\bigg|_0^2 = \frac{e^4-1}{2}$$

$$\because \{(x,y): y \le x \le 2, 0 \le y \le 2\} = \{(x,y): 0 \le x \le 2, 0 \le y \le x\}$$

$$\therefore \int_0^2\left(\int_y^2 e^{\max(x^2,y^2)}dx\right)dy = \int_0^2\int_0^x e^{\max(x^2,y^2)}dydx = \int_0^2\int_0^x e^{x^2}dydx = \int_0^2 e^{x^2}x\,dx$$

$$= \frac{e^{x^2}}{2}\bigg|_{x=0}^{x=2} = \frac{e^4-1}{2}$$

$$\therefore \int_0^1\int_0^1 e^{\max(x^2,y^2)}dxdy = e^4 - 1$$

Example 13.

(1) Find $\int_0^1\int_0^1 |x-y|dxdy = ?$ (2) Find $\int_0^2\int_0^1 \sqrt{|y-x^2|}\,dxdy = ?$

【Solution】

(1)

$$\because \int_0^1 |x-y|dx = \int_0^y |x-y|dx + \int_y^1 |x-y|dx = \int_0^y y - x\,dx + \int_y^1 x - y\,dx, \forall 0 < y < 1$$

$$\therefore \int_0^1\int_0^1 |x-y|dxdy = \int_0^1\int_0^y y - x\,dxdy + \int_0^1\int_y^1 x - y\,dxdy$$

$$\because \int_0^1\int_0^y y - x\,dxdy = \int_0^1\left(yx - \frac{x^2}{2}\right)\bigg|_{x=0}^{x=y}dy = \int_0^1\frac{y^2}{2}dy = \frac{y^3}{6}\bigg|_{y=0}^{y=1} = \frac{1}{6}$$

$$\because \{(x,y): y \le x \le 1, 0 \le y \le 1\} = \{(x,y): 0 \le x \le 1, 0 \le y \le x\}$$

$$\therefore \int_0^1\int_y^1 x - y\,dxdy = \int_0^1\int_0^x x - y\,dydx = \int_0^1\left(yx - \frac{y^2}{2}\right)\bigg|_{y=0}^{y=x}dx = \int_0^1\frac{x^2}{2}dx = \frac{x^3}{6}\bigg|_{x=0}^{x=1} = \frac{1}{6}$$

$$\therefore \int_0^1\int_0^1 |x-y|dxdy = \frac{1}{3}$$

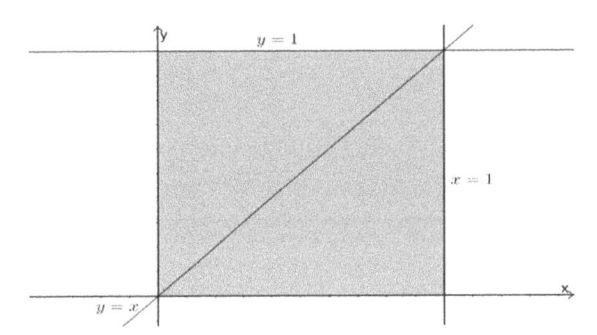

(2)

∵ $f(x,y) = \sqrt{|y - x^2|}$ is continuous on $[0,1] \times [0,2]$, by Fubini's Theorem,

$$\int_0^2 \int_0^1 \sqrt{|y - x^2|}\, dxdy = \int_0^1 \int_0^2 \sqrt{|y - x^2|}\, dydx$$

∵ $\int_0^1 \int_0^2 \sqrt{|y - x^2|}\, dydx = \int_0^1 \int_0^{x^2} \sqrt{|y - x^2|}\, dydx + \int_0^1 \int_{x^2}^2 \sqrt{|y - x^2|}\, dydx$

∵ $\int_0^1 \int_0^{x^2} \sqrt{|y - x^2|}\, dydx = \int_0^1 \int_0^{x^2} \sqrt{x^2 - y}\, dydx = \int_0^1 \left. \frac{(-2)(x^2 - y)^{\frac{3}{2}}}{3} \right|_{y=0}^{y=x^2} dx = \int_0^1 \frac{2x^3}{3} dx$

$= \left. \frac{2x^4}{12} \right|_{x=0}^{x=1} = \frac{1}{6}$

∵ $\int_0^1 \int_{x^2}^2 \sqrt{|y - x^2|}\, dydx = \int_0^1 \int_{x^2}^2 \sqrt{y - x^2}\, dydx = \int_0^1 \left. \frac{2(y - x^2)^{\frac{3}{2}}}{3} \right|_{y=x^2}^{y=2} dx$

$= \int_0^1 \frac{2(2 - x^2)^{\frac{3}{2}}}{3} dx$

Let $x = \sqrt{2}\sin\theta$ then $dx = \sqrt{2}\cos\theta\, d\theta$

∴ $\int_0^1 \frac{2(2 - x^2)^{\frac{3}{2}}}{3} dx = \int_0^{\frac{\pi}{4}} \frac{2(2 - 2\sin^2\theta)^{\frac{3}{2}}\sqrt{2}\cos\theta}{3} d\theta = \int_0^{\frac{\pi}{4}} \frac{2 \cdot 2\sqrt{2}(1 - \sin^2\theta)^{\frac{3}{2}}\sqrt{2}\cos\theta}{3} d\theta$

$= \int_0^{\frac{\pi}{4}} \frac{2 \cdot 2\sqrt{2}(\cos^2\theta)^{\frac{3}{2}}\sqrt{2}\cos\theta}{3} d\theta = \int_0^{\frac{\pi}{4}} \frac{8\cos^4\theta}{3} d\theta = \frac{8}{3}\int_0^{\frac{\pi}{4}} \left(\frac{1 + \cos 2\theta}{2}\right)^2 d\theta$

$= \frac{8}{3}\int_0^{\frac{\pi}{4}} \frac{1 + 2\cos 2\theta + \cos^2 2\theta}{4} d\theta = \frac{8}{3}\int_0^{\frac{\pi}{4}} \frac{1 + 2\cos 2\theta}{4} + \frac{1 + \cos 4\theta}{8} d\theta = \frac{\pi}{4} + \frac{2}{3}$

∴ $\int_0^1 \int_0^2 \sqrt{|y - x^2|}\, dydx = \frac{1}{6} + \frac{\pi}{4} + \frac{2}{3} = \frac{\pi}{4} + \frac{5}{6}$

8.3.2 Change the Order of Integration

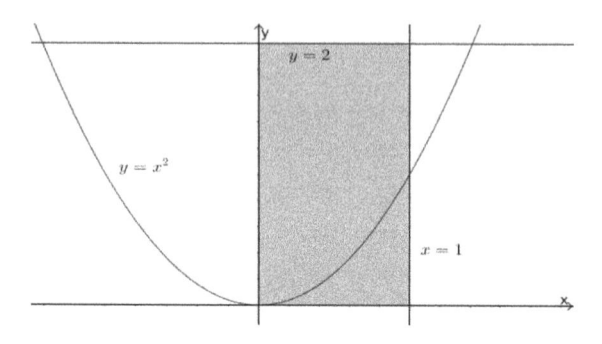

Example 14.

Find $\int_0^\pi \int_x^\pi \dfrac{\sin y}{y} \cos\dfrac{x}{y}\, dydx =?$

【Solution】

$\because \{(x,y): 0 \leq x \leq \pi, x \leq y \leq \pi\} = \{(x,y): 0 \leq x \leq y, 0 \leq y \leq \pi\}$

$\therefore \int_0^\pi \int_x^\pi \dfrac{\sin y}{y}\cos\dfrac{x}{y}\,dydx = \int_0^\pi \int_0^y \dfrac{\sin y}{y}\cos\dfrac{x}{y}\,dxdy = \int_0^\pi \left(\sin y \sin\dfrac{x}{y}\right)\Big|_{x=0}^{x=y} dy$

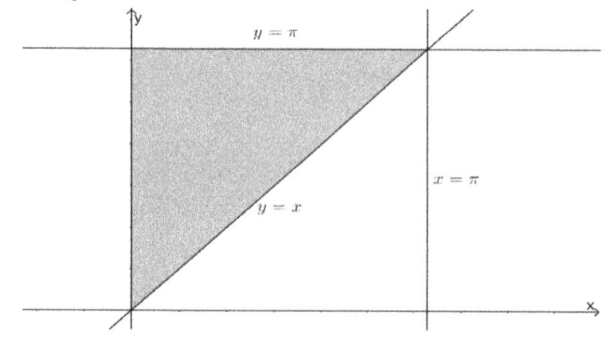

$= \int_0^\pi \sin y\,(\sin 1)\,dy = \sin 1\,(-\cos y)|_0^\pi = 2\sin 1$

Example 15.

Find $\int_0^{\frac{\pi}{2}} \int_x^{\frac{\pi}{2}} \dfrac{\cos y}{y} \sin\dfrac{x}{y}\, dydx =?$

【Solution】

$\because \{(x,y): 0 \leq x \leq \dfrac{\pi}{2}, x \leq y \leq \dfrac{\pi}{2}\} = \{(x,y): 0 \leq x \leq y, 0 \leq y \leq \dfrac{\pi}{2}\}$

$$\therefore \int_0^{\frac{\pi}{2}} \int_x^{\frac{\pi}{2}} \frac{\cos y}{y} \sin \frac{x}{y} \, dy dx = \int_0^{\frac{\pi}{2}} \int_0^y \frac{\cos y}{y} \sin \frac{x}{y} \, dx dy = \int_0^{\frac{\pi}{2}} \left(-\cos y \cos \frac{x}{y} \right) \Big|_{x=0}^{x=y} dy$$

$$= \int_0^{\frac{\pi}{2}} \cos y \, (1 - \cos 1) \, dy = (1 - \cos 1) \sin y \Big|_0^{\frac{\pi}{2}} = 1 - \cos 1$$

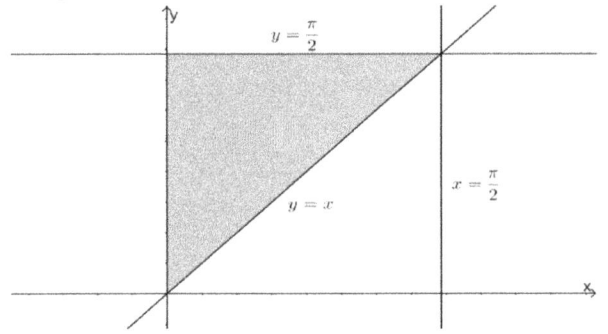

Example 16.

Find $\int_0^2 \int_x^2 \sin y^2 \, dy dx =?$

【Solution】

$\because \{(x, y): 0 \le x \le 2, x \le y \le 2\} = \{(x, y): 0 \le x \le y, 0 \le y \le 2\}$

$$\therefore \int_0^2 \int_x^2 \sin y^2 \, dy dx = \int_0^2 \int_0^y \sin y^2 \, dx dy = \int_0^2 y \sin y^2 \, dy = -\frac{\cos y^2}{2} \Big|_0^2 = \frac{1}{2}(1 - \cos 4)$$

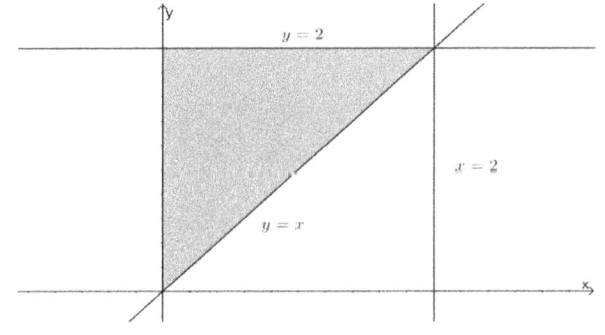

Example 17.

Find $\iint_R 1 dA =?$, $R = \{(x, y): -3x - y + 6 \le 0, 4x - x^2 - y \ge 0, y \ge 0\}$.

【Solution】

8.3.2 Change the Order of Integration

$$\iint_R 1 dA = \int_1^2 4x - x^2 - (6 - 3x)\, dx + \int_2^4 4x - x^2\, dx$$

$$= \left(\frac{7x^2}{2} - \frac{x^3}{3} - 6x\right)\Big|_1^2 + \left(2x^2 - \frac{x^3}{3}\right)\Big|_2^4 = \frac{21}{2} - \frac{7}{3} - 6 + 2\cdot 12 - \frac{56}{3} = \frac{15}{2}$$

Example 18.

Find $\iint_R \sqrt{x} - y^2\, dxdy = ?$, $R = \left\{(x,y): y \geq x^2, y \leq x^{\frac{1}{4}}\right\}$.

【Solution】

Let $y = x^2 = x^{\frac{1}{4}}$ then $x = 0$ or $1 \Rightarrow 0 \leq x \leq 1 \Rightarrow R = \{(x,y): 0 \leq x \leq 1, y \geq x^2, y \leq x^{\frac{1}{4}}\}$
∵ $f(x,y) = \sqrt{x} - y^2$ is continuous on R, by Fubini's Theorem,

$$\iint_R \sqrt{x} - y^2\, dxdy = \int_0^1 \int_{x^2}^{x^{\frac{1}{4}}} \sqrt{x} - y^2\, dydx = \int_0^1 \sqrt{x}\, y - \frac{y^3}{3}\bigg|_{y=x^2}^{y=x^{\frac{1}{4}}} dx$$

$$= \int_0^1 \sqrt{x}(x^{\frac{1}{4}}) - \frac{x^{\frac{3}{4}}}{3} - \left(\sqrt{x}(x^2) - \frac{x^6}{3}\right) dx = \int_0^1 x^{\frac{3}{4}} - \frac{x^{\frac{3}{4}}}{3} - x^{\frac{5}{2}} + \frac{x^6}{3} dx$$

$$= \left(\frac{4x^{\frac{7}{4}}}{7} - \frac{4x^{\frac{7}{4}}}{21} - \frac{2x^{\frac{7}{2}}}{7} + \frac{x^7}{21}\right)\bigg|_{x=0}^{x=1} = \frac{2}{7} - \frac{3}{21} = \frac{1}{7}$$

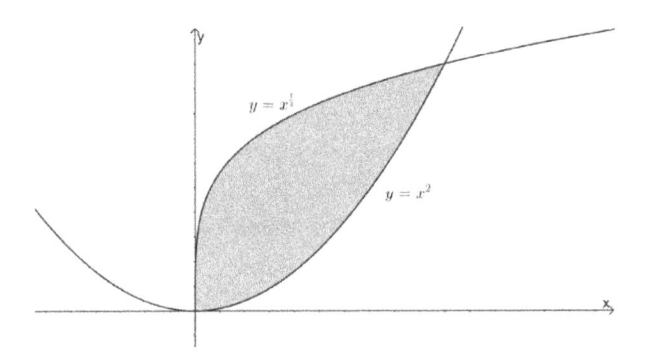

Example 19.

Find $\int_0^{\sqrt{3}} \int_y^{\sqrt{3}} \cos \frac{\pi x^2}{2} dx dy = ?$

【Solution】

$\because \{(x,y): y \le x \le \sqrt{3}, 0 \le y \le \sqrt{3}\} = \{(x,y): 0 \le x \le \sqrt{3}, 0 \le y \le x\}$

$\therefore \int_0^{\sqrt{3}} \int_y^{\sqrt{3}} \cos \frac{\pi x^2}{2} dx dy = \int_0^{\sqrt{3}} \int_0^x \cos \frac{\pi x^2}{2} dy dx = \int_0^{\sqrt{3}} x \cos \frac{\pi x^2}{2} dx = \frac{1}{\pi} \sin \frac{\pi x^2}{2} \Big|_0^{\sqrt{3}} = -\frac{1}{\pi}$

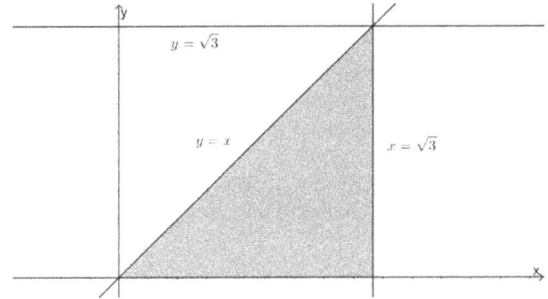

Example 20.

Find $\int_0^{\sqrt{3}} \int_y^{\sqrt{3}} \sin \frac{\pi x^2}{2} dx dy = ?$

【Solution】

$\because \{(x,y): y \le x \le \sqrt{3}, 0 \le y \le \sqrt{3}\} = \{(x,y): 0 \le x \le \sqrt{3}, 0 \le y \le x\}$

$\therefore \int_0^{\sqrt{3}} \int_y^{\sqrt{3}} \sin \frac{\pi x^2}{2} dx dy = \int_0^{\sqrt{3}} \int_0^x \sin \frac{\pi x^2}{2} dy dx = \int_0^{\sqrt{3}} x \sin \frac{\pi x^2}{2} dx = \frac{-1}{\pi} \cos \frac{\pi x^2}{2} \Big|_0^{\sqrt{3}} = \frac{1}{\pi}$

8.3.2 Change the Order of Integration

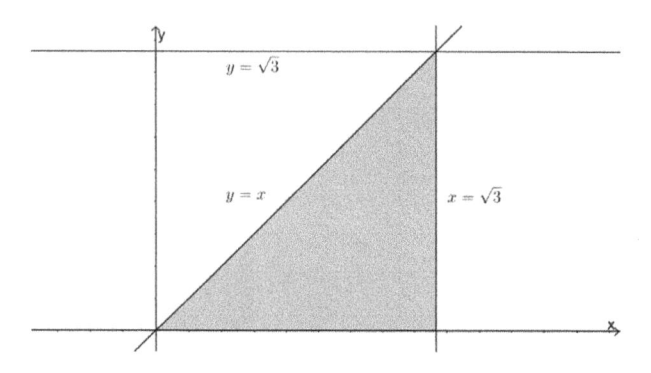

Example 21.

Find $\int_0^2 \int_x^2 \cos y^2 \, dy dx =?$

【Solution】

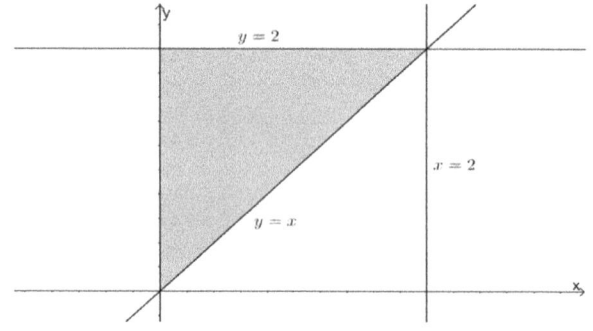

∵ $\{(x,y): 0 \le x \le 2, x \le y \le 2\} = \{(x,y): 0 \le x \le y, 0 \le y \le 2\}$

∴ $\int_0^2 \int_x^2 \cos y^2 \, dy dx = \int_0^2 \int_0^y \cos y^2 \, dx dy = \int_0^2 y \cos y^2 \, dy = \left.\frac{\sin y^2}{2}\right|_0^2 = \frac{\sin 4}{2}$

Example 22.

Find $\int_0^1 \int_{\sqrt{x}}^1 \sqrt{1+y^3} \, dy dx =?$

【Solution】

∵ $\{(x,y): 0 \le x \le 1, \sqrt{x} \le y \le 1\} = \{(x,y): 0 \le x \le y^2, 0 \le y \le 1\}$

∴ $\int_0^1 \int_{\sqrt{x}}^1 \sqrt{1+y^3} \, dy dx = \int_0^1 \int_0^{y^2} \sqrt{1+y^3} \, dx dy = \int_0^1 y^2 \sqrt{1+y^3} \, dy = \left.\frac{2}{9}(1+y^3)^{\frac{3}{2}}\right|_0^1$

$= \frac{2}{9}(2\sqrt{2} - 1)$

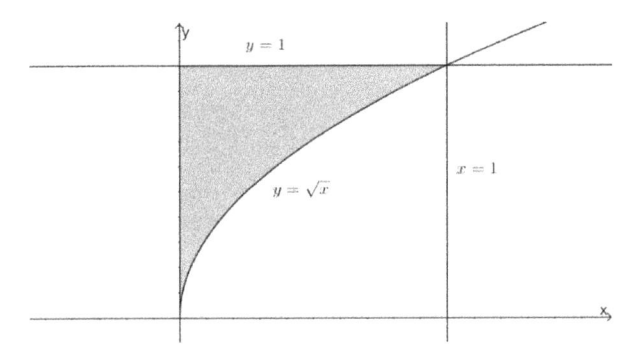

Example 23.

Find $\int_0^1 \int_{2y}^2 \cos x^2 \, dxdy =?$

【Solution】

$\because \{(x,y): 2y \leq x \leq 2, 0 \leq y \leq 1\} = \{(x,y): 0 \leq x \leq 2, 0 \leq y \leq \frac{x}{2}\}$

$\therefore \int_0^1 \int_{2y}^2 \cos x^2 \, dxdy = \int_0^2 \int_0^{\frac{x}{2}} \cos x^2 \, dydx = \int_0^2 \frac{x}{2} \cos x^2 \, dx = \frac{\sin x^2}{4}\bigg|_0^2 = \frac{\sin 4}{4}$

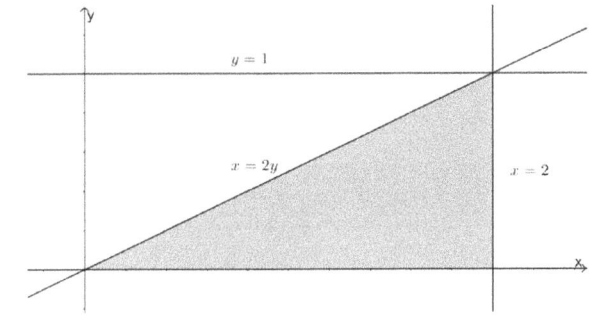

Example 24.

Find $\int_0^1 \int_{2y}^1 \sec^2 x^2 \, dxdy =?$

【Solution】

$\because \{(x,y): 2y \leq x \leq 1, 0 \leq y \leq 1\} = \{(x,y): 0 \leq x \leq 1, 0 \leq y \leq \frac{x}{2}\}$

$\therefore \int_0^1 \int_{2y}^1 \sec^2 x^2 \, dxdy = \int_0^1 \int_0^{\frac{x}{2}} \sec^2 x^2 \, dydx = \int_0^1 \frac{x}{2} \sec^2 x^2 \, dx = \frac{\tan x^2}{4}\bigg|_0^1 = \frac{\tan 1}{4}$

8.3.2 Change the Order of Integration

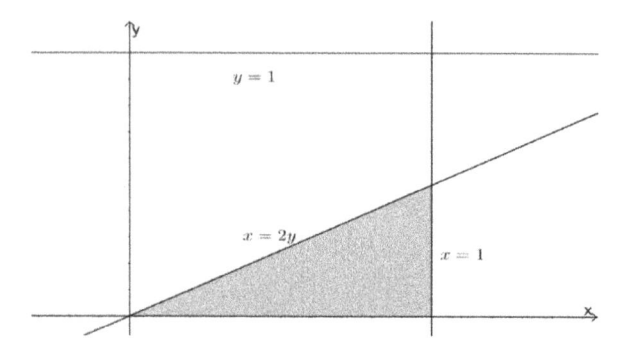

Example 25.

Find $\int_0^1 \int_y^1 e^{-x^2} dx dy = ?$

【Solution】

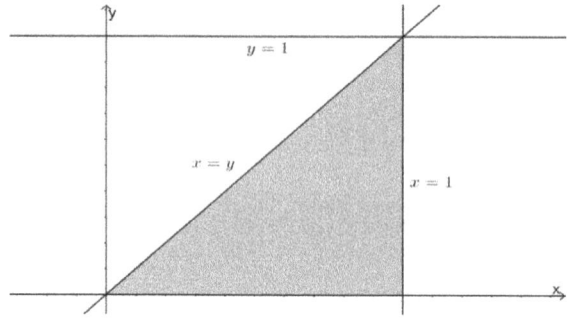

∵ $\{(x,y): y \leq x \leq 1, 0 \leq y \leq 1\} = \{(x,y): 0 \leq x \leq 1, 0 \leq y \leq x\}$

∴ $\int_0^1 \int_y^1 e^{-x^2} dx dy = \int_0^1 \int_0^x e^{-x^2} dy dx = \int_0^1 x e^{-x^2} dx = -\left.\frac{e^{-x^2}}{2}\right|_0^1 = \frac{1}{2}(1 - e^{-1})$

Example 26.

Find $\int_0^1 \int_{x^2}^1 x^3 \sin y^3 \, dy dx = ?$

【Solution】

∵ $\{(x,y): 0 \leq x \leq 1, x^2 \leq y \leq 1\} = \{(x,y): 0 \leq x \leq \sqrt{y}, 0 \leq y \leq 1\}$

∴ $\int_0^1 \int_{x^2}^1 x^3 \sin y^3 \, dy dx = \int_0^1 \int_0^{\sqrt{y}} x^3 \sin y^3 \, dx dy = \int_0^1 \frac{y^2 \sin y^3}{4} dy = -\left.\frac{\cos y^3}{12}\right|_0^1 = \frac{1 - \cos 1}{12}$

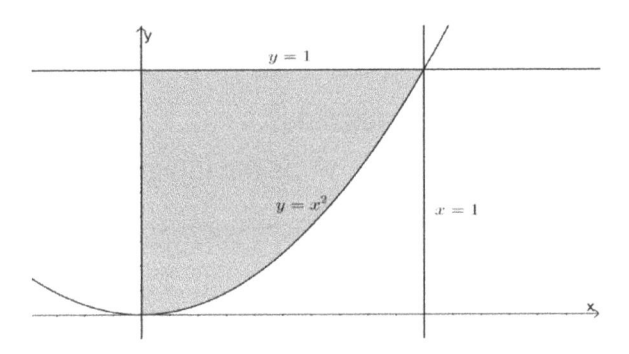

Example 27.

Find $\int_0^1 \int_{\sin^{-1} y}^{\frac{\pi}{2}} \cos x \sqrt{\cos^2 x + 1}\, dx\, dy = ?$

【Solution】

$\because \{(x,y): \sin^{-1} y \leq x \leq \frac{\pi}{2}, 0 \leq y \leq 1\} = \{(x,y): 0 \leq x \leq \frac{\pi}{2}, 0 \leq y \leq \sin x\}$

$\therefore \int_0^1 \int_{\sin^{-1} y}^{\frac{\pi}{2}} \cos x \sqrt{\cos^2 x + 1}\, dx\, dy = \int_0^{\frac{\pi}{2}} \int_0^{\sin x} \cos x \sqrt{\cos^2 x + 1}\, dy\, dx$

$= \int_0^{\frac{\pi}{2}} \sin x \cos x \sqrt{\cos^2 x + 1}\, dx = \int_0^{\frac{\pi}{2}} \frac{\sin 2x \sqrt{\frac{\cos 2x + 3}{2}}}{2}\, dx = -\left.\frac{\left(\frac{\cos 2x + 3}{2}\right)^{\frac{3}{2}}}{3}\right|_0^{\frac{\pi}{2}}$

$= \frac{2\sqrt{2} - 1}{3}$

8.3.3 Use Polar Coordinates to Compute Double Integrals

When the region of integration in a double integral involves $x^2 + y^2$ or the integrand function includes $x^2 + y^2$, it is often suitable to use polar coordinates.

Question Types:
Type 1.

Given $R = \{(x,y): 0 \leq x^2 + y^2 \leq a^2\}$, $a > 0$, find $\iint_R f(x,y)dA$.

Problem-Solving Process:
Step1.
Let $x = r\cos\theta$, $y = r\sin\theta$ then
$R = \{(x,y): 0 \leq x^2 + y^2 \leq a^2\} = \{(r,\theta): 0 \leq r \leq a, 0 \leq \theta \leq 2\pi\}$

$$dxdy = \begin{Vmatrix} \frac{\partial x}{\partial r} & \frac{\partial x}{\partial \theta} \\ \frac{\partial y}{\partial r} & \frac{\partial y}{\partial \theta} \end{Vmatrix} drd\theta = \begin{Vmatrix} \cos\theta & -r\sin\theta \\ \sin\theta & r\cos\theta \end{Vmatrix} drd\theta = rdrd\theta$$

Step2.
$$\therefore \iint_R f(x,y)dA = \int_0^{2\pi}\int_0^a f(r\cos\theta, r\sin\theta)rdrd\theta$$

Step3.
Find $\int_0^{2\pi}\int_0^a f(r\cos\theta, r\sin\theta)rdrd\theta$

Step4.
If you cannot compute $\int_0^{2\pi}\int_0^a f(r\cos\theta, r\sin\theta)rdrd\theta$, you would usually use variable substitution or integration by parts as additional methods.

Examples:

Given $R = \{(x,y): 0 \leq x^2 + y^2 \leq a^2\}$, $a > 0$, find $\iint_R f(x,y)dA = ?$

(I) If $f(x,y) = e^{x^2+y^2}$ then $\iint_R f(x,y)dA = \int_0^{2\pi}\int_0^a e^{r^2} rdrd\theta$

(II) If $f(x,y) = \ln(x^2 + y^2)$ then $\iint_R f(x,y)dA = \int_0^{2\pi} \int_0^a \ln(r^2) \, r\,dr\,d\theta$

(III) If $f(x,y) = \sin(x^2 + y^2)$ then $\iint_R f(x,y)dA = \int_0^{2\pi} \int_0^a \sin(r^2) \, r\,dr\,d\theta$

(IV) If $f(x,y) = \dfrac{1}{\sqrt{x^2+y^2}}$ then $\iint_R f(x,y)dA = \int_0^{2\pi} \int_0^a dr\,d\theta$

(V) If $f(x,y) = \dfrac{1}{\sqrt{1+x^2+y^2}}$ then $\iint_R f(x,y)dA = \int_0^{2\pi} \int_0^a \dfrac{r}{\sqrt{1+r^2}} dr\,d\theta$

(VI) If $f(x,y) = e^{-(x^2+y^2)} \cos(x^2+y^2)$ then $\iint_R f(x,y)dA = \int_0^{2\pi} \int_0^a e^{-r^2} \cos(r^2) \, r\,dr\,d\theta$

(VII) If $f(x,y) = \sin^{-1} \dfrac{y}{\sqrt{x^2+y^2}}$ then $\iint_R f(x,y)dA = \int_0^{2\pi} \int_0^a \sin^{-1}\left(\dfrac{r\sin\theta}{r}\right) r\,dr\,d\theta$

(VIII) If $f(x,y) = \sec^2(x^2+y^2)$ then $\iint_R f(x,y)dA = \int_0^{2\pi} \int_0^a \sec^2(r^2) \, r\,dr\,d\theta$

Example 1.

Find $\int_0^2 \int_0^{\sqrt{4-y^2}} e^{x^2+y^2} dx\,dy = ?$

【Solution】

Let $x = r\cos\theta$, $y = r\sin\theta$ then

$\{(x,y) : 0 \le x \le \sqrt{4-y^2}, 0 \le y \le 2\} = \{(r,\theta) : 0 \le r \le 2, 0 \le \theta \le \dfrac{\pi}{2}\}$

and $dx\,dy = \begin{Vmatrix} \dfrac{\partial x}{\partial r} & \dfrac{\partial x}{\partial \theta} \\ \dfrac{\partial y}{\partial r} & \dfrac{\partial y}{\partial \theta} \end{Vmatrix} dr\,d\theta = \begin{Vmatrix} \cos\theta & -r\sin\theta \\ \sin\theta & r\cos\theta \end{Vmatrix} dr\,d\theta = r\,dr\,d\theta$

$\therefore \int_0^2 \int_0^{\sqrt{4-y^2}} e^{x^2+y^2} dx\,dy = \int_0^{\frac{\pi}{2}} \int_0^2 e^{r^2} r\,dr\,d\theta = \int_0^{\frac{\pi}{2}} \dfrac{e^{r^2}}{2} \Big|_{r=0}^{r=2} d\theta = \int_0^{\frac{\pi}{2}} \dfrac{e^4 - 1}{2} d\theta$

$= \left(\dfrac{e^4-1}{4}\right)\pi$

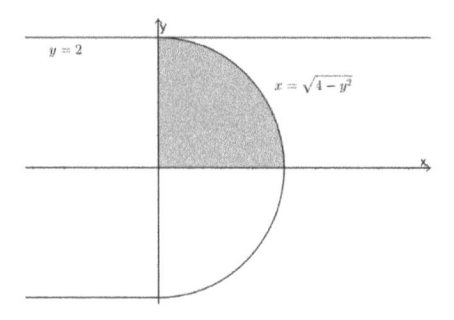

8.3.3 Use Polar Coordinates to Compute Double Integrals

Example 2.

Find $\iint_R \ln(x^2 + y^2)\, dA =?$, $R = \{(x,y): x \geq 0, y \geq 0, 0 \leq x^2 + y^2 \leq 16\}$.

【Solution】

Let $x = r\cos\theta$, $y = r\sin\theta$ then

$$\{(x,y): x \geq 0, y \geq 0, 0 \leq x^2 + y^2 \leq 16\} = \{(r,\theta): 0 \leq r \leq 4, 0 \leq \theta \leq \frac{\pi}{2}\}$$

and $dxdy = \begin{Vmatrix} \dfrac{\partial x}{\partial r} & \dfrac{\partial x}{\partial \theta} \\ \dfrac{\partial y}{\partial r} & \dfrac{\partial y}{\partial \theta} \end{Vmatrix} drd\theta = \begin{Vmatrix} \cos\theta & -r\sin\theta \\ \sin\theta & r\cos\theta \end{Vmatrix} drd\theta = rdrd\theta$

$$\therefore \iint_R \ln(x^2 + y^2)\, dA = \int_0^{\frac{\pi}{2}} \int_0^4 (\ln r^2) r\, dr d\theta = 2 \int_0^{\frac{\pi}{2}} \int_0^4 (\ln r) r\, dr d\theta$$

By integration by parts,

$$\int_0^4 (\ln r) r\, dr = \ln r \left(\frac{r^2}{2}\right)\Big|_{r=0}^{r=4} - \int_0^4 \frac{r}{2}\, dr = \ln r \left(\frac{r^2}{2}\right)\Big|_{r=0}^{r=4} - \frac{r^2}{4}\Big|_{r=0}^{r=4} = 16\ln 2 - 4$$

$$\because 2\int_0^{\frac{\pi}{2}} d\theta = \pi \quad \therefore \iint_R \ln(x^2 + y^2)\, dA = \pi(16\ln 2 - 4)$$

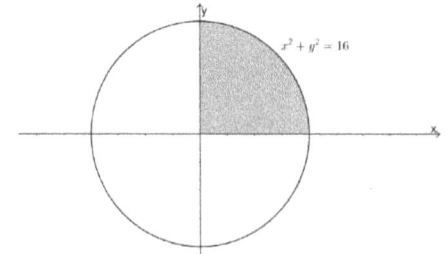

Example 3.

Find $\iint_R x^2 y \, dxdy = ?$, $R = \{(x, y): y \geq 0, \ x^2 + y^2 \leq 4\}$.

【Solution】

Let $x = r\cos\theta, \ y = r\sin\theta$ then
$$\{(x,y): y \geq 0, x^2 + y^2 \leq 4\} = \{(r,\theta): 0 \leq r \leq 2, 0 \leq \theta \leq \pi\}$$

and $dxdy = \begin{Vmatrix} \dfrac{\partial x}{\partial r} & \dfrac{\partial x}{\partial \theta} \\ \dfrac{\partial y}{\partial r} & \dfrac{\partial y}{\partial \theta} \end{Vmatrix} drd\theta = \begin{Vmatrix} \cos\theta & -r\sin\theta \\ \sin\theta & r\cos\theta \end{Vmatrix} drd\theta = rdrd\theta$

$\therefore \iint_R x^2 y \, dxdy = \int_0^\pi \int_0^2 (r\cos\theta)^2 (r\sin\theta) r \, drd\theta = \int_0^\pi \sin\theta \cos^2\theta \, d\theta \int_0^2 r^4 dr$

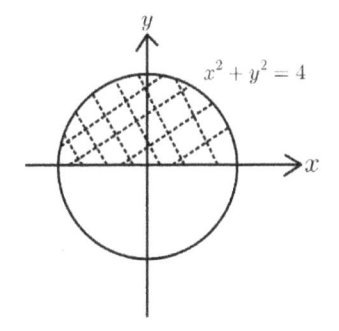

$= \dfrac{(-1)\cos^3\theta}{3}\bigg|_0^\pi \cdot \dfrac{r^5}{5}\bigg|_0^2 = \dfrac{2}{3} \cdot \dfrac{32}{5} = \dfrac{64}{15}$

Example 4.

Find $\iint_R \sqrt{x^2 + y^2} \, dA = ?$, $R = \{(x, y): x^2 + y^2 \leq 3x\}$.

【Solution】

Let $x = r\cos\theta, \ y = r\sin\theta$ then
$$\{(x,y): x^2 + y^2 \leq 3x\} = \{(r,\theta): 0 \leq r \leq 3\cos\theta, -\dfrac{\pi}{2} \leq \theta \leq \dfrac{\pi}{2}\}$$

and $dxdy = \begin{Vmatrix} \dfrac{\partial x}{\partial r} & \dfrac{\partial x}{\partial \theta} \\ \dfrac{\partial y}{\partial r} & \dfrac{\partial y}{\partial \theta} \end{Vmatrix} drd\theta = \begin{Vmatrix} \cos\theta & -r\sin\theta \\ \sin\theta & r\cos\theta \end{Vmatrix} drd\theta = rdrd\theta$

8.3.3 Use Polar Coordinates to Compute Double Integrals

$$\therefore \iint_R \sqrt{x^2+y^2}\,dA = \int_{-\frac{\pi}{2}}^{\frac{\pi}{2}} \int_0^{3\cos\theta} r^2\,dr\,d\theta = \int_{-\frac{\pi}{2}}^{\frac{\pi}{2}} \frac{r^3}{3}\bigg|_0^{3\cos\theta} d\theta = \int_{-\frac{\pi}{2}}^{\frac{\pi}{2}} 9\cos^3\theta\,d\theta$$

$$= 9\int_{-\frac{\pi}{2}}^{\frac{\pi}{2}} \cos\theta(1-\sin^2\theta)\,d\theta = 9\left(\sin\theta - \frac{\sin^3\theta}{3}\right)\bigg|_{-\frac{\pi}{2}}^{\frac{\pi}{2}} = 12$$

Example 5.

Find $\iint_R \frac{1}{\sqrt{x^2+y^2}}\,dA =?$, $R = \{(x,y): y \geq 0, 0 \leq x^2+y^2 \leq 1\}$.

【Solution】

Let $x = r\cos\theta$, $y = r\sin\theta$ then
$\{(x,y): y \geq 0, 0 \leq x^2+y^2 \leq 1\} = \{(r,\theta): 0 \leq r \leq 1, 0 \leq \theta \leq \pi\}$

and $dxdy = \begin{Vmatrix} \frac{\partial x}{\partial r} & \frac{\partial x}{\partial \theta} \\ \frac{\partial y}{\partial r} & \frac{\partial y}{\partial \theta} \end{Vmatrix} dr d\theta = \begin{Vmatrix} \cos\theta & -r\sin\theta \\ \sin\theta & r\cos\theta \end{Vmatrix} dr d\theta = r\,dr\,d\theta$

$$\therefore \iint_R \frac{1}{\sqrt{x^2+y^2}}\,dA = \int_0^\pi \int_0^1 \frac{r}{\sqrt{r^2}}\,dr\,d\theta = \pi$$

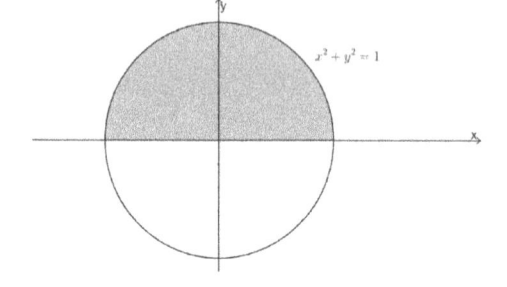

Example 6.

Find $\int_0^\infty e^{-x^2}\,dx =?$

【Solution】

$$\therefore \int_0^\infty e^{-x^2}\,dx \int_0^\infty e^{-y^2}\,dy = \int_0^\infty \int_0^\infty e^{-(x^2+y^2)}\,dy\,dx$$

Let $x = r\cos\theta$, $y = r\sin\theta$ then

8.3.3 Use Polar Coordinates to Compute Double Integrals

$$\{(x,y): y \geq 0, 0 \leq x^2 + y^2 \leq \infty\} = \{(r,\theta): 0 \leq r \leq \infty, 0 \leq \theta \leq \frac{\pi}{2}\}$$

and $dxdy = \begin{Vmatrix} \frac{\partial x}{\partial r} & \frac{\partial x}{\partial \theta} \\ \frac{\partial y}{\partial r} & \frac{\partial y}{\partial \theta} \end{Vmatrix} drd\theta = \begin{Vmatrix} \cos\theta & -r\sin\theta \\ \sin\theta & r\cos\theta \end{Vmatrix} drd\theta = rdrd\theta$

$$\therefore \int_0^\infty \int_0^\infty e^{-(x^2+y^2)} dy\, dx = \int_0^{\frac{\pi}{2}} \int_0^\infty e^{-r^2} r\, dr\, d\theta = \int_0^{\frac{\pi}{2}} d\theta \int_0^\infty e^{-r^2} r\, dr$$

$$= \frac{\pi}{2} \cdot \frac{-e^{-r^2}}{2}\Big|_{r=0}^{r=\infty} = \frac{\pi}{4} \Rightarrow \int_0^\infty e^{-x^2} dx = \sqrt{\frac{\pi}{4}}$$

Example 7.

Find $\int_0^\infty \int_0^\infty e^{-(x^2+y^2)} \cos(x^2 + y^2) dxdy = ?$

【Solution】

Let $x = r\cos\theta$, $y = r\sin\theta$ then

$$\{(x,y): y \geq 0, 0 \leq x^2 + y^2 \leq \infty\} = \{(r,\theta): 0 \leq r \leq \infty, 0 \leq \theta \leq \frac{\pi}{2}\}$$

and $dxdy = \begin{Vmatrix} \frac{\partial x}{\partial r} & \frac{\partial x}{\partial \theta} \\ \frac{\partial y}{\partial r} & \frac{\partial y}{\partial \theta} \end{Vmatrix} drd\theta = \begin{Vmatrix} \cos\theta & -r\sin\theta \\ \sin\theta & r\cos\theta \end{Vmatrix} drd\theta = rdrd\theta$

$$\therefore \int_0^\infty \int_0^\infty e^{-(x^2+y^2)} \cos(x^2+y^2) dxdy = \int_0^{\frac{\pi}{2}} \int_0^\infty e^{-r^2} \cos(r^2) r\, dr\, d\theta$$

$$= \int_0^{\frac{\pi}{2}} d\theta \int_0^\infty e^{-r^2} \cos(r^2) r\, dr = \frac{\pi}{2} \int_0^\infty e^{-r^2} \cos(r^2) r\, dr$$

Let $u = r^2$ then $du = 2rdr$ $\quad \therefore \int_0^\infty e^{-r^2} \cos(r^2) r\, dr = \frac{1}{2} \int_0^\infty e^{-u} \cos u\, du$

By integration by parts,

$$\int_0^\infty e^{-u} \cos u\, du = e^{-u} \sin u\Big|_{u=0}^{u=\infty} + \int_0^\infty e^{-u} \sin u\, du = \int_0^\infty e^{-u} \sin u\, du$$

$$= -e^{-u} \cos u\Big|_{u=0}^{u=\infty} - \int_0^\infty e^{-u} \cos u\, du$$

8.3.3 Use Polar Coordinates to Compute Double Integrals

$$\therefore \int_0^\infty e^{-u}\cos u\, du = \frac{1}{2} \Rightarrow \int_0^\infty e^{-r^2}\cos(r^2)r\, dr = \frac{1}{4}$$

$$\Rightarrow \int_0^\infty \int_0^\infty e^{-(x^2+y^2)}\cos(x^2+y^2)\, dx\, dy = \frac{\pi}{8}$$

Example 8.

Find $\iint_R \tan^{-1}\frac{y}{x}\, dA = ?$, $R = \{(x,y): x \geq 0, y \geq 0, 0 \leq x^2 + y^2 \leq a^2\}$.

【Solution】

Let $x = r\cos\theta$, $y = r\sin\theta$ then

$$\{(x,y): x \geq 0, y \geq 0, 0 \leq x^2 + y^2 \leq a^2\} = \{(r,\theta): 0 \leq r \leq a, 0 \leq \theta \leq \frac{\pi}{2}\}$$

and $dx\,dy = \begin{Vmatrix} \frac{\partial x}{\partial r} & \frac{\partial x}{\partial \theta} \\ \frac{\partial y}{\partial r} & \frac{\partial y}{\partial \theta} \end{Vmatrix} dr\,d\theta = \begin{Vmatrix} \cos\theta & -r\sin\theta \\ \sin\theta & r\cos\theta \end{Vmatrix} dr\,d\theta = r\,dr\,d\theta$

$$\therefore \iint_R \tan^{-1}\frac{y}{x}\, dA = \int_0^{\frac{\pi}{2}} \int_0^a \tan^{-1}\left(\frac{r\sin\theta}{r\cos\theta}\right) r\,dr\,d\theta = \int_0^{\frac{\pi}{2}} \int_0^a \tan^{-1}(\tan\theta)\, r\,dr\,d\theta$$

$$= \frac{\theta^2}{2}\Big|_0^{\frac{\pi}{2}} \cdot \frac{r^2}{2}\Big|_0^a = \frac{(\pi a)^2}{16}$$

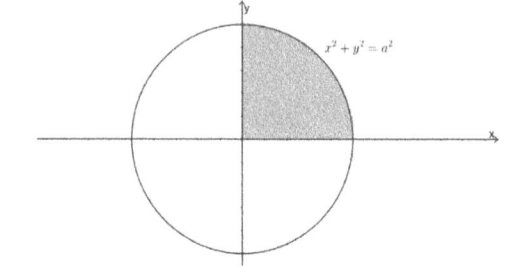

Example 9.

Find $\iint_R \sin^{-1}\frac{y}{\sqrt{x^2+y^2}}\, dA = ?$, $R = \{(x,y): x \geq 0, y \geq 0, 0 \leq x^2 + y^2 \leq a^2\}$.

【Solution】

Let $x = r\cos\theta$, $y = r\sin\theta$ then

$$\{(x,y): x \geq 0, y \geq 0, 0 \leq x^2 + y^2 \leq a^2\} = \{(r,\theta): 0 \leq r \leq a, 0 \leq \theta \leq \frac{\pi}{2}\}$$

and $dxdy = \left\| \begin{matrix} \frac{\partial x}{\partial r} & \frac{\partial x}{\partial \theta} \\ \frac{\partial y}{\partial r} & \frac{\partial y}{\partial \theta} \end{matrix} \right\| drd\theta = \left\| \begin{matrix} \cos\theta & -r\sin\theta \\ \sin\theta & r\cos\theta \end{matrix} \right\| drd\theta = rdrd\theta$

$$\therefore \iint_R \sin^{-1}\left(\frac{y}{\sqrt{x^2+y^2}}\right) dA = \int_0^{\frac{\pi}{2}} \int_0^a \sin^{-1}\left(\frac{r\sin\theta}{r}\right) rdrd\theta = \int_0^{\frac{\pi}{2}} \int_0^a \sin^{-1}(\sin\theta) rdrd\theta$$

$$= \frac{\theta^2}{2}\bigg|_0^{\frac{\pi}{2}} \cdot \frac{r^2}{2}\bigg|_0^a = \frac{(\pi a)^2}{16}$$

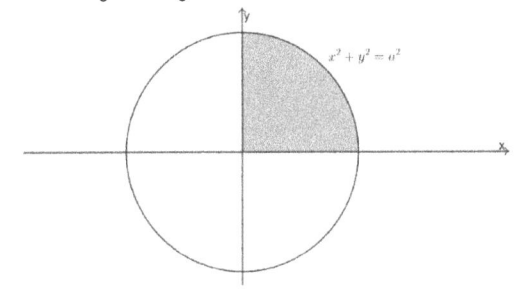

Example 10.

Find $\iint_R \cos^{-1}\frac{x}{\sqrt{x^2+y^2}} dA = ?$, $R = \{(x,y): x \geq 0, y \geq 0, 0 \leq x^2 + y^2 \leq a^2\}$.

【Solution】

Let $x = r\cos\theta$, $y = r\sin\theta$ then

$$\{(x,y): x \geq 0, y \geq 0, 0 \leq x^2 + y^2 \leq a^2\} = \{(r,\theta): 0 \leq r \leq a, 0 \leq \theta \leq \frac{\pi}{2}\}$$

and $dxdy = \left\| \begin{matrix} \frac{\partial x}{\partial r} & \frac{\partial x}{\partial \theta} \\ \frac{\partial y}{\partial r} & \frac{\partial y}{\partial \theta} \end{matrix} \right\| drd\theta = \left\| \begin{matrix} \cos\theta & -r\sin\theta \\ \sin\theta & r\cos\theta \end{matrix} \right\| drd\theta = rdrd\theta$

$$\therefore \iint_R \cos^{-1}\frac{x}{\sqrt{x^2+y^2}} dA = \int_0^{\frac{\pi}{2}} \int_0^a \cos^{-1}\left(\frac{r\cos\theta}{r}\right) rdrd\theta = \int_0^{\frac{\pi}{2}} \int_0^a \cos^{-1}(\cos\theta) rdrd\theta$$

$$= \frac{\theta^2}{2}\bigg|_0^{\frac{\pi}{2}} \cdot \frac{r^2}{2}\bigg|_0^a = \frac{(\pi a)^2}{16}$$

8.3.3 Use Polar Coordinates to Compute Double Integrals

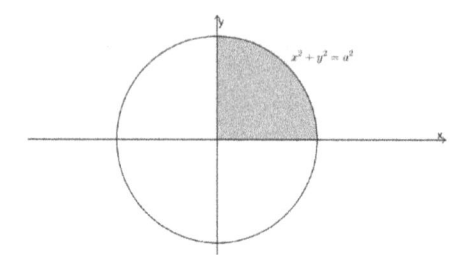

Example 11.

Find $\int_0^{\frac{1}{2}} \int_{\sqrt{3}x}^{\sqrt{1-x^2}} e^{-x^2-y^2} \, dy \, dx = ?$

【Solution】

Let $x = r\cos\theta, \ y = r\sin\theta$ then

$$\{(x,y): 0 \le x \le \frac{1}{2}, \sqrt{3}x \le y \le \sqrt{1-x^2}\} = \{(r,\theta): 0 \le r \le 1, \frac{\pi}{3} \le \theta \le \frac{\pi}{2}\}$$

and $dxdy = \begin{Vmatrix} \frac{\partial x}{\partial r} & \frac{\partial x}{\partial \theta} \\ \frac{\partial y}{\partial r} & \frac{\partial y}{\partial \theta} \end{Vmatrix} drd\theta = \begin{Vmatrix} \cos\theta & -r\sin\theta \\ \sin\theta & r\cos\theta \end{Vmatrix} drd\theta = rdrd\theta$

$$\therefore \int_0^{\frac{1}{2}} \int_{\sqrt{3}x}^{\sqrt{1-x^2}} e^{-x^2-y^2} \, dydx = \int_{\frac{\pi}{3}}^{\frac{\pi}{2}} \int_0^1 e^{-r^2} r \, dr \, d\theta = \int_0^1 e^{-r^2} r \, dr \int_{\frac{\pi}{3}}^{\frac{\pi}{2}} d\theta = -\frac{e^{-r^2}}{2} \bigg|_0^1 \cdot \frac{\pi}{6}$$

$$= \frac{\pi}{12}(1 - e^{-1})$$

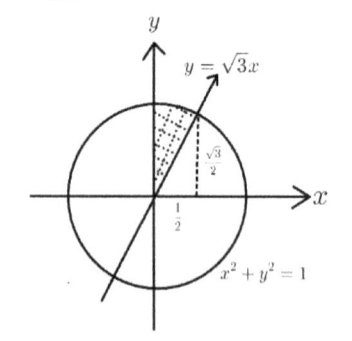

Example 12.

Find $\int_1^2 \int_0^{\sqrt{2x-x^2}} (x^2 + y^2)^{-\frac{1}{2}} \, dy \, dx = ?$

【Solution】

Let $x = r\cos\theta$, $y = r\sin\theta$ then

$$\{(x,y): 1 \leq x \leq 2, 0 \leq y \leq \sqrt{2x - x^2}\} = \{(r,\theta): \sec\theta \leq r \leq 2\cos\theta, 0 \leq \theta \leq \frac{\pi}{4}\}$$

and $dxdy = \begin{Vmatrix} \frac{\partial x}{\partial r} & \frac{\partial x}{\partial \theta} \\ \frac{\partial y}{\partial r} & \frac{\partial y}{\partial \theta} \end{Vmatrix} drd\theta = \begin{Vmatrix} \cos\theta & -r\sin\theta \\ \sin\theta & r\cos\theta \end{Vmatrix} drd\theta = rdrd\theta$

$$\therefore \int_1^2 \int_0^{\sqrt{2x-x^2}} (x^2 + y^2)^{-\frac{1}{2}} dydx = \int_0^{\frac{\pi}{4}} \int_{\sec\theta}^{2\cos\theta} (r^2)^{-\frac{1}{2}} r drd\theta = \int_0^{\frac{\pi}{4}} \int_{\sec\theta}^{2\cos\theta} drd\theta$$

$$= \int_0^{\frac{\pi}{4}} 2\cos\theta - \sec\theta\, d\theta = (2\sin\theta - \ln|\sec\theta + \tan\theta|)\Big|_0^{\frac{\pi}{4}} = \sqrt{2} - \ln(1 + \sqrt{2})$$

Example 13.

Find $\iint_R \frac{x^2}{(x^2 + y^2)^2} dxdy = ?$, $R = \{(x,y): a^2 \leq x^2 + y^2 \leq b^2, 0 < a < b\}$.

【Solution】

Let $x = r\cos\theta$, $y = r\sin\theta$ then
$R = \{(x,y): a^2 \leq x^2 + y^2 \leq b^2, 0 < a < b\} = \{(r,\theta): a \leq r \leq b, 0 \leq \theta \leq 2\pi\}$

and $dxdy = \begin{Vmatrix} \frac{\partial x}{\partial r} & \frac{\partial x}{\partial \theta} \\ \frac{\partial y}{\partial r} & \frac{\partial y}{\partial \theta} \end{Vmatrix} drd\theta = \begin{Vmatrix} \cos\theta & -r\sin\theta \\ \sin\theta & r\cos\theta \end{Vmatrix} drd\theta = rdrd\theta$

$$\therefore \iint_R \frac{x^2}{(x^2+y^2)^2} dxdy = \int_0^{2\pi} \int_a^b \left(\frac{r^2\cos^2\theta}{r^4}\right) rdrd\theta = \int_0^{2\pi} \int_a^b \frac{\cos^2\theta}{r} drd\theta$$

$$= \int_0^{2\pi} \cos^2\theta\, d\theta \int_a^b \frac{1}{r} dr = \int_0^{2\pi} \frac{1 + \cos 2\theta}{2} d\theta \int_a^b \frac{1}{r} dr = \pi \ln\frac{b}{a}$$

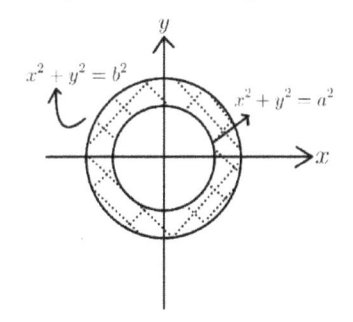

8.3.3 Use Polar Coordinates to Compute Double Integrals

Example 14.

Find $\int_0^1 \int_{x^2}^{x} (x^2+y^2)^{-\frac{1}{2}} dydx = ?$

【Solution】

Let $x = r\cos\theta$, $y = r\sin\theta$ then

$R = \{(x,y): 0 \le x \le 1, x^2 < y < x\} = \{(r,\theta): 0 \le r \le \dfrac{\sin\theta}{\cos^2\theta}, 0 \le \theta \le \dfrac{\pi}{4}\}$

and, $dxdy = \begin{Vmatrix} \dfrac{\partial x}{\partial r} & \dfrac{\partial x}{\partial \theta} \\ \dfrac{\partial y}{\partial r} & \dfrac{\partial y}{\partial \theta} \end{Vmatrix} drd\theta = \begin{Vmatrix} \cos\theta & -r\sin\theta \\ \sin\theta & r\cos\theta \end{Vmatrix} drd\theta = rdrd\theta$

$\therefore \int_0^1 \int_{x^2}^{x} (x^2+y^2)^{-\frac{1}{2}} dydx = \int_0^{\frac{\pi}{4}} \int_0^{\frac{\sin\theta}{\cos^2\theta}} \dfrac{1}{r} \cdot rdrd\theta = \int_0^{\frac{\pi}{4}} \dfrac{\sin\theta}{\cos^2\theta} d\theta = \dfrac{1}{\cos\theta}\Big|_0^{\frac{\pi}{4}} = \sqrt{2} - 1$

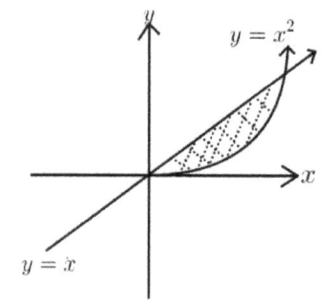

Example 15.

Find $\int_{-3}^{3} \int_{-\sqrt{9-x^2}}^{\sqrt{9-x^2}} (9 - x^2 - y^2)^{\frac{1}{2}} dydx = ?$

【Solution】

Let $x = r\cos\theta$, $y = r\sin\theta$ then

$\{(x,y): -3 \le x \le 3, -\sqrt{9-x^2} < y < \sqrt{9-x^2}\} = \{(r,\theta): 0 \le r \le 3, 0 \le \theta \le 2\pi\}$

and $dxdy = \begin{Vmatrix} \dfrac{\partial x}{\partial r} & \dfrac{\partial x}{\partial \theta} \\ \dfrac{\partial y}{\partial r} & \dfrac{\partial y}{\partial \theta} \end{Vmatrix} drd\theta = \begin{Vmatrix} \cos\theta & -r\sin\theta \\ \sin\theta & r\cos\theta \end{Vmatrix} drd\theta = rdrd\theta$

$$\therefore \int_{-3}^{3}\int_{-\sqrt{9-x^2}}^{\sqrt{9-x^2}}(9-x^2-y^2)^{\frac{1}{2}}dydx = \int_{0}^{2\pi}\int_{0}^{3}(9-r^2)^{\frac{1}{2}}\cdot r\,drd\theta = 2\pi\cdot\left.\frac{-(9-r^2)^{\frac{3}{2}}}{3}\right|_{0}^{3} = 18\pi$$

Example 16.

Find $\iint_{R}\dfrac{1}{\sqrt{1+x^2+y^2}}dA =?$, $R = \{(x,y): x^2+y^2 \leq 4\}$.

【Solution】

Let $x = r\cos\theta$, $y = r\sin\theta$ then $\{(x,y): x^2+y^2 \leq 4\} = \{(r,\theta): 0 \leq r \leq 2, 0 \leq \theta \leq 2\pi\}$

and $dxdy = \begin{Vmatrix}\dfrac{\partial x}{\partial r} & \dfrac{\partial x}{\partial \theta}\\ \dfrac{\partial y}{\partial r} & \dfrac{\partial y}{\partial \theta}\end{Vmatrix}drd\theta = \begin{Vmatrix}\cos\theta & -r\sin\theta\\ \sin\theta & r\cos\theta\end{Vmatrix}drd\theta = rdrd\theta$

$$\therefore \iint_{R}\frac{1}{\sqrt{1+x^2+y^2}}dA = \int_{0}^{2\pi}\int_{0}^{2}\frac{1}{\sqrt{1+r^2}}rdrd\theta = \int_{0}^{2\pi}d\theta\int_{0}^{2}\frac{rdr}{\sqrt{1+r^2}} = 2\pi\cdot(1+r^2)^{\frac{1}{2}}\Big|_{r=0}^{r=2}$$
$= 2\pi(\sqrt{5}-1)$

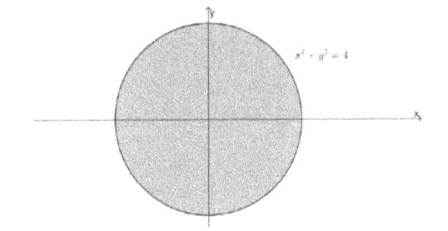

Example 17.

Find $\iint_{R}\dfrac{x^2}{\sqrt{1+(x^2+y^2)^2}}dA =?$, $R = \{(x,y): x^2+y^2 \leq 1, y > 0\}$.

【Solution】

Let $x = r\cos\theta$, $y = r\sin\theta$ then $\{(x,y): x^2+y^2 \leq 1\} = \{(r,\theta): 0 \leq r \leq 1, 0 \leq \theta \leq \pi\}$

and $dxdy = \begin{Vmatrix}\dfrac{\partial x}{\partial r} & \dfrac{\partial x}{\partial \theta}\\ \dfrac{\partial y}{\partial r} & \dfrac{\partial y}{\partial \theta}\end{Vmatrix}drd\theta = \begin{Vmatrix}\cos\theta & -r\sin\theta\\ \sin\theta & r\cos\theta\end{Vmatrix}drd\theta = rdrd\theta$

8.3.3 Use Polar Coordinates to Compute Double Integrals

$$\therefore \iint_R \frac{x^2}{\sqrt{1+(x^2+y^2)^2}} dA = \int_0^\pi \int_0^1 \frac{r^2 \sin^2 \theta}{\sqrt{1+r^4}} r \, dr d\theta = \int_0^\pi \sin^2 \theta \, d\theta \int_0^1 \frac{r^3 dr}{\sqrt{1+r^4}}$$

$$= \int_0^\pi \frac{1-\cos 2\theta}{2} d\theta \int_0^1 \frac{r^3 dr}{\sqrt{1+r^4}} = \frac{\pi}{2} \cdot \frac{1}{2}(1+r^4)^{\frac{1}{2}}\Big|_{r=0}^{r=1} = \frac{\pi}{4}(\sqrt{2}-1)$$

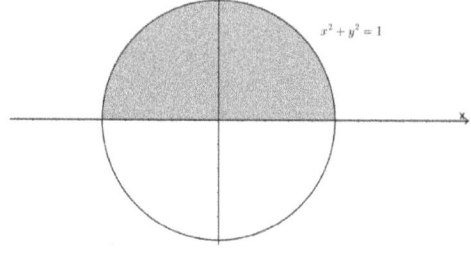

Example 18.

Find $\iint_R e^{x^2+y^2} dxdy = ?$, $R = \{(x,y): 0 \le y \le x, x^2 + y^2 \le 4\}$.

【Solution】

Let $x = r\cos\theta$, $y = r\sin\theta$ then

$$\{(x,y): 0 \le y \le x, x^2 + y^2 \le 4\} = \{(r,\theta): 0 \le r \le 2, 0 \le \theta \le \frac{\pi}{4}\}$$

and $dxdy = \begin{Vmatrix} \frac{\partial x}{\partial r} & \frac{\partial x}{\partial \theta} \\ \frac{\partial y}{\partial r} & \frac{\partial y}{\partial \theta} \end{Vmatrix} drd\theta = \begin{Vmatrix} \cos\theta & -r\sin\theta \\ \sin\theta & r\cos\theta \end{Vmatrix} drd\theta = r \, drd\theta$

$$\therefore \iint_R e^{x^2+y^2} dxdy = \int_0^{\frac{\pi}{4}} \int_0^2 e^{r^2} r \, dr d\theta = \int_0^{\frac{\pi}{4}} \frac{e^{r^2}}{2}\Big|_{r=0}^{r=2} d\theta = \int_0^{\frac{\pi}{4}} \frac{e^4 - 1}{2} d\theta = \left(\frac{e^4-1}{8}\right)\pi$$

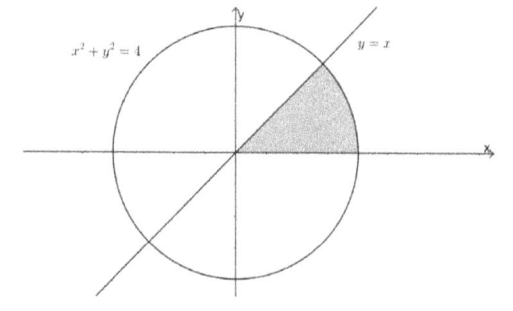

Example 19.

Find $\iint_R (x^2 + y^2)e^{(x^2+y^2)^2} dxdy = ?$, $R = \{(x,y): 0 \leq y \leq x, x^2 + y^2 \leq 4\}$.

【Solution】

Let $x = r\cos\theta$, $y = r\sin\theta$ then

$$\{(x,y): 0 \leq y \leq x, x^2 + y^2 \leq 4\} = \{(r,\theta): 0 \leq r \leq 2, 0 \leq \theta \leq \frac{\pi}{4}\}$$

and $dxdy = \begin{Vmatrix} \frac{\partial x}{\partial r} & \frac{\partial x}{\partial \theta} \\ \frac{\partial y}{\partial r} & \frac{\partial y}{\partial \theta} \end{Vmatrix} drd\theta = \begin{Vmatrix} \cos\theta & -r\sin\theta \\ \sin\theta & r\cos\theta \end{Vmatrix} drd\theta = rdrd\theta$

$\therefore \iint_R (x^2 + y^2)e^{(x^2+y^2)^2} dxdy = \int_0^{\frac{\pi}{4}} \int_0^2 e^{r^4} r^3 drd\theta = \int_0^{\frac{\pi}{4}} \frac{e^{r^4}}{4}\bigg|_{r=0}^{r=2} d\theta = \int_0^{\frac{\pi}{4}} \frac{e^{16} - 1}{4} d\theta$

$= \left(\frac{e^{16} - 1}{16}\right)\pi$

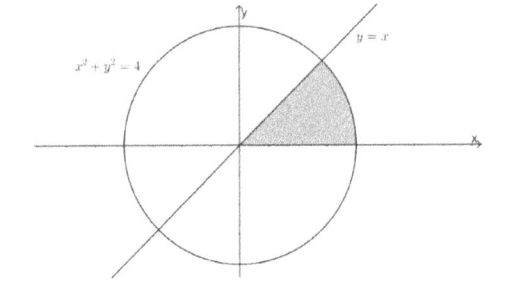

Example 20.

Find $\iint_R \sqrt{a^2 - x^2 - y^2} dA = ?$, $R = \{(x,y): x^2 + y^2 \leq a^2\}$.

【Solution】

Let $x = r\cos\theta$, $y = r\sin\theta$ then $\{(x,y): x^2 + y^2 \leq a^2\} = \{(r,\theta): 0 \leq r \leq a, 0 \leq \theta \leq 2\pi\}$

and $dxdy = \begin{Vmatrix} \frac{\partial x}{\partial r} & \frac{\partial x}{\partial \theta} \\ \frac{\partial y}{\partial r} & \frac{\partial y}{\partial \theta} \end{Vmatrix} drd\theta = \begin{Vmatrix} \cos\theta & -r\sin\theta \\ \sin\theta & r\cos\theta \end{Vmatrix} drd\theta = rdrd\theta$

$\therefore \iint_R \sqrt{a^2 - x^2 - y^2} dA = \int_0^{2\pi} \int_0^a \sqrt{a^2 - r^2} r dr d\theta = \int_0^{2\pi} d\theta \int_0^a r\sqrt{a^2 - r^2} dr$

$$= 2\pi \cdot \left.\frac{(-1)(a^2-r^2)^{\frac{3}{2}}}{3}\right|_{r=0}^{r=a} = \frac{2\pi a^3}{3}$$

Example 21.

Find $\iint_R e^{\sqrt{x^2+y^2}} dxdy =?$, $R = \{(x,y): x^2+y^2 \leq 4, 0 \leq y \leq x\}$.

【Solution】

Let $x = r\cos\theta$, $y = r\sin\theta$ then $\{(x,y): x^2+y^2 \leq 4\} = \{(r,\theta): 0 \leq r \leq 2, 0 \leq \theta \leq \frac{\pi}{4}\}$

and $dxdy = \left\|\begin{matrix}\frac{\partial x}{\partial r} & \frac{\partial x}{\partial \theta} \\ \frac{\partial y}{\partial r} & \frac{\partial y}{\partial \theta}\end{matrix}\right\| drd\theta = \left\|\begin{matrix}\cos\theta & -r\sin\theta \\ \sin\theta & r\cos\theta\end{matrix}\right\| drd\theta = rdrd\theta$

$$\therefore \iint_R e^{\sqrt{x^2+y^2}} dxdy = \int_0^{\frac{\pi}{4}} \int_0^2 e^r r\, drd\theta = \int_0^{\frac{\pi}{4}} d\theta \int_0^2 e^r r\, dr$$

By integration by parts, $\int_0^2 e^r r\, dr = 2e^2 - (e^2-1) = e^2+1$

$$\therefore \iint_R e^{\sqrt{x^2+y^2}} dxdy = \frac{\pi}{4}(e^2+1)$$

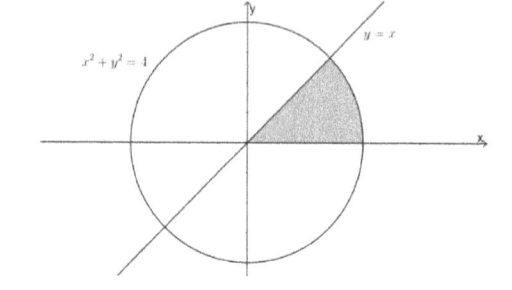

Example 22.

(1) Find $\iint_R \sin(x^2+y^2)\, dA =?$, $R = \{(x,y): \pi^2 \leq x^2+y^2 \leq 9\pi^2\}$.

(2) Find $\iint_R \sec^2(x^2+y^2)\, dA =?$, $R = \left\{(x,y): 0 \leq x^2+y^2 \leq \frac{\pi}{4}\right\}$.

【Solution】
(1)
Let $x = r\cos\theta$, $y = r\sin\theta$ then
$\{(x,y): \pi^2 \leq x^2 + y^2 \leq 9\pi^2\} = \{(r,\theta): \pi \leq r \leq 3\pi, 0 \leq \theta \leq 2\pi\}$

and $dxdy = \begin{Vmatrix} \dfrac{\partial x}{\partial r} & \dfrac{\partial x}{\partial \theta} \\ \dfrac{\partial y}{\partial r} & \dfrac{\partial y}{\partial \theta} \end{Vmatrix} drd\theta = \begin{Vmatrix} \cos\theta & -r\sin\theta \\ \sin\theta & r\cos\theta \end{Vmatrix} drd\theta = rdrd\theta$

$\therefore \iint_R \sin(x^2+y^2)\, dA = \int_0^{2\pi}\int_\pi^{3\pi} \sin(r^2)\, rdrd\theta = \int_0^{2\pi} d\theta \int_\pi^{3\pi} \sin(r^2)\, rdr$

$= 2\pi \cdot \dfrac{(-1)\cos r^2}{2}\bigg|_\pi^{3\pi} = (-\pi)(\cos 9\pi^2 - \cos \pi^2)$

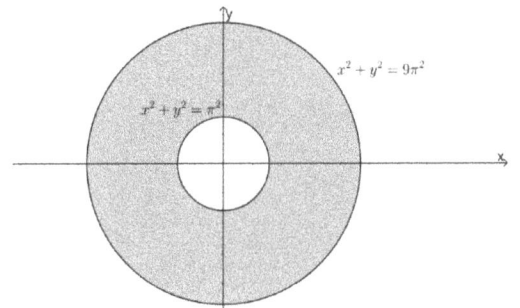

(2)
Let $x = r\cos\theta$, $y = r\sin\theta$ then
$\left\{(x,y): 0 \leq x^2 + y^2 \leq \dfrac{\pi}{4}\right\} = \left\{(r,\theta): 0 \leq r \leq \dfrac{\sqrt{\pi}}{2}, 0 \leq \theta \leq 2\pi\right\}$

and $dxdy = \begin{Vmatrix} \dfrac{\partial x}{\partial r} & \dfrac{\partial x}{\partial \theta} \\ \dfrac{\partial y}{\partial r} & \dfrac{\partial y}{\partial \theta} \end{Vmatrix} drd\theta = \begin{Vmatrix} \cos\theta & -r\sin\theta \\ \sin\theta & r\cos\theta \end{Vmatrix} drd\theta - rdrd\theta$

$\therefore \iint_R \sec^2(x^2+y^2)\, dA = \int_0^{2\pi}\int_0^{\frac{\sqrt{\pi}}{2}} \sec^2(r^2)\, rdrd\theta = \int_0^{2\pi} d\theta \int_0^{\frac{\sqrt{\pi}}{2}} \sec^2(r^2)\, rdr$

$= 2\pi \cdot \dfrac{\tan r^2}{2}\bigg|_0^{\frac{\sqrt{\pi}}{2}} = \pi$

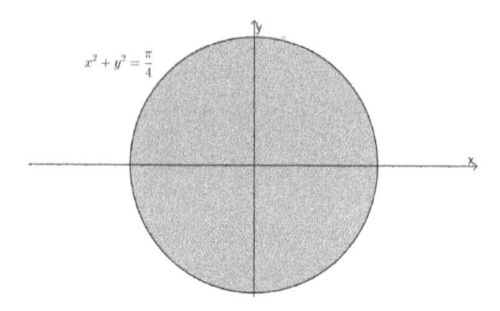

8.3.3 Use Polar Coordinates to Compute Double Integrals

$x^2 + y^2 = \frac{\pi}{4}$

Example 23.

Find $\displaystyle\int_0^a \int_0^{\sqrt{a^2-x^2}} \frac{1}{(1+x^2+y^2)^{\frac{3}{2}}} dydx = ?$

【Solution】

Let $x = r\cos\theta$, $y = r\sin\theta$ then

$\{(x,y): 0 \le x \le a, 0 < y < \sqrt{a^2 - x^2}\} = \{(r,\theta): 0 \le r \le a, 0 \le \theta \le \frac{\pi}{2}\}$

and $dxdy = \begin{Vmatrix} \frac{\partial x}{\partial r} & \frac{\partial x}{\partial \theta} \\ \frac{\partial y}{\partial r} & \frac{\partial y}{\partial \theta} \end{Vmatrix} drd\theta = \begin{Vmatrix} \cos\theta & -r\sin\theta \\ \sin\theta & r\cos\theta \end{Vmatrix} drd\theta = rdrd\theta$

$\therefore \displaystyle\int_0^a \int_0^{\sqrt{a^2-x^2}} \frac{1}{(1+x^2+y^2)^{\frac{3}{2}}} dydx = \int_0^{\frac{\pi}{2}} \int_0^a \frac{1}{(1+r^2)^{\frac{3}{2}}} \cdot rdrd\theta = -(1+r^2)^{-\frac{1}{2}} \Big|_0^a \cdot \frac{\pi}{2}$

$= \dfrac{\pi}{2}\left(1 - (1+a^2)^{-\frac{1}{2}}\right)$

Example 24.

(1) Find $\displaystyle\int_0^{4a} \int_0^{\sqrt{4ax-x^2}} x^2 + y^2 \, dydx = ?$ (2) Find $\displaystyle\int_0^{2a} \int_{-\sqrt{2ay-y^2}}^{\sqrt{2ay-y^2}} \sqrt{x^2+y^2} \, dxdy = ?$

【Solution】

(1)

Let $x = r\cos\theta$, $y = r\sin\theta$ then

$\{(x,y): 0 \le x \le 4a, 0 \le y \le \sqrt{4ax - x^2}\} = \{(r,\theta): 0 \le r \le 4a\cos\theta, 0 \le \theta \le \frac{\pi}{2}\}$

8.3.3 Use Polar Coordinates to Compute Double Integrals

and $dxdy = \begin{Vmatrix} \dfrac{\partial x}{\partial r} & \dfrac{\partial x}{\partial \theta} \\ \dfrac{\partial y}{\partial r} & \dfrac{\partial y}{\partial \theta} \end{Vmatrix} drd\theta = \begin{Vmatrix} \cos\theta & -r\sin\theta \\ \sin\theta & r\cos\theta \end{Vmatrix} drd\theta = rdrd\theta$

$\therefore \int_0^{4a} \int_0^{\sqrt{4ax-x^2}} x^2 + y^2 \, dydx = \int_0^{\frac{\pi}{2}} \int_0^{4a\cos\theta} r^3 drd\theta = \int_0^{\frac{\pi}{2}} \dfrac{r^4}{4} \Big|_0^{4a\cos\theta} d\theta = 64a^4 \int_0^{\frac{\pi}{2}} \cos^4\theta \, d\theta$

$\because \cos^4\theta = \left(\dfrac{1+\cos 2\theta}{2}\right)^2 = \dfrac{1+2\cos 2\theta + \cos^2 2\theta}{4} = \dfrac{1+2\cos 2\theta}{4} + \dfrac{1+\cos 4\theta}{8}$

$\therefore \int_0^{\frac{\pi}{2}} \cos^4\theta \, d\theta = \int_0^{\frac{\pi}{2}} \dfrac{3}{8} + \dfrac{\cos 2\theta}{2} + \dfrac{\cos 4\theta}{8} d\theta = \dfrac{3\pi}{16}$

$\therefore \int_0^{4a} \int_0^{\sqrt{4ax-x^2}} x^2 + y^2 \, dydx = 12\pi a^4$

(2)
Let $x = r\cos\theta, \ y = r\sin\theta$ then
$\{(x,y): -\sqrt{2ay-y^2} \le x \le \sqrt{2ay-y^2}, 0 \le y \le 2a\} = \{(r,\theta): 0 \le r \le 2a\sin\theta, 0 \le \theta \le \pi\}$

and $dxdy = \begin{Vmatrix} \dfrac{\partial x}{\partial r} & \dfrac{\partial x}{\partial \theta} \\ \dfrac{\partial y}{\partial r} & \dfrac{\partial y}{\partial \theta} \end{Vmatrix} drd\theta = \begin{Vmatrix} \cos\theta & -r\sin\theta \\ \sin\theta & r\cos\theta \end{Vmatrix} drd\theta = rdrd\theta$

$\therefore \int_0^{2a} \int_{-\sqrt{2ay-y^2}}^{\sqrt{2ay-y^2}} \sqrt{x^2+y^2} \, dxdy = \int_0^{\pi} \int_0^{2a\sin\theta} r^2 drd\theta = \int_0^{\pi} \dfrac{r^3}{3} \Big|_0^{2a\sin\theta} d\theta$

$= \dfrac{8a^3}{3} \int_0^{\pi} \sin^3\theta \, d\theta = \dfrac{8a^3}{3} \int_0^{\pi} \sin\theta(1-\cos^2\theta) \, d\theta = \dfrac{8a^3}{3}\left(-\cos\theta + \dfrac{\cos^3\theta}{3}\right)\Big|_0^{\pi} = \dfrac{32a^3}{9}$

8.3.3 Use Polar Coordinates to Compute Double Integrals

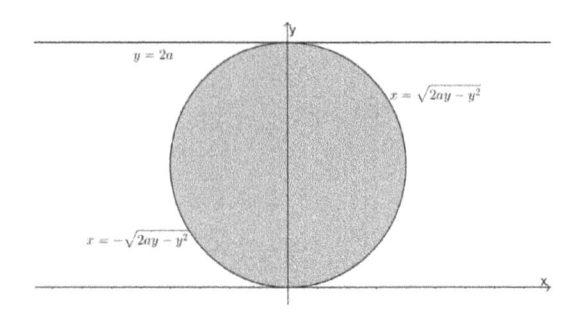

8.3.4 Use Generalized Coordinates to Compute Double Integrals

When the integration region R corresponds to the following situations, it can be considered as a case for using generalized coordinate transformations:

(i) The integration region R involves $ax + by$, $cx + dy$, where $ad - bc \neq 0$.

(ii) The integration region R or the integrand involve $ax^2 + bxy + cy^2$, where $4ac - b^2 > 0, a > 0, b > 0$.

(iii) The integration region R is bounded by $y = ax$, $y = bx$, $xy = c$, and $xy = d$, where $abcd \neq 0$.

Question Types:

Type 1.

Suppose $R = \{(x, y): 0 \leq ax + by \leq s, 0 \leq cx + dy \leq t\}$, $a, b, c, d \in R, s, t > 0$ where $ad - bc \neq 0$.

$$\text{Find } \iint_R f(ax + by, cx + dy) dA = ?$$

Problem-Solving Process:

Step1.

Let $u = ax + by$, $v = cx + dy$ then $x = \dfrac{du - bv}{ad - bc}$ and $y = \dfrac{av - cu}{ad - bc}$

$$\therefore dxdy = \begin{Vmatrix} \dfrac{\partial x}{\partial u} & \dfrac{\partial x}{\partial v} \\ \dfrac{\partial y}{\partial u} & \dfrac{\partial y}{\partial v} \end{Vmatrix} dudv = \dfrac{1}{|ad - bc|^2} \begin{Vmatrix} d & -b \\ -c & a \end{Vmatrix} dudv = \dfrac{1}{|ad - bc|} dudv$$

Step2.

$$\therefore \iint_R f(ax + by, cx + dy) dA = \dfrac{1}{|ad - bc|} \int_0^t \int_0^s f(u, v) dudv$$

Step3.

Find $\dfrac{1}{|ad-bc|}\displaystyle\int_0^t\int_0^s f(u,v)dudv =?$

Examples:

Find $\displaystyle\iint_R f(ax+by, cx+dy)dA =?, R = \{(x,y): 0 \le ax+by \le s, 0 \le cx+dy \le t\}$, $a,b,c,d \in R, s,t > 0$.

Let $u = ax+by, \; v = cx+dy$ then $x = \dfrac{du-bv}{ad-bc}$ and $y = \dfrac{av-cu}{ad-bc}$

$\because dxdy = \begin{Vmatrix} \dfrac{\partial x}{\partial u} & \dfrac{\partial x}{\partial v} \\ \dfrac{\partial y}{\partial u} & \dfrac{\partial y}{\partial v} \end{Vmatrix} dudv = \dfrac{1}{|ad-bc|^2}\begin{Vmatrix} d & -b \\ -c & a \end{Vmatrix} dudv = \dfrac{1}{|ad-bc|} dudv$

$\therefore \displaystyle\iint_R f(ax+by, cx+dy)dA = \dfrac{1}{|ad-bc|}\int_0^t\int_0^s f(u,v)dudv$

(I) If $f(ax+by, cx+dy) = (ax+by)(cx+dy)$ then

$\displaystyle\iint_R f(ax+by, cx+dy)dA = \dfrac{1}{|ad-bc|}\int_0^t\int_0^s uv\,dudv =?$

(II) If $f(ax+by, cx+dy) = e^{\frac{cx+dy}{ax+by}}$ then

$\displaystyle\iint_R f(ax+by, cx+dy)dA = \dfrac{1}{|ad-bc|}\int_0^t\int_0^s e^{\frac{v}{u}}dudv =?$

(III) If $f(ax+by, cx+dy) = (ax+by)^2 + (cx+dy)^2$ then

$\displaystyle\iint_R f(ax+by, cx+dy)dA = \dfrac{1}{|ad-bc|}\int_0^t\int_0^s u^2+v^2\,dudv =?$

Type 2.
Suppose $R = \{(x,y): ax^2+bxy+cy^2 \le d^2\}, \; 4ac-b^2 > 0, a > 0, b > 0$.

Find $\displaystyle\iint_R f(ax^2+bxy+cy^2)dA =?$

Problem-Solving Process:
Step1.
$\because ax^2+bxy+cy^2 = a(x+\dfrac{by}{2a})^2 + \dfrac{4ac-b^2}{4a}y^2$

8.3.4 Use Generalized Coordinates to Compute Double Integrals

Let $\sqrt{a}\left(x + \dfrac{by}{2a}\right) = u$, $\sqrt{\dfrac{4ac - b^2}{4a}}\, y = v$

$$\therefore dxdy = \left\|\begin{matrix}\dfrac{\partial x}{\partial u} & \dfrac{\partial x}{\partial v} \\ \dfrac{\partial y}{\partial u} & \dfrac{\partial y}{\partial v}\end{matrix}\right\| dudv = \left\|\begin{matrix}\dfrac{1}{\sqrt{a}} & 0 \\ \dfrac{2\sqrt{a}}{b} & \sqrt{\dfrac{4a}{4ac-b^2}}\end{matrix}\right\| dudv = \dfrac{2}{\sqrt{4ac - b^2}}\, dudv$$

Step2.

$$\therefore \iint_R f(ax^2 + bxy + cy^2)dA = \iint_{u^2 + v^2 \le d^2} f(u^2 + v^2)\dfrac{2}{\sqrt{4ac - b^2}}\, dudv$$

Step3.
Let $u = r\cos\theta, v = r\sin\theta$ then $\{(u,v): u^2 + v^2 \le d^2\} = \{(r,\theta): 0 \le r \le d, 0 \le \theta \le 2\pi\}$

Step4.

$$\therefore dudv = \left\|\begin{matrix}\dfrac{\partial u}{\partial r} & \dfrac{\partial u}{\partial \theta} \\ \dfrac{\partial v}{\partial r} & \dfrac{\partial v}{\partial \theta}\end{matrix}\right\| drd\theta = \left\|\begin{matrix}\cos\theta & -r\sin\theta \\ \sin\theta & r\cos\theta\end{matrix}\right\| drd\theta = rdrd\theta$$

$$\therefore \iint_{u^2+v^2\le d^2} f(u^2+v^2)\dfrac{2}{\sqrt{4ac-b^2}}\, dudv = \int_0^{2\pi}\int_0^d \dfrac{2f(r^2)}{\sqrt{4ac-b^2}}\, rdrd\theta$$

$$= \dfrac{2}{\sqrt{4ac-b^2}} \int_0^{2\pi} d\theta \int_0^d f(r^2) rdr$$

Step5.

Find $\dfrac{2}{\sqrt{4ac-b^2}} \int_0^{2\pi} d\theta \int_0^d f(r^2) rdr = ?$

Examples:

Find $\iint_R f(ax^2 + bxy + cy^2)dA = ?$, $R = \{(x,y): ax^2 + bxy + cy^2 \le d^2\}$, where $4ac - b^2 > 0, a > 0, b > 0$.

$$\therefore \iint_R f(ax^2 + bxy + cy^2)dA = \iint_{u^2+v^2\le d^2} f(u^2 + v^2)\dfrac{2}{\sqrt{4ac-b^2}}\, dudv$$

$$= \dfrac{2}{\sqrt{4ac-b^2}} \int_0^{2\pi} d\theta \int_0^d f(r^2) rdr$$

(I) If $f(t) = t$ then $\iint_R f(ax^2 + bxy + cy^2)dA = \dfrac{2}{\sqrt{4ac-b^2}}\int_0^{2\pi}d\theta\int_0^d r^3 dr$

(II) If $f(t) = e^{-t}$ then $\iint_R f(ax^2 + bxy + cy^2)dA = \dfrac{2}{\sqrt{4ac-b^2}}\int_0^{2\pi}d\theta\int_0^d e^{-r^2}rdr$

(III) If $f(t) = e^{t}$ then $\iint_R f(ax^2 + bxy + cy^2)dA = \dfrac{2}{\sqrt{4ac-b^2}}\int_0^{2\pi}d\theta\int_0^d e^{r^2}rdr$

(IV) If $f(t) = \ln(t)$ then $\iint_R f(ax^2 + bxy + cy^2)dA = \dfrac{2}{\sqrt{4ac-b^2}}\int_0^{2\pi}d\theta\int_0^d \ln(r^2)rdr$

(V) If $f(t) = \sin(t)$ then $\iint_R f(ax^2 + bxy + cy^2)dA = \dfrac{2}{\sqrt{4ac-b^2}}\int_0^{2\pi}d\theta\int_0^d \sin(r^2)rdr$

Type 3.

Assuming R is the region bounded by $y = x^\alpha$, $y = \beta x^\alpha$, $x = y^\alpha$, and $x = \gamma y^\alpha$,

where $\alpha > 1, \beta > 1, \gamma > 1$, find $\iint_R f\left(\dfrac{y}{x^\alpha}, \dfrac{x}{y^\alpha}\right)dA = ?$.

Problem-Solving Process:

Step1.

Let $u = \dfrac{y}{x^\alpha}$, $v = \dfrac{x}{y^\alpha}$ then $x = u^{\frac{\alpha}{1-\alpha^2}}v^{\frac{1}{1-\alpha^2}}$, $y = u^{\frac{1}{1-\alpha^2}}v^{\frac{\alpha}{1-\alpha^2}}$

and $dxdy = \begin{Vmatrix} \dfrac{\partial x}{\partial u} & \dfrac{\partial x}{\partial v} \\ \dfrac{\partial y}{\partial u} & \dfrac{\partial y}{\partial v} \end{Vmatrix} dudv = \begin{Vmatrix} \dfrac{\alpha u^{\left(\frac{\alpha}{1-\alpha^2}-1\right)}v^{\frac{1}{1-\alpha^2}}}{1-\alpha^2} & \dfrac{u^{\frac{\alpha}{1-\alpha^2}}v^{\left(\frac{1}{1-\alpha^2}-1\right)}}{1-\alpha^2} \\ \dfrac{u^{\left(\frac{1}{1-\alpha^2}-1\right)}v^{\frac{\alpha}{1-\alpha^2}}}{1-\alpha^2} & \dfrac{\alpha u^{\frac{1}{1-\alpha^2}}v^{\left(\frac{\alpha}{1-\alpha^2}-1\right)}}{1-\alpha^2} \end{Vmatrix} dudv$

$= \dfrac{\alpha^2 u^{\frac{\alpha-1+\alpha^2+1}{1-\alpha^2}} v^{\frac{1+\alpha-1+\alpha^2}{1-\alpha^2}} - u^{\frac{\alpha-1+\alpha^2+1}{1-\alpha^2}} v^{\frac{1+\alpha-1+\alpha^2}{1-\alpha^2}}}{(1-\alpha^2)^2} dudv = \dfrac{\alpha^2 u^{\frac{\alpha}{1-\alpha}} v^{\frac{\alpha}{1-\alpha}} - u^{\frac{\alpha}{1-\alpha}} v^{\frac{\alpha}{1-\alpha}}}{(1-\alpha^2)^2} dudv$

$= \dfrac{u^{\frac{\alpha}{1-\alpha}} v^{\frac{\alpha}{1-\alpha}}}{\alpha^2 - 1} dudv$

Step2.

Let $R = \left\{(x,y): 1 \leq \dfrac{y}{x^\alpha} \leq \beta, 1 \leq \dfrac{x}{y^\alpha} \leq \gamma \right\}$ then $R = \{(u,v): 1 \leq u \leq \beta, 1 \leq v \leq \gamma\}$

Step3.

$\therefore \iint_R f\left(\dfrac{y}{x^\alpha}, \dfrac{x}{y^\alpha}\right)dA = \int_1^\beta \int_1^\gamma f(u,v) \dfrac{u^{\frac{\alpha}{1-\alpha}} v^{\frac{\alpha}{1-\alpha}}}{\alpha^2 - 1} dvdu$

8.3.4 Use Generalized Coordinates to Compute Double Integrals

Type 4.

Find $\iint_R f\left(\frac{y}{x}, xy\right) dxdy = ?$, where R is the region bounded by $y = ax$, $y = bx$, $xy = c$, and $xy = d$ with $a < b, c < d$.

Problem-Solving Process:

Step1.

Let $\frac{y}{x} = u$, $xy = v$ then $x = \sqrt{\frac{v}{u}}$, $y = \sqrt{uv}$

and $dxdy = \begin{Vmatrix} \frac{\partial x}{\partial u} & \frac{\partial x}{\partial v} \\ \frac{\partial y}{\partial u} & \frac{\partial y}{\partial v} \end{Vmatrix} dudv = \begin{Vmatrix} \frac{1}{2}\left(\frac{v}{u}\right)^{-\frac{1}{2}}\left(-\frac{v}{u^2}\right) & \frac{1}{2}\left(\frac{v}{u}\right)^{-\frac{1}{2}}\left(\frac{1}{u}\right) \\ \frac{1}{2}(uv)^{-\frac{1}{2}}v & \frac{1}{2}(uv)^{-\frac{1}{2}}u \end{Vmatrix} dudv = \frac{1}{2u} dudv$

Step2.

Let $R = \left\{(x,y): a \leq \frac{y}{x} \leq b, c \leq xy \leq d\right\}$ then $R = \{(u,v): a \leq u \leq b, c \leq v \leq d\}$

Step3.

Find $\iint_R f\left(\frac{y}{x}, xy\right) dxdy = \int_a^b \int_c^d \frac{f(u,v)}{2u} dvdu$

Examples:

Find $\iint_R f\left(\frac{y}{x}, xy\right) dxdy = ?$, where R is the region bounded by $y = ax$, $y = bx$, $xy = c$, and $xy = d$, with $a < b$, $c < d$.

(I) If $f\left(\frac{y}{x}, xy\right) = e^{-\frac{y}{x}xy}$ then

$\iint_R f\left(\frac{y}{x}, xy\right) dxdy = \int_a^b \int_c^d \frac{f(u,v)}{2u} dvdu = \int_a^b \int_c^d \frac{e^{-uv}}{2u} dvdu$

(II) If $f\left(\frac{y}{x}, xy\right) = \left(\frac{y}{x}\right)^2 + (xy)^2$ then

$\iint_R f\left(\frac{y}{x}, xy\right) dxdy = \int_a^b \int_c^d \frac{f(u,v)}{2u} dvdu = \int_a^b \int_c^d \frac{u^2 + v^2}{2u} dvdu$

Example 1.

Find $\iint_R x^2 + y^2 dA = ?$, $R = \left\{(x,y): \dfrac{x^2}{a^2} + \dfrac{y^2}{b^2} \leq 4\right\}$.

【Solution】

Let $x = ar\cos\theta$, $y = br\sin\theta$ then $\left\{(x,y): \dfrac{x^2}{a^2} + \dfrac{y^2}{b^2} \leq 4\right\} = \{(r,\theta): 0 \leq r \leq 2, 0 \leq \theta \leq 2\pi\}$

and $dxdy = \begin{Vmatrix} \dfrac{\partial x}{\partial r} & \dfrac{\partial x}{\partial \theta} \\ \dfrac{\partial y}{\partial r} & \dfrac{\partial y}{\partial \theta} \end{Vmatrix} drd\theta = \begin{Vmatrix} a\cos\theta & -ra\sin\theta \\ b\sin\theta & rb\cos\theta \end{Vmatrix} drd\theta = abrdrd\theta$

$\therefore \iint_R x^2 + y^2 dA = \int_0^{2\pi} \int_0^2 ((ar)^2 \cos^2\theta + (br)^2 \sin^2\theta) abr \, drd\theta$

$= a^3b \int_0^{2\pi} \cos^2\theta \, d\theta \int_0^2 r^3 dr + ab^3 \int_0^{2\pi} \sin^2\theta \, d\theta \int_0^2 r^3 dr$

$\because \int_0^{2\pi} \cos^2\theta \, d\theta = \int_0^{2\pi} \dfrac{1 + \cos 2\theta}{2} d\theta = \pi,$

$\int_0^{2\pi} \sin^2\theta \, d\theta = \int_0^{2\pi} \dfrac{1 - \cos 2\theta}{2} d\theta = \pi$ and $\int_0^2 r^3 dr = \dfrac{16}{4} = 4$

$\therefore \iint_R x^2 + y^2 dA = a^3b(4\pi) + ab^3(4\pi) = 4ab\pi(a^2 + b^2)$

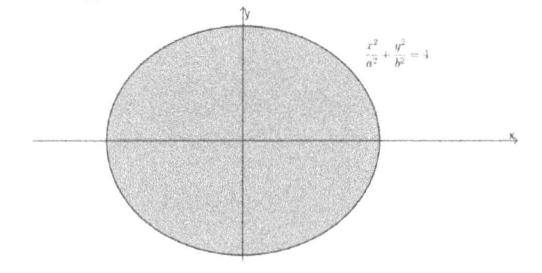

Example 2.

Find $\iint_R e^{x+y} dA = ?$, $R = \{(x,y): |x| + |y| \leq 2\}$.

【Solution】

Let $x + y = u$, $x - y = v$ then $\{(x,y): |x| + |y| \leq 2\} = \{(u,v): -2 \leq u \leq 2, -2 \leq v \leq 2\}$

8.3.4 Use Generalized Coordinates to Compute Double Integrals

and $x = \dfrac{u+v}{2}$, $y = \dfrac{u-v}{2}$, $dxdy = \begin{Vmatrix} \dfrac{\partial x}{\partial u} & \dfrac{\partial x}{\partial v} \\ \dfrac{\partial y}{\partial u} & \dfrac{\partial y}{\partial v} \end{Vmatrix} dudv = \begin{Vmatrix} \dfrac{1}{2} & \dfrac{1}{2} \\ \dfrac{1}{2} & -\dfrac{1}{2} \end{Vmatrix} dudv = |-\dfrac{1}{2}| dudv$

$\therefore \iint_R e^{x+y} dA = \int_{-2}^{2} \int_{-2}^{2} \dfrac{e^u}{2} dudv = \dfrac{1}{2} \int_{-2}^{2} dv \int_{-2}^{2} e^u \, du = 2(e^2 - e^{-2})$

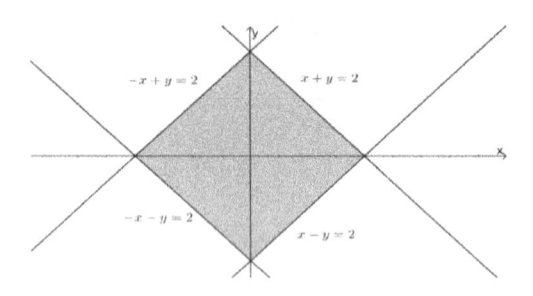

Example 3.

Find $\iint_R x^2 y \, dA = ?$, $R = \{(x,y): y \geq 0, \ (x-1)^2 + y^2 \leq 1\}$.

【Solution】

Let $x = r\cos\theta + 1$, $y = r\sin\theta$ then
$\{(x,y): (x-1)^2 + y^2 \leq 1, y \geq 0\} = \{(r,\theta): 0 \leq r \leq 1, 0 \leq \theta \leq \pi\}$

and $dxdy = \begin{Vmatrix} \dfrac{\partial x}{\partial r} & \dfrac{\partial x}{\partial \theta} \\ \dfrac{\partial y}{\partial r} & \dfrac{\partial y}{\partial \theta} \end{Vmatrix} drd\theta = \begin{Vmatrix} \cos\theta & -r\sin\theta \\ \sin\theta & r\cos\theta \end{Vmatrix} drd\theta = rdrd\theta$

$\therefore \iint_R x^2 y \, dA = \int_0^{\pi} \int_0^1 (r\cos\theta + 1)^2 r\sin\theta \, drd\theta$

$= \int_0^{\pi} \int_0^1 (r^2 \cos^2\theta + 2r\cos\theta + 1) r\sin\theta \, drd\theta$

$= \int_0^{\pi} \cos^2\theta \sin\theta \, d\theta \int_0^1 r^4 dr + \int_0^{\pi} 2\cos\theta \sin\theta \, d\theta \int_0^1 r^3 dr + \int_0^{\pi} \sin\theta \, d\theta \int_0^1 r^2 dr$

$= (-1) \dfrac{\cos^3\theta}{3}\Big|_0^{\pi} \cdot \dfrac{r^5}{5}\Big|_0^1 - \dfrac{\cos 2\theta}{2}\Big|_0^{\pi} \cdot \dfrac{r^4}{4}\Big|_0^1 - \cos\theta\Big|_0^{\pi} \cdot \dfrac{r^3}{3}\Big|_0^1 = \dfrac{4}{5}$

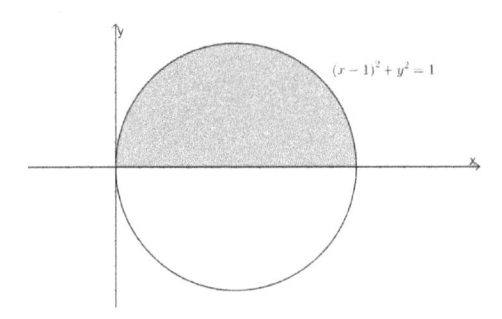

8.3.4 Use Generalized Coordinates to Compute Double Integrals

Example 4.

Find $\iint_R e^{\frac{y-x}{x+y}} dxdy = ?$, $R = \{(x,y): x \geq 0, y \geq 0, x+y \leq 3\}$.

【Solution】

Let $x + y = u$, $y - x = v$ then
$\{(x,y): x \geq 0, y \geq 0, x+y \leq 3\} = \{(u,v): 0 \leq u \leq 3, -u \leq v \leq u\}$

and $x = \dfrac{u-v}{2}$, $y = \dfrac{u+v}{2}$, $dxdy = \left\| \begin{matrix} \dfrac{\partial x}{\partial u} & \dfrac{\partial x}{\partial v} \\ \dfrac{\partial y}{\partial u} & \dfrac{\partial y}{\partial v} \end{matrix} \right\| dudv = \left\| \begin{matrix} \dfrac{1}{2} & \dfrac{-1}{2} \\ \dfrac{1}{2} & \dfrac{1}{2} \end{matrix} \right\| dudv = |\dfrac{-1}{2}| dudv$

$\therefore \iint_R e^{\frac{y-x}{x+y}} dxdy = \int_0^3 \int_{-u}^{u} e^{\frac{v}{u}} \cdot \dfrac{1}{2} dvdu = \dfrac{1}{2}\int_0^3 u e^{\frac{v}{u}} \Big|_{v=-u}^{v=u} du = \dfrac{1}{2}\int_0^3 u(e - e^{-1}) du$

$= \dfrac{1}{4}(e - e^{-1}) u^2 \big|_{u=0}^{u=3} = \dfrac{9}{4}(e - e^{-1})$

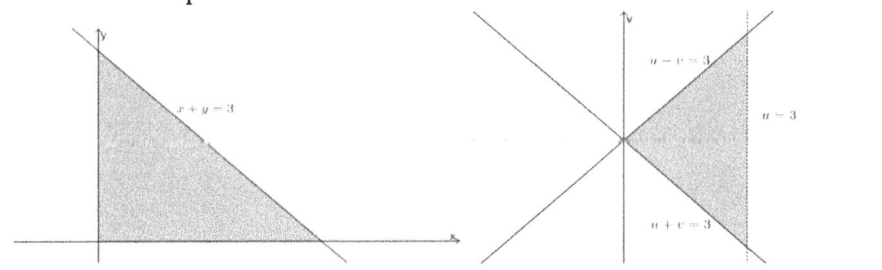

Example 5.

Prove that $\displaystyle\int_{-\infty}^{\infty}\int_{-\infty}^{\infty} e^{-(x^2+2xy+10y^2)} dxdy = \dfrac{\pi}{3}$.

【Solution】

$\because x^2 + 2xy + 10y^2 = (x+y)^2 + 9y^2$, let $(x+y) = u$, $3y = v$ then

8.3.4 Use Generalized Coordinates to Compute Double Integrals

$$dxdy = \begin{Vmatrix} \frac{\partial x}{\partial u} & \frac{\partial x}{\partial v} \\ \frac{\partial y}{\partial u} & \frac{\partial y}{\partial v} \end{Vmatrix} dudv = \begin{Vmatrix} 1 & 0 \\ 1 & \frac{1}{3} \end{Vmatrix} dudv = \frac{1}{3} dudv$$

$$\therefore \int_{-\infty}^{\infty}\int_{-\infty}^{\infty} e^{-(x^2+2xy+10y^2)} dxdy = \int_{-\infty}^{\infty}\int_{-\infty}^{\infty} e^{-(u^2+v^2)} \frac{1}{3} dudv$$

Let $u = r\cos\theta$, $v = r\sin\theta$ then

$$\{(u,v): -\infty \leq u \leq \infty, -\infty \leq v \leq \infty\} = \{(r,\theta): 0 \leq r \leq \infty, 0 \leq \theta \leq 2\pi\}$$

and $dudv = \begin{Vmatrix} \frac{\partial u}{\partial r} & \frac{\partial u}{\partial \theta} \\ \frac{\partial v}{\partial r} & \frac{\partial v}{\partial \theta} \end{Vmatrix} drd\theta = \begin{Vmatrix} \cos\theta & -r\sin\theta \\ \sin\theta & r\cos\theta \end{Vmatrix} drd\theta = r\,drd\theta$

$$\therefore \int_{-\infty}^{\infty}\int_{-\infty}^{\infty} e^{-(u^2+v^2)} du\,dv = \int_{0}^{2\pi}\int_{0}^{\infty} e^{-r^2} r\,drd\theta = \int_{0}^{2\pi} d\theta \int_{0}^{\infty} e^{-r^2} r\,dr = 2\pi \cdot \frac{-e^{-r^2}}{2}\Big|_{r=0}^{r=\infty} = \pi$$

$$\therefore \int_{-\infty}^{\infty}\int_{-\infty}^{\infty} e^{-(x^2+2xy+10y^2)} dxdy = \frac{\pi}{3}$$

Example 6.

(1) Given $a > 0$, $b > 0$ and $b^2 < 4ac$, prove that

$$\int_{-\infty}^{\infty}\int_{-\infty}^{\infty} e^{-(ax^2+bxy+cy^2)} dxdy = \frac{2\pi}{\sqrt{4ac-b^2}}.$$

(2) Given $R = \{(x,y): x^2 + 2xy + 3y^2 \leq t^2\}$, $I(t) = \iint_R e^{-(x^2+2xy+3y^2)} dA$, find

(a) $I(1) = ?$ (b) $\lim_{t\to\infty} I(t) = ?$.

【Solution】

(1)

$$\because ax^2 + bxy + cy^2 = a\left(x + \frac{by}{2a}\right)^2 + \left(\frac{4ac-b^2}{4a}\right) y^2$$

Let $\sqrt{a}\left(x + \frac{by}{2a}\right) = u$, $\sqrt{\frac{4ac-b^2}{4a}} y = v$ then

8.3.4 Use Generalized Coordinates to Compute Double Integrals

$$dxdy = \begin{Vmatrix} \dfrac{\partial x}{\partial u} & \dfrac{\partial x}{\partial v} \\ \dfrac{\partial y}{\partial u} & \dfrac{\partial y}{\partial v} \end{Vmatrix} dudv = \begin{Vmatrix} \dfrac{1}{\sqrt{a}} & 0 \\ \dfrac{2\sqrt{a}}{b} & \sqrt{\dfrac{4a}{4ac-b^2}} \end{Vmatrix} dudv = \dfrac{2}{\sqrt{4ac-b^2}} dudv$$

$$\therefore \int_{-\infty}^{\infty}\int_{-\infty}^{\infty} e^{-(ax^2+bxy+cy^2)} dxdy = \int_{-\infty}^{\infty}\int_{-\infty}^{\infty} e^{-(u^2+v^2)} \dfrac{2}{\sqrt{4ac-b^2}} dudv$$

Let $u = r\cos\theta$, $v = r\sin\theta$

then $\{(u,v): -\infty \le u \le \infty, -\infty \le v \le \infty\} = \{(r,\theta): 0 \le r \le \infty, 0 \le \theta \le 2\pi\}$

and $dudv = \begin{Vmatrix} \dfrac{\partial u}{\partial r} & \dfrac{\partial u}{\partial \theta} \\ \dfrac{\partial v}{\partial r} & \dfrac{\partial v}{\partial \theta} \end{Vmatrix} drd\theta = \begin{Vmatrix} \cos\theta & -r\sin\theta \\ \sin\theta & r\cos\theta \end{Vmatrix} drd\theta = r\, drd\theta$

$$\therefore \int_{-\infty}^{\infty}\int_{-\infty}^{\infty} e^{-(u^2+v^2)} du\, dv = \int_0^{2\pi}\int_0^{\infty} e^{-r^2} r\, drd\theta = \int_0^{2\pi} d\theta \int_0^{\infty} e^{-r^2} r\, dr = 2\pi \cdot \dfrac{-e^{-r^2}}{2}\bigg|_{r=0}^{r=\infty} = \pi$$

$$\therefore \int_{-\infty}^{\infty}\int_{-\infty}^{\infty} e^{-(u^2+v^2)} \dfrac{2}{\sqrt{4ac-b^2}} dudv = \dfrac{2\pi}{\sqrt{4ac-b^2}}$$

(2)

(a) $\because x^2 + 2xy + 3y^2 = (x+y)^2 + 2y^2$

Let $(x+y) = u$, $\sqrt{2}y = v$ then $x = u - \dfrac{v}{\sqrt{2}}$, $y = \dfrac{v}{\sqrt{2}}$

and $dxdy = \begin{Vmatrix} \dfrac{\partial x}{\partial u} & \dfrac{\partial x}{\partial v} \\ \dfrac{\partial y}{\partial u} & \dfrac{\partial y}{\partial v} \end{Vmatrix} dudv = \begin{Vmatrix} 1 & \dfrac{-1}{\sqrt{2}} \\ 0 & \dfrac{1}{\sqrt{2}} \end{Vmatrix} dudv = \sqrt{\dfrac{1}{2}} dudv$

$$\therefore \iint_R e^{-(x^2+2xy+3y^2)} dxdy = \iint_{u^2+v^2 \le 1} e^{-(u^2+v^2)} \sqrt{\dfrac{1}{2}} dudv$$

Let $u = r\cos\theta$, $v = r\sin\theta$ then $\{(u,v): u^2+v^2 \le 1\} = \{(r,\theta): 0 \le r \le 1, 0 \le \theta \le 2\pi\}$

and $dudv = \begin{Vmatrix} \dfrac{\partial u}{\partial r} & \dfrac{\partial u}{\partial \theta} \\ \dfrac{\partial v}{\partial r} & \dfrac{\partial v}{\partial \theta} \end{Vmatrix} drd\theta = \begin{Vmatrix} \cos\theta & -r\sin\theta \\ \sin\theta & r\cos\theta \end{Vmatrix} drd\theta = r\, drd\theta$

$$\therefore \iint_{u^2+v^2\le 1} e^{-(u^2+v^2)} \sqrt{\dfrac{1}{2}} dudv = \sqrt{\dfrac{1}{2}} \int_0^{2\pi}\int_0^1 e^{-r^2} r\, drd\theta = \sqrt{\dfrac{1}{2}} \int_0^{2\pi} d\theta \int_0^1 e^{-r^2} r\, dr$$

8.3.4 Use Generalized Coordinates to Compute Double Integrals

$$= \sqrt{\frac{1}{2}} \cdot 2\pi \cdot \frac{-e^{-r^2}}{2}\Big|_{r=0}^{r=1} = \frac{\pi}{\sqrt{2}}(1-e^{-1})$$

(b)

$$\therefore I(t) = \sqrt{\frac{1}{2}} \int_0^{2\pi} d\theta \int_0^t e^{-r^2} r\, dr = \sqrt{\frac{1}{2}} \cdot 2\pi \cdot \frac{-e^{-r^2}}{2}\Big|_{r=0}^{r=t} = \frac{\pi}{\sqrt{2}}(1-e^{-t})$$

$$\therefore \lim_{t\to\infty} I(t) = \lim_{t\to\infty} \frac{\pi}{\sqrt{2}}(1-e^{-t}) = \frac{\pi}{\sqrt{2}}$$

Example 7.

Find $\iint_R e^{-(x^2+xy+y^2)} dxdy = ?$, $R = \{(x,y): x^2+xy+y^2 \leq 4\}$.

【Solution】

$\because x^2 + xy + y^2 = \left(x + \frac{y}{2}\right)^2 + \frac{3}{4}y^2$, let $x + \frac{y}{2} = u$, $\sqrt{\frac{3}{4}}y = v$ then

$$dxdy = \begin{Vmatrix} \dfrac{\partial x}{\partial u} & \dfrac{\partial x}{\partial v} \\ \dfrac{\partial y}{\partial u} & \dfrac{\partial y}{\partial v} \end{Vmatrix} dudv = \begin{Vmatrix} 1 & 0 \\ 2 & \sqrt{\dfrac{4}{3}} \end{Vmatrix} dudv = \frac{2}{\sqrt{3}} dudv$$

$$\therefore \iint_R e^{-(x^2+xy+y^2)} dxdy = \iint_{u^2+v^2\leq 4} e^{-(u^2+v^2)} \frac{2}{\sqrt{3}} dudv$$

Let $u = r\cos\theta$, $v = r\sin\theta$ then

$\{(u,v): u^2+v^2 \leq 4\} = \{(r,\theta): 0 \leq r \leq 2, 0 \leq \theta \leq 2\pi\}$

$$\text{and } dudv = \begin{Vmatrix} \dfrac{\partial u}{\partial r} & \dfrac{\partial u}{\partial \theta} \\ \dfrac{\partial v}{\partial r} & \dfrac{\partial v}{\partial \theta} \end{Vmatrix} drd\theta = \begin{Vmatrix} \cos\theta & -r\sin\theta \\ \sin\theta & r\cos\theta \end{Vmatrix} drd\theta = r\,drd\theta$$

$$\therefore \iint_{u^2+v^2\leq 4} e^{-(u^2+v^2)} dudv = \int_0^{2\pi}\int_0^2 e^{-r^2} r\,drd\theta = \int_0^{2\pi} d\theta \int_0^2 e^{-r^2} r\, dr = 2\pi \cdot \frac{-e^{-r^2}}{2}\Big|_{r=0}^{r=2}$$

$= \pi(1-e^{-4})$

$$\therefore \iint_R e^{-(x^2+xy+y^2)} dxdy = \frac{2\pi(1-e^{-4})}{\sqrt{3}}$$

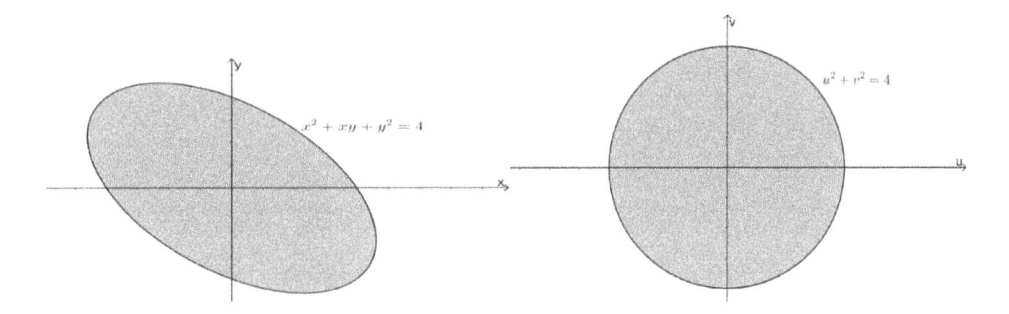

Example 8.

Find $\iint_R dA = ?$, $R = \{(x,y): ax^2 + bxy + cy^2 \leq \alpha^2\}$.

【Solution】

$\because ax^2 + bxy + cy^2 = a\left(x + \dfrac{by}{2a}\right)^2 + \dfrac{4ac - b^2}{4a}y^2$

Let $\sqrt{a}\left(x + \dfrac{by}{2a}\right) = u$, $\sqrt{\dfrac{4ac - b^2}{4a}}\, y = v$ then

$dxdy = \begin{Vmatrix} \dfrac{\partial x}{\partial u} & \dfrac{\partial x}{\partial v} \\ \dfrac{\partial y}{\partial u} & \dfrac{\partial y}{\partial v} \end{Vmatrix} dudv = \begin{Vmatrix} \dfrac{1}{\sqrt{a}} & 0 \\ \dfrac{2\sqrt{a}}{b} & \sqrt{\dfrac{4a}{4ac - b^2}} \end{Vmatrix} dudv = \dfrac{2}{\sqrt{4ac - b^2}} dudv$

$\therefore \iint_R dA = \iint_{u^2 + v^2 \leq \alpha^2} \dfrac{2}{\sqrt{4ac - b^2}} dudv$

Let $u = r\cos\theta$, $v = r\sin\theta$ then $\{(u,v): u^2 + v^2 \leq \alpha^2\} = \{(r,\theta): 0 \leq r \leq \alpha, 0 \leq \theta \leq 2\pi\}$

and $dudv = \begin{Vmatrix} \dfrac{\partial u}{\partial r} & \dfrac{\partial u}{\partial \theta} \\ \dfrac{\partial v}{\partial r} & \dfrac{\partial v}{\partial \theta} \end{Vmatrix} drd\theta = \begin{Vmatrix} \cos\theta & -r\sin\theta \\ \sin\theta & r\cos\theta \end{Vmatrix} drd\theta = rdrd\theta$

$\therefore \iint_{u^2 + v^2 \leq \alpha^2} \dfrac{2}{\sqrt{4ac - b^2}} dudv = \int_0^{2\pi} \int_0^{\alpha} \dfrac{2}{\sqrt{4ac - b^2}} rdrd\theta = \dfrac{2}{\sqrt{4ac - b^2}} \int_0^{2\pi} d\theta \int_0^{\alpha} rdr$

$= \dfrac{2\pi\alpha^2}{\sqrt{4ac - b^2}}$

8.3.4 Use Generalized Coordinates to Compute Double Integrals

Example 9.

Find $\iint_R \dfrac{x^2 - xy + y^2}{2} dA = ?$, $R = \{(x,y): x^2 - xy + y^2 \leq 2\}$

【Solution】

$\because \dfrac{x^2 - xy + y^2}{2} = \dfrac{1}{2}\left(x - \dfrac{y}{2}\right)^2 + \dfrac{3}{8}y^2$, let $\sqrt{\dfrac{1}{2}}\left(x - \dfrac{y}{2}\right) = u$, $\sqrt{\dfrac{3}{8}}y = v$ then

$dxdy = \begin{Vmatrix} \dfrac{\partial x}{\partial u} & \dfrac{\partial x}{\partial v} \\ \dfrac{\partial y}{\partial u} & \dfrac{\partial y}{\partial v} \end{Vmatrix} dudv = \begin{Vmatrix} \sqrt{2} & 0 \\ -2\sqrt{2} & \sqrt{\dfrac{8}{3}} \end{Vmatrix} dudv = \dfrac{4}{\sqrt{3}} dudv$

$\therefore \iint_R \dfrac{x^2 - xy + y^2}{2} dA = \iint_{u^2+v^2 \leq 1} (u^2 + v^2) \dfrac{4}{\sqrt{3}} dudv$

Let $u = r\cos\theta$, $v = r\sin\theta$ then $\{(u,v): u^2 + v^2 \leq 1\} = \{(r,\theta): 0 \leq r \leq 1, 0 \leq \theta \leq 2\pi\}$

and $dudv = \begin{Vmatrix} \dfrac{\partial u}{\partial r} & \dfrac{\partial u}{\partial \theta} \\ \dfrac{\partial v}{\partial r} & \dfrac{\partial v}{\partial \theta} \end{Vmatrix} drd\theta = \begin{Vmatrix} \cos\theta & -r\sin\theta \\ \sin\theta & r\cos\theta \end{Vmatrix} drd\theta = rdrd\theta$

$\therefore \iint_{u^2+v^2 \leq 1} (u^2 + v^2) \dfrac{4}{\sqrt{3}} dudv = \dfrac{4}{\sqrt{3}} \int_0^{2\pi} \int_0^1 r^2 \cdot rdrd\theta = \dfrac{4}{\sqrt{3}} \int_0^{2\pi} d\theta \int_0^1 r^3 dr$

$= \dfrac{4}{\sqrt{3}} \cdot 2\pi \cdot \dfrac{r^4}{4}\Big|_{r=0}^{r=1} = \dfrac{2\pi}{\sqrt{3}}$

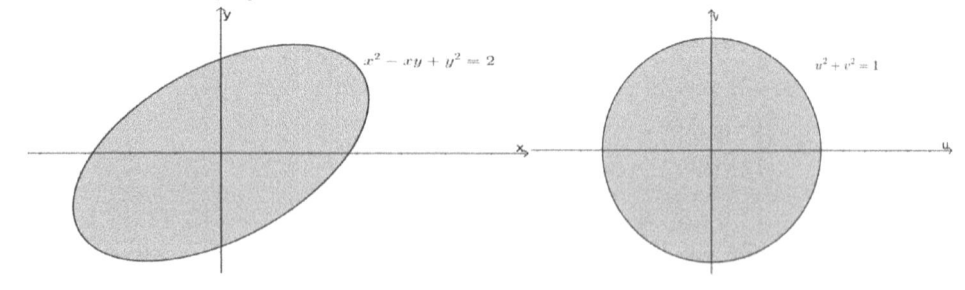

Example 10.

(1) Find the area enclosed by the lines

$x - 2y = -4$, $x - 2y = 1$, $2x - y = 0$, and $2x - y = 3$.

(2) Find the area enclosed by the lines
$$x - my = 0, \quad x - my = 1, \quad nx - y = 0, \quad \text{and} \quad nx - y = 1 \text{ where } m > n > 1.$$

(3) Find $\iint_R \dfrac{x - 2y}{3x - y} dxdy = ?$, where R is the region enclosed by the lines:
$$x - 2y = 0, x - 2y = 4, 3x - y = 1, \text{ and } 3x - y = 7.$$

【Solution】

(1)

Let $x - 2y = u$, $2x - y = v$ then $x = -\dfrac{u}{3} + \dfrac{2v}{3}$, $y = -\dfrac{2u}{3} + \dfrac{v}{3}$

and $dxdy = \begin{Vmatrix} \dfrac{\partial x}{\partial u} & \dfrac{\partial x}{\partial v} \\ \dfrac{\partial y}{\partial u} & \dfrac{\partial y}{\partial v} \end{Vmatrix} dudv = \begin{Vmatrix} \dfrac{-1}{3} & \dfrac{2}{3} \\ \dfrac{-2}{3} & \dfrac{1}{3} \end{Vmatrix} dudv = \dfrac{1}{3} dudv$

Let $R = \{(x, y): -4 \leq x - 2y \leq 1, 0 \leq 2x - y \leq 3\}$ then $R = \{(u, v): -4 \leq u \leq 1, 0 \leq v \leq 3\}$

$\therefore$ area $= \iint_R 1 dA = \int_{-4}^{1} \int_0^3 \dfrac{1}{3} dvdu = 5$

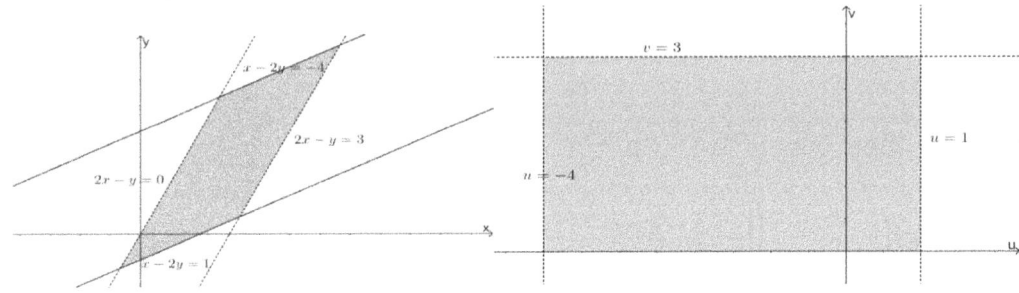

(2)

Let $x - my = u$, $nx - y = v$ then $x = \dfrac{mv - u}{mn - 1}$, $y = \dfrac{v - nu}{mn - 1}$

and $dxdy = \begin{Vmatrix} \dfrac{\partial x}{\partial u} & \dfrac{\partial x}{\partial v} \\ \dfrac{\partial y}{\partial u} & \dfrac{\partial y}{\partial v} \end{Vmatrix} dudv = \begin{Vmatrix} \dfrac{-1}{mn-1} & \dfrac{m}{mn-1} \\ \dfrac{-n}{mn-1} & \dfrac{1}{mn-1} \end{Vmatrix} dudv = \dfrac{1}{mn - 1} dudv$

Let $R = \{(x, y): 0 \leq x - my \leq 1, 0 \leq nx - y \leq 1\}$ then $R = \{(u, v): 0 \leq u \leq 1, 0 \leq v \leq 1\}$

$\therefore$ area $= \iint_R 1 dA = \int_0^1 \int_0^1 \dfrac{1}{mn - 1} dudv = \dfrac{1}{mn - 1}$

(3)

Let $x - 2y = u$, $3x - y = v$ then $x = \dfrac{-u + 2v}{5}$, $y = \dfrac{-3u + v}{5}$

8.3.4 Use Generalized Coordinates to Compute Double Integrals

and $dxdy = \begin{Vmatrix} \dfrac{\partial x}{\partial u} & \dfrac{\partial x}{\partial v} \\ \dfrac{\partial y}{\partial u} & \dfrac{\partial y}{\partial v} \end{Vmatrix} dudv = \begin{Vmatrix} \dfrac{-1}{5} & \dfrac{2}{5} \\ \dfrac{-3}{5} & \dfrac{1}{5} \end{Vmatrix} dudv = \dfrac{1}{5} dudv$

Let $R = \{(x,y): 0 \le x - 2y \le 4, 1 \le 3x - y \le 7\}$ then $R = \{(u,v): 0 \le u \le 4, 1 \le v \le 7\}$

$\therefore \iint_R \dfrac{x-2y}{3x-y} dxdy = \dfrac{1}{5}\int_1^7 \int_0^4 \dfrac{u}{v} dudv = \dfrac{8}{5}\int_1^7 \dfrac{1}{v} dv = \dfrac{8\ln 7}{5}$

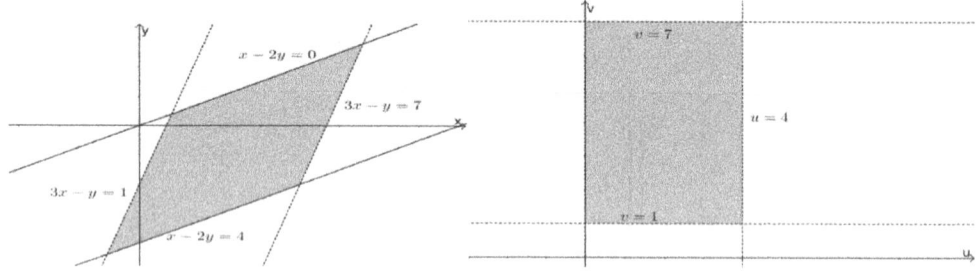

Example 11.

Find the area enclosed by the curves $y = x^2$, $y = 3x^2$, $x = y^2$, $x = 4y^2$.

【Solution】

Let $u = \dfrac{y}{x^2}$, $v = \dfrac{x}{y^2}$ then $x = u^{-\frac{2}{3}}v^{-\frac{1}{3}}$, $y = u^{-\frac{1}{3}}v^{-\frac{2}{3}}$

and $dxdy = \begin{Vmatrix} \dfrac{\partial x}{\partial u} & \dfrac{\partial x}{\partial v} \\ \dfrac{\partial y}{\partial u} & \dfrac{\partial y}{\partial v} \end{Vmatrix} dudv = \begin{Vmatrix} \dfrac{-2u^{-\frac{5}{3}}v^{-\frac{1}{3}}}{3} & \dfrac{-u^{-\frac{2}{3}}v^{-\frac{4}{3}}}{3} \\ \dfrac{-u^{-\frac{4}{3}}v^{-\frac{2}{3}}}{3} & \dfrac{-2u^{-\frac{1}{3}}v^{-\frac{5}{3}}}{3} \end{Vmatrix} dudv = \dfrac{u^{-2}v^{-2}}{3} dudv$

Let $R = \left\{(x,y): 1 \le \dfrac{y}{x^2} \le 3, 1 \le \dfrac{x}{y^2} \le 4\right\}$ then $R = \{(u,v): 1 \le u \le 3, 1 \le v \le 4\}$

$\therefore$ area $= \iint_R 1 dA = \int_1^3 \int_1^4 \dfrac{u^{-2}v^{-2}}{3} dvdu = \dfrac{1}{6}$

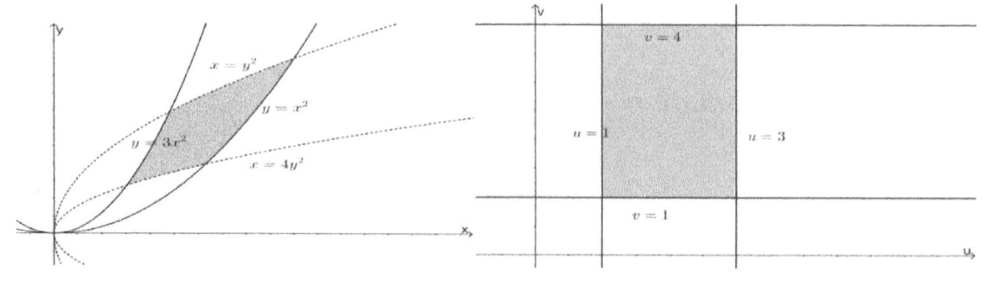

Example 12.

Find the area enclosed by: $x^2 - 4xy + 4y^2 - 2x - y - 1 = 0$ and $y = \frac{2}{5}$.

【Solution】

$\because x^2 - 4xy + 4y^2 - 2x - y - 1 = (x - 2y)^2 - (2x + y) - 1 = 0$

Let $u = x - 2y$ and $v = 2x + y$ then $x = \frac{u}{5} + \frac{2v}{5}$, $y = -\frac{2u}{5} + \frac{v}{5}$

and $dxdy = \begin{Vmatrix} \frac{\partial x}{\partial u} & \frac{\partial x}{\partial v} \\ \frac{\partial y}{\partial u} & \frac{\partial y}{\partial v} \end{Vmatrix} dudv = \begin{Vmatrix} \frac{1}{5} & \frac{2}{5} \\ -\frac{2}{5} & \frac{1}{5} \end{Vmatrix} dudv = \frac{1}{5} dudv$

Let $R = \{(u,v): -1 \leq u \leq 3, u^2 - 1 \leq v \leq 2u + 2\}$

$\therefore$ area $= \iint_R 1 dA = \int_{-1}^{3} \int_{u^2-1}^{2u+2} \frac{1}{5} dv du = \frac{32}{15}$

Example 13.

Find $\iint_R (x - y)^2 \sin^2(x + y) \, dxdy =?$, where R is the region enclosed by the points:
$(\pi, 0)$, $(2\pi, \pi)$, $(\pi, 2\pi)$, $(0, \pi)$.

【Solution】

Let $x + y = u$, $x - y = v$ then $x = \frac{u+v}{2}$, $y = \frac{u-v}{2}$

and $dxdy = \begin{Vmatrix} \frac{\partial x}{\partial u} & \frac{\partial x}{\partial v} \\ \frac{\partial y}{\partial u} & \frac{\partial y}{\partial v} \end{Vmatrix} dudv = \begin{Vmatrix} \frac{1}{2} & \frac{1}{2} \\ \frac{1}{2} & -\frac{1}{2} \end{Vmatrix} dudv = \frac{1}{2} dudv$

$\because R = \{(x,y): \pi \leq x + y \leq 3\pi, -\pi \leq x - y \leq \pi\}$ $\therefore R = \{(u,v): \pi \leq u \leq 3\pi, -\pi \leq v \leq \pi\}$

$\therefore \iint_R (x-y)^2 \sin^2(x+y) \, dxdy = \frac{1}{2} \int_{-\pi}^{\pi} \int_{\pi}^{3\pi} v^2 \sin^2 u \, dudv = \frac{1}{2} \int_{\pi}^{3\pi} \sin^2 u \, du \int_{-\pi}^{\pi} v^2 \, dv$

$= \frac{1}{2} \int_{\pi}^{3\pi} \frac{1 - \cos 2u}{2} du \int_{-\pi}^{\pi} v^2 \, dv = \frac{1}{2} \left(\frac{u}{2} - \frac{\sin 2u}{4}\right)\Big|_{\pi}^{3\pi} \cdot \frac{v^3}{3}\Big|_{-\pi}^{\pi} = \frac{\pi^4}{3}$

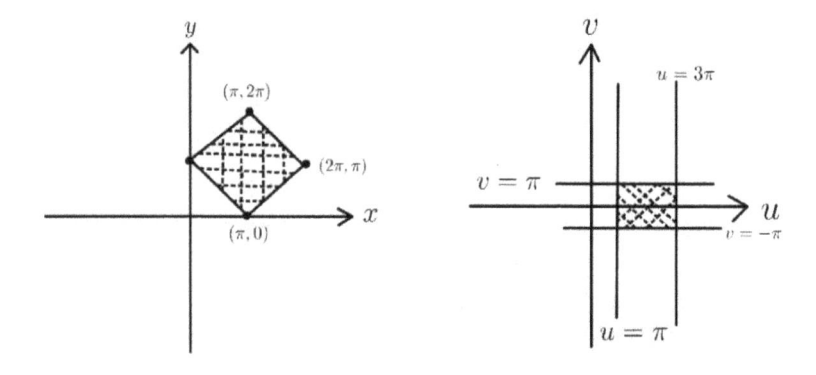

8.3.4 Use Generalized Coordinates to Compute Double Integrals

Example 14.

Find $\iint_R \sqrt{\dfrac{y-x}{x+y}}\, dxdy = ?$, where R is the region enclosed by the lines: $x = 0$, $y = x$, $x + y = 1$, and $x + y = 2$.

【Solution】

Let $y - x = u,\ x + y = v$ then $x = \dfrac{v-u}{2},\ y = \dfrac{u+v}{2}$

and $dxdy = \begin{Vmatrix} \dfrac{\partial x}{\partial u} & \dfrac{\partial x}{\partial v} \\ \dfrac{\partial y}{\partial u} & \dfrac{\partial y}{\partial v} \end{Vmatrix} dudv = \begin{Vmatrix} -\dfrac{1}{2} & \dfrac{1}{2} \\ \dfrac{1}{2} & \dfrac{1}{2} \end{Vmatrix} dudv = \dfrac{1}{2} dudv$

$\because R = \{(x,y): y \geq x, x \geq 0, 1 \leq x + y \leq 2\} \quad \therefore R = \{(x,y): 0 \leq u \leq v, 1 \leq v \leq 2\}$

$\therefore \iint_R \sqrt{\dfrac{y-x}{x+y}}\, dxdy = \dfrac{1}{2}\int_1^2 \int_0^v \sqrt{\dfrac{u}{v}}\, dudv = \dfrac{1}{2}\int_1^2 \sqrt{\dfrac{1}{v}} \cdot \dfrac{2u^{\frac{3}{2}}}{3}\bigg|_0^v dv = \dfrac{1}{3}\int_1^2 v\, dv = \dfrac{1}{2}$

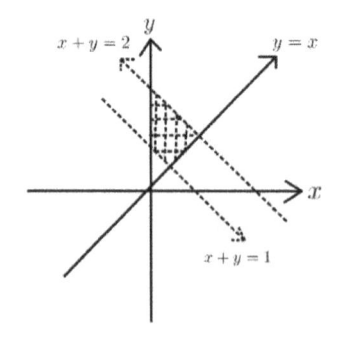

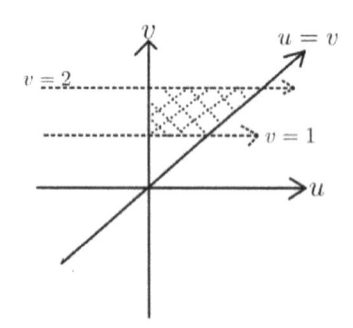

Example 15.

Find $\iint_R (x+y)^4 \, dxdy = ?$, where R is the region enclosed by the points $(1,0)$, $(1,3)$, $(2,2)$, and $(0,1)$.

【Solution】

Let $x + y = u$, $2x - y = v$ then $x = \dfrac{u+v}{3}$, $y = \dfrac{2u-v}{3}$

and $dxdy = \begin{Vmatrix} \dfrac{\partial x}{\partial u} & \dfrac{\partial x}{\partial v} \\ \dfrac{\partial y}{\partial u} & \dfrac{\partial y}{\partial v} \end{Vmatrix} dudv = \begin{Vmatrix} \dfrac{1}{3} & \dfrac{1}{3} \\ \dfrac{2}{3} & -\dfrac{1}{3} \end{Vmatrix} dudv = \dfrac{1}{3} dudv$

$\because R = \{(x,y): 1 \leq x+y \leq 4, -1 \leq 2x - y \leq 2\}$ $\therefore R = \{(u,v): 1 \leq u \leq 4, -1 \leq v \leq 2\}$

$\therefore \iint_R (x+y)^4 \, dxdy = \dfrac{1}{3} \int_{-1}^{2} \int_{1}^{4} u^4 \, dudv = \dfrac{u^5}{5}\bigg|_1^4 = \dfrac{1023}{5}$

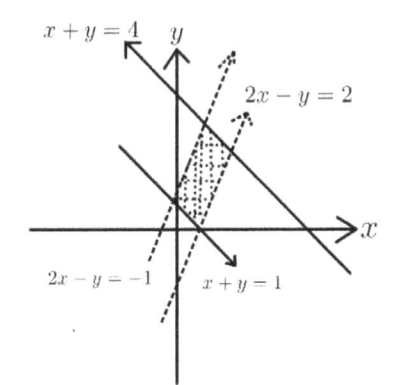

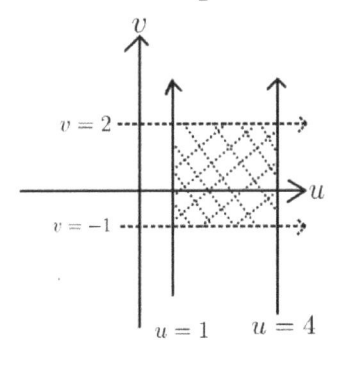

Example 16.

Find $\iint_R (\sqrt{x} + \sqrt{y}) \, dxdy = ?$, where R is the region enclosed by $\sqrt{x} + \sqrt{y} = 1$, $x = 0$, and $y = 0$.

【Solution】

Let $x = r\cos^4\theta$, $y = r\sin^4\theta$ then

8.3.4 Use Generalized Coordinates to Compute Double Integrals

$$dxdy = \begin{Vmatrix} \dfrac{\partial x}{\partial r} & \dfrac{\partial x}{\partial \theta} \\ \dfrac{\partial y}{\partial r} & \dfrac{\partial y}{\partial \theta} \end{Vmatrix} drd\theta = \begin{Vmatrix} \cos^4\theta & -4r\cos^3\theta\sin\theta \\ \sin^4\theta & 4r\sin^3\theta\cos\theta \end{Vmatrix} drd\theta = 4r\sin^3\theta\cos^3\theta\, drd\theta$$

$\because R = \{(x,y): x \geq 0,\ y \geq 0, \sqrt{x}+\sqrt{y} \leq 1\} = \{(r,\theta): 0 \leq r \leq 1, 0 \leq \theta \leq \dfrac{\pi}{2}\}$

$\therefore \iint_R (\sqrt{x}+\sqrt{y})\, dxdy = \int_0^{\frac{\pi}{2}} \int_0^1 \sqrt{r}(4r\sin^3\theta\cos^3\theta)\, drd\theta = \int_0^1 4r^{\frac{3}{2}}\, dr \int_0^{\frac{\pi}{2}} \sin^3\theta\cos^3\theta\, d\theta$

$= \int_0^1 4r^{\frac{3}{2}}\, dr \int_0^{\frac{\pi}{2}} \left(\dfrac{\sin 2\theta}{2}\right)^3 d\theta = \dfrac{2}{15}$

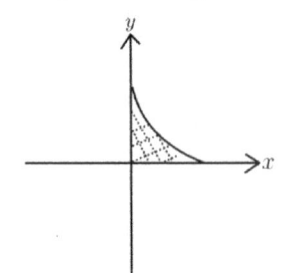

Example 17.

Find $\iint_R x^2+y^2\, dxdy = ?$, where R is the region enclosed by $x^2-y^2 = 1$, $x^2-y^2 = 9$, $xy = 2$, and $xy = 4$.

【Solution】

Let $x^2-y^2 = u,\ 2xy = v$ then

$$dudv = \begin{Vmatrix} \dfrac{\partial u}{\partial x} & \dfrac{\partial u}{\partial y} \\ \dfrac{\partial v}{\partial x} & \dfrac{\partial v}{\partial y} \end{Vmatrix} dxdy = \begin{Vmatrix} 2x & -2y \\ 2y & 2x \end{Vmatrix} dxdy = 4(x^2+y^2)dxdy$$

$\because (x^2+y^2)^2 = (x^2-y^2)^2 + (2xy)^2 = u^2+v^2\quad \therefore x^2+y^2 = (u^2+v^2)^{\frac{1}{2}}$

$\therefore dxdy = \dfrac{dudv}{4(u^2+v^2)^{\frac{1}{2}}}$

$\because R = \{(x,y): 1 \leq x^2-y^2 \leq 9, 4 \leq 2xy \leq 8\}\quad \therefore R = \{(u,v): 1 \leq u \leq 9, 4 \leq v \leq 8\}$

$$\therefore \iint_R x^2 + y^2 \, dxdy = \int_4^8 \int_1^9 \frac{(u^2+v^2)^{\frac{1}{2}}}{4(u^2+v^2)^{\frac{1}{2}}} dudv = 8$$

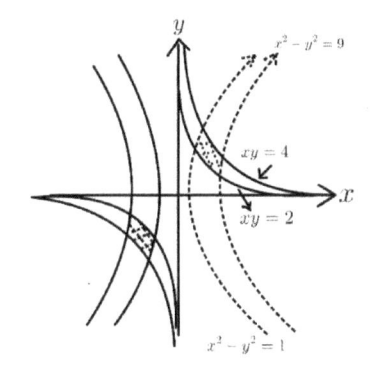

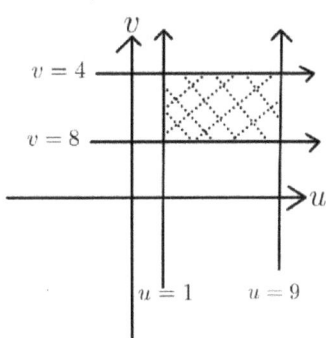

Example 18.

(1) Find $\iint_R x^2 + y^2 dA = ?$, where R is the region enclosed by the lines $-x = y$, $y = -x + 3$, $x - 3 = y$, and $y = x$.

(2) Find $\iint_R xy \, dA = ?$, where R is the region enclosed in the first quadrant by the curves $x^2 + y^2 = a$, $x^2 + y^2 = b$, $x^2 - y^2 = c$, $x^2 - y^2 = d$, $(b > a > 0, d > c > 0)$.

【Solution】

(1)

Let $x + y = u$, $x - y = v$ then $x = \dfrac{u+v}{2}$, $y = \dfrac{u-v}{2}$

and $dxdy = \begin{Vmatrix} \dfrac{\partial x}{\partial u} & \dfrac{\partial x}{\partial v} \\ \dfrac{\partial y}{\partial u} & \dfrac{\partial y}{\partial v} \end{Vmatrix} dudv = \begin{Vmatrix} \dfrac{1}{2} & \dfrac{1}{2} \\ \dfrac{1}{2} & \dfrac{-1}{2} \end{Vmatrix} dudv = \dfrac{1}{2} dudv$

$\because R = \{(x,y): 0 \le x + y \le 3, 0 \le x - y \le 3\}$ $\therefore R = \{(u,v): 0 \le u \le 3, 0 \le v \le 3\}$

$\therefore \iint_R x^2 + y^2 dA = \int_0^3 \int_0^3 \left(\left(\dfrac{u+v}{2}\right)^2 + \left(\dfrac{u-v}{2}\right)^2\right) \dfrac{1}{2} dudv = \dfrac{1}{4}\int_0^3 \int_0^3 u^2 + v^2 dudv = \dfrac{27}{2}$

8.3.4 Use Generalized Coordinates to Compute Double Integrals

(2)

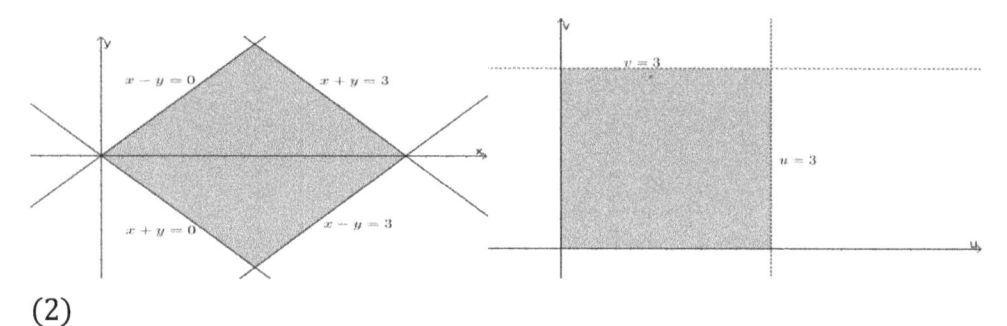

Let $x^2 + y^2 = u$, $x^2 - y^2 = v$ then $x = \dfrac{(u+v)^{\frac{1}{2}}}{\sqrt{2}}$, $y = \dfrac{(u-v)^{\frac{1}{2}}}{\sqrt{2}}$

and $dxdy = \begin{Vmatrix} \dfrac{\partial x}{\partial u} & \dfrac{\partial x}{\partial v} \\ \dfrac{\partial y}{\partial u} & \dfrac{\partial y}{\partial v} \end{Vmatrix} dudv = \begin{Vmatrix} \dfrac{(u+v)^{-\frac{1}{2}}}{2\sqrt{2}} & \dfrac{(u+v)^{-\frac{1}{2}}}{2\sqrt{2}} \\ \dfrac{(u-v)^{-\frac{1}{2}}}{2\sqrt{2}} & \dfrac{(u-v)^{-\frac{1}{2}}}{2\sqrt{2}} \end{Vmatrix} dudv = \dfrac{(u^2-v^2)^{-\frac{1}{2}}}{4} dudv$

$\because R = \{(x,y): a \le x^2+y^2 \le b, c \le x^2-y^2 \le d\}$ $\therefore R = \{(u,v): a \le u \le b, c \le v \le d\}$

$\therefore \iint_R xy\,dA = \int_c^d \int_a^b \dfrac{(u^2-v^2)^{\frac{1}{2}}}{2} \cdot \dfrac{(u^2-v^2)^{-\frac{1}{2}}}{4} dudv = \dfrac{(b-a)(c-d)}{8}$

Example 19.

Find $\displaystyle\int_0^{\frac{1}{2}} \int_0^{1-2y} e^{\frac{x}{x+2y}} dxdy = ?$

【Solution】

Let $x = u$, $x + 2y = v$ then $x = u$, $y = \dfrac{-u+v}{2}$

$dxdy = \begin{Vmatrix} \dfrac{\partial x}{\partial u} & \dfrac{\partial x}{\partial v} \\ \dfrac{\partial y}{\partial u} & \dfrac{\partial y}{\partial v} \end{Vmatrix} dudv = \begin{Vmatrix} 1 & 0 \\ -\dfrac{1}{2} & \dfrac{1}{2} \end{Vmatrix} dudv = \dfrac{1}{2} dudv$

Let $R = \{(x,y): 0 \le x \le 1-2y, 0 \le y \le \dfrac{1}{2}\}$ then $R = \{(u,v): 0 \le u \le v, 0 \le v \le 1\}$

$\therefore \displaystyle\int_0^{\frac{1}{2}} \int_0^{1-2y} e^{\frac{x}{x+2y}} dxdy = \dfrac{1}{2}\int_0^1 \int_0^v e^{\frac{u}{v}} dudv = \dfrac{1}{2}\int_0^1 v e^{\frac{u}{v}}\Big|_{u=0}^{u=v} dv = \dfrac{1}{2}\int_0^1 v(e-1)\,dv = \dfrac{e-1}{4}$

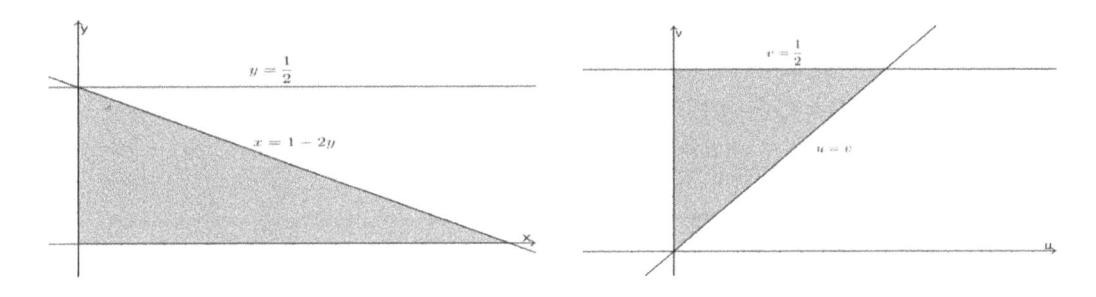

8.3.4 Use Generalized Coordinates to Compute Double Integrals

Example 20.

Find $\iint_R 3xy\,dA = ?$, where R is the region enclosed by the lines $x - 2y = 0$, $x - 2y = -4$, $x + y = 4$, and $x + y = 1$.

【Solution】

Let $x - 2y = u$, $x + y = v$ then $x = \dfrac{u + 2v}{3}$, $y = \dfrac{-u + v}{3}$

and $dxdy = \begin{Vmatrix} \dfrac{\partial x}{\partial u} & \dfrac{\partial x}{\partial v} \\ \dfrac{\partial y}{\partial u} & \dfrac{\partial y}{\partial v} \end{Vmatrix} dudv = \begin{Vmatrix} \dfrac{1}{3} & \dfrac{2}{3} \\ \dfrac{-1}{3} & \dfrac{1}{3} \end{Vmatrix} dudv = \dfrac{1}{3}dudv$

$\because R = \{(x, y): -4 \le x - 2y \le 0, 1 \le x + y \le 4\}$ $\therefore R = \{(u, v): -4 \le u \le 0, 1 \le v \le 4\}$

$\therefore \iint_R 3xy\,dxdy = \int_1^4 \int_{-4}^0 \left(\dfrac{u + 2v}{3}\right)\left(\dfrac{-u + v}{3}\right) dudv = \dfrac{164}{9}$

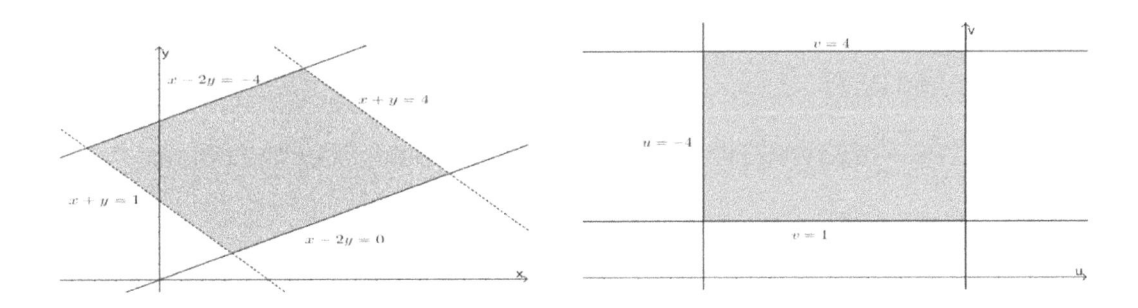

Example 21.

Find $\iint_R (2x + y)dxdy = ?$, where R is the region enclosed by the curves $x^2 - 2xy + y^2 + x + y = 0$ and $x + y + 4 = 0$.

8.3.4 Use Generalized Coordinates to Compute Double Integrals

【Solution】

$\because x^2 - 2xy + y^2 + x + y = 0 \Leftrightarrow (x-y)^2 = -(x+y)$

Let $x + y = u$, $x - y = v$ then $x = \dfrac{u+v}{2}$, $y = \dfrac{u-v}{2}$

and $dxdy = \left\| \begin{matrix} \dfrac{\partial x}{\partial u} & \dfrac{\partial x}{\partial v} \\ \dfrac{\partial y}{\partial u} & \dfrac{\partial y}{\partial v} \end{matrix} \right\| dudv = \left\| \begin{matrix} \dfrac{1}{2} & \dfrac{1}{2} \\ \dfrac{1}{2} & \dfrac{1}{2} \end{matrix} \right\| dudv = \dfrac{1}{2} dudv$

$\because R = \{(u,v): -v^2 \le u \le 0, -2 \le v \le 2\}$

$\therefore \iint_R (2x + y) dxdy = \int_{-2}^{2} \int_{-v^2}^{0} 3u + v \, dudv = -\dfrac{96}{5}$

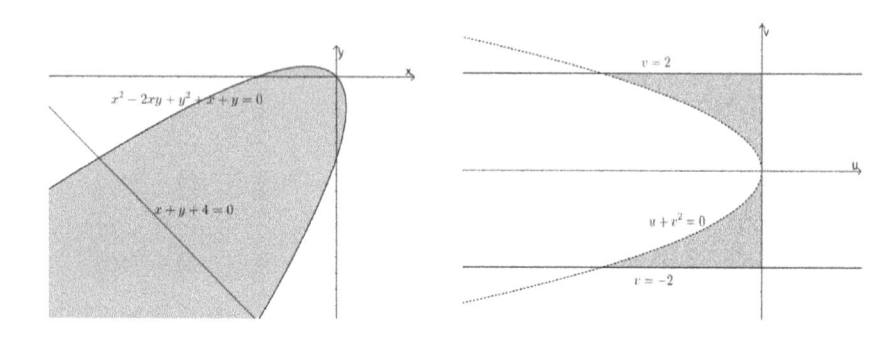

Example 22.

(1) Find $\iint_R \cos \dfrac{y-x}{x+y} dxdy =?$, where R is the region enclosed by the points
(1,0), (3,0), (0,1), and (0,3).

(2) Find $\iint_R \sec^2 \dfrac{y-x}{x+y} dxdy =?$, where R is the region enclosed by the points
(1,0), (3,0), (0,1), and (0,3).

【Solution】

(1)

Let $y - x = u$, $x + y = v$ then $x = \dfrac{-u+v}{2}$, $y = \dfrac{u+v}{2}$

and $dxdy = \begin{Vmatrix} \dfrac{\partial x}{\partial u} & \dfrac{\partial x}{\partial v} \\ \dfrac{\partial y}{\partial u} & \dfrac{\partial y}{\partial v} \end{Vmatrix} dudv = \begin{Vmatrix} -\dfrac{1}{2} & \dfrac{1}{2} \\ \dfrac{1}{2} & \dfrac{1}{2} \end{Vmatrix} dudv = \dfrac{1}{2} dudv$

$\because R = \{(x,y): x \geq 0, y \geq 0, 1 \leq x + y \leq 3\}$ $\therefore R = \{(u,v): -v \leq u \leq v, 1 \leq v \leq 3\}$

$\therefore \iint_R \cos\dfrac{y-x}{x+y} dxdy = \dfrac{1}{2}\int_1^3 \int_{-v}^{v} \cos\dfrac{u}{v} dudv = \dfrac{1}{2}\int_1^3 v\sin\dfrac{u}{v}\Big|_{u=-v}^{u=v} dv = \sin 1 \int_1^3 v\, dv$
$= 4\sin 1$

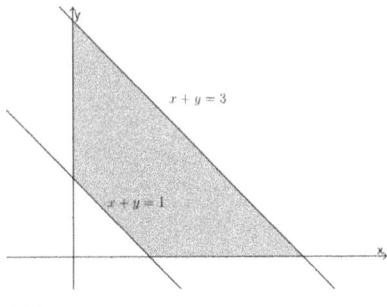

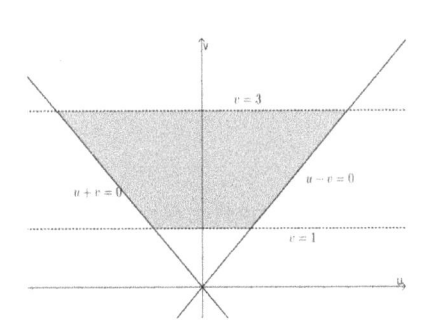

(2)

Let $y - x = u$, $x + y = v$ then $x = \dfrac{-u+v}{2}$, $y = \dfrac{u+v}{2}$

and $dxdy = \begin{Vmatrix} \dfrac{\partial x}{\partial u} & \dfrac{\partial x}{\partial v} \\ \dfrac{\partial y}{\partial u} & \dfrac{\partial y}{\partial v} \end{Vmatrix} dudv = \begin{Vmatrix} -\dfrac{1}{2} & \dfrac{1}{2} \\ \dfrac{1}{2} & \dfrac{1}{2} \end{Vmatrix} dudv = \dfrac{1}{2} dudv$

$\because R = \{(x,y): x \geq 0, y \geq 0, 1 \leq x + y \leq 3\}$ $\therefore R = \{(u,v): -v \leq u \leq v, 1 \leq v \leq 3\}$

$\therefore \iint_R \sec^2\dfrac{y-x}{x+y} dxdy = \dfrac{1}{2}\int_1^3 \int_{-v}^{v} \sec^2\dfrac{u}{v} dudv = \dfrac{1}{2}\int_1^3 v\tan\dfrac{u}{v}\Big|_{u=-v}^{u=v} dv = \tan 1 \int_1^3 v\, dv$
$= 4\tan 1$

Example 23.

(1) Find $\iint_R \left(1 - \dfrac{x^2}{25} - \dfrac{y^2}{4}\right)^{\frac{3}{2}} dxdy = ?$, $R = \left\{(x,y): \dfrac{x^2}{25} + \dfrac{y^2}{4} \leq 1, y \geq 0\right\}$.

(2) Find $\iint_R \cos(25x^2 + 4y^2)\, dxdy = ?$, $R = \left\{(x,y): \dfrac{x^2}{4} + \dfrac{y^2}{25} \leq 1\right\}$.

【Solution】
(1)

8.3.4 Use Generalized Coordinates to Compute Double Integrals

Let $x = 5r\cos\theta$, $y = 2r\sin\theta$ then $R = \{(r,\theta): 0 \leq r \leq 1, 0 \leq \theta \leq \pi\}$

and $dxdy = \left\| \begin{matrix} \dfrac{\partial x}{\partial r} & \dfrac{\partial x}{\partial \theta} \\ \dfrac{\partial y}{\partial r} & \dfrac{\partial y}{\partial \theta} \end{matrix} \right\| drd\theta = \left\| \begin{matrix} 5\cos\theta & -5r\sin\theta \\ 2\sin\theta & 2r\cos\theta \end{matrix} \right\| drd\theta = 10rdrd\theta$

$\therefore \iint_R \left(1 - \dfrac{x^2}{25} - \dfrac{y^2}{4}\right)^{\frac{3}{2}} dxdy = \int_0^\pi \int_0^1 (1-r^2)^{\frac{3}{2}} 10 r dr d\theta = 10 \int_0^\pi d\theta \int_0^1 (1-r^2)^{\frac{3}{2}} r dr$

$= 10\pi \cdot \left. \dfrac{-(1-r^2)^{\frac{5}{2}}}{5} \right|_0^1 = 2\pi$

(2)

Let $x = 2r\cos\theta$, $y = 5r\sin\theta$ then $R = \{(r,\theta): 0 \leq r \leq 1, 0 \leq \theta \leq 2\pi\}$

and $dxdy = \left\| \begin{matrix} \dfrac{\partial x}{\partial r} & \dfrac{\partial x}{\partial \theta} \\ \dfrac{\partial y}{\partial r} & \dfrac{\partial y}{\partial \theta} \end{matrix} \right\| drd\theta = \left\| \begin{matrix} 2\cos\theta & -2r\sin\theta \\ 5\sin\theta & 5r\cos\theta \end{matrix} \right\| drd\theta = 10rdrd\theta$

$\therefore \iint_R \cos(25x^2 + 4y^2) dxdy = 10 \int_0^{2\pi} \int_0^1 \cos(100r^2) r dr d\theta$

$= 10 \int_0^{2\pi} d\theta \int_0^1 \cos(100r^2) r dr = 10(2\pi) \cdot \left. \dfrac{\sin(100r^2)}{200} \right|_0^1 = \dfrac{\pi \sin 100}{10}$

Example 24.

Find $\int_1^2 \int_0^x \dfrac{1}{\sqrt{x^2+y^2}} dy dx = ?$

【Solution】

Let $x = r\cos\theta$, $y = r\sin\theta$ then

$\{(x,y): 1 \leq x \leq 2, 0 \leq y \leq x\} = \{(r,\theta): \dfrac{1}{\cos\theta} \leq r \leq \dfrac{2}{\cos\theta}, 0 \leq \theta \leq \dfrac{\pi}{4}\}$

and $dxdy = \left\| \begin{matrix} \dfrac{\partial x}{\partial r} & \dfrac{\partial x}{\partial \theta} \\ \dfrac{\partial y}{\partial r} & \dfrac{\partial y}{\partial \theta} \end{matrix} \right\| drd\theta = \left\| \begin{matrix} \cos\theta & -r\sin\theta \\ \sin\theta & r\cos\theta \end{matrix} \right\| drd\theta = rdrd\theta$

$$\therefore \int_1^2 \int_0^x \frac{1}{\sqrt{x^2+y^2}} dy dx = \int_0^{\frac{\pi}{4}} \int_{\frac{1}{\cos\theta}}^{\frac{2}{\cos\theta}} \frac{r}{\sqrt{r^2}} dr d\theta = \int_0^{\frac{\pi}{4}} \frac{1}{\cos\theta} d\theta = \ln(\sec\theta + \tan\theta)|_0^{\frac{\pi}{4}}$$
$$= \ln(\sqrt{2}+1)$$

Example 25.

Find $\iint_R e^{-xy} dxdy =?$, where R is the region enclosed by $y=x$, $y=2x$, $xy=1$, and $xy=4$.

【Solution】

Let $\frac{y}{x} = u$, $xy = v$ then $x = \sqrt{\frac{v}{u}}$, $y = \sqrt{uv}$

and $dxdy = \begin{Vmatrix} \frac{\partial x}{\partial u} & \frac{\partial x}{\partial v} \\ \frac{\partial y}{\partial u} & \frac{\partial y}{\partial v} \end{Vmatrix} dudv = \begin{Vmatrix} \frac{1}{2}\left(\frac{v}{u}\right)^{-\frac{1}{2}}\left(-\frac{v}{u^2}\right) & \frac{1}{2}\left(\frac{v}{u}\right)^{-\frac{1}{2}}\left(\frac{1}{u}\right) \\ \frac{1}{2}(uv)^{-\frac{1}{2}}v & \frac{1}{2}(uv)^{-\frac{1}{2}}u \end{Vmatrix} dudv = \frac{1}{2u} dudv$

$\because R = \left\{(x,y): 1 \leq \frac{y}{x} \leq 2, 1 \leq xy \leq 4\right\}$ $\therefore R = \{(u,v): 1 \leq u \leq 2, 1 \leq v \leq 4\}$

$\therefore \iint_R e^{-xy} dxdy = 2\int_1^4 \int_1^2 \frac{e^{-v}}{2u} dudv = \int_1^4 e^{-v} dv \int_1^2 \frac{1}{u} du = \left(\frac{1}{e} - \frac{1}{e^4}\right) \ln 2$

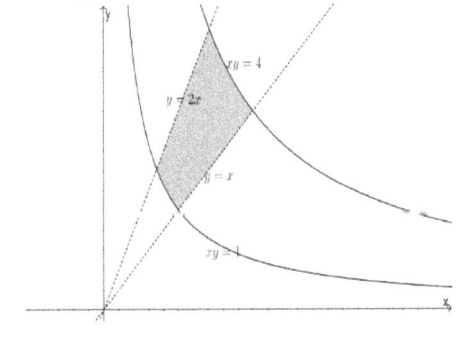

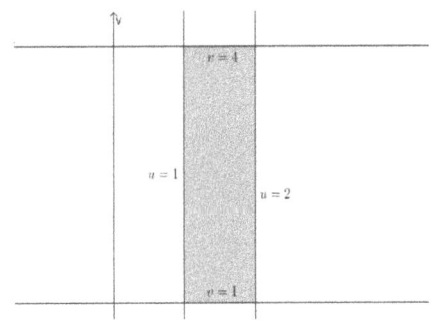

Example 26.

Find $\iint_R x^2+y^2 dxdy =?$, where $R = \left\{(x,y), 1 \leq xy \leq 4, 1 \leq \frac{y}{x} \leq 4\right\}$.

【Solution】

8.3.4 Use Generalized Coordinates to Compute Double Integrals

Let $\dfrac{y}{x} = u$, $xy = v$ then $x = \sqrt{\dfrac{v}{u}}$, $y = \sqrt{uv}$

and $dxdy = \begin{Vmatrix} \dfrac{\partial x}{\partial u} & \dfrac{\partial x}{\partial v} \\ \dfrac{\partial y}{\partial u} & \dfrac{\partial y}{\partial v} \end{Vmatrix} dudv = \begin{Vmatrix} \dfrac{1}{2}\left(\dfrac{v}{u}\right)^{-\frac{1}{2}}\left(-\dfrac{v}{u^2}\right) & \dfrac{1}{2}\left(\dfrac{v}{u}\right)^{-\frac{1}{2}}\left(\dfrac{1}{u}\right) \\ \dfrac{1}{2}(uv)^{-\frac{1}{2}}v & \dfrac{1}{2}(uv)^{-\frac{1}{2}}u \end{Vmatrix} dudv = \dfrac{1}{2u} dudv$

$\because R = \left\{(x,y): 1 \le \dfrac{y}{x} \le 4, 1 \le xy \le 4\right\}$ $\therefore R = \{(u,v): 1 \le u \le 4, 1 \le v \le 4\}$

$\iint_R x^2 + y^2 dxdy = 2\int_1^4 \int_1^4 \left(\dfrac{v}{u^2}+v\right)\dfrac{1}{2} dudv = \int_1^4\int_1^4 \left(\dfrac{v}{u^2}+v\right) dudv = \int_1^4 \dfrac{15v}{4} dv = \dfrac{225}{8}$

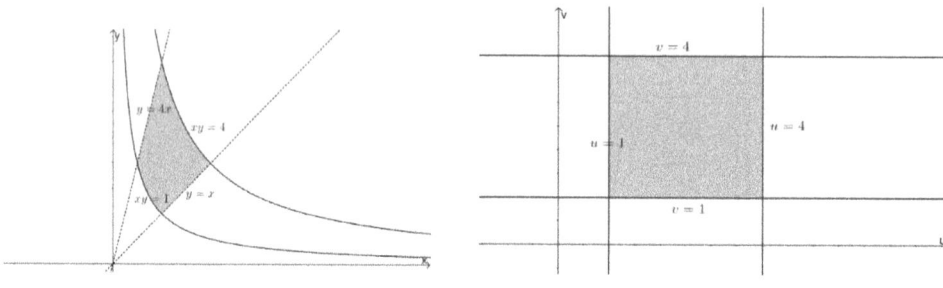

Example 27.

Find $\displaystyle\int_{\frac{1}{2}}^{1} \int_{1-x}^{x} \dfrac{1}{\sqrt{x^2+y^2}} dydx = ?$

【Solution】

Let $x = r\cos\theta$, $y = r\sin\theta$ then

$\left\{(x,y): 1-x \le y \le x, \dfrac{1}{2} \le x \le 1\right\} = \left\{(r,\theta): \dfrac{1}{\cos\theta+\sin\theta} \le r \le \dfrac{1}{\cos\theta}, 0 \le \theta \le \dfrac{\pi}{4}\right\}$

and $dxdy = \begin{Vmatrix} \dfrac{\partial x}{\partial r} & \dfrac{\partial x}{\partial \theta} \\ \dfrac{\partial y}{\partial r} & \dfrac{\partial y}{\partial \theta} \end{Vmatrix} drd\theta = \begin{Vmatrix} \cos\theta & -r\sin\theta \\ \sin\theta & r\cos\theta \end{Vmatrix} drd\theta = rdrd\theta$

$\therefore \displaystyle\int_{\frac{1}{2}}^{1}\int_{1-x}^{x} \dfrac{1}{\sqrt{x^2+y^2}} dydx = \int_0^{\frac{\pi}{4}} \int_{\frac{1}{\cos\theta+\sin\theta}}^{\frac{1}{\cos\theta}} \dfrac{r}{\sqrt{r^2}} drd\theta = \int_0^{\frac{\pi}{4}} \dfrac{1}{\cos\theta} - \dfrac{1}{\cos\theta+\sin\theta} d\theta$

$= \displaystyle\int_0^{\frac{\pi}{4}} \dfrac{1}{\cos\theta} - \dfrac{1}{\sqrt{2}\sin(\theta+\frac{\pi}{4})} d\theta$

$$= \ln(\sec\theta + \tan\theta)|_0^{\frac{\pi}{4}} + \frac{1}{\sqrt{2}}\ln(|\csc\left(\theta + \frac{\pi}{4}\right) + \cot\left(\theta + \frac{\pi}{4}\right)|)\Big|_0^{\frac{\pi}{4}} = (1 - \frac{1}{\sqrt{2}})\ln(1 + \sqrt{2})$$

Example 28.

Find $\iint_R (x+y)^2 \, dxdy = ?$, where R is the region enclosed by the lines $x+y=0$, $x+y=1$, $2x-y=0$, and $2x-y=3$.

【Solution】

Let $x+y=u$, $2x-y=v$ then $x = \dfrac{u+v}{3}$, $y = \dfrac{2u-v}{3}$

and $dxdy = \begin{Vmatrix} \dfrac{\partial x}{\partial u} & \dfrac{\partial x}{\partial v} \\ \dfrac{\partial y}{\partial u} & \dfrac{\partial y}{\partial v} \end{Vmatrix} dudv = \begin{Vmatrix} \dfrac{1}{3} & \dfrac{1}{3} \\ \dfrac{2}{3} & \dfrac{-1}{3} \end{Vmatrix} dudv = \dfrac{1}{3} dudv$

∵ $R = \{(x,y): 0 \leq x+y \leq 1, 0 \leq 2x-y \leq 3\}$ ∴ $R = \{(u,v): 0 \leq u \leq 1, 0 \leq v \leq 3\}$

∴ $\iint_R (x+y)^2 \, dxdy = \dfrac{1}{3}\int_0^3 \int_0^1 u^2 \, dudv = \dfrac{u^3}{3}\Big|_0^1 = \dfrac{1}{3}$

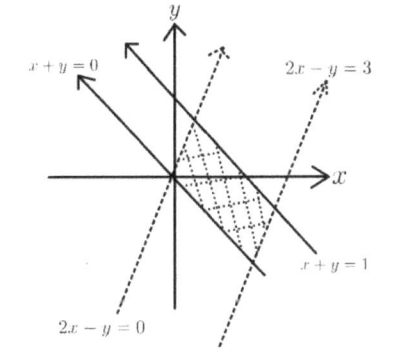

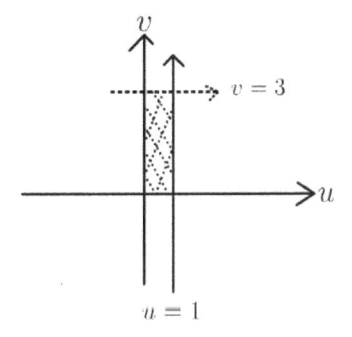

8.4 Methods for Solving Triple Integrals

The types of exam questions for triple integrals include cases where the order of integration does not need to be changed and can be directly converted into double integrals before turning into definite integrals, transforming into double integrals that involve $x^2 + y^2$ and then using polar

8.4.1 Directly Compute the Iterated Integrals without Changing the Order

coordinates, and using spherical coordinates to compute triple integrals, where the order of integration does not need to be changed for direct evaluation of iterated integrals.

8.4.1 Directly Compute the Iterated Integrals without Changing the Order

Question Types:

Type 1.

Given $f_1(x), f_2(x), g_1(x, y)$, and $g_2(x, y)$, assume $R = \{(x, y): a \leq x \leq b, f_1(x) \leq y \leq f_2(x)\}$ and $g_1(x, y) \leq z \leq g_2(x, y)$. Find $\iint_R \int_{g_1(x,y)}^{g_2(x,y)} f(x, y) dz dA = ?$

Problem-Solving Process:

Step1.

Convert the triple integral into a double integral:

$$\iint_R \int_{g_1(x,y)}^{g_2(x,y)} f(x, y) dz dA = \iint_R f(x, y)(g_2(x, y) - g_1(x, y)) dA$$

Step2.

By using Fubini's Theorem,

$$\iint_R f(x, y)(g_2(x, y) - g_1(x, y)) dA = \int_a^b \int_{f_1(x)}^{f_2(x)} f(x, y)(g_2(x, y) - g_1(x, y)) dy dx$$

Step3.

Evaluate $\int_{f_1(x)}^{f_2(x)} f(x, y)(g_2(x, y) - g_1(x, y)) dy = ?$

Assume the computation gives $h(x) = \int_{f_1(x)}^{f_2(x)} f(x, y)(g_2(x, y) - g_1(x, y)) dy$

then $\int_a^b \int_{f_1(x)}^{f_2(x)} f(x, y)(g_2(x, y) - g_1(x, y)) dy dx = \int_a^b h(x) dx.$ Find $\int_a^b h(x) dx = ?$

Type 2.

Given $f_1(x), f_2(x), g_1(x, y)$, and $g_2(x, y)$, assume $R = \{(x, y): c \leq y \leq d, f_1(y) \leq x \leq f_2(y)\}$ and $g_1(x, y) \leq z \leq g_2(x, y)$. Find $\iint_R \int_{g_1(x,y)}^{g_2(x,y)} f(x, y, z) dz dA = ?$

Problem-Solving Process:

Step1.

Convert the triple integral into a double integral:

8.4.1 Directly Compute the Iterated Integrals without Changing the Order

$$\iint_R \int_{g_1(x,y)}^{g_2(x,y)} f(x,y,z)dzdA = \iint_R f(x,y)(g_2(x,y) - g_1(x,y))dA$$

Step2.
By using Fubini's Theorem,

$$\iint_R f(x,y)(g_2(x,y) - g_1(x,y))dA = \int_c^d \int_{f_1(y)}^{f_2(y)} f(x,y)(g_2(x,y) - g_1(x,y))dxdy$$

Step3.

Evaluate $\int_{f_1(y)}^{f_2(y)} f(x,y)(g_2(x,y) - g_1(x,y))dx =?$

Assume the computation gives $h(y) = \int_{f_1(y)}^{f_2(y)} f(x,y)(g_2(x,y) - g_1(x,y))dx$

then $\int_c^d \int_{f_1(y)}^{f_2(y)} f(x,y)(g_2(x,y) - g_1(x,y))dxdy = \int_c^d h(y)\,dy.$ Find $\int_c^d h(y)\,dy =?$

Example 1.

(1) Find the volume of the region bounded by $y = 2x$ and $y = x^2$, between xy-plane and the surface $z = x^3 + 2y$.

(2) Suppose V is the region bounded by $z > 0$, $z = x^3 + 2y$, $y = 2x$, and $y = x^2$.

Find $\iiint_V x\,dV = ?$

【Solution】

(1)

Let $R = \{(x,y): 0 \le x \le 2, x^2 \le y \le 2x\}$ then

$$\text{volume} = \iint_R \int_0^{x^3+2y} dzdA = \int_0^2 \int_{x^2}^{2x} x^3 + 2y\,dydx = \int_0^2 x^3 y + y^2 |_{y=x^2}^{y=2x} dx$$

$$= \int_0^2 2x^4 - x^5 + 4x^2 - x^4 dx = \frac{32}{5}$$

(2)

Let $R = \{(x,y): 0 \le x \le 2, x^2 \le y \le 2x\}$ then

$$\iiint_V x\,dV = \iint_R \int_0^{x^3+2y} x\,dzdA = \int_0^2 \int_{x^2}^{2x} x(x^3 + 2y)\,dydx = \int_0^2 x\left(x^3 y + y^2 |_{y=x^2}^{y=2x}\right) dx$$

8.4.1 Directly Compute the Iterated Integrals without Changing the Order

$$= \int_0^2 2x^5 - x^6 + 4x^3 - x^5 dx = \frac{x^6}{6} - \frac{x^7}{7} + x^4 \Big|_0^2 = \frac{224 - 384 + 336}{21} = \frac{176}{21}$$

Example 2.

Find $\displaystyle\int_0^{\frac{\pi}{2}} \int_{\sin 2z}^{0} \int_0^{2yz} \sin\frac{x}{y} dxdydz = ?$

【Solution】

$$\int_0^{\frac{\pi}{2}} \int_{\sin 2z}^{0} \int_0^{2yz} \sin\frac{x}{y} dxdydz = \int_0^{\frac{\pi}{2}} \int_{\sin 2z}^{0} -y\cos\frac{x}{y}\Big|_{x=0}^{x=2yz} dydz = \int_0^{\frac{\pi}{2}} \int_{\sin 2z}^{0} y - y\cos 2z\, dydz$$

$$= \int_0^{\frac{\pi}{2}} \left(\frac{y^2}{2} - \frac{y^2 \cos 2z}{2}\right)\Big|_{y=\sin 2z}^{y=0} dz = \int_0^{\frac{\pi}{2}} -\frac{\sin^2 2z}{2} + \frac{\sin^2 2z \cos 2z}{2} dz$$

$$= \int_0^{\frac{\pi}{2}} -\frac{1-\cos 4z}{4} + \frac{\sin^2 2z \cos 2z}{2} dz = -\frac{\pi}{8}$$

Example 3.

Find $\displaystyle\int_0^1 \int_0^{\sqrt{\ln 2}} \int_0^1 \frac{xyze^{x^2}}{1+y^2} dydxdz = ?$

【Solution】

$$\int_0^1 \int_0^{\sqrt{\ln 2}} \int_0^1 \frac{xyze^{x^2}}{1+y^2} dydxdz = \int_0^1 \frac{y}{1+y^2} dy \int_0^{\sqrt{\ln 2}} xe^{x^2} dx \int_0^1 zdz$$

$$= \frac{\ln(1+y^2)}{2}\Big|_0^1 \cdot \frac{e^{x^2}}{2}\Big|_0^{\sqrt{\ln 2}} \cdot \frac{z^2}{2}\Big|_0^1 = \frac{\ln 2}{8}$$

Example 4.

(1) Find the volume of the region bounded by $2y = x$, $x = 0$, between xy-plane and the plane $x + 2y + z = 2$.

(2) Suppose V is the region bounded by $z > 0$, $x + 2y + z = 2$, $2y = x$, and $x = 0$. Find $\iiint_V x dV = ?$

【Solution】

(1)

Let $R = \left\{(x, y): 0 \leq x \leq 1, \frac{x}{2} \leq y \leq 1 - \frac{x}{2}\right\}$ then

$$\text{volume} = \iint_R \int_0^{2-x-2y} dz dA = \int_0^1 \int_{\frac{x}{2}}^{1-\frac{x}{2}} 2 - x - 2y \, dy dx = \int_0^1 2y - xy - y^2 \Big|_{y=\frac{x}{2}}^{y=1-\frac{x}{2}} dx$$

$$= \int_0^1 1 - 2x + x^2 dx = \frac{1}{3}$$

(2)

Let $R = \left\{(x, y): 0 \leq x \leq 1, \frac{x}{2} \leq y \leq 1 - \frac{x}{2}\right\}$ then

$$\iiint_V x dV = \iint_R \int_0^{2-x-2y} x dz dA = \int_0^1 \int_{\frac{x}{2}}^{1-\frac{x}{2}} x(2 - x - 2y) dy dx$$

$$= \int_0^1 x\left(2y - xy - y^2 \Big|_{y=\frac{x}{2}}^{y=1-\frac{x}{2}}\right) dx = \int_0^1 x - 2x^2 + x^3 dx = \frac{x^2}{2} - \frac{2x^3}{3} + \frac{x^4}{4}\Big|_0^1 = \frac{1}{12}$$

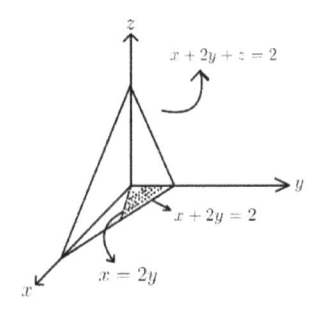

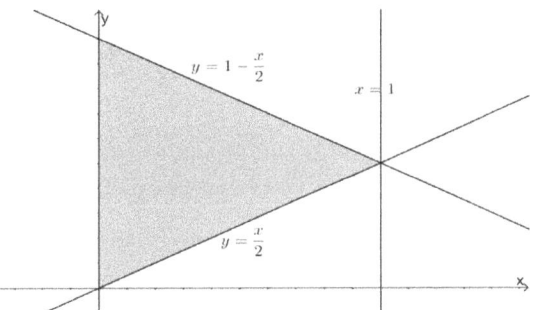

Example 5.

Find the volume of the region bounded by $y = x, y = 0, z = 0, 6x + 2y + 3z = 6$.

【Solution】

Let $R = \{(x, y): 0 \leq x \leq 1, 0 \leq y \leq x\}$ then

8.4.1 Directly Compute the Iterated Integrals without Changing the Order 218

$$V = \iint_R \frac{6-6x-2y}{3} dydx = \int_0^1 \int_0^x 2 - 2x - \frac{2y}{3} dydx = \int_0^1 (2-2x)y - \frac{y^2}{3}\bigg|_{y=0}^{y=x} dx$$

$$= \int_0^1 2x - \frac{7x^2}{3} dx = x^2 - \frac{7x^3}{9}\bigg|_0^1 = \frac{2}{9}$$

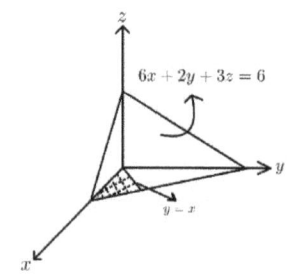

Example 6.

Find $\int_0^{\frac{\pi}{2}} \int_0^z \int_0^y \sin(x+y+z)dxdydz = ?$

【Solution】

$$\int_0^{\frac{\pi}{2}} \int_0^z \int_0^y \sin(x+y+z)dxdydz = \int_0^{\frac{\pi}{2}} \int_0^z -\cos(x+y+z)\big|_0^y dydz$$

$$= \int_0^{\frac{\pi}{2}} \int_0^z \cos(y+z) - \cos(2y+z) \, dydz = \int_0^{\frac{\pi}{2}} \left(\sin(y+z) - \frac{\sin(2y+z)}{2}\right)\bigg|_{y=0}^{y=z} dz$$

$$= \int_0^{\frac{\pi}{2}} \sin(2z) - \frac{\sin z}{2} - \frac{\sin(3z)}{2} dz = -\frac{\cos 2z}{2} + \frac{\cos z}{2} + \frac{\cos(3z)}{6}\bigg|_0^{\frac{\pi}{2}} = \frac{1}{3}$$

Example 7.

Find $\int_0^5 \int_{-2}^4 \int_1^2 6xy^2z^3 dxdydz = ?$

【Solution】

$$\int_0^5 \int_{-2}^4 \int_1^2 6xy^2z^3 dxdydz = \int_0^5 \int_{-2}^4 3x^2y^2z^3\big|_{x=1}^{x=2} dydz = \int_0^5 \int_{-2}^4 9y^2z^3 dydz$$

$$= \int_0^5 3y^3z^3\big|_{y=-2}^{y=4} dz = 216 \int_0^5 z^3 \, dz = 33750$$

Example 8.

(1) Find the volume of the region bounded by $x = 0, z = 0, y = x^2$ and $y + z = 9$.

(2) Suppose V is the region bounded by $x = 0, z = 0, y = x^2$, and $y + z = 9$.

Find $\iiint_V x dV = ?$

【Solution】

(1)

Let $R = \{(x, y): 0 \leq x \leq 3, x^2 \leq y \leq 9\}$ then

$$\text{volume} = \iint_R \int_0^{9-y} dz dA = \int_0^3 \int_{x^2}^9 9 - y dy dx = \int_0^3 9y - \frac{y^2}{2}\Big|_{y=x^2}^{y=9} dx$$

$$= \int_0^3 9(9 - x^2) - \frac{81 - x^4}{2} dx = \int_0^3 \frac{81}{2} - 9x^2 + \frac{x^4}{2} dx = \left(\frac{81x}{2} - 3x^3 + \frac{x^5}{10}\right)\Big|_0^3 = \frac{324}{5}$$

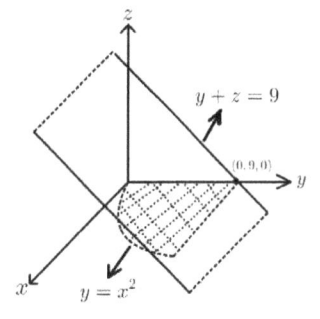

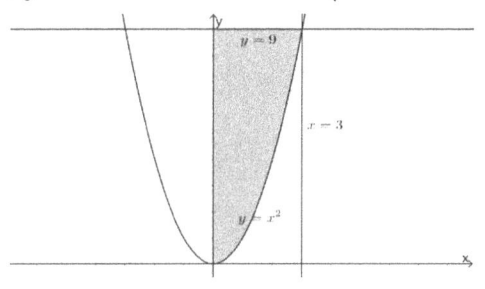

(2)

Let $R = \{(x, y): 0 \leq x \leq 3, x^2 \leq y \leq 9\}$ then

$$\iiint_V x dV = \iint_R \int_0^{9-y} x dz dA = \int_0^3 \int_{x^2}^9 x(9 - y) dy dx = \int_0^3 x\left(9y - \frac{y^2}{2}\Big|_{y=x^2}^{y=9}\right) dx$$

$$= \int_0^3 x\left(9(9 - x^2) - \frac{81 - x^4}{2}\right) dx = \int_0^3 \frac{81x}{2} - 9x^3 + \frac{x^5}{2} dx = \left(\frac{81x^2}{4} - \frac{9x^4}{4} + \frac{x^6}{12}\right)\Big|_0^3$$

$$= \frac{81}{4} \cdot 9 - \frac{9}{4} \cdot 81 + \frac{9^3}{12} = \frac{243}{4}$$

Example 9.

(1) Find the volume of the region bounded by $3x + 2y + z = 6$, $x = 0$, $y = 0$ and $z = 0$.

(2) Suppose V is the region bounded by $3x + 2y + z = 6$, $x = 0$, $y = 0$, and $z = 0$.

8.4.1 Directly Compute the Iterated Integrals without Changing the Order

Find $\iiint_V x\,dV = ?$

【Solution】

(1)

Let $R = \left\{(x,y): 0 \leq x \leq 2, 0 \leq y \leq 3 - \dfrac{3x}{2}\right\}$ then

$$\text{volume} = \iint_R \int_0^{6-3x-2y} dz\,dA = \int_0^2 \int_0^{3-\frac{3x}{2}} 6 - 3x - 2y\,dy\,dx$$

$$= \int_0^2 (6y - 3xy - y^2)\Big|_{y=0}^{y=3-\frac{3x}{2}} dx = \int_0^2 9 - 9x + \dfrac{9x^2}{4}\,dx = \left(9x - \dfrac{9}{2}x^2 + \dfrac{3x^3}{4}\right)\Big|_0^2 = 6$$

(2)

Let $R = \left\{(x,y): 0 \leq x \leq 2, 0 \leq y \leq 3 - \dfrac{3x}{2}\right\}$ then

$$\iiint_V x\,dV = \iint_R \int_0^{6-3x-2y} x\,dz\,dA = \int_0^2 \int_0^{3-\frac{3x}{2}} x(6 - 3x - 2y)\,dy\,dx$$

$$= \int_0^2 x\left((6y - 3xy - y^2)\Big|_{y=0}^{y=3-\frac{3x}{2}}\right) dx = \int_0^2 9x - 9x^2 + \dfrac{9x^3}{4}\,dx = \left(\dfrac{9x^2}{2} - 3x^3 + \dfrac{9x^4}{16}\right)\Big|_0^2 = 3$$

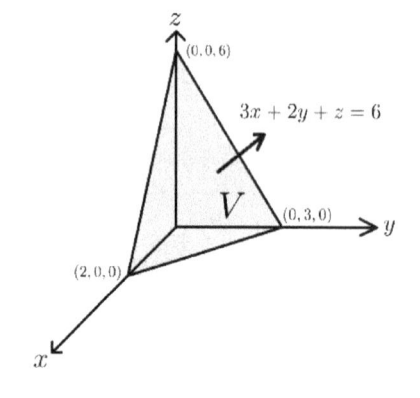

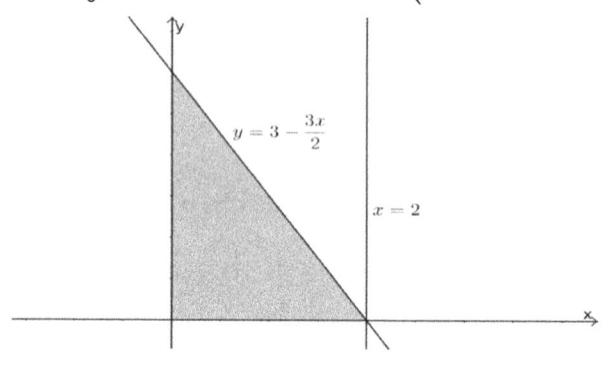

Example 10.

Suppose V is the tetrahedron bounded by $x + y + z = 1$, $x = 0$, $y = 0$, and $z = 0$.

Find $\iiint_V \dfrac{24}{(1+x+y+z)^4}\,dV = ?$

【Solution】

Let $R = \{(x,y): 0 \leq x \leq 1, 0 \leq y \leq 1 - x\}$ then

8.4.1 Directly Compute the Iterated Integrals without Changing the Order

$$\iiint_V \frac{24}{(1+x+y+z)^4} dV = \iint_R \int_0^{1-x-y} \frac{24}{(1+x+y+z)^4} dz dA$$

$$= \int_0^1 \int_0^{1-x} -8(1+x+y+z)^{-3}|_{z=0}^{z=1-x-y} dy dx = \int_0^1 \int_0^{1-x} -1 + 8(1+x+y)^{-3} dy dx$$

$$= \int_0^1 (-y - 4(1+x+y)^{-2})|_{y=0}^{y=1-x} dx = \int_0^1 4(1+x)^{-2} + x - 2 dx$$

$$= (-4)(1+x)^{-1} + \frac{x^2}{2} - 2x \bigg|_0^1 = \frac{1}{2}$$

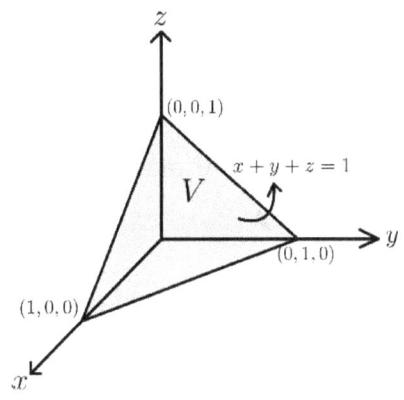

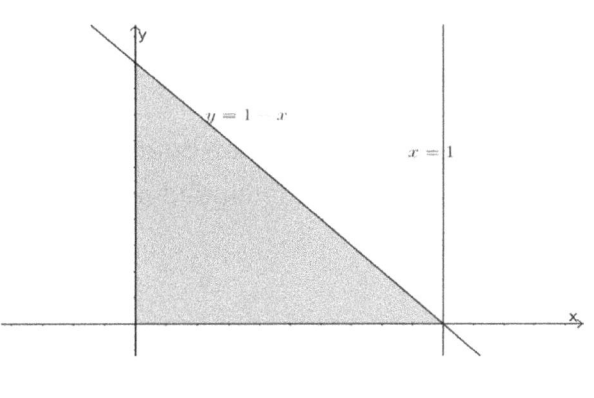

Example 11.

Suppose V is the tetrahedron bounded by $4x + 4y + z = 4$, $x = 0$, $y = 0$, and $z = 0$.

Find $\iiint_V 2x dV = ?$

【Solution】

Let $R = \{(x,y): 0 \leq x \leq 1, 0 \leq y \leq 1-x\}$ then

$$\iiint_V 2x dV = \iint_R \int_0^{4-4x-4y} 2x dz dA = \int_0^1 \int_0^{1-x} 2xz|_{z=0}^{z=4-4x-4y} dy dx$$

$$= \int_0^1 \int_0^{1-x} 8x - 8x^2 - 8xy dy dx = \int_0^1 (8xy - 8x^2 y - 4xy^2)|_{y=0}^{y=1-x} dx$$

$$= \int_0^1 8x(1-x) - 8x^2(1-x) - 4x(1-x)^2 dx = 2\left(x^2 - \frac{4x^3}{3} + \frac{x^4}{2}\right)\bigg|_0^1 = \frac{1}{3}$$

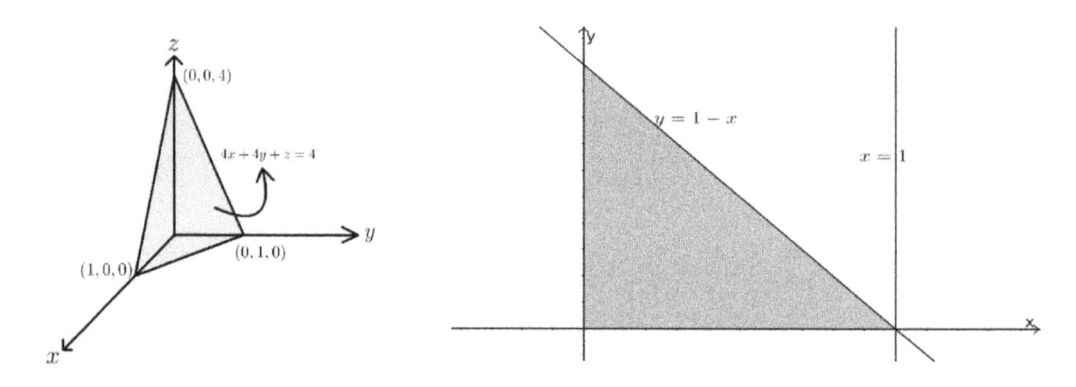

8.4.1 Directly Compute the Iterated Integrals without Changing the Order 222

Example 12.

(1) Find the volume of the region bounded by $z = 0$, $z = 1 - x - y$, $x = 0$, and $y = 0$.

(2) Suppose V is the region bounded by $z = 0$, $z = 1 - x - y$, $x = 0$, and $y = 0$.

Find $\iiint_V x\,dV = ?$

【Solution】

(1)

Let $R = \left\{(x,y): 0 \leq x \leq \dfrac{1}{2}, 0 \leq y \leq \dfrac{1}{2} - x\right\}$ then

$$\text{volume} = \iint_R \int_0^{1-x-y} dz\,dA = \int_0^{\frac{1}{2}} \int_0^{\frac{1}{2}-x} 1 - x - y\,dy\,dx = \int_0^{\frac{1}{2}} \left(y - xy - \dfrac{y^2}{2}\right)\bigg|_0^{\frac{1}{2}-x} dx$$

$$= \int_0^{\frac{1}{2}} \left(\dfrac{1}{2} - x\right) - x\left(\dfrac{1}{2} - x\right) - \dfrac{(\frac{1}{2} - x)^2}{2}\,dx = \int_0^{\frac{1}{2}} \dfrac{x^2}{2} - x + \dfrac{3}{8}\,dx = \left(\dfrac{3x}{8} - \dfrac{x^2}{2} + \dfrac{x^3}{6}\right)\bigg|_0^{\frac{1}{2}} = \dfrac{1}{12}$$

(2)

Let $R = \left\{(x,y): 0 \leq x \leq \dfrac{1}{2}, 0 \leq y \leq \dfrac{1}{2} - x\right\}$ then

$$\iiint_V x\,dV = \iint_R \int_0^{1-x-y} x\,dz\,dA = \int_0^{\frac{1}{2}} \int_0^{\frac{1}{2}-x} x(1 - x - y)\,dy\,dx$$

$$= \int_0^{\frac{1}{2}} x\left(y - xy - \dfrac{y^2}{2}\right)\bigg|_0^{\frac{1}{2}-x} dx = \int_0^{\frac{1}{2}} x\left(\left(\dfrac{1}{2} - x\right) - x\left(\dfrac{1}{2} - x\right) - \dfrac{(\frac{1}{2} - x)^2}{2}\right)dx$$

$$= \int_0^{\frac{1}{2}} \dfrac{x^3}{2} - x^2 + \dfrac{3x}{8}\,dx = \left(\dfrac{x^4}{8} - \dfrac{x^3}{3} + \dfrac{3x^2}{16}\right)\bigg|_0^{\frac{1}{2}} = \dfrac{1}{8}\left(\dfrac{1}{16} - \dfrac{1}{3} + \dfrac{3}{8}\right) = \dfrac{5}{384}$$

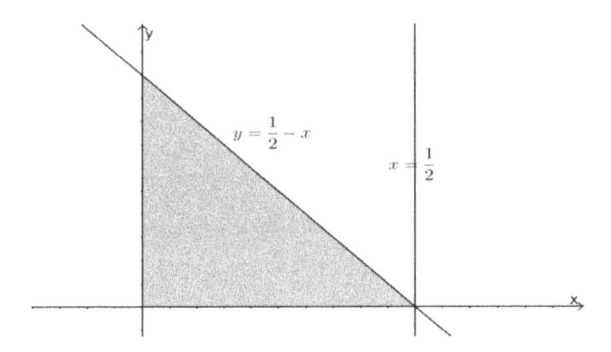

Example 13.

(1) Find the volume of the region bounded by $y + 2z = 2$, $z = x^2$, and $y = 0$.
(2) Suppose V is the region bounded by $y + 2z = 2$, $z = x^2$, and $y = 0$.

Find $\iiint_V x^2 dV =?$

【Solution】

(1)
Let $R = \{(x,z): -1 \leq x \leq 1, x^2 \leq z \leq 1\}$ then

$$\text{volume} = \iint_R \int_0^{2-2z} dy dA = 2\int_{-1}^{1}\int_{x^2}^{1} 1 - z \, dz dx = 2\int_{-1}^{1} \left(z - \frac{z^2}{2}\right)\Big|_{x^2}^{1} dx$$

$$= 2\int_{-1}^{1}\left(\frac{1}{2} - x^2 + \frac{x^4}{2}\right) dx = \frac{16}{15}$$

(2)
Let $R = \{(x,z): -1 \leq x \leq 1, x^2 \leq z \leq 1\}$ then

$$\iiint_V x^2 dV = \iint_R \int_0^{2-2z} x^2 dy dA = 2\int_{-1}^{1}\int_{x^2}^{1} x^2(1-z) dz dx = 2\int_{-1}^{1} x^2\left(\left(z - \frac{z^2}{2}\right)\Big|_{x^2}^{1}\right) dx$$

$$= 2\int_{-1}^{1}\left(\frac{x^2}{2} - x^4 + \frac{x^6}{2}\right) dx = 2\left(\frac{x^3}{6} - \frac{x^5}{5} + \frac{x^7}{14}\right)\Big|_{-1}^{1} = \frac{16}{105}$$

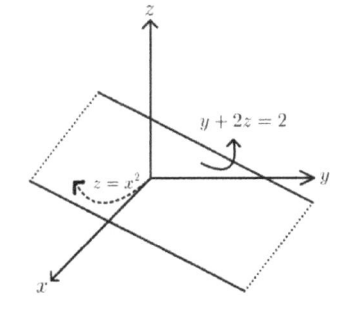

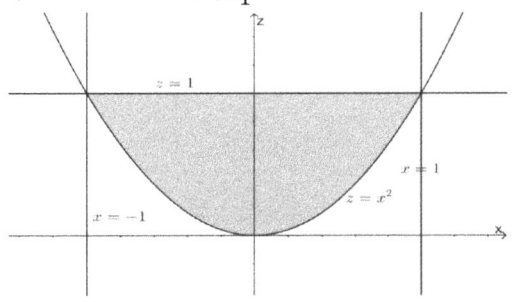

8.4.1 Directly Compute the Iterated Integrals without Changing the Order

Example 14.

Suppose V is the tetrahedron bounded by $x + y + z = d(d > 0)$, $x = 0$, $y = 0$, and $z = 0$. Find $\iiint_V \dfrac{24}{(d+x+y+z)^4} dV = ?$

【Solution】

Let $R = \{(x,y): 0 \le x \le d, 0 \le y \le d - x\}$ then

$$\iiint_V \frac{24}{(d+x+y+z)^4} dV = \iint_R \int_0^{d-x-y} \frac{24}{(d+x+y+z)^4} dz dA$$

$$= \int_0^d \int_0^{d-x} -8(d+x+y+z)^{-3}|_{z=0}^{z=d-x-y} dy dx = \int_0^d \int_0^{d-x} -d^{-3} + 8(d+x+y)^{-3} dy dx$$

$$= \int_0^d (-d^{-3}y - 4(d+x+y)^{-2})|_{y=0}^{y=d-x} dx = \int_0^d 4(d+x)^{-2} + d^{-3}x - 2d^{-2} dx$$

$$= -4(d+x)^{-1} + d^{-3} \cdot \frac{x^2}{2} - 2d^{-2}x \Big|_0^d = \frac{1}{2d}$$

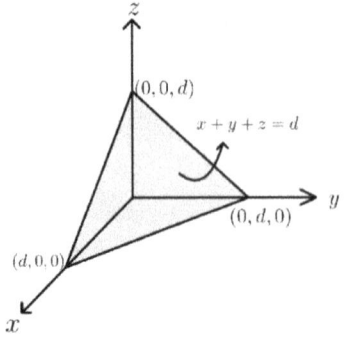

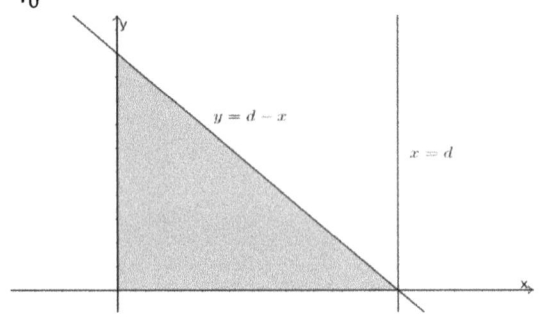

Example 15.

Suppose V is the tetrahedron bounded by $dx + dy + z = d(d > 0)$, $x = 0$, $y = 0$, and $z = 0$. Find $\iiint_V 2x dV =?$

【Solution】

Let $R = \{(x,y): 0 \le x \le 1, 0 \le y \le 1 - x\}$ then

$$\iiint_V 2x dV = \iint_R \int_0^{d-dx-dy} 2x dz dA = \int_0^1 \int_0^{1-x} 2xz|_{z=0}^{z=d-dx-dy} dy dx$$

$$= \int_0^1 \int_0^{1-x} 2d(x - x^2 - xy)dydx = 2d\int_0^1 (xy - x^2y - \frac{xy^2}{2})\Big|_{y=0}^{y=1-x} dx$$

$$= 2d\int_0^1 x(1-x) - x^2(1-x) - \frac{x(1-x)^2}{2} dx = d\left(\frac{x^2}{2} - \frac{2x^3}{3} + \frac{x^4}{4}\right)\Big|_0^1 = \frac{d}{12}$$

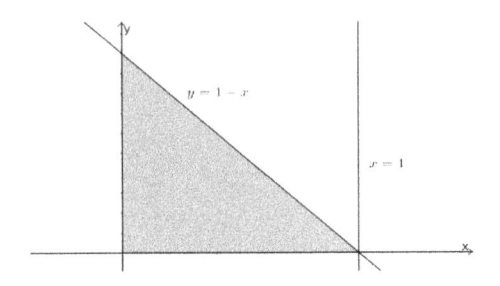

Example 16.

(1) Find the volume of the region bounded by $z = 3x^2$, $z = 4 - x^2$, $y + 2z = 12$, and $y = 0$.

(2) Suppose V is the region bounded by $z = 3x^2$, $z = 4 - x^2$, $y + 2z = 12$, and $y = 0$. Find $\iiint_V x^2 dV = ?$

【Solution】

(1)
Let $R = \{(x,z): -1 \le x \le 1, 3x^2 \le z \le 4 - x^2\}$ then

$$\text{volume} = \iint_R \int_0^{12-2z} dydA = 2\int_{-1}^1 \int_{3x^2}^{4-x^2} 6 - zdzdx = 2\int_{-1}^1 (6z - \frac{z^2}{2})\Big|_{3x^2}^{4-x^2} dx$$

$$= 8\int_{-1}^1 x^4 - 5x^2 + 4dx = \frac{608}{15}$$

(2)
Let $R = \{(x,z): -1 \le x \le 1, 3x^2 \le z \le 4 - x^2\}$ then

$$\iiint_V x^2 dV = \iint_R \int_0^{12-2z} x^2 dydA = 2\int_{-1}^1 \int_{3x^2}^{4-x^2} x^2(6-z)dzdx$$

$$= 2\int_{-1}^1 x^2\left((6z - \frac{z^2}{2})\Big|_{3x^2}^{4-x^2}\right)dx = 8\int_{-1}^1 x^6 - 5x^4 + 4x^2 dx = 8\left(\frac{x^7}{7} - x^5 + \frac{4x^3}{3}\right)\Big|_{-1}^1$$

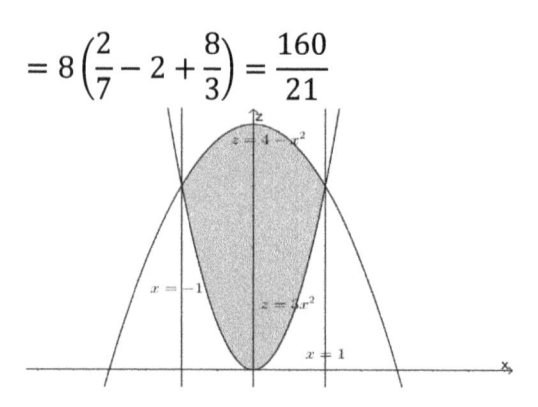

8.4.1 Directly Compute the Iterated Integrals without Changing the Order

$$= 8\left(\frac{2}{7} - 2 + \frac{8}{3}\right) = \frac{160}{21}$$

Example 17.

(1) Find the volume of the region bounded by $z = x^2$, $z = 1 - y$, $x = 0$, and $y = 0$.

(2) Suppose V is the region bounded by $z = x^2$, $z = 1 - y$, $x = 0$, and $y = 0$.

Find $\iiint_V x dV = ?$

【Solution】

(1)

Let $R = \{(x, y): 0 \le x \le 1, 0 \le y \le 1 - x^2\}$ then

$$\text{volume} = \iint_R \int_{x^2}^{1-y} dz dA = \int_0^1 \int_0^{1-x^2} 1 - y - x^2 dy dx = \int_0^1 \left(y - \frac{y^2}{2} - x^2 y\right)\Big|_0^{1-x^2} dx$$

$$= \int_0^1 \frac{x^4}{2} - x^2 + \frac{1}{2} dx = \frac{4}{15}$$

(2)

Let $R = \{(x, y): 0 \le x \le 1, 0 \le y \le 1 - x^2\}$ then

$$\iiint_V x dV = \iint_R \int_{x^2}^{1-y} x dz dA = \int_0^1 \int_0^{1-x^2} x(1 - y - x^2) dy dx$$

$$= \int_0^1 x\left((y - \frac{y^2}{2} - x^2 y)\Big|_0^{1-x^2}\right) dx = \int_0^1 \frac{x^5}{2} - x^3 + \frac{x}{2} dx = \frac{x^6}{12} - \frac{x^4}{4} + \frac{x^2}{4}\Big|_0^1 = \frac{1}{12}$$

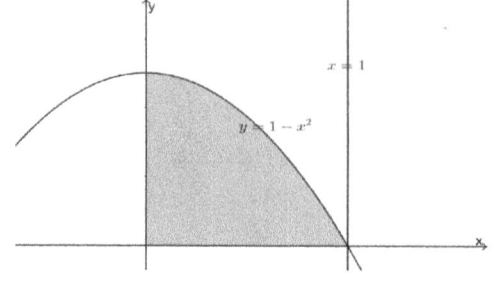

Example 18.

Suppose V is the tetrahedron bounded by $2x + 3y + z = 6, x = 0, y = 0,$ and $z = 0$.

Find $\iiint_V y^2 dx dy dz =?$

【Solution】

Let $R = \left\{(x,y): 0 \le x \le 3, 0 \le y \le 2 - \frac{2x}{3}\right\}$ then

$$\iiint_V y^2 dx dy dz = \iint_R \int_0^{6-2x-3y} y^2 dz dA = \int_0^3 \int_0^{2-\frac{2x}{3}} y^2 z \Big|_{z=0}^{z=6-2x-3y} dy dx$$

$$= \int_0^3 \int_0^{2-\frac{2x}{3}} y^2 (6 - 2x - 3y) dy dx = \int_0^3 \left(\frac{y^3(6-2x)}{3} - \frac{3y^4}{4}\right) \Big|_{y=0}^{y=2-\frac{2x}{3}} dx$$

$$= \frac{1}{4} \int_0^3 (2 - \frac{2x}{3})^4 dx = \frac{1}{20} \left(-\frac{3}{2}\right)(2-\frac{2x}{3})^5 \Big|_0^3 = \frac{12}{5}$$

Example 19.

Suppose V is the tetrahedron bounded by $x + y + z = 1$, $x = 0$, $y = 0$, and $z = 0$.

Find $\iiint_V z dV = ?$

【Solution】

Let $R = \{(x,y): 0 \le x \le 1, 0 \le y \le 1 - x\}$ then

8.4.1 Directly Compute the Iterated Integrals without Changing the Order 228

$$\iiint_V z\,dV = \iint_R \int_0^{1-x-y} z\,dz\,dA = \int_0^1 \int_0^{1-x} \left.\frac{z^2}{2}\right|_{z=0}^{z=1-x-y} dy\,dx$$

$$= \int_0^1 \int_0^{1-x} \frac{(1-x-y)^2}{2} dy\,dx = \frac{-1}{2}\int_0^1 \left.\frac{(1-x-y)^3}{3}\right|_{y=0}^{y=1-x} dx = \frac{1}{6}\int_0^1 (1-x)^3\,dx$$

$$= \frac{-1}{6}\left.\left(\frac{(1-x)^4}{4}\right)\right|_0^1 = \frac{1}{24}$$

Example 20.

Suppose V is the tetrahedron bounded by $x + y + z = d\,(d > 0)$, $x = 0$, $y = 0$, and $z = 0$. Find $\iiint_V z\,dV =?$

【Solution】

Let $R = \{(x,y): 0 \le x \le d, 0 \le y \le d-x\}$ then

$$\iiint_V z\,dV = \iint_R \int_0^{d-x-y} z\,dz\,dA = \int_0^d \int_0^{d-x} \left.\frac{z^2}{2}\right|_{z=0}^{z=d-x-y} dy\,dx$$

$$= \int_0^d \int_0^{d-x} \frac{(d-x-y)^2}{2} dy\,dx = \frac{-1}{2}\int_0^d \left.\frac{(d-x-y)^3}{3}\right|_{y=0}^{y=d-x} dx = \frac{1}{6}\int_0^d (d-x)^3\,dx$$

$$= \frac{-1}{6}\left.\left(\frac{(d-x)^4}{4}\right)\right|_0^d = \frac{d^4}{24}$$

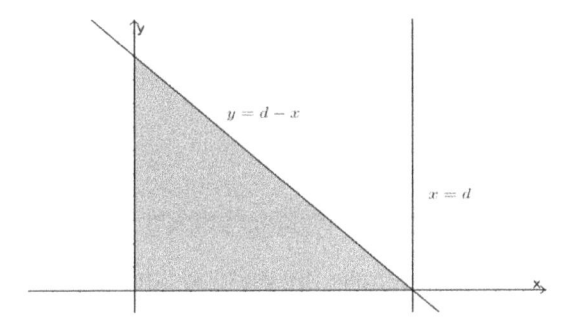

[Figure: shaded triangular region bounded by $y = d-x$, $x = d$, and the axes]

8.4.2 Use Polar Coordinates to Compute Triple Integrals

Question Types:

Type 1.

Given $f_1(x,y)$, $f_2(x,y)$, assume $R = \{(x,y): x^2 + y^2 \leq a\}$, $a > 0$ and $f_1(x,y) \leq z \leq f_2(x,y)$. Find $V = \iint_R \int_{f_2(x,y)}^{f_1(x,y)} f(x,y) dz dA = ?$

Problem-Solving Process:

Step1.
$$V = \iint_R \int_{f_2(x,y)}^{f_1(x,y)} f(x,y) dz dA = \iint_R f(x,y)\big(f_1(x,y) - f_2(x,y)\big) dx dy$$

Step2.
Let $x = r \cos\theta$, $y = r \sin\theta$ then
$R = \{(x,y): x^2 + y^2 \leq a\} = \{(r,\theta): 0 \leq r \leq a, 0 \leq \theta \leq 2\pi\}$

and $dx dy = \begin{Vmatrix} \dfrac{\partial x}{\partial r} & \dfrac{\partial x}{\partial \theta} \\ \dfrac{\partial y}{\partial r} & \dfrac{\partial y}{\partial \theta} \end{Vmatrix} dr d\theta = \begin{Vmatrix} \cos\theta & -r\sin\theta \\ \sin\theta & r\cos\theta \end{Vmatrix} dr d\theta = r dr d\theta$

Step3.
$$V = \iint_R f(x,y)\big(f_1(x,y) - f_2(x,y)\big) dx dy$$
$$= \int_0^{2\pi} \int_0^a f(r\cos\theta, r\sin\theta)\big(f_1(r\cos\theta, r\sin\theta) - f_2(r\cos\theta, r\sin\theta)\big) r dr d\theta$$

Step4.

Find $\int_0^{2\pi} \int_0^a f(r\cos\theta, r\sin\theta)\big(f_1(r\cos\theta, r\sin\theta) - f_2(r\cos\theta, r\sin\theta)\big) r dr d\theta$.

8.4.2 Use Polar Coordinates to Compute Triple Integrals

This may also involve using variable substitution or integration by parts to find the integral value.

Type 2.

In space, many quadratic surfaces, when intersected pairwise, enclose a closed region whose projection onto the xy-plane often appears as $\frac{x^2}{a^2} + \frac{y^2}{b^2} \leq 1$. We can use the methods mentioned above to find the value, and below is an organized list of surfaces that can enclose $\frac{x^2}{a^2} + \frac{y^2}{b^2} \leq 1$.

First, let's review common quadratic surfaces:

Cylinder: $x^2 + y^2 = a^2$, $f_1(x,y) \leq z \leq f_2(x,y)$

Ellipsoid: $\frac{x^2}{a^2} + \frac{y^2}{b^2} + \frac{z^2}{c^2} = 1$

Elliptic Paraboloid: $\frac{x^2}{a^2} + \frac{y^2}{b^2} = \frac{z}{c}$

Elliptic Cone: $\frac{x^2}{a^2} + \frac{y^2}{b^2} = \frac{z^2}{c^2}$

Hyperboloid: $\frac{x^2}{a^2} + \frac{y^2}{b^2} - \frac{z^2}{c^2} = 1$

Question Types:

Find the closed region enclosed by the cylinder: $x^2 + y^2 = a^2$, $f_1(x,y) \leq z \leq f_2(x,y)$.

Find the closed region enclosed by the elliptic paraboloid: $\frac{x^2}{a^2} + \frac{y^2}{b^2} = \frac{z}{c}$ and $z = f_1(x,y)$.

Find the closed region enclosed by two elliptic paraboloids.

Find the closed region enclosed by an elliptic paraboloid and an elliptic cone.

Find the closed region enclosed by the ellipsoid $\frac{x^2}{a^2} + \frac{y^2}{b^2} + \frac{z^2}{c^2} = 1$ and the plane $z = f_1(x,y)$.

Find the closed region enclosed by the ellipsoid $\frac{x^2}{a^2} + \frac{y^2}{b^2} + \frac{z^2}{c^2} = 1$ and an elliptic paraboloid.

Find the closed region enclosed by the ellipsoid $\frac{x^2}{a^2} + \frac{y^2}{b^2} + \frac{z^2}{c^2} = 1$ and the cylinder $x^2 + y^2 = r^2$.

Problem-Solving Process:

Step1.

Find α, $f_1(x,y)$, $f_2(x,y)$ such that $V = \iint_R \int_{f_2(x,y)}^{f_1(x,y)} dz dA = \iint_R f_1(x,y) - f_2(x,y) dxdy$

where $R = \{(x,y): x^2 + y^2 \leq \alpha^2\}$, $\alpha > 0$.

Step2.

8.4.2 Use Polar Coordinates to Compute Triple Integrals

Use the polar coordinate transformation to evaluate $\iint_R f_1(x,y) - f_2(x,y)dxdy =?$

Let $x = r\cos\theta$, $y = r\sin\theta$ then
$R = \{(x,y): x^2 + y^2 \leq a\} = \{(r,\theta): 0 \leq r \leq a, 0 \leq \theta \leq 2\pi\}$

and $dxdy = \begin{Vmatrix} \frac{\partial x}{\partial r} & \frac{\partial x}{\partial \theta} \\ \frac{\partial y}{\partial r} & \frac{\partial y}{\partial \theta} \end{Vmatrix} drd\theta = \begin{Vmatrix} \cos\theta & -r\sin\theta \\ \sin\theta & r\cos\theta \end{Vmatrix} drd\theta = rdrd\theta$

$\iint_R f_1(x,y) - f_2(x,y)dxdy = \int_0^{2\pi}\int_0^a (f_1(r\cos\theta, r\sin\theta) - f_2(r\cos\theta, r\sin\theta))rdrd\theta$

Example 1.
(1) Find the volume of the region between the xy-plane and the paraboloid $z = x^2 + y^2$ within the cylinder $x^2 + y^2 = a^2$.
(2) Find the volume of the cylinder $x^2 + y^2 = 9$ within the region bounded by the plane $2x + z = 8$ and the plane $z = 0$.

【Solution】

(1)

Let $R = \{(x,y): x^2 + y^2 \leq a^2\}$ then $V = \iint_R \int_0^{x^2+y^2} dzdA = \iint_{x^2+y^2 \leq a^2} x^2 + y^2 dxdy$

Let $x = r\cos\theta$, $y = r\sin\theta$ then $\{(x,y): x^2 + y^2 \leq a^2\} = \{(r,\theta): 0 \leq r \leq a, 0 \leq \theta \leq 2\pi\}$

and $dxdy = \begin{Vmatrix} \frac{\partial x}{\partial r} & \frac{\partial x}{\partial \theta} \\ \frac{\partial y}{\partial r} & \frac{\partial y}{\partial \theta} \end{Vmatrix} drd\theta = \begin{Vmatrix} \cos\theta & -r\sin\theta \\ \sin\theta & r\cos\theta \end{Vmatrix} drd\theta = rdrd\theta$

$\therefore \iint_{x^2+y^2 \leq a^2} x^2 + y^2 dxdy = \int_0^{2\pi}\int_0^a r^3 drd\theta = \int_0^{2\pi} d\theta \int_0^a r^3 dr = \frac{\pi a^4}{2}$

8.4.2 Use Polar Coordinates to Compute Triple Integrals

(2)

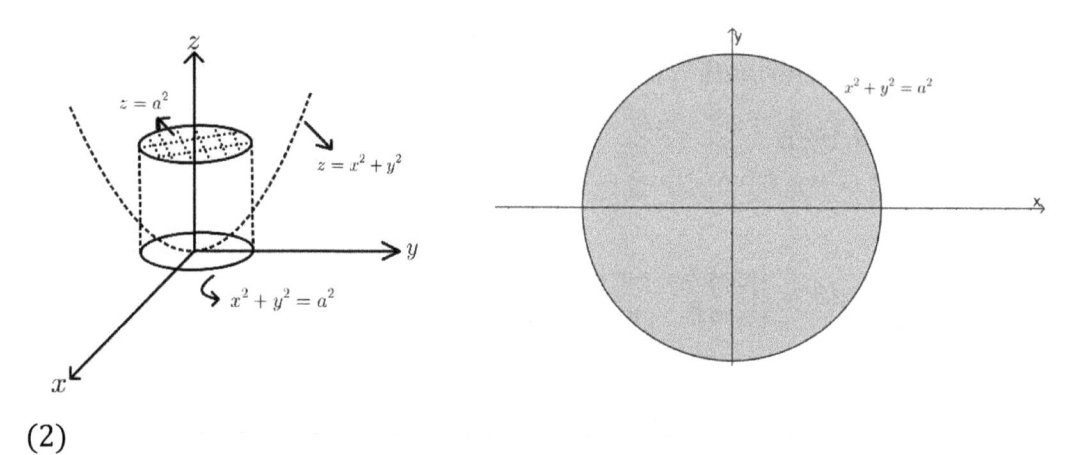

Let $R = \{(x, y): x^2 + y^2 \leq 9\}$ then $V = \iint_R \int_0^{8-2x} dz dA = 2 \iint_{x^2+y^2 \leq 9} 4 - x \, dx dy$

Let $x = r \cos \theta$, $y = r \sin \theta$ then $\{(x, y): x^2 + y^2 \leq 9\} = \{(r, \theta): 0 \leq r \leq 3, 0 \leq \theta \leq 2\pi\}$

and $dxdy = \begin{Vmatrix} \frac{\partial x}{\partial r} & \frac{\partial x}{\partial \theta} \\ \frac{\partial y}{\partial r} & \frac{\partial y}{\partial \theta} \end{Vmatrix} drd\theta = \begin{Vmatrix} \cos \theta & -r \sin \theta \\ \sin \theta & r \cos \theta \end{Vmatrix} drd\theta = rdrd\theta$

$\therefore 2 \iint_{x^2+y^2 \leq 9} 4 - x \, dxdy = 2 \int_0^{2\pi} \int_0^3 (4 - r \cos \theta) r \, drd\theta = 2 \int_0^{2\pi} 2r^2|_0^3 - \frac{r^3 \cos \theta}{3} \Big|_0^3 d\theta$

$= 72\pi$

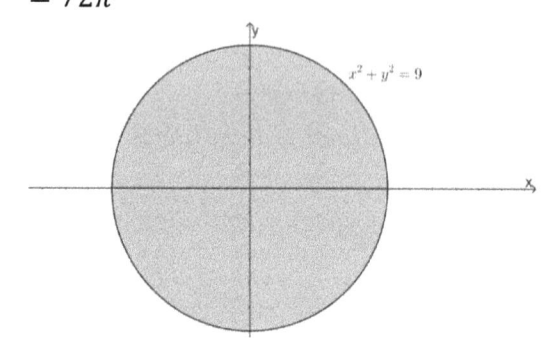

Example 2.

Assume V is the region bounded by the surfaces $x^2 + y^2 = 4z$, $x^2 + y^2 = 8y$, and $z = 0$. Find $\iiint_V dxdydz = ?$

【Solution】

Let $R = \{(x, y): x^2 + (y - 4)^2 \leq 16\}$ then

$$\iiint_V dxdydz = \iint_R \int_0^{\frac{x^2+y^2}{4}} dzdA = \iint_{x^2+(y-4)^2\le 16} \frac{x^2+y^2}{4} dxdy$$

Let $x = 4r\cos\theta$, $y = 4r\sin\theta$ then
$\{(x,y): x^2 + (y-4)^2 \le 16, y \ge 0\} = \{(r,\theta): 0 \le r \le 2\sin\theta, 0 \le \theta \le \pi\}$

and $dxdy = \begin{Vmatrix} \frac{\partial x}{\partial r} & \frac{\partial x}{\partial \theta} \\ \frac{\partial y}{\partial r} & \frac{\partial y}{\partial \theta} \end{Vmatrix} drd\theta = \begin{Vmatrix} 4\cos\theta & -4r\sin\theta \\ 4\sin\theta & 4r\cos\theta \end{Vmatrix} drd\theta = 16r drd\theta$

$\therefore \iint_{x^2+(y-4)^2\le 16} \frac{x^2+y^2}{4} dxdy = \int_0^\pi \int_0^{2\sin\theta} 4r^2 \cdot 16r drd\theta = 64 \int_0^\pi \frac{r^4}{4} \Big|_0^{2\sin\theta} d\theta$

$= 256 \int_0^\pi \sin^4\theta\, d\theta = 256 \cdot \frac{3}{4} \int_0^\pi \sin^2\theta\, d\theta = 256 \cdot \frac{3}{4} \cdot \frac{\pi}{2} = 96\pi$

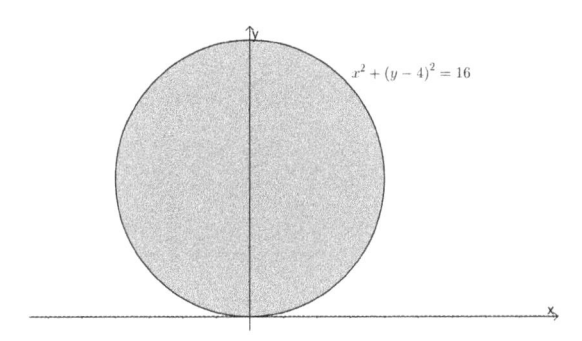

$x^2 + (y-4)^2 = 16$

Example 3.
Assume V is the region bounded by the surfaces $x^2 + y^2 = 4$, $2(x^2 + y^2) = z$, and the xy-plane. Find $\iiint_V dxdydz = ?$

【Solution】
Let $R = \{(x,y): x^2 + y^2 \le 4\}$ then

$$\iiint_V dxdydz = \iint_R \int_0^{2(x^2+y^2)} dzdA = \iint_{x^2+y^2\le 4} 2(x^2+y^2)dxdy$$

Let $x = r\cos\theta$, $y = r\sin\theta$ then $\{(x,y): x^2 + y^2 \le 4\} = \{(r,\theta): 0 \le r \le 2, 0 \le \theta \le 2\pi\}$

and $dxdy = \begin{Vmatrix} \frac{\partial x}{\partial r} & \frac{\partial x}{\partial \theta} \\ \frac{\partial y}{\partial r} & \frac{\partial y}{\partial \theta} \end{Vmatrix} drd\theta = \begin{Vmatrix} \cos\theta & -r\sin\theta \\ \sin\theta & r\cos\theta \end{Vmatrix} drd\theta = rdrd\theta$

8.4.2 Use Polar Coordinates to Compute Triple Integrals

$$\therefore \iint_{x^2+y^2 \leq 4} 2(x^2+y^2)dxdy = \int_0^{2\pi} \int_0^2 2r^3 dr d\theta = 16\pi$$

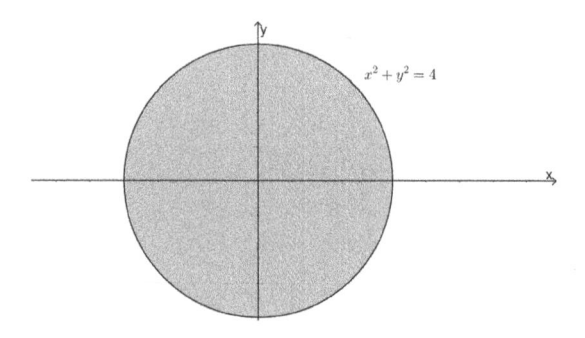

Example 4.

(1) Find the volume of the paraboloid $z = 2x^2 + 2y^2$ within the region $0 \leq z \leq 8$.

(2) Assume V is the region bounded by $z = 2x^2 + 2y^2, x > 0, y > 0, z = 8$.

Find $\iiint_V xdxdydz = ?$

【Solution】

(1)

Let $R = \{(x,y): x^2 + y^2 \leq 4\}$ then $V = \iint_R \int_{2x^2+2y^2}^8 dzdA = 2\iint_{x^2+y^2 \leq 4} 4-(x^2+y^2)dxdy$

Let $x = r\cos\theta$, $y = r\sin\theta$ then $\{(x,y): x^2 + y^2 \leq 4\} = \{(r,\theta): 0 \leq r \leq 2, 0 \leq \theta \leq 2\pi\}$

and $dxdy = \begin{Vmatrix} \dfrac{\partial x}{\partial r} & \dfrac{\partial x}{\partial \theta} \\ \dfrac{\partial y}{\partial r} & \dfrac{\partial y}{\partial \theta} \end{Vmatrix} drd\theta = \begin{Vmatrix} \cos\theta & -r\sin\theta \\ \sin\theta & r\cos\theta \end{Vmatrix} drd\theta = rdrd\theta$

$$\therefore 2\iint_{x^2+y^2 \leq 4} 4-(x^2+y^2)dxdy = 2\int_0^{2\pi}\int_0^2 4r - r^3 drd\theta = 2\int_0^{2\pi} d\theta \int_0^2 4r - r^3 dr$$

$$= 4\pi \left(2r^2 - \frac{r^4}{4}\right)\Big|_0^2 = 16\pi$$

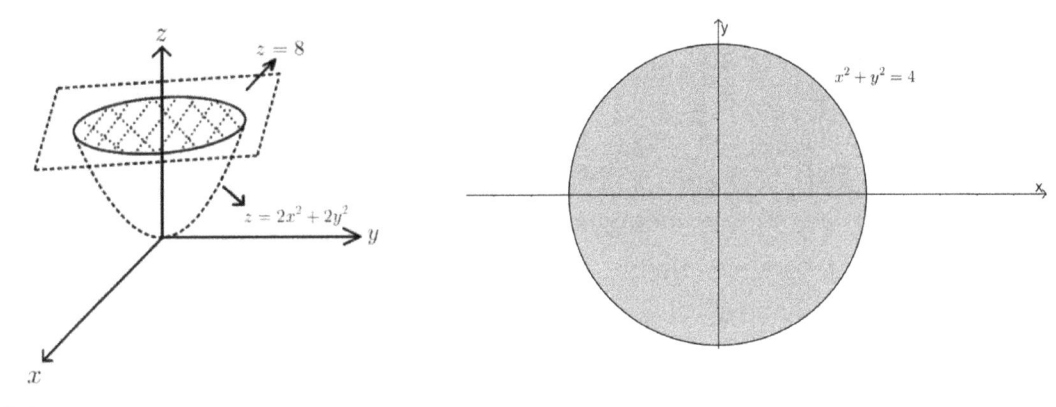

(2)
Let $R = \{(x,y): x^2 + y^2 \leq 4, x > 0, y > 0\}$ then

$$\iiint_V xdxdydz = \iint_R \int_{2x^2+2y^2}^{8} xdzdA = 2\iint_{x^2+y^2\leq 4} x(4-(x^2+y^2))dxdy$$

Let $x = r\cos\theta$, $y = r\sin\theta$ then

$$\{(x,y): x^2+y^2 \leq 4, x > 0, y > 0\} = \{(r,\theta): 0 \leq r \leq 2, 0 \leq \theta \leq \frac{\pi}{2}\}$$

and $dxdy = \begin{Vmatrix} \dfrac{\partial x}{\partial r} & \dfrac{\partial x}{\partial \theta} \\ \dfrac{\partial y}{\partial r} & \dfrac{\partial y}{\partial \theta} \end{Vmatrix} drd\theta = \begin{Vmatrix} \cos\theta & -r\sin\theta \\ \sin\theta & r\cos\theta \end{Vmatrix} drd\theta = rdrd\theta$

$$\therefore 2\iint_{x^2+y^2\leq 4} x(4-(x^2+y^2))dxdy = 2\int_0^{\frac{\pi}{2}} \cos\theta\, d\theta \int_0^2 4r^2 - r^4 dr = 2\left(\frac{4r^3}{3} - \frac{r^5}{5}\right)\Big|_0^2 = \frac{128}{15}$$

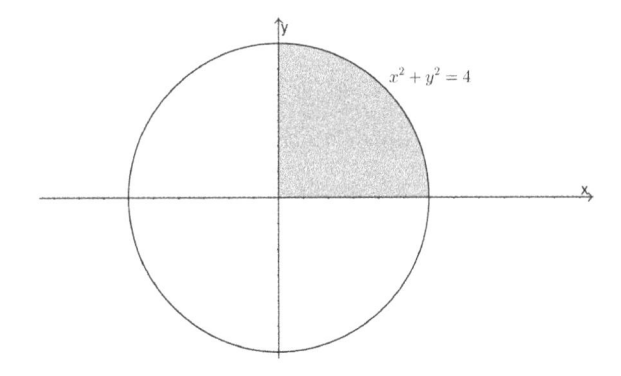

Example 5.
　　Find the volume of the region bounded below by the surface $z = 1 - x^2 - y^2$ and above by the surface $z = 1 - y$.

【Solution】

8.4.2 Use Polar Coordinates to Compute Triple Integrals

Let $R = \left\{(x,y): x^2 + \left(y - \dfrac{1}{2}\right)^2 \leq \dfrac{1}{4}\right\}$ then

$$V = \iint_R \int_{1-y}^{1-x^2-y^2} dzdA = \iint_{x^2+\left(y-\frac{1}{2}\right)^2 \leq \frac{1}{4}} \dfrac{1}{4} - x^2 - \left(y - \dfrac{1}{2}\right)^2 dxdy$$

Let $x = \dfrac{1}{2}r\cos\theta$, $y = \dfrac{1}{2} + \dfrac{1}{2}r\sin\theta$ then

$$\left\{(x,y): x^2 + \left(y - \dfrac{1}{2}\right)^2 \leq \dfrac{1}{4}\right\} = \{(r,\theta): 0 \leq r \leq 1, 0 \leq \theta \leq 2\pi\}$$

and $dxdy = \left\| \begin{matrix} \dfrac{\partial x}{\partial r} & \dfrac{\partial x}{\partial \theta} \\ \dfrac{\partial y}{\partial r} & \dfrac{\partial y}{\partial \theta} \end{matrix} \right\| drd\theta = \left\| \begin{matrix} \dfrac{1}{2}\cos\theta & \dfrac{1}{2}r\sin\theta \\ \dfrac{1}{2}\sin\theta & \dfrac{1}{2}r\cos\theta \end{matrix} \right\| drd\theta = \dfrac{1}{4}rdrd\theta$

$$\therefore \iint_{x^2+\left(y-\frac{1}{2}\right)^2 \leq \frac{1}{4}} \dfrac{1}{4} - x^2 - \left(y - \dfrac{1}{2}\right)^2 dxdy = \int_0^{2\pi} \int_0^1 \left(\dfrac{1}{4} - \dfrac{1}{4}r^2\right)\dfrac{1}{4}rdrd\theta$$

$$= \dfrac{1}{16} \int_0^{2\pi} d\theta \int_0^1 r - r^3 dr = \dfrac{\pi}{32}$$

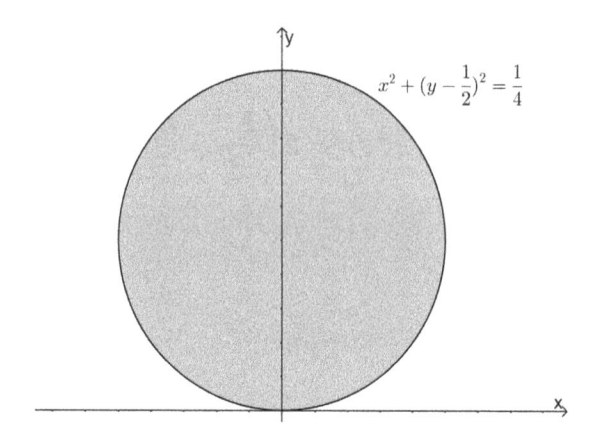

Example 6.

Assume V is the region bounded by $z \leq 1 - x^2 - y^2$, $z \geq 1 - y$, $x > 0, y > 0$, and $z > 0$. Find $\iiint_V xdxdydz = ?$

【Solution】

Let $R = \left\{(x,y): x^2 + \left(y - \dfrac{1}{2}\right)^2 \leq \dfrac{1}{4}, x > 0, y > 0\right\}$ then

$$\iiint_V xdxdydz = \iint_R \int_{1-y}^{1-x^2-y^2} dzdA = \iint_R x\left(\frac{1}{4}-x^2-\left(y-\frac{1}{2}\right)^2\right)dxdy$$

Let $x = \frac{1}{2}r\cos\theta$, $y = \frac{1}{2}+\frac{1}{2}r\sin\theta$ then

$$\{(x,y): x^2+\left(y-\frac{1}{2}\right)^2 \le \frac{1}{4}, x > 0, y > 0\} = \{(r,\theta): 0 \le r \le 1, 0 \le \theta \le \frac{\pi}{2}\}$$

and $dxdy = \begin{Vmatrix} \frac{\partial x}{\partial r} & \frac{\partial x}{\partial \theta} \\ \frac{\partial y}{\partial r} & \frac{\partial y}{\partial \theta} \end{Vmatrix} drd\theta = \begin{Vmatrix} \frac{1}{2}\cos\theta & \frac{1}{2}r\sin\theta \\ \frac{1}{2}\sin\theta & \frac{1}{2}r\cos\theta \end{Vmatrix} drd\theta = \frac{1}{4}rdrd\theta$

$$\therefore \iint_R x\left(\frac{1}{4}-x^2-\left(y-\frac{1}{2}\right)^2\right)dxdy = \int_0^{\frac{\pi}{2}} \frac{1}{2}r\cos\theta \int_0^1 \left(\frac{1}{4}-\frac{1}{4}r^2\right)\frac{1}{4}rdrd\theta$$

$$= \frac{1}{32}\int_0^{\frac{\pi}{2}}\cos\theta \, d\theta \int_0^1 r^2-r^4 dr = \frac{1}{240}$$

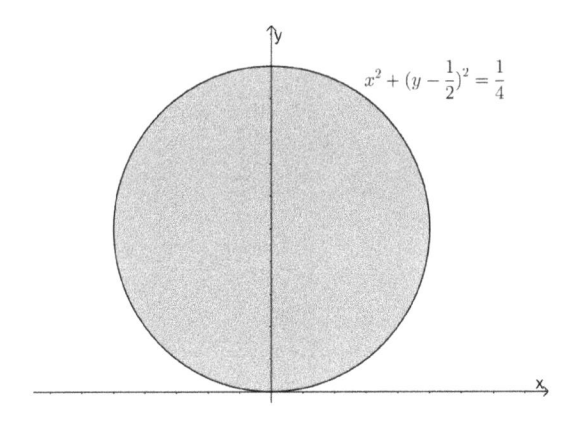

$x^2+(y-\frac{1}{2})^2 = \frac{1}{4}$

Example 7.

(1) Find the volume of the region bounded by the paraboloid $6x^2+6y^2+z = 20$ and the plane $z = 8$.

(2) Assume V is the region bounded by $6x^2+6y^2+z = 20$, $x > 0$, $y > 0$, $z > 0$, and $z = 8$. Find $\iiint_V xdxdydz = ?$

【Solution】

(1)

Let $R = \{(x,y): x^2+y^2 \le 2\}$ then

8.4.2 Use Polar Coordinates to Compute Triple Integrals

$$V = \iint_R \int_8^{20-6x^2-6y^2} dz\,dA = 2\iint_{x^2+y^2 \leq 2} 6 - 3x^2 - 3y^2\,dxdy$$

Let $x = r\cos\theta$, $y = r\sin\theta$ then $\{(x,y): x^2 + y^2 \leq 2\} = \{(r,\theta): 0 \leq r \leq \sqrt{2}, 0 \leq \theta \leq 2\pi\}$

and $dxdy = \left\| \begin{matrix} \frac{\partial x}{\partial r} & \frac{\partial x}{\partial \theta} \\ \frac{\partial y}{\partial r} & \frac{\partial y}{\partial \theta} \end{matrix} \right\| dr d\theta = \left\| \begin{matrix} \cos\theta & -r\sin\theta \\ \sin\theta & r\cos\theta \end{matrix} \right\| drd\theta = rdrd\theta$

$$\therefore 2\iint_{x^2+y^2 \leq 2} 6 - 3x^2 - 3y^2\,dxdy = 2\int_0^{2\pi}\int_0^{\sqrt{2}} 6r - 3r^3\,drd\theta = 2\int_0^{2\pi} d\theta \int_0^{\sqrt{2}} 6r - 3r^3\,dr$$

$$= 4\pi \left(3r^2 - \frac{3r^4}{4} \right)\Big|_0^{\sqrt{2}} = 12\pi$$

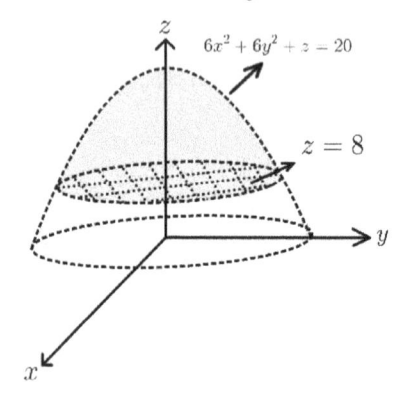

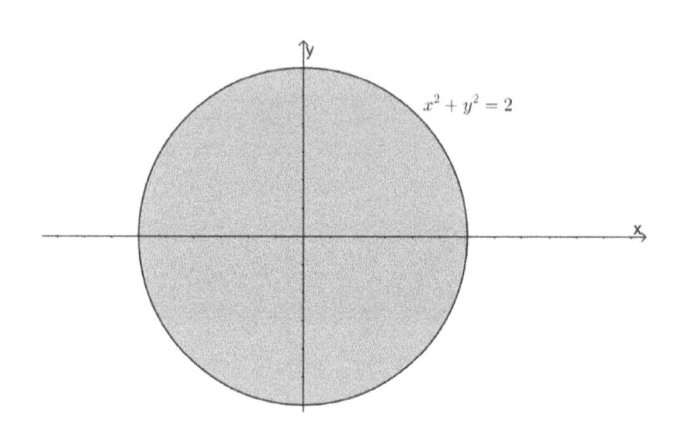

(2)
Let $R = \{(x,y): x^2 + y^2 \leq 2, x > 0, y > 0\}$ then

$$V = \iint_R \int_8^{20-6x^2-6y^2} x\,dz\,dA = 2\iint_{x^2+y^2 \leq 2, x>0, y>0} x(6 - 3x^2 - 3y^2)dxdy$$

Let $x = r\cos\theta$, $y = r\sin\theta$ then

$$\{(x,y): x^2 + y^2 \leq 2, x > 0, y > 0\} = \{(r,\theta): 0 \leq r \leq \sqrt{2}, 0 \leq \theta \leq \frac{\pi}{2}\}$$

and $dxdy = \left\| \begin{matrix} \frac{\partial x}{\partial r} & \frac{\partial x}{\partial \theta} \\ \frac{\partial y}{\partial r} & \frac{\partial y}{\partial \theta} \end{matrix} \right\| dr d\theta = \left\| \begin{matrix} \cos\theta & -r\sin\theta \\ \sin\theta & r\cos\theta \end{matrix} \right\| drd\theta = rdrd\theta$

$$\therefore \iiint_V x\,dxdydz = 2\iint_{x^2+y^2 \leq 2, x>0, y>0} x(6 - 3x^2 - 3y^2)dxdy$$

$$= 2\int_0^{\frac{\pi}{2}} \cos\theta \int_0^{\sqrt{2}} 6r^2 - 3r^4\,drd\theta = 2\int_0^{\frac{\pi}{2}} \cos\theta\,d\theta \int_0^{\sqrt{2}} 6r^2 - 3r^4\,dr = 2\left(2r^3 - \frac{3r^5}{5} \right)\Big|_0^{\sqrt{2}}$$

$$= \frac{16\sqrt{2}}{5}$$

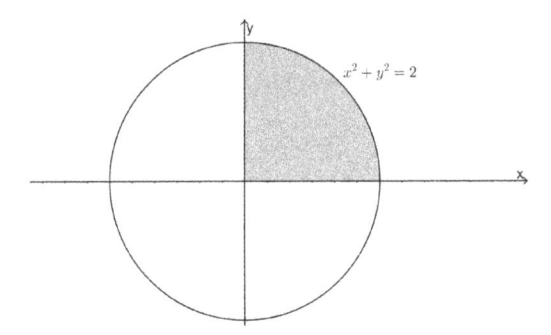

Example 8.
Find the volume enclosed by the two paraboloids $z = 5x^2 + 5y^2$ and $z = 6 - 7x^2 - y^2$.

【Solution】

Let $R = \{(x, y): 2x^2 + y^2 \leq 1\}$ then

$$V = \iint_R \int_{5x^2+5y^2}^{6-7x^2-y^2} dz dA = \iint_R z\Big|_{5x^2+5y^2}^{6-7x^2-y^2} dA = \iint_R 6 - 12x^2 - 6y^2 dA$$

Let $x = \dfrac{u}{\sqrt{2}}$, $y = v$ then $dxdy = \begin{Vmatrix} \dfrac{\partial x}{\partial u} & \dfrac{\partial x}{\partial v} \\ \dfrac{\partial y}{\partial u} & \dfrac{\partial y}{\partial v} \end{Vmatrix} dudv = \begin{Vmatrix} \dfrac{1}{\sqrt{2}} & -0 \\ 0 & 1 \end{Vmatrix} dudv = \dfrac{1}{\sqrt{2}} dudv$

and $R = \{(u, v): u^2 + v^2 \leq 1\}$

$$\therefore \iint_R 6 - 12x^2 - 6y^2 dA = \frac{6}{\sqrt{2}} \iint_{u^2+v^2 \leq 1} 1 - u^2 - v^2 dudv$$

Let $u = r\cos\theta$, $v = r\sin\theta$ then $\{(u,v): u^2+v^2 \leq 1\} = \{(r,\theta): 0 \leq r \leq 1, 0 \leq \theta \leq 2\pi\}$

and $dudv = \begin{Vmatrix} \dfrac{\partial u}{\partial r} & \dfrac{\partial u}{\partial \theta} \\ \dfrac{\partial v}{\partial r} & \dfrac{\partial v}{\partial \theta} \end{Vmatrix} drd\theta = \begin{Vmatrix} \cos\theta & -r\sin\theta \\ \sin\theta & r\cos\theta \end{Vmatrix} drd\theta = rdrd\theta$

$$\therefore V = \frac{6}{\sqrt{2}} \iint_{u^2+v^2 \leq 1} 1 - u^2 - v^2 dudv = \frac{6}{\sqrt{2}} \int_0^1 \int_0^{2\pi} (1-r^2) r d\theta dr = \frac{3\sqrt{2}}{2}$$

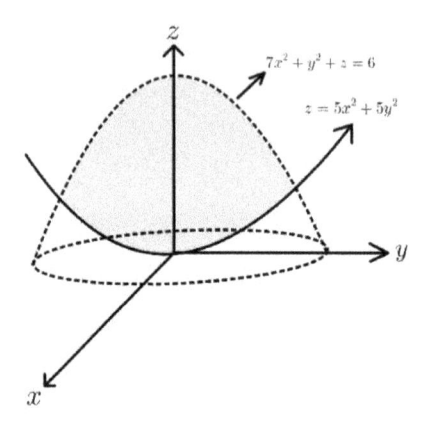

8.4.2 Use Polar Coordinates to Compute Triple Integrals

Example 9.

Calculate the volume of the region enclosed by the sphere $x^2 + y^2 + z^2 \leq 6$ and the paraboloid $z \geq x^2 + y^2$.

【Solution】

Let $R = \{(x,y): x^2 + y^2 \leq 2\}$ then

$$V = \iint_R \int_{x^2+y^2}^{\sqrt{6-x^2-y^2}} dz dA = \iint_{x^2+y^2 \leq 2} \sqrt{6-x^2-y^2} - x^2 - y^2 dxdy$$

Let $x = r\cos\theta, y = r\sin\theta$ then $\{(x,y): x^2 + y^2 \leq 2\} = \{(r,\theta): 0 \leq r \leq \sqrt{2}, 0 \leq \theta \leq 2\pi\}$

and $dxdy = \begin{Vmatrix} \dfrac{\partial x}{\partial r} & \dfrac{\partial x}{\partial \theta} \\ \dfrac{\partial y}{\partial r} & \dfrac{\partial y}{\partial \theta} \end{Vmatrix} drd\theta = \begin{Vmatrix} \cos\theta & -r\sin\theta \\ \sin\theta & r\cos\theta \end{Vmatrix} drd\theta = rdrd\theta$

$$\therefore \iint_{x^2+y^2 \leq 2} \sqrt{6-x^2-y^2} - x^2 - y^2 dxdy = \int_0^{2\pi} \int_0^{\sqrt{2}} (\sqrt{6-r^2} - r^2) r dr d\theta$$

$$= (2\pi) \cdot \left(\dfrac{(-1)(6-r^2)^{\frac{3}{2}}}{3} - \dfrac{r^4}{4} \right) \Big|_{r=0}^{r=\sqrt{2}} = (2\pi) \cdot \left(\dfrac{-8+6\sqrt{6}}{3} - 1 \right) = \dfrac{2\pi}{3}(6\sqrt{6} - 11)$$

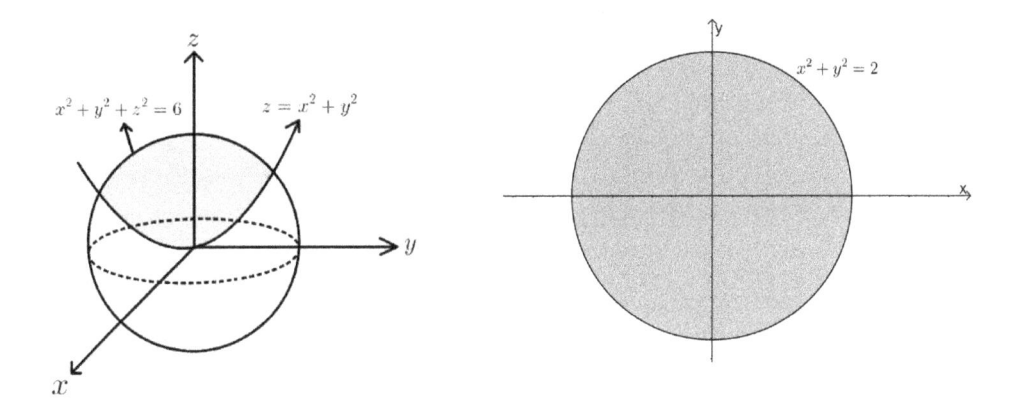

8.4.2 Use Polar Coordinates to Compute Triple Integrals

Example 10.

Calculate the volume of the region enclosed by the sphere $x^2 + y^2 + z^2 = 4a^2$ and the cylinder $x^2 + (y - a)^2 = a^2$.

【Solution】

Let $R = \{(x, y): x^2 + (y - a)^2 \leq a^2\}$ then

$$V = \iint_R \int_{-\sqrt{4a^2-x^2-y^2}}^{\sqrt{4a^2-x^2-y^2}} dz dA = \iint_{x^2+(y-a)^2 \leq a^2} 2\sqrt{4a^2 - x^2 - y^2} dx dy$$

Let $x = ar\cos\theta, y = ar\sin\theta$ then
$\{(x, y): x^2 + (y - a)^2 \leq a^2\} = \{(r, \theta): 0 \leq r \leq 2\sin\theta, 0 \leq \theta \leq \pi\}$

and $dxdy = \begin{Vmatrix} \dfrac{\partial x}{\partial r} & \dfrac{\partial x}{\partial \theta} \\ \dfrac{\partial y}{\partial r} & \dfrac{\partial y}{\partial \theta} \end{Vmatrix} drd\theta = \begin{Vmatrix} a\cos\theta & ar\sin\theta \\ a\sin\theta & ar\cos\theta \end{Vmatrix} drd\theta = a^2 r drd\theta$

$$\therefore \iint_{x^2+(y-a)^2 \leq a^2} 2\sqrt{4a^2 - x^2 - y^2} dx dy = 2\int_0^{\pi} \int_0^{2\sin\theta} (\sqrt{4a^2 - a^2 r^2}) a^2 r drd\theta$$

$$= 4a^3 \int_0^{\frac{\pi}{2}} \int_0^{2\sin\theta} (\sqrt{4 - r^2}) r drd\theta = 4a^3 \int_0^{\frac{\pi}{2}} \left(\dfrac{(-1)(4-r^2)^{\frac{3}{2}}}{3} \right) \Bigg|_0^{2\sin\theta} d\theta$$

$$= 32a^3 \int_0^{\frac{\pi}{2}} \dfrac{1 - \cos^3\theta}{3} d\theta = 32a^3 \left(\dfrac{\pi}{6} - \dfrac{2}{9} \right)$$

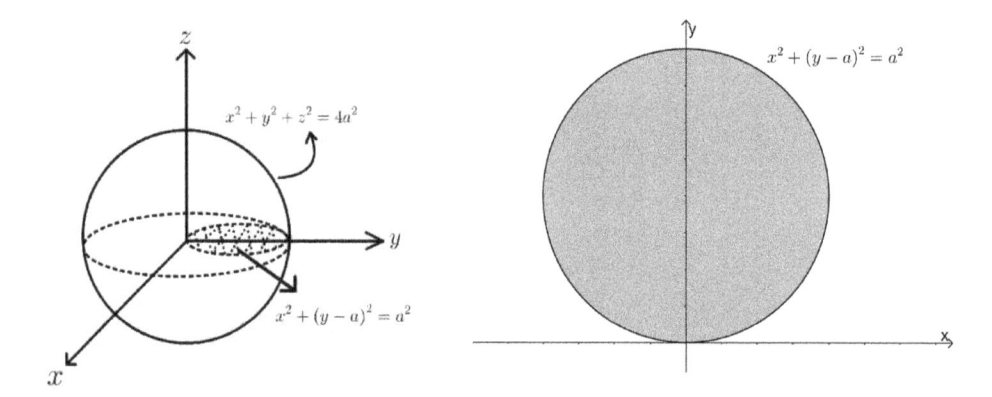

8.4.2 Use Polar Coordinates to Compute Triple Integrals

Example 11.

Find the volume of the region enclosed by $z^2 = x^2 + y^2$ and $nz = x^2 + y^2$, $n \in N$.

【Solution】

Let $R = \{(x,y): x^2 + y^2 \leq n^2\}$ then

$$V = \iint_R \int_{\frac{x^2+y^2}{n}}^{\sqrt{x^2+y^2}} dz dA = \iint_{x^2+y^2 \leq n^2} \sqrt{x^2+y^2} - \left(\frac{x^2+y^2}{n}\right) dxdy$$

Let $x = r\cos\theta, y = r\sin\theta$ then $\{(x,y): x^2+y^2 \leq n^2\} = \{(r,\theta): 0 \leq r \leq n, 0 \leq \theta \leq 2\pi\}$

and $dxdy = \begin{Vmatrix} \dfrac{\partial x}{\partial r} & \dfrac{\partial x}{\partial \theta} \\ \dfrac{\partial y}{\partial r} & \dfrac{\partial y}{\partial \theta} \end{Vmatrix} drd\theta = \begin{Vmatrix} \cos\theta & -r\sin\theta \\ \sin\theta & r\cos\theta \end{Vmatrix} drd\theta = rdrd\theta$

$$\therefore \iint_{x^2+y^2 \leq n^2} \sqrt{x^2+y^2} - \left(\frac{x^2+y^2}{n}\right) dxdy = \int_0^{2\pi} \int_0^n r^2 - \frac{r^3}{n} drd\theta = 2\pi\left(\frac{n^3}{3} - \frac{n^3}{4}\right) = \frac{\pi n^3}{6}$$

Example 12.

Find $\displaystyle\int_0^3 \int_0^{\sqrt{9-x^2}} \int_0^2 \sqrt{x^2+y^2}\, dzdydx = ?$

【Solution】

Let $x = r\cos\theta, y = r\sin\theta$ then

$\{(x,y): 0 \leq x \leq 3, 0 \leq y \leq \sqrt{9-x^2}\} = \{(r,\theta): 0 \leq r \leq 3, 0 \leq \theta \leq \dfrac{\pi}{2}\}$

and $dxdy = \begin{Vmatrix} \dfrac{\partial x}{\partial r} & \dfrac{\partial x}{\partial \theta} \\ \dfrac{\partial y}{\partial r} & \dfrac{\partial y}{\partial \theta} \end{Vmatrix} drd\theta = \begin{Vmatrix} \cos\theta & -r\sin\theta \\ \sin\theta & r\cos\theta \end{Vmatrix} drd\theta = rdrd\theta$

$$\therefore \int_0^3 \int_0^{\sqrt{9-x^2}} \int_0^2 \sqrt{x^2+y^2}\, dzdydx = \int_0^3 \int_0^{\frac{\pi}{2}} 2r^2 d\theta dr = \left.\frac{2r^3}{3}\right|_0^3 \left.\frac{\pi}{2}\right|_0 = 9\pi$$

Example 13.
  (1) Find the volume of the region enclosed by $z = x^2 + 3y^2 + 1$ and $z - 9 + x^2 + y^2 = 0$.
  (2) Assume V is the region bounded by $z = x^2 + 3y^2 + 1, z - 9 + x^2 + y^2 = 0$, $x > 0,\ y > 0,\ \text{and } z > 0$. Find $\iiint_V xdxdydz =?$

【Solution】

(1)
Let $R = \{(x,y): x^2 + 2y^2 \leq 4\}$ then
$$V = \iint_R \int_{x^2+3y^2+1}^{9-x^2-y^2} dzdA = \iint_{x^2+2y^2\leq 4} 8 - 2x^2 - 4y^2 dxdy$$

Let $x = 2r\cos\theta,\ y = \sqrt{2}r\sin\theta$ then
$\{(x,y): x^2 + 2y^2 \leq 4\} = \{(r,\theta): 0 \leq r \leq 1, 0 \leq \theta \leq 2\pi\}$

and $dxdy = \begin{Vmatrix} \frac{\partial x}{\partial r} & \frac{\partial x}{\partial \theta} \\ \frac{\partial y}{\partial r} & \frac{\partial y}{\partial \theta} \end{Vmatrix} drd\theta = \begin{Vmatrix} 2\cos\theta & -2r\sin\theta \\ \sqrt{2}\sin\theta & \sqrt{2}r\cos\theta \end{Vmatrix} drd\theta = 2\sqrt{2}rdrd\theta$

$$\therefore \iint_{x^2+2y^2\leq 4} 8 - 2x^2 - 4y^2 dxdy = 2\sqrt{2}\int_0^{2\pi}\int_0^1 8r - 8r^3 drd\theta = 16\sqrt{2}\int_0^{2\pi} d\theta \int_0^1 r - r^3 dr$$
$$= 8\sqrt{2}\pi$$

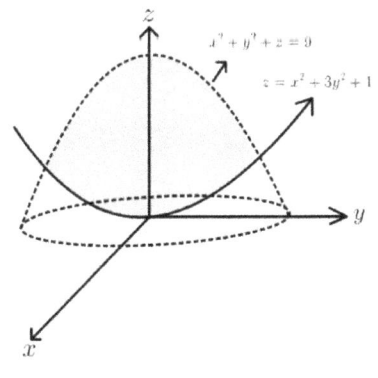

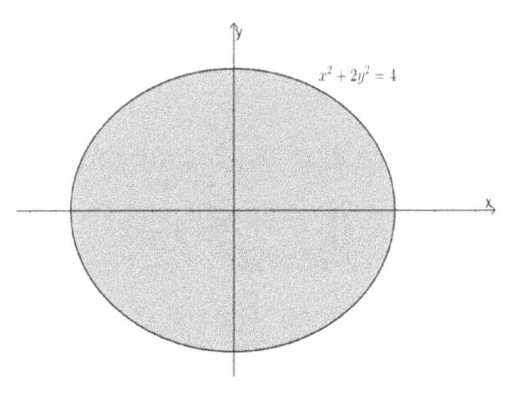

(2)
Let $R = \{(x,y): x^2 + 2y^2 \leq 4, x > 0, y > 0\ \}$ then

8.4.2 Use Polar Coordinates to Compute Triple Integrals

$$\iiint_V x\,dxdydz = \iint_R \int_{x^2+3y^2+1}^{9-x^2-y^2} x\,dz\,dA = \iint_{x^2+2y^2\le 4, x>0, y>0} x(8-2x^2-4y^2)\,dxdy$$

Let $x = 2r\cos\theta$, $y = \sqrt{2}r\sin\theta$ then

$$\{(x,y): x^2 + 2y^2 \le 4\} = \{(r,\theta): 0 \le r \le 1, 0 \le \theta \le \frac{\pi}{2}\}$$

and $dxdy = \left\| \begin{matrix} \frac{\partial x}{\partial r} & \frac{\partial x}{\partial \theta} \\ \frac{\partial y}{\partial r} & \frac{\partial y}{\partial \theta} \end{matrix} \right\| drd\theta = \left\| \begin{matrix} 2\cos\theta & -2r\sin\theta \\ \sqrt{2}\sin\theta & \sqrt{2}r\cos\theta \end{matrix} \right\| drd\theta = 2\sqrt{2}rdrd\theta$

$$\therefore \iint_{x^2+2y^2\le 4, x>0, y>0} x(8-2x^2-4y^2)\,dxdy = 2\sqrt{2}\int_0^{\frac{\pi}{2}} 2r\cos\theta \int_0^1 8r^2 - 8r^4\,drd\theta$$

$$= 32\sqrt{2}\int_0^{\frac{\pi}{2}} \cos\theta\,d\theta \int_0^1 r^2 - r^4\,dr = \frac{64\sqrt{2}}{15}$$

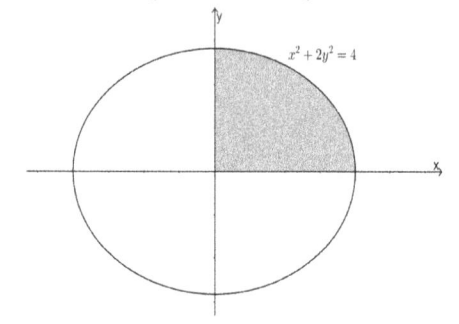

Example 14.

Assume V is the region bounded by $z = x^2 + y^2$, $x = 0$, $y = 0$, and $z = 4$.

Find $\iiint_V 3x\,dxdydz = ?$

【Solution】

Let $R = \{(x,y): x \ge 0, y \ge 0, x^2 + y^2 \le 4\}$ then

$$V = \iint_R \int_{x^2+y^2}^{4} 3x\,dzdA = \iint_{x^2+y^2\le 4} 3x(4-x^2-y^2)\,dxdy$$

Let $x = r\cos\theta, y = r\sin\theta$ then $\{(x,y): x^2+y^2 \le 4\} = \{(r,\theta): 0 \le r \le 2, 0 \le \theta \le \frac{\pi}{2}\}$

and $dxdy = \left\| \begin{matrix} \frac{\partial x}{\partial r} & \frac{\partial x}{\partial \theta} \\ \frac{\partial y}{\partial r} & \frac{\partial y}{\partial \theta} \end{matrix} \right\| drd\theta = \left\| \begin{matrix} \cos\theta & -r\sin\theta \\ \sin\theta & r\cos\theta \end{matrix} \right\| drd\theta = rdrd\theta$

$$\therefore \iint_{x^2+y^2\leq 4} 3x(4-x^2-y^2)dxdy = \int_0^{\frac{\pi}{2}}\int_0^2 3r\cos\theta\,(4-r^2)rdrd\theta$$

$$= \int_0^{\frac{\pi}{2}}\cos\theta d\theta \int_0^2 (12r^2 - 3r^4)dr = \frac{64}{5}$$

Example 15.

Find $\int_0^4 \int_0^{\sqrt{16-x^2}} \int_0^1 \ln\sqrt{x^2+y^2}\,dzdydx = ?$

【Solution】

Let $x = r\cos\theta$, $y = r\sin\theta$ then

$\{(x,y): 0 \leq x \leq 4, 0 \leq y \leq \sqrt{16-x^2}\,\} = \{(r,\theta): 0 \leq r \leq 4, 0 \leq \theta \leq \frac{\pi}{2}\}$

and $dxdy = \begin{Vmatrix} \frac{\partial x}{\partial r} & \frac{\partial x}{\partial \theta} \\ \frac{\partial y}{\partial r} & \frac{\partial y}{\partial \theta} \end{Vmatrix} drd\theta = \begin{Vmatrix} \cos\theta & -r\sin\theta \\ \sin\theta & r\cos\theta \end{Vmatrix} drd\theta = rdrd\theta$

$$\therefore \int_0^4 \int_0^{\sqrt{16-x^2}} \int_0^1 \ln\sqrt{x^2+y^2}\,dzdydx = \int_0^4 \int_0^{\frac{\pi}{2}} r\ln r\,d\theta dr = \int_0^{\frac{\pi}{2}}d\theta \int_0^4 r\ln r\,dr$$

$$= \frac{\pi}{2}\cdot \left(\frac{r^2\ln r}{2} - \frac{r^2}{4}\right)\Big|_0^4 = \pi(8\ln 2 - 2)$$

Example 16.

(1) Find the volume of the region enclosed by the xy-plane, $z = x^2 - 2$ and $x^2 + y^2 = 1$.

(2) Find the volume of the region enclosed by $z = 1 - 2x^2 - 2y^2$ and

8.4.2 Use Polar Coordinates to Compute Triple Integrals

$$z = (x^2 + y^2)^2 - 2.$$

【Solution】

(1)

Let $R = \{(x,y): x^2 + y^2 \leq 1\}$ then $V = \iint_R \int_{x^2-2}^{0} dz dA = \iint_{x^2+y^2 \leq 1} 2 - x^2 dxdy$

Let $x = r\cos\theta, y = r\sin\theta$ then $\{(x,y): x^2 + y^2 \leq 1\} = \{(r,\theta): 0 \leq r \leq 1, 0 \leq \theta \leq 2\pi\}$

and $dxdy = \begin{Vmatrix} \dfrac{\partial x}{\partial r} & \dfrac{\partial x}{\partial \theta} \\ \dfrac{\partial y}{\partial r} & \dfrac{\partial y}{\partial \theta} \end{Vmatrix} drd\theta = \begin{Vmatrix} \cos\theta & -r\sin\theta \\ \sin\theta & r\cos\theta \end{Vmatrix} drd\theta = rdrd\theta$

$\therefore \iint_{x^2+y^2\leq 1} 2 - x^2 dxdy = \int_0^{2\pi} \int_0^1 (2 - r^2\cos^2\theta) r dr d\theta = \int_0^{2\pi} 1 - \dfrac{\cos^2\theta}{4} d\theta$

$= \int_0^{2\pi} 1 - \dfrac{1}{4}\left(\dfrac{1 + \cos 2\theta}{2}\right) d\theta = \dfrac{7\pi}{4}$

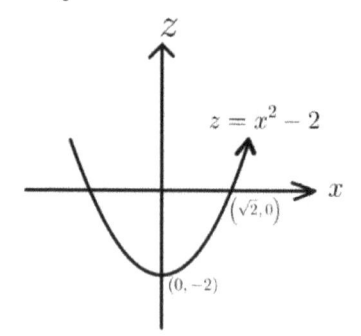

(2)

Let $R = \{(x,y): x^2 + y^2 \leq 1\}$ then

$V = \iint_R \int_{(x^2+y^2)^2-2}^{1-2x^2-2y^2} dz dA = \iint_{x^2+y^2 \leq 1} 1 - 2x^2 - 2y^2 - ((x^2+y^2)^2 - 2) dxdy$

Let $x = r\cos\theta, y = r\sin\theta$ then $\{(x,y): x^2 + y^2 \leq 1\} = \{(r,\theta): 0 \leq r \leq 1, 0 \leq \theta \leq 2\pi\}$

and $dxdy = \begin{Vmatrix} \dfrac{\partial x}{\partial r} & \dfrac{\partial x}{\partial \theta} \\ \dfrac{\partial y}{\partial r} & \dfrac{\partial y}{\partial \theta} \end{Vmatrix} drd\theta = \begin{Vmatrix} \cos\theta & -r\sin\theta \\ \sin\theta & r\cos\theta \end{Vmatrix} drd\theta = rdrd\theta$

$\therefore \iint_{x^2+y^2\leq 1} 1 - 2x^2 - 2y^2 - ((x^2+y^2)^2 - 2) dxdy = \int_0^{2\pi} \int_0^1 (-r^4 - 2r^2 + 3) r dr d\theta$

$= \int_0^{2\pi} d\theta \int_0^1 (-r^5 - 2r^3 + 3r) dr = 2\pi \left(-\dfrac{1}{6} - \dfrac{2}{4} + \dfrac{3}{2}\right) = \dfrac{5\pi}{3}$

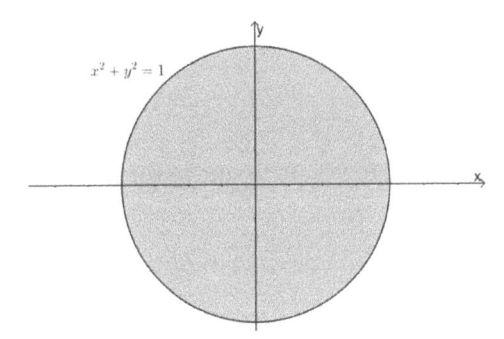

8.4.2 Use Polar Coordinates to Compute Triple Integrals

Example 17.

Find $\int_0^1 \int_0^{\sqrt{1-x^2}} \int_0^1 e^{\sqrt{x^2+y^2}} dz\,dy\,dx = ?$

【Solution】

Let $x = r\cos\theta$, $y = r\sin\theta$ then

$\{(x,y): 0 \le x \le 1, 0 \le y \le \sqrt{1-x^2}\} = \{(r,\theta): 0 \le r \le 1, 0 \le \theta \le \frac{\pi}{2}\}$

and $dxdy = \begin{Vmatrix} \frac{\partial x}{\partial r} & \frac{\partial x}{\partial \theta} \\ \frac{\partial y}{\partial r} & \frac{\partial y}{\partial \theta} \end{Vmatrix} dr d\theta = \begin{Vmatrix} \cos\theta & -r\sin\theta \\ \sin\theta & r\cos\theta \end{Vmatrix} dr d\theta = r\,dr\,d\theta$

$\therefore \int_0^1 \int_0^{\sqrt{1-x^2}} \int_0^1 e^{\sqrt{x^2+y^2}} dz\,dy\,dx = \int_0^1 \int_0^{\frac{\pi}{2}} re^r d\theta dr = \int_0^{\frac{\pi}{2}} d\theta \int_0^1 e^r r\,dr = \frac{\pi}{2} \cdot (re^r - e^r)|_0^1 = \frac{\pi}{2}$

Example 18.

(1) Calculate the volume of the sphere $x^2 + y^2 + z^2 = 25$ after being truncated by the cylinder $x^2 + y^2 = 16$.

(2) Calculate the remaining volume of the sphere $x^2 + y^2 + z^2 = 25$ after being truncated by the cylinder $x^2 + y^2 = 16$.

【Solution】

(1)
Let $R = \{(x,y): x^2 + y^2 \le 16\}$ then

$V = \iint_R \int_{-\sqrt{25-x^2-y^2}}^{\sqrt{25-x^2-y^2}} dz\,dA = \iint_{x^2+y^2 \le 16} 2\sqrt{25-x^2-y^2}\,dxdy$

Let $x = r\cos\theta, y = r\sin\theta$ then $\{(x,y): x^2 + y^2 \le 16\} = \{(r,\theta): 0 \le r \le 4, 0 \le \theta \le 2\pi\}$

8.4.2 Use Polar Coordinates to Compute Triple Integrals

and $dxdy = \begin{Vmatrix} \dfrac{\partial x}{\partial r} & \dfrac{\partial x}{\partial \theta} \\ \dfrac{\partial y}{\partial r} & \dfrac{\partial y}{\partial \theta} \end{Vmatrix} drd\theta = \begin{Vmatrix} \cos\theta & -r\sin\theta \\ \sin\theta & r\cos\theta \end{Vmatrix} drd\theta = rdrd\theta$

$\therefore \iint_{x^2+y^2 \leq 16} 2\sqrt{25-x^2-y^2}\, dxdy = 2\int_0^{2\pi} d\theta \int_0^4 \sqrt{25-r^2}\, rdr$

$= 2(2\pi) \cdot \dfrac{(-1)(25-r^2)^{\frac{3}{2}}}{3} \Big|_0^4 = \dfrac{392\pi}{3}$

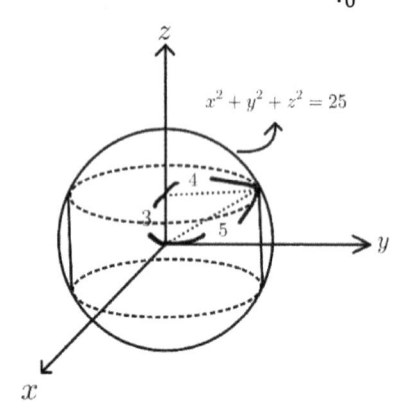

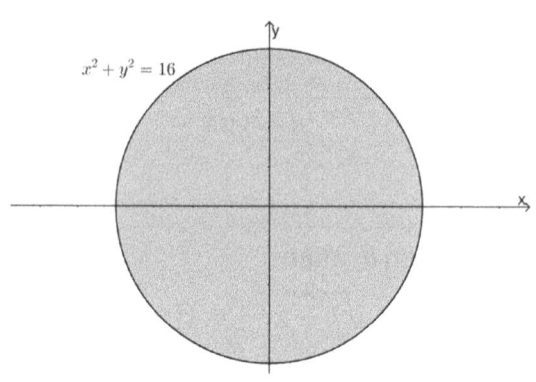

(2)
Let $R = \{(x,y): 16 \leq x^2 + y^2 \leq 25\}$ then

$V = \iint_R \int_{-\sqrt{25-x^2-y^2}}^{\sqrt{25-x^2-y^2}} dzdA = \iint_{16 \leq x^2+y^2 \leq 25} 2\sqrt{25-x^2-y^2}\, dxdy$

Let $x = r\cos\theta, y = r\sin\theta$ then $\{(x,y): x^2+y^2 \leq 1\} = \{(r,\theta): 4 \leq r \leq 5, 0 \leq \theta \leq 2\pi\}$

and $dxdy = \begin{Vmatrix} \dfrac{\partial x}{\partial r} & \dfrac{\partial x}{\partial \theta} \\ \dfrac{\partial y}{\partial r} & \dfrac{\partial y}{\partial \theta} \end{Vmatrix} drd\theta = \begin{Vmatrix} \cos\theta & -r\sin\theta \\ \sin\theta & r\cos\theta \end{Vmatrix} drd\theta = rdrd\theta$

$\therefore \iint_{16 \leq x^2+y^2 \leq 25} 2\sqrt{25-x^2-y^2}\, dxdy = 2\int_0^{2\pi} d\theta \int_4^5 \sqrt{25-r^2}\, rdr$

$= 2(2\pi) \cdot \dfrac{(-1)(25-r^2)^{\frac{3}{2}}}{3} \Big|_4^5 = 36\pi$

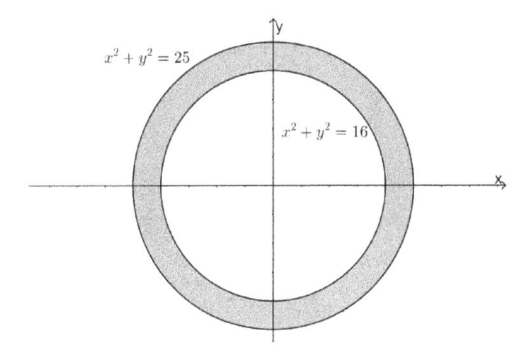

8.4.2 Use Polar Coordinates to Compute Triple Integrals

Example 19.

Calculate the volume of the sphere $x^2 + y^2 + z^2 = 4$ after being truncated by the cylinder $x^2 + y^2 = 2x$.

【Solution】

Let $R = \{(x,y): (x-1)^2 + y^2 \leq 1\}$ then

$$V = \iint_R \int_{-\sqrt{4-x^2-y^2}}^{\sqrt{4-x^2-y^2}} dz dA = \iint_{(x-1)^2+y^2\leq 1} 2\sqrt{4-x^2-y^2} dxdy$$

Let $x = r\cos\theta$, $y = r\sin\theta$ then

$\{(x,y): (x-1)^2 + y^2 \leq 1\} = \{(r,\theta): 0 \leq r \leq 2\cos\theta, 0 \leq \theta \leq \pi\}$

and $dxdy = \begin{Vmatrix} \frac{\partial x}{\partial r} & \frac{\partial x}{\partial \theta} \\ \frac{\partial y}{\partial r} & \frac{\partial y}{\partial \theta} \end{Vmatrix} drd\theta = \begin{Vmatrix} a\cos\theta & ar\sin\theta \\ a\sin\theta & ar\cos\theta \end{Vmatrix} drd\theta = rdrd\theta$

$$\therefore \iint_{(x-1)^2+y^2\leq 1} 2\sqrt{4-x^2-y^2} dxdy = 2\int_0^\pi \int_0^{2\cos\theta} (\sqrt{4-r^2}) r dr d\theta$$

$$= 4\int_0^{\frac{\pi}{2}} \int_0^{2\cos\theta} (\sqrt{4-r^2}) r dr d\theta = 4\int_0^{\frac{\pi}{2}} \left(\frac{(-1)(4-r^2)^{\frac{3}{2}}}{3} \right)\Bigg|_0^{2\cos\theta} d\theta = 32\int_0^{\frac{\pi}{2}} \frac{1-\sin^3\theta}{3} d\theta$$

$$= 32\left(\frac{\pi}{6} - \frac{2}{9}\right)$$

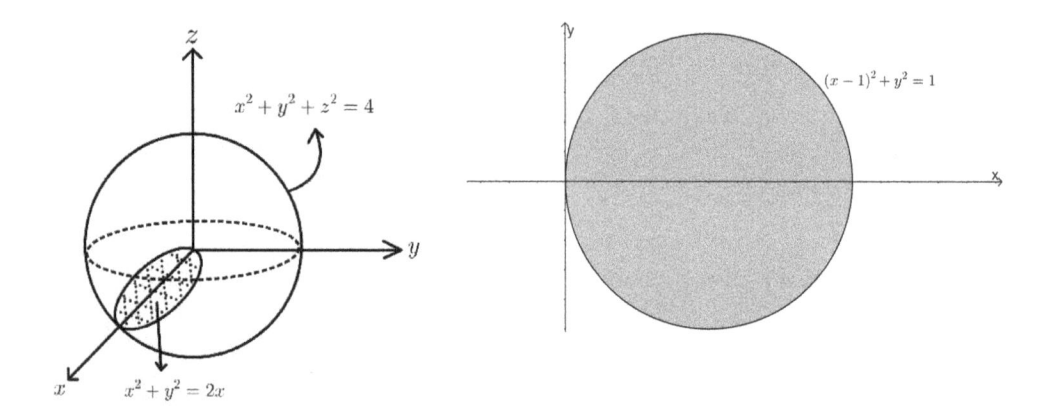

8.4.2 Use Polar Coordinates to Compute Triple Integrals

Example 20.

(1) Calculate the volume of the sphere $x^2 + y^2 + z^2 = 16$ after being truncated by the cylinder $r = 4\cos\theta$.

(2) Calculate the volume of the sphere $x^2 + y^2 + z^2 = 4$ after being truncated by the cylinder $r = 2\sin\theta$.

【Solution】

(1)
Let $R = \{(x, y): (x-2)^2 + y^2 \le 4, x \ge 0\}$ then

$$V = \iint_R \int_{-\sqrt{16-x^2-y^2}}^{\sqrt{16-x^2-y^2}} dz\, dA = \iint_{(x-2)^2+y^2 \le 4} 2\sqrt{16-x^2-y^2}\, dx\, dy$$

Let $x = r\cos\theta, y = r\sin\theta$ then

$$\{(x,y): (x-2)^2 + y^2 \le 4, x \ge 0\} = \{(r, \theta): 0 \le r \le 4\cos\theta, \frac{-\pi}{2} \le \theta \le \frac{\pi}{2}\}$$

and $dx\, dy = \begin{Vmatrix} \frac{\partial x}{\partial r} & \frac{\partial x}{\partial \theta} \\ \frac{\partial y}{\partial r} & \frac{\partial y}{\partial \theta} \end{Vmatrix} dr\, d\theta = \begin{Vmatrix} \cos\theta & -r\sin\theta \\ \sin\theta & r\cos\theta \end{Vmatrix} dr\, d\theta = r\, dr\, d\theta$

$$\therefore \iint_{(x-2)^2+y^2 \le 4} 2\sqrt{16-x^2-y^2}\, dx\, dy = 2\int_{-\frac{\pi}{2}}^{\frac{\pi}{2}} \int_0^{4\cos\theta} \sqrt{16-r^2}\, r\, dr\, d\theta$$

$$= 2\int_{-\frac{\pi}{2}}^{\frac{\pi}{2}} \frac{(-1)(16-r^2)^{\frac{3}{2}}}{3} \Big|_{r=0}^{r=4\cos\theta} d\theta = \frac{256}{3}\int_0^{\frac{\pi}{2}} 1-(1-\cos^2\theta)^{\frac{3}{2}} d\theta = \frac{256}{3}\int_0^{\frac{\pi}{2}} 1-\sin^3\theta\, d\theta$$

$$= \left(\frac{\pi}{2} - \frac{2}{3}\right)\frac{256}{3}$$

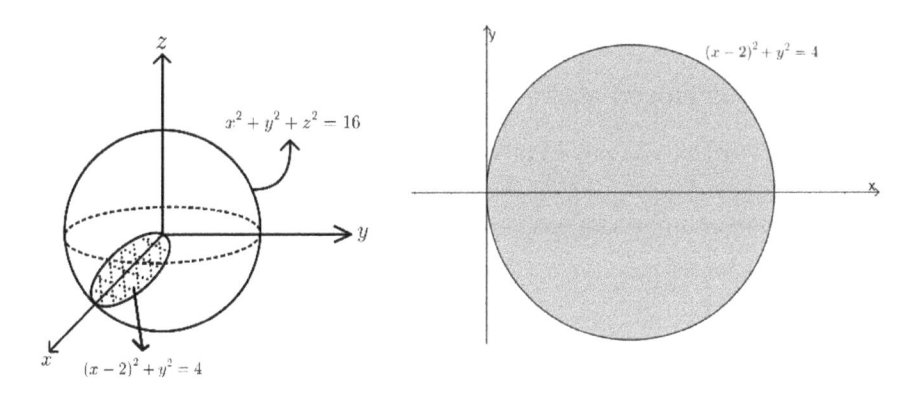

(2)
Let $R = \{(x,y): x^2 + (y-1)^2 \leq 1, y \geq 0\}$ then

$$V = \iint_R \int_{-\sqrt{4-x^2-y^2}}^{\sqrt{4-x^2-y^2}} dz dA = \iint_{x^2+(y-1)^2 \leq 1} 2\sqrt{4-x^2-y^2} dxdy$$

Let $x = r\cos\theta, y = r\sin\theta$ then
$\{(x,y): x^2 + (y-1)^2 \leq 1, y \geq 0\} = \{(r,\theta): 0 \leq r \leq 2\sin\theta, 0 \leq \theta \leq \pi\}$

and $dxdy = \left\|\begin{matrix}\dfrac{\partial x}{\partial r} & \dfrac{\partial x}{\partial \theta}\\ \dfrac{\partial y}{\partial r} & \dfrac{\partial y}{\partial \theta}\end{matrix}\right\| drd\theta = \left\|\begin{matrix}\cos\theta & -r\sin\theta\\ \sin\theta & r\cos\theta\end{matrix}\right\| drd\theta = rdrd\theta$

$\therefore \iint_{x^2+(y-1)^2 \leq 1} 2\sqrt{4-x^2-y^2} dxdy = 2\int_0^\pi \int_0^{2\sin\theta} \sqrt{4-r^2} rdrd\theta$

$= 2\int_0^\pi \left.\dfrac{(-1)(4-r^2)^{\frac{3}{2}}}{3}\right|_{r=0}^{r=2\sin\theta} d\theta = \dfrac{32}{3}\int_0^{\frac{\pi}{2}} 1-(1-\sin^2\theta)^{\frac{3}{2}} d\theta = \dfrac{32}{3}\int_0^{\frac{\pi}{2}} 1-\cos^3\theta\, d\theta$

$= \left(\dfrac{\pi}{2} - \dfrac{2}{3}\right)\dfrac{32}{3}$

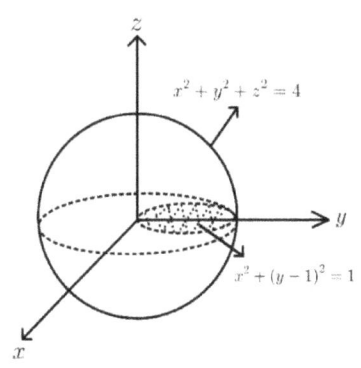

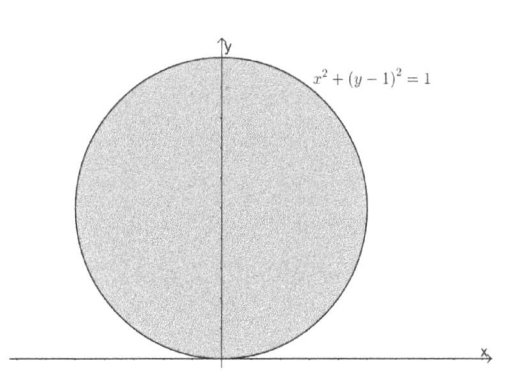

8.4.2 Use Polar Coordinates to Compute Triple Integrals

Example 21.
Find the volume of the region enclosed above by the surface $z = x^2 + y^2$ and below by the surface $z = y$.

【Solution】

Let $R = \left\{(x, y): x^2 + \left(y - \dfrac{1}{2}\right)^2 \leq \dfrac{1}{4}\right\}$ then

$$V = \iint_R \int_{x^2+y^2}^{y} dzdA = \iint_R \dfrac{1}{4} - x^2 - \left(y - \dfrac{1}{2}\right)^2 dxdy$$

Let $x = \dfrac{1}{2}r\cos\theta$, $y = \dfrac{1}{2} + \dfrac{1}{2}r\sin\theta$ then

$\left\{(x,y): x^2 + \left(y - \dfrac{1}{2}\right)^2 \leq \dfrac{1}{4}\right\} = \{(r, \theta): 0 \leq r \leq 1, 0 \leq \theta \leq 2\pi\}$

and $dxdy = \begin{Vmatrix} \dfrac{\partial x}{\partial r} & \dfrac{\partial x}{\partial \theta} \\ \dfrac{\partial y}{\partial r} & \dfrac{\partial y}{\partial \theta} \end{Vmatrix} drd\theta = \begin{Vmatrix} \dfrac{1}{2}\cos\theta & \dfrac{1}{2}r\sin\theta \\ \dfrac{1}{2}\sin\theta & \dfrac{1}{2}r\cos\theta \end{Vmatrix} drd\theta = \dfrac{1}{4}rdrd\theta$

$\therefore \iint_R \dfrac{1}{4} - x^2 - \left(y - \dfrac{1}{2}\right)^2 dxdy = \int_0^{2\pi} \int_0^1 \left(\dfrac{1}{4} - \dfrac{1}{4}r^2\right) \dfrac{1}{4} rdrd\theta = \dfrac{1}{16}\int_0^{2\pi} d\theta \int_0^1 r - r^3 dr$

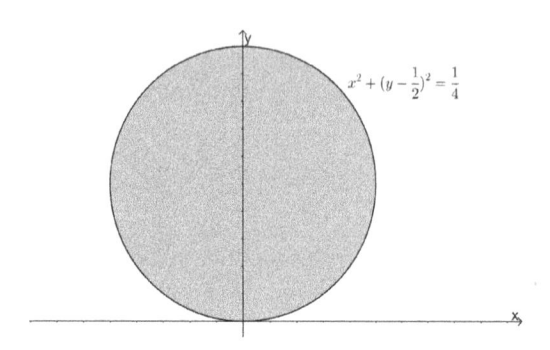

$= \dfrac{\pi}{32}$

$x^2 + (y - \tfrac{1}{2})^2 = \tfrac{1}{4}$

Example 22.
Assume V is the region bounded by $z \geq x^2 + y^2$, $z \leq y$, $x > 0$, $y > 0$, and $z > 0$.
Find $\iiint_V xdxdydz = ?$

【Solution】

Let $R = \left\{(x,y): x^2 + \left(y - \dfrac{1}{2}\right)^2 \leq \dfrac{1}{4}, x > 0, y > 0\right\}$ then

$$\iiint_V x\,dx\,dy\,dz = \iint_R \int_{x^2+y^2}^{y} x\,dz\,dA = \iint_R x\left(\dfrac{1}{4} - x^2 - \left(y - \dfrac{1}{2}\right)^2\right)dx\,dy$$

Let $x = \dfrac{1}{2}r\cos\theta$, $y = \dfrac{1}{2} + \dfrac{1}{2}r\sin\theta$ then

$\{(x,y): x^2 + \left(y - \dfrac{1}{2}\right)^2 \leq \dfrac{1}{4}, x > 0, y > 0\} = \{(r,\theta): 0 \leq r \leq 1, 0 \leq \theta \leq \dfrac{\pi}{2}\}$

and $dx\,dy = \begin{Vmatrix} \dfrac{\partial x}{\partial r} & \dfrac{\partial x}{\partial \theta} \\ \dfrac{\partial y}{\partial r} & \dfrac{\partial y}{\partial \theta} \end{Vmatrix} dr\,d\theta = \begin{Vmatrix} \dfrac{1}{2}\cos\theta & \dfrac{1}{2}r\sin\theta \\ \dfrac{1}{2}\sin\theta & \dfrac{1}{2}r\cos\theta \end{Vmatrix} dr\,d\theta = \dfrac{1}{4}r\,dr\,d\theta$

$\therefore \iint_R x\left(\dfrac{1}{4} - x^2 - \left(y - \dfrac{1}{2}\right)^2\right)dx\,dy$

$= \displaystyle\int_0^{\frac{\pi}{2}} \dfrac{1}{2}r\cos\theta \int_0^1 \left(\dfrac{1}{4} - \dfrac{1}{4}r^2\right)\dfrac{1}{4}r\,dr\,d\theta = \dfrac{1}{32}\int_0^{\frac{\pi}{2}}\cos\theta\,d\theta\int_0^1 r^2 - r^4\,dr = \dfrac{1}{240}$

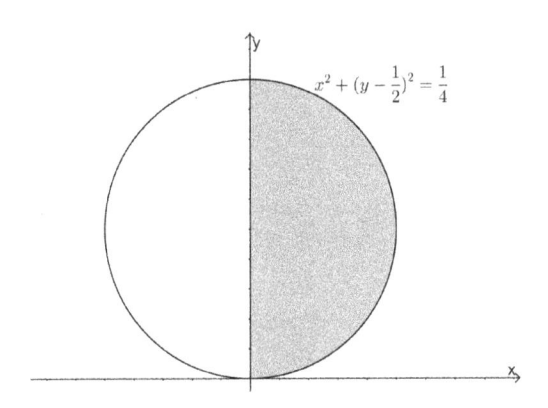

Example 23.

(1) Calculate the volume of the region enclosed by $x^2 + y^2 = 1$ and $4x^2 + 4y^2 + z^2 = 64$.

(2) Find the volume of the region enclosed above by the surface $z = \sqrt{4 - 3x^2 - 3y^2}$ and below by $z = \sqrt{x^2 + y^2}$.

【Solution】

(1)

Let $R = \{(x,y): x^2 + y^2 \leq 1\}$ then

8.4.2 Use Polar Coordinates to Compute Triple Integrals

$$V = \iint_R \int_{-\sqrt{64-4x^2-4y^2}}^{\sqrt{64-4x^2-4y^2}} dz\, dA = \iint_{x^2+y^2 \leq 1} 2\sqrt{64 - 4x^2 - 4y^2}\, dx\, dy$$

Let $x = r\cos\theta, y = r\sin\theta$ then $\{(x,y): x^2 + y^2 \leq 1\} = \{(r,\theta): 0 \leq r \leq 1, 0 \leq \theta \leq 2\pi\}$

and $dx\,dy = \begin{Vmatrix} \dfrac{\partial x}{\partial r} & \dfrac{\partial x}{\partial \theta} \\ \dfrac{\partial y}{\partial r} & \dfrac{\partial y}{\partial \theta} \end{Vmatrix} dr\,d\theta = \begin{Vmatrix} \cos\theta & -r\sin\theta \\ \sin\theta & r\cos\theta \end{Vmatrix} dr\,d\theta = r\,dr\,d\theta$

$\therefore \iint_{x^2+y^2 \leq 1} 2\sqrt{64 - 4x^2 - 4y^2}\, dx\, dy = 2\int_0^{2\pi} \int_0^1 \sqrt{64 - 4r^2}\, r\, dr\, d\theta$

$= 4 \cdot (2\pi) \cdot \left. \dfrac{(-1)(16-r^2)^{\frac{3}{2}}}{3} \right|_{r=0}^{r=1} = \dfrac{8\pi}{3}(64 - 15\sqrt{15})$

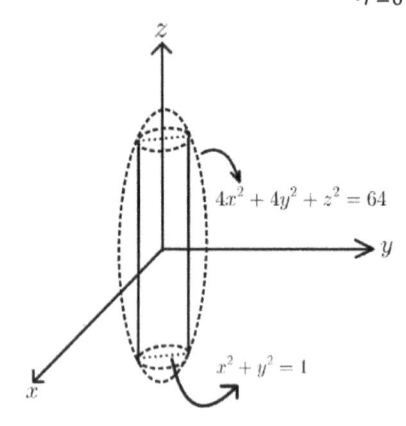

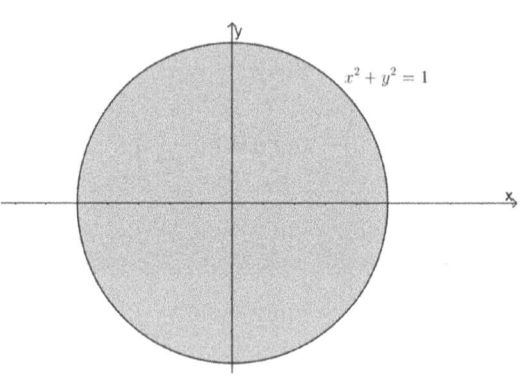

(2)
Let $R = \{(x,y): x^2 + y^2 \leq 1\}$ then

$$V = \iint_R \int_{\sqrt{x^2+y^2}}^{\sqrt{4-3x^2-3y^2}} dz\, dA = \iint_{x^2+y^2 \leq 1} \sqrt{4 - 3x^2 - 3y^2} - \sqrt{x^2 + y^2}\, dx\, dy$$

Let $x = r\cos\theta, y = r\sin\theta$ then $\{(x,y): x^2 + y^2 \leq 1\} = \{(r,\theta): 0 \leq r \leq 1, 0 \leq \theta \leq 2\pi\}$

and $dx\,dy = \begin{Vmatrix} \dfrac{\partial x}{\partial r} & \dfrac{\partial x}{\partial \theta} \\ \dfrac{\partial y}{\partial r} & \dfrac{\partial y}{\partial \theta} \end{Vmatrix} dr\,d\theta = \begin{Vmatrix} \cos\theta & -r\sin\theta \\ \sin\theta & r\cos\theta \end{Vmatrix} dr\,d\theta = r\,dr\,d\theta$

$\therefore \iint_{x^2+y^2 \leq 1} \sqrt{4 - 3x^2 - 3y^2} - \sqrt{x^2 + y^2}\, dx\, dy = \int_0^{2\pi} \int_0^1 (\sqrt{4 - 3r^2} - r) r\, dr\, d\theta$

$$= (2\pi) \cdot \left(\frac{(-1)(4-3r^2)^{\frac{3}{2}}}{9} - \frac{r^3}{3} \right) \Bigg|_{r=0}^{r=1} = \frac{8\pi}{9}$$

Example 24.

(1) Calculate the smaller portion of the sphere $x^2 + y^2 + z^2 \leq 100$ after being truncated by the plane $z = 6$.

(2) Calculate the smaller portion of the sphere $x^2 + y^2 + z^2 \leq 100$ after being truncated by the plane $z = 8$.

【Solution】

(1)
Let $R = \{(x,y): x^2 + y^2 \leq 64\}$ then

$$V = \iint_R \int_6^{\sqrt{100-x^2-y^2}} dz dA = \iint_{x^2+y^2 \leq 64} \sqrt{100 - x^2 - y^2} - 6 \, dxdy$$

Let $x = r\cos\theta, y = r\sin\theta$ then $\{(x,y): x^2 + y^2 \leq 64\} = \{(r,\theta): 0 \leq r \leq 8, 0 \leq \theta \leq 2\pi\}$

and $dxdy = \begin{Vmatrix} \frac{\partial x}{\partial r} & \frac{\partial x}{\partial \theta} \\ \frac{\partial y}{\partial r} & \frac{\partial y}{\partial \theta} \end{Vmatrix} drd\theta = \begin{Vmatrix} \cos\theta & -r\sin\theta \\ \sin\theta & r\cos\theta \end{Vmatrix} drd\theta = rdrd\theta$

$$\therefore \iint_{x^2+y^2 \leq 64} \sqrt{100 - x^2 - y^2} - 6 \, dxdy = \int_0^{2\pi} \int_0^8 (\sqrt{100 - r^2} - 6) r \, drd\theta$$

$$= (2\pi) \cdot \left(\frac{(-1)(100-r^2)^{\frac{3}{2}}}{3} - \frac{6r^2}{2} \right) \Bigg|_{r=0}^{r=8} = \frac{356\pi}{3}$$

8.4.2 Use Polar Coordinates to Compute Triple Integrals

(2)
Let $R = \{(x,y): x^2 + y^2 \leq 36\}$ then

$$V = \iint_R \int_8^{\sqrt{100-x^2-y^2}} dzdA = \iint_{x^2+y^2 \leq 36} \sqrt{100 - x^2 - y^2} - 8 dxdy$$

Let $x = r\cos\theta, y = r\sin\theta$ then $\{(x,y): x^2 + y^2 \leq 36\} = \{(r,\theta): 0 \leq r \leq 6, 0 \leq \theta \leq 2\pi\}$

and $dxdy = \begin{Vmatrix} \dfrac{\partial x}{\partial r} & \dfrac{\partial x}{\partial \theta} \\ \dfrac{\partial y}{\partial r} & \dfrac{\partial y}{\partial \theta} \end{Vmatrix} drd\theta = \begin{Vmatrix} \cos\theta & -r\sin\theta \\ \sin\theta & r\cos\theta \end{Vmatrix} drd\theta = rdrd\theta$

$\therefore \iint_{x^2+y^2 \leq 36} \sqrt{100 - x^2 - y^2} - 8 dxdy = \int_0^{2\pi} \int_0^6 (\sqrt{100 - r^2} - 8) r drd\theta$

$= (2\pi) \cdot \left(\dfrac{(-1)(100-r^2)^{\frac{3}{2}}}{3} - \dfrac{8r^2}{2} \right) \Bigg|_{r=0}^{r=6} = \dfrac{112\pi}{3}$

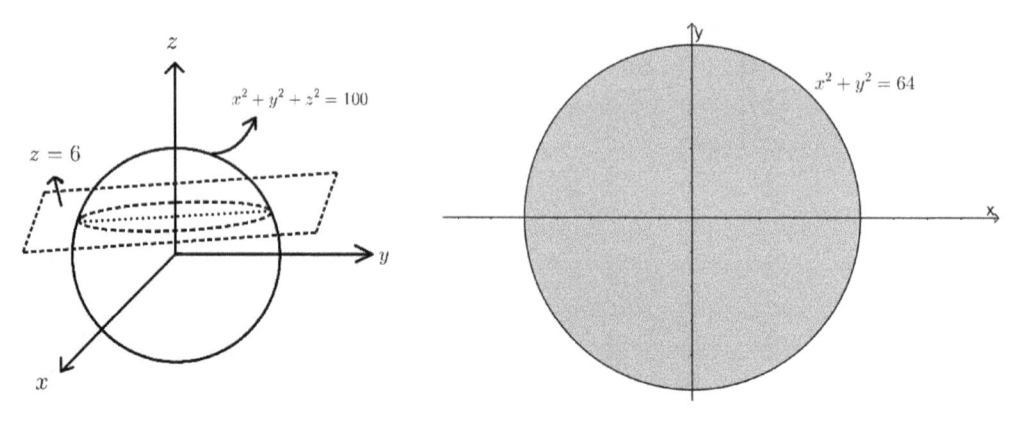

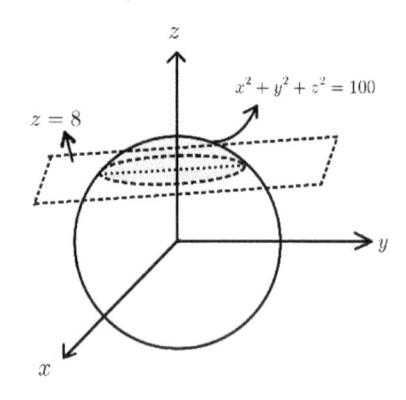

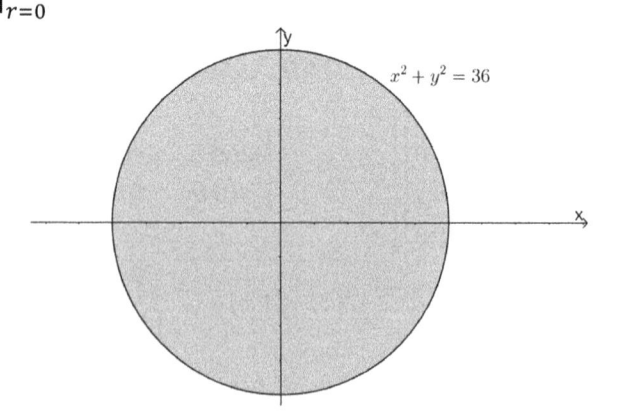

Example 25.

8.4.2 Use Polar Coordinates to Compute Triple Integrals

Find $\int_0^3 \int_0^{\sqrt{9-x^2}} \int_0^2 \sqrt{x^2+y^2} \cos \sqrt{x^2+y^2}\, dzdydx =?$

【Solution】

Let $x = r\cos\theta, y = r\sin\theta$ then

$\{(x,y): 0 \leq x \leq 3, 0 \leq y \leq \sqrt{9-x^2}\} = \{(r,\theta): 0 \leq r \leq 3, 0 \leq \theta \leq \frac{\pi}{2}\}$

and $dxdy = \begin{Vmatrix} \frac{\partial x}{\partial r} & \frac{\partial x}{\partial \theta} \\ \frac{\partial y}{\partial r} & \frac{\partial y}{\partial \theta} \end{Vmatrix} drd\theta = \begin{Vmatrix} \cos\theta & -r\sin\theta \\ \sin\theta & r\cos\theta \end{Vmatrix} drd\theta = rdrd\theta$

$\therefore \int_0^3 \int_0^{\sqrt{9-x^2}} \int_0^2 \sqrt{x^2+y^2} \cos\sqrt{x^2+y^2}\, dzdydx = 2\int_0^3 \int_0^{\frac{\pi}{2}} r^2 \cos r\, d\theta dr$

$= 2\int_0^{\frac{\pi}{2}} d\theta \int_0^3 r^2 \cos r\, dr$

Let $u = r^2$, $dv = \cos r\, dr$ then $du = 2rdr$, $v = \sin r$, by integration by parts,

then $\int r^2 \cos r\, dr = r^2 \sin r - 2 \int r \sin r\, dr + c$

Let $s = r$, $dt = \sin r\, dr$ then $ds = dr$, $t = -\cos r$, by integration by parts,

then $\int r \sin r\, dr = -r\cos r + \int \cos r dr = -r\cos r + \sin r$

$\therefore \int r^2 \cos r\, dr = r^2 \sin r - 2\int r\sin r\, dr + c = r^2 \sin r + 2r\cos r - 2\sin r + c$

Let $a, b \in R$ then

$\int_a^b r^2 \cos r\, dx = b^2 \sin b + 2b\cos b - 2\sin b - (a^2 \sin a + 2a \cos a - 2\sin a)$

$\therefore \int_0^3 \int_0^{\sqrt{9-x^2}} \int_0^2 \sqrt{x^2+y^2} \cos\sqrt{x^2+y^2}\, dzdydx = 2\int_0^{\frac{\pi}{2}} d\theta \int_0^3 r^2 \cos r\, dr$

$= \pi(7\sin 3 + 6\cos 3)$

Example 26.

(1) Calculate the volume of the intersection between the sphere $x^2 + y^2 + z^2 \leq 1$ and $z \geq \sqrt{3x^2 + 3y^2}$.

(2) Calculate the volume of the intersection between the sphere $x^2 + y^2 + z^2 \leq 1$ and $\sqrt{3}z \geq \sqrt{x^2 + y^2}$.

8.4.2 Use Polar Coordinates to Compute Triple Integrals

【Solution】

(1)

Let $R = \{(x,y): x^2 + y^2 \le \frac{1}{4}\}$ then

$$V = \iint_R \int_{\sqrt{3x^2+3y^2}}^{\sqrt{1-x^2-y^2}} dz\,dA = \iint_{x^2+y^2 \le \frac{1}{4}} \sqrt{1-x^2-y^2} - \sqrt{3x^2+3y^2}\,dxdy$$

Let $x = r\cos\theta, y = r\sin\theta$ then $\{(x,y): x^2 + y^2 \le \frac{1}{4}\} = \{(r,\theta): 0 \le r \le \frac{1}{2}, 0 \le \theta \le 2\pi\}$

and $dxdy = \begin{Vmatrix} \frac{\partial x}{\partial r} & \frac{\partial x}{\partial \theta} \\ \frac{\partial y}{\partial r} & \frac{\partial y}{\partial \theta} \end{Vmatrix} drd\theta = \begin{Vmatrix} \cos\theta & -r\sin\theta \\ \sin\theta & r\cos\theta \end{Vmatrix} drd\theta = r\,drd\theta$

$$\therefore \iint_{x^2+y^2 \le \frac{1}{4}} \sqrt{1-x^2-y^2} - \sqrt{3x^2+3y^2}\,dxdy = \int_0^{2\pi} \int_0^{\frac{1}{2}} (\sqrt{1-r^2} - \sqrt{3r^2})r\,drd\theta$$

$$= (2\pi) \cdot \left(\frac{(-1)(1-r^2)^{\frac{3}{2}}}{3} - \frac{\sqrt{3}r^3}{3} \right) \Big|_{r=0}^{r=\frac{1}{2}} = 2\pi \left(\frac{2-\sqrt{3}}{6} \right) = \pi \left(\frac{2-\sqrt{3}}{3} \right)$$

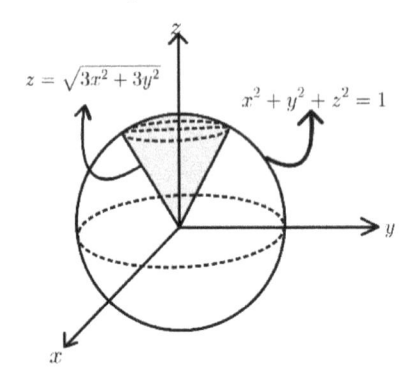

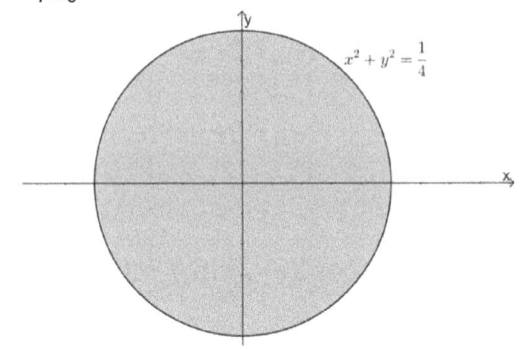

(2)

Let $R = \{(x,y): x^2 + y^2 \le \frac{3}{4}\}$ then

$$V = \iint_R \int_{\sqrt{\frac{x^2+y^2}{3}}}^{\sqrt{1-x^2-y^2}} dz\,dA = \iint_{x^2+y^2 \le \frac{3}{4}} \sqrt{1-x^2-y^2} - \sqrt{\frac{x^2+y^2}{3}}\,dxdy$$

Let $x = r\cos\theta, y = r\sin\theta$ then $\{(x,y): x^2 + y^2 \le \frac{3}{4}\} = \{(r,\theta): 0 \le r \le \sqrt{\frac{3}{4}}, 0 \le \theta \le 2\pi\}$

and $dxdy = \begin{Vmatrix} \frac{\partial x}{\partial r} & \frac{\partial x}{\partial \theta} \\ \frac{\partial y}{\partial r} & \frac{\partial y}{\partial \theta} \end{Vmatrix} drd\theta = \begin{Vmatrix} \cos\theta & -r\sin\theta \\ \sin\theta & r\cos\theta \end{Vmatrix} drd\theta = rdrd\theta$

$$\therefore \iint_{x^2+y^2\leq\frac{3}{4}} \sqrt{1-x^2-y^2} - \sqrt{\frac{x^2+y^2}{3}} dxdy = \int_0^{2\pi} \int_0^{\frac{\sqrt{3}}{2}} \left(\sqrt{1-r^2} - \sqrt{\frac{r^2}{3}}\right) rdrd\theta$$

$$= (2\pi) \cdot \left(\frac{(-1)(1-r^2)^{\frac{3}{2}}}{3} - \frac{r^3}{3\sqrt{3}}\right) \Bigg|_{r=0}^{r=\frac{\sqrt{3}}{2}} = 2\pi \left(\frac{1}{6}\right) = \frac{\pi}{3}$$

Example 27.

(1) Calculate the volume of the region enclosed by the sphere $x^2+y^2+(z-2)^2 \leq 4$ and $z \geq \sqrt{x^2+y^2}$.

(2) Calculate the volume of the region enclosed by the sphere $x^2+y^2+z^2 \leq z$ and $z \geq \sqrt{x^2+y^2}$.

【Solution】

(1)
Let $R = \{(x,y): x^2+y^2 \leq 4\}$ then

$$V = \iint_R \int_{\sqrt{x^2+y^2}}^{2+\sqrt{4-x^2-y^2}} dzdA = \iint_{x^2+y^2\leq 4} 2 + \sqrt{4-x^2-y^2} - \sqrt{x^2+y^2} dxdy$$

Let $x = r\cos\theta, y = r\sin\theta$ then $\{(x,y): x^2+y^2 \leq 4\} = \{(r,\theta): 0 \leq r \leq 2, 0 \leq \theta \leq 2\pi\}$

8.4.2 Use Polar Coordinates to Compute Triple Integrals

and $dxdy = \begin{Vmatrix} \frac{\partial x}{\partial r} & \frac{\partial x}{\partial \theta} \\ \frac{\partial y}{\partial r} & \frac{\partial y}{\partial \theta} \end{Vmatrix} drd\theta = \begin{Vmatrix} \cos\theta & -r\sin\theta \\ \sin\theta & r\cos\theta \end{Vmatrix} drd\theta = rdrd\theta$

$\therefore \iint_{x^2+y^2\leq 4} 2 + \sqrt{4-x^2-y^2} - \sqrt{x^2+y^2} dxdy = \int_0^{2\pi}\int_0^2 (2+\sqrt{4-r^2}-\sqrt{r^2})rdrd\theta$

$= (2\pi)\cdot\left(r^2 + \frac{(-1)(4-r^2)^{\frac{3}{2}}}{3} - \frac{r^3}{3}\right)\Big|_{r=0}^{r=2} = 8\pi$

(2)

Let $R = \left\{(x,y): x^2+y^2 \leq \frac{1}{4}\right\}$ then

$V = \iint_R \int_{\sqrt{x^2+y^2}}^{\frac{1}{2}+\sqrt{\frac{1}{4}-x^2-y^2}} dzdA = \iint_{x^2+y^2\leq\frac{1}{4}} \frac{1}{2} + \sqrt{\frac{1}{4}-x^2-y^2} - \sqrt{x^2+y^2} dxdy$

Let $x = r\cos\theta, y = r\sin\theta$ then $\{(x,y): x^2+y^2 \leq \frac{1}{4}\} = \{(r,\theta): 0\leq r\leq\frac{1}{2}, 0\leq\theta\leq 2\pi\}$

and $dxdy = \begin{Vmatrix} \frac{\partial x}{\partial r} & \frac{\partial x}{\partial \theta} \\ \frac{\partial y}{\partial r} & \frac{\partial y}{\partial \theta} \end{Vmatrix} drd\theta = \begin{Vmatrix} \cos\theta & -r\sin\theta \\ \sin\theta & r\cos\theta \end{Vmatrix} drd\theta = rdrd\theta$

$\therefore \iint_{x^2+y^2\leq\frac{1}{4}} \frac{1}{2} + \sqrt{\frac{1}{4}-x^2-y^2} - \sqrt{x^2+y^2} dxdy = \int_0^{2\pi}\int_0^{\frac{1}{2}}\left(\frac{1}{2}+\sqrt{\frac{1}{4}-r^2}-r\right)rdrd\theta$

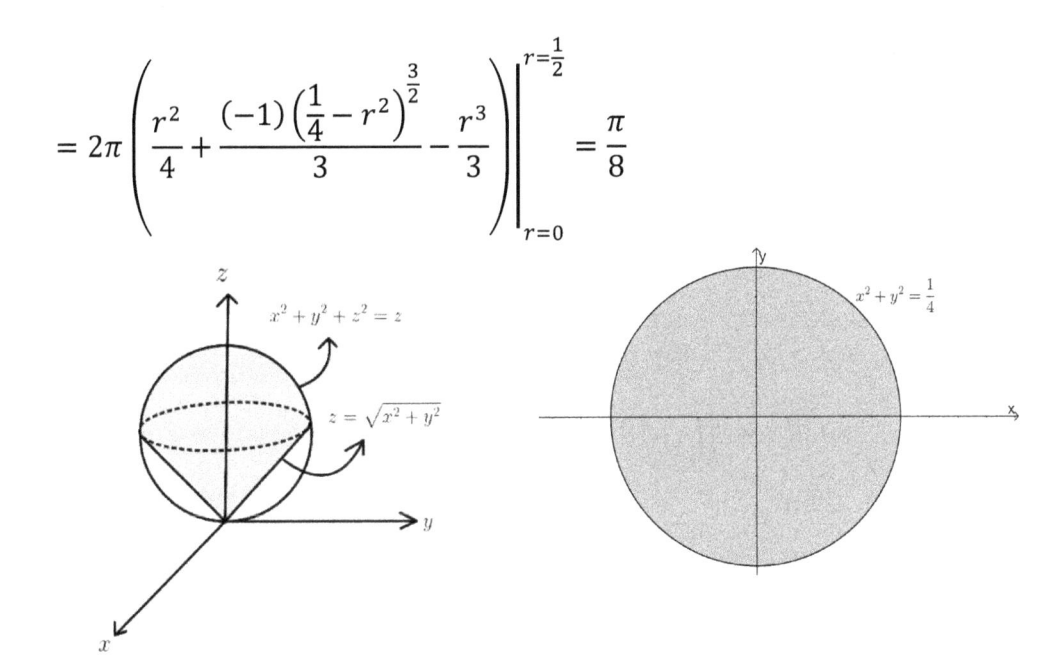

$$= 2\pi \left(\frac{r^2}{4} + \frac{(-1)\left(\frac{1}{4} - r^2\right)^{\frac{3}{2}}}{3} - \frac{r^3}{3} \right) \Bigg|_{r=0}^{r=\frac{1}{2}} = \frac{\pi}{8}$$

Example 28.

Find the volume of the region enclosed above by $z = x^2 + 8y^2$ and below $z = 9 - y^2$.

【Solution】

Let $R = \{(x, y): x^2 + 9y^2 \leq 9\}$ then $V = \iint_R \int_{x^2+8y^2}^{9-y^2} dz dA = \iint_R 9 - x^2 - 9y^2 dxdy$

Let $x = 3r \cos \theta, y = r \sin \theta$ then $\{(x, y): x^2 + 9y^2 \leq 9\} = \{(r, \theta): 0 \leq r \leq 1, 0 \leq \theta \leq 2\pi\}$

and $dxdy = \begin{Vmatrix} \frac{\partial x}{\partial r} & \frac{\partial x}{\partial \theta} \\ \frac{\partial y}{\partial r} & \frac{\partial y}{\partial \theta} \end{Vmatrix} drd\theta = \begin{Vmatrix} 3\cos \theta & -3r \sin \theta \\ \sin \theta & r \cos \theta \end{Vmatrix} drd\theta = 3rdrd\theta$

$\therefore \iint_R 9 - x^2 - 9y^2 dxdy = \int_0^{2\pi} \int_0^1 (9 - 9r^2) 3r dr d\theta = 27 \int_0^{2\pi} d\theta \int_0^1 r - r^3 dr = \frac{27\pi}{2}$

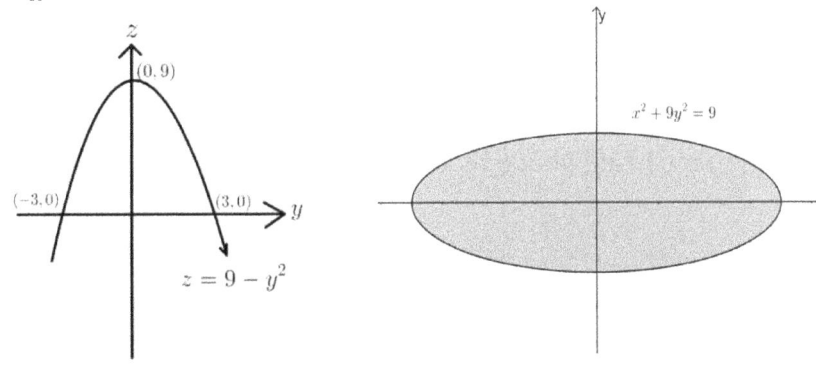

Example 29.

Assume V is the region bounded by $z \geq x^2 + 8y^2, z \leq 9 - y^2, x > 0, y > 0,$ and $z > 0$.

Find $\iiint_V x\,dx\,dy\,dz = ?$

【Solution】

Let $R = \{(x,y): x^2 + 9y^2 \leq 9, x > 0, y > 0\}$ then

$$\iiint_V x\,dx\,dy\,dz = \iint_R \int_{x^2+8y^2}^{9-y^2} x\,dz\,dA = \iint_R x(9 - x^2 - 9y^2)\,dx\,dy$$

Let $x = 3r\cos\theta, y = r\sin\theta$ then

$$\{(x,y): x^2 + 9y^2 \leq 9, x > 0, y > 0\} = \{(r,\theta): 0 \leq r \leq 1, 0 \leq \theta \leq \frac{\pi}{2}\}$$

and $dx\,dy = \begin{Vmatrix} \frac{\partial x}{\partial r} & \frac{\partial x}{\partial \theta} \\ \frac{\partial y}{\partial r} & \frac{\partial y}{\partial \theta} \end{Vmatrix} dr\,d\theta = \begin{Vmatrix} 2\cos\theta & -2r\sin\theta \\ \sin\theta & r\cos\theta \end{Vmatrix} dr\,d\theta = 2r\,dr\,d\theta$

$$\therefore \iint_R x(9 - x^2 - 9y^2)\,dx\,dy = \int_0^{\frac{\pi}{2}} 3r\cos\theta \int_0^1 (9 - 9r^2) 2r\,dr\,d\theta$$

$$= 54 \int_0^{\frac{\pi}{2}} \cos\theta\,d\theta \int_0^1 r^2 - r^4\,dr = \frac{36}{5}$$

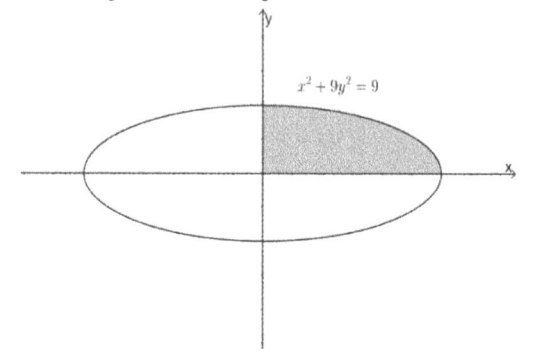

Example 30.

Find the volume of the region enclosed above by $z = 8 - 2y$ and below by $z = x^2 + y^2$.

【Solution】

Let $R = \{(x,y): x^2 + (y+1)^2 \leq 9\}$ then

$$V = \iint_R \int_{x^2+y^2}^{8-2y} dzdA = \iint_{x^2+(y+1)^2\leq 9} 9 - x^2 - (y+1)^2 dxdy$$

Let $x = 3r\cos\theta, y = -1 + 3r\sin\theta$ then
$\{(x,y): x^2 + (y+1)^2 \leq 9\} = \{(r,\theta): 0 \leq r \leq 1, 0 \leq \theta \leq 2\pi\}$

and $dxdy = \begin{Vmatrix} \dfrac{\partial x}{\partial r} & \dfrac{\partial x}{\partial \theta} \\ \dfrac{\partial y}{\partial r} & \dfrac{\partial y}{\partial \theta} \end{Vmatrix} drd\theta = \begin{Vmatrix} 3\cos\theta & 3r\sin\theta \\ 3\sin\theta & 3r\cos\theta \end{Vmatrix} drd\theta = 9rdrd\theta$

$$\therefore \iint_{x^2+(y+1)^2\leq 9} 9 - x^2 - (y+1)^2 dxdy = \int_0^{2\pi}\int_0^1 (9 - 9r^2)9rdrd\theta = 81\int_0^{2\pi} d\theta \int_0^1 r - r^3 dr$$
$$= \frac{81\pi}{2}$$

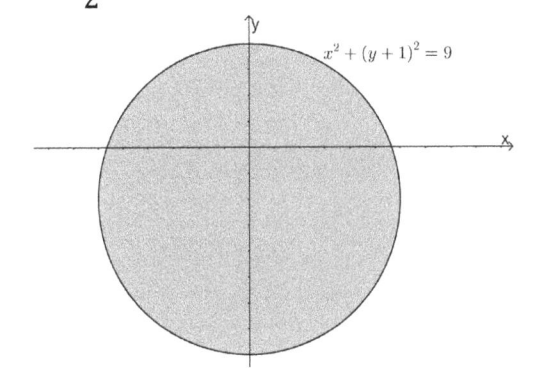

Example 31.
    Assume V is the region bounded by $z \leq 8 - 2y$, $z \geq x^2 + y^2$, $x > 0$, $y > 0$, $z > 0$.

Find $\iiint_V xdxdydz = ?$

【Solution】
Let $R = \{(x,y): x^2 + (y+1)^2 \leq 9, x > 0, y > 0\}$ then

$$\iiint_V xdxdydz = \iint_R \int_{x^2+y^2}^{8-2y} dzdA = \iint_R x(9 - x^2 - (y+1)^2)dxdy$$

Let $x = 3r\cos\theta, y = -1 + 3r\sin\theta$ then
$\{(x,y): x^2 + (y+1)^2 \leq 9\} = \{(r,\theta): 0 \leq r \leq 1, 0 \leq \theta \leq \dfrac{\pi}{2}\}$

8.4.2 Use Polar Coordinates to Compute Triple Integrals

and $dxdy = \begin{Vmatrix} \dfrac{\partial x}{\partial r} & \dfrac{\partial x}{\partial \theta} \\ \dfrac{\partial y}{\partial r} & \dfrac{\partial y}{\partial \theta} \end{Vmatrix} drd\theta = \begin{Vmatrix} 3\cos\theta & 3r\sin\theta \\ 3\sin\theta & 3r\cos\theta \end{Vmatrix} drd\theta = 9rdrd\theta$

$\therefore \iint_{x^2+(y+1)^2 \le 9, x>0, y>0} x(9 - x^2 - (y+1)^2) dxdy = \int_0^{\frac{\pi}{2}} 3r\cos\theta \int_0^1 (9 - 9r^2) 9rdrd\theta$

$= 243 \int_0^{\frac{\pi}{2}} \cos\theta \, d\theta \int_0^1 r^2 - r^4 dr = \dfrac{162}{5}$

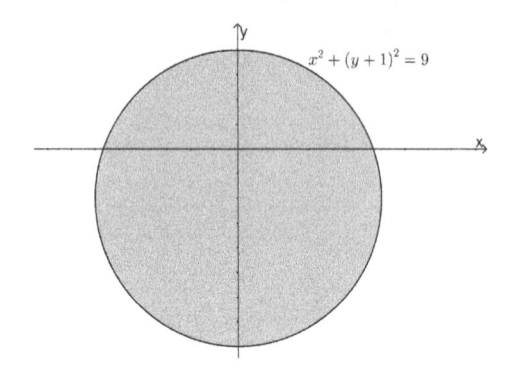

Example 32.

Find $\displaystyle\int_0^1 \int_0^{\sqrt{1-x^2}} \int_0^1 \tan^{-1}\sqrt{x^2+y^2} \, dzdydx = ?$

【Solution】

Let $x = r\cos\theta, y = r\sin\theta$ then

$\{(x,y): 0 \le x \le 1, 0 \le y \le \sqrt{1-x^2}\} = \{(r,\theta): 0 \le r \le 1, 0 \le \theta \le \dfrac{\pi}{2}\}$

and $dxdy = \begin{Vmatrix} \dfrac{\partial x}{\partial r} & \dfrac{\partial x}{\partial \theta} \\ \dfrac{\partial y}{\partial r} & \dfrac{\partial y}{\partial \theta} \end{Vmatrix} drd\theta = \begin{Vmatrix} \cos\theta & -r\sin\theta \\ \sin\theta & r\cos\theta \end{Vmatrix} drd\theta = rdrd\theta$

$\therefore \displaystyle\int_0^1 \int_0^{\sqrt{1-x^2}} \int_0^1 \tan^{-1}\sqrt{x^2+y^2}\, dzdydx = \int_0^1 \int_0^{\frac{\pi}{2}} r\tan^{-1} r\, d\theta dr = \int_0^{\frac{\pi}{2}} d\theta \int_0^1 r\tan^{-1} r\, dr$

Let $u = \tan^{-1} r$, $dv = rdr$ then $du = \dfrac{dr}{1+r^2}$, $v = \dfrac{r^2}{2}$, by integration by parts,

$\int r\tan^{-1} r \, dr = \dfrac{r^2}{2}\tan^{-1} r - \dfrac{1}{2}\int \dfrac{r^2 dr}{1+r^2} = \dfrac{r^2}{2}\tan^{-1} r - \dfrac{1}{2}\int \dfrac{1 + r^2 - 1 dr}{1+r^2}$

$$= \frac{r^2}{2}\tan^{-1} r - \frac{r}{2} + \frac{1}{2}\tan^{-1} r + c$$

$$\therefore \int_0^1 \int_0^{\sqrt{1-x^2}} \int_0^1 \tan^{-1}\sqrt{x^2+y^2}\, dzdydx = \int_0^{\frac{\pi}{2}} d\theta \int_0^1 r\tan^{-1} r\, dr = \frac{\pi}{2}\left(\frac{\pi}{4} - \frac{1}{2}\right)$$

8.4.3 Use Spherical Coordinates to Compute Triple Integrals

The appropriate time to use spherical coordinate transformations is when the integration region V in three-dimensional space involves the expression $x^2 + y^2 + z^2$.

Question Types:

Type 1.

Assume $V = \{(x,y,z): x^2 + y^2 + z^2 \leq a\},\ a > 0$. Find $\iiint_V f(x,y,z)dV = ?$

Problem-Solving Process:

Step1.

Let $x = \rho\sin\varphi\cos\theta,\ y = \rho\sin\varphi\sin\theta,\ z = \rho\cos\varphi$ then

$$dxdydz = \begin{Vmatrix} \dfrac{\partial x}{\partial \rho} & \dfrac{\partial x}{\partial \varphi} & \dfrac{\partial x}{\partial \theta} \\ \dfrac{\partial y}{\partial \rho} & \dfrac{\partial y}{\partial \varphi} & \dfrac{\partial y}{\partial \theta} \\ \dfrac{\partial z}{\partial \rho} & \dfrac{\partial z}{\partial \varphi} & \dfrac{\partial z}{\partial \theta} \end{Vmatrix} d\rho d\varphi d\theta$$

$$= \begin{Vmatrix} \sin\varphi\cos\theta & \rho\cos\varphi\cos\theta & -\rho\sin\varphi\sin\theta \\ \sin\varphi\sin\theta & \rho\cos\varphi\sin\theta & -\rho\sin\varphi\cos\theta \\ \cos\varphi & -\rho\sin\varphi & 0 \end{Vmatrix} d\rho d\varphi d\theta = \rho^2 \sin\varphi\, d\rho d\varphi d\theta$$

where $\{(\rho,\varphi,\theta): 0 \leq \rho \leq a, 0 \leq \varphi \leq \pi, 0 \leq \theta \leq 2\pi\}$

Step2.

$$\iiint_V f(x,y,z)dV = \int_0^{\pi}\int_0^{2\pi}\int_0^a f(\rho\sin\varphi\cos\theta, \rho\sin\varphi\sin\theta, \rho\cos\varphi)\rho^2\sin\varphi\, d\rho d\theta d\varphi$$

Step3.

Find $\int_0^{\pi}\int_0^{2\pi}\int_0^a f(\rho\sin\varphi\cos\theta, \rho\sin\varphi\sin\theta, \rho\cos\varphi)\rho^2\sin\varphi\, d\rho d\theta d\varphi = ?$

Examples:

8.4.3 Use Spherical Coordinates to Compute Triple Integrals

Find $\iiint_V f(x,y,z)dV = ?$, $V = \{(x,y,z): x^2 + y^2 + z^2 \leq a\}$, $a > 0$.

Let $x = \rho \sin\varphi \cos\theta$, $y = \rho \sin\varphi \sin\theta$, $z = \rho \cos\varphi$ then $dxdydz = \rho^2 \sin\varphi \, d\rho d\varphi d\theta$ where $0 \leq \rho \leq a$, $0 \leq \varphi \leq \pi$, $0 \leq \theta \leq 2\pi$.

$$\iiint_V f(x,y,z)dV = \int_0^\pi \int_0^{2\pi} \int_0^a f(\rho\sin\varphi\cos\theta, \rho\sin\varphi\sin\theta, \rho\cos\varphi)\rho^2 \sin\varphi \, d\rho d\theta d\varphi$$

(I) If $f(x,y,z) = \dfrac{1}{\sqrt{x^2+y^2+z^2}}$, then

$$\iiint_V \frac{1}{\sqrt{x^2+y^2+z^2}} dV = \int_0^\pi \int_0^{2\pi} \int_0^a \rho \sin\varphi \, d\rho d\theta d\varphi$$

(II) If $f(x,y,z) = \dfrac{z^2}{\sqrt{x^2+y^2+z^2}}$, then

$$\iiint_V \frac{z^2}{\sqrt{x^2+y^2+z^2}} dV = \int_0^\pi \int_0^{2\pi} \int_0^a \frac{\rho^2 \cos^2\varphi \, \rho^2 \sin\varphi}{\rho} d\rho d\theta d\varphi$$

(III) If $f(x,y,z) = \sqrt{1-(x^2+y^2+z^2)}$, then

$$\iiint_V \sqrt{1-(x^2+y^2+z^2)} dV = \int_0^\pi \int_0^{2\pi} \int_0^a \sqrt{1-\rho^2} \rho^2 \sin\varphi \, d\rho d\theta d\varphi$$

(IV) If $f(x,y,z) = \dfrac{1}{\sqrt{1-(x^2+y^2+z^2)}}$, then

$$\iiint_V \frac{1}{\sqrt{1-(x^2+y^2+z^2)}} dV = \int_0^\pi \int_0^{2\pi} \int_0^a \frac{\rho^2}{\sqrt{1-\rho^2}} \sin\varphi \, d\rho d\theta d\varphi$$

(V) If $f(x,y,z) = xyz$, then

$$\iiint_V xyz \, dV = \int_0^\pi \sin^3\varphi \cos\varphi \, d\varphi \int_0^{2\pi} \cos\theta \sin\theta \, d\theta \int_0^a \rho^5 d\rho$$

(VI) If $f(x,y,z) = (x^2+y^2+z^2)^{\frac{3}{2}}$, then

$$\iiint_V (x^2+y^2+z^2)^{\frac{3}{2}} dV = \int_0^\pi \int_0^{2\pi} \int_0^a (\rho^2)^{\frac{3}{2}} \rho^2 \sin\varphi \, d\rho d\theta d\varphi$$

(VII) If $f(x,y,z) = \dfrac{\cos\sqrt{x^2+y^2+z^2}}{x^2+y^2+z^2}$, then

$$\iiint_V \frac{\cos\sqrt{x^2+y^2+z^2}}{x^2+y^2+z^2} dV = \int_0^\pi \int_0^{2\pi} \int_0^a \frac{\cos\rho}{\rho^2} \rho^2 \sin\varphi \, d\rho d\theta d\varphi$$

8.4.3 Use Spherical Coordinates to Compute Triple Integrals

Example 1.

(1) Find the volume of the region enclosed by the surface $x^{\frac{2}{3}} + y^{\frac{2}{3}} + z^{\frac{2}{3}} = r^2$.

(2) Find the volume of the region enclosed by the surface $x^2 + 2y^2 + 3z^2 = 4$.

【Solution】

(1)

Let $x = u^3$, $y = v^3$, $z = s^3$ then

$$dxdydz = \begin{Vmatrix} \frac{\partial x}{\partial u} & \frac{\partial x}{\partial v} & \frac{\partial x}{\partial s} \\ \frac{\partial y}{\partial u} & \frac{\partial y}{\partial v} & \frac{\partial y}{\partial s} \\ \frac{\partial z}{\partial u} & \frac{\partial z}{\partial v} & \frac{\partial z}{\partial s} \end{Vmatrix} dudvds = \begin{Vmatrix} 3u^2 & 0 & 0 \\ 0 & 3v^2 & 0 \\ 0 & 0 & 3s^2 \end{Vmatrix} = 27u^2v^2s^2 dudvds$$

$$\therefore V = \iiint_{x^{\frac{2}{3}}+y^{\frac{2}{3}}+z^{\frac{2}{3}}\leq r^2} dxdydz = \iiint_{u^2+v^2+s^2\leq r^2} 27u^2v^2s^2 dudvds$$

Let $u = \rho \sin \varphi \cos \theta$, $v = \rho \sin \varphi \sin \theta$, $s = \rho \cos \varphi$ then

$$dudvds = \begin{Vmatrix} \frac{\partial u}{\partial \rho} & \frac{\partial u}{\partial \varphi} & \frac{\partial u}{\partial \theta} \\ \frac{\partial v}{\partial \rho} & \frac{\partial v}{\partial \varphi} & \frac{\partial v}{\partial \theta} \\ \frac{\partial s}{\partial \rho} & \frac{\partial s}{\partial \varphi} & \frac{\partial s}{\partial \theta} \end{Vmatrix} d\rho d\varphi d\theta$$

$$= \begin{Vmatrix} \sin \varphi \cos \theta & \rho \cos \varphi \cos \theta & -\rho \sin \varphi \sin \theta \\ \sin \varphi \sin \theta & \rho \cos \varphi \sin \theta & -\rho \sin \varphi \cos \theta \\ \cos \varphi & -\rho \sin \varphi & 0 \end{Vmatrix} d\rho d\varphi d\theta = \rho^2 \sin \varphi \, d\rho d\varphi d\theta$$

and $\{(u, v, s): u^2 + v^2 + s^2 \leq r^2\} = \{(\rho, \varphi, \theta): 0 \leq \rho \leq r, 0 \leq \varphi \leq \pi, 0 \leq \theta \leq 2\pi\}$

$$\therefore \iiint_{u^2+v^2+s^2\leq r^2} 27u^2v^2s^2 dudvds$$

$$= \int_0^\pi \int_0^{2\pi} \int_0^r 27(\rho \sin \varphi \cos \theta)^2 (\rho \sin \varphi \sin \theta)^2 (\rho \cos \varphi)^2 \rho^2 \sin \varphi \, d\rho d\theta d\varphi$$

$$= 27 \int_0^\pi \sin^5 \varphi \cos^2 \varphi \, d\varphi \int_0^{2\pi} \cos^2 \theta \sin^2 \theta d\theta \int_0^r \rho^8 d\rho$$

$$\because \int_0^\pi \sin^5 \varphi \cos^2 \varphi \, d\varphi = \int_0^\pi \sin \varphi \, (1 - \cos^2 \varphi)^2 \cos^2 \varphi \, d\varphi$$

$$= \left(\frac{(-1)\cos^3 \varphi}{3} + \frac{2\cos^5 \varphi}{5} + \frac{(-1)\cos^7 \varphi}{7} \right) \Big|_0^\pi = \frac{16}{105}$$

8.4.3 Use Spherical Coordinates to Compute Triple Integrals

and $\int_0^{2\pi} \cos^2\theta \sin^2\theta \, d\theta = \int_0^{2\pi} \left(\frac{\sin 2\theta}{2}\right)^2 d\theta = \frac{1}{4}\int_0^{2\pi} \frac{1-\cos 4\theta}{2} d\theta = \frac{1}{4}\left(\frac{\theta}{2} - \frac{\sin 4\theta}{8}\right)\Big|_0^{2\pi} = \frac{\pi}{4}$

$\therefore V = 27 \int_0^{\pi} \sin^5\varphi \cos^2\varphi \, d\varphi \int_0^{2\pi} \cos^2\theta \sin^2\theta \, d\theta \int_0^r \rho^8 d\rho = 27 \cdot \frac{16}{105} \cdot \frac{\pi}{4} \cdot \frac{r^9}{9} = \frac{4\pi r^9}{35}$

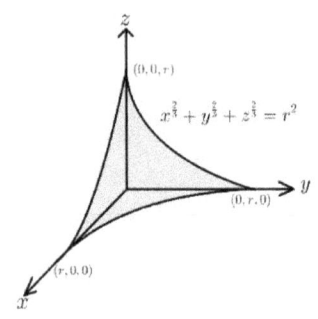

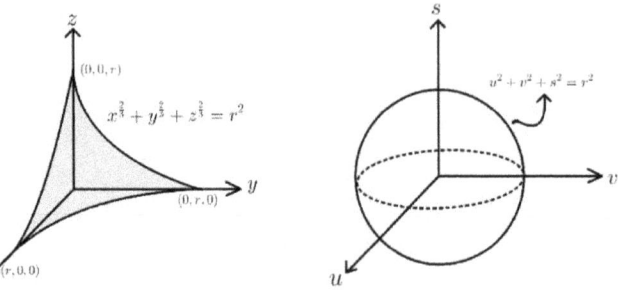

(2)

Let $x = u$, $y = \dfrac{v}{\sqrt{2}}$, $z = \dfrac{s}{\sqrt{3}}$ then

$dxdydz = \begin{Vmatrix} \dfrac{\partial x}{\partial u} & \dfrac{\partial x}{\partial v} & \dfrac{\partial x}{\partial s} \\ \dfrac{\partial y}{\partial u} & \dfrac{\partial y}{\partial v} & \dfrac{\partial y}{\partial s} \\ \dfrac{\partial z}{\partial u} & \dfrac{\partial z}{\partial v} & \dfrac{\partial z}{\partial s} \end{Vmatrix} dudvds = \begin{Vmatrix} 1 & 0 & 0 \\ 0 & \dfrac{1}{\sqrt{2}} & 0 \\ 0 & 0 & \dfrac{1}{\sqrt{3}} \end{Vmatrix} = \frac{1}{\sqrt{6}} dudvds$

$\therefore \iiint_{x^2+2y^2+3z^2 \leq 4} dxdydz = \iiint_{u^2+v^2+s^2 \leq 4} \frac{1}{\sqrt{6}} dudvds$

Let $u = \rho \sin\varphi \cos\theta$, $v = \rho \sin\varphi \sin\theta$, $s = \rho \cos\varphi$ then

$dudvds = \begin{Vmatrix} \dfrac{\partial u}{\partial \rho} & \dfrac{\partial u}{\partial \varphi} & \dfrac{\partial u}{\partial \theta} \\ \dfrac{\partial v}{\partial \rho} & \dfrac{\partial v}{\partial \varphi} & \dfrac{\partial v}{\partial \theta} \\ \dfrac{\partial s}{\partial \rho} & \dfrac{\partial s}{\partial \varphi} & \dfrac{\partial s}{\partial \theta} \end{Vmatrix} d\rho d\varphi d\theta$

$= \begin{Vmatrix} \sin\varphi \cos\theta & \rho\cos\varphi \cos\theta & -\rho\sin\varphi \sin\theta \\ \sin\varphi \sin\theta & \rho\cos\varphi \sin\theta & -\rho\sin\varphi \cos\theta \\ \cos\varphi & -\rho\sin\varphi & 0 \end{Vmatrix} d\rho d\varphi d\theta = \rho^2 \sin\varphi \, d\rho d\varphi d\theta$

and $\{(u,v,s): u^2 + v^2 + s^2 \leq 4\} = \{(\rho,\varphi,\theta): 0 \leq \rho \leq 2, 0 \leq \varphi \leq \pi, 0 \leq \theta \leq 2\pi\}$

$\therefore \iiint_{u^2+v^2+s^2 \leq 4} \frac{1}{\sqrt{6}} dudvds = \frac{1}{\sqrt{6}} \int_0^{\pi} \sin\varphi \, d\varphi \int_0^{2\pi} d\theta \int_0^2 \rho^2 d\rho = \frac{32\pi}{3\sqrt{6}}$

Example 2.

$$\text{Find } \int_0^1 \int_0^{\sqrt{1-x^2}} \int_0^{\sqrt{1-(x^2+y^2)}} x^2+y^2+z^2 \, dz\,dy\,dx = ?$$

【Solution】

Let $x = \rho \sin\varphi \cos\theta$, $y = \rho \sin\varphi \sin\theta$, $z = \rho \cos\varphi$ then

$$dxdydz = \begin{Vmatrix} \dfrac{\partial x}{\partial \rho} & \dfrac{\partial x}{\partial \varphi} & \dfrac{\partial x}{\partial \theta} \\ \dfrac{\partial y}{\partial \rho} & \dfrac{\partial y}{\partial \varphi} & \dfrac{\partial y}{\partial \theta} \\ \dfrac{\partial z}{\partial \rho} & \dfrac{\partial z}{\partial \varphi} & \dfrac{\partial z}{\partial \theta} \end{Vmatrix} d\rho d\varphi d\theta$$

$$= \begin{Vmatrix} \sin\varphi\cos\theta & \rho\cos\varphi\cos\theta & -\rho\sin\varphi\sin\theta \\ \sin\varphi\sin\theta & \rho\cos\varphi\sin\theta & -\rho\sin\varphi\cos\theta \\ \cos\varphi & -\rho\sin\varphi & 0 \end{Vmatrix} d\rho d\varphi d\theta = \rho^2 \sin\varphi \, d\rho d\varphi d\theta$$

Let $R = \left\{(\rho, \varphi, \theta): 0 \leq \rho \leq 1, 0 \leq \varphi \leq \dfrac{\pi}{2}, 0 \leq \theta \leq \dfrac{\pi}{2}\right\}$ then

$$\int_0^1 \int_0^{\sqrt{1-x^2}} \int_0^{\sqrt{1-(x^2+y^2)}} x^2+y^2+z^2 \, dz\,dy\,dx = \iiint_R \rho^4 \sin\varphi \, dV = \left(\dfrac{\rho^5}{5}\bigg|_0^1\right) \int_0^{\frac{\pi}{2}} \sin\varphi \, d\varphi \int_0^{\frac{\pi}{2}} d\theta$$

$$= \dfrac{\pi}{10}$$

Example 3.

$$\text{Find } \int_{-2}^2 \int_{-\sqrt{4-x^2}}^{\sqrt{4-x^2}} \int_0^{\sqrt{4-(x^2+y^2)}} z^2\sqrt{x^2+y^2+z^2} \, dz\,dy\,dx = ?$$

【Solution】

Let $x = \rho \sin\varphi \cos\theta$, $y = \rho \sin\varphi \sin\theta$, $z = \rho \cos\varphi$ then

$$dxdydz = \begin{Vmatrix} \dfrac{\partial x}{\partial \rho} & \dfrac{\partial x}{\partial \varphi} & \dfrac{\partial x}{\partial \theta} \\ \dfrac{\partial y}{\partial \rho} & \dfrac{\partial y}{\partial \varphi} & \dfrac{\partial y}{\partial \theta} \\ \dfrac{\partial z}{\partial \rho} & \dfrac{\partial z}{\partial \varphi} & \dfrac{\partial z}{\partial \theta} \end{Vmatrix} d\rho d\varphi d\theta$$

8.4.3 Use Spherical Coordinates to Compute Triple Integrals

$$= \begin{Vmatrix} \sin\varphi\cos\theta & \rho\cos\varphi\cos\theta & -\rho\sin\varphi\sin\theta \\ \sin\varphi\sin\theta & \rho\cos\varphi\sin\theta & -\rho\sin\varphi\cos\theta \\ \cos\varphi & -\rho\sin\varphi & 0 \end{Vmatrix} d\rho d\varphi d\theta = \rho^2 \sin\varphi \, d\rho d\varphi d\theta$$

Let $R = \left\{(\rho, \varphi, \theta): 0 \le \rho \le 2, 0 \le \varphi \le \dfrac{\pi}{2}, 0 \le \theta \le 2\pi\right\}$ then

$$\int_{-2}^{2}\int_{-\sqrt{4-x^2}}^{\sqrt{4-x^2}}\int_{0}^{\sqrt{4-(x^2+y^2)}} z^2\sqrt{x^2+y^2+z^2}\,dzdydx = \iiint_R \rho^3 \cos^2\varphi\,\rho^2 \sin\varphi\,dV$$

$$= \left(\dfrac{\rho^6}{6}\Big|_0^2\right)\int_0^{\frac{\pi}{2}} \cos^2\varphi \sin\varphi \, d\varphi \int_0^{2\pi} d\theta = \dfrac{64}{6}\cdot\dfrac{1}{3}\cdot 2\pi = \dfrac{64\pi}{9}$$

Example 4.

Let $V = \left\{(x,y,z): \dfrac{x^2}{a^2} + \dfrac{y^2}{b^2} + \dfrac{z^2}{c^2} \le 1\right\}$. Find $\iiint_V x^2 y^2 \, dxdydz =?$

【Solution】

Let $x = a\rho\sin\varphi\cos\theta$, $y = b\rho\sin\varphi\sin\theta$, $z = c\rho\cos\varphi$ then

$$dxdydz = \begin{Vmatrix} \dfrac{\partial x}{\partial \rho} & \dfrac{\partial x}{\partial \varphi} & \dfrac{\partial x}{\partial \theta} \\ \dfrac{\partial y}{\partial \rho} & \dfrac{\partial y}{\partial \varphi} & \dfrac{\partial y}{\partial \theta} \\ \dfrac{\partial z}{\partial \rho} & \dfrac{\partial z}{\partial \varphi} & \dfrac{\partial z}{\partial \theta} \end{Vmatrix} d\rho d\varphi d\theta$$

$$= abc \begin{Vmatrix} \sin\varphi\cos\theta & \rho\cos\varphi\cos\theta & -\rho\sin\varphi\sin\theta \\ \sin\varphi\sin\theta & \rho\cos\varphi\sin\theta & -\rho\sin\varphi\cos\theta \\ \cos\varphi & -\rho\sin\varphi & 0 \end{Vmatrix} d\rho d\varphi d\theta = abc\rho^2 \sin\varphi \, d\rho d\varphi d\theta$$

Let $R = \{(\rho, \varphi, \theta): 0 \le \rho \le 1, 0 \le \varphi \le \pi, 0 \le \theta \le 2\pi\}$ then

$$\iiint_V x^2 y^2 \, dxdydz = a^3 b^3 c \iiint_R \rho^6 \sin^5\varphi \sin^2\theta \cos^2\theta \, dV$$

$$= a^3 b^3 c \int_0^{\pi}\int_0^{2\pi}\int_0^1 \rho^6 \sin^5\varphi \sin^2\theta \cos^2\theta \, d\rho d\theta d\varphi$$

$$= a^3 b^3 c \left(\dfrac{\rho^7}{7}\Big|_0^1\right)\int_0^{\pi} \sin^5\varphi \, d\varphi \int_0^{2\pi} \sin^2\theta \cos^2\theta \, d\theta$$

$$\because \int_0^{\pi}\sin^5\varphi \, d\varphi = \dfrac{4}{5}\cdot\dfrac{2}{3}\cdot\int_0^{\pi}\sin\varphi \, d\varphi = \dfrac{16}{15}$$

$$\int_0^{2\pi} \sin^2\theta \cos^2\theta \, d\theta = \int_0^{2\pi} \left(\frac{\sin 2\theta}{2}\right)^2 d\theta = \int_0^{2\pi} \frac{1 - \cos 4\theta}{8} d\theta = \frac{\pi}{4}$$

$$\therefore \iiint_V x^2 y^2 \, dxdydz = \frac{4\pi a^3 b^3 c}{105}$$

Example 5.

Given R as the region enclosed by $x^2 + 2y^2 + 3z^2 = 4$ and $z > 0$, find $\iiint_R z \, dV = ?$

【Solution】

Let $x = u, y = \dfrac{v}{\sqrt{2}}, z = \dfrac{s}{\sqrt{3}}$ then

$$dxdydz = \begin{Vmatrix} \dfrac{\partial x}{\partial u} & \dfrac{\partial x}{\partial v} & \dfrac{\partial x}{\partial s} \\ \dfrac{\partial y}{\partial u} & \dfrac{\partial y}{\partial v} & \dfrac{\partial y}{\partial s} \\ \dfrac{\partial z}{\partial u} & \dfrac{\partial z}{\partial v} & \dfrac{\partial z}{\partial s} \end{Vmatrix} dudvds = \begin{Vmatrix} 1 & 0 & 0 \\ 0 & \dfrac{1}{\sqrt{2}} & 0 \\ 0 & 0 & \dfrac{1}{\sqrt{3}} \end{Vmatrix} = \frac{1}{\sqrt{6}} dudvds$$

$$\therefore \iiint_{x^2+2y^2+3z^2 \leq 4} z \, dxdydz = \frac{1}{3\sqrt{2}} \iiint_{u^2+v^2+s^2 \leq 4} s \, dudvds$$

Let $u = \rho \sin\varphi \cos\theta, v = \rho \sin\varphi \sin\theta, s = \rho \cos\varphi$ then

$$dudvds = \begin{Vmatrix} \dfrac{\partial u}{\partial \rho} & \dfrac{\partial u}{\partial \varphi} & \dfrac{\partial u}{\partial \theta} \\ \dfrac{\partial v}{\partial \rho} & \dfrac{\partial v}{\partial \varphi} & \dfrac{\partial v}{\partial \theta} \\ \dfrac{\partial s}{\partial \rho} & \dfrac{\partial s}{\partial \varphi} & \dfrac{\partial s}{\partial \theta} \end{Vmatrix} d\rho d\varphi d\theta$$

$$= \begin{Vmatrix} \sin\varphi \cos\theta & \rho\cos\varphi \cos\theta & -\rho\sin\varphi \sin\theta \\ \sin\varphi \sin\theta & \rho\cos\varphi \sin\theta & -\rho\sin\varphi \cos\theta \\ \cos\varphi & -\rho\sin\varphi & 0 \end{Vmatrix} d\rho d\varphi d\theta = \rho^2 \sin\varphi \, d\rho d\varphi d\theta$$

and $\{(u,v,s): u^2 + v^2 + s^2 \leq 4, s > 0\} = \{(\rho, \varphi, \theta): 0 \leq \rho \leq 2, 0 \leq \varphi \leq \dfrac{\pi}{2}, 0 \leq \theta \leq 2\pi\}$

$$\therefore \frac{1}{3\sqrt{2}} \iiint_{u^2+v^2+s^2 \leq 4} s \, dudvds = \frac{1}{\sqrt{6}} \int_0^{\frac{\pi}{2}} \sin\varphi \cos\varphi \, d\varphi \int_0^{2\pi} d\theta \int_0^2 \rho^3 d\rho$$

$$= \frac{\pi}{\sqrt{6}} \int_0^{\frac{\pi}{2}} \sin 2\varphi \, d\varphi \cdot \frac{\rho^4}{4}\Big|_0^2 = \frac{\pi}{2\sqrt{6}} \cdot (-\cos 2\varphi)\Big|_0^{\frac{\pi}{2}} \cdot 4 = \frac{4\pi}{\sqrt{6}}$$

8.4.3 Use Spherical Coordinates to Compute Triple Integrals

Example 6.

Find $\int_{-\infty}^{\infty}\int_{-\infty}^{\infty}\int_{-\infty}^{\infty} e^{-(x^2+y^2+z^2)} dxdydz = ?$

【Solution】

Let $x = \rho \sin\varphi \cos\theta, y = \rho \sin\varphi \sin\theta, z = \rho \cos\varphi$ then

$$dxdydz = \begin{Vmatrix} \dfrac{\partial x}{\partial \rho} & \dfrac{\partial x}{\partial \varphi} & \dfrac{\partial x}{\partial \theta} \\ \dfrac{\partial y}{\partial \rho} & \dfrac{\partial y}{\partial \varphi} & \dfrac{\partial y}{\partial \theta} \\ \dfrac{\partial z}{\partial \rho} & \dfrac{\partial z}{\partial \varphi} & \dfrac{\partial z}{\partial \theta} \end{Vmatrix} d\rho d\varphi d\theta$$

$$= \begin{Vmatrix} \sin\varphi \cos\theta & \rho\cos\varphi \cos\theta & -\rho\sin\varphi \sin\theta \\ \sin\varphi \sin\theta & \rho\cos\varphi \sin\theta & -\rho\sin\varphi \cos\theta \\ \cos\varphi & -\rho\sin\varphi & 0 \end{Vmatrix} d\rho d\varphi d\theta = \rho^2 \sin\varphi \, d\rho d\varphi d\theta$$

Let $R = \{(\rho, \varphi, \theta) : 0 \le \rho \le 1, 0 \le \varphi \le \pi, 0 \le \theta \le 2\pi\}$ then

$$\int_{-\infty}^{\infty}\int_{-\infty}^{\infty}\int_{-\infty}^{\infty} e^{-(x^2+y^2+z^2)} dxdydz = \iiint_R e^{-\rho^2} \rho^2 \sin\varphi \, d\rho d\theta d\varphi$$

$$= \int_0^{\pi}\int_0^{2\pi}\int_0^{\infty} e^{-\rho^2} \rho^2 \sin\varphi \, d\rho d\theta d\varphi = \int_0^{\pi} \sin\varphi \int_0^{2\pi} d\theta \int_0^{\infty} e^{-\rho^2} \rho^2 d\rho$$

By integration by parts,

$$\int_0^{\infty} e^{-\rho^2} \rho^2 d\rho = \dfrac{-\rho}{2} e^{-\rho^2} \Big|_0^{\infty} + \dfrac{1}{2}\int_0^{\infty} e^{-\rho^2} d\rho = \dfrac{\sqrt{\pi}}{4}$$

$$\therefore \int_{-\infty}^{\infty}\int_{-\infty}^{\infty}\int_{-\infty}^{\infty} e^{-(x^2+y^2+z^2)} dxdydz = \int_0^{\pi} \sin\varphi \int_0^{2\pi} d\theta \int_0^{\infty} e^{-\rho^2} \rho^2 d\rho = 2(2\pi)\dfrac{\sqrt{\pi}}{4} = \pi\sqrt{\pi}$$

Example 7.

Find $\int_{-\infty}^{\infty}\int_{-\infty}^{\infty}\int_{-\infty}^{\infty} e^{-(x^2+2y^2+3z^2)} dxdydz = ?$

【Solution】

Let $x = \rho \sin\varphi \cos\theta, y = \dfrac{\rho \sin\varphi \sin\theta}{\sqrt{2}}, z = \dfrac{\rho \cos\varphi}{\sqrt{3}}$ then

8.4.3 Use Spherical Coordinates to Compute Triple Integrals

$$dxdydz = \begin{Vmatrix} \dfrac{\partial x}{\partial \rho} & \dfrac{\partial x}{\partial \varphi} & \dfrac{\partial x}{\partial \theta} \\ \dfrac{\partial y}{\partial \rho} & \dfrac{\partial y}{\partial \varphi} & \dfrac{\partial y}{\partial \theta} \\ \dfrac{\partial z}{\partial \rho} & \dfrac{\partial z}{\partial \varphi} & \dfrac{\partial z}{\partial \theta} \end{Vmatrix} d\rho d\varphi d\theta$$

$$= \dfrac{1}{\sqrt{6}} \begin{Vmatrix} \sin\varphi\cos\theta & \rho\cos\varphi\cos\theta & -\rho\sin\varphi\sin\theta \\ \sin\varphi\sin\theta & \rho\cos\varphi\sin\theta & -\rho\sin\varphi\cos\theta \\ \cos\varphi & -\rho\sin\varphi & 0 \end{Vmatrix} d\rho d\varphi d\theta = \dfrac{\rho^2 \sin\varphi \, d\rho d\varphi d\theta}{\sqrt{6}}$$

Let $R = \{(\rho, \varphi, \theta): 0 \le \rho \le 1, 0 \le \varphi \le \pi, 0 \le \theta \le 2\pi\}$ then

$$\int_{-\infty}^{\infty}\int_{-\infty}^{\infty}\int_{-\infty}^{\infty} e^{-(x^2+2y^2+3z^2)} dxdydz = \dfrac{1}{\sqrt{6}}\iiint_R e^{-\rho^2}\rho^2 \sin\varphi \, d\rho d\theta d\varphi$$

$$= \dfrac{1}{\sqrt{6}}\int_0^{\pi}\int_0^{2\pi}\int_0^{\infty} e^{-\rho^2}\rho^2 \sin\varphi \, d\rho d\theta d\varphi = \dfrac{1}{\sqrt{6}}\int_0^{\pi} \sin\varphi \int_0^{2\pi} d\theta \int_0^{\infty} e^{-\rho^2}\rho^2 d\rho$$

By integration by parts,

$$\int_0^{\infty} e^{-\rho^2}\rho^2 d\rho = \dfrac{-\rho}{2}e^{-\rho^2}\Big|_0^{\infty} + \dfrac{1}{2}\int_0^{\infty} e^{-\rho^2} d\rho = \dfrac{\sqrt{\pi}}{4}$$

$$\therefore \dfrac{1}{\sqrt{6}}\int_0^{\pi} \sin\varphi \int_0^{2\pi} d\theta \int_0^{\infty} e^{-\rho^2}\rho^2 d\rho = \dfrac{1}{\sqrt{6}} \cdot 2(2\pi) \dfrac{\sqrt{\pi}}{4} = \pi\sqrt{\dfrac{\pi}{6}}$$

Example 8.

(1) Let $V = \{(x,y,z): x^2+y^2+z^2 \le 1\}$. Find $\iiint_V \dfrac{dV}{9-(x^2+y^2+z^2)} = ?$

(2) Let $V = \{(x,y,z): 1 \le x^2+y^2+z^2 \le 4, x \ge 0, y \ge 0\}$. Find $\iiint_V (3x+2z) dV = ?$

【Solution】

(1)
Let $x = \rho\sin\varphi\cos\theta$, $y = \rho\sin\varphi\sin\theta$, $z = \rho\cos\varphi$ then

$$dxdydz = \begin{Vmatrix} \dfrac{\partial x}{\partial \rho} & \dfrac{\partial x}{\partial \varphi} & \dfrac{\partial x}{\partial \theta} \\ \dfrac{\partial y}{\partial \rho} & \dfrac{\partial y}{\partial \varphi} & \dfrac{\partial y}{\partial \theta} \\ \dfrac{\partial z}{\partial \rho} & \dfrac{\partial z}{\partial \varphi} & \dfrac{\partial z}{\partial \theta} \end{Vmatrix} d\rho d\varphi d\theta$$

8.4.3 Use Spherical Coordinates to Compute Triple Integrals 274

$$= \begin{Vmatrix} \sin\varphi\cos\theta & \rho\cos\varphi\cos\theta & -\rho\sin\varphi\sin\theta \\ \sin\varphi\sin\theta & \rho\cos\varphi\sin\theta & -\rho\sin\varphi\cos\theta \\ \cos\varphi & -\rho\sin\varphi & 0 \end{Vmatrix} d\rho d\varphi d\theta = \rho^2\sin\varphi\, d\rho d\varphi d\theta$$

Let $R = \{(\rho,\varphi,\theta): 0 \le \rho \le 1, 0 \le \varphi \le \pi, 0 \le \theta \le 2\pi\}$ then

$$\iiint_V \frac{dV}{9-(x^2+y^2+z^2)} = \iiint_R \frac{\rho^2\sin\varphi}{9-\rho^2} d\rho d\theta d\varphi = \int_0^\pi \int_0^{2\pi} \int_0^1 \frac{\rho^2\sin\varphi}{9-\rho^2} d\rho d\theta d\varphi$$

$$= \int_0^\pi \sin\varphi d\varphi \int_0^{2\pi} d\theta \int_0^1 \frac{\rho^2}{9-\rho^2} d\rho = 2(2\pi)\left(\frac{3}{2}\ln\left|\frac{3+\rho}{3-\rho}\right| - \rho\right)\Big|_0^1 = 4\pi\left(\frac{3\ln 2}{2} - 1\right)$$

(2)

Let $x = \rho\sin\varphi\cos\theta, y = \rho\sin\varphi\sin\theta, z = \rho\cos\varphi$ then

$$dxdydz = \begin{Vmatrix} \frac{\partial x}{\partial \rho} & \frac{\partial x}{\partial \varphi} & \frac{\partial x}{\partial \theta} \\ \frac{\partial y}{\partial \rho} & \frac{\partial y}{\partial \varphi} & \frac{\partial y}{\partial \theta} \\ \frac{\partial z}{\partial \rho} & \frac{\partial z}{\partial \varphi} & \frac{\partial z}{\partial \theta} \end{Vmatrix} d\rho d\varphi d\theta$$

$$= \begin{Vmatrix} \sin\varphi\cos\theta & \rho\cos\varphi\cos\theta & -\rho\sin\varphi\sin\theta \\ \sin\varphi\sin\theta & \rho\cos\varphi\sin\theta & -\rho\sin\varphi\cos\theta \\ \cos\varphi & -\rho\sin\varphi & 0 \end{Vmatrix} d\rho d\varphi d\theta = \rho^2\sin\varphi\, d\rho d\varphi d\theta$$

Let $R = \{(\rho,\varphi,\theta): 1 \le \rho \le 2, 0 \le \varphi \le \pi, 0 \le \theta \le \frac{\pi}{2}\}$ then

$$\iiint_V (3x+2z)dV = \iiint_R (3\rho\sin\varphi\cos\theta + 2\rho\cos\varphi)\rho^2\sin\varphi\, d\rho d\theta d\varphi$$

$$= \int_0^\pi \int_0^{\frac{\pi}{2}} \int_1^2 (3\rho\sin\varphi\cos\theta + 2\rho\cos\varphi)\rho^2\sin\varphi\, d\rho d\theta d\varphi$$

$$= \int_0^\pi \int_0^{\frac{\pi}{2}} \int_1^2 (3\rho^3\sin^2\varphi\cos\theta + 2\rho^3\cos\varphi\sin\varphi)d\rho d\theta d\varphi$$

$$= \int_0^\pi \int_0^{\frac{\pi}{2}} \sin^2\varphi\cos\theta \frac{3\rho^4}{4}\Big|_1^2 + \cos\varphi\sin\varphi \frac{\rho^4}{2}\Big|_1^2 d\rho d\theta d\varphi$$

$$= \int_0^\pi \int_0^{\frac{\pi}{2}} \sin^2\varphi\cos\theta \left(\frac{45}{4}\right) + \cos\varphi\sin\varphi \left(\frac{15}{2}\right) d\theta d\varphi$$

$$= \int_0^\pi \sin^2\varphi\sin\theta\Big|_{\theta=0}^{\theta=\frac{\pi}{2}} \left(\frac{45}{4}\right) + \left(\frac{\pi}{2}\right)\cos\varphi\sin\varphi \left(\frac{15}{2}\right) d\varphi$$

$$= \int_0^\pi \sin^2\varphi \left(\frac{45}{4}\right) + \left(\frac{\pi}{2}\right)\cos\varphi\sin\varphi \left(\frac{15}{2}\right) d\varphi = \frac{45}{4}\int_0^\pi \frac{1-\cos 2\varphi}{2} d\varphi + \frac{15\pi}{4}\int_0^\pi \cos\varphi\sin\varphi\, d\varphi$$

$$= \frac{45\pi}{8}$$

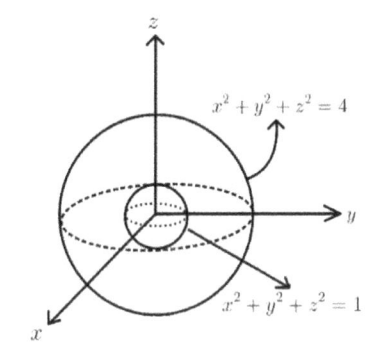

Example 9.

(1) Let $V = \{(x,y,z): 1 \leq x^2 + y^2 + z^2 \leq 9, x \geq 0, y \geq 0, z \geq 0\}$. Find $\iiint_V z\, dV = ?$

(2) Let $V = \{(x,y,z): 1 \leq x^2 + y^2 + z^2 \leq 3\}$. Find $\iiint_V \frac{\cos\sqrt{x^2+y^2+z^2}\, dxdydz}{x^2+y^2+z^2} = ?$

【Solution】

(1)
Let $x = \rho \sin\varphi \cos\theta$, $y = \rho \sin\varphi \sin\theta$, $z = \rho \cos\varphi$ then

$$dxdydz = \begin{Vmatrix} \frac{\partial x}{\partial \rho} & \frac{\partial x}{\partial \varphi} & \frac{\partial x}{\partial \theta} \\ \frac{\partial y}{\partial \rho} & \frac{\partial y}{\partial \varphi} & \frac{\partial y}{\partial \theta} \\ \frac{\partial z}{\partial \rho} & \frac{\partial z}{\partial \varphi} & \frac{\partial z}{\partial \theta} \end{Vmatrix} d\rho d\varphi d\theta$$

$$= \begin{Vmatrix} \sin\varphi\cos\theta & \rho\cos\varphi\cos\theta & -\rho\sin\varphi\sin\theta \\ \sin\varphi\sin\theta & \rho\cos\varphi\sin\theta & -\rho\sin\varphi\cos\theta \\ \cos\varphi & -\rho\sin\varphi & 0 \end{Vmatrix} d\rho d\varphi d\theta = \rho^2 \sin\varphi\, d\rho d\varphi d\theta$$

Let $R = \{(\rho,\varphi,\theta): 1 \leq \rho \leq 3, 0 \leq \varphi \leq \frac{\pi}{2}, 0 \leq \theta \leq \frac{\pi}{2}\}$ then

$$\iiint_V z\, dV = \iiint_R \rho\cos\varphi\, \rho^2 \sin\varphi\, d\rho d\theta d\varphi = \int_0^{\frac{\pi}{2}} \int_0^{\frac{\pi}{2}} \int_1^3 \rho\cos\varphi\, \rho^2 \sin\varphi\, d\rho d\theta d\varphi$$

$$= \int_0^{\frac{\pi}{2}} \int_0^{\frac{\pi}{2}} \int_1^3 \rho^3 \cos\varphi \sin\varphi\, d\rho d\theta d\varphi = \int_0^{\frac{\pi}{2}} \cos\varphi \sin\varphi\, d\varphi \int_0^{\frac{\pi}{2}} d\theta \int_1^3 \rho^3 d\rho$$

$$= \left(-\frac{\cos 2\varphi}{4}\Big|_0^{\frac{\pi}{2}}\right)\left(\frac{\pi}{2}\right)20 = 5\pi$$

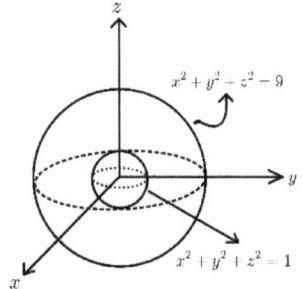

(2)
Let $x = \rho \sin \varphi \cos \theta$, $y = \rho \sin \varphi \sin \theta$, $z = \rho \cos \varphi$ then

$$dxdydz = \begin{Vmatrix} \frac{\partial x}{\partial \rho} & \frac{\partial x}{\partial \varphi} & \frac{\partial x}{\partial \theta} \\ \frac{\partial y}{\partial \rho} & \frac{\partial y}{\partial \varphi} & \frac{\partial y}{\partial \theta} \\ \frac{\partial z}{\partial \rho} & \frac{\partial z}{\partial \varphi} & \frac{\partial z}{\partial \theta} \end{Vmatrix} d\rho d\varphi d\theta$$

$$= \begin{Vmatrix} \sin \varphi \cos \theta & \rho \cos \varphi \cos \theta & -\rho \sin \varphi \sin \theta \\ \sin \varphi \sin \theta & \rho \cos \varphi \sin \theta & -\rho \sin \varphi \cos \theta \\ \cos \varphi & -\rho \sin \varphi & 0 \end{Vmatrix} d\rho d\varphi d\theta = \rho^2 \sin \varphi \, d\rho d\varphi d\theta$$

Let $R = \{(\rho, \varphi, \theta): 1 \leq \rho \leq \sqrt{3}, 0 \leq \varphi \leq \pi, 0 \leq \theta \leq 2\pi\}$ then

$$\iiint_{1 \leq x^2+y^2+z^2 \leq 3} \frac{\cos\sqrt{x^2+y^2+z^2}\,dxdydz}{x^2+y^2+z^2} dV = \iiint_R \frac{\cos \rho}{\rho^2} \rho^2 \sin \varphi \, d\rho d\theta d\varphi$$

$$= \int_0^\pi \int_0^{2\pi} \int_1^{\sqrt{3}} \frac{\cos \rho}{\rho^2} \rho^2 \sin \varphi \, d\rho d\theta d\varphi = \int_0^\pi \sin \varphi \, d\varphi \int_0^{2\pi} d\theta \int_1^{\sqrt{3}} \cos \rho \, d\rho$$

$$= (-\cos \varphi|_0^\pi)(2\pi)\left(\sin \rho|_1^{\sqrt{3}}\right) = 4\pi(\sin \sqrt{3} - \sin 1)$$

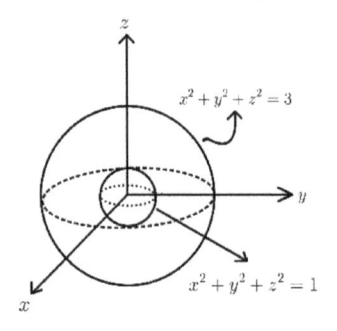

Example 10.

Let $V = \{(x, y, z): x^2 + y^2 + z^2 \leq 1\}$. Find $\iiint_V \dfrac{z^2 dxdydz}{\sqrt{1 - x^2 - y^2 - z^2}} = ?$

【Solution】

Let $x = \rho \sin \varphi \cos \theta$, $y = \rho \sin \varphi \sin \theta$, $z = \rho \cos \varphi$ then

$$dxdydz = \begin{Vmatrix} \dfrac{\partial x}{\partial \rho} & \dfrac{\partial x}{\partial \varphi} & \dfrac{\partial x}{\partial \theta} \\ \dfrac{\partial y}{\partial \rho} & \dfrac{\partial y}{\partial \varphi} & \dfrac{\partial y}{\partial \theta} \\ \dfrac{\partial z}{\partial \rho} & \dfrac{\partial z}{\partial \varphi} & \dfrac{\partial z}{\partial \theta} \end{Vmatrix} d\rho d\varphi d\theta$$

$$= \begin{Vmatrix} \sin \varphi \cos \theta & \rho \cos \varphi \cos \theta & -\rho \sin \varphi \sin \theta \\ \sin \varphi \sin \theta & \rho \cos \varphi \sin \theta & -\rho \sin \varphi \cos \theta \\ \cos \varphi & -\rho \sin \varphi & 0 \end{Vmatrix} d\rho d\varphi d\theta = \rho^2 \sin \varphi \, d\rho d\varphi d\theta$$

Let $R = \{(\rho, \varphi, \theta): 0 \leq \rho \leq 1, 0 \leq \varphi \leq \pi, 0 \leq \theta \leq 2\pi\}$ then

$$\iiint_V \dfrac{z^2 dxdydz}{\sqrt{1 - x^2 - y^2 - z^2}} = \iiint_R \dfrac{\rho^4 \sin \varphi \cos^2 \varphi}{\sqrt{1 - \rho^2}} d\rho d\theta d\varphi$$

$$= \int_0^\pi \int_0^{2\pi} \int_0^1 \dfrac{\rho^4 \sin \varphi \cos^2 \varphi}{\sqrt{1 - \rho^2}} d\rho d\theta d\varphi = \int_0^\pi \cos^2 \varphi \sin \varphi d\varphi \int_0^{2\pi} d\theta \int_0^1 \dfrac{\rho^4}{\sqrt{1 - \rho^2}} d\rho$$

$$= -\dfrac{\cos^3 \varphi}{3} \bigg|_0^\pi (2\pi) \left(\int_0^1 \dfrac{\rho^4}{\sqrt{1 - \rho^2}} d\rho \right) = \dfrac{4\pi}{3} \left(\int_0^1 \dfrac{\rho^4}{\sqrt{1 - \rho^2}} d\rho \right)$$

Let $\rho = \sin \theta$ then $d\rho = \cos \theta \, d\theta$

$$\therefore \int_0^1 \dfrac{\rho^4}{\sqrt{1 - \rho^2}} d\rho = \int_0^{\frac{\pi}{2}} \dfrac{\sin^4 \theta \cos \theta}{\sqrt{1 - \sin^2 \theta}} d\theta = \int_0^{\frac{\pi}{2}} \sin^4 \theta \, d\theta = \dfrac{1}{4} \left(\dfrac{3\theta}{2} - \sin 2\theta + \dfrac{\sin 4\theta}{8} \right) \bigg|_0^{\frac{\pi}{2}} = \dfrac{3\pi}{16}$$

$$\therefore \iiint_V \dfrac{z^2 dxdydz}{\sqrt{1 - x^2 - y^2 - z^2}} = \dfrac{4\pi}{3} \left(\int_0^1 \dfrac{\rho^4}{\sqrt{1 - \rho^2}} d\rho \right) = \dfrac{\pi^2}{4}$$

Example 11.

Let $V = \{(x, y, z): x^2 + y^2 + z^2 \leq 1, z > 0\}$. Find $\iiint_V \dfrac{z dxdydz}{\sqrt{1 - x^2 - y^2 - z^2}} = ?$

【Solution】

Let $x = \rho \sin \varphi \cos \theta$, $y = \rho \sin \varphi \sin \theta$, $z = \rho \cos \varphi$ then

8.4.3 Use Spherical Coordinates to Compute Triple Integrals

$$dxdydz = \begin{Vmatrix} \dfrac{\partial x}{\partial \rho} & \dfrac{\partial x}{\partial \varphi} & \dfrac{\partial x}{\partial \theta} \\ \dfrac{\partial y}{\partial \rho} & \dfrac{\partial y}{\partial \varphi} & \dfrac{\partial y}{\partial \theta} \\ \dfrac{\partial z}{\partial \rho} & \dfrac{\partial z}{\partial \varphi} & \dfrac{\partial z}{\partial \theta} \end{Vmatrix} d\rho d\varphi d\theta$$

$$= \begin{Vmatrix} \sin\varphi\cos\theta & \rho\cos\varphi\cos\theta & -\rho\sin\varphi\sin\theta \\ \sin\varphi\sin\theta & \rho\cos\varphi\sin\theta & -\rho\sin\varphi\cos\theta \\ \cos\varphi & -\rho\sin\varphi & 0 \end{Vmatrix} d\rho d\varphi d\theta = \rho^2 \sin\varphi \, d\rho d\varphi d\theta$$

Let $R = \left\{(\rho, \varphi, \theta) : 0 \le \rho \le 1, 0 \le \varphi \le \dfrac{\pi}{2}, 0 \le \theta \le 2\pi\right\}$ then

$$\iiint_V \frac{z\,dxdydz}{\sqrt{1-x^2-y^2-z^2}} = \iiint_R \frac{\rho^3 \sin\varphi \cos\varphi}{\sqrt{1-\rho^2}} d\rho d\theta d\varphi$$

$$= \int_0^{\frac{\pi}{2}} \int_0^{2\pi} \int_0^1 \frac{\rho^3 \sin\varphi \cos\varphi}{\sqrt{1-\rho^2}} d\rho d\theta d\varphi = \int_0^{\frac{\pi}{2}} \cos\varphi \sin\varphi d\varphi \int_0^{2\pi} d\theta \int_0^1 \frac{\rho^3}{\sqrt{1-\rho^2}} d\rho$$

$$= -\frac{\cos 2\varphi}{4}\bigg|_0^{\frac{\pi}{2}} (2\pi) \left(\int_0^1 \frac{\rho^3}{\sqrt{1-\rho^2}} d\rho\right) = \pi \left(\int_0^1 \frac{\rho^3}{\sqrt{1-\rho^2}} d\rho\right)$$

Let $\rho = \sin\theta$ then $d\rho = \cos\theta \, d\theta$

$$\therefore \int_0^1 \frac{\rho^3}{\sqrt{1-\rho^2}} d\rho = \int_0^{\frac{\pi}{2}} \frac{\sin^3\theta \cos\theta}{\sqrt{1-\sin^2\theta}} d\theta = \int_0^{\frac{\pi}{2}} \sin^3\theta \, d\theta = -\cos\theta + \frac{\cos^3\theta}{3}\bigg|_0^{\frac{\pi}{2}} = \frac{2}{3}$$

$$\therefore \iiint_V \frac{z\,dxdydz}{\sqrt{1-x^2-y^2-z^2}} = \frac{2\pi}{3}$$

Example 12.

Let $V = \{(x, y, z) : x^2 + y^2 + z^2 \le r^2\}$. Find $\iiint_V (x^2 + y^2 + z^2)^{\frac{3}{2}} dxdydz = ?$

【Solution】

Let $x = \rho \sin\varphi \cos\theta$, $y = \rho \sin\varphi \sin\theta$, $z = \rho \cos\varphi$ then

8.4.3 Use Spherical Coordinates to Compute Triple Integrals

$$dxdydz = \begin{Vmatrix} \dfrac{\partial x}{\partial \rho} & \dfrac{\partial x}{\partial \varphi} & \dfrac{\partial x}{\partial \theta} \\ \dfrac{\partial y}{\partial \rho} & \dfrac{\partial y}{\partial \varphi} & \dfrac{\partial y}{\partial \theta} \\ \dfrac{\partial z}{\partial \rho} & \dfrac{\partial z}{\partial \varphi} & \dfrac{\partial z}{\partial \theta} \end{Vmatrix} d\rho d\varphi d\theta$$

$$= \begin{Vmatrix} \sin\varphi\cos\theta & \rho\cos\varphi\cos\theta & -\rho\sin\varphi\sin\theta \\ \sin\varphi\sin\theta & \rho\cos\varphi\sin\theta & -\rho\sin\varphi\cos\theta \\ \cos\varphi & -\rho\sin\varphi & 0 \end{Vmatrix} d\rho d\varphi d\theta = \rho^2 \sin\varphi \, d\rho d\varphi d\theta$$

Let $R = \{(\rho, \varphi, \theta): 0 \leq \rho \leq r, 0 \leq \varphi \leq \pi, 0 \leq \theta \leq 2\pi\}$ then

$$\iiint_V (x^2 + y^2 + z^2)^{\frac{3}{2}} dxdydz = \iiint_R (\rho^2)^{\frac{3}{2}} \rho^2 \sin\varphi \, d\rho d\theta d\varphi$$

$$= \int_0^\pi \int_0^{2\pi} \int_0^r (\rho^2)^{\frac{3}{2}} \rho^2 \sin\varphi \, d\rho d\theta d\varphi = \int_0^\pi \sin\varphi d\varphi \int_0^{2\pi} d\theta \int_0^r \rho^5 d\rho = 2(2\pi)\left(\frac{r^6}{6}\right) = \frac{2\pi r^6}{3}$$

Example 13.

Find $\displaystyle\int_{-3}^{3} \int_0^{\sqrt{9-y^2}} \int_{-\sqrt{9-x^2-y^2}}^{\sqrt{9-x^2-y^2}} y^2\sqrt{x^2+y^2+z^2} \, dzdxdy = ?$

【Solution】

Let $x = \rho\sin\varphi\cos\theta$, $y = \rho\sin\varphi\sin\theta$, $z = \rho\cos\varphi$ then

$$dxdydz = \begin{Vmatrix} \dfrac{\partial x}{\partial \rho} & \dfrac{\partial x}{\partial \varphi} & \dfrac{\partial x}{\partial \theta} \\ \dfrac{\partial y}{\partial \rho} & \dfrac{\partial y}{\partial \varphi} & \dfrac{\partial y}{\partial \theta} \\ \dfrac{\partial z}{\partial \rho} & \dfrac{\partial z}{\partial \varphi} & \dfrac{\partial z}{\partial \theta} \end{Vmatrix} d\rho d\varphi d\theta$$

$$= \begin{Vmatrix} \sin\varphi\cos\theta & \rho\cos\varphi\cos\theta & -\rho\sin\varphi\sin\theta \\ \sin\varphi\sin\theta & \rho\cos\varphi\sin\theta & -\rho\sin\varphi\cos\theta \\ \cos\varphi & -\rho\sin\varphi & 0 \end{Vmatrix} d\rho d\varphi d\theta = \rho^2 \sin\varphi \, d\rho d\varphi d\theta$$

Let $R = \left\{(\rho, \varphi, \theta): 0 \leq \rho \leq 3, 0 \leq \varphi \leq \pi, -\dfrac{\pi}{2} \leq \theta \leq \dfrac{\pi}{2}\right\}$ then

$$\int_{-3}^{3} \int_0^{\sqrt{9-y^2}} \int_{-\sqrt{9-x^2-y^2}}^{\sqrt{9-x^2-y^2}} y^2\sqrt{x^2+y^2+z^2} \, dzdxdy = \iiint_R (\rho\sin\varphi\sin\theta)^2 \rho^3 \sin\varphi \, d\rho d\theta d\varphi$$

$$= \int_0^\pi \int_{-\frac{\pi}{2}}^{\frac{\pi}{2}} \int_0^3 (\rho \sin\varphi \sin\theta)^2 \rho^3 \sin\varphi \, d\rho d\theta d\varphi = \int_0^\pi \sin^3\varphi \, d\varphi \int_{-\frac{\pi}{2}}^{\frac{\pi}{2}} \sin^2\theta \, d\theta \int_0^3 \rho^5 d\rho$$

$$= \frac{4}{3} \cdot \frac{\pi}{2} \cdot \frac{243}{2} = 81\pi$$

Example 14.

Find $\int_0^r \int_0^{\sqrt{r^2-x^2}} \int_0^{\sqrt{r^2-x^2-y^2}} \frac{1}{x^2+y^2+z^2} dzdydx = ?$, $r > 0$

【Solution】

Let $x = \rho \sin\varphi \cos\theta$, $y = \rho \sin\varphi \sin\theta$, $z = \rho \cos\varphi$ then

$$dxdydz = \begin{Vmatrix} \frac{\partial x}{\partial \rho} & \frac{\partial x}{\partial \varphi} & \frac{\partial x}{\partial \theta} \\ \frac{\partial y}{\partial \rho} & \frac{\partial y}{\partial \varphi} & \frac{\partial y}{\partial \theta} \\ \frac{\partial z}{\partial \rho} & \frac{\partial z}{\partial \varphi} & \frac{\partial z}{\partial \theta} \end{Vmatrix} d\rho d\varphi d\theta$$

$$= \begin{Vmatrix} \sin\varphi \cos\theta & \rho\cos\varphi \cos\theta & -\rho\sin\varphi \sin\theta \\ \sin\varphi \sin\theta & \rho\cos\varphi \sin\theta & -\rho\sin\varphi \cos\theta \\ \cos\varphi & -\rho\sin\varphi & 0 \end{Vmatrix} d\rho d\varphi d\theta = \rho^2 \sin\varphi \, d\rho d\varphi d\theta$$

Let $R = \left\{(\rho, \varphi, \theta): 0 \leq \rho \leq r, 0 \leq \varphi \leq \frac{\pi}{2}, 0 \leq \theta \leq \frac{\pi}{2}\right\}$ then

$$\int_0^r \int_0^{\sqrt{r^2-x^2}} \int_0^{\sqrt{r^2-x^2-y^2}} \frac{1}{x^2+y^2+z^2} dzdydx = \iiint_R \frac{1}{\rho^2} \rho^2 \sin\varphi \, d\rho d\theta d\varphi$$

$$= \int_0^{\frac{\pi}{2}} \int_0^{\frac{\pi}{2}} \int_0^r \frac{1}{\rho^2} \rho^2 \sin\varphi \, d\rho d\theta d\varphi = \int_0^{\frac{\pi}{2}} \sin\varphi \, d\varphi \int_0^{\frac{\pi}{2}} d\theta \int_0^r 1 d\rho = \frac{\pi r}{2}$$

Example 15.

Let $V = \{(x, y, z): x^2 + y^2 + z^2 \leq 1, x \geq 0, y \geq 0, z \geq 0\}$. Find $\iiint_V xyz \, dxdydz = ?$

【Solution】

Let $x = \rho \sin\varphi \cos\theta$, $y = \rho \sin\varphi \sin\theta$, $z = \rho \cos\varphi$ then

$$dxdydz = \begin{Vmatrix} \dfrac{\partial x}{\partial \rho} & \dfrac{\partial x}{\partial \varphi} & \dfrac{\partial x}{\partial \theta} \\ \dfrac{\partial y}{\partial \rho} & \dfrac{\partial y}{\partial \varphi} & \dfrac{\partial y}{\partial \theta} \\ \dfrac{\partial z}{\partial \rho} & \dfrac{\partial z}{\partial \varphi} & \dfrac{\partial z}{\partial \theta} \end{Vmatrix} d\rho d\varphi d\theta$$

$$= \begin{Vmatrix} \sin\varphi\cos\theta & \rho\cos\varphi\cos\theta & -\rho\sin\varphi\sin\theta \\ \sin\varphi\sin\theta & \rho\cos\varphi\sin\theta & -\rho\sin\varphi\cos\theta \\ \cos\varphi & -\rho\sin\varphi & 0 \end{Vmatrix} d\rho d\varphi d\theta = \rho^2 \sin\varphi\, d\rho d\varphi d\theta$$

Let $R = \left\{(\rho, \varphi, \theta): 0 \leq \rho \leq 1, 0 \leq \varphi \leq \dfrac{\pi}{2}, 0 \leq \theta \leq \dfrac{\pi}{2}\right\}$ then

$$\iiint_V xyz\, dxdydz = \iiint_R (\rho\sin\varphi\cos\theta \cdot \rho\sin\varphi\sin\theta \cdot \rho\cos\varphi)\rho^2 \sin\varphi\, d\rho d\theta d\varphi$$

$$= \int_0^{\frac{\pi}{2}} \int_0^{\frac{\pi}{2}} \int_0^1 (\rho\sin\varphi\cos\theta \cdot \rho\sin\varphi\sin\theta \cdot \rho\cos\varphi)\rho^2 \sin\varphi\, d\rho d\theta d\varphi$$

$$= \left(\dfrac{\rho^6}{6}\bigg|_0^1\right) \int_0^{\frac{\pi}{2}} \sin^3\varphi\cos\varphi\, d\varphi \int_0^{\frac{\pi}{2}} \cos\theta\sin\theta\, d\theta$$

$$\because \int_0^{\frac{\pi}{2}} \cos\theta\sin\theta\, d\theta = \dfrac{1}{2}\int_0^{\frac{\pi}{2}} \sin 2\theta\, d\theta = \dfrac{-1}{4}(\cos 2\theta)\bigg|_0^{\frac{\pi}{2}} = \dfrac{1}{2}$$

$$\int_0^{\frac{\pi}{2}} \sin^3\varphi\cos\varphi\, d\varphi = \dfrac{\sin^4\varphi}{4}\bigg|_0^{\frac{\pi}{2}} = \dfrac{1}{4}$$

$$\therefore \iiint_V xyz\, dxdydz = \dfrac{1}{6} \cdot \dfrac{1}{4} \cdot \dfrac{1}{2} = \dfrac{1}{48}$$

Example 16.

Let $V = \{(x, y, z): x^2 + y^2 + z^2 \leq 1\}$. Find $\iiint_V y^2\, dV = ?$

【Solution】

Let $x = \rho\sin\varphi\cos\theta$, $y = \rho\sin\varphi\sin\theta$, $z = \rho\cos\varphi$ then

8.4.3 Use Spherical Coordinates to Compute Triple Integrals

$$dxdydz = \begin{Vmatrix} \dfrac{\partial x}{\partial \rho} & \dfrac{\partial x}{\partial \varphi} & \dfrac{\partial x}{\partial \theta} \\ \dfrac{\partial y}{\partial \rho} & \dfrac{\partial y}{\partial \varphi} & \dfrac{\partial y}{\partial \theta} \\ \dfrac{\partial z}{\partial \rho} & \dfrac{\partial z}{\partial \varphi} & \dfrac{\partial z}{\partial \theta} \end{Vmatrix} d\rho d\varphi d\theta$$

$$= \begin{Vmatrix} \sin\varphi\cos\theta & \rho\cos\varphi\cos\theta & -\rho\sin\varphi\sin\theta \\ \sin\varphi\sin\theta & \rho\cos\varphi\sin\theta & -\rho\sin\varphi\cos\theta \\ \cos\varphi & -\rho\sin\varphi & 0 \end{Vmatrix} d\rho d\varphi d\theta = \rho^2 \sin\varphi \, d\rho d\varphi d\theta$$

Let $R = \{(\rho, \varphi, \theta): 0 \le \rho \le 1, 0 \le \varphi \le \pi, 0 \le \theta \le 2\pi\}$ then

$$\iiint_V y^2 \, dV = \iiint_R \rho^4 \sin^3\varphi \sin^2\theta \, d\rho d\theta d\varphi = \int_0^\pi \int_0^{2\pi} \int_0^1 \rho^4 \sin^3\varphi \sin^2\theta \, d\rho d\theta d\varphi$$

$$= \left(\dfrac{\rho^5}{5}\Big|_0^1\right) \int_0^\pi \sin^3\varphi \, d\varphi \int_0^{2\pi} \sin^2\theta \, d\theta = \dfrac{1}{5} \cdot \dfrac{4}{3} \cdot \pi = \dfrac{4\pi}{15}$$

Example 17.

Find the volume of the region enclosed by the surface $(x^2 + y^2 + z^2)^2 = 2z(x^2 + y^2)$.

【Solution】

Let $x = \rho\sin\varphi\cos\theta$, $y = \rho\sin\varphi\sin\theta$, $z = \rho\cos\varphi$ then

$$dxdydz = \begin{Vmatrix} \dfrac{\partial x}{\partial \rho} & \dfrac{\partial x}{\partial \varphi} & \dfrac{\partial x}{\partial \theta} \\ \dfrac{\partial y}{\partial \rho} & \dfrac{\partial y}{\partial \varphi} & \dfrac{\partial y}{\partial \theta} \\ \dfrac{\partial z}{\partial \rho} & \dfrac{\partial z}{\partial \varphi} & \dfrac{\partial z}{\partial \theta} \end{Vmatrix} d\rho d\varphi d\theta$$

$$= \begin{Vmatrix} \sin\varphi\cos\theta & \rho\cos\varphi\cos\theta & -\rho\sin\varphi\sin\theta \\ \sin\varphi\sin\theta & \rho\cos\varphi\sin\theta & -\rho\sin\varphi\cos\theta \\ \cos\varphi & -\rho\sin\varphi & 0 \end{Vmatrix} d\rho d\varphi d\theta = \rho^2 \sin\varphi \, d\rho d\varphi d\theta$$

Let $R = \left\{(\rho, \varphi, \theta): 0 \le \rho \le 2\cos\varphi\sin^2\varphi, 0 \le \varphi \le \dfrac{\pi}{2}, 0 \le \theta \le 2\pi\right\}$ then

$$V = \int_0^{2\pi} \int_0^{\pi/2} \int_0^{2\cos\varphi\sin^2\varphi} \rho^2 \sin\varphi \, d\rho d\varphi d\theta = 2\pi \int_0^{\pi/2} \dfrac{\rho^3}{3}\Big|_0^{2\cos\varphi\sin^2\varphi} \sin\varphi \, d\varphi$$

$$= \dfrac{16\pi}{3} \int_0^{\pi/2} \cos^3\varphi \sin^7\varphi \, d\varphi = \dfrac{16\pi}{3} \int_0^{\pi/2} \cos\varphi (1 - \sin^2\varphi) \sin^7\varphi \, d\varphi$$

$$= \frac{16\pi}{3}\left(\frac{\sin^8 \varphi}{8} - \frac{\sin^{10} \varphi}{10}\right)\bigg|_0^{\frac{\pi}{2}} = \frac{2\pi}{15}$$

Example 18.

(1) Calculate the volume of the sphere $x^2 + y^2 + z^2 \leq 100$ after being cut by the plane $z = 4$, leaving the smaller portion.

(2) Calculate the volume of the sphere $x^2 + y^2 + z^2 \leq 100$ after being cut by the plane $z = 2$, leaving the smaller portion.

【Solution】

(1)
Let $x = \rho \sin \varphi \cos \theta$, $y = \rho \sin \varphi \sin \theta$, $z = \rho \cos \varphi$ then

$$dxdydz = \begin{Vmatrix} \frac{\partial x}{\partial \rho} & \frac{\partial x}{\partial \varphi} & \frac{\partial x}{\partial \theta} \\ \frac{\partial y}{\partial \rho} & \frac{\partial y}{\partial \varphi} & \frac{\partial y}{\partial \theta} \\ \frac{\partial z}{\partial \rho} & \frac{\partial z}{\partial \varphi} & \frac{\partial z}{\partial \theta} \end{Vmatrix} d\rho d\varphi d\theta$$

$$= \begin{Vmatrix} \sin \varphi \cos \theta & \rho\cos \varphi \cos \theta & -\rho\sin \varphi \sin \theta \\ \sin \varphi \sin \theta & \rho\cos \varphi \sin \theta & -\rho\sin \varphi \cos \theta \\ \cos \varphi & -\rho\sin \varphi & 0 \end{Vmatrix} d\rho d\varphi d\theta = \rho^2 \sin \varphi \, d\rho d\varphi d\theta$$

Let $R = \left\{(\rho, \varphi, \theta): \dfrac{4}{\cos \varphi} \leq \rho \leq 10, 0 \leq \varphi \leq \cos^{-1}\dfrac{2}{5}, 0 \leq \theta \leq 2\pi\right\}$ then

$$V = \iiint_R \rho^2 \sin \varphi \, d\rho d\theta d\varphi = \int_0^{\cos^{-1}\frac{2}{5}} \int_0^{2\pi} \int_{\frac{4}{\cos \varphi}}^{10} \rho^2 \sin \varphi \, d\rho d\theta d\varphi$$

$$= \int_0^{\cos^{-1}\frac{2}{5}} \int_0^{2\pi} \frac{\rho^3}{3}\bigg|_{\frac{4}{\cos \varphi}}^{10} \sin \varphi \, d\theta d\varphi = \int_0^{\cos^{-1}\frac{2}{5}} \frac{1}{3}\left(1000 - \left(\frac{4}{\cos \varphi}\right)^3\right) \sin \varphi \, d\varphi \int_0^{2\pi} d\theta$$

$$= \frac{2\pi}{3}\left(-1000 \cdot \cos \varphi\bigg|_0^{\cos^{-1}\frac{2}{5}} - \frac{4^3}{2} \cdot \cos^{-2} \varphi\bigg|_0^{\cos^{-1}\frac{2}{5}}\right)$$

$$= \frac{2\pi}{3}\left(-1000 \cdot \left(\frac{2}{5} - 1\right) - \frac{4^3}{2} \cdot \left(\frac{25}{4} - 1\right)\right) = \frac{2\pi}{3}\left(1000 \cdot \left(\frac{3}{5}\right) - \frac{16}{2} \cdot 21\right) = 288\pi$$

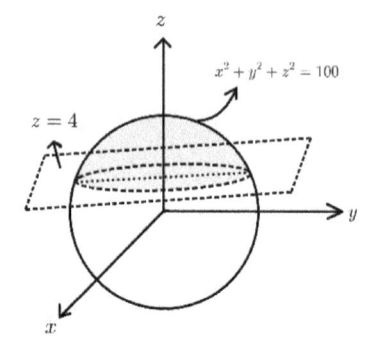

8.4.3 Use Spherical Coordinates to Compute Triple Integrals 284

(2)
Let $x = \rho \sin\varphi \cos\theta$, $y = \rho \sin\varphi \sin\theta$, $z = \rho \cos\varphi$ then

$$dxdydz = \begin{Vmatrix} \dfrac{\partial x}{\partial \rho} & \dfrac{\partial x}{\partial \varphi} & \dfrac{\partial x}{\partial \theta} \\ \dfrac{\partial y}{\partial \rho} & \dfrac{\partial y}{\partial \varphi} & \dfrac{\partial y}{\partial \theta} \\ \dfrac{\partial z}{\partial \rho} & \dfrac{\partial z}{\partial \varphi} & \dfrac{\partial z}{\partial \theta} \end{Vmatrix} d\rho d\varphi d\theta$$

$$= \begin{Vmatrix} \sin\varphi \cos\theta & \rho\cos\varphi \cos\theta & -\rho\sin\varphi \sin\theta \\ \sin\varphi \sin\theta & \rho\cos\varphi \sin\theta & -\rho\sin\varphi \cos\theta \\ \cos\varphi & -\rho\sin\varphi & 0 \end{Vmatrix} d\rho d\varphi d\theta = \rho^2 \sin\varphi \, d\rho d\varphi d\theta$$

Let $R = \left\{ (\rho, \varphi, \theta) : \dfrac{2}{\cos\varphi} \leq \rho \leq 10, 0 \leq \varphi \leq \cos^{-1}\dfrac{1}{5}, 0 \leq \theta \leq 2\pi \right\}$ then

$$V = \iiint_R \rho^2 \sin\varphi \, d\rho d\theta d\varphi = \int_0^{\cos^{-1}\frac{1}{5}} \int_0^{2\pi} \int_{\frac{2}{\cos\varphi}}^{10} \rho^2 \sin\varphi \, d\rho d\theta d\varphi$$

$$= \int_0^{\cos^{-1}\frac{1}{5}} \int_0^{2\pi} \dfrac{\rho^3}{3} \Big|_{\frac{2}{\cos\varphi}}^{10} \sin\varphi \, d\theta d\varphi = \int_0^{\cos^{-1}\frac{1}{5}} \dfrac{1}{3}\left(1000 - \left(\dfrac{2}{\cos\varphi}\right)^3\right) \sin\varphi \, d\varphi \int_0^{2\pi} d\theta$$

$$= \dfrac{2\pi}{3}\left(-1000 \cdot \cos\varphi\Big|_0^{\cos^{-1}\frac{1}{5}} - \dfrac{2^3}{2} \cdot \cos^{-2}\varphi\Big|_0^{\cos^{-1}\frac{1}{5}}\right)$$

$$= \dfrac{2\pi}{3}\left(-1000 \cdot \left(\dfrac{1}{5} - 1\right) - \dfrac{2^3}{2} \cdot (25 - 1)\right) = \dfrac{2\pi}{3}\left(1000 \cdot \left(\dfrac{4}{5}\right) - 96\right) = \dfrac{1408\pi}{3}$$

Example 19.
(1)Calculate the volume of the intersection between the sphere $x^2 + y^2 + z^2 \leq r^2$ and $z \geq \sqrt{3x^2 + 3y^2}$.
(2)Calculate the volume of the intersection between the sphere $x^2 + y^2 + z^2 \leq 1$

and $\sqrt{3}z \geq \sqrt{x^2 + y^2}$.

【Solution】

(1)
Let $x = \rho \sin\varphi \cos\theta$, $y = \rho \sin\varphi \sin\theta$, $z = \rho \cos\varphi$ then

$$dxdydz = \begin{Vmatrix} \dfrac{\partial x}{\partial \rho} & \dfrac{\partial x}{\partial \varphi} & \dfrac{\partial x}{\partial \theta} \\ \dfrac{\partial y}{\partial \rho} & \dfrac{\partial y}{\partial \varphi} & \dfrac{\partial y}{\partial \theta} \\ \dfrac{\partial z}{\partial \rho} & \dfrac{\partial z}{\partial \varphi} & \dfrac{\partial z}{\partial \theta} \end{Vmatrix} d\rho d\varphi d\theta$$

$$= \begin{Vmatrix} \sin\varphi \cos\theta & \rho\cos\varphi \cos\theta & -\rho\sin\varphi \sin\theta \\ \sin\varphi \sin\theta & \rho\cos\varphi \sin\theta & -\rho\sin\varphi \cos\theta \\ \cos\varphi & -\rho\sin\varphi & 0 \end{Vmatrix} d\rho d\varphi d\theta = \rho^2 \sin\varphi \, d\rho d\varphi d\theta$$

Let $R = \left\{(\rho, \varphi, \theta): 0 \leq \rho \leq r, 0 \leq \varphi \leq \dfrac{\pi}{6}, 0 \leq \theta \leq 2\pi\right\}$ then

$$V = \iiint_R \rho^2 \sin\varphi \, d\rho d\theta d\varphi = \int_0^{\frac{\pi}{6}} \int_0^{2\pi} \int_0^r \rho^2 \sin\varphi \, d\rho d\theta d\varphi = \int_0^{\frac{\pi}{6}} \sin\varphi \, d\varphi \int_0^{2\pi} d\theta \int_0^r \rho^2 d\rho$$

$$= \left(\dfrac{1}{3} - \dfrac{\sqrt{3}}{6}\right) 2\pi r^3$$

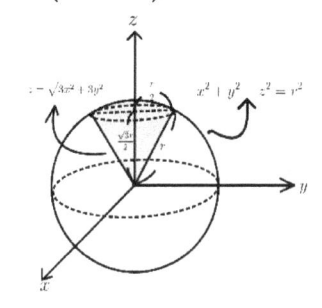

(2)
Let $x = \rho \sin\varphi \cos\theta$, $y = \rho \sin\varphi \sin\theta$, $z = \rho \cos\varphi$ then

$$dxdydz = \begin{Vmatrix} \dfrac{\partial x}{\partial \rho} & \dfrac{\partial x}{\partial \varphi} & \dfrac{\partial x}{\partial \theta} \\ \dfrac{\partial y}{\partial \rho} & \dfrac{\partial y}{\partial \varphi} & \dfrac{\partial y}{\partial \theta} \\ \dfrac{\partial z}{\partial \rho} & \dfrac{\partial z}{\partial \varphi} & \dfrac{\partial z}{\partial \theta} \end{Vmatrix} d\rho d\varphi d\theta$$

8.4.3 Use Spherical Coordinates to Compute Triple Integrals 286

$$= \begin{Vmatrix} \sin\varphi\cos\theta & \rho\cos\varphi\cos\theta & -\rho\sin\varphi\sin\theta \\ \sin\varphi\sin\theta & \rho\cos\varphi\sin\theta & -\rho\sin\varphi\cos\theta \\ \cos\varphi & -\rho\sin\varphi & 0 \end{Vmatrix} d\rho d\varphi d\theta = \rho^2 \sin\varphi \, d\rho d\varphi d\theta$$

Let $R = \{(\rho, \varphi, \theta): 0 \leq \rho \leq 1, 0 \leq \varphi \leq \frac{\pi}{3}, 0 \leq \theta \leq 2\pi\}$ then

$$V = \iiint_R \rho^2 \sin\varphi \, d\rho d\theta d\varphi = \int_0^{\frac{\pi}{3}} \int_0^{2\pi} \int_0^1 \rho^2 \sin\varphi \, d\rho d\theta d\varphi = \int_0^{\frac{\pi}{3}} \sin\varphi \, d\varphi \int_0^{2\pi} d\theta \int_0^1 \rho^2 d\rho$$

$$= \frac{\pi}{3}$$

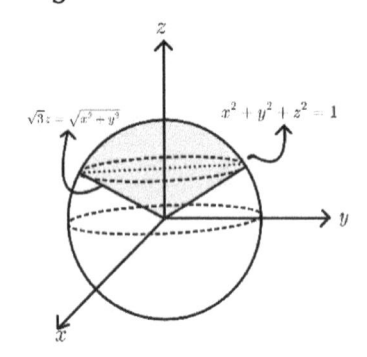

Example 20.

(1) Calculate the volume of the intersection between the sphere $x^2 + y^2 + (z-1)^2 \leq 1$ and $\sqrt{3}\, z \geq \sqrt{x^2 + y^2}$.

(2) Calculate the volume of the intersection between the sphere $x^2 + y^2 + z^2 \leq z$ and $\sqrt{3}\, z \geq \sqrt{x^2 + y^2}$.

【Solution】

(1)

Let $x = \rho\sin\varphi\cos\theta$, $y = \rho\sin\varphi\sin\theta$, $z = \rho\cos\varphi$ then

$$dxdydz = \begin{Vmatrix} \dfrac{\partial x}{\partial \rho} & \dfrac{\partial x}{\partial \varphi} & \dfrac{\partial x}{\partial \theta} \\ \dfrac{\partial y}{\partial \rho} & \dfrac{\partial y}{\partial \varphi} & \dfrac{\partial y}{\partial \theta} \\ \dfrac{\partial z}{\partial \rho} & \dfrac{\partial z}{\partial \varphi} & \dfrac{\partial z}{\partial \theta} \end{Vmatrix} d\rho d\varphi d\theta$$

$$= \begin{Vmatrix} \sin\varphi\cos\theta & \rho\cos\varphi\cos\theta & -\rho\sin\varphi\sin\theta \\ \sin\varphi\sin\theta & \rho\cos\varphi\sin\theta & -\rho\sin\varphi\cos\theta \\ \cos\varphi & -\rho\sin\varphi & 0 \end{Vmatrix} d\rho d\varphi d\theta = \rho^2 \sin\varphi \, d\rho d\varphi d\theta$$

Let $R = \{(\rho, \varphi, \theta): 0 \leq \rho \leq 2\cos\varphi, 0 \leq \varphi \leq \frac{\pi}{3}, 0 \leq \theta \leq 2\pi\}$ then

$$V = \iiint_R \rho^2 \sin\varphi \, d\rho d\theta d\varphi = \int_0^{\frac{\pi}{3}} \int_0^{2\pi} \int_0^{2\cos\varphi} \rho^2 \sin\varphi \, d\rho d\theta d\varphi$$

$$= 2\pi \int_0^{\frac{\pi}{3}} \int_0^{2\cos\varphi} \rho^2 \sin\varphi \, d\rho d\varphi = 2\pi \int_0^{\frac{\pi}{3}} \sin\varphi \int_0^{2\cos\varphi} \rho^2 \, d\rho d\varphi$$

$$\because \int_0^{2\cos\varphi} \rho^2 \, d\rho = \frac{\rho^3}{3}\Big|_0^{2\cos\varphi} = \frac{8\cos^3\varphi}{3}$$

$$\therefore 2\pi \int_0^{\frac{\pi}{3}} \sin\varphi \int_0^{2\cos\varphi} \rho^2 \, d\rho d\varphi = \frac{16\pi}{3} \int_0^{\frac{\pi}{3}} \sin\varphi \cos^3\varphi \, d\varphi = \frac{-16\pi}{3} \cdot \frac{\cos^4\varphi}{4}\Big|_0^{\frac{\pi}{3}}$$

$$= \frac{-4\pi}{3}\left(\frac{1}{16} - 1\right) = \frac{5\pi}{4}$$

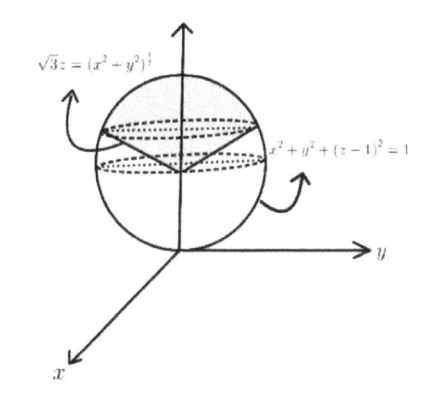

(2)
Let $x = \rho \sin\varphi \cos\theta$, $y = \rho \sin\varphi \sin\theta$, $z = \rho \cos\varphi$ then

$$dxdydz = \begin{Vmatrix} \frac{\partial x}{\partial \rho} & \frac{\partial x}{\partial \varphi} & \frac{\partial x}{\partial \theta} \\ \frac{\partial y}{\partial \rho} & \frac{\partial y}{\partial \varphi} & \frac{\partial y}{\partial \theta} \\ \frac{\partial z}{\partial \rho} & \frac{\partial z}{\partial \varphi} & \frac{\partial z}{\partial \theta} \end{Vmatrix} d\rho d\varphi d\theta$$

$$= \begin{Vmatrix} \sin\varphi \cos\theta & \rho\cos\varphi \cos\theta & -\rho\sin\varphi \sin\theta \\ \sin\varphi \sin\theta & \rho\cos\varphi \sin\theta & -\rho\sin\varphi \cos\theta \\ \cos\varphi & -\rho\sin\varphi & 0 \end{Vmatrix} d\rho d\varphi d\theta = \rho^2 \sin\varphi \, d\rho d\varphi d\theta$$

Let $R = \left\{(\rho, \varphi, \theta): 0 \leq \rho \leq \cos\varphi, 0 \leq \varphi \leq \frac{\pi}{3}, 0 \leq \theta \leq 2\pi\right\}$ then

$$V = \iiint_R \rho^2 \sin\varphi \, d\rho d\theta d\varphi = \int_0^{\frac{\pi}{3}} \int_0^{2\pi} \int_0^{\cos\varphi} \rho^2 \sin\varphi \, d\rho d\theta d\varphi$$

8.4.3 Use Spherical Coordinates to Compute Triple Integrals

$$= 2\pi \int_0^{\frac{\pi}{3}} \int_0^{\cos\varphi} \rho^2 \sin\varphi \, d\rho d\varphi = 2\pi \int_0^{\frac{\pi}{3}} \sin\varphi \int_0^{\cos\varphi} \rho^2 \, d\rho d\varphi$$

$$\because \int_0^{\cos\varphi} \rho^2 \, d\rho = \frac{\rho^3}{3}\Big|_0^{\cos\varphi} = \frac{\cos^3\varphi}{3}$$

$$\therefore 2\pi \int_0^{\frac{\pi}{3}} \sin\varphi \int_0^{\cos\varphi} \rho^2 \, d\rho d\varphi = \frac{2\pi}{3} \int_0^{\frac{\pi}{3}} \sin\varphi \cos^3\varphi \, d\varphi = -\frac{2\pi}{3} \cdot \frac{\cos^4\varphi}{4}\Big|_0^{\frac{\pi}{3}} = -\frac{\pi}{6}\left(\frac{1}{16} - 1\right)$$

$$= \frac{5\pi}{32}$$

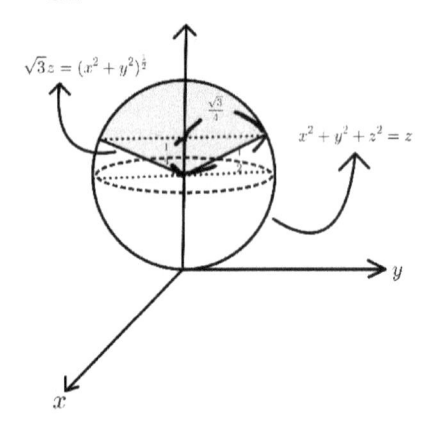

Example 21.

(1) Using a change of variables, find the volume of the region enclosed by

$$\left(\frac{x^2}{a^2} + \frac{y^2}{b^2} + \frac{z^2}{c^2}\right)^{\frac{3}{2}} = \frac{x^2}{a^2} + \frac{y^2}{b^2}.$$

(2) Given V as the region enclosed by $\left(\frac{x^2}{a^2} + \frac{y^2}{b^2} + \frac{z^2}{c^2}\right)^{\frac{3}{2}} = \frac{x^2}{a^2} + \frac{y^2}{b^2}$ and $z > 0$,

Find $\iiint_V z \, dx dy dz = ?$

【Solution】

(1)

Let $x = a\rho \sin\varphi \cos\theta$, $y = b\rho \sin\varphi \sin\theta$, $z = c\rho \cos\varphi$ then

$$dxdydz = \begin{Vmatrix} \dfrac{\partial x}{\partial \rho} & \dfrac{\partial x}{\partial \varphi} & \dfrac{\partial x}{\partial \theta} \\ \dfrac{\partial y}{\partial \rho} & \dfrac{\partial y}{\partial \varphi} & \dfrac{\partial y}{\partial \theta} \\ \dfrac{\partial z}{\partial \rho} & \dfrac{\partial z}{\partial \varphi} & \dfrac{\partial z}{\partial \theta} \end{Vmatrix} d\rho d\varphi d\theta$$

$$= abc \begin{Vmatrix} \sin\varphi\cos\theta & \rho\cos\varphi\cos\theta & -\rho\sin\varphi\sin\theta \\ \sin\varphi\sin\theta & \rho\cos\varphi\sin\theta & -\rho\sin\varphi\cos\theta \\ \cos\varphi & -\rho\sin\varphi & 0 \end{Vmatrix} d\rho d\varphi d\theta = abc\rho^2 \sin\varphi \, d\rho d\varphi d\theta$$

Let $R = \{(\rho, \varphi, \theta): 0 \leq \rho \leq \sin^2\varphi, 0 \leq \varphi \leq \pi, 0 \leq \theta \leq 2\pi\}$ then

$$V = abc \iiint_R \rho^2 \sin\varphi \, d\rho d\theta d\varphi = abc \int_0^\pi \int_0^{2\pi} \int_0^{\sin^2\varphi} \rho^2 \sin\varphi \, d\rho d\theta d\varphi$$

$$= abc \int_0^\pi \int_0^{2\pi} \dfrac{\rho^3}{3}\bigg|_0^{\sin^2\varphi} \sin\varphi \, d\theta d\varphi = \dfrac{abc}{3} \cdot 2\pi \cdot \int_0^\pi \sin^7\varphi \, d\varphi = \dfrac{64abc\pi}{105}$$

(2)
Let $x = a\rho\sin\varphi\cos\theta, y = b\rho\sin\varphi\sin\theta, z = c\rho\cos\varphi$ then

$$dxdydz = \begin{Vmatrix} \dfrac{\partial x}{\partial \rho} & \dfrac{\partial x}{\partial \varphi} & \dfrac{\partial x}{\partial \theta} \\ \dfrac{\partial y}{\partial \rho} & \dfrac{\partial y}{\partial \varphi} & \dfrac{\partial y}{\partial \theta} \\ \dfrac{\partial z}{\partial \rho} & \dfrac{\partial z}{\partial \varphi} & \dfrac{\partial z}{\partial \theta} \end{Vmatrix} d\rho d\varphi d\theta$$

$$= abc \begin{Vmatrix} \sin\varphi\cos\theta & \rho\cos\varphi\cos\theta & -\rho\sin\varphi\sin\theta \\ \sin\varphi\sin\theta & \rho\cos\varphi\sin\theta & -\rho\sin\varphi\cos\theta \\ \cos\varphi & -\rho\sin\varphi & 0 \end{Vmatrix} d\rho d\varphi d\theta$$

$$= abc\rho^2 \sin\varphi \, d\rho d\varphi d\theta$$

Let $R = \left\{(\rho, \varphi, \theta): 0 \leq \rho \leq \sin^2\varphi, 0 \leq \varphi \leq \dfrac{\pi}{2}, 0 \leq \theta \leq 2\pi\right\}$ then

$$\iiint_V z \, dxdydz = abc^2 \iiint_R \rho^3 \sin\varphi \cos\varphi \, d\rho d\theta d\varphi$$

$$= abc^2 \int_0^{\frac{\pi}{2}} \int_0^{2\pi} \int_0^{\sin^2\varphi} \rho^3 \sin\varphi \cos\varphi \, d\rho d\theta d\varphi = abc^2 \int_0^{\frac{\pi}{2}} \int_0^{2\pi} \dfrac{\rho^4}{4}\bigg|_0^{\sin^2\varphi} \sin\varphi \cos\varphi \, d\theta d\varphi$$

$$= \dfrac{abc^2}{4} \cdot 2\pi \cdot \int_0^{\frac{\pi}{2}} \sin^9\varphi \cos\varphi \, d\varphi = \dfrac{abc^2\pi}{20}$$

8.4.3 Use Spherical Coordinates to Compute Triple Integrals

Example 22.

Let $V = \{(x, y, z): x^2 + y^2 + z^2 \leq 1\}$. Find $\displaystyle\iiint_V \frac{dxdydz}{\sqrt{1-x^2-y^2-z^2}} = ?$

【Solution】

Let $x = \rho \sin\varphi \cos\theta$, $y = \rho \sin\varphi \sin\theta$, $z = \rho \cos\varphi$ then

$$dxdydz = \begin{Vmatrix} \dfrac{\partial x}{\partial \rho} & \dfrac{\partial x}{\partial \varphi} & \dfrac{\partial x}{\partial \theta} \\ \dfrac{\partial y}{\partial \rho} & \dfrac{\partial y}{\partial \varphi} & \dfrac{\partial y}{\partial \theta} \\ \dfrac{\partial z}{\partial \rho} & \dfrac{\partial z}{\partial \varphi} & \dfrac{\partial z}{\partial \theta} \end{Vmatrix} d\rho d\varphi d\theta$$

$$= \begin{Vmatrix} \sin\varphi\cos\theta & \rho\cos\varphi\cos\theta & -\rho\sin\varphi\sin\theta \\ \sin\varphi\sin\theta & \rho\cos\varphi\sin\theta & -\rho\sin\varphi\cos\theta \\ \cos\varphi & -\rho\sin\varphi & 0 \end{Vmatrix} d\rho d\varphi d\theta = \rho^2 \sin\varphi\, d\rho d\varphi d\theta$$

Let $R = \{(\rho, \varphi, \theta): 0 \leq \rho \leq 1, 0 \leq \varphi \leq \pi, 0 \leq \theta \leq 2\pi\}$ then

$$\iiint_V \frac{dxdydz}{\sqrt{1-x^2-y^2-z^2}} = \iiint_R \frac{\rho^2 \sin\varphi}{\sqrt{1-\rho^2}} d\rho d\theta d\varphi = \int_0^\pi \int_0^{2\pi} \int_0^1 \frac{\rho^2 \sin\varphi}{\sqrt{1-\rho^2}} d\rho d\theta d\varphi$$

$$= \int_0^\pi \sin\varphi d\varphi \int_0^{2\pi} d\theta \int_0^1 \frac{\rho^2}{\sqrt{1-\rho^2}} d\rho = 2(2\pi)\left(\int_0^1 \frac{\rho^2}{\sqrt{1-\rho^2}} d\rho\right)$$

Let $\rho = \sin\theta$ then $d\rho = \cos\theta\, d\theta$

$$\therefore \int_0^1 \frac{\rho^2}{\sqrt{1-\rho^2}} d\rho = \int_0^{\frac{\pi}{2}} \frac{\sin^2\theta \cos\theta}{\sqrt{1-\sin^2\theta}} d\theta = \int_0^{\frac{\pi}{2}} \sin^2\theta\, d\theta = \int_0^{\frac{\pi}{2}} \frac{1-\cos 2\theta}{2} d\theta$$

$$= \frac{1}{2}\left(\frac{\pi}{2} - \frac{\sin 2\theta}{2}\Big|_0^{\frac{\pi}{2}}\right) = \frac{\pi}{4}$$

$$\therefore \iiint_V \frac{dxdydz}{\sqrt{1-x^2-y^2-z^2}} = 2(2\pi)\left(\int_0^1 \frac{\rho^2}{\sqrt{1-\rho^2}} d\rho\right) = \pi^2$$

Example 23.

Let $V = \{(x, y, z): x^2 + y^2 + z^2 \leq r^2\}$. Find $\displaystyle\iiint_V z^2(x^2+y^2+z^2)^{\frac{3}{2}} dxdydz = ?$

【Solution】

Let $x = \rho \sin\varphi \cos\theta$, $y = \rho \sin\varphi \sin\theta$, $z = \rho \cos\varphi$ then

$$dxdydz = \begin{Vmatrix} \frac{\partial x}{\partial \rho} & \frac{\partial x}{\partial \varphi} & \frac{\partial x}{\partial \theta} \\ \frac{\partial y}{\partial \rho} & \frac{\partial y}{\partial \varphi} & \frac{\partial y}{\partial \theta} \\ \frac{\partial z}{\partial \rho} & \frac{\partial z}{\partial \varphi} & \frac{\partial z}{\partial \theta} \end{Vmatrix} d\rho d\varphi d\theta$$

$$= \begin{Vmatrix} \sin\varphi\cos\theta & \rho\cos\varphi\cos\theta & -\rho\sin\varphi\sin\theta \\ \sin\varphi\sin\theta & \rho\cos\varphi\sin\theta & -\rho\sin\varphi\cos\theta \\ \cos\varphi & -\rho\sin\varphi & 0 \end{Vmatrix} d\rho d\varphi d\theta = \rho^2 \sin\varphi \, d\rho d\varphi d\theta$$

Let $R = \{(\rho, \varphi, \theta): 0 \leq \rho \leq r, 0 \leq \varphi \leq \pi, 0 \leq \theta \leq 2\pi\}$ then

$$\iiint_V z^2(x^2 + y^2 + z^2)^{\frac{3}{2}} dxdydz = \iiint_R \rho^2 \cos^2\varphi \, (\rho^2)^{\frac{3}{2}} \rho^2 \sin\varphi \, d\rho d\theta d\varphi$$

$$= \int_0^\pi \int_0^{2\pi} \int_0^r \rho^2 \cos^2\varphi \, (\rho^2)^{\frac{3}{2}} \rho^2 \sin\varphi \, d\rho d\theta d\varphi = \int_0^\pi \cos^2\varphi \sin\varphi d\varphi \int_0^{2\pi} d\theta \int_0^r \rho^7 d\rho$$

$$= -\frac{\cos^3\varphi}{3}\Big|_0^\pi (2\pi) \left(\frac{r^8}{8}\right) = \frac{\pi r^8}{6}$$

Example 24.

Let $V = \{(x, y, z): x^2 + y^2 + z^2 \leq r^2, z > 0\}$. Find $\iiint_V z(x^2 + y^2 + z^2)^{\frac{3}{2}} dxdydz =?$

【Solution】

Let $x = \rho \sin\varphi \cos\theta$, $y = \rho \sin\varphi \sin\theta$, $z = \rho \cos\varphi$ then

$$dxdydz = \begin{Vmatrix} \frac{\partial x}{\partial \rho} & \frac{\partial x}{\partial \varphi} & \frac{\partial x}{\partial \theta} \\ \frac{\partial y}{\partial \rho} & \frac{\partial y}{\partial \varphi} & \frac{\partial y}{\partial \theta} \\ \frac{\partial z}{\partial \rho} & \frac{\partial z}{\partial \varphi} & \frac{\partial z}{\partial \theta} \end{Vmatrix} d\rho d\varphi d\theta$$

$$= \begin{Vmatrix} \sin\varphi\cos\theta & \rho\cos\varphi\cos\theta & -\rho\sin\varphi\sin\theta \\ \sin\varphi\sin\theta & \rho\cos\varphi\sin\theta & -\rho\sin\varphi\cos\theta \\ \cos\varphi & -\rho\sin\varphi & 0 \end{Vmatrix} d\rho d\varphi d\theta = \rho^2 \sin\varphi \, d\rho d\varphi d\theta$$

Let $R = \left\{(\rho, \varphi, \theta): 0 \leq \rho \leq r, 0 \leq \varphi \leq \frac{\pi}{2}, 0 \leq \theta \leq 2\pi\right\}$ then

$$\iiint_V z(x^2 + y^2 + z^2)^{\frac{3}{2}} dxdydz = \iiint_R \rho \cos\varphi \, (\rho^2)^{\frac{3}{2}} \rho^2 \sin\varphi \, d\rho d\theta d\varphi$$

8.4.3 Use Spherical Coordinates to Compute Triple Integrals   292

$$= \int_0^{\frac{\pi}{2}} \int_0^{2\pi} \int_0^r \rho \cos\varphi \, (\rho^2)^{\frac{3}{2}} \rho^2 \sin\varphi \, d\rho d\theta d\varphi = \int_0^{\frac{\pi}{2}} \cos\varphi \sin\varphi d\varphi \int_0^{2\pi} d\theta \int_0^r \rho^6 d\rho$$

$$= -\frac{\cos 2\varphi}{4}\Big|_0^{\frac{\pi}{2}} (2\pi) \left(\frac{r^7}{7}\right) = \frac{\pi r^7}{7}$$

Example 25.

Let $V = \left\{(x, y, z): \dfrac{x^2}{a^2} + \dfrac{y^2}{b^2} + \dfrac{z^2}{c^2} \leq r^2\right\}$. Find $\iiint_V xyz \, dV = ?$

【Solution】

Let $x = a\rho \sin\varphi \cos\theta$, $y = b\rho \sin\varphi \sin\theta$, $z = c\rho \cos\varphi$ then

$$dxdydz = \begin{Vmatrix} \dfrac{\partial x}{\partial \rho} & \dfrac{\partial x}{\partial \varphi} & \dfrac{\partial x}{\partial \theta} \\ \dfrac{\partial y}{\partial \rho} & \dfrac{\partial y}{\partial \varphi} & \dfrac{\partial y}{\partial \theta} \\ \dfrac{\partial z}{\partial \rho} & \dfrac{\partial z}{\partial \varphi} & \dfrac{\partial z}{\partial \theta} \end{Vmatrix} d\rho d\varphi d\theta$$

$$= abc \begin{Vmatrix} \sin\varphi \cos\theta & \rho\cos\varphi \cos\theta & -\rho\sin\varphi \sin\theta \\ \sin\varphi \sin\theta & \rho\cos\varphi \sin\theta & -\rho\sin\varphi \cos\theta \\ \cos\varphi & -\rho\sin\varphi & 0 \end{Vmatrix} d\rho d\varphi d\theta = abc\rho^2 \sin\varphi \, d\rho d\varphi d\theta$$

Let $R = \{(\rho, \varphi, \theta): 0 \leq \rho \leq r, 0 \leq \varphi \leq \pi, 0 \leq \theta \leq 2\pi\}$ then

$$\iiint_V xyz \, dV = a^2b^2c^2 \iiint_R (\rho \sin\varphi \cos\theta \cdot \rho \sin\varphi \sin\theta \cdot \rho \cos\varphi)\rho^2 \sin\varphi \, d\rho d\theta d\varphi$$

$$= a^2b^2c^2 \int_0^{\pi} \int_0^{2\pi} \int_0^r (\rho \sin\varphi \cos\theta \cdot \rho \sin\varphi \sin\theta \cdot \rho \cos\varphi)\rho^2 \sin\varphi \, d\rho d\theta d\varphi$$

$$= a^2b^2c^2 \left(\frac{\rho^6}{6}\Big|_0^r\right) \int_0^{\pi} \sin^3\varphi \cos\varphi \, d\varphi \int_0^{2\pi} \cos\theta \sin\theta \, d\theta$$

$$\because \int_0^{2\pi} \cos\theta \sin\theta \, d\theta = \frac{1}{2}\int_0^{2\pi} \sin 2\theta \, d\theta = \frac{-1}{2}(\cos 2\theta)\Big|_0^{2\pi} = 0 \quad \therefore \iiint_V xyz \, dV = 0$$

Example 26.

Let $V = \left\{(x, y, z): \dfrac{x^2}{a^2} + \dfrac{y^2}{b^2} + \dfrac{z^2}{c^2} \leq r^2\right\}$. Find $\iiint_V \sqrt{r^2 - \left(\dfrac{x^2}{a^2} + \dfrac{y^2}{b^2} + \dfrac{z^2}{c^2}\right)} \, dxdydz = ?$

【Solution】

Let $x = a\rho \sin \varphi \cos \theta$, $y = b\rho \sin \varphi \sin \theta$, $z = c\rho \cos \varphi$ then

$$dxdydz = \begin{Vmatrix} \dfrac{\partial x}{\partial \rho} & \dfrac{\partial x}{\partial \varphi} & \dfrac{\partial x}{\partial \theta} \\ \dfrac{\partial y}{\partial \rho} & \dfrac{\partial y}{\partial \varphi} & \dfrac{\partial y}{\partial \theta} \\ \dfrac{\partial z}{\partial \rho} & \dfrac{\partial z}{\partial \varphi} & \dfrac{\partial z}{\partial \theta} \end{Vmatrix} d\rho d\varphi d\theta$$

$$= abc \begin{Vmatrix} \sin \varphi \cos \theta & \rho \cos \varphi \cos \theta & -\rho \sin \varphi \sin \theta \\ \sin \varphi \sin \theta & \rho \cos \varphi \sin \theta & -\rho \sin \varphi \cos \theta \\ \cos \varphi & -\rho \sin \varphi & 0 \end{Vmatrix} d\rho d\varphi d\theta = abc\rho^2 \sin \varphi \, d\rho d\varphi d\theta$$

Let $R = \{(\rho, \varphi, \theta): 0 \le \rho \le r, 0 \le \varphi \le \pi, 0 \le \theta \le 2\pi\}$ then

$$\iiint_V \sqrt{r^2 - \left(\dfrac{x^2}{a^2} + \dfrac{y^2}{b^2} + \dfrac{z^2}{c^2}\right)} \, dxdydz = abc \iiint_R \sqrt{r^2 - \rho^2} \rho^2 \sin \varphi \, d\rho d\theta d\varphi$$

$$= abc \int_0^\pi \int_0^{2\pi} \int_0^r \sqrt{r^2 - \rho^2} \rho^2 \sin \varphi \, d\rho d\theta d\varphi = abc \int_0^\pi \sin \varphi d\varphi \int_0^{2\pi} d\theta \int_0^r \sqrt{r^2 - \rho^2} \rho^2 d\rho$$

$$= abc(2)(2\pi) \int_0^r \sqrt{r^2 - \rho^2} \rho^2 d\rho$$

Let $\rho = r \sin \theta$ then $d\rho = r \cos \theta \, d\theta$

$$\therefore \int_0^r \sqrt{r^2 - \rho^2} \rho^2 d\rho = r^4 \int_0^{\frac{\pi}{2}} \sin^2 \theta \cos^2 \theta \, d\theta = r^4 \int_0^{\frac{\pi}{2}} \left(\dfrac{\sin 2\theta}{2}\right)^2 d\theta = \dfrac{r^4}{4} \int_0^{\frac{\pi}{2}} \dfrac{1 - \cos 4\theta}{2} d\theta$$

$$= \dfrac{r^4}{8} \left(\dfrac{\pi}{2} - \dfrac{\sin 4\theta}{4}\bigg|_0^{\frac{\pi}{2}}\right) = \dfrac{\pi r^4}{16}$$

$$\therefore \iiint_V \sqrt{r^2 - \left(\dfrac{x^2}{a^2} + \dfrac{y^2}{b^2} + \dfrac{z^2}{c^2}\right)} dxdydz = \dfrac{abc\pi^2 r^4}{4}$$

Example 27.
   Find the volume of the region enclosed by the surface $(x^2 + y^2 + z^2)^3 = 27a^3 xyz$ (for $a > 0$) in the first octant.

【Solution】

Let $x = \rho \sin \varphi \cos \theta$, $y = \rho \sin \varphi \sin \theta$, $z = \rho \cos \varphi$ then

8.4.3 Use Spherical Coordinates to Compute Triple Integrals

$$dxdydz = \begin{Vmatrix} \dfrac{\partial x}{\partial \rho} & \dfrac{\partial x}{\partial \varphi} & \dfrac{\partial x}{\partial \theta} \\ \dfrac{\partial y}{\partial \rho} & \dfrac{\partial y}{\partial \varphi} & \dfrac{\partial y}{\partial \theta} \\ \dfrac{\partial z}{\partial \rho} & \dfrac{\partial z}{\partial \varphi} & \dfrac{\partial z}{\partial \theta} \end{Vmatrix} d\rho d\varphi d\theta$$

$$= \begin{Vmatrix} \sin\varphi\cos\theta & \rho\cos\varphi\cos\theta & -\rho\sin\varphi\sin\theta \\ \sin\varphi\sin\theta & \rho\cos\varphi\sin\theta & -\rho\sin\varphi\cos\theta \\ \cos\varphi & -\rho\sin\varphi & 0 \end{Vmatrix} d\rho d\varphi d\theta = \rho^2 \sin\varphi \, d\rho d\varphi d\theta$$

Let $R = \left\{(\rho, \varphi, \theta): 0 \leq \rho \leq 3a\sqrt[3]{\sin^2\varphi \cos\varphi \cos\theta \sin\theta}, 0 \leq \varphi \leq \dfrac{\pi}{2}, 0 \leq \theta \leq \dfrac{\pi}{2}\right\}$ then

$$V = \iiint_R \rho^2 \sin\varphi \, d\rho d\theta d\varphi = \int_0^{\frac{\pi}{2}} \int_0^{\frac{\pi}{2}} \int_0^{3a\sqrt[3]{\sin^2\varphi \cos\varphi \cos\theta \sin\theta}} \rho^2 \sin\varphi \, d\rho d\theta d\varphi$$

$$= \int_0^{\frac{\pi}{2}} \int_0^{\frac{\pi}{2}} 9a^3 \sin^3\varphi \cos\varphi \cos\theta \sin\theta \, d\theta d\varphi = 9a^3 \left(\dfrac{\sin^4\varphi}{4}\right)\bigg|_0^{\frac{\pi}{2}} \left(\dfrac{-\cos 2\theta}{4}\right)\bigg|_0^{\frac{\pi}{2}}$$

$$= 9a^3 \cdot \dfrac{1}{4} \cdot \dfrac{1}{2} = \dfrac{9a^3}{8}$$

Example 28.

Let $V = \{(x, y, z): 1 \leq x^2 + y^2 + z^2 \leq 9, z > 0\}$. Find $\iiint_V \dfrac{z^2 dV}{\sqrt{x^2 + y^2 + z^2}} = ?$

【Solution】

Let $x = \rho\sin\varphi\cos\theta$, $y = \rho\sin\varphi\sin\theta$, $z = \rho\cos\varphi$ then

$$dxdydz = \begin{Vmatrix} \dfrac{\partial x}{\partial \rho} & \dfrac{\partial x}{\partial \varphi} & \dfrac{\partial x}{\partial \theta} \\ \dfrac{\partial y}{\partial \rho} & \dfrac{\partial y}{\partial \varphi} & \dfrac{\partial y}{\partial \theta} \\ \dfrac{\partial z}{\partial \rho} & \dfrac{\partial z}{\partial \varphi} & \dfrac{\partial z}{\partial \theta} \end{Vmatrix} d\rho d\varphi d\theta$$

$$= \begin{Vmatrix} \sin\varphi\cos\theta & \rho\cos\varphi\cos\theta & -\rho\sin\varphi\sin\theta \\ \sin\varphi\sin\theta & \rho\cos\varphi\sin\theta & -\rho\sin\varphi\cos\theta \\ \cos\varphi & -\rho\sin\varphi & 0 \end{Vmatrix} d\rho d\varphi d\theta = \rho^2 \sin\varphi \, d\rho d\varphi d\theta$$

Let $R = \left\{(\rho, \varphi, \theta): 1 \leq \rho \leq 3, 0 \leq \varphi \leq \dfrac{\pi}{2}, 0 \leq \theta \leq 2\pi\right\}$ then

$$\iiint_V \frac{z^2 dV}{\sqrt{x^2+y^2+z^2}} = \iiint_R \frac{\rho^2 \cos^2 \varphi \, \rho^2 \sin \varphi}{\rho} d\rho d\theta d\varphi$$

$$= \int_0^{\frac{\pi}{2}} \int_0^{2\pi} \int_1^3 \frac{\rho^2 \cos^2 \varphi \, \rho^2 \sin \varphi}{\rho} d\rho d\theta d\varphi = \int_0^{\frac{\pi}{2}} \sin \varphi \cos^2 \varphi \, d\varphi \int_0^{2\pi} d\theta \int_1^3 \rho^3 d\rho$$

$$= \frac{-\cos^3 \varphi}{3} \Big|_0^{\frac{\pi}{2}} (2\pi)(20) = \frac{40\pi}{3}$$

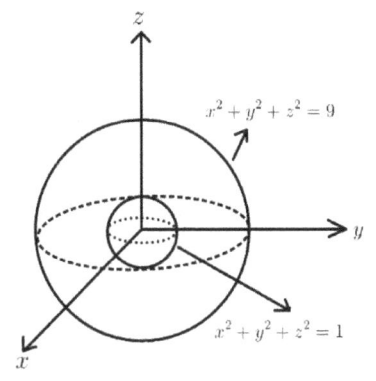

Example 29.

Let $V = \{(x,y,z): 1 \leq x^2 + y^2 + z^2 \leq 9, z > 0\}$. Find $\iiint_V \frac{zdV}{\sqrt{x^2+y^2+z^2}} = ?$

【Solution】

Let $x = \rho \sin \varphi \cos \theta, y = \rho \sin \varphi \sin \theta, z = \rho \cos \varphi$ then

$$dxdydz = \begin{Vmatrix} \frac{\partial x}{\partial \rho} & \frac{\partial x}{\partial \varphi} & \frac{\partial x}{\partial \theta} \\ \frac{\partial y}{\partial \rho} & \frac{\partial y}{\partial \varphi} & \frac{\partial y}{\partial \theta} \\ \frac{\partial z}{\partial \rho} & \frac{\partial z}{\partial \varphi} & \frac{\partial z}{\partial \theta} \end{Vmatrix} d\rho d\varphi d\theta$$

$$= \begin{Vmatrix} \sin \varphi \cos \theta & \rho \cos \varphi \cos \theta & -\rho \sin \varphi \sin \theta \\ \sin \varphi \sin \theta & \rho \cos \varphi \sin \theta & -\rho \sin \varphi \cos \theta \\ \cos \varphi & -\rho \sin \varphi & 0 \end{Vmatrix} d\rho d\varphi d\theta = \rho^2 \sin \varphi \, d\rho d\varphi d\theta$$

Let $R = \{(\rho, \varphi, \theta): 1 \leq \rho \leq 3, 0 \leq \varphi \leq \frac{\pi}{2}, 0 \leq \theta \leq 2\pi\}$ then

$$\iiint_V \frac{zdV}{\sqrt{x^2+y^2+z^2}} = \iiint_R \frac{\rho \cos \varphi \, \rho^2 \sin \varphi}{\rho} d\rho d\theta d\varphi$$

8.4.3 Use Spherical Coordinates to Compute Triple Integrals

$$= \int_0^{\frac{\pi}{2}} \int_0^{2\pi} \int_1^3 \frac{\rho \cos\varphi \, \rho^2 \sin\varphi}{\rho} d\rho d\theta d\varphi = \int_0^{\frac{\pi}{2}} \sin\varphi \cos\varphi \, d\varphi \int_0^{2\pi} d\theta \int_1^3 \rho^2 d\rho$$

$$= \frac{1}{2} \int_0^{\frac{\pi}{2}} \sin 2\varphi \, d\varphi \int_0^{2\pi} d\theta \int_1^3 \rho^2 d\rho = \frac{-\cos 2\varphi}{4} \Big|_0^{\frac{\pi}{2}} (2\pi) \frac{26}{3} = \frac{26\pi}{3}$$

Example 30.

Let $V = \{(x, y, z): x^2 + y^2 + z^2 \le 1\}$. Find $\iiint_V \frac{z^2 dV}{9 - (x^2 + y^2 + z^2)} = ?$

【Solution】

Let $x = \rho \sin\varphi \cos\theta$, $y = \rho \sin\varphi \sin\theta$, $z = \rho \cos\varphi$ then

$$dxdydz = \begin{Vmatrix} \frac{\partial x}{\partial \rho} & \frac{\partial x}{\partial \varphi} & \frac{\partial x}{\partial \theta} \\ \frac{\partial y}{\partial \rho} & \frac{\partial y}{\partial \varphi} & \frac{\partial y}{\partial \theta} \\ \frac{\partial z}{\partial \rho} & \frac{\partial z}{\partial \varphi} & \frac{\partial z}{\partial \theta} \end{Vmatrix} d\rho d\varphi d\theta$$

Let $R = \{(\rho, \varphi, \theta): 0 \le \rho \le 1, 0 \le \varphi \le \pi, 0 \le \theta \le 2\pi\}$ then

$$\iiint_V \frac{z^2 dV}{9 - (x^2 + y^2 + z^2)} = \iiint_R \frac{\rho^4 \sin\varphi \cos^2\varphi}{9 - \rho^2} d\rho d\theta d\varphi$$

$$= \int_0^\pi \int_0^{2\pi} \int_0^1 \frac{\rho^4 \sin\varphi \cos^2\varphi}{9 - \rho^2} d\rho d\theta d\varphi$$

$$= \int_0^\pi \cos^2\varphi \sin\varphi \, d\varphi \int_0^{2\pi} d\theta \int_0^1 \frac{\rho^2(\rho^2 - 9) + 9(\rho^2 - 9) + 81}{9 - \rho^2} d\rho$$

$$= \frac{-\cos^3\varphi}{3} \Big|_0^\pi \cdot \int_0^{2\pi} d\theta \int_0^1 -\rho^2 - 9 + \frac{81}{9 - \rho^2} d\rho = \frac{4\pi}{3} \cdot \left(-\frac{\rho^3}{3} - 9\rho + \frac{81}{6} \ln \frac{3+\rho}{3-\rho} \right) \Big|_0^1$$

$$= \frac{4\pi}{3} \cdot \left(\frac{28}{3} + \frac{81}{6} \ln 2 \right)$$

Example 31.

Let $V = \{(x, y, z): x^2 + y^2 + z^2 \le 1, z > 0\}$. Find $\iiint_V \frac{zdV}{9 - (x^2 + y^2 + z^2)}$.

【Solution】

Let $x = \rho \sin\varphi \cos\theta$, $y = \rho \sin\varphi \sin\theta$, $z = \rho \cos\varphi$ then

$$dxdydz = \begin{Vmatrix} \dfrac{\partial x}{\partial \rho} & \dfrac{\partial x}{\partial \varphi} & \dfrac{\partial x}{\partial \theta} \\ \dfrac{\partial y}{\partial \rho} & \dfrac{\partial y}{\partial \varphi} & \dfrac{\partial y}{\partial \theta} \\ \dfrac{\partial z}{\partial \rho} & \dfrac{\partial z}{\partial \varphi} & \dfrac{\partial z}{\partial \theta} \end{Vmatrix} d\rho d\varphi d\theta$$

Let $R = \left\{(\rho, \varphi, \theta): 0 \le \rho \le 1, 0 \le \varphi \le \dfrac{\pi}{2}, 0 \le \theta \le 2\pi\right\}$ then

$$\iiint_V \frac{zdV}{9-(x^2+y^2+z^2)} = \iiint_R \frac{\rho^3 \sin\varphi \cos\varphi}{9-\rho^2} d\rho d\theta d\varphi$$

$$= \int_0^{\frac{\pi}{2}} \int_0^{2\pi} \int_0^1 \frac{\rho^3 \sin\varphi \cos\varphi}{9-\rho^2} d\rho d\theta d\varphi = \int_0^{\frac{\pi}{2}} \cos\varphi \sin\varphi d\varphi \int_0^{2\pi} d\theta \int_0^1 \frac{\rho(\rho^2-9)+9\rho}{9-\rho^2} d\rho$$

$$= \frac{1}{2}\int_0^{\frac{\pi}{2}} \sin 2\varphi d\varphi \int_0^{2\pi} d\theta \int_0^1 -\rho + \frac{9\rho}{9-\rho^2} d\rho = -\left.\frac{\cos 2\varphi}{4}\right|_0^{\frac{\pi}{2}} \cdot 2\pi \cdot \left.\left(-\frac{\rho^2}{2} - \frac{9}{2}\ln(9-\rho^2)\right)\right|_0^1$$

$$= \left(\frac{1}{2}\right)(2\pi)\left(-\frac{1}{2} - \frac{9}{2}\ln\frac{8}{9}\right) = \pi\left(-\frac{1}{2} + \frac{9}{2}(2\ln 3 - 3\ln 2)\right)$$

Example 32.

Let $V = \{(x, y, z): 1 \le x^2 + y^2 + z^2 \le 3\}$. Find $\displaystyle\iiint_V \frac{z^2 \cos\sqrt{x^2+y^2+z^2}\, dxdydz}{(x^2+y^2+z^2)^2} = ?$

【Solution】

Let $x = \rho \sin\varphi \cos\theta$, $y = \rho \sin\varphi \sin\theta$, $z = \rho \cos\varphi$ then

$$dxdydz = \begin{Vmatrix} \dfrac{\partial x}{\partial \rho} & \dfrac{\partial x}{\partial \varphi} & \dfrac{\partial x}{\partial \theta} \\ \dfrac{\partial y}{\partial \rho} & \dfrac{\partial y}{\partial \varphi} & \dfrac{\partial y}{\partial \theta} \\ \dfrac{\partial z}{\partial \rho} & \dfrac{\partial z}{\partial \varphi} & \dfrac{\partial z}{\partial \theta} \end{Vmatrix} d\rho d\varphi d\theta$$

$$= \begin{Vmatrix} \sin\varphi \cos\theta & \rho\cos\varphi\cos\theta & -\rho\sin\varphi\sin\theta \\ \sin\varphi \sin\theta & \rho\cos\varphi\sin\theta & -\rho\sin\varphi\cos\theta \\ \cos\varphi & -\rho\sin\varphi & 0 \end{Vmatrix} d\rho d\varphi d\theta = \rho^2 \sin\varphi\, d\rho d\varphi d\theta$$

Let $R = \{(\rho, \varphi, \theta): 1 \le \rho \le \sqrt{3}, 0 \le \varphi \le \pi, 0 \le \theta \le 2\pi\}$ then

8.4.3 Use Spherical Coordinates to Compute Triple Integrals

$$\iiint_V \frac{z^2 \cos\sqrt{x^2+y^2+z^2}\,dxdydz}{(x^2+y^2+z^2)^2}\,dV = \iiint_R \frac{\cos\rho}{\rho^4}\rho^4 \sin\varphi \cos^2\varphi\,d\rho d\theta d\varphi$$

$$= \int_0^\pi \int_0^{2\pi} \int_1^{\sqrt{3}} \frac{\cos\rho}{\rho^4}\rho^4 \sin\varphi \cos^2\varphi\,d\rho d\theta d\varphi = \left.\frac{-\cos^3\varphi}{3}\right|_0^\pi \cdot 2\pi \cdot \int_1^{\sqrt{3}} \cos\rho\,d\rho$$

$$= \frac{4\pi}{3}\left(\sin\rho\big|_1^{\sqrt{3}}\right) = \frac{4\pi}{3}(\sin\sqrt{3}-\sin 1)$$

Example 33.

Let $V = \{(x,y,z): 1 \leq x^2+y^2+z^2 \leq 3, z > 0\}$.

Find $\iiint_V \dfrac{z\cos\sqrt{x^2+y^2+z^2}\,dxdydz}{(x^2+y^2+z^2)^{\frac{3}{2}}} = ?$

【Solution】

Let $x = \rho\sin\varphi\cos\theta, y = \rho\sin\varphi\sin\theta, z = \rho\cos\varphi$ then

$$dxdydz = \begin{Vmatrix} \dfrac{\partial x}{\partial \rho} & \dfrac{\partial x}{\partial \varphi} & \dfrac{\partial x}{\partial \theta} \\ \dfrac{\partial y}{\partial \rho} & \dfrac{\partial y}{\partial \varphi} & \dfrac{\partial y}{\partial \theta} \\ \dfrac{\partial z}{\partial \rho} & \dfrac{\partial z}{\partial \varphi} & \dfrac{\partial z}{\partial \theta} \end{Vmatrix} d\rho d\varphi d\theta$$

$$= \begin{Vmatrix} \sin\varphi\cos\theta & \rho\cos\varphi\cos\theta & -\rho\sin\varphi\sin\theta \\ \sin\varphi\sin\theta & \rho\cos\varphi\sin\theta & -\rho\sin\varphi\cos\theta \\ \cos\varphi & -\rho\sin\varphi & 0 \end{Vmatrix} d\rho d\varphi d\theta = \rho^2\sin\varphi\,d\rho d\varphi d\theta$$

Let $R = \left\{(\rho,\varphi,\theta): 1 \leq \rho \leq \sqrt{3}, 0 \leq \varphi \leq \dfrac{\pi}{2}, 0 \leq \theta \leq 2\pi\right\}$ then

$$\iiint_V \frac{z\cos\sqrt{x^2+y^2+z^2}\,dxdydz}{(x^2+y^2+z^2)^{\frac{3}{2}}} = \iiint_R \frac{\cos\rho}{\rho^3}\rho^3 \sin\varphi\cos\varphi\,d\rho d\theta d\varphi$$

$$= \int_0^{\frac{\pi}{2}} \int_0^{2\pi} \int_1^{\sqrt{3}} \frac{\cos\rho}{\rho^3}\rho^3 \sin\varphi\cos\varphi\,d\rho d\theta d\varphi = \left.\frac{-\cos 2\varphi}{4}\right|_0^{\frac{\pi}{2}} \cdot 2\pi \cdot \int_1^{\sqrt{3}} \cos\rho\,d\rho = \pi\left(\sin\rho\big|_1^{\sqrt{3}}\right)$$

$$= \pi(\sin\sqrt{3} - \sin 1)$$

Example 34.

8.4.3 Use Spherical Coordinates to Compute Triple Integrals

Find $\int_0^1 \int_0^{\sqrt{1-x^2}} \int_{\sqrt{x^2+y^2}}^{\sqrt{2-x^2-y^2}} dz\,dy\,dx = ?$

【Solution】

Let $x = \rho \sin\varphi \cos\theta$, $y = \rho \sin\varphi \sin\theta$, $z = \rho \cos\varphi$ then

$$dxdydz = \begin{Vmatrix} \dfrac{\partial x}{\partial \rho} & \dfrac{\partial x}{\partial \varphi} & \dfrac{\partial x}{\partial \theta} \\ \dfrac{\partial y}{\partial \rho} & \dfrac{\partial y}{\partial \varphi} & \dfrac{\partial y}{\partial \theta} \\ \dfrac{\partial z}{\partial \rho} & \dfrac{\partial z}{\partial \varphi} & \dfrac{\partial z}{\partial \theta} \end{Vmatrix} d\rho d\varphi d\theta$$

$$= \begin{Vmatrix} \sin\varphi \cos\theta & \rho\cos\varphi \cos\theta & -\rho\sin\varphi \sin\theta \\ \sin\varphi \sin\theta & \rho\cos\varphi \sin\theta & -\rho\sin\varphi \cos\theta \\ \cos\varphi & -\rho\sin\varphi & 0 \end{Vmatrix} d\rho d\varphi d\theta = \rho^2 \sin\varphi \, d\rho d\varphi d\theta$$

Let $R = \left\{(\rho,\varphi,\theta): 0 \leq \rho \leq \sqrt{2}, 0 \leq \varphi \leq \dfrac{\pi}{4}, 0 \leq \theta \leq \dfrac{\pi}{2}\right\}$ then

$$\int_0^1 \int_0^{\sqrt{1-x^2}} \int_{\sqrt{x^2+y^2}}^{\sqrt{2-x^2-y^2}} dz\,dy\,dx = \iiint_R \rho^2 \sin\varphi \, d\rho d\theta d\varphi = \int_0^{\pi/4} \int_0^{\pi/2} \int_0^{\sqrt{2}} \rho^2 \sin\varphi \, d\rho d\theta d\varphi$$

$$= \dfrac{(\sqrt{2}-1)\pi}{3}$$

Example 35.

Let $V = \left\{(x,y,z): \dfrac{x^2}{a^2} + \dfrac{y^2}{b^2} + \dfrac{z^2}{c^2} \leq 1\right\}$. Find $\iiint_V dxdydz = ?$

【Solution】

Let $x = a\rho \sin\varphi \cos\theta$, $y = b\rho \sin\varphi \sin\theta$, $z = c\rho \cos\varphi$ then

$$dxdydz = \begin{Vmatrix} \dfrac{\partial x}{\partial \rho} & \dfrac{\partial x}{\partial \varphi} & \dfrac{\partial x}{\partial \theta} \\ \dfrac{\partial y}{\partial \rho} & \dfrac{\partial y}{\partial \varphi} & \dfrac{\partial y}{\partial \theta} \\ \dfrac{\partial z}{\partial \rho} & \dfrac{\partial z}{\partial \varphi} & \dfrac{\partial z}{\partial \theta} \end{Vmatrix} d\rho d\varphi d\theta$$

$$= abc \begin{Vmatrix} \sin\varphi \cos\theta & \rho\cos\varphi \cos\theta & -\rho\sin\varphi \sin\theta \\ \sin\varphi \sin\theta & \rho\cos\varphi \sin\theta & -\rho\sin\varphi \cos\theta \\ \cos\varphi & -\rho\sin\varphi & 0 \end{Vmatrix} d\rho d\varphi d\theta = abc\rho^2 \sin\varphi \, d\rho d\varphi d\theta$$

8.4.3 Use Spherical Coordinates to Compute Triple Integrals

Let $R = \{(\rho, \varphi, \theta): 0 \leq \rho \leq 1, 0 \leq \varphi \leq \pi, 0 \leq \theta \leq 2\pi\}$ then

$$\iiint_V dxdydz = abc \iiint_R \rho^2 \sin\varphi \, d\rho d\theta d\varphi = abc \int_0^\pi \int_0^{2\pi} \int_0^1 \rho^2 \sin\varphi \, d\rho d\theta d\varphi$$
$$= \frac{4\pi abc}{3}$$

Example 36.

Find $\iiint_V \left(3x + \dfrac{1}{1+x^2+y^2+z^2}\right) dV = ?$, where $V = \{(x,y,z): x^2 + y^2 + z^2 \leq 4\}$.

【Solution】

Let $x = \rho \sin\varphi \cos\theta$, $y = \rho \sin\varphi \sin\theta$, $z = \rho \cos\varphi$ then

$$dxdydz = \begin{Vmatrix} \dfrac{\partial x}{\partial \rho} & \dfrac{\partial x}{\partial \varphi} & \dfrac{\partial x}{\partial \theta} \\ \dfrac{\partial y}{\partial \rho} & \dfrac{\partial y}{\partial \varphi} & \dfrac{\partial y}{\partial \theta} \\ \dfrac{\partial z}{\partial \rho} & \dfrac{\partial z}{\partial \varphi} & \dfrac{\partial z}{\partial \theta} \end{Vmatrix} d\rho d\varphi d\theta = \rho^2 \sin\varphi \, d\rho d\varphi d\theta$$

Let $R = \{(\rho, \varphi, \theta): 0 \leq \rho \leq 2, 0 \leq \varphi \leq \pi, 0 \leq \theta \leq 2\pi\}$ then

$$\iiint_V \left(3x + \frac{1}{1+x^2+y^2+z^2}\right) dV = \iiint_R \left(3\rho \sin\varphi \cos\theta + \frac{1}{1+\rho^2}\right) \rho^2 \sin\varphi \, d\rho d\theta d\varphi$$

$$= \int_0^\pi \int_0^{2\pi} \int_0^2 \left(3\rho \sin\varphi \cos\theta + \frac{1}{1+\rho^2}\right) \rho^2 \sin\varphi \, d\rho d\theta d\varphi$$

$$= \int_0^\pi \int_0^{2\pi} \int_0^2 3\rho^3 \sin^2\varphi \cos\theta \, d\rho d\theta d\varphi + \int_0^\pi \int_0^{2\pi} \int_0^2 \frac{\rho^2 \sin\varphi}{1+\rho^2} \, d\rho d\theta d\varphi$$

$$= \int_0^\pi \sin^2\varphi \, d\varphi \int_0^{2\pi} \cos\theta \, d\theta \int_0^2 3\rho^3 d\rho + 2\pi \int_0^\pi \sin\varphi \, d\varphi \int_0^2 \frac{\rho^2}{1+\rho^2} d\rho$$

$$= 2\pi \int_0^\pi \sin\varphi \, d\varphi \int_0^2 \frac{\rho^2}{1+\rho^2} d\rho = 4\pi \int_0^2 1 - \frac{1}{1+\rho^2} d\rho = 4\pi(2 - \tan^{-1} 2)$$

Example 37.

Find the mass of the solid bounded below by the half-cone $z = \sqrt{x^2 + y^2}$ and above by the spherical surface $x^2 + y^2 + z^2 = 1$ given that the density function

$$f(x,y,z) = e^{(x^2+y^2+z^2)^{\frac{3}{2}}}.$$

【Solution】

$\because$ mass $= \iiint_V f(x,y,z)\, dxdydz$

Let $x = \rho \sin\varphi \cos\theta$, $y = \rho \sin\varphi \sin\theta$, $z = \rho \cos\varphi$ then

$$dxdydz = \begin{Vmatrix} \dfrac{\partial x}{\partial \rho} & \dfrac{\partial x}{\partial \varphi} & \dfrac{\partial x}{\partial \theta} \\ \dfrac{\partial y}{\partial \rho} & \dfrac{\partial y}{\partial \varphi} & \dfrac{\partial y}{\partial \theta} \\ \dfrac{\partial z}{\partial \rho} & \dfrac{\partial z}{\partial \varphi} & \dfrac{\partial z}{\partial \theta} \end{Vmatrix} d\rho d\varphi d\theta = \rho^2 \sin\varphi\, d\rho d\varphi d\theta$$

Let $R = \left\{(\rho,\varphi,\theta): 0 \leq \rho \leq 1, \dfrac{\pi}{4} \leq \varphi \leq \pi, 0 \leq \theta \leq 2\pi\right\}$ then

$$\iiint_V f(x,y,z)\, dxdydz = \iiint_V e^{(x^2+y^2+z^2)^{\frac{3}{2}}}\, dxdydz = \iiint_R e^{\rho^3} \rho^2 \sin\varphi\, d\rho d\theta d\varphi$$

$$= \int_{\frac{\pi}{4}}^{\pi} \int_0^{2\pi} \int_0^1 e^{\rho^3} \rho^2 \sin\varphi\, d\rho d\theta d\varphi = 2\pi \int_0^1 e^{\rho^3} \rho^2 d\rho \int_{\frac{\pi}{4}}^{\pi} \sin\varphi\, d\varphi = 2\pi \cdot \left.\dfrac{e^{\rho^3}}{3}\right|_0^1 \cdot (-\cos\varphi)\big|_{\frac{\pi}{4}}^{\pi}$$

$$= \dfrac{2\pi(e-1)\left(1 + \dfrac{1}{\sqrt{2}}\right)}{3}$$

CHAPTER 9 VECTOR CALCULUS

Next, we will sequentially introduce line integrals, Green's Theorem, surface integrals, Stokes' Theorem, and Gauss's Theorem (The Divergence Theorem). The integrand in line integrals can be either scalar functions or vector functions, and the curves may be two-dimensional planar curves or three-dimensional space curves. In either case, they can be transformed into definite integrals of single-variable functions.

When the first-order partial derivatives of the integrand exist and are continuous, Green's Theorem serves as an important tool for computing integrals over closed curves in the two-dimensional plane and double integrals. Additionally, Green's Theorem is a form that extends the Fundamental Theorem of Calculus to two dimensions. This theorem connects the line integral along a simple closed curve in the plane with the double integral over the region it encloses, enabling a bidirectional conversion between line integrals and double integrals. If calculating the line integral is difficult, one can use this theorem to convert it into a double integral; conversely, if the double integral is challenging to compute, it can be transformed into a line integral using this theorem. Green's Theorem requires that the region D in the double integral is simply connected. It is important to note that the union of several disjoint simple connected sets is not a simple connected set; therefore, Green's Theorem can sometimes be extended to apply to non-simply connected regions. Readers should be familiar with various applications of Green's Theorem. When the path of the line integral is complex or when the integrand is complicated, one should try to use Green's Theorem to transform the problem into a double integral, which is usually evaluated using Fubini's Theorem. Moreover, it may also involve transformations to polar coordinates or generalized coordinates for evaluation. In any case, accurately identifying the boundaries of the integrals is crucial. When the integrand does not satisfy the conditions of having first-order partial derivatives existing and being continuous within the enclosed region, one must construct another closed region that approximates the original region and within which the original integrand meets the conditions for applying Green's Theorem. Only then can Green's Theorem be used to compute the value, which is then approximated back to the original closed region. Conversely, when given a closed region and asked to find its area, one can also apply Green's Theorem to transform the problem into a line integral. Therefore, the ability to compute line integrals is essential.

Surface integrals can be categorized into three types: calculating the surface area of a closed region in space, integrating a scalar function over a surface S, and determining the flux of a vector function through the surface. Ultimately, all three types are transformed into calculations involving double integrals. Given the parameterization of the surface S as $\vec{r}(s, t)$, the calculations for these three types are related to the cross product of the two partial derivatives $\vec{r}_s \times \vec{r}_t$ and its magnitude $|\vec{r}_s \times \vec{r}_t|$. Therefore, efficiently and accurately determining these two components is crucial. The surfaces involved are typically cylinders, elliptical cylinders, cones, parabolas, etc. Regardless of the combination, the problem of computing surface integrals can be transformed into a problem of double integrals using projection methods, and it is also essential to accurately and quickly determine the boundaries of the projected integrals. Additionally, the projected integrals may require transformations to polar coordinates or generalized coordinates.

Stokes' Theorem is not only a generalization of the Fundamental Theorem of Calculus to higher dimensions but also a higher-dimensional extension of Green's Theorem. While Green's Theorem connects the double integral over a planar region to the line integral around a closed curve in the plane, Stokes' Theorem relates the line integral along the boundary curve C of a non-closed surface S to the surface integral of the curl of $\vec{F}$ over that surface. Stokes' Theorem shows that the line integral of $\vec{F}$ along the boundary curve C of the non-closed surface S can be converted into the flux of the curl of $\vec{F}$ across the surface S. Conversely, one can also obtain the flux of the curl of $\vec{F}$ over the surface S by calculating the line integral along the boundary curve C. Stokes' Theorem has several types of examination problems, with one important condition being that the surface must be non-closed and its boundary must be a closed curve in space. Furthermore, the orientation of the surface S and whether the boundary curve C is traversed counterclockwise are critically related. Additionally, when the integrand of the line integral is complex, Stokes' Theorem can be utilized to transform the problem into a surface integral, which can subsequently be projected into a double integral problem. Thus, developing the ability to compute double integrals is essential.

In summary, Gauss's Theorem (The Divergence Theorem), Stokes' Theorem, and Green's Theorem are all extensions of the Fundamental Theorem of Calculus. Moreover, Gauss's Theorem allows us to transform the flux of a vector field $\vec{F}$ through a closed surface S (a surface integral) into the triple integral of the divergence of $\vec{F}$ over a closed volume V. For example, if a closed surface is composed of surfaces S_1 and S_2, the original method for calculating the flux through the two surfaces involves computing their individual fluxes and adding them together. However, by applying Gauss's Theorem, this problem can be converted into a volume integral of a scalar function. Conversely, one can also obtain the triple integral of the divergence of $\vec{F}$ over the volume V by computing the flux of $\vec{F}$ across the closed surface S. When the first-order partial derivatives of the vector function exist and are continuous within the region enclosed by a smooth closed surface, if the integrand is complex, Gauss's Theorem can be used to transform the surface integral problem into a triple integral problem. Therefore, the ability to compute triple integrals is crucial, and this may also require transformations to polar coordinates or generalized coordinates for evaluation.

9.1 Parametric Equations

【Definition】

Assume that x and y are continuous functions of t, where $t \in I$. Then the equations
$$x = x(t) \text{ and } y = y(t)$$
are called parametric equations. As t varies within the interval I, the set of points (x, y) obtained is referred to as a parametric curve, denoted by the symbol C.

To familiarize yourself with and understand the graph of a curve represented parametrically, it is essential to practice rewriting these two equations into a single equation involving the variables x and y. On the other hand, when a single equation for the variables x and y is given, the corresponding parametric equations are not unique.

Example 1.
Try to express the following parametric equations as a single equation.
(1) $x(t) = t^2 - 5, \ y(t) = 2t + 1, \ -2 \leq t \leq 3$.
(2) $x(t) = \sqrt{2t + 3}, \ y(t) = 2t + 1, -2 \leq t \leq 6$.
(3) $x(t) = 2\cos t, \ y(t) = \sqrt{3} \sin t, -2 \leq t \leq 7$.

【Solution】

(1)
$$\because t = \frac{y-1}{2} \quad \therefore x = \left(\frac{y-1}{2}\right)^2 - 5 = \frac{y^2 - 2y + 1}{4} - 5 = \frac{y^2 - 2y - 19}{4}$$

$$\therefore x = \frac{y^2 - 2y - 19}{4}, \quad \forall -1 \le x \le 4$$

(2)

$\because x = \sqrt{2t+3} \quad \therefore x^2 = 2t+3 \Rightarrow t = \dfrac{x^2-3}{2}$

$\because y = 2t+5 \quad \therefore y = 2\left(\dfrac{x^2-3}{2}\right) + 5 = x^2 + 2, \quad \forall 0 \le x \le \sqrt{15}$

(3)

$\because x(t) = 2\cos t \quad \therefore \cos t = \dfrac{x}{2}, \quad \because y(t) = \sqrt{3}\sin t \quad \therefore \sin t = \dfrac{y}{\sqrt{3}}$

$\because \cos^2 t + \sin^2 t = 1 \quad \therefore \left(\dfrac{x}{2}\right)^2 + \left(\dfrac{y}{\sqrt{3}}\right)^2 = 1 \Rightarrow \dfrac{x^2}{4} + \dfrac{y^2}{3} = 1$

Example 2.

Try to find two different sets of parametric equations corresponding to the following single equation.

(1) $y = 3x^2 - 1$ (2) $y = \ln x$ (3) $x = y^2 - 6y + 8$

【Solution】

(1)
Let $x(t) = t$ then $y(t) = 3t^2 - 1$
Let $x(t) = 2t - 1$ then $y(t) = 3x^2 - 1 = 3(2t-1)^2 - 1 = 12t^2 - 12t + 2$

(2)
Let $x(t) = t$ then $y(t) = \ln t$
Let $x(t) = t^2$ then $y(t) = \ln t^2 = 2\ln t$

(3)
Let $y(t) = t$ then $x(t) = t^2 - 6t + 8$
Let $y(t) = t+2$ then $x(t) = y^2 - 6y + 8 = (t+2)^2 - 6(t+2) + 8 = t^2 - 2t$

Example 3.

Try to express the following parametric equations as a single equation.

(1) $x(t) = t^5, \quad y(t) = 5\ln t$

(2) $x = e^{3t}, y = t + 3$

(3) $x(t) = 2t - 5, \quad y(t) = 4t - 7$

(4) $x = \sqrt{t}, \quad y = 2 - t$

(5) $x(t) = t^2 - 2, \quad y(t) = \dfrac{t}{3}, -2 \le t \le 6$.

(6) $x = t^2 - 2t, \ y = t + 2, 0 \le t \le 5$
(7) $x(t) = 2\cos t, \ y(t) = 3\sin t, 0 \le t \le 2\pi$
(8) $x(t) = 2 + \cos t, \ y(t) = 4 - \sin t$
(9) $x = a\cos t, y = a\sin t, \ 0 \le t \le 2\pi$
(10) $x(t) = \sin t, \ y(t) = \cos 2t$
(11) $x(t) = \csc t, \ y(t) = \cot t, \ 0 < t < \dfrac{\pi}{2}$
(12) $x = \cos t, \ y = \sec t, \ 0 < t < \dfrac{\pi}{2}$
(13) $x = \sinh t, y = \cosh t$
(14) $x(t) = 3\cos 7t, \ y(t) = 3\sin 7t$
(15) $x(t) = 3\cosh 4t, \ y(t) = 4\sinh 4t$
(16) $x = \cos t, \ y = \cos^2 t$

【Solution】

(1)

$\because x(t) = t^5 \ \therefore 5\ln t = \ln x \Rightarrow y = \ln x, \ \forall x \in [0, \infty)$

(2)

$\because x = e^{3t}$ and $y = t + 3 \ \therefore t = \dfrac{\ln x}{3} = y - 3 \Rightarrow y = \dfrac{\ln x}{3} + 3$

$\therefore$ the curve given parametric equations is $y = \dfrac{\ln x}{3} + 3$

(3)

$\because t = \dfrac{x+5}{2} \ \therefore y = 4t - 7 = 4\left(\dfrac{x+5}{2}\right) - 7 = 2x + 3 \Rightarrow y = 2x + 3, \ \forall x \in R$

(4)

$\because x^2 = t$ and $t = 2 - y$

$\therefore$ the curve given parametric equations is the parabola $y = 2 - x^2, \ \forall x \ge 0$

(5)

$\because t = 3y \ \therefore x = t^2 - 2 = (3y)^2 - 2 = 9y^2 - 2 \Rightarrow x = 9y^2 - 2$ where $-2 \le x < \infty$

(6)

$\because t = y - 2, \ \therefore x = t^2 - 2t = (y-2)^2 - 2(y-2) = y^2 - 6y + 8$

$\therefore$ the curve given parametric equations is the parabola $x = y^2 - 6y + 8, \ -2 \le y \le 3$

(7)

$\because x(t) = 2\cos t \ \therefore \cos t = \dfrac{x}{2}, \ \because y(t) = 3\sin t \ \therefore \sin t = \dfrac{y}{3}$

$\because \cos^2 t + \sin^2 t = 1 \ \therefore \left(\dfrac{x}{2}\right)^2 + \left(\dfrac{y}{3}\right)^2 = 1 \Rightarrow \dfrac{x^2}{4} + \dfrac{y^2}{9} = 1$ where $-2 \le x \le 2$

(8)
$\because x(t) = 2 + \cos t \quad \therefore \cos t = x - 2, \quad \because y(t) = 4 - \sin t \quad \therefore \sin t = 4 - y$
$\because \cos^2 t + \sin^2 t = 1 \quad \therefore (x-2)^2 + (y-4)^2 = 1, \text{ where } 1 \leq x \leq 3$

(9)
$\because x^2 + y^2 = a^2 \cos^2 t + a^2 \sin^2 t = a^2$
$\therefore$ the curve given parametric equations is the circle with radius a, $x^2 + y^2 = a^2$

(10)
$\because x(t) = \sin t, \quad y(t) = \cos 2t$
$\because \sin^2 t = \dfrac{1 - \cos 2t}{2} \quad \therefore x^2 = \dfrac{1-y}{2} \Rightarrow y = 1 - x^2, \text{ where } -1 \leq x \leq 1$

(11)
$\because x(t) = \csc t, \quad y(t) = \cot t$
$\because \csc^2 t = \cot^2 t + 1 \quad \therefore x^2 = y^2 + 1 \Rightarrow y = \sqrt{x^2 - 1}, \text{ where } x > 1$

(12)
$\because x = \cos t, \quad y = \sec t \quad \therefore y = \dfrac{1}{x}, \quad \forall y > 1$
$\therefore$ the curve given parametric equations is $y = \dfrac{1}{x}, \quad \forall y > 1$

(13)
$\because \cosh^2 t - \sinh^2 t = 1, \quad x = \sinh t \text{ and } y = \cosh t$
$\therefore$ the curve given parametric equations is $y^2 - x^2 = 1$

(14)
$\because x(t) = 3 \cos 7t, \quad y(t) = 3 \sin 7t,$
$\because \cos^2 7t + \sin^2 7t = 1 \quad \therefore \left(\dfrac{x}{3}\right)^2 + \left(\dfrac{y}{3}\right)^2 = 1 \quad \therefore x^2 + y^2 = 9, \text{ where } -3 \leq x \leq 3$

(15)
$\because x(t) = 3 \cosh 4t, \quad y(t) = 4 \sinh 4t,$
$\because \cosh^2 4t - \sinh^2 4t = 1 \quad \therefore \left(\dfrac{x}{3}\right)^2 - \left(\dfrac{y}{4}\right)^2 = 1, \text{ where } -3 \leq x \leq 3$
$\therefore$ the curve is hyperbola

(16)
$\because x = \cos t, \quad y = \cos^2 t, \quad \therefore y = \cos^2 t = x^2, \quad \forall -1 \leq x \leq 1$
$\therefore$ the curve given parametric equations is the parabola $y = x^2, \quad \forall -1 \leq x \leq 1$

9.2 Given the Vector Representation of a Curve, Find Its Derivative

9.2 Given the Vector Representation of a Curve, Find Its Derivative

【Definition】

A vector-valued function is a function of the form
$$\vec{r}(t) = (f(t), g(t)) \quad \text{or} \quad \vec{r}(t) = (f(t), g(t), h(t))$$
where the component functions f, g and h are real-valued functions of the parameter t.

In the above definition, the former is a two-dimensional vector-valued function, while the latter is a three-dimensional vector-valued function.

【Definition】

Given a vector-valued function $\vec{r}(t)$, if
$$\lim_{\Delta t \to 0} \frac{\vec{r}(t + \Delta t) - \vec{r}(t)}{\Delta t} \quad \text{exists,}$$
then $\vec{r}(t)$ is said to be differentiable at t, and we denote
$$\vec{r}'(t) = \lim_{\Delta t \to 0} \frac{\vec{r}(t + \Delta t) - \vec{r}(t)}{\Delta t}.$$
Furthermore, if $\vec{r}'(t)$ exists, $\forall t \in (a, b)$ then $\vec{r}(t)$ is said to be differentiable on the interval (a, b).

【Theorem】

Assume $\vec{r}(t) = (f(t), g(t), h(t))$ where $f(t), g(t)$ and $h(t)$ are differentiable functions. Then
$$\vec{r}'(t) = (f'(t), g'(t), h'(t)).$$

Proof:

$\because f(t), g(t), h(t)$ are differentiable functions

$$\therefore \vec{r}'(t) = \lim_{\Delta t \to 0} \frac{\vec{r}(t + \Delta t) - \vec{r}(t)}{\Delta t} = \lim_{\Delta t \to 0} \frac{\big(f(t+\Delta t), g(t+\Delta t), h(t+\Delta t)\big) - \big(f(t), g(t), h(t)\big)}{\Delta t}$$

$$= \lim_{\Delta t \to 0} \left(\frac{f(t+\Delta t) - f(t)}{\Delta t}, \frac{g(t+\Delta t) - g(t)}{\Delta t}, \frac{h(t+\Delta t) - h(t)}{\Delta t} \right)$$

$$= \left(\lim_{\Delta t \to 0} \frac{f(t+\Delta t) - f(t)}{\Delta t}, \lim_{\Delta t \to 0} \frac{g(t+\Delta t) - g(t)}{\Delta t}, \lim_{\Delta t \to 0} \frac{h(t+\Delta t) - h(t)}{\Delta t} \right)$$

$$= (f'(t), g'(t), h'(t))$$

【Definition】

Given a vector-valued function $\vec{r}(t)$, if $\vec{r}'(t)$ is continuous and $\vec{r}'(t) \neq 0, \forall t \in I$, then $\vec{r}(t)$ is called a smooth vector-valued function on I. If a curve is described by a smooth vector-valued function, then the curve is referred to as a smooth curve.

Next, we will list the important formulas related to the line integrals of scalar and vector functions. Detailed derivations can be found in the respective sections. The first calculation focus is on finding the line integral of a scalar function f over a given curve C. The formula is

$$\int_C f(x, y, z) ds = \int_a^b f(\vec{r}(t))|\vec{r}'(t)| dt.$$

The second calculation focus is on finding the line integral of a vector function $\vec{F}$ over a given curve C. The formula is

$$\int_C \vec{F} \cdot d\vec{r} = \int_a^b \vec{F}(\vec{r}(t)) \cdot \vec{r}'(t) dt,$$

where $C: \vec{r}(t) = (x(t), y(t), z(t))$, $\forall a \leq t \leq b$.

From the above observations, it is important to efficiently and quickly determine $\vec{r}'(t)$ and $|\vec{r}'(t)|$ for a given curve C: $\vec{r}(t) = (x(t), y(t), z(t))$.

Example 1.
Assume $\vec{r}(t) = (\sqrt{t}, 2 - t)$. Find $\vec{r}'(t)$ and $\vec{r}'(1)$.

【Solution】

$$\vec{r}'(t) = \left(\frac{1}{2\sqrt{t}}, -1\right) \text{ and } \vec{r}'(1) = \left(\frac{1}{2\sqrt{2}}, -1\right)$$

Example 2.
Try to find the derivatives of the following vector-valued functions, i. e., $\vec{r}'(t) =?$
(1) $\vec{r}(t) = (7t + 10, 5t^2 + 2t - 3)$ (2) $\vec{r}(t) = (5\cos t, 7\sin t)$
(3) $\vec{r}(t) = (e^t \sin t, e^{-t} \cos t, -e^{3t})$ (4) $\vec{r}(t) = (t \ln t, 7e^t \sin t, \cos t + \sin t)$
(5) $\vec{r}(t) = (5 + t^4, te^{-t}, \cos 2t)$

【Solution】

(1)
$\vec{r}'(t) = (7, 10t + 2)$
(2)
$\vec{r}'(t) = (-5 \sin t, 7 \cos t)$

(3)
$\vec{r}'(t) = (e^t(\sin t + \cos t), -e^{-t}(\cos t + \sin t), -3e^{3t})$
(4)
$\vec{r}'(t) = (1 + \ln t, 7e^t(\sin t + \cos t), -\sin t + \cos t)$
(5)
$\vec{r}'(t) = (4t^3, (1-t)e^{-t}, -2\sin 2t)$

Example 3.
Find $\vec{r}'(t)$ and $|\vec{r}'(t)|$ for the following vector-valued functions.
(1)$\vec{r}(t) = (t, at), \ a > 0$ (2)$\vec{r}(t) = (t, 4t, 3t)$ (3)$\vec{r}(t) = (\cos t, \sin t, t)$
(4)$\vec{r}(t) = (e^t, e^{at}, e^{-t})$ (5)$\vec{r}(t) = (0, 3\sin t, \sin^2 t)$ (6)$\vec{r}(t) = (e^t \sin t, e^t \cos t)$
(7)$\vec{r}(t) = (r \cos t, r \sin t), \ r > 0$

【Solution】
(1)
$\vec{r}'(t) = (1, a), \ |\vec{r}'(t)| = \sqrt{1 + a^2}$
(2)
$\vec{r}'(t) = (1,4,3), \ |\vec{r}'(t)| = \sqrt{26}$
(3)
$\vec{r}'(t) = (-\sin t, \cos t, 1), \ |\vec{r}'(t)| = \sqrt{\cos^2 t + \sin^2 t + 1} = \sqrt{2}$
(4)
$\vec{r}'(t) = (e^t, ae^t, -e^{-t}), \ |\vec{r}'(t)| = \sqrt{e^{2t}(1 + a^2) + e^{-2t}}$
(5)
$\vec{r}'(t) = (0, 3\cos t, 2\sin t \cos t), \ |\vec{r}'(t)| = \sqrt{9\cos^2 t + 4\sin^2 t \cos^2 t}$
(6)
$\vec{r}'(t) = (e^t(\sin t + \cos t), e^t(\cos t - \sin t)), \ |\vec{r}'(t)| = \sqrt{2e^{2t}}$
(7)
$\vec{r}'(t) = (-r \sin t, r \cos t), \ |\vec{r}'(t)| = r\sqrt{\cos^2 t + \sin^2 t} = r$

Example 4.
The curve C is given by the vector-valued function $\vec{r}(t) = (t, t^2), \forall t \in R$. Is C smooth?
【Solution】
$\because \vec{r}'(t) = (1, 2t) \ \therefore \vec{r}'(t) \neq \vec{0}, \forall t \in R$ and $\vec{r}'(t)$ is continuous, $\forall t \ \therefore C$ is a smooth curve

Example 5.
The curve C is given by the vector-valued function $\vec{r}(t) = (t^4, t^5), \forall t \in R$. Is C smooth?

【Solution】
$\because \vec{r}'(t) = (4t^3, 5t^4), \quad \therefore \vec{r}'(0) = (0,0) \quad \therefore C$ is not a smooth curve

Example 6.
The curve C is given by parametric equations $\vec{r}(t) = (t, |t|), \forall t \in R$. Is C smooth?

【Solution】
$\because \vec{r}'(0)$ doesn't exist $\therefore C$ is not a smooth curve

Example 7.
Determine whether the following curves are smooth curves.
(1) $\vec{r}(t) = (1 + t^3, te^{-t}, \sin 2t), \forall t \in R$
(2) $\vec{r}(t) = (3 \cos t, 4 \sin t), \forall t \in R$
(3) $\vec{r}(t) = (e^t \sin t, e^t \cos t, -e^{2t}), \forall t \in R$
(4) $\vec{r}(t) = (t \ln t, 5e^t \cos t, \cos t - \sin t), \forall t \in R$
(5) $\vec{r}(t) = (0, 3 \sin t, \sin^2 t), \forall t \in R$
(6) $\vec{r}(t) = (e^t \sin t, e^{-t} \cos t)$
(7) $\vec{r}(t) = (t^3 - 12t + 17, \ t^2 - 4t + 8), \forall t \in R$
(8) $\vec{r}(t) = \left(t, \ t^2 \sin \dfrac{1}{t}\right), \forall t \neq 0$ and $\vec{r}(0) = \vec{0}$.

【Solution】
(1)
$\because \vec{r}'(t) = (3t^2, (1-t)e^{-t}, 2\cos 2t) \neq \vec{0}$ and $\vec{r}'(t)$ is continuous, $\forall t \in R \quad \therefore C$ is smooth
(2)
$\because \vec{r}'(t) = (-3\sin t, 4\cos t) \neq \vec{0}$ and $\vec{r}'(t)$ is continuous, $\forall t \in R \quad \therefore C$ is smooth
(3)
$\because \vec{r}'(t) = (e^t(\sin t + \cos t), e^t(\cos t - \sin t), -2e^{2t}) \neq \vec{0}$ and $\vec{r}'(t)$ is continuous, $\forall t \in R$
$\therefore C$ is smooth
(4)
$\because \vec{r}'(t) = (1 + \ln t, 5e^t(\cos t - \sin t), -\sin t - \cos t) \neq \vec{0}$ and $\vec{r}'(t)$ is continuous, $\forall t \in R$
$\therefore C$ is smooth
(5)

$\because \vec{r}'(t) = (0, 3\cos t, 2\sin t \cos t)$ $\quad \therefore \vec{r}'\left(\dfrac{\pi}{2}\right) = \vec{0}$ $\quad \therefore C$ is not smooth

(6)

$\because \vec{r}'(t) = \left(e^t(\sin t + \cos t), -e^t(\sin t + \cos t)\right) = \left(e^t \sin\left(t + \dfrac{\pi}{4}\right), -e^t \sin\left(t + \dfrac{\pi}{4}\right)\right)$

$\therefore \vec{r}'\left(-\dfrac{\pi}{4}\right) = \vec{0}$ $\quad \therefore C$ is not smooth

(7)

$\because x(t) = t^3 - 12t + 17, \quad y(t) = t^2 - 4t + 8$

$\therefore x'(t) = 3t^2 - 12, \quad y'(t) = 2t - 4 \Rightarrow x'(2) = y'(2) = 0$

$\therefore C$ is not a smooth curve

(8)

$\because y(t) = t^2 \sin\dfrac{1}{t}$ $\quad \therefore y'(t) = -\cos\dfrac{1}{t} + 2t \sin\dfrac{1}{t}, \quad \forall t \neq 0$

$\because y'(0) = 0$ and $\lim\limits_{t \to 0} y'(t) = \lim\limits_{t \to 0} -\cos\dfrac{1}{t} + 2t \sin\dfrac{1}{t} \neq 0$

$\therefore y'(t)$ isn't continuous at $t = 0$ $\quad \therefore \vec{r}'(t)$ isn't continuous at $t = 0$

$\therefore C$ is not smooth

9.3 Line Integrals

The integrand in a line integral can be either a scalar function or a vector function, and the curve C can be a curve in a two-dimensional plane or three-dimensional space. The types usually include parameterized curves, circles, arcs, ellipses, etc. Regardless of the case, the key step is to convert it into a single-variable integral, so it is important to clearly understand the form of the integrand after conversion. The curve C can be further classified as closed or open based on its appearance. When the curve is a closed curve in a two-dimensional plane, sometimes Green's Theorem can be used to convert the line integral into a double integral. When the curve is a closed curve in three-dimensional space, Stokes' Theorem can sometimes be used to convert the line integral into a surface integral.

9.3.1 For Scalar Functions, Find the Line Integral along Curve C

Consider a spatial curve C, described by the parametric equations
$$x = x(t), \quad y = y(t), \quad z = z(t), \quad \forall a \leq t \leq b$$
or represented by a vector-valued function
$$\vec{r}(t) = (x(t), y(t), z(t)), \quad \forall a \leq t \leq b.$$

9.3.1 For Scalar Functions, Find the Line Integral along Curve C

Assume C is a smooth spatial curve (where $\vec{r}'(t)$ is continuous and $\vec{r}'(t) \neq 0$). We divide the interval $[a, b]$ into n equal-width subintervals, where $x_i = x(t_i)$, $y_i = y(t_i)$, $z_i = z(t_i)$, $\forall 0 \leq i \leq n$. These corresponding points (x_i, y_i, z_i) divide C into n subarcs C_i with lengths $\Delta s_1, \Delta s_1 \ldots \Delta s_n$ (the lengths $\Delta s_1, \Delta s_2, \ldots, \Delta s_n$ do not necessarily have to be equal). For each subarc, we choose an arbitrary point (x_i^*, y_i^*, z_i^*). If the domain of the function f includes the spatial curve C, then the value of the function f at the point (x_i^*, y_i^*, z_i^*) is multiplied by the length of the subarc Δs_i. Summing these products yields the Riemann sum of the form

$$\sum_{i=1}^{n} f(x_i^*, y_i^*, z_i^*)\Delta s_i.$$

Through the limit of this Riemann sum, we can define the line integral of a scalar function.

【Definition】

Assume the function f is defined on a smooth spatial curve C, where

$$C: x = x(t),\ y = y(t),\ z = z(t),\quad \forall a \leq t \leq b.$$

If the limit

$$\lim_{n \to \infty} \sum_{i=1}^{n} f(x_i^*, y_i^*, z_i^*)\Delta s_i \text{ exists,}$$

we denote it as $\int_C f(x, y, z)ds$ and refer to it as the line integral of the function f along the spatial curve C.

The line integral for a smooth planar curve is similar. Consider a two-dimensional curve C, described by the parametric equations

$$x = x(t),\ y = y(t),\quad \forall a \leq t \leq b$$

or represented by a vector-valued function

$$\vec{r}(t) = (x(t), y(t)),\quad \forall a \leq t \leq b.$$

The method of subdividing the planar curve C is similar to that in three-dimensional space (as shown in the diagram below). The value of the function f at the point (x_i^*, y_i^*) is multiplied by the length of the subarc Δs_i, and summing these yields the Riemann sum of the form

$$\sum_{i=1}^{n} f(x_i^*, y_i^*)\Delta s_i,$$

9.3.1 For Scalar Functions, Find the Line Integral along Curve C

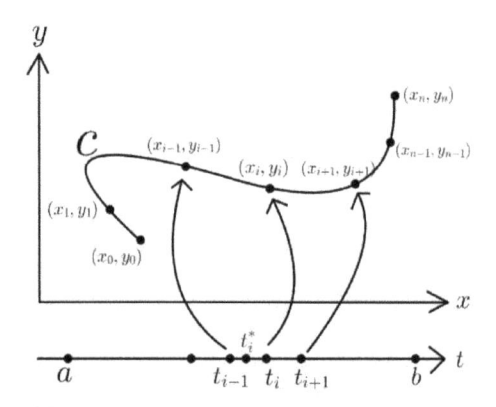

Through the limit of this Riemann sum, we can define the line integral of a scalar function.

【Definition】

Assume the function f is defined on a smooth planar curve C, where
$$C: x = x(t), \quad y = y(t), \quad \forall a \le t \le b.$$

If the limit
$$\lim_{n \to \infty} \sum_{i=1}^{n} f(x_i^*, y_i^*) \Delta s_i \text{ exists,}$$

we denote it as $\int_C f(x, y) ds$ and refer to it as the line integral of the function f along the planar curve C.

It is worth noting that any continuous function defined on a closed interval is Riemann integrable. Therefore, if the function f defined above is continuous, then the limit of the aforementioned Riemann sum must exist. Additionally, if the curve $C(t)$ is smooth, $\forall t \in [a, b]$, thrn $\vec{r}'(t)$ is continuous, $\forall a \le t \le b$. By the arc length formula,
$$\Delta s_i = \int_{t_{i-1}}^{t_i} |\vec{r}'(t)| dt.$$

Let $\Delta t_i = t_i - t_{i-1}$ approach zero, then
$$\Delta s_i = \int_{t_{i-1}}^{t_i} |\vec{r}'(t)| dt \approx |\vec{r}'(t_i^*)| \Delta t_i, \text{ where } t_i^* \in [t_i, t_{i-1}].$$

Thus, we have
$$\sum_{i=1}^{n} f(x_i^*, y_i^*) \Delta s_i = \sum_{i=1}^{n} f(\vec{r}(t_i^*)) \Delta s_i \approx \sum_{i=1}^{n} f(\vec{r}(t_i^*)) |\vec{r}'(t_i^*)| \Delta t_i.$$

Therefore,

$$\int_C f(x,y)ds = \lim_{n\to\infty}\sum_{i=1}^n f(x_i^*,y_i^*)\Delta s_i = \lim_{n\to\infty}\sum_{i=1}^n f(\vec{r}(t_i^*))|\vec{r}'(t_i^*)|\Delta t_i = \int_a^b f(\vec{r}(t))|\vec{r}'(t)|dt.$$

The above formula illustrates that computing the line integral of the function f along the curve C is equivalent to calculating the Riemann integral of a single-variable function. Notably, the integrand for the single-variable Riemann integral is

$$f(\vec{r}(t))|\vec{r}'(t)|.$$

Thus, finding the line integral along the curve involves first determining the integrand and then computing the single-variable Riemann integral. Below, we will provide a preliminary explanation of the different types of line integrals after converting them to single-variable Riemann integrals.

To grasp the key differences, it is essential to be familiar with the characteristics of the curves, which can be broadly classified into two categories: curves in a two-dimensional plane and curves in three-dimensional space. Furthermore, these can be subdivided into circular and non-circular curves, as well as whether the curves are expressed in parametric form. In these different contexts, there will be a clear corresponding computation formula for

$$\int_a^b f(\vec{r}(t))|\vec{r}'(t)|dt.$$

Assuming $\vec{r}(t) = (x(t), y(t))$, we have

$$\vec{r}'(t) = (x'(t), y'(t)) \Rightarrow |\vec{r}'(t)| = \sqrt{(x'(t))^2 + (y'(t))^2}.$$

Therefore, the line integral of the function f along the curve C can be rewritten as

$$\int_C f(x,y)ds = \int_a^b f(\vec{r}(t))|\vec{r}'(t)|dt = \int_a^b f(x(t), y(t))\sqrt{(x'(t))^2 + (y'(t))^2}dt.$$

Additionally, when given a two-dimensional curve $C: y = g(x), \forall x \in [a,b]$, we have
$$\vec{r}(x) = (x, g(x)) \Rightarrow \vec{r}'(x) = (1, g'(x))$$
and
$$|\vec{r}'(x)| = \sqrt{1 + (g'(x))^2}.$$

Thus, the line integral of the scalar function f along the curve C can be expressed as

$$\int_C f(x,y)ds = \int_a^b f(\vec{r}(x))|\vec{r}'(x)|dx = \int_a^b f(x, g(x))\sqrt{1 + (g'(x))^2}dx.$$

When considering the circle $C: x^2 + y^2 = r^2, r > 0$, we can parametrize it using polar coordinates

9.3.1 For Scalar Functions, Find the Line Integral along Curve C

Since
$$x = r\cos\theta, y = r\sin\theta \text{ then } \vec{r}(\theta) = (r\cos\theta, r\sin\theta), \ 0 \le \theta \le 2\pi.$$

$$\vec{r}'(\theta) = (-r\sin\theta, r\cos\theta) \text{ and } |\vec{r}'(\theta)| = r\sqrt{\cos^2\theta + \sin^2\theta} = r,$$

the line integral can be rewritten as

$$\int_C f(x,y)ds = \int_0^{2\pi} f(\vec{r}(\theta))|\vec{r}'(\theta)|d\theta = \int_0^{2\pi} f(r\cos\theta, r\sin\theta)rd\theta.$$

The above discusses the methods for calculating line integrals of two-dimensional curves under different circumstances. Next, we will explain the case of three-dimensionalcurves. Consider a parameterized smooth spatial curve

$$C: \vec{r}(t) = (x(t), y(t), z(t)), \ \forall a \le t \le b.$$

If f is a continuous function defined on the curve C in three variables, we can define the line integral of f along the spatial curve C using a method similar to that used for planar curves

$$\int_C f(x,y,z)ds = \lim_{n\to\infty} \sum_{i=1}^{n} f(x_i^*, y_i^*, z_i^*)\Delta s_i.$$

Using a formula analogous to that applied to planar curves, we compute the line integral of the spatial curve C as follows,

$$\int_C f(x,y,z)ds = \int_a^b f(\vec{r}(t))|\vec{r}'(t)|dt.$$

Since

$$\vec{r}'(t) = (x'(t), y'(t), z'(t)) \text{ and } |\vec{r}'(t)| = \sqrt{(x'(t))^2 + (y'(t))^2 + (z'(t))^2},$$

the line integral can be rewritten using the following formula:

$$\int_C f(x,y,z)ds = \int_a^b f(\vec{r}(t))|\vec{r}'(t)|dt$$
$$= \int_a^b f(x(t), y(t), z(t))\sqrt{(x'(t))^2 + (y'(t))^2 + (z'(t))^2}dt.$$

The above provides a clear method for calculating the line integral for a single smooth curve. In terms of exam types, the curve C may be a piecewise smooth curve (as shown in the diagram below). Next, we will explain how to define the line integral when the curve C is a piecewise smooth curve. Assume C is a piecewise smooth curve, meaning it is composed of a finite number of smooth segments $C_1, \ldots, C_n$, where the endpoint of C_i coincides with the starting point of C_{i+1}. Then, the line integral of f along C is the sum of the line integrals of f along each smooth segment C_j, $\forall 1 \leq j \leq n$, i.e.,

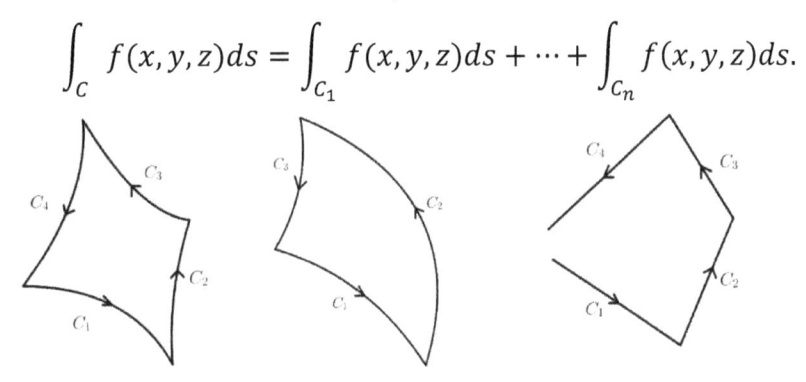

$$\int_C f(x,y,z)ds = \int_{C_1} f(x,y,z)ds + \cdots + \int_{C_n} f(x,y,z)ds.$$

Question Types:

Type 1.

Find $\int_C f(x,y)ds =?$, where the curve $C: y = g(x)$, $\forall x \in [a,b]$.

Problem-Solving Process:

Step1.

Let $\vec{r}(x) = (x, g(x))$ then $\vec{r}'(x) = (1, g'(x))$ and $|\vec{r}'(x)| = \sqrt{1 + (g'(x))^2}$

Step2.

$$\int_C f(x,y)ds = \int_a^b f(\vec{r}(x))|\vec{r}'(x)|dx = \int_a^b f(x, g(x))\sqrt{1 + (g'(x))^2}\,dx$$

Type 2.

$$\text{Find } \int_C f(x,y)ds =?,$$

where $C: (x(t), y(t))$, $\forall a \leq t \leq b$ is a line segment in two-dimensional space.

$$\text{Find } \int_C f(x,y,z)ds =?,$$

where $C: (x(t), y(t), z(t))$, $\forall a \leq t \leq b$ is a line segment in three-dimensional space.

Problem-Solving Process:

Step1.

If the curve C is a line segment in two-dimensional space:

Let $C: \vec{r}(t) = (x(t), y(t))$, $\forall a \leq t \leq b$ then

$\vec{r}'(t) = (x'(t), y'(t))$ and $|\vec{r}'(t)| = \sqrt{(x'(t))^2 + (y'(t))^2}$

9.3.1 For Scalar Functions, Find the Line Integral along Curve C 318

$$\therefore \int_C f(x,y)ds = \int_a^b f(\vec{r}(t))|\vec{r}'(t)|dt = \int_a^b f(x(t),y(t))\sqrt{(x'(t))^2 + (y'(t))^2}dt$$

If the curve C is a line segment in three-dimensional space:

Let $C: \vec{r}(t) = (x(t), y(t), z(t))$, $\forall a \leq t \leq b$ then

$$\vec{r}'(t) = (x'(t), y'(t), z'(t)) \text{ and } |\vec{r}'(t)| = \sqrt{(x'(t))^2 + (y'(t))^2 + (z'(t))^2}$$

$$\therefore \int_C f(x,y,z)ds = \int_a^b f(\vec{r}(t))|\vec{r}'(t)|dt$$

$$= \int_a^b f(x(t),y(t),z(t))\sqrt{(x'(t))^2 + (y'(t))^2 + (z'(t))^2}dt$$

Step2.

If the curve C is a line segment in two-dimensional space, compute

$$\int_a^b f(x(t),y(t))\sqrt{(x'(t))^2 + (y'(t))^2}dt$$

If the curve C is a line segment in three-dimensional space, compute

$$\int_a^b f(x(t),y(t),z(t))\sqrt{(x'(t))^2 + (y'(t))^2 + (z'(t))^2}dt$$

Examples:

(I) Find $\int_C f(x,y)ds =?$, where $f(x,y) = xy^3$, $C: \vec{r}(t) = (t, at)$, $-1 \leq t \leq 1$.

$\because \vec{r}(t) = (t, at) \quad \therefore \vec{r}'(t) = (1, a) \text{ and } |\vec{r}'(t)| = \sqrt{1 + a^2}$

$$\int_C f(x,y)ds = \int_{-1}^1 f(\vec{r}(t))|\vec{r}'(t)|dt = \int_{-1}^1 t(at)^3 \cdot \sqrt{1 + a^2}dt = \frac{2a^3\sqrt{1+a^2}}{5}$$

Type 3.

Find $\int_C f(x,y)ds =?$, where $C: x^2 + y^2 = r^2$, $r > 0$.

Problem-Solving Process:

Step1.

Let $x = r\cos\theta$, $y = r\sin\theta$, $0 \leq \theta \leq 2\pi$ then $\vec{r}'(\theta) = (-r\sin\theta, r\cos\theta)$

$\therefore |\vec{r}'(\theta)| = r\sqrt{\cos^2\theta + \sin^2\theta} = r$

$$\therefore \int_C f(x,y)ds = \int_0^{2\pi} f(\vec{r}(\theta))|\vec{r}'(\theta)|d\theta = \int_0^{2\pi} f(r\cos\theta, r\sin\theta)rd\theta$$

Step2.

Find $\int_0^{2\pi} f(r\cos\theta, r\sin\theta)rd\theta = ?$

Examples:

Find $\int_C f(x,y)ds = ?$, where $C: x^2 + y^2 = r^2$, $r > 0$.

(I) If $f(x,y) = e^{x^2+y^2}$ then $\int_C e^{x^2+y^2} ds = \int_0^{2\pi} e^{r^2} rd\theta$

(II) If $f(x,y) = \ln(x^2+y^2)$ then $\int_C \ln(x^2+y^2) ds = \int_0^{2\pi} \ln(r^2) rd\theta$

(III) If $f(x,y) = \sin(x^2+y^2)$ then $\int_C \sin(x^2+y^2) ds = \int_0^{2\pi} \sin(r^2) rd\theta$

(IV) If $f(x,y) = \sin^{-1}(x^2+y^2)$ then $\int_C \sin^{-1}(x^2+y^2) ds = \int_0^{2\pi} \sin^{-1}(r^2) rd\theta$

(V) If $f(x,y) = \tan^{-1}(x^2+y^2)$ then $\int_C \tan^{-1}(x^2+y^2) ds = \int_0^{2\pi} \tan^{-1}(r^2) rd\theta$

(VI) If $f(x,y) = \dfrac{1}{\sqrt{x^2+y^2}}$ then $\int_C \dfrac{1}{\sqrt{x^2+y^2}} ds = \int_0^{2\pi} \int_0^a drd\theta$

(VII) If $f(x,y) = \dfrac{1}{\sqrt{1+x^2+y^2}}$ then $\int_C \dfrac{1}{\sqrt{1+x^2+y^2}} ds = \int_0^{2\pi} \dfrac{r}{\sqrt{1+r^2}} d\theta$

Example 1.

Find $\int_C xy^2 dt = ?$, $C: x = 4t$, $y = e^t$, $0 \le t \le 2$.

【Solution】

$\int_C xy^2 dt = \int_0^2 4t \cdot e^{2t} dt = 4\left(\dfrac{e^{2t}t}{2}\bigg|_0^2 - \dfrac{1}{2}\int_0^2 e^{2t} dt\right) = 4\left(e^4 - \dfrac{e^4-1}{4}\right) = 3e^4 + 1$

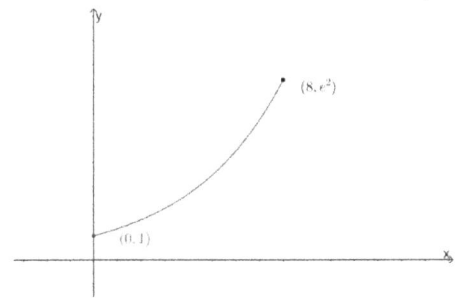

9.3.1 For Scalar Functions, Find the Line Integral along Curve C

Example 2.

Suppose $C: y = ax$ and $f(x, y) = xy^3$, find the line integral of f along C from point $(-1, -a)$ to point $(1, a)$, where $a > 0$, i.e., $\int_C f(x, y)ds =?$.

【Solution】

Let $\vec{r}(t) = (t, at), -1 \leq t \leq 1$ then $\vec{r}'(t) = (1, a)$ and $|\vec{r}'(t)| = \sqrt{1 + a^2}$

$$\therefore \int_C f(x,y)ds = \int_{-1}^{1} f(\vec{r}(t))|\vec{r}'(t)|dt = \int_{-1}^{1} t(at)^3 \cdot \sqrt{1 + a^2} dt = \frac{2a^3\sqrt{1 + a^2}}{5}$$

Example 3.

Find $\int_C 2 + x^2 y \, ds =?$, $C: x^2 + y^2 = r^2$, $y \geq 0$.

【Solution】

Let $\vec{r}(\theta) = (r\cos\theta, r\sin\theta)$, $0 \leq \theta \leq \pi$ then $\vec{r}'(\theta) = (-r\sin\theta, r\cos\theta)$ and $|\vec{r}'(\theta)| = r$

$$\int_C 2 + x^2 y \, ds = \int_0^{\pi} (2 + r^3 \cos^2\theta \sin\theta) r \, d\theta = 2\pi r - r^4 \cdot \left.\frac{\cos^3\theta}{3}\right|_0^{\pi} = 2\pi r + \frac{2r^4}{3}$$

Example 4.

Find $\int_C x - 3y^2 + z \, ds =?$, where C is the line segment from $(0,0,0)$ to $(1,4,3)$.

【Solution】

Let $\vec{r}(t) = (t, 4t, 3t)$, $0 \leq t \leq 1$ then $\vec{r}'(t) = (1,4,3)$ and $|\vec{r}'(t)| = \sqrt{26}$

$$\therefore \int_C x - 3y^2 + z \, ds = \int_0^1 (t - 3(4t)^2 + 3t) \cdot \sqrt{26} \, dt = -14\sqrt{26}$$

Example 5.

Find $\int_C 2x + y \, ds =?$, where C is the arc of the circle $x^2 + y^2 = 25$ from $(4,3)$ to $(3,4)$.

【Solution】

Let $\vec{r}(\theta) = (5\cos\theta, 5\sin\theta)$ then $\cos^{-1}\frac{4}{5} \leq \theta \leq \cos^{-1}\frac{3}{5}$ and $\sin^{-1}\frac{3}{5} \leq \theta \leq \sin^{-1}\frac{4}{5}$

$\therefore \vec{r}'(\theta) = (-5\sin\theta, 5\cos\theta)$ and $|\vec{r}'(\theta)| = 5\sqrt{\cos^2\theta + \sin^2\theta} \, d\theta = 5$

$$\int_C 2x + y \, ds = \int_{\sin^{-1}\frac{3}{5}}^{\sin^{-1}\frac{4}{5}} 2 \cdot 5\cos\theta \cdot |\vec{r}'(\theta)| d\theta + \int_{\cos^{-1}\frac{4}{5}}^{\cos^{-1}\frac{3}{5}} 5\sin\theta \cdot |\vec{r}'(\theta)| d\theta$$

$$= 5\int_{\sin^{-1}\frac{3}{5}}^{\sin^{-1}\frac{4}{5}} 10\cos\theta \, d\theta + 5\int_{\cos^{-1}\frac{4}{5}}^{\cos^{-1}\frac{3}{5}} 5\sin\theta \, d\theta = -25\cos\theta \Big|_{\cos^{-1}\frac{4}{5}}^{\cos^{-1}\frac{3}{5}} + 50\sin\theta \Big|_{\sin^{-1}\frac{3}{5}}^{\sin^{-1}\frac{4}{5}} = 15$$

Example 6.

Find $\int_C 3 + x^2 y \, ds = ?$, where C is the arc of the unit circle $x^2 + y^2 = 1$ in the first quadrant.

【Solution】

Let $\vec{r}(t) = (\cos t, \sin t)$ then

$\vec{r}'(t) = (-\sin t, \cos t)$ and $|\vec{r}'(t)| = \sqrt{\sin^2 t + \cos^2 t} = 1, 0 \leq t \leq \frac{\pi}{2}$

$$\therefore \int_C 3 + x^2 y \, ds = \int_0^{\frac{\pi}{2}} (3 + \cos^2 t \sin t) |\vec{r}'(t)| dt = \int_0^{\frac{\pi}{2}} (3 + \cos^2 t \sin t) dt = \left(3t - \frac{\cos^3 t}{3}\right)\Big|_0^{\frac{\pi}{2}}$$

$$= \frac{3\pi}{2} + \frac{1}{3}$$

Example 7.

Find $\int_C 2x \, ds = ?$, where C includes: C_1, the parabola $y = x^2$ from $(0,0)$ to $(2,4)$ and C_2, the vertical line from $(2,4)$ to $(2,5)$.

【Solution】

9.3.1 For Scalar Functions, Find the Line Integral along Curve C

∵ C_1: $y = x^2$, $0 \leq x \leq 2$

∴ $\int_{C_1} 2x \, ds = \int_0^2 2x \sqrt{\left(\dfrac{dx}{dx}\right)^2 + \left(\dfrac{dy}{dx}\right)^2} \, dx = \int_0^2 2x\sqrt{1+4x^2} \, dx = \dfrac{17\sqrt{17}-1}{6}$

∵ C_2: $x = 2$, $4 \leq y \leq 5$ ∴ $\int_{C_2} 2x \, ds = \int_4^5 2\sqrt{\left(\dfrac{dx}{dy}\right)^2 + \left(\dfrac{dy}{dy}\right)^2} \, dy = 2$

∴ $\int_C 2x \, ds = \int_{C_1} 2x \, ds + \int_{C_2} 2x \, ds = \dfrac{17\sqrt{17}-1}{6} + 2$

Example 8.

Find $\int_C y^2 \, dx + x \, dy$, (a) $C = C_1$: C_1, the line segment from $(-5,-3)$ to $(0,2)$

(b) $C = C_2$: C_2, the parabola $x = 4 - y^2$ from $(-5,-3)$ to $(0,2)$.

【Solution】

(a)

∵ C_1: $x(t) = 5t - 5$, $y(t) = 5t - 3$, $\forall 0 \leq t \leq 1$

∴ $\int_{C_1} y^2 \, dx + x \, dy = \int_0^1 (5t-3)^2 5 \, dt + (5t-5) 5 \, dt = -\dfrac{5}{6}$

(b)

∵ C_2: $x = 4 - y^2$, $\forall -3 \leq y \leq 2$

∴ $\int_{C_2} y^2 \, dx + x \, dy = \int_{-3}^2 y^2(-2y) \, dy + (4 - y^2) \, dy = \dfrac{245}{6}$

Example 9.

Find $\int_C xy \, dx + y \, dy$, where C: $x = 2t$, $y = 10t$, $0 \leq t \leq 2$.

【Solution】

Let $C: \vec{r}(t) = (2t, 10t)$, $\forall 0 \leq t \leq 2$ then

$\int_C xy \, dx + y \, dy = \int_0^2 20t^2 \cdot 2 \, dt + 10t \cdot 10 \, dt = \left(\dfrac{40t^3}{3} + 50t^2\right)\Big|_0^2 = \dfrac{320}{3} + 200 = \dfrac{920}{3}$

Example 10.

Find $\int_C \dfrac{y}{2x^2-y^2}\,ds$, where $C(t)=(t,t),\ 1\le t\le 3$.

【Solution】

Let $C:\vec{r}(t)=(t,t),\ \forall\,0\le t\le 3$ then $\vec{r}'(t)=(1,1)$ and $|\vec{r}'(t)|=\sqrt{2}$

$$\therefore \int_C \dfrac{y}{2x^2-y^2}ds = \int_1^3 \dfrac{y(t)|\vec{r}'(t)|}{2x^2(t)-y^2(t)}dt = \sqrt{2}\int_1^3 \dfrac{t}{t^2}dt = \sqrt{2}\int_1^3 \dfrac{1}{t}dt = \sqrt{2}\ln t\Big|_1^3 = \sqrt{2}\ln 3$$

Example 11.

Find $\int_C xyz\,ds = ?$, where $C: x=2\sin t,\ y=t,\ z=-2\cos t,\ \forall\,0\le t\le \pi$.

【Solution】

Let $C:\vec{r}(t)=(2\sin t,\ t,\ -2\cos t),\ \forall\,0\le t\le \pi$ then

$\vec{r}'(t)=(2\cos t,\ 1,\ 2\sin t)$ and $|\vec{r}'(t)|=\sqrt{(2\cos t)^2+(1)^2+(2\sin t)^2}=\sqrt{5}$

$$\therefore \int_C xyz\,ds = -4\int_0^\pi t\sin t\cos t\cdot|\vec{r}'(t)|dt = -4\sqrt{5}\int_0^\pi t\sin t\cos t\,dt = -2\sqrt{5}\int_0^\pi t\sin 2t\,dt$$

$$= 2\sqrt{5}\cdot\dfrac{t\cos 2t}{2}\Big|_0^\pi = \sqrt{5}\pi$$

Example 12.

Evaluate $\int_C x\cos z\,ds = ?$, where

(1) C is the circular helix given by the equation $x=\cos t,\ y=\sin t,\ z=t,\ 0\le t\le 2\pi$.
(2) C is the circular helix given by the equation $x=\cos t,\ y=\sin t,\ z=0,\ 0\le t\le 2\pi$.

【Solution】

(1)

$$\int_C x\cos z\,ds = \int_0^{2\pi} \cos t\cos t\sqrt{(x'(t))^2+(y'(t))^2+(z'(t))^2}\,dt$$

$$= \int_0^{2\pi} \cos^2 t\sqrt{\sin^2 t+\cos^2 t+1}\,dt = \sqrt{2}\int_0^{2\pi}\dfrac{1+\cos 2t}{2}dt = \sqrt{2}\pi$$

(2)

$$\int_C x\cos z\,ds = \int_0^{2\pi} \cos t\sqrt{(x'(t))^2+(y'(t))^2+(z'(t))^2}\,dt = \int_0^{2\pi} \cos t\sqrt{\sin^2 t+\cos^2 t}\,dt$$

9.3.1 For Scalar Functions, Find the Line Integral along Curve C

= 0

Example 13.

Find $\int_C ydx + zdy + xdz$, where C consists of: C_1, the line segment from $(3,0,0)$ to $(4,4,5)$ and C_2, the vertical line segment from $(4,4,5)$ to $(4,4,0)$.

【Solution】

$\because C_1 : \vec{r}(t) = (1-t)(3,0,0) + t(4,4,5) = (3+t, 4t, 5t), 0 \le t \le 1$

$\therefore \int_{C_1} ydx + zdy + xdz = \int_0^1 4tdt + 5t(4)dt + (3+t)5dt = \dfrac{59}{2}$

$\because C_2 : \vec{r}(t) = (1-t)(4,4,5) + t(4,4,0) = (4,4,5-5t), 0 \le t \le 1$

$\therefore \int_{C_2} ydx + zdy + xdz = \int_0^1 -20dt = -20$

$\therefore \int_C ydx + zdy + xdz = \dfrac{59}{2} - 20 = \dfrac{19}{2}$

Example 14.

Find $\int_C x - yds = ?$, where $\vec{r}(t) = (4t, 3t), \forall 0 \le t \le 4$.

【Solution】

$\because C : \vec{r}(t) = (4t, 3t), \forall 0 \le t \le 4 \quad \therefore \vec{r}'(t) = (4,3)$ and $|\vec{r}'(t)| = 5, \forall 0 \le t \le 4$

$\therefore \int_C x - yds = \int_0^4 (x(t) - y(t))|\vec{r}'(t)|dt = \int_0^4 t \cdot 5dt = \dfrac{5t^2}{2}\bigg|_0^4 = 40$

Example 15.

Find $\int_C xy^4 ds = ?$, where C is the right half of the circle $x^2 + y^2 = 16$, oriented counterclockwise.

【Solution】

Let $\vec{r}(t) = (4\cos t, 4\sin t)$ then $\vec{r}'(t) = (-4\sin t, 4\cos t)$ and $|\vec{r}'(t)| = 4, -\dfrac{\pi}{2} \le t \le \dfrac{\pi}{2}$

$$\therefore \int_C xy^4 ds = \int_{-\frac{\pi}{2}}^{\frac{\pi}{2}} x(t)y^4(t)|\vec{r}'(t)|dt = 4^5 \int_{-\frac{\pi}{2}}^{\frac{\pi}{2}} (\cos t \sin^4 t) \cdot 4 dt = 4^6 \int_{-\frac{\pi}{2}}^{\frac{\pi}{2}} \cos t \sin^4 t \, dt$$

$$= 4^6 \left(\frac{\sin^5 t}{5}\right)\Big|_{-\frac{\pi}{2}}^{\frac{\pi}{2}} = \frac{8192}{5}$$

Example 16.

Find $\int_C yzdx + xzdy + xydz = ?$, where C is the line segment from $(1,1,1)$ to $(3,2,0)$.

【Solution】

Let $C: \vec{r}(t) = (1-t)(1,1,1) + t(3,2,0) = (1+2t, t+1, 1-t), \ 0 \le t \le 1$

$$\therefore \int_C yzdx + xzdy + xydz = \int_0^1 (t+1)(1-t)2 + (1+2t)(1-t) - (1+2t)(t+1)dt$$

$$= \int_0^1 2 - 2t - 6t^2 dt = 2t - t^2 - 2t^3 \big|_0^1 = -1$$

Example 17.

Evaluate $\int_C x+y+z \, ds = ?$, where C is parameterized by $(\cos t, \sin t, t), t \in [0, 2\pi]$.

【Solution】

Let $\vec{r}(t) = (x(t), y(t), z(t)) = (\cos t, \sin t, t), \ \forall 0 \le t \le 2\pi$ then
$\vec{r}'(t) = (-\sin t, \cos t, 1)$ and $|\vec{r}'(t)| = \sqrt{2}, \ \forall 0 \le t \le 2\pi$

$$\therefore \int_C x+y+z \, ds = \int_0^{2\pi} (x(t)+y(t)+z(t))|\vec{r}'(t)| \, dt = \int_0^{2\pi} (\cos t + \sin t + t)\sqrt{2} \, dt$$

$$= 2\sqrt{2}\pi^2$$

Example 18.

Find $\int_C (x^2+y^2+z^2)^2 ds = ?$, where C is the space helix $(\cos t, \sin t, 3t)$, from $(1,0,0)$ to $(1,0,6\pi)$.

【Solution】

Let $\vec{r}(t) = (x(t), y(t), z(t)) = (\cos t, \sin t, 3t), \ \forall 0 \le t \le 2\pi$ then

$\vec{r}'(t) = (-\sin t, \cos t, 3)$ and $|\vec{r}'(t)| = \sqrt{10}$, $\forall 0 \le t \le 2\pi$

$$\therefore \int_C (x^2 + y^2 + z^2)^2 ds = \int_0^{2\pi} (x^2(t) + y^2(t) + z^2(t))^2 |\vec{r}'(t)| dt$$

$$= \int_0^{2\pi} ((\cos t)^2 + (\sin t)^2 + (3t)^2)^2 \sqrt{10} \, dt = \sqrt{10} \int_0^{2\pi} (1 + 9t^2)^2 \, dt$$

$$= \sqrt{10} \left(2\pi + 48\pi^3 + \frac{2592\pi^5}{5} \right)$$

Example 19.

Find $\int_C xe^{yz} ds = ?$, where C is the line segment from $(0,0,0)$ to $(1,2,3)$.

【Solution】

Let $C: \vec{r}(t) = (1-t)(0,0,0) + t(1,2,3) = (t, 2t, 3t)$, $0 \le t \le 1$ then
$\vec{r}'(t) = (1,2,3)$ and $|\vec{r}'(t)| = \sqrt{14}$

$$\therefore \int_C xe^{yz} ds = \int_0^1 x(t)e^{y(t)z(t)} |\vec{r}'(t)| dt = \int_0^1 te^{6t^2} \sqrt{14} dt = \frac{\sqrt{14}e^{6t^2}}{12}\bigg|_0^1 = \frac{\sqrt{14}(e^6 - 1)}{12}$$

Example 20.

Find $\int_C z^2 dx + x^2 dy + y^2 dz = ?$, where C is the line segment from $(1,0,0)$ to $(4,1,2)$.

【Solution】

Let $C: \vec{r}(t) = (1-t)(1,0,0) + t(4,1,2) = (1 + 3t, t, 2t)$, $0 \le t \le 1$ then

$$\int_C z^2 dx + x^2 dy + y^2 dz = \int_0^1 4t^2 \cdot 3 + (1 + 3t)^2 + t^2 \cdot 2 \, dt$$

$$= 4t^3 + (t + 3t^2 + 3t^3) + \frac{2t^3}{3}\bigg|_0^1 = \frac{35}{3}$$

9.3.2 For Vector Functions, Find the Line Integral along Curve C

Assume $\vec{F}(x, y, z) = (f_1(x, y, z), f_2(x, y, z), f_3(x, y, z))$ is a continuous vector field in three-dimensional space, and its domain includes a smooth spatial curve
$$C: \vec{r}(t) = (x(t), y(t), z(t)), \quad \forall t \in [a, b].$$
Divide the parameter interval $[a, b]$ into n equal-width subintervals, where $x_i = x(t_i)$, $y_i =$

$y(t_i)$, and $z_i = z(t_i)$, $\forall 0 \leq i \leq n$. These corresponding points (x_i, y_i, z_i) divide C into n sub-arcs C_i with lengths $\Delta s_1, \Delta s_1 \ldots \Delta s_n$ (where $\Delta s_1, \Delta s_2, \ldots, \Delta s_n$ need not be equal). At the i-th sub-arc, choose a point (x_i^*, y_i^*, z_i^*) corresponding to the parameter value t_i^*. From a physical perspective, as a particle moves along the curve C_i, if Δs_i is very small, it will be very close to the direction of the unit tangent vector $\vec{u}(t_i^*)$ (as shown in the figure), Therefore, the work done by the force $\vec{F}$ in moving the particle along the curve C_i from the starting point to the endpoint can be approximated as

$$\vec{F}(x_i^*, y_i^*, z_i^*) \cdot \Delta s_i \vec{u}(t_i^*) = \vec{F}(\vec{r}(t_i^*)) \cdot \Delta s_i \vec{u}(t_i^*).$$

Since the tangent vector of the curve is $\vec{r}'(t) = \dfrac{d\vec{r}(t)}{dt}$, the unit tangent vector is given by $\vec{u}(t) = \dfrac{\vec{r}'(t)}{|\vec{r}'(t)|}$. Thus,

$$\vec{F}(\vec{r}(t_i^*)) \cdot \Delta s_i \vec{u}(t_i^*) = \vec{F}(\vec{r}(t_i^*)) \cdot \Delta s_i \left(\dfrac{\vec{r}'(t_i^*)}{|\vec{r}'(t_i^*)|} \right).$$

Hence, the total work done in moving the particle along the curve can be approximated by

$$\sum_{i=1}^{n} \vec{F}(\vec{r}(t_i^*)) \cdot \Delta s_i \left(\dfrac{\vec{r}'(t_i^*)}{|\vec{r}'(t_i^*)|} \right).$$

Since $\Delta s_i = \displaystyle\int_{t_{i-1}}^{t_i} |\vec{r}'(t)| dt$, when $\Delta t_i = t_i - t_{i-1}$ approach zero,

$$\Delta s_i = \int_{t_{i-1}}^{t_i} |\vec{r}'(t)| dt \approx |\vec{r}'(t_i^*)| \Delta t_i.$$

Thus, the total work done in moving the particle along the curve can be approximated as

$$\sum_{i=1}^{n} \vec{F}(\vec{r}(t_i^*)) \cdot |\vec{r}'(t_i^*)| \Delta t_i \left(\dfrac{\vec{r}'(t_i^*)}{|\vec{r}'(t_i^*)|} \right) = \sum_{i-1}^{n} \left(\vec{F}(\vec{r}(t_i^*)) \cdot \vec{r}'(t_i^*) \right) \Delta t_i.$$

Therefore, the work W done by the force field $\vec{F}$ is defined as the limit of the above Riemann sum

$$W = \lim_{n \to \infty} \sum_{i=1}^{n} \left(\vec{F}(\vec{r}(t_i^*)) \cdot \vec{r}'(t_i^*) \right) \Delta t_i = \int_a^b \vec{F}(\vec{r}(t)) \cdot \vec{r}'(t) dt.$$

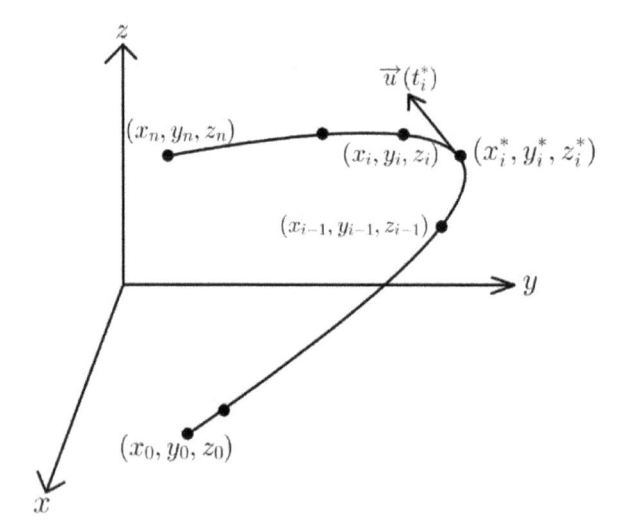

9.3.2 For Vector Functions, Find the Line Integral along Curve C

Based on the observations above, the following explains the definition of the line integral of a vector function.

【Definition】

Assume $\vec{F}$ is a continuous vector field defined along a smooth curve $C: \vec{r}(t), \forall t \in [a, b]$. The line integral of $\vec{F}$ along the curve C is expressed as

$$\int_C \vec{F} \cdot d\vec{r} = \int_a^b \vec{F}(\vec{r}(t)) \cdot \vec{r}'(t) dt.$$

The above formula shows that calculating the line integral of a vector function is equivalent to evaluating the Riemann integral of a single-variable function. Specifically, the integrand of the single-variable Riemann integral is

$$\vec{F}(\vec{r}(t)) \cdot \vec{r}'(t).$$

Therefore, to compute the line integral of a vector function, one must first determine the integrand and then calculate the single-variable Riemann integral.

The following will provide a preliminary explanation of the different types of single-variable Riemann integrals that arise after transforming the line integral of the vector function along a curve. It is important to note the relationship between the line integral of a vector function and that of a scalar function. Assume

$$\vec{F} = (f_1, f_2, f_3) \text{ and curve } C: \vec{r}(t) = (x, y, z).$$

Then

$$d\vec{r} = (dx, dy, dz),$$

which leads to,

9.3.2 For Vector Functions, Find the Line Integral along Curve C

$$\int_C \vec{F} \cdot d\vec{r} = \int_C (f_1, f_2, f_3) \cdot (dx, dy, dz) = \int_C f_1 dx + f_2 dy + f_3 dz.$$

Understanding the key aspects of the problem is crucial, similar to the previous section. one needs to be familiar with the shape of the curve. Broadly, curves can be divided into two categories: those in two-dimensional planes and those in three-dimensional space. Furthermore, they can be subdivided into circular and non-circular, as well as whether they are expressed using parametric equations. In these different contexts, there needs to be a clear corresponding formula for calculating $\int_a^b \vec{F}(\vec{r}(t)) \cdot \vec{r}'(t) dt$.

Given the vector function $\vec{F} = (f_1(x,y), f_2(x,y))$ and the curve C defined by the two-dimensional curve $y = g(x)$, $\forall a \le x \le b$, the goal is to compute the line integral

$$\int_C \vec{F} \cdot d\vec{r}.$$

First, we parameterize the curve. Let

$$\vec{r}(x) = (x, g(x))$$

then

$$\vec{r}'(x) = (1, g'(x)).$$

Therefore, the line integral can be expressed as

$$\int_C \vec{F} \cdot d\vec{r} = \int_a^b \vec{F}(\vec{r}(x)) \cdot \vec{r}'(x) dx = \int_a^b \vec{F}(\vec{r}(x)) \cdot (1, g'(x)) dx$$
$$= \int_a^b f_1(x, g(x)) dx + \int_a^b f_2(x, g(x)) g'(x) dx$$

In particular, if C is a circle defined by $x^2 + y^2 = r^2$, we can use polar coordinates. Let

$$x = r \cos \theta, \quad y = r \sin \theta$$

then

$$\vec{r}(\theta) = (r \cos \theta, r \sin \theta), \quad \forall \, 0 \le \theta \le 2\pi.$$

Since

$$\vec{r}'(\theta) = (-r \sin \theta, r \cos \theta),$$

the line integral $\int_C \vec{F} \cdot d\vec{r}$ can be expressed as the sum of two integrals

$$\int_C \vec{F} \cdot d\vec{r} = \int_0^{2\pi} \vec{F}(\vec{r}(\theta)) \cdot \vec{r}'(\theta) d\theta = \int_0^{2\pi} \left(f_1(\vec{r}(\theta)), f_2(\vec{r}(\theta)) \right) \cdot (-r \sin \theta, r \cos \theta) d\theta$$
$$= \int_0^{2\pi} f_1(r \cos \theta, r \sin \theta)(-r \sin \theta) d\theta + \int_0^{2\pi} f_2(r \cos \theta, r \sin \theta)(r \cos \theta) d\theta$$

9.3.2 For Vector Functions, Find the Line Integral along Curve C

Given the vector function in three-dimensional space
$$\vec{F} = (f_1(x,y,z), f_2(x,y,z), f_3(x,y,z)),$$
the goal is to compute the line integral
$$\int_C \vec{F} \cdot d\vec{r}$$
where the vector function of the curve C is given by $\vec{r}(t) = (x(t), y(t), z(t))$, $\forall a \leq t \leq b$.
Since
$$\vec{r}'(t) = (x'(t), y'(t), z'(t)),$$
we have
$$\int_C \vec{F} \cdot d\vec{r} = \int_a^b \vec{F}(\vec{r}(t)) \cdot \vec{r}'(t) dt$$
$$= \int_a^b f_1(\vec{r}(t)) x'(t) dt + \int_a^b f_2(\vec{r}(t)) y'(t) dt + \int_a^b f_3(\vec{r}(t)) z'(t) dt.$$

Given the vector-valued function $\vec{F} = (f_1(x,y,z), f_2(x,y,z), f_3(x,y,z))$ and the closed elliptical curve in three-dimensional space $C: \frac{x^2}{a^2} + \frac{y^2}{b^2} = 1$ with $z = c$, the curve can be expressed as $C = \left\{(x,y,z): \frac{x^2}{a^2} + \frac{y^2}{b^2} = 1, z = c\right\}$. By using polar coordinates, we set
$$x = a\cos\theta, \quad y = b\sin\theta$$
then
$$\vec{r}(\theta) = (a\cos\theta, b\sin\theta, c) \text{ and } \vec{r}'(\theta) = (-a\sin\theta, b\cos\theta, 0).$$
Thus, the line integral can be expressed as
$$\oint_C \vec{F} \cdot d\vec{r} = \int_0^{2\pi} \vec{F}(\vec{r}(\theta)) \cdot \vec{r}'(\theta) d\theta$$
$$= \int_0^{2\pi} f_1(a\cos\theta, b\sin\theta)(-a\sin\theta) d\theta + \int_0^{2\pi} f_2(a\cos\theta, b\sin\theta) b\cos\theta \, d\theta$$

The above provides the formulas for calculating the line integral when $\vec{F}$ is a vector-valued function, under different circumstances for the curve C. However, for different types of exam questions, it is essential to provide a detailed problem-solving process.

Question Types:

Type 1.

Given the vector-valued function $\vec{F} = (f_1(x,y), f_2(x,y))$, suppose the curve C is a line segment in the two-dimensional plane along $y = g(x)$, where $a \leq x \leq b$.

9.3.2 For Vector Functions, Find the Line Integral along Curve C

$$\text{Find} \int_C \vec{F} \cdot d\vec{r} = ?$$

Problem-Solving Process:

Step1.

Let $\vec{r}(x) = (x, g(x))$ then $\vec{r}'(x) = (1, g'(x))$

$$\therefore \int_C \vec{F} \cdot d\vec{r} = \int_a^b \vec{F}(\vec{r}(x)) \cdot \vec{r}'(x)dx = \int_a^b f_1(x, g(x))dx + \int_a^b f_2(x, g(x))g'(x)dx$$

Step2.

Find $\int_a^b f_1(x, g(x))dx + \int_a^b f_2(x, g(x))g'(x)dx = ?$

Examples:

(I) Find $\int_C \vec{F} \cdot d\vec{r} = ?$,

where $\vec{F} = (y \cos x, x \sin y)$ and the curve $C = \{(x, y): 0 \leq x \leq a, y = x\}$.

$$\int_C \vec{F} \cdot d\vec{r} = \int_C y \cos x \, dx + x \sin y \, dy = \int_a^0 x \cos x \, dx + x \sin x \, dx$$
$$= -(a+1)\sin a + (a-1)\cos a + 1$$

Type 2.

Given the vector function $\vec{F} = (f_1(x, y, z), f_2(x, y, z), f_3(x, y, z))$, find $\int_C \vec{F} \cdot d\vec{r} = ?$,

where the curve C is definde as: $\vec{r}(t) = (x(t), y(t), z(t))$, $\forall a \leq t \leq b$.

Problem-Solving Process:

Step1.

$$\int_C \vec{F} \cdot d\vec{r} = \int_a^b f_1(\vec{r}(t))x'(t)dt + \int_a^b f_2(\vec{r}(t))y'(t)dt + \int_a^b f_3(\vec{r}(t))z'(t)dt$$

Step2.

Find $\int_a^b f_1(\vec{r}(t))x'(t)dt + \int_a^b f_2(\vec{r}(t))y'(t)dt + \int_a^b f_3(\vec{r}(t))z'(t)dt = ?$

Examples:

Find $\int_C \vec{F} \cdot d\vec{r} = ?$

(I)If $\vec{F} = (x^2 y, x - z, xyz)$ and $\vec{r}(t) = (t, t^2, 2)$, $0 \leq t \leq 2$

9.3.2 For Vector Functions, Find the Line Integral along Curve C 332

then $\int_C \vec{F} \cdot d\vec{r} = \int_0^2 \vec{F}(\vec{r}(t)) \cdot \vec{r}'(t)dt = \int_0^2 (t^2 \cdot t^2 + (t-2) \cdot 2t + 2t^3 \cdot 0) \, dt = \dfrac{56}{15}$

(II) If $\vec{F} = (x^3, 3zy^2, x^2y)$ and $\vec{r}(t) = (3t, 2t, t)$, $-1 \le t \le 0$

then $\int_C \vec{F} \cdot d\vec{r} = \int_{-1}^0 \vec{F}(\vec{r}(t)) \cdot \vec{r}'(t)dt = \int_{-1}^0 (27t^3 \cdot 3 + 3t(2t)^2 \cdot 2 + 9t^2 \cdot 2t) \, dt = \dfrac{-123}{4}$

(III) If $\vec{F} = (xy, yz, xz)$ and $\vec{r}(t) = (t, t^2, t^3)$, $0 \le t \le 1$

then $\int_C \vec{F} \cdot d\vec{r} = \int_0^1 \vec{F}(\vec{r}(t)) \cdot \vec{r}'(t)dt = \int_{-1}^0 t^3 + 5t^6 \, dt = \dfrac{27}{28}$

(IV) If $\vec{F} = (x^2, -xy)$ and $\vec{r}(t) = (\cos t, \sin t), 0 \le t \le \dfrac{\pi}{2}$

then $\int_C \vec{F} \cdot d\vec{r} = \int_0^{\frac{\pi}{2}} \vec{F}(\vec{r}(t)) \cdot \vec{r}'(t)dt = \int_0^{\frac{\pi}{2}} (\cos^2 t, -\cos t \sin t) \cdot (-\sin t, \cos t)dt$

$= -2 \int_0^{\frac{\pi}{2}} \sin t \cos^2 t \, dt = -\dfrac{2}{3}$

Type 3.

Find $\oint_C \vec{F} \cdot d\vec{r} = ?$, where $\vec{F} = (f_1(x,y), f_2(x,y))$, and the curve C is a circle of radius r: $x^2 + y^2 = r^2$.

Problem-Solving Process:

Step1.

Let $x = r \cos \theta$, $y = r \sin \theta$, $0 \le \theta \le 2\pi$ then $\oint_C \vec{F} \cdot d\vec{r} = \int_C f_1(x,y)dx + f_2(x,y)dy$

Step2.

$\int_C f_1(x,y)dx = \int_0^{2\pi} f_1(r \cos \theta, r \sin \theta)(-r \sin \theta)d\theta$

$\int_C f_2(x,y)dy = \int_0^{2\pi} f_2(r \cos \theta, r \sin \theta)(r \cos \theta)d\theta$

Step3.

Find $\int_0^{2\pi} f_1(r \cos \theta, r \sin \theta)(-r \sin \theta)d\theta + \int_0^{2\pi} f_2(r \cos \theta, r \sin \theta)(r \cos \theta)d\theta = ?$

<u>Examples:</u>

Find $\oint_C \vec{F} \cdot d\vec{r} = ?$, where $\vec{F} = (f_1(x,y), f_2(x,y))$ and the curve $C: x^2 + y^2 = r^2$.

(I) If $f_1(x,y) = \dfrac{-y}{x^2 + y^2}$, $f_2(x,y) = \dfrac{x}{x^2 + y^2}$ then

$$\oint_C \vec{F} \cdot d\vec{r} = \int_0^{2\pi} \cos^2\theta + \sin^2\theta \, d\theta = 2\pi$$

(II) If $f_1(x,y) = \dfrac{x^2}{x^2 + y^2}$, $f_2(x,y) = \dfrac{y^2}{x^2 + y^2}$ then

$$\oint_C \vec{F} \cdot d\vec{r} = r \int_0^{2\pi} -\cos^2\theta \sin\theta + \sin^2\theta \cos\theta \, d\theta = 0$$

(III) If $f_1(x,y) = y^3$, $f_2(x,y) = -x^3$ then

$$\oint_C \vec{F} \cdot d\vec{r} = \int_0^{2\pi} r^3 \sin^3\theta \cdot (-r\sin\theta) - r^3 \cos^3\theta \cdot (r\cos\theta) d\theta = -\dfrac{3r^4\pi}{2}$$

Type 4.

Find $\oint_C \vec{F} \cdot d\vec{r} = ?$, where $\vec{F} = (f_1(x,y,z), f_2(x,y,z), f_3(x,y,z))$ and the curve C is a closed circle in three-dimensional space: $x^2 + y^2 = r^2$, $z = c$.

Problem-Solving Process:

Step 1.
Let $C = \{(x,y,z): x^2 + y^2 = r^2, z = c\}$
Let $x = r\cos\theta$, $y = r\sin\theta$ then
$\vec{r}(\theta) = (r\cos\theta, r\sin\theta, c)$ ∴ $\vec{r}'(\theta) = (-a\sin\theta, b\cos\theta, 0)$

Step 2.

$$\oint_C \vec{F} \cdot d\vec{r} = \int_0^{2\pi} \vec{F}(\vec{r}(\theta)) \cdot \vec{r}'(\theta) d\theta = \int_C (f_1, f_2, f_3) \cdot (-a\sin\theta, b\cos\theta, 0) d\theta$$

$$= \int_0^{2\pi} f_1(r\cos\theta, r\sin\theta)(-r\sin\theta) d\theta + \int_0^{2\pi} f_2(r\cos\theta, r\sin\theta) r\cos\theta \, d\theta$$

Step 3.

Find $\displaystyle\int_0^{2\pi} f_1(r\cos\theta, r\sin\theta)(-r\sin\theta) d\theta + \int_0^{2\pi} f_2(r\cos\theta, r\sin\theta) r\cos\theta \, d\theta = ?$

Examples:

Find $\oint_C \vec{F} \cdot d\vec{r} = ?$, where $\vec{F} = (f_1(x,y,z), f_2(x,y,z), f_3(x,y,z))$ and the curve C is a closed

9.3.2 For Vector Functions, Find the Line Integral along Curve C 334

circle in three-dimensional space: $x^2 + y^2 = r^2$, $z = c$.

(I)If $\vec{F} = (-y^3 \cos z, x^3 e^z, -e^z)$ then

$$\oint_C \vec{F} \cdot d\vec{r} = \int_0^{2\pi} \vec{F}(\vec{r}(\theta)) \cdot \vec{r}'(\theta) d\theta$$

$$= \int_0^{2\pi} (-(a \sin \theta)^3, (a \cos \theta)^3, 1) \cdot (-a\sin \theta, a \cos \theta, 0) d\theta = \frac{3a^4 \pi}{2}$$

Type 5.

Find $\oint_C \vec{F} \cdot d\vec{r} =?$, where $\vec{F} = (f_1(x,y,z), f_2(x,y,z), f_3(x,y,z))$ and the curve C is a closed ellipse in three-dimensional space: $\frac{x^2}{a^2} + \frac{y^2}{b^2} = 1$, $z = c$.

Problem-Solving Process:

Step1.

Let $C = \{(x,y,z): \frac{x^2}{a^2} + \frac{y^2}{b^2} = 1, z = c\}$

Let $x = a\cos \theta$, $y = b \sin \theta$ then $\vec{r}(\theta) = (a\cos \theta, b \sin \theta, c)$

$\therefore \vec{r}'(\theta) = (-a\sin \theta, b \cos \theta, 0)$

Step2.

$$\oint_C \vec{F} \cdot d\vec{r} = \int_0^{2\pi} \vec{F}(\vec{r}(\theta)) \cdot \vec{r}'(\theta) d\theta = \int_C (f_1, f_2, f_3) \cdot (-a\sin \theta, b \cos \theta, 0) d\theta$$

$$= \int_0^{2\pi} f_1(a\cos \theta, b \sin \theta)(-a\sin \theta) d\theta + \int_0^{2\pi} f_2(a\cos \theta, b \sin \theta) b\cos \theta \, d\theta$$

Step3.

Find $\int_0^{2\pi} f_1(a\cos \theta, b \sin \theta)(-a\sin \theta) d\theta + \int_0^{2\pi} f_2(a\cos \theta, b \sin \theta) b\cos \theta \, d\theta =?$

Examples:

Find $\oint_C \vec{F} \cdot d\vec{r} =?$, where $\vec{F} = (f_1(x,y,z), f_2(x,y,z), f_3(x,y,z))$ and the curve C is a closed ellipse in three-dimensional space: $\frac{x^2}{a^2} + \frac{y^2}{b^2} = 1$, $z = c$.

(I)If $\vec{F} = (2x - y, 2y + z, xyz)$ then

$$\oint_C \vec{F} \cdot d\vec{r} = \int_0^{2\pi} \vec{F}(\vec{r}(\theta)) \cdot \vec{r}'(\theta) d\theta$$

$$= \int_0^{2\pi} (2a\cos\theta - b\sin\theta, 2b\sin\theta, abc\cos\theta\sin\theta) \cdot (-a\sin\theta, b\cos\theta, 0) d\theta = ab\pi$$

Type 6.
The intersection of a quadric surface with the plane $z = c$ in space often results in an ellipse or a circle. Below is a summary of several cases where the intersection forms a circle or ellipse:

(i) C is the intersection of an ellipsoid $\dfrac{x^2}{a^2} + \dfrac{y^2}{b^2} + \dfrac{z^2}{c^2} = 1$ and $z = 0 \Rightarrow C: \dfrac{x^2}{a^2} + \dfrac{y^2}{b^2} = 1, z = 0$

(ii) C is the intersection of an ellipsoid $\dfrac{x^2}{a^2} + \dfrac{y^2}{b^2} + \dfrac{z^2}{c^2} = c_1$ and $\dfrac{x^2}{a^2} + \dfrac{y^2}{b^2} = c_2$ $(c_1 > c_2 > 0)$

$\Rightarrow C: \dfrac{x^2}{a^2} + \dfrac{y^2}{b^2} = c_2, \ z = c\sqrt{c_1 - c_2}$

(iii) C is the intersection of an elliptic paraboloid $\dfrac{x^2}{a^2} + \dfrac{y^2}{b^2} = \dfrac{z}{c}$ and $z = c$

$\Rightarrow C: \dfrac{x^2}{a^2} + \dfrac{y^2}{b^2} = 1, \ z = c$

(iv) C is the intersection of an elliptic cone $\dfrac{x^2}{a^2} + \dfrac{y^2}{b^2} = \dfrac{z^2}{c^2}$ and $z = c$

$\Rightarrow C: \dfrac{x^2}{a^2} + \dfrac{y^2}{b^2} = 1, \ z = c$

(v) C is the intersection of a hyperboloid $\dfrac{x^2}{a^2} + \dfrac{y^2}{b^2} - \dfrac{z^2}{c^2} = 1$ and $z = 0$

$\Rightarrow C: \dfrac{x^2}{a^2} + \dfrac{y^2}{b^2} = 1, \ z = 0$

Find $\oint_C \vec{F} \cdot d\vec{r} = ?$, where $\vec{F} = (f_1(x,y,z), f_2(x,y,z), f_3(x,y,z))$ and the curve C is the closed ellipse in three-dimensional space described above. "

Example 1.

$$\text{Find } \oint (6y + x) dx + (y + 2x) dy = ?, \ C: (x-2)^2 + (y-3)^2 = r^2.$$

【Solution】
Let $x = 2 + r\cos\theta, y = 3 + r\sin\theta, \ 0 \le \theta \le 2\pi$ then

$$\oint (6y + x) dx + (y + 2x) dy$$

9.3.2 For Vector Functions, Find the Line Integral along Curve C

$$= \int_0^{2\pi} (6(3 + r\sin\theta) + 2 + r\cos\theta)(-r\sin\theta) + (3 + r\sin\theta + 2(2 + r\cos\theta))r\cos\theta\, d\theta$$

$$= \int_0^{2\pi} -6r^2 \sin^2\theta + 2r^2 \cos^2\theta\, d\theta = \int_0^{2\pi} -6r^2\sin^2\theta + 2r^2(1 - \sin^2\theta)d\theta$$

$$= \int_0^{2\pi} -8r^2\sin^2\theta + 2r^2 d\theta = 4\pi r^2 - 8r^2 \int_0^{2\pi} \frac{1 - \cos 2\theta}{2} d\theta = -4\pi r^2$$

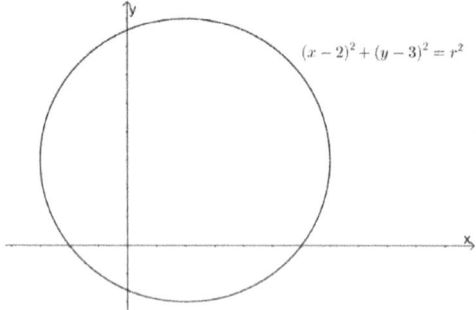

$(x-2)^2 + (y-3)^2 = r^2$

Example 2.

Find $\int_C x^2 y\, dx + (x - z)dy + xyz\, dz = ?$, where C follows $y = x^2$, $z = 2$, from point $(0,0,2)$ to point $(2,4,2)$.

【Solution】

Let $\vec{r}(t) = (t, t^2, 2), 0 \le t \le 2$ then $\vec{r}'(t) = (1, 2t, 0)$

$$\therefore \int_C x^2 y\, dx + (x - z)dy + xyz\, dz = \int_0^2 (t^2 \cdot t^2, t - 2, 2t^3) \cdot (1, 2t, 0)dt = \int_0^2 t^4 + 2t^2 - 4t\, dt$$

$$= \frac{t^5}{5} + \frac{2t^3}{3} - 2t^2 \Big|_0^2 = \frac{56}{15}$$

Example 3.

Find $\int_C x^3 dx + 3zy^2 dy + x^2 y\, dz = ?$, where C is the line segment from $(-3, -2, -1)$ to $(0,0,0)$.

【Solution】

Let $\vec{r}(t) = (3t, 2t, t)$, $-1 \le t \le 0$ then $\vec{r}'(t) = (3, 2, 1)$

$$\therefore \int_C x^3 dx + 3zy^2 dy + x^2 y\, dz = \int_{-1}^0 27t^3 \cdot 3dt + 3t(2t)^2 \cdot 2dt + 9t^2 \cdot 2t dt = \frac{-123}{4}$$

9.3.2 For Vector Functions, Find the Line Integral along Curve C

Example 4.

Find $\oint y^3 dx - x^3 dy = ?$, $C: x^2 + y^2 = r^2$.

【Solution】

Let $x = r\cos\theta, y = r\sin\theta, 0 \leq \theta \leq 2\pi$ then

$$\oint y^3 dx - x^3 dy = \int_0^{2\pi} r^3 \sin^3\theta \cdot (-r\sin\theta) - r^3 \cos^3\theta \cdot (r\cos\theta) d\theta$$

$$= -r^4 \int_0^{2\pi} \sin^4\theta + \cos^4\theta \, d\theta = -r^4 \int_0^{2\pi} (\sin^2\theta + \cos^2\theta)^2 - 2\sin^2\theta \cos^2\theta \, d\theta$$

$$= -r^4 \int_0^{2\pi} 1 - 2\left(\frac{\sin 2\theta}{2}\right)^2 d\theta = -r^4 \int_0^{2\pi} 1 - \frac{\sin^2 2\theta}{2} d\theta = -r^4 \int_0^{2\pi} 1 - \frac{1-\cos 4\theta}{4} d\theta$$

$$= -r^4 \cdot \frac{3}{4} \cdot 2\pi = -\frac{3r^4\pi}{2}$$

Example 5.

Suppose the curve C is defined by $\vec{r}(t) = (\cos t, \sin t)$, $0 \leq t \leq \frac{\pi}{2}$. Find the work done by the force field $\vec{F}(x, y) = (x, -y^2)$ along C.

【Solution】

$\because \vec{r}(t) = (\cos t, \sin t) \quad \therefore \vec{r}'(t) = (-\sin t, \cos t), \quad 0 \leq t \leq \frac{\pi}{2}$

$\therefore \int_C \vec{F} \cdot d\vec{r} = \int_0^{\frac{\pi}{2}} (\cos t, -\sin^2 t) \cdot \vec{r}'(t) dt = \int_0^{\frac{\pi}{2}} \cos t (-\sin t) - \sin^2 t \cdot \cos t \, dt = -\frac{1}{2} - \frac{1}{3}$

$= -\frac{5}{6}$

Example 6.

Find $\int_C \frac{-y}{x^2 + y^2} dx + \frac{x}{x^2 + y^2} dy = ?$, where C is the circle $x^2 + y^2 = r^2$.

【Solution】

Let $x = r\cos\theta$, $y = r\sin\theta$, $0 \leq \theta \leq 2\pi$ then

$$\int_C \frac{-y}{x^2 + y^2} dx + \frac{x}{x^2 + y^2} dy = \int_0^{2\pi} (-\sin\theta)(-\sin\theta) + \cos\theta \cdot \cos\theta \, d\theta$$

$$= \int_0^{2\pi} \sin^2\theta + \cos^2\theta \, d\theta = \int_0^{2\pi} d\theta = 2\pi$$

Example 7.

Find $\oint 2y\,dx + (x^2+y^2)\,dy = ?$, $C: x^2+(y-3)^2 = r^2$.

【Solution】

Let $x = r\cos\theta$, $y = 3+r\sin\theta$, $0 \le \theta \le 2\pi$ then

$$\oint 2y\,dx + (x^2+y^2)\,dy = \int_0^{2\pi} 2(3+r\sin\theta)(-r\sin\theta) + \big((9+r^2)+6r\sin\theta\big)r\cos\theta \, d\theta$$

$$= \int_0^{2\pi} -6r\sin\theta - 2r^2\sin^2\theta + (9+r^2)r\cos\theta + 6r^2\sin\theta\cos\theta \, d\theta$$

$$= \int_0^{2\pi} -2r^2\sin^2\theta + 6r^2\sin\theta\cos\theta \, d\theta = \int_0^{2\pi} -2r^2\left(\frac{1-\cos 2\theta}{2}\right) + 3r^2\sin 2\theta \, d\theta = -2r^2\pi$$

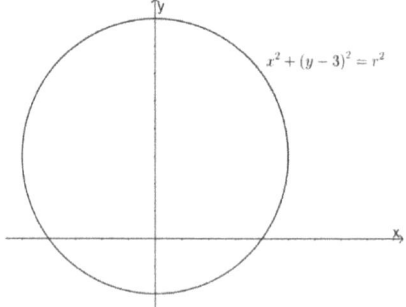

$x^2+(y-3)^2 = r^2$

Example 8.

Let $\vec{F} = \left(-z, x, \dfrac{y^2 z}{2}\right)$, $C: z = a, x^2+y^2 = a^2$. Find $\oint_C \vec{F} \cdot d\vec{r} = ?$

【Solution】

Let $\vec{r}(t) = (a\cos t, a\sin t, a)$, $0 \le t \le 2\pi$ then $\vec{r}'(t) = (-a\sin t, a\cos t, 0)$

$$\oint_C \vec{F}\cdot d\vec{r} = \int_0^{2\pi} \vec{F}(\vec{r}(t))\cdot \vec{r}'(t)\,dt = \int_0^{2\pi} \left(-a, a\cos t, \frac{a^3\sin^2 t}{2}\right)\cdot(-a\sin t, a\cos t, 0)\,dt$$

$$= \int_0^{2\pi} a^2\sin t + a^2\cos^2 t \, dt = 2\pi\cdot\frac{a^2}{2} = \pi a^2$$

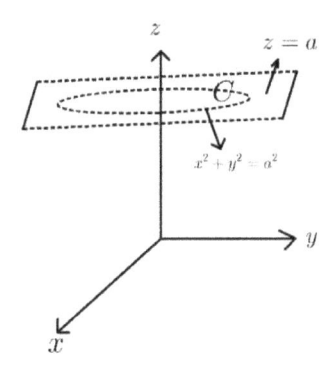

Example 9.

Let $\vec{F} = (0, -xz, -xy)$, $C: x^2 + z^2 \le 1$, $y = 1$. Find $\oint_C \vec{F} \cdot d\vec{r} = ?$

【Solution】

Let $\vec{r}(t) = (\cos t, 1, \sin t)$, $0 \le t \le 2\pi$ then $\vec{r}'(t) = (-\sin t, 0, \cos t)$

$$\therefore \oint_C \vec{F} \cdot d\vec{r} = \int_0^{2\pi} \vec{F}(\vec{r}(t)) \cdot \vec{r}'(t) \, dt = \int_0^{2\pi} (0, -\cos t \sin t, -\cos t) \cdot (-\sin t, 0, \cos t) \, dt$$

$$= -\int_0^{2\pi} \cos^2 t \, dt = -\pi$$

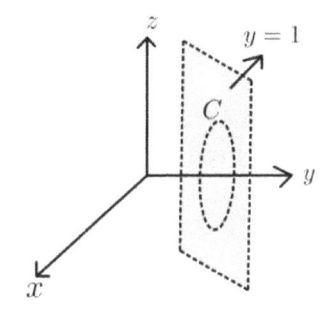

Example 10.

Find $\int_C \dfrac{-y}{x^2+y^2} dx + \dfrac{x}{x^2+y^2} dy = ?$, where C is the arc of the circle $x^2 + y^2 = 4$ from $(0,2)$ to $(-\sqrt{2}, -\sqrt{2})$.

【Solution】

Let $x = 2\cos\theta$, $y = 2\sin\theta$, $\dfrac{\pi}{2} \le \theta \le \dfrac{5\pi}{4}$ then

9.3.2 For Vector Functions, Find the Line Integral along Curve C

$$\int_C \frac{-y}{x^2+y^2} dx + \frac{x}{x^2+y^2} dy = \int_{\frac{\pi}{2}}^{\frac{5\pi}{4}} \frac{-2\sin\theta}{4} \cdot (-2\sin\theta) + \frac{2\cos\theta}{4} \cdot 2\cos\theta \, d\theta$$

$$= \int_{\frac{\pi}{2}}^{\frac{5\pi}{4}} \cos^2\theta + \sin^2\theta \, d\theta = \frac{3\pi}{4}$$

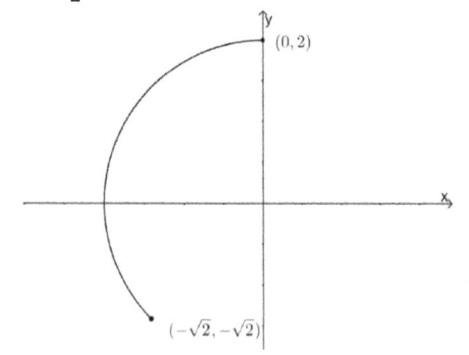

Example 11.

Find $\int_C \frac{x^2}{x^2+y^2} dx + \frac{y^2}{x^2+y^2} dy = ?$, where C is the arc of the circle $x^2+y^2=r^2$ from $(r, 0)$ to $\left(-\frac{r}{\sqrt{2}}, \frac{r}{\sqrt{2}}\right)$, $r > 0$.

【Solution】

Let $x = r\cos\theta$, $y = r\sin\theta$, $0 \le \theta \le \frac{3\pi}{4}$ then

$$\int_C \frac{x^2}{x^2+y^2} dx + \frac{y^2}{x^2+y^2} dy = \int_0^{\frac{3\pi}{4}} \frac{r^2\cos^2\theta}{r^2} \cdot (-r\sin\theta) + \frac{r^2\sin^2\theta}{r^2} \cdot r\cos\theta \, d\theta$$

$$= r\int_0^{\frac{3\pi}{4}} -\cos^2\theta\sin\theta + \sin^2\theta\cos\theta \, d\theta = \frac{r}{3}(\cos^3\theta + \sin^3\theta)\Big|_0^{\frac{3\pi}{4}} = -\frac{r}{3}$$

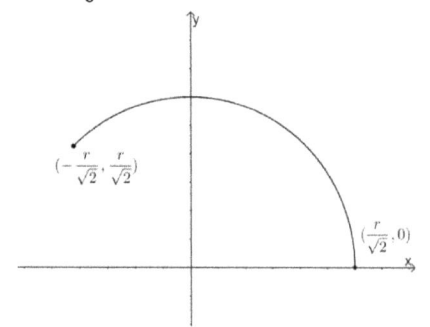

Example 12.

Find $\int_C yzdx - xzdy + xydz = ?$, $C: \vec{r}(t) = (e^t, e^{at}, e^{-t})$, $0 \leq t \leq 1$.

【Solution】

$\because \vec{r}(t) = (e^t, e^{at}, e^{-t})$ $\therefore \vec{r}'(t) = (e^t, ae^{at}, -e^{-t})$

$\therefore \int_C yzdx - xzdy + xydz = \int_0^1 e^{(a-1)t} \cdot e^t - ae^{at} - e^{at} dt = 1 - e^a$

Example 13.

Find $\int_C \vec{F} \cdot d\vec{r} = ?$, $\vec{F} = (z, x, y)$, $C: \vec{r}(t) = (0, 3\sin t, \sin^2 t)$, $0 \leq t \leq \dfrac{\pi}{2}$.

【Solution】

$\because \vec{r}(t) = (0, 3\sin t, \sin^2 t)$ $\therefore \vec{r}'(t) = (0, 3\cos t, 2\sin t \cos t)$

$\therefore \int_C \vec{F} \cdot d\vec{r} = \int_0^{\frac{\pi}{2}} (\sin^2 t, 0, 3\sin t) \cdot (0, 3\cos t, 2\sin t \cos t) dt = \int_0^{\frac{\pi}{2}} 6\sin^2 t \cos t \, dt$

$= 2\sin^3 t \big|_0^{\frac{\pi}{2}} = 2$

Example 14.

$\vec{F} = (3x - 4y + 2z, 4x + 2y - 3z^2, 2xz - 4y^2 + z^3)$, the elliptical curve

$C: \dfrac{x^2}{16} + \dfrac{y^2}{9} = 1$, $z = 0$. Find the work done along the upper half of the ellipse C,

$\int_C \vec{F} \cdot d\vec{r} = ?$

【Solution】

Let $\vec{r}(t) = (4\cos t, 3\sin t, 0)$, $0 \leq t \leq \pi$ then $\vec{r}'(t) = (-4\sin t, 3\cos t, 0)$

$\therefore \int_C \vec{F} \cdot d\vec{r} = \int_0^{\pi} \vec{F}(t) \cdot \vec{r}'(t) dt$

$= \int_0^{\pi} (12\cos t - 12\sin t, 16\cos t + 6\sin t)(-4\sin t, 3\cos t) dt$

$= \int_0^{\pi} (12\cos t - 12\sin t)(-4\sin t) + (16\cos t + 6\sin t)(3\cos t) dt$

$= 48 \int_0^{\pi} \sin^2 t + \cos^2 t \, dt = 48\pi$

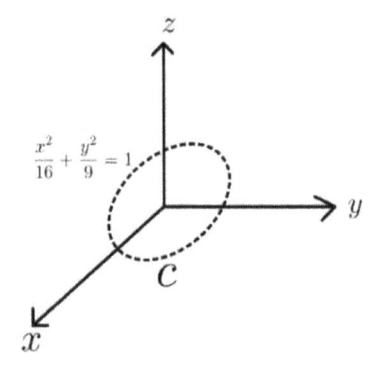

9.3.2 For Vector Functions, Find the Line Integral along Curve C

Example 15.

Let $\vec{F} = \left(0, \dfrac{2x^3}{3}, 2y^3\right)$. The surface S is defined by $x + y + z = 1$ in the first quadrant, and C is the counterclockwise boundary of S. Find $\oint_C \vec{F} \cdot d\vec{r} = ?$

【Solution】

Let $C_1: \vec{r}_1(t) = (1-t, t, 0)$, $C_2: \vec{r}_2(t) = (0, 1-t, t)$, $C_3: \vec{r}_3(t) = (t, 0, 1-t)$, $0 \le t \le 1$

then $\vec{r}_1'(t) = (-1, 1, 0)$, $\vec{r}_2'(t) = (0, -1, 1)$, $\vec{r}_3'(t) = (1, 0, -1)$

$$\therefore \oint_C \vec{F} \cdot d\vec{r} = \int_{C_1} \vec{F}(\vec{r}_1(t)) \cdot \vec{r}_1'(t) dt + \int_{C_2} \vec{F}(\vec{r}_2(t)) \cdot \vec{r}_2'(t) dt + \int_{C_3} \vec{F}(\vec{r}_3(t)) \cdot \vec{r}_3'(t) dt$$

$$= \dfrac{2}{3}\int_0^1 (1-t)^3 \, dt + 2\int_0^1 (1-t)^3 \, dt = \dfrac{2}{3}$$

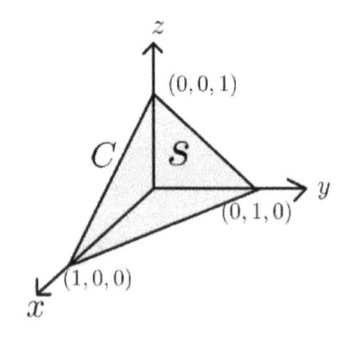

Example 16.

Let $\vec{F} = \left(\dfrac{z^2}{2}, \dfrac{x^2}{2}, \dfrac{y^2}{2}\right)$. The surface S is formed by points $(a, 0,0), (0, a, 0), (0,0, a)$ in the first quadrant, and C is the counterclockwise boundary of S. Find $\oint_C \vec{F} \cdot d\vec{r} = ?$

【Solution】

Let $C_1: \vec{r}_1(t) = (a - at, at, 0), C_2: \vec{r}_2(t) = (0, a - at, at),$ and $C_3: \vec{r}_3(t) = (at, 0, a - at),$

$0 \le t \le 1$ then $\vec{r}_1'(t) = (-a, a, 0), \vec{r}_2'(t) = (0, -a, a),$ and $\vec{r}_3'(t) = (a, 0, -a)$

$$\therefore \oint_C \vec{F} \cdot d\vec{r} = \int_{C_1} \vec{F}(\vec{r}_1(t)) \cdot \vec{r}_1'(t)dt + \int_{C_2} \vec{F}(\vec{r}_2(t)) \cdot \vec{r}_2'(t)dt + \int_{C_3} \vec{F}(\vec{r}_3(t)) \cdot \vec{r}_3'(t)dt$$

$$= \dfrac{a}{2}\int_0^1 (a - at)^3 \, dt \cdot 3 = \dfrac{a^3}{2}$$

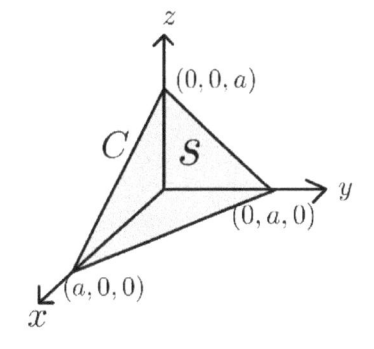

Example 17.

Suppose $\vec{F} = (0, e^x + 16xy, e^y + 6y^2),$ and the surface S is defined by $x + y + z = 6$ in the first quadrant, and C is the counterclockwise boundary of S. Find $\oint_C \vec{F} \cdot d\vec{r} = ?$

【Solution】

Let $C_1: \vec{r}_1(t) = (6 - 6t, 6t, 0), C_2: \vec{r}_2(t) = (0, 6 - 6t, 6t), C_3: \vec{r}_3(t) = (6t, 0, 6 - 6t)$

then $\vec{r}_1'(t) = (-6, 6, 0), \vec{r}_2'(t) = (0, -6, 6),$ and $\vec{r}_3'(t) = (6, 0, -6), 0 \le t \le 1$

$$\therefore \oint_C \vec{F} \cdot d\vec{r} = \int_{C_1} \vec{F}(\vec{r}_1(t)) \cdot \vec{r}_1'(t)dt + \int_{C_2} \vec{F}(\vec{r}_2(t)) \cdot \vec{r}_2'(t)dt + \int_{C_3} \vec{F}(\vec{r}_3(t)) \cdot \vec{r}_3'(t)dt$$

$$\therefore \int_{C_1} \vec{F}(\vec{r}_1(t)) \cdot \vec{r}_1'(t)dt = 6\int_0^1 e^{6-6t} + 96t(6-6t)\,dt = e^6 - 1 + 576$$

$$\int_{C_2} \vec{F}(\vec{r}_2(t)) \cdot \vec{r}_2'(t)dt = -6\int_0^1 dt + 6\int_0^1 e^{6-6t} + 6(6-6t)^2\,dt = -6 + e^6 - 1 + 72$$

$$= e^6 + 65$$

and $\int_{C_3} \vec{F}(\vec{r}_3(t)) \cdot \vec{r}_3'(t)dt = -6\int_0^1 dt = -6$

$$\therefore \oint_C \vec{F} \cdot d\vec{r} = 634 + 2e^6$$

Example 18.

Find $\int_C y\cos x\,dx + x\sin y\,dy = ?$, where C is the triangle formed by points $(0,0)$, $(a,0), (a,a)$, traversed in a counterclockwise direction, $a > 0$.

【Solution】

Let $C_1 = \{(x,y): 0 \leq x \leq a, y = 0\}$, $C_2 = \{(x,y): 0 \leq y \leq a, x = a\}$
and $C_3 = \{(x,y): 0 \leq x \leq a, y = x\}$ then

$$\int_C y\cos x\,dx + x\sin y\,dy$$

$$= \int_{C_1} y\cos x\,dx + x\sin y\,dy + \int_{C_2} y\cos x\,dx + x\sin y\,dy + \int_{C_3} y\cos x\,dx + x\sin y\,dy$$

$$\therefore \int_{C_1} y\cos x\,dx + x\sin y\,dy = 0,$$

$$\int_{C_2} y\cos x\,dx + x\sin y\,dy = a\int_0^a \sin y\,dy = a(-\cos a + 1),$$

$$\int_{C_3} y\cos x\,dx + x\sin y\,dy = \int_a^0 x\cos x\,dx + x\sin x\,dx = -(a+1)\sin a + (a-1)\cos a + 1$$

$$\therefore \oint y\cos x\,dx + x\sin y\,dy = -(a+1)\sin a - \cos a + a + 1$$

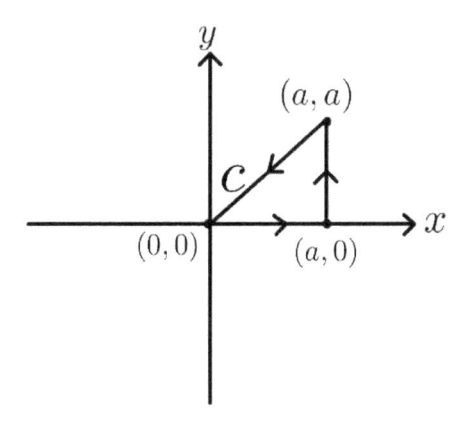

Example 19.

Find $\int_C -y^4 dx + xy^2 dy = ?$, where C is the triangle formed by points $(0,0)$, $(1,0)$, $(1,1)$, traversed in a counterclockwise direction.

【Solution】

Let $C_1 = \{(x, y): 0 \le x \le 1, y = 0\}$, $C_2 = \{(x, y): 0 \le y \le 1, x = 1\}$, and $C_3 = \{(x, y): 0 \le x \le 1, y = x\}$ then

$$\int_C -y^4 dx + xy^2 dy = \int_{C_1} -y^4 dx + xy^2 dy + \int_{C_2} -y^4 dx + xy^2 dy + \int_{C_3} -y^4 dx + xy^2 dy$$

$$\because \int_{C_1} -y^4 dx + xy^2 dy = 0, \quad \int_{C_2} -y^4 dx + xy^2 dy = \int_0^1 y^2 dy = \frac{1}{3}$$

and $\int_{C_3} -y^4 dx + xy^2 dy = \int_1^0 -x^4 dx + x^3 dy = \frac{-1}{20}$

$$\therefore \oint -y^4 dx + xy^2 dy = \frac{1}{3} - \frac{1}{20} = \frac{17}{60}$$

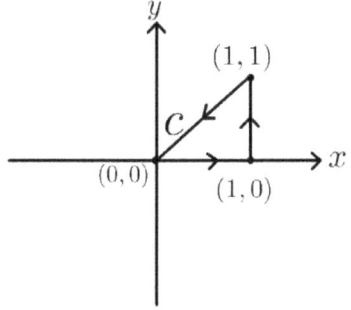

9.3.2 For Vector Functions, Find the Line Integral along Curve C

Example 20.

Suppose $\vec{F} = \left(x(x^2+y^2)^{-\frac{3}{2}}, y(x^2+y^2)^{-\frac{3}{2}}\right)$, and the curve C is defined by

$\vec{r}(t) = (e^t \sin t, e^t \cos t)$, $0 \le t \le 2$. Find $\int_C \vec{F} \cdot d\vec{r} = ?$

【Solution】

$\because \vec{r}(t) = (e^t \sin t, e^t \cos t) \quad \therefore \vec{r}'(t) = (e^t(\sin t + \cos t), e^t(\cos t - \sin t))$

$\because \vec{F} = \left(x(x^2+y^2)^{-\frac{3}{2}}, y(x^2+y^2)^{-\frac{3}{2}}\right)$ and $\int_C \vec{F} \cdot d\vec{r} = \int_0^2 \vec{F}(\vec{r}(t)) \cdot \vec{r}'(t) dt$

$\because \vec{F}(\vec{r}(t)) \cdot \vec{r}'(t) = e^t \sin t \cdot e^{-3t} \cdot e^t(\sin t + \cos t) + e^t \cos t \cdot e^{-3t} \cdot e^t(\cos t - \sin t) = e^{-t}$

$\therefore \int_0^2 \vec{F}(\vec{r}(t)) \cdot \vec{r}'(t) dt = \int_0^2 e^{-t} dt = 1 - e^{-2}$

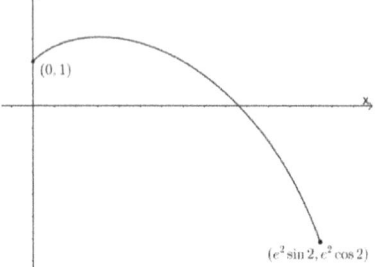

(0, 1)

$(e^2 \sin 2, e^2 \cos 2)$

Example 21.

Suppose the curve C is defined by $\vec{r}(t) = (t, t^2)$, $0 \le t \le 1$. Find the work done by the force field $\vec{F}(x, y) = (x^2, -y)$ along C.

【Solution】

$\because \vec{r}(t) = (t, t^2) \quad \therefore \vec{r}'(t) = (1, 2t)$

$\therefore \int_C \vec{F} \cdot d\vec{r} = \int_0^2 \vec{F}(\vec{r}(t)) \cdot \vec{r}'(t) dt = \int_0^2 (t^2, -t^2) \cdot (1, 2t) dt = \int_0^1 t^2 - 2t^3 \, dt = \left(\frac{t^3}{3} - \frac{t^4}{2}\right)\Big|_0^1$

$= -\frac{1}{6}$

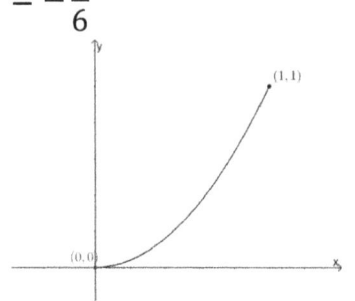

(1, 1)

(0, 0)

Example 22.

Suppose $\vec{F}(x,y,z) = (3x^2 + 6y, -14yz, 20xz^2)$. Find the line integral from $(0,0,0)$ to $(1,1,1)$ along the following paths: (1)$C: x = t, y = t^2, z = t^3$ (2)C: the straight line from $(0,0,0)$ to $(1,1,1)$.

【Solution】

(1)

Let $\vec{r}(t) = (t, t^2, t^3)$ then $\vec{r}'(t) = (1, 2t, 3t^2)$

$$\therefore \int_C \vec{F} \cdot d\vec{r} = \int_0^1 \vec{F}(\vec{r}(t)) \cdot \vec{r}'(t)dt = \int_C \vec{F}(t, t^2, t^3) \cdot (1, 2t, 3t^2)dt$$

$$= \int_0^1 (3t^2 + 6t^2) - 14t^5 \cdot 2t + 20t^7 \cdot 3t^2 dt = \int_0^1 9t^2 - 28t^6 + 60t^9 dt = 5$$

(2)

Let $\vec{r}(t) = (t, t, t)$ then $\vec{r}'(t) = (1,1,1)$

$$\int_C \vec{F} \cdot d\vec{r} = \int_0^1 \vec{F}(\vec{r}(t)) \cdot \vec{r}'(t)dt = \int_C \vec{F}(t,t,t) \cdot (1,1,1)dt = \int_0^1 (3t^2 + 6t) - 14t^2 + 20t^3 dt$$

$$= \int_0^1 20t^3 - 11t^2 + 6t\, dt = \frac{13}{3}$$

Example 23.

Find $\int_C (x^2 + y^2)dx - xdy = ?$, where C is the arc of the circle $x^2 + y^2 = r^2$, from $(r, 0)$ to $(0, r)$.

【Solution】

Let $x = r\cos\theta$, $y = r\sin\theta$, $0 \leq \theta \leq \frac{\pi}{2}$ then $x'(\theta) = -r\sin\theta$, $y'(\theta) = r\cos\theta$

$$\therefore \int_C (x^2 + y^2)dx - xdy = \int_0^{\frac{\pi}{2}} r^2(\cos^2\theta + \sin^2\theta)(-r\sin\theta) - r^2\cos^2\theta\, d\theta$$

$$= \int_0^{\frac{\pi}{2}} -r^3 \sin\theta - r^2 \cos^2\theta\, d\theta = -r^3 - \frac{\pi r^2}{4}$$

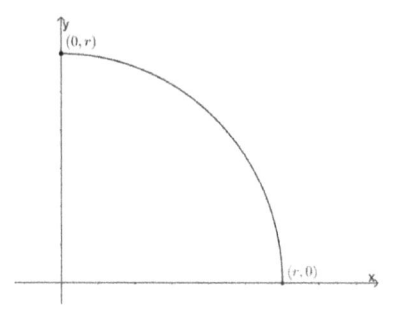

9.3.2 For Vector Functions, Find the Line Integral along Curve C

Example 24.

Find $\int_C x\,dx - z\,dy + y\,dz = ?$, where the curve C is defined by $\vec{r}(t) = (t^2, -t, t^2)$, $\forall\ 1 \le t \le 3$.

【Solution】

Let $\vec{r}(t) = (t^2, -t, t^2)$ then $\vec{r}'(t) = (2t, -1, 2t)$

$$\therefore \int_C x\,dx - z\,dy + y\,dz = \int_1^3 t^2 \cdot 2t\,dt - t^2 \cdot (-1)\,dt - t \cdot 2t\,dt = \left(\frac{t^4}{2} - \frac{t^3}{3}\right)\Big|_1^3 = \frac{81}{2} - 9 - \frac{1}{6}$$

$$= \frac{94}{3}$$

Example 25.

Find $\int_C yz^2\,dx + (xz^2 + ze^{yz})\,dy + \left(2xyz + ye^{yz} + \frac{1}{1+z}\right)dz = ?$, where the curve C is defined by $\vec{r}(t) = (t, t^2, t^3), 0 \le t \le 2$.

【Solution】

Let $\vec{r}(t) = (t, t^2, t^3)$ then $\vec{r}'(t) = (1, 2t, 3t^2)$

$$\therefore \int_C yz^2\,dx + (xz^2 + ze^{yz})\,dy + \left(2xyz + ye^{yz} + \frac{1}{1+z}\right)dz$$

$$= \int_0^2 t^8 + (t^7 + t^3 e^{t^5})\,2t + \left(2t^6 + t^2 e^{t^5} + \frac{1}{1+t^3}\right)3t^2\,dt$$

$$= \int_0^2 9t^8\,dt + 5\int_0^2 t^4 e^{t^5}\,dt + \int_0^2 \frac{3t^2}{1+t^3}\,dt = \left(t^9 + e^{t^5} + \ln(1+t^3)\right)\Big|_0^2 = 511 + e^{32} + 2\ln 3$$

Example 26.

Find $\int_C x^2 y dx + xy dy = ?$, where C is the segment of the curve $x = \sqrt{1-y^2}$ from $(1,0)$ to $(0,1)$.

【Solution】

$\because x = \sqrt{1-y^2} \quad \therefore dx = \frac{1}{2}(1-y^2)^{-\frac{1}{2}}(-2y)dy$

$\therefore \int_C x^2 y dx + xy dy = \int_0^1 (1-y^2) y \cdot \frac{1}{2}(1-y^2)^{-\frac{1}{2}}(-2y) + y\sqrt{1-y^2} dy$

$= \int_0^1 -y^2 \cdot (1-y^2)^{\frac{1}{2}} + y\sqrt{1-y^2} dy$

$\because \int_0^1 y\sqrt{1-y^2} dy = \left. \frac{-(1-y^2)^{\frac{3}{2}}}{3} \right|_0^1 = \frac{1}{3}$

Let $y = \sin\theta$ then $dy = \cos\theta\, d\theta$

$\therefore \int_0^1 -y^2 \cdot (1-y^2)^{\frac{1}{2}} dy = -\int_0^{\frac{\pi}{2}} \sin^2\theta \cdot \cos^2\theta\, d\theta = -\int_0^{\frac{\pi}{2}} \left(\frac{\sin 2\theta}{2}\right)^2 d\theta = -\frac{1}{4}\int_0^{\frac{\pi}{2}} \sin^2 2\theta\, d\theta$

$= -\frac{1}{4}\int_0^{\frac{\pi}{2}} \frac{1-\cos 4\theta}{2} d\theta = -\frac{\pi}{16}$

$\therefore \int_C x^2 y dx + xy dy = -\frac{\pi}{16} + \frac{1}{3}$

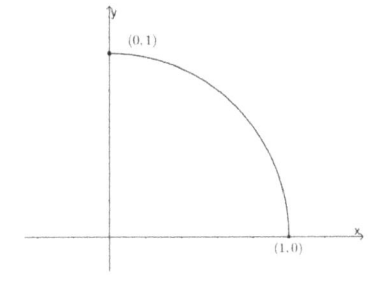

Example 27.

Let $\vec{F} = (-y^3 \cos z, x^3 e^z, -e^z)$. The curve C is the intersection of the surface $S: x^2 + y^2 + (z-a)^2 = 2a^2$ with $z = 0$. Find $\oint_C \vec{F} \cdot d\vec{r} = ?$

【Solution】

Let curve C be the counterclockwise closed curve formed by the intersection of surface S

with $z = 0$. Then $C = \{(x, y, z): x^2 + y^2 = a^2, z = 0\}$

Let $\vec{r}(\theta) = (a\cos\theta, a\sin\theta, 0)$ then $\vec{r}'(\theta) = (-a\sin\theta, a\cos\theta, 0)$

$$\therefore \oint_C \vec{F} \cdot d\vec{r} = \int_0^{2\pi} \vec{F}(\vec{r}(\theta)) \cdot \vec{r}'(\theta) \, d\theta$$

$$= \int_0^{2\pi} (-(a\sin\theta)^3, (a\cos\theta)^3, -1) \cdot (-a\sin\theta, a\cos\theta, 0) \, d\theta = a^4 \int_0^{2\pi} \sin^4\theta + \cos^4\theta \, d\theta$$

$\because \sin^4\theta + \cos^4\theta = (\sin^2\theta + \cos^2\theta)^2 - 2\sin^2\theta\cos^2\theta = 1 - 2(\sin\theta\cos\theta)^2$

$= 1 - 2\left(\dfrac{\sin 2\theta}{2}\right)^2 = 1 - \dfrac{\sin^2 2\theta}{2} = 1 - \dfrac{1 - \cos 4\theta}{4} = \dfrac{3 + \cos 4\theta}{4}$

$$\therefore \int_0^{2\pi} \sin^4\theta + \cos^4\theta \, d\theta = \int_0^{2\pi} \dfrac{3 + \cos 4\theta}{4} \, d\theta = \dfrac{3\pi}{2}$$

$$\therefore \oint_C \vec{F} \cdot d\vec{r} = \dfrac{3a^4\pi}{2}$$

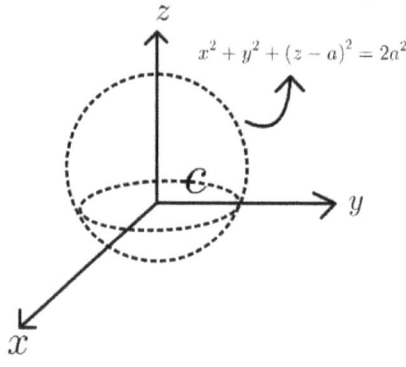

Example 28.

Let $\vec{F} = (2x - y, 2y + z, xyz)$. The curve C is the counterclockwise closed curve formed by the intersection of the surface S: $\dfrac{x^2}{a^2} + \dfrac{y^2}{b^2} + \dfrac{z^2}{c^2} = 1$ with $z = 0$.

Find $\oint_C \vec{F} \cdot d\vec{r} = ?$

【Solution】

Let curve C be the counterclockwise closed curve formed by the intersection of surface S with $z = 0$. Then, $C = \{(x, y, z): \dfrac{x^2}{a^2} + \dfrac{y^2}{b^2} = 1, z = 0\}$

Let $\vec{r}(\theta) = (a\cos\theta, b\sin\theta, 0)$ then $\vec{r}'(\theta) = (-a\sin\theta, b\cos\theta, 0)$

$$\oint_C \vec{F} \cdot d\vec{r} = \int_0^{2\pi} \vec{F}(\vec{r}(\theta)) \cdot \vec{r}'(\theta) \, d\theta$$

$$= \int_0^{2\pi} (2a \cos\theta - b \sin\theta, 2b \sin\theta, 0) \cdot (-a\sin\theta, b\cos\theta, 0) \, d\theta$$

$$= \int_0^{2\pi} -2a^2 \sin\theta \cos\theta + ab \sin^2\theta + 2b^2 \sin\theta \cos\theta \, d\theta$$

$$= \int_0^{2\pi} 2(b^2 - a^2) \sin\theta \cos\theta + ab \sin^2\theta \, d\theta = \int_0^{2\pi} (b^2 - a^2) \sin 2\theta + ab \left(\frac{1 - \cos 2\theta}{2} \right) d\theta$$

$$= ab\pi$$

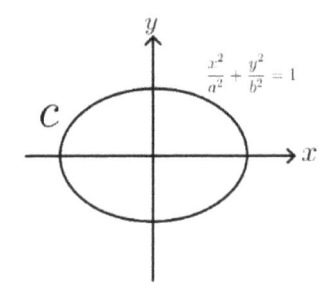

Example 29.

Let $\vec{F} = (xz, yz, xy)$. The curve C is the counterclockwise closed curve formed by the intersection of the surface $S: x^2 + y^2 + z^2 = 4$ $(z > 0)$ and $x^2 + y^2 = 1$.

Find $\oint_C \vec{F} \cdot d\vec{r} = ?$

【Solution】

Let C be the closed curve formed by the intersection of surface S and $x^2 + y^2 = 1$. Then
$C = \{(x, y, z): x^2 + y^2 = 1, z = \sqrt{3}\}$
Let $\vec{r}(\theta) = (\cos\theta, \sin\theta, \sqrt{3})$ then $\vec{r}'(\theta) = (-\sin\theta, \cos\theta, 0)$

$$\therefore \oint_C \vec{F} \cdot d\vec{r} = \int_0^{2\pi} \vec{F}(\vec{r}(\theta)) \cdot \vec{r}'(\theta) \, d\theta$$

$$= \int_0^{2\pi} (\sqrt{3} \cos\theta, \sqrt{3} \sin\theta, \sin\theta \cos\theta) \cdot (-\sin\theta, \cos\theta, 0) \, d\theta = \sqrt{3} \int_0^{2\pi} 0 \, d\theta = 0$$

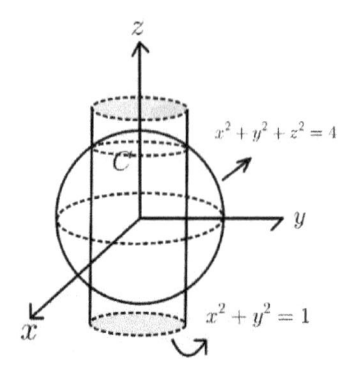

9.3.2 For Vector Functions, Find the Line Integral along Curve C

Example 30.

Let $\vec{F} = (2x - y, -yz^2, -zy^2)$. The curve C is the counterclockwise closed curve formed by the intersection of the sphere $x^2 + y^2 + z^2 = 1$ and the plane $z = 0$.

Find $\oint_C \vec{F} \cdot d\vec{r} = ?$

【Solution】

Let C be the closed curve formed by the intersection of the sphere and the plane $z = 0$.
Then, $C = \{(x, y, z): x^2 + y^2 = 1, z = 0\}$
Let $\vec{r}(\theta) = (\cos\theta, \sin\theta, 0)$ then $\vec{r}'(\theta) = (-\sin\theta, \cos\theta, 0)$

$$\therefore \oint_C \vec{F} \cdot d\vec{r} = \int_0^{2\pi} \vec{F}(\vec{r}(\theta)) \cdot \vec{r}'(\theta) \, d\theta = \int_0^{2\pi} (2\cos\theta - \sin\theta, 0, 0) \cdot (-\sin\theta, \cos\theta, 0) \, d\theta$$

$$= \int_0^{2\pi} -2\sin\theta\cos\theta + \sin^2\theta \, d\theta = \int_0^{2\pi} -2\sin\theta\cos\theta + \frac{1-\cos 2\theta}{2} \, d\theta = \pi$$

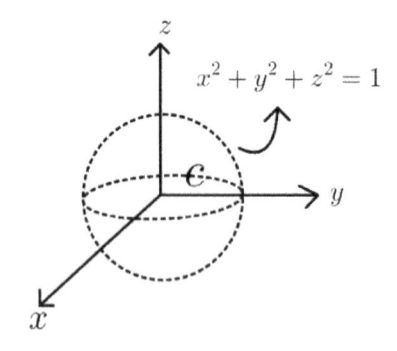

Example 31.

Let $\vec{F} = (y^2, x^2, 0)$. The curve C is the counterclockwise closed curve formed by the intersection of the surface $S: x^2 + y^2 \leq 1$ and the plane $z = 0$. Find $\oint_C \vec{F} \cdot d\vec{r} = ?$

【Solution】

Let C be the closed curve formed by the intersection of the surface S and the plane $z = 0$.
Then, $C = \{(x, y, z): x^2 + y^2 = 1, z = 0\}$
Let $\vec{r}(\theta) = (\cos\theta, \sin\theta, 0)$ then $\vec{r}'(\theta) = (-\sin\theta, \cos\theta, 0)$

$$\oint_C \vec{F} \cdot d\vec{r} = \int_0^{2\pi} \vec{F}(\vec{r}(\theta)) \cdot \vec{r}'(\theta)\, d\theta = \int_0^{2\pi} (\sin^2, \cos^2, 0) \cdot (-\sin\theta, \cos\theta, 0)$$

$$= \int_0^{2\pi} -\sin^3\theta + \cos^3\theta\, d\theta$$

$\because -\sin^3\theta + \cos^3\theta = (\cos\theta - \sin\theta)(\cos^2\theta + \sin\theta\cos\theta + \sin^2\theta)$
$= (\cos\theta - \sin\theta)(1 + \sin\theta\cos\theta) = (\cos\theta - \sin\theta) - \sin^2\theta\cos\theta + \sin\theta\cos^2\theta$

$$\therefore \int_0^{2\pi} -\sin^3\theta + \cos^3\theta\, d\theta = \int_0^{2\pi} (\cos\theta - \sin\theta) - \sin^2\theta\cos\theta + \sin\theta\cos^2\theta\, d\theta$$

$$= -\frac{\sin^3\theta}{3} + \frac{\cos^3\theta}{3}\Big|_0^{2\pi} = 0$$

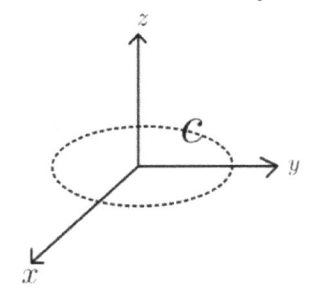

9.4 Green's Theorem

Green's Theorem is related to simply connected sets. Below, we first introduce the definition of a simply connected set.

【Definition】

(i) If the curve $\vec{r}(t)$, $\forall a \leq t \leq b$ satisfies $\vec{r}(t_1) \neq \vec{r}(t_2)$, $\forall a < t_1 < t_2 < b$ then it is called a simple curve.

(ii) Furthermore, if the simple curve satisfies $\vec{r}(a) = \vec{r}(b)$, then it is called a simple closed curve.

closed, simple

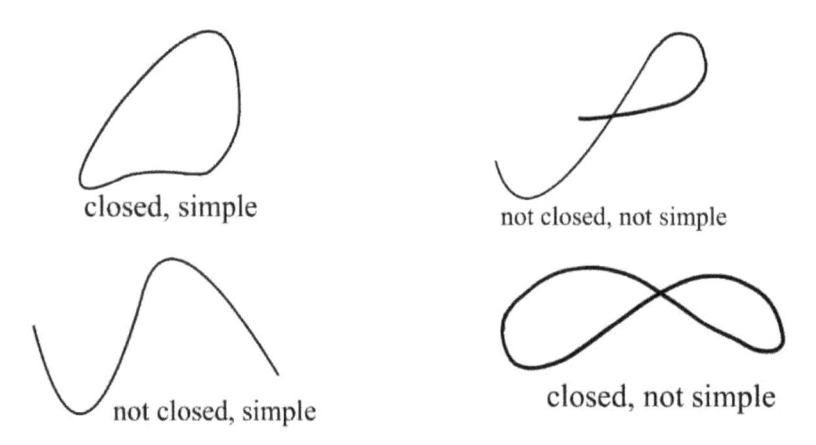

not closed, not simple

not closed, simple

closed, not simple

【Definition】

If every region enclosed by an arbitrary simple closed curve within a connected set D is contained entirely within D, then D is simply connected.

From the above definition, we can infer that the union of two disjoint open sets is not simply connected, or if there exists an open set within a set D that is not part of D, then D is not simply connected, as illustrated in the diagram below.

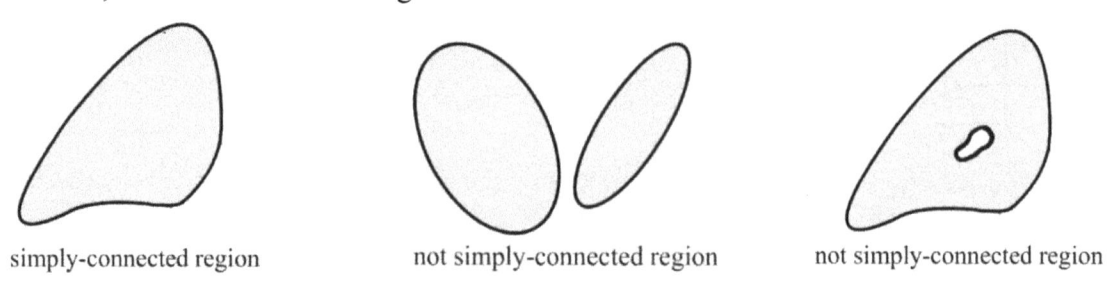

simply-connected region

not simply-connected region

not simply-connected region

    Next, we introduce Green's Theorem. When the first-order partial derivatives of the integrand exist and are continuous, Green's Theorem becomes an important tool for calculating line integrals along two-dimensional closed curves and double integrals. Moreover, Green's Theorem is a two-dimensional extension of the Fundamental Theorem of Calculus. This theorem links the line integral along a simple closed plane curve C to the double integral over the region R enclosed by the curve (as shown in the diagram below), facilitating the conversion between line integrals and double integrals. When the calculation of a line integral is difficult, this theorem allows it to be transformed into a double integral, and conversely, when double integrals are challenging to compute, they can be converted into line integrals using this theorem. Green's Theorem requires that the region D involved in the double integral be simply connected.

Importantly, the union of several disjoint simply connected sets is not simply connected. Therefore, Green's Theorem can be extended to apply to non-simply connected sets by first dividing the region into several simply connected subsets, applying Green's Theorem to each one individually, and then summing the results. Green's Theorem comes in two forms: the **curl form** and the **divergence form**. The rewritten expressions of both forms provide intuitive explanations for **Stokes' Theorem** and the **Divergence Theorem**. Additionally, in Green's Theorem, the positive orientation of a closed curve refers to the counterclockwise direction.

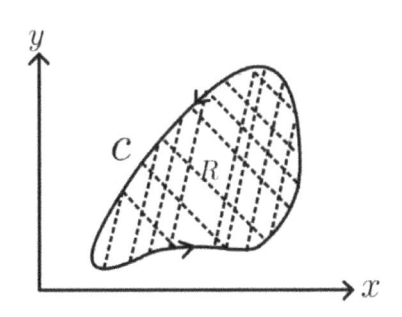

The types of exam questions related to Green's Theorem include: demonstrating that line integrals are path-independent, finding the area of a given closed region R, converting a closed line integral into a surface integral, converting a closed line integral into a surface integral combined with a coordinate transformation, converting a surface integral into a closed line integral, and constructing a closed region that does not contain singularities and then applying Green's Theorem. It is worth noting that, except for the first case, in all other types,

$$\frac{\partial g}{\partial x}(x, y) - \frac{\partial f}{\partial y}(x, y) \neq 0.$$

In general, if the problem involves a two-dimensional closed curve, it is likely an opportunity to apply Green's Theorem. One of the conditions for Green's Theorem to hold is that the functions f and g must have continuous first-order partial derivatives in the enclosed region R. If this condition is not met, such as when the first-order partial derivatives of f and g do not exist at the origin, we typically first construct a region R' where f and g have continuous first-order partial derivatives, and then let R' approach R, as illustrated in the diagram below.

9.4 Green's Theorem

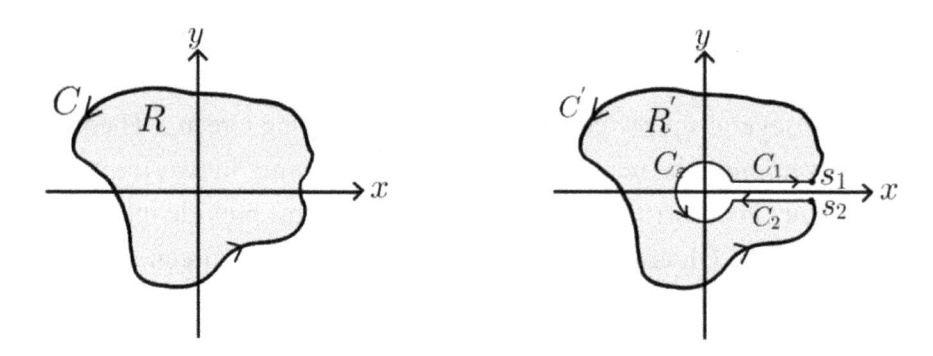

【Theorem】 Green's Theorem

Assume
1. C is a counterclockwise-oriented closed curve in two-dimensional space, enclosing a region R, and R is simply connected.
2. $f(x,y)$ and $g(x,y)$ are defined on an open set B, $R \subseteq B$ and the first-order partial derivatives of f and g exist and are continuous in B.

Then,
$$\iint_R \frac{\partial g}{\partial x}(x,y) - \frac{\partial f}{\partial y}(x,y) \, dxdy = \oint_C f(x,y) \, dx + g(x,y) \, dy.$$

Proof:

Let C be made up of C_1 and C_2, where $C_1: y = f_1(x)$, and $C_2: y = f_2(x)$, $a \leq x \leq b$

then $\oint_C f(x,y) \, dx = \oint_{C_1} f(x,y) \, dx - \oint_{C_2} f(x,y) \, dx$

Claim: $\iint_R \frac{\partial f}{\partial y}(x,y) \, dxdy = -\oint_C f(x,y) \, dx$

$\because \iint_R \frac{\partial f}{\partial y}(x,y) \, dxdy = \int_a^b \int_{f_1(x)}^{f_2(x)} \frac{\partial f}{\partial y}(x,y) \, dydx$

By Fundamental Theorem of Calculus,

$\int_{f_1(x)}^{f_2(x)} \frac{\partial f}{\partial y}(x,y) \, dy = f(x, f_2(x)) - f(x, f_1(x))$

$\therefore \iint_R \frac{\partial f}{\partial y}(x,y) \, dxdy = \int_a^b \int_{f_1(x)}^{f_2(x)} \frac{\partial f}{\partial y}(x,y) \, dydx = \int_a^b f(x, f_2(x)) - f(x, f_1(x)) \, dx$

$= \int_a^b f(x, f_2(x)) \, dx - \int_a^b f(x, f_1(x)) \, dx = -\left(\oint_{C_1} f(x,y) \, dx - \oint_{C_2} f(x,y) \, dx \right)$

$$= -\oint_C f(x,y)\, dx$$

Let C be made up of C_3 and C_4, where $C_3: y = g_1(y)$ and $C_4: y = g_2(y)$, $c \leq y \leq d$

then $\oint_C g(x,y)\, dx = \oint_{C_3} g(x,y)\, dx - \oint_{C_4} g(x,y)\, dx$

Claim: $\iint_R \dfrac{\partial g}{\partial x}(x,y)\, dxdy = \oint_C g(x,y)\, dy$

$\because \iint_R \dfrac{\partial g}{\partial x}(x,y)\, dxdy = \int_c^d \int_{g_2(y)}^{g_1(y)} \dfrac{\partial g}{\partial x}(x,y)\, dxdy$

By Fundamental Theorem of Calculus,

$\int_{g_2(y)}^{g_1(y)} \dfrac{\partial g}{\partial x}(x,y)\, dx = g(g_1(y), y) - g(g_2(y), y)$

$\therefore \iint_R \dfrac{\partial g}{\partial x}(x,y)\, dxdy = \int_c^d \int_{g_2(y)}^{g_1(y)} \dfrac{\partial g}{\partial x}(x,y)\, dxdy = \int_c^d g(g_1(y), y) - g(g_2(y), y)\, dy$

$= \int_c^d g(g_1(y), y)\, dy - \int_c^d g(g_2(y), y)\, dy = \oint_{C_3} g(x,y)\, dy - \oint_{C_4} g(x,y)\, dy = \oint_C g(x,y)\, dy$

Therefore,

$$\iint_R \dfrac{\partial g}{\partial x}(x,y) - \dfrac{\partial f}{\partial y}(x,y)\, dxdy = \oint_C f(x,y)\, dx + g(x,y) dy.$$

Green's Theorem can be rewritten in terms of curl ($\nabla \times \vec{F}$) and divergence(div $\vec{F}$). The reformulated expressions provide intuitive explanations for **Stokes' Theorem** and the **Divergence Theorem**. Suppose the closed curve in the plane is C, and the enclosed region is R, where the partial derivatives of the functions f and g exist and are continuous. Consider the vector field $\vec{F} = (f, g)$.

$\because \operatorname{curl} \vec{F} = \nabla \times \vec{F} = \begin{Vmatrix} i & j & k \\ \dfrac{\partial}{\partial x} & \dfrac{\partial}{\partial y} & \dfrac{\partial}{\partial z} \\ f & g & 0 \end{Vmatrix} = \left(\dfrac{\partial g}{\partial x} - \dfrac{\partial f}{\partial y} \right) k \qquad \therefore \dfrac{\partial g}{\partial x} - \dfrac{\partial f}{\partial y} = \left(\operatorname{curl} \vec{F} \right) \cdot k$

On the other hand, the line integral can be expressed as

$$\oint_C f dx + g dy = \oint_C \vec{F} \cdot d\vec{r}, \text{ where } \vec{F} = (f(x,y), g(x,y), 0).$$

By Green's Theorem,

9.4 Green's Theorem

$$\oint_C \vec{F} \cdot d\vec{r} = \oint_C f\,dx + g\,dy = \iint_R \frac{\partial g}{\partial x}(x,y) - \frac{\partial f}{\partial y}(x,y)\,dx\,dy = \iint_R (\nabla \times \vec{F}) \cdot \mathbf{k}\,dA.$$

Thus, the vector form of Green's Theorem is

$$\oint_C \vec{F} \cdot d\vec{r} = \iint_R (\nabla \times \vec{F}) \cdot \mathbf{k}\,dA.$$

For three-dimensional space, the intuitive extension is:

$$\oint_C \vec{F} \cdot d\vec{r} = \iint_S (\nabla \times \vec{F}) \cdot \vec{n}\,dA.$$

where S is an open surface in space and C is the boundary of S (a closed curve in space).

Next, we explain the divergence form of Green's Theorem. Suppose the curve C is represented by the vector equation

$$\vec{r}(t) = (x(t), y(t)), \quad a \leq t \leq b.$$

∵ the unit tangent vector $= \left(\dfrac{x'(t)}{|\vec{r}'(t)|}, \dfrac{y'(t)}{|\vec{r}'(t)|} \right)$

∴ the outward unit normal vector $\vec{n} = \left(\dfrac{y'(t)}{|\vec{r}'(t)|}, -\dfrac{x'(t)}{|\vec{r}'(t)|} \right)$

Therefore,

$$\oint_C \vec{F} \cdot \vec{n}\,ds = \int_a^b (\vec{F} \cdot \vec{n})(t)|\vec{r}'(t)|\,dt = \int_a^b \left(\frac{f(x(t),y(t))y'(t)}{|\vec{r}'(t)|} - \frac{g(x(t),y(t))x'(t)}{|\vec{r}'(t)|} \right) |\vec{r}'(t)|\,dt$$

$$= \int_a^b f(x(t),y(t))y'(t) - g(x(t),y(t))x'(t)\,dt = \int_C f\,dy - g\,dx$$

According to Green's Theorem,

$$\int_C f\,dy - g\,dx = \iint_R \frac{\partial f}{\partial x} + \frac{\partial g}{\partial y}\,dA = \iint_R \operatorname{div} \vec{F}\,dA.$$

Thus,

$$\oint_C \vec{F} \cdot \vec{n}\,ds = \iint_R \operatorname{div} \vec{F}\,dA.$$

For three-dimensional space, the intuitive extension is

$$\oiint_S \vec{F} \cdot \vec{n}\,dA = \iiint_V \nabla \cdot \vec{F}\,dV$$

where S is a closed surface in space, and V is the volume enclosed by S.

9.4.1 The Line Integral Is Path-Independent

The subsequent explanations of Stokes' Theorem and the Divergence Theorem are based on the rewritten form of Green's Theorem described above.

9.4.1 The Line Integral Is Path-Independent

In Green's Theorem, the functions f and g satisfy the condition that their first partial derivatives exist and are continuous. Using

$$\frac{\partial g}{\partial x}(x,y) - \frac{\partial f}{\partial y}(x,y) = 0,$$

we can prove that the line integral is independent of the path.

Question Types:

Type 1.

Given the functions $f(x,y), g(x,y),$ and two points (x_0, y_0) and $(x_1, y_1),$ prove that

$$\int_{(x_0,y_0)}^{(x_1,y_1)} f(x,y)dx + g(x,y)dy$$

is independent of the path and compute this integral.

Problem-Solving Process:

Step1.

Let C_1 and C_2 be any two curves connecting (x_0, y_0) to (x_1, y_1).
Let C be the closed curve formed by C_1 and C_2 in a counterclockwise direction, and let R be the enclosed region.

Step2.

Claim: $\frac{\partial g}{\partial x}(x,y) - \frac{\partial f}{\partial y}(x,y) = 0$

Step3.

By Green's Theorem, $\oint_C fdx + gdy = \iint_R \frac{\partial g}{\partial x}(x,y) - \frac{\partial f}{\partial y}(x,y)dxdy = 0$

$\therefore \int_{C_1} fdx + gdy - \int_{C_2} fdx + gdy = 0 \Rightarrow \int_{C_1} fdx + gdy = \int_{C_2} fdx + gdy$

$\therefore \int_{(x_0,y_0)}^{(x_1,y_1)} f(x,y)dx + g(x,y)dy$ is independent of the path.

Step4.

$\because \int_{(x_0,y_0)}^{(x_1,y_1)} f(x,y)dx + g(x,y)dy$ is path-independent

$\therefore \exists G(x,y)$ such that $\nabla G(x,y,z) = (f(x,y), g(x,y))$

9.4.1 The Line Integral Is Path-Independent

and $\int_{(x_0,y_0)}^{(x_1,y_1)} f dx + g dy = \int_{(x_0,y_0)}^{(x_1,y_1)} \nabla G \cdot (dx, dy) = \int_{(x_0,y_0)}^{(x_1,y_1)} dG = G(x_1, y_1) - G(x_0, y_0)$

Step 5.
Find $G(x, y) = ?$ and compute $G(x_1, y_1) - G(x_0, y_0) = ?$

Examples:
Given two points (x_0, y_0) and (x_1, y_1) on the plane,

prove that $\int_{(x_0,y_0)}^{(x_1,y_1)} f(x,y)dx + g(x,y)dy$ is independent of the path.

Let C_1 and C_2 be any two curves from (x_0, y_0) to (x_1, y_1).
Let C be the closed curve formed by C_1 and C_2 in a counterclockwise direction, and let R be the enclosed region.

(I) If $f(x,y) = 6xy^2 - y^3$, $g(x,y) = 6x^2y - 3xy^2$ then
$\frac{\partial g}{\partial x}(x,y) - \frac{\partial f}{\partial y}(x,y) = 12xy - 3y^2 - (12xy - 3y^2) = 0$

(II) if $f(x,y) = 2xy - y^4 + 3$, $g(x,y) = x^2 - 4xy^3$ then
$\frac{\partial g}{\partial x}(x,y) - \frac{\partial f}{\partial y}(x,y) = 2x - 4y^3 - (2x - 4y^3) = 0$

(III) If $f(x,y) = 4x^3 y$, $g(x,y) = x^4$ then
$\frac{\partial g}{\partial x}(x,y) - \frac{\partial f}{\partial y}(x,y) = 4x^3 - (4x^3) = 0$

By Green's Theorem, $\oint_C f dx + g dy = \iint_R \frac{\partial g}{\partial x}(x,y) - \frac{\partial f}{\partial y}(x,y) dx dy = 0$

$\therefore \int_{C_1} f dx + g dy - \int_{C_2} f dx + g dy = 0 \Rightarrow \int_{C_1} f dx + g dy = \int_{C_2} f dx + g dy$

$\therefore \int_{(x_0,y_0)}^{(x_1,y_1)} f(x,y)dx + g(x,y)dy$ is independent of the path.

Example 1.

Prove that the integral $\int_{(1,1)}^{(2,2)} (6xy^2 - y^3)dx + (6x^2y - 3xy^2)dy$ is path-independent, and calculate the value of this integral.

【Solution】

Let C_1 and C_2 be any two curves from $(1,1)$ to $(2,2)$.

Let C be the counterclockwise closed curve formed by C_1 and C_2, and R be the enclosed region formed by this curve.

Let $f(x,y) = 6xy^2 - y^3$, $g(x,y) = 6x^2y - 3xy^2$ then

$$\frac{\partial g}{\partial x}(x,y) - \frac{\partial f}{\partial y}(x,y) = 12xy - 3y^2 - (12xy - 3y^2) = 0$$

∵ the first-order partial derivatives of f and g exist and are continuous

By Green's Theorem, $\oint_C fdx + gdy = \iint_R \frac{\partial g}{\partial x}(x,y) - \frac{\partial f}{\partial y}(x,y)dxdy = 0$

∴ $\int_{C_1} fdx + gdy - \int_{C_2} fdx + gdy = 0 \Rightarrow \int_{C_1} fdx + gdy = \int_{C_2} fdx + gdy$

∴ $\int_{(1,1)}^{(2,2)} (6xy^2 - y^3)dx + (6x^2y - 3xy^2)dy$ is independent of the path

∴ $\exists G(x,y)$ s.t. $\nabla G(x,y) = (f(x,y), g(x,y))$

and $\int_{(1,1)}^{(2,2)} fdx + gdy = \int_{(1,1)}^{(2,2)} \nabla G \cdot (dx, dy) = \int_{(1,1)}^{(2,2)} dG = G(3,4) - G(1,2)$

∴ $\begin{cases} \frac{\partial G}{\partial x} = 6xy^2 - y^3 \\ \frac{\partial G}{\partial y} = 6x^2y - 3xy^2 \end{cases}$ ∴ $\begin{cases} G(x,y,z) = 3x^2y^2 - xy^3 + c \\ G(x,y,z) = 3x^2y^2 - xy^3 + c \end{cases}$

∴ $G(x,y,z) = 3x^2y^2 - xy^3 + c$

∴ $\int_{(1,1)}^{(2,2)} (6xy^2 - y^3)dx + (6x^2y - 3xy^2)dy = 3x^2y^2 - xy^3|_{(1,1)}^{(2,2)} = 30$

Example 2.

Find $\oint_C (x^2y \cos x + 2xy \sin x - y^2 e^x)dx + (x^2 \sin x - 2ye^x)dy = ?$, where

$C: x^{\frac{2}{3}} + y^{\frac{2}{3}} = a^{\frac{2}{3}}$ is a counterclockwise closed curve.

【Solution】

Let $f(x,y) = x^2y \cos x + 2xy \sin x - y^2 e^x$, $g(x,y) = x^2 \sin x - 2ye^x$ then

$$\frac{\partial g}{\partial x}(x,y) - \frac{\partial f}{\partial y}(x,y) = 2x \sin x + x^2 \cos x - 2ye^x - (x^2 \cos x + 2x \sin x - 2ye^x) = 0$$

∵ the first-order partial derivatives of f and g exist and are continuous

By Green's Theorem, $\oint_C fdx + gdy = \iint_R \frac{\partial g}{\partial x}(x,y) - \frac{\partial f}{\partial y}(x,y)dxdy = 0$

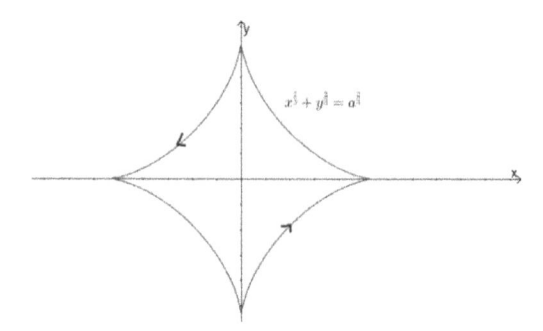

9.4.1 The Line Integral Is Path-Independent

$x^{\frac{2}{3}} + y^{\frac{2}{3}} = a^{\frac{2}{3}}$

Example 3.

Prove that the integral $\int_{(0,0)}^{(3,1)} (2xy - y^4 + 3)dx + (x^2 - 4xy^3)dy$ is path-independent, and calculate the value of this integral.

【Solution】

Let C_1 and C_2 be any two curves from $(0,0)$ to $(3,1)$.
Let C be the counterclockwise closed curve formed by C_1 and C_2, and R be the enclosed region formed by this curve.
Let $f(x,y) = 2xy - y^4 + 3$, $g(x,y) = x^2 - 4xy^3$ then

$$\frac{\partial g}{\partial x}(x,y) - \frac{\partial f}{\partial y}(x,y) = 2x - 4y^3 - (2x - 4y^3) = 0$$

∵ the first-order partial derivatives of f and g exist and are continuous

By Green's Theorem, $\oint_C f\,dx + g\,dy = \iint_R \frac{\partial g}{\partial x}(x,y) - \frac{\partial f}{\partial y}(x,y)\,dxdy = 0$

∴ $\int_{C_1} f\,dx + g\,dy - \int_{C_2} f\,dx + g\,dy = 0 \Rightarrow \int_{C_1} f\,dx + g\,dy = \int_{C_2} f\,dx + g\,dy$

∴ $\int_{(0,0)}^{(3,1)} (2xy - y^4 + 3)dx + (x^2 - 4xy^3)dy$ is independent of the path

∴ $\exists G(x,y)$ s. t. $\nabla G(x,y) = (f(x,y), g(x,y))$

and $\int_{(0,0)}^{(3,1)} f\,dx + g\,dy = \int_{(0,0)}^{(3,1)} \nabla G \cdot (dx, dy) = \int_{(0,0)}^{(3,1)} dG = G(3,1) - G(0,0)$

∵ $\begin{cases} \dfrac{\partial G}{\partial x} = 2xy - y^4 + 3 \\ \dfrac{\partial G}{\partial y} = x^2 - 4xy^3 \end{cases}$ ∴ $\begin{cases} G(x,y,z) = x^2y - xy^4 + 3x + c \\ G(x,y,z) = x^2y - xy^4 + c \end{cases}$

∴ $G(x,y,z) = x^2y - xy^4 + 3x + c$

$$\therefore \int_{(0,0)}^{(3,1)} (2xy - y^4 + 3)dx + (x^2 - 4xy^3)dy = x^2y - xy^4 + 3x\big|_{(0,0)}^{(3,1)} = 15$$

Example 4.

Prove that the integral $\int_{(0,0)}^{(a,b)} 4x^3 y dx + x^4 dy$ is path-independent, and calculate the value of this integral.

【Solution】

Let C_1 and C_2 be any two curves from $(0,0)$ to (a, b).
Let C be the counterclockwise closed curve formed by C_1 and C_2, and R be the enclosed region formed by this curve.

Let $f(x, y) = 4x^3 y$, $g(x, y) = x^4$ then $\dfrac{\partial g}{\partial x}(x, y) - \dfrac{\partial f}{\partial y}(x, y) = 4x^3 - (4x^3) = 0$

∵ the first-order partial derivatives of f and g exist and are continuous

By Green's Theorem, $\oint_C f dx + g dy = \iint_R \dfrac{\partial g}{\partial x}(x, y) - \dfrac{\partial f}{\partial y}(x, y) dx dy = 0$

$$\therefore \int_{C_1} f dx + g dy - \int_{C_2} f dx + g dy = 0 \Rightarrow \int_{C_1} f dx + g dy = \int_{C_2} f dx + g dy$$

$$\therefore \int_{(0,0)}^{(a,b)} 4x^3 y dx + x^4 dy \text{ is independent of the path}$$

∴ ∃$G(x, y)$ s.t. $\nabla G(x, y) = (f(x, y), g(x, y))$

and $\int_{(0,0)}^{(a,b)} f dx + g dy = \int_{(0,0)}^{(a,b)} \nabla G \cdot (dx, dy) = \int_{(0,0)}^{(a,b)} dG = G(a, b) - G(0,0)$

∵ $\begin{cases} \dfrac{\partial G}{\partial x} = 4x^3 y \\ \dfrac{\partial G}{\partial y} = x^4 \end{cases}$ ∴ $\begin{cases} G(x, y, z) = x^4 y + c \\ G(x, y, z) = x^4 y + c \end{cases}$

$\therefore G(x, y, z) = x^4 y + c$ ∴ $\int_{(0,0)}^{(a,b)} 4x^3 y dx + x^4 dy = x^4 y\big|_{(0,0)}^{(a,b)} = a^4 b$

Example 5.

Find $\int_{(0,0)}^{(-2,-1)} (10x^4 - 2xy^3)dx - 3x^2 y^2 dy = ?$, where the integration path is defined by $x^4 - 6xy^3 = 4y^2$.

9.4.1 The Line Integral Is Path-Independent

【Solution】

Let C_1 and C_2 be any two curves from $(0,0)$ to $(-2,-1)$.

Let C be the counterclockwise closed curve formed by C_1 and C_2, and R be the enclosed region formed by this curve.

Let $f(x,y) = 10x^4 - 2xy^3, g(x,y) = -3x^2y^2$ then

$$\frac{\partial g}{\partial x}(x,y) - \frac{\partial f}{\partial y}(x,y) = -6xy^2 - (-6xy^2) = 0$$

∵ the first-order partial derivatives of f and g exist and are continuous

By Green's Theorem, $\oint_C fdx + gdy = \iint_R \frac{\partial g}{\partial x}(x,y) - \frac{\partial f}{\partial y}(x,y) dxdy = 0$

$\therefore \int_{C_1} fdx + gdy - \int_{C_2} fdx + gdy = 0 \Rightarrow \int_{C_1} fdx + gdy = \int_{C_2} fdx + gdy$

$\therefore \int_{(0,0)}^{(-2,-1)} (10x^4 - 2xy^3)dx - 3x^2y^2dy$ is independent of the path

$\therefore \exists G(x,y)$ s.t. $\nabla G(x,y) = (f(x,y), g(x,y))$ and

$\int_{(0,0)}^{(-2,-1)} fdx + gdy = \int_{(0,0)}^{(-2,-1)} \nabla G \cdot (dx, dy) = \int_{(0,0)}^{(-2,-1)} dG = G(-2,-1) - G(0,0)$

$\therefore \begin{cases} \frac{\partial G}{\partial x} = 10x^4 - 2xy^3 \\ \frac{\partial G}{\partial y} = -3x^2y^2 \end{cases} \quad \therefore \begin{cases} G(x,y,z) = 2x^5 - x^2y^3 + c \\ G(x,y,z) = -x^2y^3 + c \end{cases} \quad \therefore G(x,y,z) = 2x^5 - x^2y^3 + c$

$\therefore \int_{(0,0)}^{(-2,-1)} (10x^4 - 2xy^3)dx - (3x^2y^2)dy = 2x^5 - x^2y^3 \big|_{(0,0)}^{(-2,-1)} = -60$

Example 6.

Find $\int_{(0,0)}^{(\frac{\pi}{2},1)} (x^2 + 6xy - 2y^2)dx + (3x^2 - 4xy + 2y)dy = ?$, where the integration path is defined by $y = \sin x$.

【Solution】

Let C_1 and C_2 be any two curves from $(0,0)$ to $\left(\frac{\pi}{2}, 1\right)$.

Let C be the counterclockwise closed curve formed by C_1 and C_2, and R be the enclosed region formed by this curve.

Let $f(x,y) = x^2 + 6xy - 2y^2$, $g(x,y) = 3x^2 - 4xy + 2y$ then

$$\frac{\partial g}{\partial x}(x,y) - \frac{\partial f}{\partial y}(x,y) = 6x - 4y - (6x - 4y) = 0$$

∵ the first-order partial derivatives of f and g exist and are continuous

By Green's Theorem, $\oint_C fdx + gdy = \iint_R \frac{\partial g}{\partial x}(x,y) - \frac{\partial f}{\partial y}(x,y)dxdy = 0$

$$\therefore \int_{C_1} fdx + gdy - \int_{C_2} fdx + gdy = 0 \Rightarrow \int_{C_1} fdx + gdy = \int_{C_2} fdx + gdy$$

$$\therefore \int_{(0,0)}^{(\frac{\pi}{2},1)} (x^2 + 6xy - 2y^2)dx + (3x^2 - 4xy + 2y)dy \text{ is independent of the path}$$

∴ ∃$G(x,y)$ s.t. $\nabla G(x,y) = (f(x,y), g(x,y))$

and $\int_{(0,0)}^{(\frac{\pi}{2},1)} fdx + gdy = \int_{(0,0)}^{(\frac{\pi}{2},1)} \nabla G \cdot (dx, dy) = \int_{(0,0)}^{(\frac{\pi}{2},1)} dG = G\left(\frac{\pi}{2}, 1\right) - G(0,0)$

$$\therefore \begin{cases} \frac{\partial G}{\partial x} = x^2 + 6xy - 2y^2 \\ \frac{\partial G}{\partial y} = 3x^2 - 4xy + 2y \end{cases} \quad \therefore \begin{cases} G(x,y,z) = \frac{x^3}{3} + 3x^2y - 2xy^2 + c \\ G(x,y,z) = 3x^2y - 2xy^2 + y^2 + c \end{cases}$$

$$\therefore G(x,y,z) = 3x^2y - 2xy^2 + y^2 + \frac{x^3}{3} + c$$

$$\therefore \int_{(0,0)}^{(\frac{\pi}{2},1)} (x^2 + 6xy - 2y^2)dx + (3x^2 - 4xy + 2y)dy = 3x^2y - 2xy^2 + y^2 + \frac{x^3}{3}\bigg|_{(0,0)}^{(\frac{\pi}{2},1)}$$

$$= 3\left(\frac{\pi}{2}\right)^2 - 2\left(\frac{\pi}{2}\right) + 1 + \frac{\left(\frac{\pi}{2}\right)^3}{3}$$

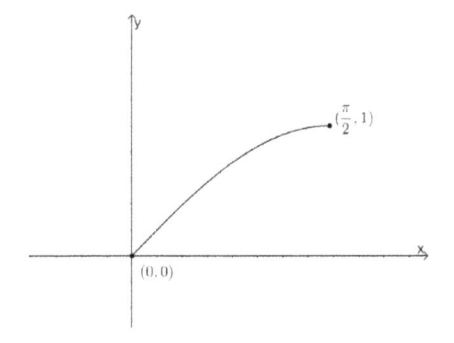

9.4.2 Find the Area of the Closed Region R

Question Types:

9.4.2 Find the Area of the Closed Region R

Type 1.
Assume C is the boundary of the closed region R, and we aim to find the area enclosed by R.

Problem-Solving Process:

Let $f(x,y) = -y$, $g(x,y) = x$. By Green's Theorem, the area enclosed by R is given by

$$\text{Area of } R = \iint_R dxdy = \frac{1}{2}\iint_R \frac{\partial x}{\partial x} - \frac{\partial(-y)}{\partial y} dxdy = \frac{1}{2}\oint_C -y\,dx + x\,dy$$

Calculate $\dfrac{1}{2}\oint_C -y\,dx + x\,dy = ?$

Remark:

The problem of finding the area of the closed region R is transformed into calculating

$\dfrac{1}{2}\oint_C -y\,dx + x\,dy = ?$, where C is the boundary of R.

Examples:

Assume R is a closed region, and C is its boundary. To find the area enclosed by R, we compute $\iint_R dxdy = ?$

Let $f(x,y) = -y$, $g(x,y) = x$. By Green's Theorem,

$$\iint_R dxdy = \frac{1}{2}\iint_R \frac{\partial x}{\partial x} - \frac{\partial(-y)}{\partial y} dxdy = \frac{1}{2}\oint_C -y\,dx + x\,dy$$

(I) If C: $\dfrac{x^2}{a^2} + \dfrac{y^2}{b^2} = 1$ then

$$\frac{1}{2}\oint_C -y\,dx + x\,dy = \frac{1}{2}\int_0^{2\pi}(-b\sin\theta)(-a\sin\theta) + (a\cos\theta)(b\cos\theta)d\theta = ab\pi$$

(II) If C: $r(\theta) = a(1 \pm \cos\theta)$ then

$$\frac{1}{2}\oint_C -y\,dx + x\,dy = \frac{1}{2}\oint_C r^2(\theta)\,d\theta = \frac{1}{2}\left(\int_0^{2\pi} a^2(1+\cos\theta)^2 d\theta\right) = \frac{3a^2\pi}{2}$$

(III) If C: $r(\theta) = a(1 \pm \sin\theta)$ then

$$\frac{1}{2}\oint_C -y\,dx + x\,dy = \frac{1}{2}\oint_C r^2(\theta)\,d\theta = \frac{1}{2}\left(\int_0^{2\pi} a^2(1 \pm \sin\theta)^2 d\theta\right) = \frac{3a^2\pi}{2}$$

Example 1.

Assume R is a closed region and C is the counterclockwise boundary of R. Use Green's Theorem to find the area enclosed by R.

【Solution】

Let $f(x,y) = -y$, $g(x,y) = x$,

∵ the first-order partial derivatives of f and g exist and are continuous

By Green's Theorem,

the area enclosed by R = $\iint_R dxdy = \frac{1}{2}\iint_R \frac{\partial x}{\partial x} - \frac{\partial(-y)}{\partial y} dxdy = \frac{1}{2}\oint_C -y\,dx + x\,dy$

Example 2.

Use Green's Theorem to find the area enclosed by $\frac{x^2}{a^2} + \frac{y^2}{b^2} = 1$.

【Solution】

Let R: $\frac{x^2}{a^2} + \frac{y^2}{b^2} \leq 1$, and let C be the counterclockwise boundary of R.

The area enclosed by $R = \frac{1}{2}\oint_C -y\,dx + x\,dy$

Let $x = a\cos\theta$, $y = b\sin\theta$ and $0 \leq \theta \leq 2\pi$ then

$\frac{1}{2}\oint_C -y\,dx + x\,dy = \frac{1}{2}\int_0^{2\pi}(-b\sin\theta)(-a\sin\theta) + a\cos\theta\, b\cos\theta\, d\theta = ab\pi$

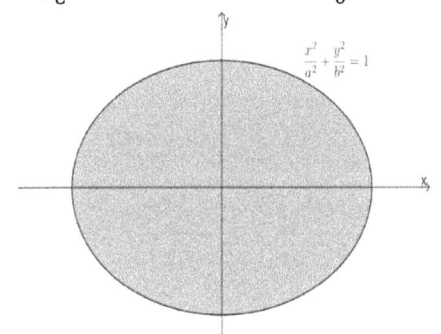

Example 3.

Use Green's Theorem to find the area enclosed by $y = x^r$, $x = y^r (r > 1)$.

【Solution】

Let $C_1: y = x^r$, $C_2: x = y^r$ and $0 \leq x, y \leq 1$

9.4.2 Find the Area of the Closed Region R

Let C be the counterclockwise closed curve enclosed by C_1 and C_2, and let R be the area enclosed by C. By Green's Theorem,

the area R $= \dfrac{1}{2}\oint_C -y\,dx + x\,dy = \dfrac{1}{2}\int_{C_1} -y dx + x dy + \dfrac{1}{2}\int_{C_2} -y dx + x dy$

$= \dfrac{1}{2}\int_0^1 -x^r dx + x(rx^{r-1})dx + \dfrac{1}{2}\int_1^0 -y(ry^{r-1})dy + y^r dy$

$= \dfrac{1}{2}\int_0^1 (r-1)x^r dx + \dfrac{1}{2}\int_1^0 (1-r)y^r dy = \dfrac{1}{2}\left(\dfrac{r-1}{r+1}\cdot x^{r+1}\big|_0^1 + \dfrac{1-r}{r+1}\cdot y^{r+1}\big|_1^0\right)$

$= \dfrac{1}{2}\cdot\dfrac{r-1-(1-r)}{r+1} = \dfrac{r-1}{r+1}$

Example 4.

Use Green's Theorem to find tha area enclosed by $r(\theta) = a(1+\cos\theta)$.

【Solution】

The enclosed area $= \dfrac{1}{2}\oint_C -y\,dx + x\,dy = \dfrac{1}{2}\oint_C r^2(\theta)\,d\theta = \dfrac{1}{2}\int_0^{2\pi} a^2(1+\cos\theta)^2\,d\theta$

$= \dfrac{a^2}{2}\int_0^{2\pi} 1 + \dfrac{1+\cos 2\theta}{2} + 2\cos\theta\,d\theta = \dfrac{a^2}{2}\left(\theta + 2\sin\theta + \dfrac{\theta}{2} + \dfrac{\sin 2\theta}{4}\right)\bigg|_0^{2\pi} = \dfrac{3a^2\pi}{2}$

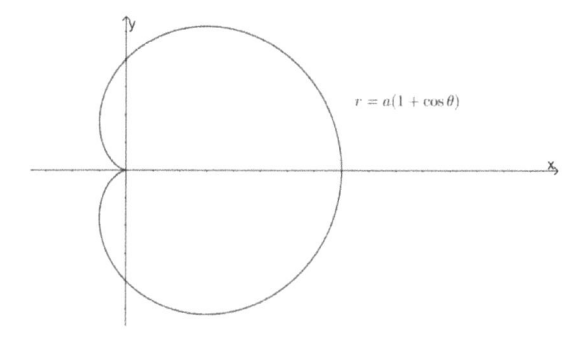

Example 5.

Using Green's Theorem, find the area enclosed by $\vec{r}(t) = (\cos^3 t, \sin^3 t)$, $0 \le t \le 2\pi$.

【Solution】

Let C be the counterclockwise closed boundary defined by $x = \cos^3 t, y = \sin^3 t$, $0 \le t \le 2\pi$. The region enclosed is denoted as R. By Green's Theorem,

$$\iint_R dxdy = \frac{1}{2}\oint_C -y\,dx + x\,dy$$

$\because dx = 3\cos^2 t\,(-\sin t)dt, \quad dy = 3\sin^2 t \cos t\,dt$

$$\therefore \frac{1}{2}\oint_C -y\,dx + x\,dy = \frac{1}{2}\int_0^{2\pi}(-\sin^3 t)\,3\cos^2 t\,(-\sin t) + (\cos^3 t)(3\sin^2 t \cos t)dt$$

$$= \frac{3}{2}\int_0^{2\pi}(\sin^2 t + \cos^2 t)\cos^2 t \sin^2 t\,dt = \frac{3}{2}\int_0^{2\pi}\cos^2 t \sin^2 t\,dt = \frac{3}{2}\int_0^{2\pi}\left(\frac{\sin 2t}{2}\right)^2 dt$$

$$= \frac{3}{2}\int_0^{2\pi}\frac{1 - \cos 4t}{8}dt = \frac{3\pi}{8}$$

9.4.3 Convert Closed Line Integrals into Surface Integrals

When the integrand of a line integral is relatively complex, using Green's Theorem to convert the closed line integral into a surface integral can simplify the integrand. According to Green's Theorem,

$$\oint_C f(x,y)\,dx + g(x,y)dy = \iint_R \frac{\partial g}{\partial x}(x,y) - \frac{\partial f}{\partial y}(x,y)dxdy.$$

Question Types:

9.4.3 Convert Closed Line Integrals into Surface Integrals

Type 1.

Find $\oint_C f(x,y)\,dx + g(x,y)\,dy = ?$, where $\frac{\partial g}{\partial x}(x,y) - \frac{\partial f}{\partial y}(x,y) = c$, c is a constant, and the curve C is a counterclockwise closed circle $x^2 + y^2 = r^2$. Additionally, f and g have continuous first-order partial derivatives.

Problem-Solving Process:

Let $R = \{(x,y): x^2 + y^2 \le r^2\}$. By Green's Theorem,

$$\oint_C f(x,y)\,dx + g(x,y)\,dy = \iint_R \frac{\partial g}{\partial x}(x,y) - \frac{\partial f}{\partial y}(x,y)\,dxdy = \iint_{x^2+y^2 \le r^2} c\,dxdy = c\pi r^2$$

Examples:

Find $\oint_C f(x,y)\,dx + g(x,y)\,dy = ?$, where $C: x^2 + y^2 = r^2, r > 0$

(I) If $f(x,y) = 3x^2y + \cos x + e^y$, $g(x,y) = x^3 + xe^y + x$ then $\frac{\partial g}{\partial x}(x,y) - \frac{\partial f}{\partial y}(x,y) = 1$

$\therefore \oint_C f(x,y)\,dx + g(x,y)\,dy = \iint_R \frac{\partial g}{\partial x}(x,y) - \frac{\partial f}{\partial y}(x,y)\,dxdy = \iint_{x^2+y^2 \le r^2} dxdy = \pi r^2$

(II) If $f(x,y) = x + 6y$, $g(x,y) = 2x + y$ then $\frac{\partial g}{\partial x}(x,y) - \frac{\partial f}{\partial y}(x,y) = 2 - 6 = -4$

$\therefore \oint_C f(x,y)\,dx + g(x,y)\,dy = \iint_R \frac{\partial g}{\partial x}(x,y) - \frac{\partial f}{\partial y}(x,y)\,dxdy = \iint_{x^2+y^2 \le r^2} -4\,dxdy$
$= -4\pi r^2$

Remark

Additional explanation: Sometimes the region of integration is not exactly a circle, but an ellipse instead. The method is similar.

Type 2.

Using Green's Theorem and Fubini's Theorem, determine $\oint_C f(x,y)\,dx + g(x,y)\,dy = ?$, where C is the counterclockwise closed boundary of $R = [a,b] \times [c,d]$, and the first-order partial derivatives of f and g exist and are continuous.

Problem-Solving Process:

Step 1.

By Green's Theorem,

$$\oint_C f(x,y)\,dx + g(x,y)dy = \iint_R \frac{\partial g}{\partial x}(x,y) - \frac{\partial f}{\partial y}(x,y)dxdy = \iint_R h(x,y)dxdy$$

where $\frac{\partial g}{\partial x}(x,y) - \frac{\partial f}{\partial y}(x,y) = h(x,y)$

Step2.

By Fubini'sTheorem, $\iint_R h(x,y)dxdy = \int_a^b \int_c^d h(x,y)dydx$

Step3.

Find $\int_a^b \int_c^d h(x,y)dydx =?$

Examples:

Find $\oint_C f(x,y)\,dx + g(x,y)dy =?$

(I) If $f(x,y) = x^2$, $g(x,y) = 2xy$, C is the rectangular boundary with vertices at $(0,0)$, $(0,5), (12,5)$, and $(12,0)$, oriented clockwise.

Using Green's Theorem,

$$-\oint_C x^2\,dx + 2xydy = -\int_0^5 \int_0^{12} 2y\,dxdy = -\int_0^5 24y\,dy = -12y^2|_0^5 = -300$$

Remark

Sometimes the region of integration is not necessarily a rectangle, but the method is similar.

Example 1.

Using Green's Theorem, find $\oint_C (3x^2y + \cos x + e^y)\,dx + (x^3 + xe^y + x)dy =?$,

where C is the counterclockwise closed circle defined by $x^2 + y^2 = 1$.

【Solution】

By Green's Theorem,

$$\oint_C (3x^2y + \cos x + e^y)\,dx + (x^3 + xe^y + x)dy = \iint_{x^2+y^2 \le 1} dxdy = \pi$$

Example 2.

9.4.3 Convert Closed Line Integrals into Surface Integrals

Using Green's Theorem, find $\oint_C (6y + x)dx + (y + 2x)dy = ?$, where C is the counterclockwise closed circle defined by $(x - 2)^2 + (y - 3)^2 = r^2$.

【Solution】

Let $R = \{(x, y): (x - 2)^2 + (y - 3)^2 \leq r^2\}$

By Green's Theorem, $\oint_C (6y + x)dx + (y + 2x)dy = \iint_R 2 - 6 \, dxdy = -4\pi r^2$

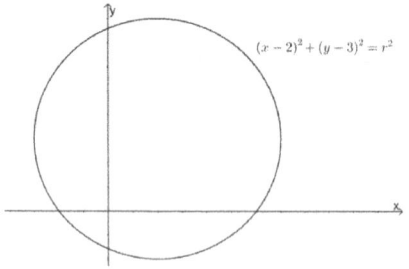

Example 3.

Using Green's Theorem, find $\oint_C 3x - 4y\,dx + (4x + 2y)dy = ?$, where C is the counterclockwise closed boundary of the upper half of the ellipse defined by $\frac{x^2}{16} + \frac{y^2}{9} = 1, y \geq 0$.

【Solution】

Let $R = \{(x, y): \frac{x^2}{16} + \frac{x^2}{9} \leq 1, y \geq 0\}$

By Green's Theorem, $\oint_C 3x - 4y\,dx + (4x + 2y)dy = \iint_R 8 \, dxdy = \frac{8 \cdot 4 \cdot 3 \cdot \pi}{2} = 48\pi$

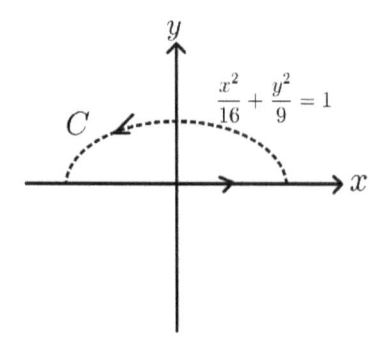

Example 4.

Using Green's Theorem, find $\oint_C x^4 \, dx + xy \, dy = ?$, where C is the counterclockwise triangle defined by the points $(0,0)$, $(2,0)$ and $(0,2)$.

【Solution】

Let $R = \{(x,y): 0 \leq x \leq 2, 0 \leq y \leq x\}$

By Green's Theorem, $\oint_C x^4 \, dx + xy \, dy = \iint_R y \, dx \, dy$

By Fubini's Theorem, $\iint_R y \, dx \, dy = \int_0^2 \int_0^x y \, dy \, dx = \int_0^2 \frac{y^2}{2}\Big|_0^x dx = \int_0^2 \frac{x^2}{2} dx = \frac{x^3}{6}\Big|_0^2 = \frac{4}{3}$

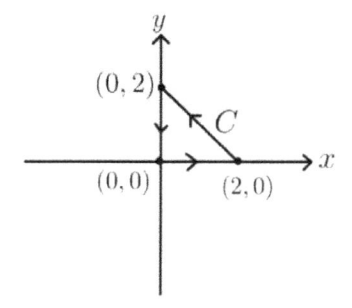

Example 5.

Using Green's Theorem, find $\oint_C (x + 6y) \, dx + (2x + y) \, dy = ?$, where C is the counterclockwise closed circle defined by $(x - 2)^2 + (y - 3)^2 = 9$.

【Solution】

By Green's Theorem,

$\oint_C (x + 6y) \, dx + (2x + y) \, dy = \iint_{(x-2)^2+(y-3)^2 \leq 9} 2 - 6 \, dx \, dy = -36\pi$

Example 6.

Using Green's Theorem, find $\oint_C x^2 \, dx + 2xy \, dy = ?$, where C is the clockwise rectangle defined by the points $(0,0)$, $(0,5)$, $(12,5)$, and $(12,0)$.

【Solution】

By Green's Theorem,

9.4.3 Convert Closed Line Integrals into Surface Integrals

$$-\oint_C x^2\, dx + 2xy\, dy = -\int_0^5 \int_0^{12} 2y\, dx dy = -\int_0^5 24y\, dy = -12y^2 \big|_0^5 = -300$$

(0,5) (12,5)
(0,0) (12,0)

Example 7.

Using Green's Theorem, find $\oint_C (x^2 - xy^3)\, dx + (y^2 - 2xy)\, dy = ?$, where C is the counterclockwise closed curve surrounding the points $(0,0), (r, 0), (r, r),$ and $(0, r)$ with $r > 0$.

【Solution】

Let $R = \{(x, y): 0 \le x \le r, 0 \le y \le r\}$

By Green's Theorem, $\oint_C (x^2 - xy^3)\, dx + (y^2 - 2xy)\, dy = \iint_R -2y + 3xy^2\, dx dy$

By Fubini's Theorem,

$$\iint_R -2y + 3xy^2\, dx dy = \int_0^r \int_0^r -2y + 3xy^2\, dy dx = \int_0^r -y^2 + xy^3 \big|_0^r\, dx$$

$$= \int_0^r -r^2 + xr^3\, dx = -r^3 + \frac{r^5}{2}$$

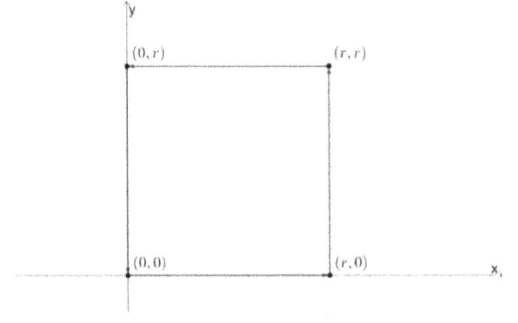

(0,r) (r,r)
(0,0) (r,0)

Example 8.

Using Green's Theorem, find $\oint_C \left(\dfrac{y^2}{2} - 4y\right) dx + \left(-\dfrac{x^2}{2}\right) dy = ?$, where C is the counterclockwise closed boundary of $R = \{(x, y): 0 \leq x \leq 2, 0 \leq y \leq 2\}$.

【Solution】

By Green's Theorem, $\oint_C \left(\dfrac{y^2}{2} - 4y\right) dx + \left(-\dfrac{x^2}{2}\right) dy = \iint_R 4 - x - y \, dA$

$\because f(x,y) = 4 - x - y$ is continuous on R, by Fubini's Theorem,

$\iint_R 4 - x - y \, dA = \int_0^2 \int_0^2 4 - x - y \, dy \, dx = \int_0^2 4y - xy - \dfrac{y^2}{2} \Big|_{y=0}^{y=2} dx = \int_0^2 8 - 2x - 2 \, dx$

$= (6x - x^2)\big|_{x=0}^{x=2} = 12 - 4 = 8$

Example 9.

Using Green's Theorem, find $\oint_C xy^2 \, dx + (5x^2 y + x) dy = ?$, where C is the counterclockwise closed boundary of $R = \{(x, y): 1 \leq x \leq 2, 0 \leq y \leq 2\}$.

【Solution】

By Green's Theorem, $\oint_C xy^2 \, dx + (5x^2 y + x) dy = \iint_R 1 + 8xy \, dA$

$\because f(x,y) = 1 + 8xy$ is continuous on R, by Fubini's Theorem,

$\iint_R 1 + 8xy \, dA = \int_0^2 \int_1^2 1 + 8xy \, dx \, dy = \int_0^2 (x + 4x^2 y)\big|_{x=1}^{x=2} dy = \int_0^2 1 + 12y \, dy$

$= (y + 6y^2)\big|_{y=0}^{y=2} = 2 + 6 \cdot 4 = 26$

Example 10.

Find $\oint_C (y - \sin x) \, dx + \left(\cos x + \sqrt{2}\right) dy = ?$, where C is the counterclockwise triangle defined by the points $(0,0)$, $\left(\dfrac{\pi}{2}, 0\right)$, and $\left(\dfrac{\pi}{2}, 1\right)$: (1)Direct computation (2) Using Green's Theorem.

【Solution】

(1)

9.4.3 Convert Closed Line Integrals into Surface Integrals

Let $C_1 = \{(x,y): 0 \le x \le \frac{\pi}{2}, y = 0\}$, $C_2 = \{(x,y): 0 \le y \le 1, x = \frac{\pi}{2}\}$

$C_3 = \{(x,y): y = \frac{2x}{\pi}, 0 \le x \le \frac{\pi}{2}\}$ then $C = C_1 \cup C_2 \cup C_3$

$\because \int_{C_1} (y - \sin x)\,dx + (\cos x + \sqrt{2})\,dy = \int_0^{\frac{\pi}{2}} -\sin x\,dx = -1$

$\int_{C_2} (y - \sin x)\,dx + (\cos x + \sqrt{2})\,dy = \int_0^1 \cos\frac{\pi}{2} + \sqrt{2}\,dy = \sqrt{2}$

$\int_{C_3} (y - \sin x)\,dx + (\cos x + \sqrt{2})\,dy = \int_{\frac{\pi}{2}}^0 \left(\frac{2x}{\pi} - \sin x\right) + \frac{2}{\pi}(\cos x + \sqrt{2})\,dx$

$= \left(\frac{x^2}{\pi} + \cos x + \frac{2}{\pi}\sin x + \frac{2\sqrt{2}}{\pi}x\right)\Bigg|_{\frac{\pi}{2}}^0 = 1 - \frac{\pi}{4} - \frac{2}{\pi} - \sqrt{2}$

$\therefore \oint_C (y - \sin x)\,dx + \cos x\,dy = -\frac{\pi}{4} - \frac{2}{\pi}$

(2)

Let $R = \{(x,y): 0 \le x \le \frac{\pi}{2}, 0 \le y \le \frac{2x}{\pi}\}$

By Green's Theorem, $\oint_C (y - \sin x)\,dx + (\cos x + \sqrt{2})\,dy = \iint_R -\sin x - 1\,dxdy$

By Fubini's Theorem,

$\iint_R -\sin x - 1\,dxdy = \int_0^{\frac{\pi}{2}} \int_0^{\frac{2x}{\pi}} -\sin x - 1\,dydx = \int_0^{\frac{\pi}{2}} (-y\sin x - y)\Big|_{y=0}^{y=\frac{2x}{\pi}} dx$

$= -\frac{2}{\pi}\left(\int_0^{\frac{\pi}{2}} x\sin x + x\,dx\right)$

By integration by parts, $\int_0^{\frac{\pi}{2}} x\sin x + x\,dx = \left(-x\cos x + \sin x + \frac{x^2}{2}\right)\Big|_0^{\frac{\pi}{2}} = 1 + \frac{\pi^2}{8}$

$\therefore \oint_C (y - \sin x)\,dx + (\cos x + \sqrt{2})\,dy = -\frac{2}{\pi}\left(1 + \frac{\pi^2}{8}\right) = -\frac{2}{\pi} - \frac{\pi}{4}$

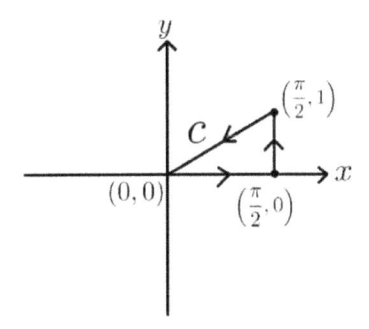

Example 11.

Using Green's Theorem, find $\oint_C y \cos x \, dx + x \sin y \, dy = ?$, where C is the counterclockwise closed path with vertices at $(0,0)$, $(a, 0)$, and (a, a) with $a > 0$.

【Solution】

Let $R = \{(x, y): 0 \leq x \leq a, 0 \leq y \leq x\}$

By Green's Theorem, $\oint_C y \cos x \, dx + x \sin y \, dy = \iint_R \sin y - \cos x \, dxdy$

By Fubini's Theorem,

$$\iint_R \sin y - \cos x \, dxdy = \int_0^a \int_0^x \sin y - \cos x \, dydx = \int_0^a (-\cos y)|_0^x - x \cos x \, dx$$

$$= \int_0^a 1 - \cos x - x \cos x \, dx = a - \sin a - a \sin a - \cos a + 1$$

$$= -(a + 1) \sin a - \cos a + a + 1$$

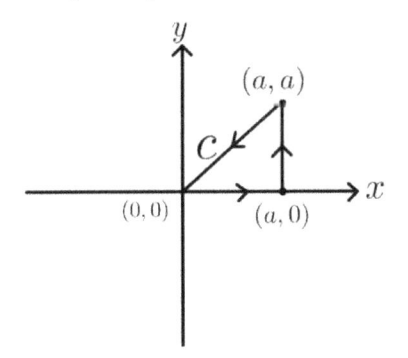

Example 12.

Using Green's Theorem, find $\oint_C -y^4 dx + xy^2 dy = ?$, where C is the

9.4.3 Convert Closed Line Integrals into Surface Integrals

counterclockwise closed path defined by the vertices $(0,0), (1,0),$ and $(1,1)$.

【Solution】

Let $R = \{(x,y): 0 \leq x \leq 1, 0 \leq y \leq x\}$

By Green's Theorem, $\oint_C -y^4 dx + xy^2 dy = \iint_R y^2 + 4y^3 dxdy$

By Fubini's Theorem,

$$\iint_R y^2 + 4y^3 dxdy = \int_0^1 \int_0^x y^2 + 4y^3 \, dydx = \int_0^1 \frac{x^3}{3} + x^4 dx = \frac{x^4}{12} + \frac{x^5}{5} = \frac{17}{60}$$

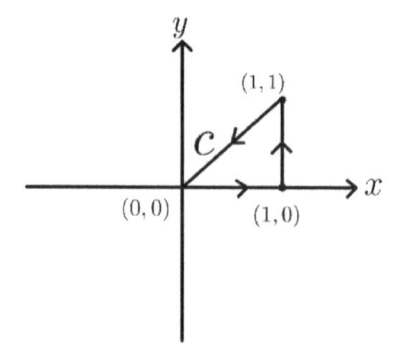

Example 13.

Using Green's Theorem, find $\oint_C (2xy - x^2) dx + (x + y^2) dy = ?$, where C is the closed curve from $(0,0)$ along $y = x^2$ to $(1,1)$ and back along $x = y^2$ to $(0,0)$.

【Solution】

Let $R = \{(x,y): 0 \leq x \leq 1, x^2 \leq y \leq \sqrt{x}\}$

By Green's Theorem, $\oint_C (2xy - x^2) dx + (x + y^2) dy = \iint_R 1 - 2x \, dxdy$

By Fubini's Theorem,

$$\iint_R 1 - 2x \, dxdy = \int_0^1 \int_{x^2}^{\sqrt{x}} 1 - 2x \, dydx = \int_0^1 x^{\frac{1}{2}} - 2x^{\frac{3}{2}} - x^2 + 2x^3 dx = \frac{1}{30}$$

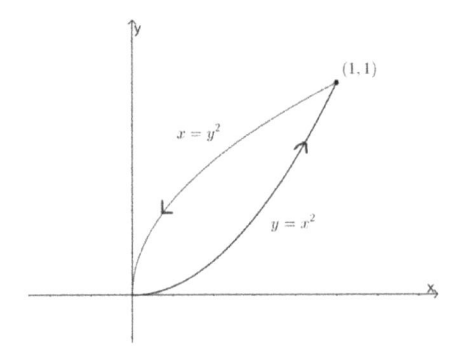

Example 14.

Let C be the counterclockwise closed path from $(0,0)$ along $y^2 = 2x$ to $(2,2)$, then along $y^3 = 4x$ back to $(0,0)$. Find $\oint_C y\,dx + x^2 y\,dy = ?$: (1) Direct computation (2) Using Green's Theorem.

【Solution】

(1)
Let $C_1 = \{(x,y): 0 \le x \le 2, y^2 = 2x\}$, $C_2 = \{(x,y): 0 \le x \le 2, y^3 = 4x\}$ then $C = C_1 \cup C_2$

$$\because \int_{C_1} y\,dx + x^2 y\,dy = \int_0^2 \sqrt{2x}\,dx + x^2\,dx = \frac{16}{3}$$

$$\int_{C_2} y\,dx + x^2 y\,dy = \int_2^0 (4x)^{\frac{1}{3}}dx + \frac{4}{3}(4x)^{\frac{-1}{3}}dx = -3 - 2 = -5$$

$$\therefore \oint_C y\,dx + x^2 y\,dy = \frac{16}{3} - 5 = \frac{1}{3}$$

(2)

Let $R = \{(x,y): \frac{y^3}{4} \le x \le \frac{y^2}{2}, 0 \le y \le 2\}$

By Green's Theorem, $\oint_C y\,dx + x^2 y\,dy = \iint_R 2xy - 1\,dxdy$

By Fubini's Theorem,

$$\iint_R 2xy - 1\,dxdy = \int_0^2 \int_{\frac{y^3}{4}}^{\frac{y^2}{2}} 2xy - 1\,dxdy = \int_0^2 (x^2 y - x)\Big|_{\frac{y^3}{4}}^{\frac{y^2}{2}}dy$$

$$= \int_0^2 \frac{y^5}{4} - \frac{y^2}{2} - \left(\frac{y^7}{16} - \frac{y^3}{4}\right)dy = \frac{1}{3}$$

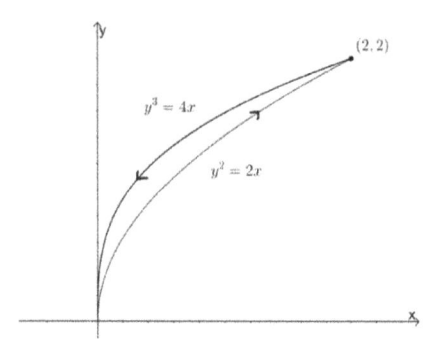

9.4.3 Convert Closed Line Integrals into Surface Integrals 380

Example 15.

Using Green's Theorem, find $\oint_C \frac{xy^2}{2}dx + (x^2y)dy = ?$, where C is the counterclockwise closed boundary of $R = \{(x,y): 0 \leq x \leq 2, 2x - 4 \leq y \leq 0\}$.

【Solution】

By Green's Theorem, $\oint_C \frac{xy^2}{2}dx + (x^2y)dy = \iint_R xy\,dA$

∵ $f(x,y) = xy$ is continuous on R, by Fubini's Theorem,

$$\iint_R xy\,dA = \int_0^2 \int_{2x-4}^0 xy\,dy\,dx = \int_0^2 \frac{xy^2}{2}\bigg|_{y=2x-4}^{y=0} dx = -\int_0^2 \frac{x(2x-4)^2}{2}dx$$

$$= -2\int_0^2 x(x-2)^2 dx = -2\int_0^2 x^3 - 4x^2 + 4x\,dx = -2\left(\frac{x^4}{4} - \frac{4x^3}{3} + 2x^2\right)\bigg|_{x=0}^{x=2}$$

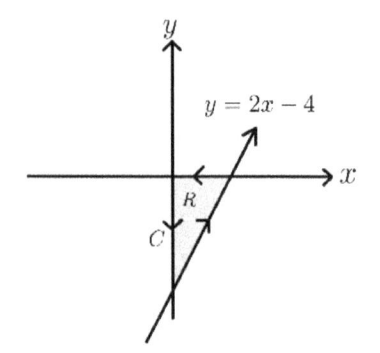

$$= -2\left(4 - \frac{32}{3} + 8\right) = \frac{-8}{3}$$

Example 16.

Using Green's Theorem, find $\oint_C \frac{y^3}{3} dx + x^2 dy =?$, where C is the counterclockwise closed boundary of $R = \{(x,y): x - y + 1 \leq 0, y \leq 2, x + y - 1 \geq 0\}$.

【Solution】

By Green's Theorem, $\oint_C \frac{y^3}{3} dx + x^2 dy = \iint_R 2x - y^2 dA$

$\because f(x,y) = 2x - y^2$ is continuous on R, by Fubinis Theorem,

$\iint_R 2x - y^2 dA = \int_1^2 \int_{1-y}^{y-1} 2x - y^2 dx dy = \int_1^2 x^2 - y^2 x \big|_{x=1-y}^{x=y-1} dy$

$= \int_1^2 (y-1)^2 - y^2(y-1) - ((1-y)^2 - y^2(1-y)) dy = \int_1^2 -2y^3 + 2y^2 dy$

$= \frac{-y^4}{2} \bigg|_{y=1}^{y=2} + \frac{2y^3}{3} \bigg|_{y=1}^{y=2} = \frac{-17}{6}$

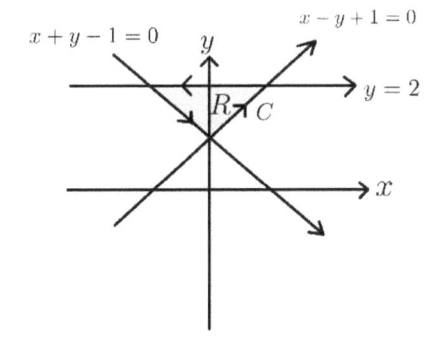

Example 17.

Using Green's Theorem, find $\oint_C y\, dx + 2x dy =?$, where C is the counterclockwise closed boundary of $R = \{(x,y): -3x - y + 6 \leq 0, 4x - x^2 - y \geq 0, y \geq 0\}$.

【Solution】

By Green's Theorem,

$\oint_C y\, dx + 2x dy = \iint_R 1 dA = \int_1^2 4x - x^2 - (6 - 3x) dx + \int_2^4 4x - x^2 dx$

$= \left(\frac{7x^2}{2} - \frac{x^3}{3} - 6x\right)\bigg|_1^2 + \left(2x^2 - \frac{x^3}{3}\right)\bigg|_2^4 = \frac{21}{2} - \frac{7}{3} - 6 + 2 \cdot 12 - \frac{56}{3} = \frac{15}{2}$

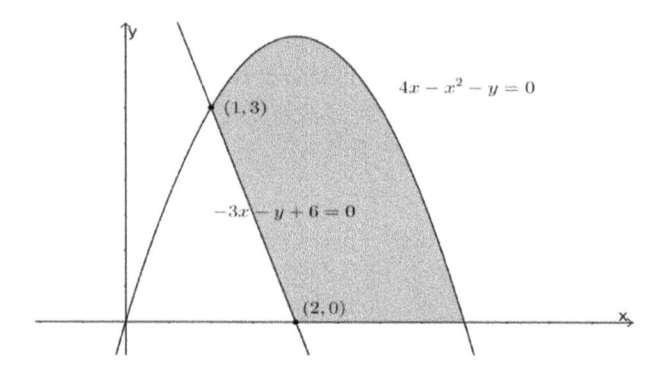

9.4.3 Convert Closed Line Integrals into Surface Integrals

Example 18.

Using Green's Theorem, find $\oint_C \dfrac{y^3}{3} dx + \dfrac{2x^{\frac{3}{2}}}{3} dy = ?$, where C is the counterclockwise closed boundary of $R = \left\{(x,y): y \geq x^2, y \leq x^{\frac{1}{4}}\right\}$.

【Solution】

By Green's Theorem, $\oint_C \dfrac{y^3}{3} dx + \dfrac{2x^{\frac{3}{2}}}{3} dy = \iint_R \sqrt{x} - y^2 \, dxdy$

Let $x^2 = x^{\frac{1}{4}}$ then $x = 0$ or $1 \Rightarrow 0 \leq x \leq 1 \Rightarrow R = \{(x,y): 0 \leq x \leq 1, y \geq x^2, y \leq x^{\frac{1}{4}}\}$

∵ $f(x,y) = \sqrt{x} - y^2$ is continuous on R, by Fubinis Theorem,

$$\iint_R \sqrt{x} - y^2 \, dxdy = \int_0^1 \int_{x^2}^{x^{\frac{1}{4}}} \sqrt{x} - y^2 \, dydx = \int_0^1 \sqrt{x}y - \dfrac{y^3}{3} \Big|_{y=x^2}^{y=x^{\frac{1}{4}}} dx$$

$$= \int_0^1 \sqrt{x}(x^{\frac{1}{4}}) - \dfrac{x^{\frac{3}{4}}}{3} - \left(\sqrt{x}(x^2) - \dfrac{x^6}{3}\right) dx = \int_0^1 x^{\frac{3}{4}} - \dfrac{x^{\frac{3}{4}}}{3} - x^{\frac{5}{2}} + \dfrac{x^6}{3} dx$$

$$= \left(\dfrac{4x^{\frac{7}{4}}}{7} - \dfrac{4x^{\frac{7}{4}}}{21} - \dfrac{2x^{\frac{7}{2}}}{7} + \dfrac{x^7}{21}\right)\Big|_{x=0}^{x=1} = \dfrac{2}{7} - \dfrac{3}{21} = \dfrac{1}{7}$$

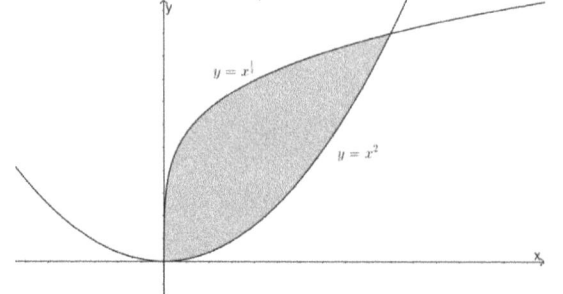

9.4.4 Use Coordinate Transformation to Compute Surface Integrals

If the integrand of a closed line integral is complex and using Green's Theorem to convert the line integral into a surface integral does not allow for direct evaluation, you may need to use polar coordinate transformations or generalized coordinate transformations.

Question Types:
Type 1.

Find $\oint_C f(x,y)\,dx + g(x,y)\,dy = ?$, where C is a counterclockwise closed circle $x^2 + y^2 = a^2$, and the first partial derivatives of f and g exist and are continuous.

Problem-Solving Process:

Step1.

Let $R = \{(x,y): x^2 + y^2 \leq a^2\}$. By Green's Theorem,

$$\oint_C f(x,y)\,dx + g(x,y)\,dy = \iint_R \frac{\partial g}{\partial x}(x,y) - \frac{\partial f}{\partial y}(x,y)\,dxdy = \iint_R h(x,y)\,dxdy$$

where $\frac{\partial g}{\partial x}(x,y) - \frac{\partial f}{\partial y}(x,y) = h(x,y)$

Step2.

Let $x = r\cos\theta$, $y = r\sin\theta$ then $\{(x,y): x^2 + y^2 \leq a^2\} = \{(r,\theta): 0 \leq r \leq a, 0 \leq \theta \leq 2\pi\}$

and $dxdy = \left\| \begin{matrix} \frac{\partial x}{\partial r} & \frac{\partial x}{\partial \theta} \\ \frac{\partial y}{\partial r} & \frac{\partial y}{\partial \theta} \end{matrix} \right\| drd\theta = \left\| \begin{matrix} \cos\theta & -r\sin\theta \\ \sin\theta & r\cos\theta \end{matrix} \right\| drd\theta = rdrd\theta$

Step3.

$$\iint_R h(x,y)\,dxdy = \int_0^{2\pi}\int_0^a h(r\cos\theta, r\sin\theta)\,rdrd\theta$$

Step4.

Find $\int_0^{2\pi}\int_0^a h(r\cos\theta, r\sin\theta)\,rdrd\theta = ?$

Examples:

(I) $f(x,y) = -\frac{y^3}{3}$, $g(x,y) = \frac{x^3}{3}$, and C is the counterclockwise closed circle $x^2 + y^2 = a^2$.

9.4.4 Use Coordinate Transformation to Compute Surface Integrals

Find $\oint_C f(x,y)\,dx + g(x,y)\,dy = ?$

Let $R = \{(x,y): x^2 + y^2 \leq a^2\}$. By Green's Theorem,

$$\oint_C f(x,y)\,dx + g(x,y)\,dy = \iint_R \frac{\partial g}{\partial x}(x,y) - \frac{\partial f}{\partial y}(x,y)\,dxdy = \iint_R x^2 + y^2\,dxdy$$

Let $x = r\cos\theta$, $y = r\sin\theta$ then $\{(x,y): x^2 + y^2 \leq a^2\} = \{(r,\theta): 0 \leq r \leq a, 0 \leq \theta \leq 2\pi\}$

and $dxdy = \left\| \begin{matrix} \frac{\partial x}{\partial r} & \frac{\partial x}{\partial \theta} \\ \frac{\partial y}{\partial r} & \frac{\partial y}{\partial \theta} \end{matrix} \right\| drd\theta = \left\| \begin{matrix} \cos\theta & -r\sin\theta \\ \sin\theta & r\cos\theta \end{matrix} \right\| drd\theta = rdrd\theta$

$$\iint_R x^2 + y^2\,dxdy = \int_0^{2\pi}\int_0^a r^3\,drd\theta = \frac{a^4\pi}{2}$$

Remark:
Sometimes the region of integration is an ellipse, and the method is similar.

Example 1.

Use Green's Theorem to compute $\oint_C y^3\,dx - x^3\,dy = ?$, where C is a counterclockwise closed circle $x^2 + y^2 = 25$.

【Solution】

Let $R = \{(x,y): x^2 + y^2 \leq 25\}$

By Green's Theorem, $\oint_C y^3\,dx - x^3\,dy = -\iint_R 3x^2 + 3y^2\,dxdy$

Let $x = r\cos\theta$, $y = r\sin\theta$ then $\{(x,y): x^2 + y^2 \leq 25\} = \{(r,\theta): 0 \leq r \leq 5, 0 \leq \theta \leq 2\pi\}$

and $dxdy = \left\| \begin{matrix} \frac{\partial x}{\partial r} & \frac{\partial x}{\partial \theta} \\ \frac{\partial y}{\partial r} & \frac{\partial y}{\partial \theta} \end{matrix} \right\| drd\theta = \left\| \begin{matrix} \cos\theta & -r\sin\theta \\ \sin\theta & r\cos\theta \end{matrix} \right\| drd\theta = rdrd\theta$

$\therefore -3\iint_R x^2 + y^2\,dxdy = -3\int_0^{2\pi}\int_0^5 r^3\,drd\theta = -6\pi \cdot \frac{r^4}{4}\bigg|_0^5 = -\frac{3\pi}{2} \cdot 625 = -\frac{1875\pi}{2}$

Example 2.

Use Green's Theorem to compute $\oint_C (6y + x) dx + (y + 2x)dy$, where C is a counterclockwise closed circle: $(x - 2)^2 + (y - 3)^2 = 9$.

【Solution】

Let $R = \{(x, y): (x - 2)^2 + (y - 3)^2 \leq 9\}$

By Green's Theorem, $\oint_C (6y + x) dx + (y + 2x)dy = \iint_R 2 - 6 dxdy$

Let $x = 2 + r\cos\theta, y = 3 + r\sin\theta$ then
$\{(x, y): (x - 2)^2 + (y - 3)^2 \leq 9\} = \{(r, \theta): 0 \leq r \leq 3, 0 \leq \theta \leq 2\pi\}$

and $dxdy = \begin{Vmatrix} \frac{\partial x}{\partial r} & \frac{\partial x}{\partial \theta} \\ \frac{\partial y}{\partial r} & \frac{\partial y}{\partial \theta} \end{Vmatrix} drd\theta = \begin{Vmatrix} \cos\theta & -r\sin\theta \\ \sin\theta & r\cos\theta \end{Vmatrix} drd\theta = rdrd\theta$

$\therefore -4\iint_R dxdy = -4\int_0^{2\pi}\int_0^3 r\,drd\theta = -8\pi \cdot \frac{r^2}{2}\Big|_0^3 = -36\pi$

Example 3.

Use Green's Theorem to compute $\oint_C 2y\, dx + (x^2 + y^2)dy = ?$, where C is a counterclockwise closed circle: $x^2 + (y - 3)^2 = 16$.

【Solution】

Let $R = \{(x, y): x^2 + (y - 3)^2 \leq 16\}$

By Green's Theorem, $\oint_C 2y\, dx + (x^2 + y^2)dy - \iint_R 2x - 2 dxdy$

Let $x = r\cos\theta, y = 3 + r\sin\theta$ then
$\{(x, y): x^2 + (y - 3)^2 \leq 16\} = \{(r, \theta): 0 \leq r \leq 4, 0 \leq \theta \leq 2\pi\}$

and $dxdy = \begin{Vmatrix} \frac{\partial x}{\partial r} & \frac{\partial x}{\partial \theta} \\ \frac{\partial y}{\partial r} & \frac{\partial y}{\partial \theta} \end{Vmatrix} drd\theta = \begin{Vmatrix} \cos\theta & -r\sin\theta \\ \sin\theta & r\cos\theta \end{Vmatrix} drd\theta = rdrd\theta$

$\therefore \iint_R 2x - 2 dxdy = 2\int_0^{2\pi}\int_0^4 r(r\cos\theta - 1)\,drd\theta = 2\int_0^{2\pi}\left(\frac{r^3}{3}\cos\theta - \frac{r^2}{2}\right)\Big|_0^4 d\theta = -32\pi$

9.4.4 Use Coordinate Transformation to Compute Surface Integrals

Example 4.

Use Green's Theorem to compute $\oint_C 2ydx + (x^2 + y^2)dy = ?$, where C is a counterclockwise closed circle: $x^2 + (y - 3)^2 = a^2, (a > 3)$.

【Solution】

Let $R = \{(x, y): x^2 + (y - 3)^2 \le a^2\}$

By Green's Theorem, $\oint_C 2ydx + (x^2 + y^2)dy = \iint_R 2x - 2dxdy$

Let $x = r\cos\theta, y = r\sin\theta, 0 \le \theta \le 2\pi$ then

$$\iint_R 2x - 2dxdy = \int_0^{2\pi}\int_0^a 2r^2\cos\theta - 2r\, drd\theta = \int_0^{2\pi}\int_0^a -2r\, drd\theta = -2a^2\pi$$

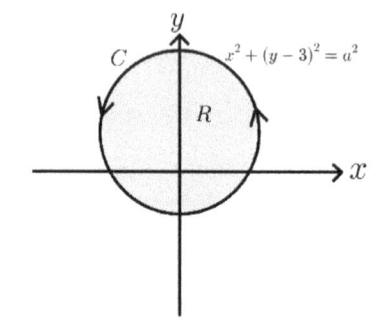

Example 5.

Assume C is a counterclockwise closed boundary $R = \{(x, y): y \ge 0, x^2 + y^2 \le 4\}$.

Use Green's Theorem to compute $\oint_C x^2y^2\, dx + x^3y\, dy = ?$

【Solution】

By Green's Theorem, $\oint_C x^2y^2\, dx + x^3y\, dy = \iint_R x^2y\, dxdy$

Let $x = r\cos\theta, y = r\sin\theta$ then

$\{(x, y): y \ge 0, x^2 + y^2 \le 4\} = \{(r, \theta): 0 \le r \le 2, 0 \le \theta \le \pi\}$

and $dxdy = \begin{Vmatrix} \dfrac{\partial x}{\partial r} & \dfrac{\partial x}{\partial \theta} \\ \dfrac{\partial y}{\partial r} & \dfrac{\partial y}{\partial \theta} \end{Vmatrix} drd\theta = \begin{Vmatrix} \cos\theta & -r\sin\theta \\ \sin\theta & r\cos\theta \end{Vmatrix} drd\theta = rdrd\theta$

$$\therefore \iint_R x^2 y \, dxdy = \int_0^\pi \int_0^2 (r\cos\theta)^2 (r\sin\theta) r \, drd\theta = \int_0^\pi \sin\theta \cos^2\theta \, d\theta \int_0^2 r^4 dr$$

$$= \frac{(-1)\cos^3\theta}{3}\bigg|_0^\pi \cdot \frac{r^5}{5}\bigg|_0^2 = \frac{2}{3} \cdot \frac{32}{5} = \frac{64}{15}$$

Example 6.

Use Green's Theorem to compute $\oint_C -\frac{y^3}{3} dx + \frac{x^3}{3} dy = ?$, where C is the counterclockwise closed curve $R = \left\{(x,y): \frac{x^2}{a^2} + \frac{y^2}{b^2} \leq 4\right\}$.

【Solution】

By Green's Theorem, $\oint_C -\frac{y^3}{3} dx + \frac{x^3}{3} dy = \iint_R x^2 + y^2 \, dA$

Let $x = ar\cos\theta$, $y = br\sin\theta$ then

$$\left\{(x,y): \frac{x^2}{a^2} + \frac{y^2}{b^2} \leq 4\right\} = \{(r,\theta): 0 \leq r \leq 2, 0 \leq \theta \leq 2\pi\}$$

and $dxdy = \begin{Vmatrix} \frac{\partial x}{\partial r} & \frac{\partial x}{\partial \theta} \\ \frac{\partial y}{\partial r} & \frac{\partial y}{\partial \theta} \end{Vmatrix} drd\theta = \begin{Vmatrix} a\cos\theta & -ra\sin\theta \\ b\sin\theta & rb\cos\theta \end{Vmatrix} drd\theta = abr \, drd\theta$

$$\therefore \iint_R x^2 + y^2 \, dA = \int_0^{2\pi} \int_0^2 ((ar)^2 \cos^2\theta + (br)^2 \sin^2\theta) abr \, drd\theta$$

$$= a^3 b \int_0^{2\pi} \cos^2\theta \, d\theta \int_0^2 r^3 dr + ab^3 \int_0^{2\pi} \sin^2\theta \, d\theta \int_0^2 r^3 dr$$

$$\because \int_0^{2\pi} \cos^2\theta \, d\theta = \int_0^{2\pi} \frac{1+\cos 2\theta}{2} d\theta = \pi$$

$$\because \int_0^{2\pi} \sin^2\theta \, d\theta = \int_0^{2\pi} \frac{1-\cos 2\theta}{2} d\theta = \pi \text{ and } \int_0^2 r^3 dr = \frac{16}{4} = 4$$

$$\therefore \iint_R x^2 + y^2 \, dA = a^3 b (4\pi) + ab^3 (4\pi) = 4ab\pi(a^2 + b^2)$$

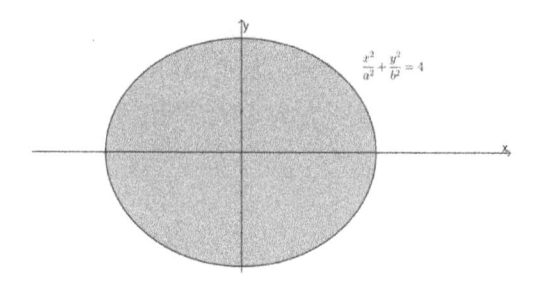

9.4.4 Use Coordinate Transformation to Compute Surface Integrals

Example 7.

Use Green's Theorem to compute $\oint_C (y^3 - 2y)\,dx - (2x^3 + x)\,dy = ?$, where C is a counterclockwise closed circle $6x^2 + 3y^2 = 1$.

【Solution】

Let $R = \{(x,y): 6x^2 + 3y^2 \leq 1\}$

By Green's Theorem, $\oint_C (y^3 - 2y)\,dx - (2x^3 + x)\,dy = \iint_R 1 - 6x^2 - 3y^2\,dxdy$

Let $x = \dfrac{r\cos\theta}{\sqrt{6}}, y = \dfrac{r\sin\theta}{\sqrt{3}}$ then $\{(x,y): 6x^2 + 3x^2 \leq 1\} = \{(r,\theta): 0 \leq r \leq 1, 0 \leq \theta \leq 2\pi\}$

and $dxdy = \left\|\begin{matrix} \dfrac{\partial x}{\partial r} & \dfrac{\partial x}{\partial \theta} \\ \dfrac{\partial y}{\partial r} & \dfrac{\partial y}{\partial \theta} \end{matrix}\right\| drd\theta = \left\|\begin{matrix} \dfrac{\cos\theta}{\sqrt{6}} & \dfrac{-r\sin\theta}{\sqrt{6}} \\ \dfrac{\sin\theta}{\sqrt{3}} & \dfrac{r\cos\theta}{\sqrt{3}} \end{matrix}\right\| drd\theta = \dfrac{1}{3\sqrt{2}} r\,drd\theta$

$\therefore \iint_R 1 - 6x^2 - 3y^2\,dxdy = \int_0^{2\pi}\int_0^1 (1 - r^2)\dfrac{1}{3\sqrt{2}} r\,drd\theta = \dfrac{\pi}{6\sqrt{2}}$

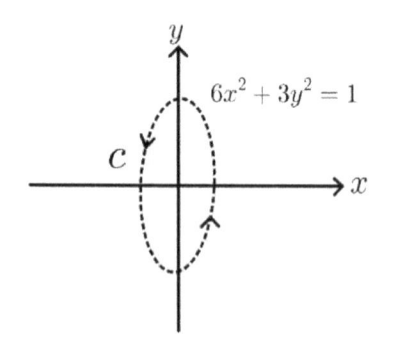

Example 8.

Use Green's Theorem to find $\oint_C (y^3 + e^{-x^2} + e^x) dx + (e^{-y^2} - x^3 + y^2 + 6xy)dy$, where C is a counterclockwise closed circle $x^2 + (y-1)^2 = 1$.

【Solution】

Let $R = \{(x,y): x^2 + (y-1)^2 \leq 1\}$, by Green's Theorem,

$$\oint_C (y^3 + e^{-x^2} + e^x) dx + (e^{-y^2} - x^3 + y^2 + 6xy)dy = \iint_R -3x^2 + 6y - 3y^2 dxdy$$

$$= 3\iint_R 1 - x^2 - (y-1)^2 dxdy$$

Let $x = r\cos\theta, y = r\sin\theta + 1$ then
$\{(x,y): x^2 + (y-1)^2 \leq 1\} = \{(r,\theta): 0 \leq r \leq 1, 0 \leq \theta \leq 2\pi\}$

and $dxdy = \begin{Vmatrix} \dfrac{\partial x}{\partial r} & \dfrac{\partial x}{\partial \theta} \\ \dfrac{\partial y}{\partial r} & \dfrac{\partial y}{\partial \theta} \end{Vmatrix} drd\theta = \begin{Vmatrix} \cos\theta & -r\sin\theta \\ \sin\theta & r\cos\theta \end{Vmatrix} drd\theta = rdrd\theta$

$$\therefore 3\iint_R 1 - x^2 - (y-1)^2 dxdy = 3\int_0^{2\pi}\int_0^1 (1-r^2)r\, drd\theta = \frac{3\pi}{2}$$

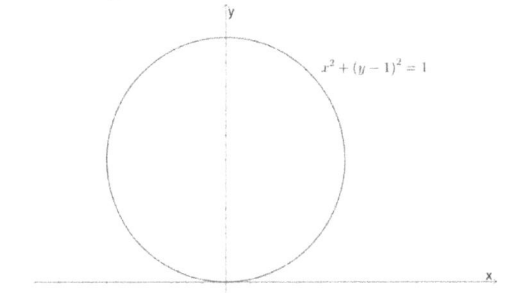

$x^2 + (y-1)^2 = 1$

Example 9.

Use Green's Theorem to find $\oint_C x^2y^2 dx + x^3y dy = ?$, where C is the counterclockwise closed boundary of $R = \{(x,y): y \geq 0, (x-1)^2 + y^2 \leq 1\}$.

【Solution】

By Green's Theorem, $\oint_C x^2y^2 dx + x^3y dy = \iint_R x^2y\, dA$

9.4.4 Use Coordinate Transformation to Compute Surface Integrals

Let $x = r\cos\theta + 1$, $y = r\sin\theta$ then
$\{(x,y): (x-1)^2 + y^2 \leq 1, y \geq 0\} = \{(r,\theta): 0 \leq r \leq 1, 0 \leq \theta \leq \pi\}$

and $dxdy = \begin{Vmatrix} \dfrac{\partial x}{\partial r} & \dfrac{\partial x}{\partial \theta} \\ \dfrac{\partial y}{\partial r} & \dfrac{\partial y}{\partial \theta} \end{Vmatrix} drd\theta = \begin{Vmatrix} \cos\theta & -r\sin\theta \\ \sin\theta & r\cos\theta \end{Vmatrix} drd\theta = rdrd\theta$

$\therefore \iint_R x^2 y \, dA = \int_0^\pi \int_0^1 (r\cos\theta + 1)^2 r\sin\theta \, rdrd\theta$

$= \int_0^\pi \int_0^1 (r^2 \cos^2\theta + 2r\cos\theta + 1) r\sin\theta \, rdrd\theta$

$= \int_0^\pi \cos^2\theta \sin\theta \, d\theta \int_0^1 r^4 dr + \int_0^\pi 2\cos\theta \sin\theta \, d\theta \int_0^1 r^3 dr + \int_0^\pi \sin\theta \, d\theta \int_0^1 r^2 dr$

$= (-1)\dfrac{\cos^3\theta}{3}\Big|_0^\pi \cdot \dfrac{r^5}{5}\Big|_0^1 - \dfrac{\cos 2\theta}{2}\Big|_0^\pi \cdot \dfrac{r^4}{4}\Big|_0^1 - \cos\theta\Big|_0^\pi \cdot \dfrac{r^3}{3}\Big|_0^1 = \dfrac{4}{5}$

Example 10.

Assume $C = \{(x,y): x = 2\cos t, y = 2\sin t\}$, $0 \leq t \leq 2\pi$. Use Green's Theorem to compute $\oint_C e^x y - y^3 dx + (e^x + x^3) dy$.

【Solution】

Let $R = \{(x,y): x^2 + y^2 \leq 4\}$

By Green's Theorem, $\oint_C e^x y - y^3 dx + (e^x + x^3) dy = \iint_R 3x^2 + 3y^2 dxdy$

Let $x = r\cos\theta$, $y = r\sin\theta$ then $\{(x,y): x^2 + y^2 \leq 4\} = \{(r,\theta): 0 \leq r \leq 2, 0 \leq \theta \leq 2\pi\}$

and $dxdy = \begin{Vmatrix} \dfrac{\partial x}{\partial r} & \dfrac{\partial x}{\partial \theta} \\ \dfrac{\partial y}{\partial r} & \dfrac{\partial y}{\partial \theta} \end{Vmatrix} drd\theta = \begin{Vmatrix} \cos\theta & -r\sin\theta \\ \sin\theta & r\cos\theta \end{Vmatrix} drd\theta = rdrd\theta$

$\therefore 3\iint_R x^2 + y^2 dxdy = 3\int_0^{2\pi}\int_0^2 r^3\, drd\theta = 24\pi$

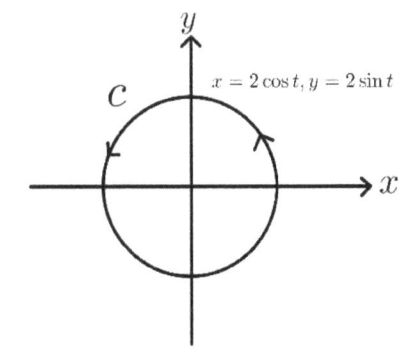

$x = 2\cos t, y = 2\sin t$

Example 11.

Use Green's Theorem to compute $\oint_C -\dfrac{2y^{\frac{3}{2}}}{3}dx + \dfrac{2x^{\frac{3}{2}}}{3}dy =?$, where R is the region enclosed by $\sqrt{x} + \sqrt{y} = 1$, $x = 0$, $y = 0$, and C is the counterclockwise boundary of R.

【Solution】

By Green's Theorem, $\oint_C -\dfrac{2y^{\frac{3}{2}}}{3}dx + \dfrac{2x^{\frac{3}{2}}}{3}dy = \iint_R \sqrt{x} + \sqrt{y}\, dxdy$

Let $x = r\cos^4\theta$, $y = r\sin^4\theta$ then

$dxdy = \begin{Vmatrix} \dfrac{\partial x}{\partial r} & \dfrac{\partial x}{\partial \theta} \\ \dfrac{\partial y}{\partial r} & \dfrac{\partial y}{\partial \theta} \end{Vmatrix} drd\theta = \begin{Vmatrix} \cos^4\theta & -4r\cos^3\theta\sin\theta \\ \sin^4\theta & 4r\sin^3\theta\cos\theta \end{Vmatrix} drd\theta = 4r\sin^3\theta\cos^3\theta\, drd\theta$

$\because R = \{(x,y): x \geq 0,\ y \geq 0, \sqrt{x} + \sqrt{y} \leq 1\} = \{(r,\theta): 0 \leq r \leq 1, 0 \leq \theta \leq \dfrac{\pi}{2}\}$

$\therefore \iint_R \sqrt{x} + \sqrt{y}\, dxdy = \int_0^{\frac{\pi}{2}}\int_0^1 \sqrt{r}(4r\sin^3\theta\cos^3\theta)\, drd\theta = \int_0^1 4r^{\frac{3}{2}}dr \int_0^{\frac{\pi}{2}}\sin^3\theta\cos^3\theta\, d\theta$

$= \int_0^1 4r^{\frac{3}{2}}dr \int_0^{\frac{\pi}{2}}\left(\dfrac{\sin 2\theta}{2}\right)^3 d\theta = \dfrac{2}{15}$

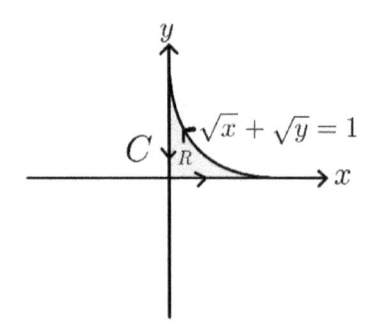

9.4.4 Use Coordinate Transformation to Compute Surface Integrals

[Figure: curve $\sqrt{x}+\sqrt{y}=1$ with region R and curve C in the first quadrant]

Example 12.

Use Green's Theorem to compute $\oint_C (6y + x)\, dx + (y + 2x)\, dy$, where C is a counterclockwise closed circle $(x - x_0)^2 + (y - y_0)^2 = a^2$.

【Solution】

Let $R = \{(x, y): (x - x_0)^2 + (y - y_0)^2 \leq a^2\}$

By Green's Theorem, $\oint_C (6y + x)\, dx + (y + 2x)\, dy = \iint_R 2 - 6\, dxdy$

Let $x = x_0 + r\cos\theta$, $y = y_0 + r\sin\theta$ then
$\{(x, y): (x - x_0)^2 + (y - y_0)^2 \leq a^2\} = \{(r, \theta): 0 \leq r \leq a, 0 \leq \theta \leq 2\pi\}$

and $dxdy = \begin{Vmatrix} \dfrac{\partial x}{\partial r} & \dfrac{\partial x}{\partial \theta} \\ \dfrac{\partial y}{\partial r} & \dfrac{\partial y}{\partial \theta} \end{Vmatrix} drd\theta = \begin{Vmatrix} \cos\theta & -r\sin\theta \\ \sin\theta & r\cos\theta \end{Vmatrix} drd\theta = rdrd\theta$

$\therefore -4 \iint_R dxdy = -4 \int_0^{2\pi}\int_0^a r\, drd\theta = -4a^2\pi$

Example 13.

Assume C is a counterclockwise closed ellipse $\dfrac{x^2}{a^2} + \dfrac{y^2}{b^2} = 1$, and f, g satisfy

$\dfrac{\partial g}{\partial x}(x, y) - \dfrac{\partial f}{\partial y}(x, y) = 1 - \left(\dfrac{x^2}{a^2} + \dfrac{y^2}{b^2}\right)$. Use Green's Theorem to compute

$\oint_C f(x, y)\, dx + g(x, y)\, dy = ?$

【Solution】

Let $R = \{(x,y): \dfrac{x^2}{a^2} + \dfrac{y^2}{b^2} \leq 1\}$. By Green's Theorem,

$$\oint_C f(x,y)\,dx + g(x,y)\,dy = \iint_R \dfrac{\partial g}{\partial x}(x,y) - \dfrac{\partial f}{\partial y}(x,y)\,dxdy = \iint_R 1 - \left(\dfrac{x^2}{a^2} + \dfrac{y^2}{b^2}\right) dxdy$$

Let $x = ra\cos\theta,\ y = rb\sin\theta$ then $\{(x,y): \dfrac{x^2}{a^2} + \dfrac{y^2}{b^2} \leq 1\} = \{(r,\theta): 0 \leq r \leq 1, 0 \leq \theta \leq 2\pi\}$

and $dxdy = \begin{Vmatrix} \dfrac{\partial x}{\partial r} & \dfrac{\partial x}{\partial \theta} \\ \dfrac{\partial y}{\partial r} & \dfrac{\partial y}{\partial \theta} \end{Vmatrix} drd\theta = \begin{Vmatrix} a\cos\theta & -ra\sin\theta \\ b\sin\theta & rb\cos\theta \end{Vmatrix} drd\theta = abr\,drd\theta$

$\therefore \iint_R 1 - \left(\dfrac{x^2}{a^2} + \dfrac{y^2}{b^2}\right) dxdy = \int_0^{2\pi}\int_0^1 (1 - r^2)abr\,drd\theta = \dfrac{ab\pi}{2}$

Example 14.

Use Green's Theorem to compute $\displaystyle\oint_C \dfrac{e^{x+y}}{3}\,dx + \dfrac{4e^{x+y}}{3}\,dy = ?$, where C is the counterclockwise closed boundary of $R = \{(x,y): |x| + |y| \leq 2\}$.

【Solution】

By Green's Theorem, $\displaystyle\oint_C \dfrac{e^{x+y}}{3}\,dx + \dfrac{4e^{x+y}}{3}\,dy = \iint_R e^{x+y}\,dA$

Let $x + y = u,\ x - y = v$ then $\{(x,y): |x| + |y| \leq 2\} = \{(u,v): -2 \leq u \leq 2, -2 \leq v \leq 2\}$

and $x = \dfrac{u+v}{2},\ y = \dfrac{u-v}{2},\ dxdy = \begin{Vmatrix} \dfrac{\partial x}{\partial u} & \dfrac{\partial x}{\partial v} \\ \dfrac{\partial y}{\partial u} & \dfrac{\partial y}{\partial v} \end{Vmatrix} dudv = \begin{Vmatrix} \dfrac{1}{2} & \dfrac{1}{2} \\ \dfrac{1}{2} & \dfrac{-1}{2} \end{Vmatrix} dudv = |\dfrac{-1}{2}|dudv$

$\therefore \iint_R e^{x+y}\,dA = \int_{-2}^{2}\int_{-2}^{2} \dfrac{e^u}{2}\,dudv = \dfrac{1}{2}\int_{-2}^{2} dv \int_{-2}^{2} e^u\,du = 2(e^2 - e^{-2})$

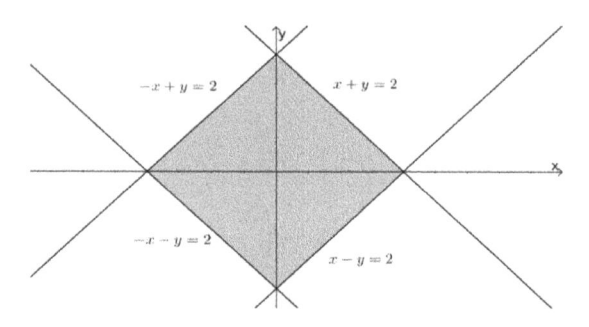

9.4.4 Use Coordinate Transformation to Compute Surface Integrals

Example 15.

Use Green's Theorem to compute $\oint_C \dfrac{y}{2} dx + \dfrac{3x}{2} dy = ?$, where C is the counterclockwise closed boundary of $R = \{(x, y): ax^2 + bxy + cy^2 \leq \alpha^2\}$.

【Solution】

By Green's Theorem, $\oint_C \dfrac{y}{2} dx + \dfrac{3x}{2} dy = \iint_R dA$

$\because ax^2 + bxy + cy^2 = a\left(x + \dfrac{by}{2a}\right)^2 + \dfrac{4ac - b^2}{4a} y^2$

Let $\sqrt{a}\left(x + \dfrac{by}{2a}\right) = u$, $\sqrt{\dfrac{4ac - b^2}{4a}} y = v$ then

$dxdy = \begin{Vmatrix} \dfrac{\partial x}{\partial u} & \dfrac{\partial x}{\partial v} \\ \dfrac{\partial y}{\partial u} & \dfrac{\partial y}{\partial v} \end{Vmatrix} dudv = \begin{Vmatrix} \dfrac{1}{\sqrt{a}} & 0 \\ \dfrac{2\sqrt{a}}{b} & \sqrt{\dfrac{4a}{4ac - b^2}} \end{Vmatrix} dudv = \dfrac{2}{\sqrt{4ac - b^2}} dudv$

$\therefore \iint_R dA = \iint_{u^2 + v^2 \leq \alpha^2} \dfrac{2}{\sqrt{4ac - b^2}} dudv$

Let $u = r\cos\theta, v = r\sin\theta$ then $\{(u, v): u^2 + v^2 \leq \alpha^2\} = \{(r, \theta): 0 \leq r \leq \alpha, 0 \leq \theta \leq 2\pi\}$

and $dudv = \begin{Vmatrix} \dfrac{\partial u}{\partial r} & \dfrac{\partial u}{\partial \theta} \\ \dfrac{\partial v}{\partial r} & \dfrac{\partial v}{\partial \theta} \end{Vmatrix} drd\theta = \begin{Vmatrix} \cos\theta & -r\sin\theta \\ \sin\theta & r\cos\theta \end{Vmatrix} drd\theta = rdrd\theta$

$\therefore \iint_{u^2 + v^2 \leq \alpha^2} \dfrac{2}{\sqrt{4ac - b^2}} dudv = \int_0^{2\pi} \int_0^\alpha \dfrac{2}{\sqrt{4ac - b^2}} rdrd\theta = \dfrac{2}{\sqrt{4ac - b^2}} \int_0^{2\pi} d\theta \int_0^\alpha rdr$

$$= \frac{2\pi a^2}{\sqrt{4ac - b^2}}$$

Example 16.

Use Green's Theorem to compute $\oint_C \left(\frac{xy^2}{4} - \frac{y^3}{6}\right) dx + \left(\frac{x^3}{6}\right) dy = ?$, where C is the counterclockwise closed boundary of $R = \{(x, y): x^2 - xy + y^2 \leq 2\}$.

【Solution】

By Green's Theorem, $\oint_C \left(\frac{xy^2}{4} - \frac{y^3}{6}\right) dx + \left(\frac{x^3}{6}\right) dy = \iint_R \frac{x^2 - xy + y^2}{2} dA$

$\because \dfrac{x^2 - xy + y^2}{2} = \dfrac{1}{2}\left(x - \dfrac{y}{2}\right)^2 + \dfrac{3}{8}y^2,$

Let $\sqrt{\dfrac{1}{2}}\left(x - \dfrac{y}{2}\right) = u,\ \sqrt{\dfrac{3}{8}}y = v$ then

$$dxdy = \begin{Vmatrix} \dfrac{\partial x}{\partial u} & \dfrac{\partial x}{\partial v} \\ \dfrac{\partial y}{\partial u} & \dfrac{\partial y}{\partial v} \end{Vmatrix} dudv = \begin{Vmatrix} \sqrt{2} & 0 \\ -2\sqrt{2} & \sqrt{\dfrac{8}{3}} \end{Vmatrix} dudv = \dfrac{4}{\sqrt{3}} dudv$$

$\therefore \iint_R \dfrac{x^2 - xy + y^2}{2} dA = \iint_{u^2+v^2 \leq 1} (u^2 + v^2)\dfrac{4}{\sqrt{3}} dudv$

Let $u = r\cos\theta,\ v = r\sin\theta$ then $\{(u,v): u^2 + v^2 \leq 1\} = \{(r, \theta): 0 \leq r \leq 1, 0 \leq \theta \leq 2\pi\}$

and $dudv = \begin{Vmatrix} \dfrac{\partial u}{\partial r} & \dfrac{\partial u}{\partial \theta} \\ \dfrac{\partial v}{\partial r} & \dfrac{\partial v}{\partial \theta} \end{Vmatrix} drd\theta = \begin{Vmatrix} \cos\theta & -r\sin\theta \\ \sin\theta & r\cos\theta \end{Vmatrix} drd\theta = r\, drd\theta$

$\therefore \iint_{u^2+v^2 \leq 1} (u^2 + v^2)\dfrac{4}{\sqrt{3}} dudv = \dfrac{4}{\sqrt{3}} \int_0^{2\pi} \int_0^1 r^2 \cdot r\, drd\theta = \dfrac{4}{\sqrt{3}} \int_0^{2\pi} d\theta \int_0^1 r^3 dr$

$= \dfrac{4}{\sqrt{3}} \cdot 2\pi \cdot \dfrac{r^4}{4}\bigg|_{r=0}^{r=1} = \dfrac{2\pi}{\sqrt{3}}$

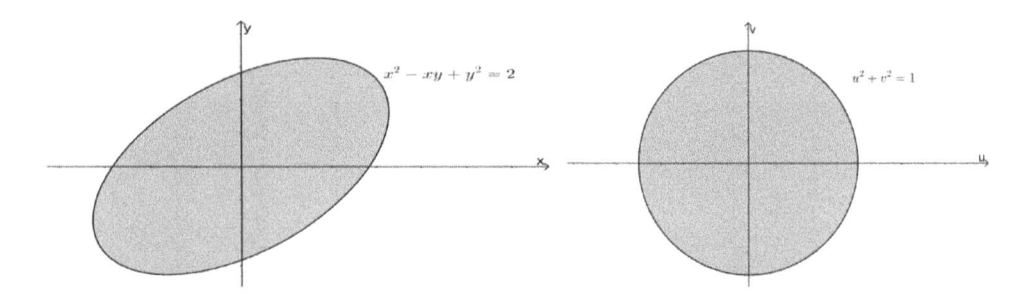

Example 17.

Use Green's Theorem to compute $\oint_C \dfrac{y}{2} dx + \dfrac{3x}{2} dy = ?$, where R is the region bounded by $y = x^2, y = 3x^2, x = y^2, x = 4y^2$, and C is the counterclockwise boundary of R.

【Solution】

By Green's Theorem, $\oint_C \dfrac{y}{2} dx + \dfrac{3x}{2} dy = \iint_R dA$

Let $u = \dfrac{y}{x^2}, \ v = \dfrac{x}{y^2}$ then $x = u^{-\frac{2}{3}} v^{-\frac{1}{3}}, \ y = u^{-\frac{1}{3}} v^{-\frac{2}{3}}$

and $dxdy = \begin{Vmatrix} \dfrac{\partial x}{\partial u} & \dfrac{\partial x}{\partial v} \\ \dfrac{\partial y}{\partial u} & \dfrac{\partial y}{\partial v} \end{Vmatrix} dudv = \begin{Vmatrix} \dfrac{-2u^{-\frac{5}{3}}v^{-\frac{1}{3}}}{3} & \dfrac{-u^{-\frac{2}{3}}v^{-\frac{4}{3}}}{3} \\ \dfrac{-u^{-\frac{4}{3}}v^{-\frac{2}{3}}}{3} & \dfrac{-2u^{-\frac{1}{3}}v^{-\frac{5}{3}}}{3} \end{Vmatrix} dudv = \dfrac{u^{-2}v^{-2}}{3} dudv$

Let $R = \left\{(x,y) : 1 \le \dfrac{y}{x^2} \le 3, 1 \le \dfrac{x}{y^2} \le 4\right\}$ then $R = \{(u,v) : 1 \le u \le 3, 1 \le v \le 4\}$

$\therefore \iint_R 1 dA = \int_1^3 \int_1^4 \dfrac{u^{-2} v^{-2}}{3} dvdu = \dfrac{1}{6}$

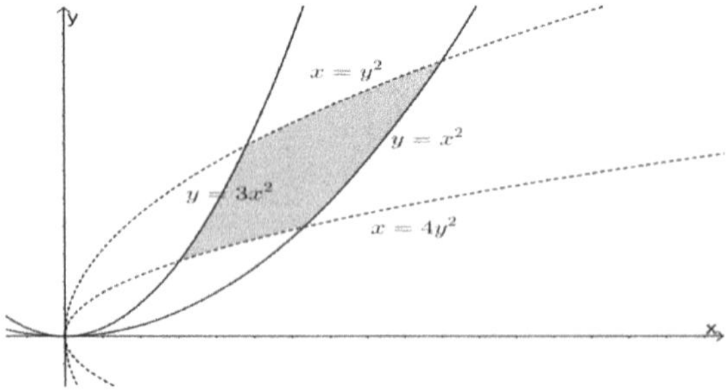

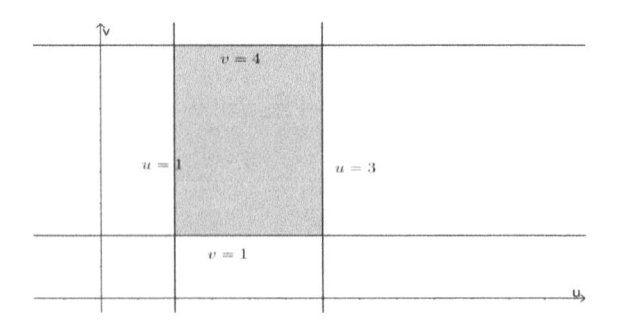

Example 18.

Use Green's Theorem to compute $\oint_C -\dfrac{y^3}{3}dx + \dfrac{x^3}{3}dy = ?$, where R is the region enclosed by $x^2 - y^2 = 1$, $x^2 - y^2 = 9$, $xy = 2$, $xy = 4$, and C is the counterclockwise boundary of R.

【Solution】

By Green's Theorem, $\oint_C -\dfrac{y^3}{3}dx + \dfrac{x^3}{3}dy = \iint_R x^2 + y^2\, dxdy$

Let $x^2 - y^2 = u$, $2xy = v$ then

$$dudv = \begin{Vmatrix} \dfrac{\partial u}{\partial x} & \dfrac{\partial u}{\partial y} \\ \dfrac{\partial v}{\partial x} & \dfrac{\partial v}{\partial y} \end{Vmatrix} dxdy = \begin{Vmatrix} 2x & -2y \\ 2y & 2x \end{Vmatrix} dxdy = 4(x^2 + y^2)dxdy$$

$\because (x^2 + y^2)^2 = (x^2 - y^2)^2 + (2xy)^2 = u^2 + v^2 \quad \therefore x^2 + y^2 = (u^2 + v^2)^{\frac{1}{2}}$

$\therefore dxdy = \dfrac{dudv}{4(u^2 + v^2)^{\frac{1}{2}}}$

$\because R = \{(x,y): 1 \le x^2 - y^2 \le 9, 4 \le 2xy \le 8\} \quad \therefore R = \{(u,v): 1 \le u \le 9, 4 \le v \le 8\}$

$\therefore \iint_R x^2 + y^2\, dxdy = \int_4^8 \int_1^9 \dfrac{(u^2 + v^2)^{\frac{1}{2}}}{4(u^2 + v^2)^{\frac{1}{2}}} dudv = 8$

Example 19.

(1) Use Green's Theorem to compute $\oint_C -\dfrac{y^3}{3}dx + \dfrac{x^3}{3}dy = ?$, where R is the region enclosed by $-x = y$, $y = -x + 3$, $x - 3 = y$, $y = x$, and C is the counterclockwise boundary of R.

(2) Use Green's Theorem to compute $\oint_C \dfrac{xy^2}{2}dx + x^2y\,dy = ?$, where R is the region enclosed by $x^2 + y^2 = a$, $x^2 + y^2 = b$, $x^2 - y^2 = c$, $x^2 - y^2 = d$, $(b > a > 0, d > c > 0)$ and C is the counterclockwise boundary of R in the first quadrant.

【Solution】

(1)

9.4.4 Use Coordinate Transformation to Compute Surface Integrals

By Green's Theorem, $\oint_C -\dfrac{y^3}{3}dx + \dfrac{x^3}{3}dy = \iint_R x^2 + y^2\, dxdy$

Let $x + y = u$, $x - y = v$ then $x = \dfrac{u+v}{2}$, $y = \dfrac{u-v}{2}$

and $dxdy = \begin{Vmatrix} \dfrac{\partial x}{\partial u} & \dfrac{\partial x}{\partial v} \\ \dfrac{\partial y}{\partial u} & \dfrac{\partial y}{\partial v} \end{Vmatrix} dudv = \begin{Vmatrix} \dfrac{1}{2} & \dfrac{1}{2} \\ \dfrac{1}{2} & \dfrac{-1}{2} \end{Vmatrix} dudv = \dfrac{1}{2} dudv$

∵ $R = \{(x, y): 0 \le x + y \le 3, 0 \le x - y \le 3\}$ ∴ $R = \{(u, v): 0 \le u \le 3, 0 \le v \le 3\}$

∴ $\iint_R x^2 + y^2 dA = \int_0^3 \int_0^3 \left(\left(\dfrac{u+v}{2}\right)^2 + \left(\dfrac{u-v}{2}\right)^2\right)\dfrac{1}{2} dudv = \dfrac{1}{4}\int_0^3 \int_0^3 u^2 + v^2 dudv = \dfrac{27}{2}$

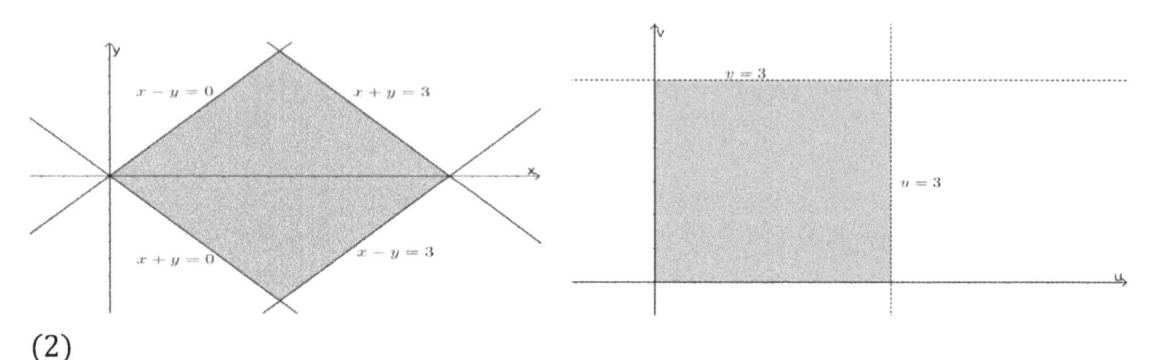

(2)

By Green's Theorem, $\oint_C \dfrac{xy^2}{2}dx + x^2y\, dy = \iint_R xy\, dxdy$

Let $x^2 + y^2 = u$, $x^2 - y^2 = v$ then $x = \dfrac{(u+v)^{\frac{1}{2}}}{\sqrt{2}}$, $y = \dfrac{(u-v)^{\frac{1}{2}}}{\sqrt{2}}$

and $dxdy = \begin{Vmatrix} \dfrac{\partial x}{\partial u} & \dfrac{\partial x}{\partial v} \\ \dfrac{\partial y}{\partial u} & \dfrac{\partial y}{\partial v} \end{Vmatrix} dudv = \begin{Vmatrix} \dfrac{(u+v)^{-\frac{1}{2}}}{2\sqrt{2}} & \dfrac{(u+v)^{-\frac{1}{2}}}{2\sqrt{2}} \\ \dfrac{(u-v)^{-\frac{1}{2}}}{2\sqrt{2}} & \dfrac{(u-v)^{-\frac{1}{2}}}{2\sqrt{2}} \end{Vmatrix} dudv = \dfrac{(u^2-v^2)^{-\frac{1}{2}}}{4} dudv$

∵ $R = \{(x, y): a \le x^2 + y^2 \le b, c \le x^2 - y^2 \le d\}$ ∴ $R = \{(u, v): a \le u \le b, c \le v \le d\}$

∴ $\iint_R xy\, dA = \int_c^d \int_a^b \dfrac{(u^2-v^2)^{\frac{1}{2}}}{2} \cdot \dfrac{(u^2-v^2)^{-\frac{1}{2}}}{4} dudv = \dfrac{(b-a)(c-d)}{8}$

Example 20.

Use Green's Theorem to compute $\oint_C -\dfrac{y^2}{2}dx + x^2 dy = ?$, where R is the region enclosed by $x^2 - 2xy + y^2 + x + y = 0$ and $x + y + 4 = 0$, and C is the counterclockwise boundary of R.

【Solution】

By Green's Theorem, $\oint_C -\dfrac{y^2}{2}dx + x^2 dy = \iint_R 2x + y \, dxdy$

$\because x^2 - 2xy + y^2 + x + y = 0 \Leftrightarrow (x-y)^2 = -(x+y)$

Let $x + y = u$, $x - y = v$ then $x = \dfrac{u+v}{2}$, $y = \dfrac{u-v}{2}$

and $dxdy = \left\| \begin{matrix} \dfrac{\partial x}{\partial u} & \dfrac{\partial x}{\partial v} \\ \dfrac{\partial y}{\partial u} & \dfrac{\partial y}{\partial v} \end{matrix} \right\| dudv = \left\| \begin{matrix} \dfrac{1}{2} & \dfrac{1}{2} \\ \dfrac{1}{2} & \dfrac{1}{2} \end{matrix} \right\| dudv = \dfrac{1}{2} dudv$

$\because R = \{(u,v): -v^2 \le u \le 0, -2 \le v \le 2\}$

$\therefore \iint_R 2x + y \, dxdy = \int_{-2}^{2}\int_{-v^2}^{0} 3u + v \, dudv = -\dfrac{96}{5}$

9.4.5 Convert Surface Integrals into Closed Line Integrals

If the integrand of a surface integral is complex, then use Green's Theorem to convert the surface integral into a closed line integral.

Type 1.

Use Green's Theorem to find $\iint_R \dfrac{\partial g}{\partial x}(x,y) - \dfrac{\partial f}{\partial y}(x,y) \, dxdy = ?$, where

$R = \left\{ (x,y): \dfrac{x^2}{a^2} + \dfrac{y^2}{b^2} \le r^2 \right\}$, C is the counterclockwise closed boundary of R, and

9.4.5 Convert Surface Integrals into Closed Line Integrals

the first partial derivatives of f and g exist and are continuous.

Problem-Solving Process:

Step1.

By Green's Theorem, $\iint_R \frac{\partial g}{\partial x}(x,y) - \frac{\partial f}{\partial y}(x,y) dxdy = \oint_C f(x,y)\,dx + g(x,y)dy$

Step2.

Let $x = ra\cos\theta$, $y = rb\sin\theta$ then

$$\oint_C f(x,y)\,dx + g(x,y)dy$$
$$= \int_0^{2\pi} f(ra\cos\theta, rb\sin\theta)(-ra\sin\theta)\,d\theta + \int_0^{2\pi} g(ra\cos\theta, rb\sin\theta)(rb\cos\theta)\,d\theta$$

Type 2.

Use Green's Theorem to find $\iint_R \frac{\partial g}{\partial x}(x,y) - \frac{\partial f}{\partial y}(x,y)\,dxdy =?$, where C is the counterclockwise closed boundary of R, and $C = \bigcup_{i=1}^{k} C_i$, $k \in N$, C_i is a smooth line segment, and the first partial derivatives of f and g exist and are continuous.

Problem-Solving Process:

Step1.

By Green's Theorem, $\iint_R \frac{\partial g}{\partial x}(x,y) - \frac{\partial f}{\partial y}(x,y) dxdy = \oint_C f(x,y)\,dx + g(x,y)dy$

Step2.

$$\oint_C f(x,y)\,dx + g(x,y)dy = \sum_{i=1}^{k} \int_{C_i} f(x,y)\,dx + g(x,y)dy$$

Example 1.

Use Green's Theorem to find $\iint_R \frac{\partial g}{\partial x}(x,y) - \frac{\partial f}{\partial y}(x,y)\,dxdy =?$, where

$f(x,y) = \frac{x^2 y^2}{2}$, $g(x,y) = \frac{2x^3 y}{3}$ and $R = \{(x,y): y \geq 0,\ x^2 + y^2 \leq 4\}$.

【Solution】

Let $C_1: \{(x,y): x^2 + y^2 = 4, y > 0\}$, $C_2: \{(x,y): -2 \leq x \leq 2, y = 0\}$ and $C = C_1 \cup C_2$

By Green's Theorem, $\iint_R \frac{\partial g}{\partial x}(x,y) - \frac{\partial f}{\partial y}(x,y) \, dxdy = \oint_C \frac{x^2 y^2}{2} dx + \frac{2x^3 y}{3} dy$

Let $x = 2\cos\theta$, $y = 2\sin\theta$, $0 \leq \theta \leq \pi$ then

$\int_{C_1} f(x,y)dx + g(x,y)dy = \int_{C_1} \frac{x^2 y^2}{2} dx + \frac{2x^3 y}{3} dy$

$= \int_0^\pi \frac{16 \cdot \cos^2\theta \cdot \sin^2\theta}{2}(-2\sin\theta)d\theta + \frac{2}{3}\int_0^\pi 16 \cdot \cos^3\theta \cdot \sin\theta \,(2\cos\theta)\, d\theta$

$= -16 \int_0^\pi (1 - \sin^2\theta) \cdot \sin^3\theta \, d\theta + \frac{64}{3}\int_0^\pi \cos^4\theta \cdot \sin\theta \, d\theta$

$= -16\left(-\cos\theta + \frac{\cos^3\theta}{3} + \frac{\cos^5\theta}{5} - \frac{2\cos^3\theta}{3} + \cos\theta\right)\Big|_0^\pi - \frac{64}{3} \cdot \frac{\cos^5\theta}{5}\Big|_0^\pi$

$= -16\left(2 - \frac{2}{3} - \frac{2}{5} + \frac{4}{3} - 2\right) - \frac{64}{3} \cdot \left(\frac{-2}{5}\right) = \frac{64}{15}$

$\because \int_{C_2} \frac{x^2 y^2}{2} dx + \frac{2x^3 y}{3} dy = 0 \quad \therefore \oint_C \frac{x^2 y^2}{2} dx + \frac{2x^3 y}{3} dy = \frac{64}{15}$

Example 2.

Use Green's Theorem to compute $\iint_R \frac{\partial g}{\partial x}(x,y) - \frac{\partial f}{\partial y}(x,y) \, dxdy = ?$, where

$f(x,y) = -\frac{y^3}{3}$, $g(x,y) = \frac{x^3}{3}$, $R = \left\{(x,y): \frac{x^2}{a^2} + \frac{y^2}{b^2} \leq 4\right\}$.

【Solution】

Let C be the counterclockwise closed boundary of R

By Green's Theorem, $\iint_R \frac{\partial g}{\partial x}(x,y) - \frac{\partial f}{\partial y}(x,y) \, dxdy = \oint_C -\frac{y^3}{3} dx + \frac{x^3}{3} dy$

Let $x = 2a\cos\theta$, $y = 2b\sin\theta$ then

$\oint_C -\frac{y^3}{3} dx + \frac{x^3}{3} dy = \frac{-1}{3}\int_0^{2\pi} 8b^3 \cdot \sin^3\theta \cdot (-2a\sin\theta) d\theta + \frac{1}{3}\int_0^{2\pi} 8a^3 \cdot \cos^3\theta \cdot 2b\cos\theta \, d\theta$

$= \frac{16b^3 a}{3} \int_0^{2\pi} \sin^4\theta \, d\theta + \frac{16a^3 b}{3}\int_0^{2\pi} \cos^4\theta \, d\theta = \frac{16b^3 a}{3} \cdot \frac{3\pi}{4} + \frac{16a^3 b}{3} \cdot \frac{3\pi}{4} = 4ab\pi(a^2 + b^2)$

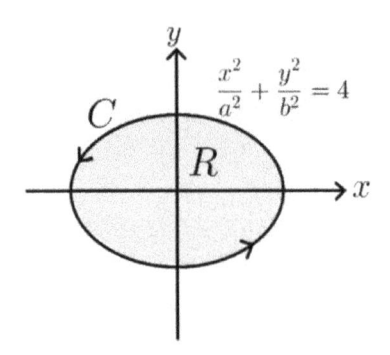

9.4.5 Convert Surface Integrals into Closed Line Integrals

Example 3.

Use Green's Theorem to compute $\iint_R \frac{\partial g}{\partial x}(x,y) - \frac{\partial f}{\partial y}(x,y)\, dxdy = ?$ where $f(x,y) = e^{x+y}$, $g(x,y) = 2e^{x+y}$, $R = \{(x,y): |x| + |y| \leq 2\}$.

【Solution】

Let $C_1: \{(x,y): x - y = 2\}, C_2: \{(x,y): x + y = 2\}, C_3: \{(x,y): -x + y = 2\}$,
$C_4: \{(x,y): -x - y = 2\}$ and $C = C_1 \cup C_2 \cup C_3 \cup C_4$.

By Green's Theorem, $\iint_R \frac{\partial g}{\partial x}(x,y) - \frac{\partial f}{\partial y}(x,y)\, dxdy = \oint_C e^{x+y}\, dx + 2e^{x+y}\, dy$

$\because \int_{C_1} f(x,y)dx + g(x,y)dy = \int_{C_1} e^{x+y} dx + 2e^{x+y} dy = \int_0^2 e^{x+y} dx + 2e^{x+y} dx$

$= 3 \int_0^2 e^{2x-2}\, dx = \frac{3}{2}(e^2 - e^{-2})$

$\int_{C_2} f(x,y)dx + g(x,y)dy = \int_2^0 e^2 - 2e^2\, dx = 2e^2$

$\int_{C_4} f(x,y)dx + g(x,y)dy = \int_{-2}^0 e^{x+y} - 2e^{x+y}\, dx = -\int_{-2}^0 e^{-2}\, dx = -2e^2$

$\int_{C_3} f(x,y)dx + g(x,y)dy = \int_{C_3} e^{x+y} dx + 2e^{x+y} dy = \int_0^{-2} e^{x+y} dx + 2e^{x+y} dx$

$= 3 \int_0^{-2} e^{2x+2}\, dx = \frac{3}{2}(e^{-2} - e^2)$

$\therefore \oint_C f\, dx + g\, dy = \frac{3}{2}(e^2 - e^{-2}) + 2e^2 + \frac{3}{2}(e^{-2} - e^2) - 2e^2 = 2(e^2 - e^{-2})$

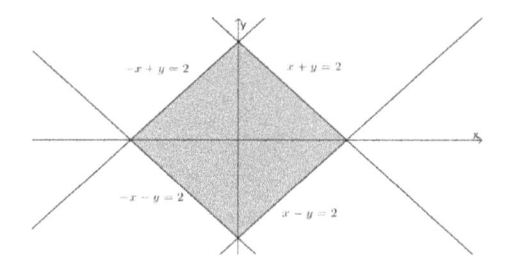

Example 4.

Use Green's Theorem to compute $\iint_R \frac{\partial g}{\partial x}(x,y) - \frac{\partial f}{\partial y}(x,y)\,dxdy = ?$, where

$f(x,y) = -\frac{2y^{\frac{3}{2}}}{3}$, $g(x,y) = \frac{2x^{\frac{3}{2}}}{3}$, R is the region enclosed by $\sqrt{x}+\sqrt{y}=1$, $x=0$, and $y=0$.

【Solution】

Let $C_1: \{(x,y): \sqrt{x}+\sqrt{y}=1, 0 \le x \le 1\}$, $C_2: \{(x,y): x=0, 0 \le y \le 2\}$,
$C_3: \{(x,y): y=0, 0 \le x \le 1\}$ and $C = C_1 \cup C_2 \cup C_3$

By Green's Theorem,

$$\iint_R \frac{\partial g}{\partial x}(x,y) - \frac{\partial f}{\partial y}(x,y)\,dxdy = \iint_R (\sqrt{x}+\sqrt{y})\,dxdy = \oint_C -\frac{2y^{\frac{3}{2}}}{3}dx + \frac{2x^{\frac{3}{2}}}{3}dy$$

$$\because \int_{C_1} -\frac{2y^{\frac{3}{2}}}{3}dx + \frac{2x^{\frac{3}{2}}}{3}dy = \int_1^0 -\frac{2}{3}\left(1-t^{\frac{1}{2}}\right)^3 dt + \frac{2}{3}\int_1^0 t^{\frac{3}{2}}\left(1-t^{-\frac{1}{2}}\right)dt$$

Let $u = 1 - t^{\frac{1}{2}}$ then

$$\int_1^0 -\frac{2}{3}\left(1-t^{\frac{1}{2}}\right)^3 dt = \frac{4}{3}\int_0^1 u^3(1-u)\,du = \frac{1}{15} \quad \text{and} \quad \frac{2}{3}\int_1^0 t^{\frac{3}{2}}\left(1-t^{-\frac{1}{2}}\right)dt = \frac{1}{15}$$

$$\because \int_{C_2} -\frac{2y^{\frac{3}{2}}}{3}dx + \frac{2x^{\frac{3}{2}}}{3}dy = 0 \quad \text{and} \quad \int_{C_3} -\frac{2y^{\frac{3}{2}}}{3}dx + \frac{2x^{\frac{3}{2}}}{3}dy = 0$$

$$\therefore \oint_C -\frac{2y^{\frac{3}{2}}}{3}dx + \frac{2x^{\frac{3}{2}}}{3}dy = \frac{2}{15}$$

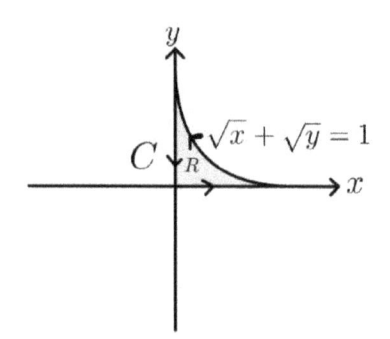

9.4.5 Convert Surface Integrals into Closed Line Integrals

Example 5.

Let $f(x, y) = \left(\dfrac{y^2}{2} - 4y\right)$, $g(x, y) = -\dfrac{x^2}{2}$, and $R = \{(x, y): 0 \leq x \leq 2, 0 \leq y \leq 2\}$.

Use Green's Theorem to compute $\iint_R \dfrac{\partial g}{\partial x}(x, y) - \dfrac{\partial f}{\partial y}(x, y)\, dxdy = ?$

【Solution】

Let $C_1: \{(x, y): 0 \leq x \leq 2, y = 0\}$, $C_2: \{(x, y): x = 2, 0 \leq y \leq 2\}$,
$C_3: \{(x, y): 0 \leq x \leq 2, y = 2\}$, $C_4: \{(x, y): x = 0, 0 \leq y \leq 2\}$ and $C = C_1 \cup C_2 \cup C_3 \cup C_4$

By Green's Theorem,

$$\iint_R \dfrac{\partial g}{\partial x}(x, y) - \dfrac{\partial f}{\partial y}(x, y)\, dxdy = \iint_R 4 - x - y\, dA = \oint_C \left(\dfrac{y^2}{2} - 4y\right) dx + \left(-\dfrac{x^2}{2}\right) dy$$

$\because \int_{C_1} \left(\dfrac{y^2}{2} - 4y\right) dx + \left(-\dfrac{x^2}{2}\right) dy = 0,\quad \int_{C_2} \left(\dfrac{y^2}{2} - 4y\right) dx + \left(-\dfrac{x^2}{2}\right) dy = \int_0^2 -2\, dx = -4$

$\int_{C_3} \left(\dfrac{y^2}{2} - 4y\right) dx + \left(-\dfrac{x^2}{2}\right) dy = \int_2^0 2 - 8\, dx = 12$ and $\int_{C_4} \left(\dfrac{y^2}{2} - 4y\right) dx + \left(-\dfrac{x^2}{2}\right) dy = 0$

$\therefore \oint_C \left(\dfrac{y^2}{2} - 4y\right) dx + \left(-\dfrac{x^2}{2}\right) dy = 8$

Example 6.

Use Green's Theorem to compute $\iint_R \dfrac{\partial g}{\partial x}(x, y) - \dfrac{\partial f}{\partial y}(x, y)\, dxdy = ?$, where

$f(x, y) = \dfrac{xy^2}{2}$, $g(x, y) = x^2 y$, and $R = \{(x, y): 0 \leq x \leq 2, 2x - 4 \leq y \leq 0\}$.

【Solution】
Let $C_1: \{(x,y): 0 \leq x \leq 2, y = 2x - 4\}$, $C_2: \{(x,y): 0 \leq x \leq 2, y = 0\}$,
$C_3: \{(x,y): x = 0, -4 \leq y \leq 0\}$ and $C = C_1 \cup C_2 \cup C_3$
By Green's Theorem,

$$\iint_R \frac{\partial g}{\partial x}(x,y) - \frac{\partial f}{\partial y}(x,y)\, dxdy = \iint_R xy\, dA = \oint_C \frac{xy^2}{2} dx + (x^2 y) dy$$

$$\because \int_{C_1} \frac{xy^2}{2} dx + (x^2 y) dy = \int_0^2 \frac{x(2x-4)^2}{2} dx + 2x^2(2x-4) dx = \int_0^2 (2x^2 - 4x)(3x-2) dx$$

$$= \int_0^2 6x^3 - 16x^2 + 8x\, dx = \frac{-8}{3},$$

$$\int_{C_2} \frac{xy^2}{2} dx + (x^2 y) dy = 0 \text{ and } \int_{C_3} \frac{xy^2}{2} dx + (x^2 y) dy = 0$$

$$\therefore \oint_C \frac{xy^2}{2} dx + (x^2 y) dy = \frac{-8}{3}$$

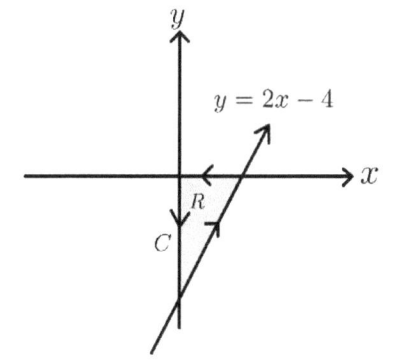

Example 7.

Let $f(x, y) = \frac{y^3}{3}$, $g(x, y) = x^2$, and $R = \{(x,y): x - y + 1 \leq 0, y \leq 2, x + y - 1 \geq 0\}$.

Use Green's Theorem to compute $\iint_R \frac{\partial g}{\partial x}(x,y) - \frac{\partial f}{\partial y}(x,y)\, dxdy = ?$

【Solution】
Let $C_1: \{(x,y): 0 \leq x \leq 1, x - y + 1 = 0\}$, $C_2: \{(x,y): -1 \leq x \leq 1, y = 2\}$,
$C_3: \{(x,y): x + y - 1 = 0, -1 \leq x \leq 0\}$ and $C = C_1 \cup C_2 \cup C_3$
By Green's Theorem,

9.4.5 Convert Surface Integrals into Closed Line Integrals

$$\iint_R \frac{\partial g}{\partial x}(x,y) - \frac{\partial f}{\partial y}(x,y)\, dxdy = \iint_R 2x - y^2\, dA = \oint_C \frac{y^3}{3} dx + x^2 dy$$

$$\therefore \int_{C_1} \frac{y^3}{3} dx + x^2 dy = \int_0^1 \frac{(x+1)^3}{3} dx + x^2 dx = \int_0^1 \frac{x^3 + 6x^2 + 3x + 1}{3} dx = \frac{19}{12},$$

$$\int_{C_2} \frac{y^3}{3} dx + x^2 dy = \int_1^{-1} \frac{8}{3} dx = -\frac{16}{3},$$

$$\int_{C_3} \frac{y^3}{3} dx + x^2 dy = \int_{-1}^0 \frac{(1-x)^3}{3} dx - x^2 dx = \int_{-1}^0 \frac{1 - 3x - x^3}{3} dx = \frac{11}{12}$$

$$\therefore \oint_C \frac{y^3}{3} dx + x^2 dy = \frac{19}{12} - \frac{16}{3} + \frac{11}{12} = \frac{-17}{6}$$

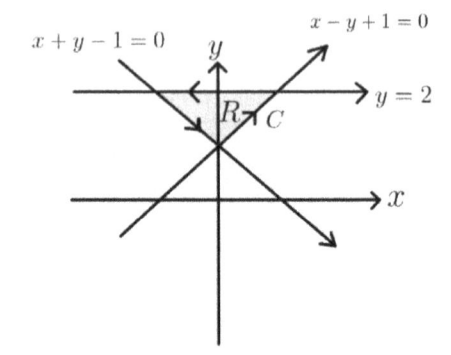

Example 8.

Let $f(x,y) = y$, $g(x,y) = 2x$, $R = \{(x,y): -3x - y + 6 \leq 0, 4x - x^2 - y \geq 0, y \geq 0\}$.

Use Green's Theorem to compute $\iint_R \frac{\partial g}{\partial x}(x,y) - \frac{\partial f}{\partial y}(x,y)\, dxdy =?$

【Solution】

Let $C_1: \{(x,y): 1 \leq x \leq 4, 4x - x^2 - y = 0\}$, $C_2: \{(x,y): 1 \leq x \leq 2, -3x - y + 6 = 0\}$, $C_3: \{(x,y): 2 \leq x \leq 4, y = 0\}$ and $C = C_1 \cup C_2 \cup C_3$

By Green's Theorem, $\iint_R \frac{\partial g}{\partial x}(x,y) - \frac{\partial f}{\partial y}(x,y)\, dxdy = \iint_R dA = \oint_C y\, dx + 2x\, dy$

$$\therefore \int_{C_1} y\, dx + 2x\, dy = \int_4^1 4x - x^2\, dx + 2x(4 - 2x)\, dx = \int_4^1 -5x^2 + 12x\, dx = \left. \frac{-5x^3}{3} + 6x^2 \right|_4^1$$

$= 15$,

$$\int_{C_2} ydx + 2xdy = \int_1^2 6 - 3xdx + 2x(-3)dx = \int_1^2 -9x + 6dx = \left.\frac{-9x^2}{2} + 6x\right|_1^2 = -\frac{15}{2}$$

and $\int_{C_3} ydx + 2xdy = 0$

$\therefore \oint_C y\,dx + 2xdy = \dfrac{15}{2}$

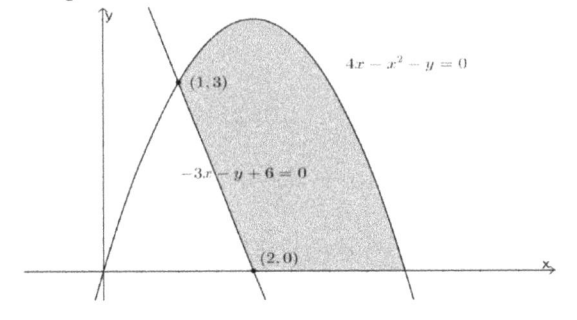

Example 9.

Use Green's Theorem to compute $\iint_R \dfrac{\partial g}{\partial x}(x,y) - \dfrac{\partial f}{\partial y}(x,y)\,dxdy = ?$, where

$f(x,y) = \dfrac{y^3}{3}, g(x,y) = \dfrac{2x^{\frac{3}{2}}}{3}$, and $R = \left\{(x,y): y \geq x^2, y \leq x^{\frac{1}{4}}\right\}$.

【Solution】

Let $C_1: \{(x,y): 0 \leq x \leq 1, y = x^2\}$, $C_2: \left\{(x,y): 0 \leq x \leq 1, y = x^{\frac{1}{4}}\right\}$ and $C = C_1 \cup C_2$

By Green's Theorem,

$$\iint_R \frac{\partial g}{\partial x}(x,y) - \frac{\partial f}{\partial y}(x,y)\,dxdy = \iint_R \sqrt{x} - y^2\,dxdy = \oint_C \frac{y^3}{3}dx + \frac{2x^{\frac{3}{2}}}{3}dy$$

$\therefore \int_{C_1} \dfrac{y^3}{3}dx + \dfrac{2x^{\frac{3}{2}}}{3}dy = \int_0^1 \dfrac{x^6}{3}dx + \dfrac{2x^{\frac{3}{2}}}{3} \cdot 2xdx = \dfrac{3}{7}$,

$\int_{C_2} \dfrac{y^3}{3}dx + \dfrac{2x^{\frac{3}{2}}}{3}dy = \int_1^0 \dfrac{x^{\frac{3}{4}}}{3}dx + \dfrac{x^{\frac{3}{4}}}{6}dx = -\dfrac{2}{7}$

$\therefore \oint_C \dfrac{y^3}{3}dx + \dfrac{2x^{\frac{3}{2}}}{3}dy = \dfrac{1}{7}$

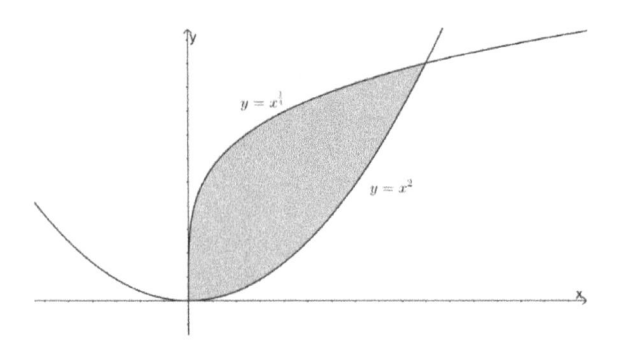

9.4.5 Convert Surface Integrals into Closed Line Integrals

Example 10.

Let $f(x,y) = 6y + x$, $g(x,y) = y + 2x$, and $R = \{(x,y): (x-2)^2 + (y-3)^2 \leq 9\}$.

Use Green's Theorem to compute $\iint_R \frac{\partial g}{\partial x}(x,y) - \frac{\partial f}{\partial y}(x,y)\, dxdy = ?$

【Solution】

Let C be a counterclockwise closed circle: $\{(x,y): (x-2)^2 + (y-3)^2 = 9\}$

By Green's Theorem,

$$\iint_R \frac{\partial g}{\partial x}(x,y) - \frac{\partial f}{\partial y}(x,y)\, dxdy = \iint_R 2 - 6\, dxdy = \oint_C (6y+x)\, dx + (y+2x)\, dy$$

Let $x = 2 + 3\cos\theta$, $y = 3 + 3\sin\theta$ then

$$\oint_C (6y + x)\, dx + (y + 2x)\, dy$$

$$= \int_0^{2\pi} (6(3 + 3\sin\theta) + (2 + 3\cos\theta))(-3\sin\theta)\, d\theta$$

$$+ \int_0^{2\pi} (3 + 3\sin\theta + (4 + 6\cos\theta))3\cos\theta\, d\theta$$

$$= \int_0^{2\pi} -\frac{54}{2} + 18 \cdot \frac{1}{2}\, d\theta = 2\pi \cdot (-18) = -36\pi$$

Example 11.

Let $f(x,y) = y^3 - 2y$, $g(x,y) = -(2x^3 + x)$, and $R = \{(x,y): 6x^2 + 3y^2 \leq 1\}$.

Use Green's Theorem to compute $\iint_R \frac{\partial g}{\partial x}(x,y) - \frac{\partial f}{\partial y}(x,y)\, dxdy = ?$

【Solution】

Let C be a counterclockwise closed circle: $\{(x,y): 6x^2 + 3y^2 = 1\}$
By Green's Theorem,

$$\iint_R \frac{\partial g}{\partial x}(x,y) - \frac{\partial f}{\partial y}(x,y)\,dxdy = \iint_R 1 - 6x^2 - 3y^2\,dxdy = \oint_C y^3 - 2y\,dx - (2x^3 + x)dy$$

Let $x = \dfrac{\cos\theta}{\sqrt{6}},\ y = \dfrac{\sin\theta}{\sqrt{3}}$ then

$$\oint_C (y^3 - 2y)\,dx - (2x^3 + x)dy$$

$$= \int_0^{2\pi}\left(\frac{\sin^3\theta}{3\sqrt{3}} - \frac{2\sin\theta}{\sqrt{3}}\right)\left(-\frac{\sin\theta}{\sqrt{6}}\right)d\theta - \int_0^{2\pi}\left(\frac{\cos^3\theta}{3\sqrt{6}} + \frac{\cos\theta}{\sqrt{6}}\right)\left(\frac{\cos\theta}{\sqrt{3}}\right)d\theta$$

$$= -\frac{1}{9\sqrt{2}}\int_0^{2\pi}\sin^4\theta\,d\theta + \frac{\sqrt{2}}{3}\int_0^{2\pi}\sin^2\theta\,d\theta - \frac{1}{9\sqrt{2}}\int_0^{2\pi}\cos^4\theta\,d\theta - \frac{1}{3\sqrt{2}}\int_0^{2\pi}\cos^2\theta\,d\theta$$

$$= -\frac{1}{9\sqrt{2}}\cdot\frac{3}{8}\cdot 2\pi + \frac{\sqrt{2}}{3}\cdot\frac{1}{\sqrt{2}}\cdot 2\pi - \frac{1}{9\sqrt{2}}\cdot\frac{3}{8}\cdot 2\pi - \frac{1}{3\sqrt{2}}\cdot\frac{1}{2}\cdot 2\pi = \frac{\pi}{6\sqrt{2}}$$

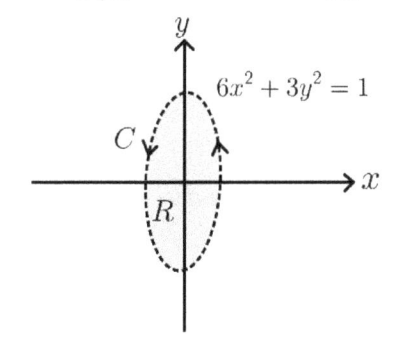

Example 12.

Let $f(x,y) = y^3,\ g(x,y) = -x^3$, and $R = \{(x,y): x^2 + y^2 \leq r^2\}$. Use Green's Theorem to compute $\iint_R \dfrac{\partial g}{\partial x}(x,y) - \dfrac{\partial f}{\partial y}(x,y)\,dxdy = ?$

【Solution】

Let C be a counterclockwise closed circle: $\{(x,y): x^2 + y^2 = r^2\}$
By Green's Theorem,

$$\iint_R \frac{\partial g}{\partial x}(x,y) - \frac{\partial f}{\partial y}(x,y)\,dxdy = \iint_R -3x^2 - 3y^2\,dxdy = \oint_C y^3 dx - x^3\,dy$$

Let $x = r\cos\theta,\ y = r\sin\theta,\ 0 \leq \theta \leq 2\pi$ then

9.4.5 Convert Surface Integrals into Closed Line Integrals

$$\oint_C y^3 dx - x^3 dy = \int_0^{2\pi} r^3 \sin^3 \theta \cdot (-r \sin \theta) - r^3 \cos^3 \theta \cdot (r \cos \theta) d\theta$$

$$= -r^4 \int_0^{2\pi} \sin^4 \theta + \cos^4 \theta \, d\theta = -r^4 \int_0^{2\pi} (\sin^2 \theta + \cos^2 \theta)^2 - 2 \sin^2 \theta \cos^2 \theta \, d\theta$$

$$= -r^4 \int_0^{2\pi} 1 - 2 \left(\frac{\sin 2\theta}{2}\right)^2 d\theta = -r^4 \int_0^{2\pi} 1 - \frac{\sin^2 2\theta}{2} d\theta = -r^4 \int_0^{2\pi} 1 - \frac{1 - \cos 4\theta}{4} d\theta$$

$$= -r^4 \cdot \frac{3}{4} \cdot 2\pi = -\frac{3r^4 \pi}{2}$$

Example 13.

Let $f(x, y) = 2y$, $g(x, y) = x^2 + y^2$, and $R = \{(x, y): x^2 + (y - 3)^2 \leq r^2\}$, $r > 3$.

Use Green's Theorem to compute $\iint_R \frac{\partial g}{\partial x}(x, y) - \frac{\partial f}{\partial y}(x, y) \, dxdy = ?$

【Solution】

Let C be a counterclockwise closed circle: $\{(x, y): x^2 + (y - 3)^2 = r^2\}$

By Green's Theorem,

$$\iint_R \frac{\partial g}{\partial x}(x, y) - \frac{\partial f}{\partial y}(x, y) \, dxdy = \iint_R 2x - 2 dxdy = \int_C 2y dx + (x^2 + y^2) dy$$

Let $x = r\cos \theta$, $y = 3 + r \sin \theta$, $0 \leq \theta \leq 2\pi$ then

$$\int_C 2y dx + (x^2 + y^2) dy = \int_0^{2\pi} 2(3 + r \sin \theta)(-r \sin \theta) + ((9 + r^2) + 6r \sin \theta) r \cos \theta \, d\theta$$

$$= \int_0^{2\pi} -6r \sin \theta - 2r^2 \sin^2 \theta + (9 + r^2) r \cos \theta + 6r^2 \sin \theta \cos \theta \, d\theta$$

$$= \int_0^{2\pi} -2r^2 \sin^2 \theta + 6r^2 \sin \theta \cos \theta \, d\theta = \int_0^{2\pi} -2r^2 \left(\frac{1 - \cos 2\theta}{2}\right) + 3r^2 \sin 2\theta \, d\theta = -2r^2 \pi$$

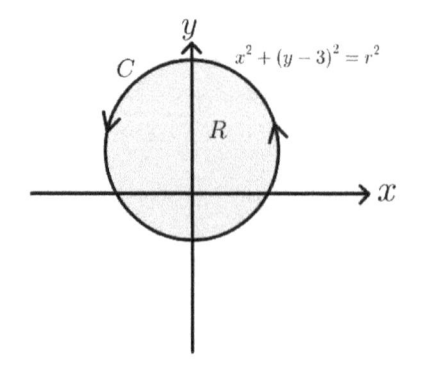

Example 14.

Let $f(x,y) = y\cos x$, $g(x,y) = x\sin y$, and R is the counterclockwise closed triangular region enclosed by the points $(0,0), (a,0),$ and $(a,a),$ where $a > 0.$ Use Green's Theorem to compute $\iint_R \frac{\partial g}{\partial x}(x,y) - \frac{\partial f}{\partial y}(x,y)\, dxdy = ?$

【Solution】

Let $C_1 = \{(x,y): 0 \le x \le a, y = 0\}$, $C_2 = \{(x,y): 0 \le y \le a, x = a\}$,
$C_3 = \{(x,y): 0 \le x \le a, y = x\}$ and $C = C_1 \cup C_2 \cup C_3$
By Green's Theorem,

$$\iint_R \frac{\partial g}{\partial x}(x,y) - \frac{\partial f}{\partial y}(x,y)\, dxdy = \iint_R \sin y - \cos x \, dxdy = \oint_C y\cos x\, dx + x\sin y\, dy$$

$$\because \oint_C y\cos x\, dx + x\sin y\, dy$$

$$= \int_{C_1} y\cos x\, dx + x\sin y\, dy + \int_{C_2} y\cos x\, dx + x\sin y\, dy + \int_{C_3} y\cos x\, dx + x\sin y\, dy$$

$$\because \int_{C_1} y\cos x\, dx + x\sin y\, dy = 0$$

$$\int_{C_2} y\cos x\, dx + x\sin y\, dy = a\int_0^a \sin y\, dy = a(-\cos a + 1)$$

$$\int_{C_3} y\cos x\, dx + x\sin y\, dy = \int_a^0 x\cos x\, dx + x\sin x\, dx = -(a+1)\sin a + (a-1)\cos a + 1$$

$$\therefore \oint_C y\cos x\, dx + x\sin y\, dy = -(a+1)\sin a - \cos a + a + 1$$

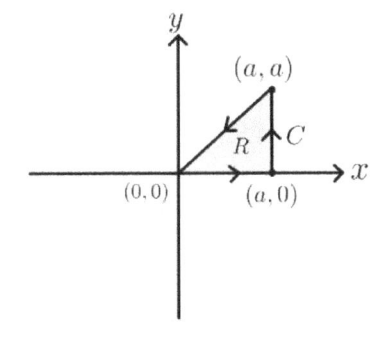

Example 15.

Let $f(x,y) = -y^4$, $g(x,y) = xy^2$, and R is the counterclockwise closed triangular region enclosed by the points $(0,0), (1,0), (1,1)$. Use Green's Theorem to compute

$$\iint_R \frac{\partial g}{\partial x}(x,y) - \frac{\partial f}{\partial y}(x,y) \, dxdy = ?$$

【Solution】

Let $C_1 = \{(x,y): 0 \leq x \leq 1, y = 0\}$, $C_2 = \{(x,y): 0 \leq y \leq 1, x = 1\}$
$C_3 = \{(x,y): 0 \leq x \leq 1, y = x\}$ and $C = C_1 \cup C_2 \cup C_3$
By Green's Theorem,

$$\iint_R \frac{\partial g}{\partial x}(x,y) - \frac{\partial f}{\partial y}(x,y) \, dxdy = \iint_R 2xy + 4y^3 \, dxdy = \oint_C -y^4 dx + xy^2 dy$$

$$\therefore \oint_C -y^4 dx + xy^2 dy = \sum_{j=1}^3 \int_{C_j} -y^4 dx + xy^2 dy$$

$$\therefore \int_{C_1} -y^4 dx + xy^2 dy = 0, \quad \int_{C_2} -y^4 dx + xy^2 dy = \int_0^1 y^2 dy = \frac{1}{3}$$

$$\int_{C_3} -y^4 dx + xy^2 dy = \int_1^0 -x^4 dx + x^3 dy = \frac{-1}{20}$$

$$\therefore \oint_C -y^4 dx + xy^2 dy = \frac{1}{3} - \frac{1}{20} = \frac{17}{60}$$

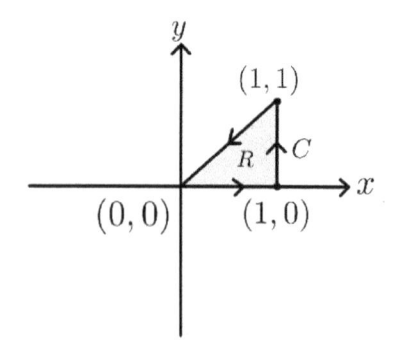

9.4.6 Create a Closed Region without Singularities

If the functions f and g in Green's Theorem do not satisfy the condition of having first partial derivatives that exist and are continuous within a closed region, then Green's Theorem cannot be

9.4.6 Create a Closed Region without Singularities

directly applied. Instead, a closed region that does not contain singular points (where the functions f and g are discontinuous or non-differentiable) must first be created before using Green's Theorem.

Question Types:
Type 1.

Find $\oint_C f(x,y)\,dx + g(x,y)\,dy = ?$, where C is any closed curve that includes the origin (as shown in the diagram), and the first partial derivatives of f and g do not exist or are discontinuous at the origin.

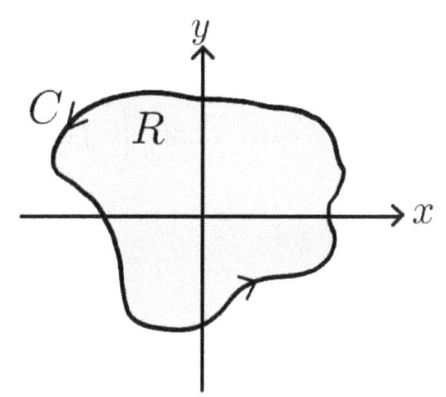

Remark:
Common closed curves include rectangles, squares, ellipses, circles, and irregular shapes. Regardless of the type of curve, the proof structure is similar.

Problem-Solving Process:

Step1.

Let O_ε be a circle of radius ε that contains the origin, and choose two points on the circle: $(\varepsilon\cos\theta, \varepsilon\sin\theta)(\varepsilon\cos(-\theta), \varepsilon\sin(-\theta))$, where $\theta > 0$.

Assume the horizontal line passing through $(\varepsilon\cos\theta, \varepsilon\sin\theta)$ intersect C at s_1 and the horizontal line passing through $(\varepsilon\cos(-\theta), \varepsilon\sin(-\theta))$ intersect C at s_2.

Let C_ε be the counterclockwise arc on O_ε from $(\varepsilon\cos\theta, \varepsilon\sin\theta)$ to $(\varepsilon\cos(-\theta), \varepsilon\sin(-\theta))$.

Let C_1 be the line segment from $(\varepsilon\cos\theta, \varepsilon\sin\theta)$ to s_1.

Let C_2 be the line segment from s_2 to $(\varepsilon\cos(-\theta), \varepsilon\sin(-\theta))$.

Let C' be the counterclockwise arc of the curve from s_1 to s_2.

Let R' be the closed region enclosed by C_ε、C_1、C_2 and C' that does not contain the origin.

Step2.

By Green's Theorem,

9.4.6 Create a Closed Region without Singularities

$$\int_{C'} fdx + gdy + \int_{C_1} fdx + gdy - \int_{C_\varepsilon} fdx + gdy + \int_{C_2} fdx + gdy$$

$$= \iint_{R'} \frac{\partial g}{\partial x}(x,y) - \frac{\partial f}{\partial y}(x,y) dxdy$$

Step3.

Claim: $\lim_{\theta \to 0} \int_{C'} fdx + gdy = \lim_{\theta \to 0} \int_{C_\varepsilon} fdx + gdy$

$\because \frac{\partial g}{\partial x}(x,y) - \frac{\partial f}{\partial y}(x,y) = 0 \quad \therefore \iint_{R'} \frac{\partial g}{\partial x}(x,y) - \frac{\partial f}{\partial y}(x,y) dxdy = 0$

$$\Rightarrow \int_{C'} fdx + gdy + \int_{C_1} fdx + gdy - \int_{C_\varepsilon} fdx + gdy + \int_{C_2} fdx + gdy = 0$$

$\because \lim_{\theta \to 0} \int_{C_1} fdx + gdy + \int_{C_2} fdx + gdy = 0 \quad \therefore \lim_{\theta \to 0} \int_{C'} fdx + gdy - \int_{C_\varepsilon} fdx + gdy = 0$

$\therefore \lim_{\theta \to 0} \int_{C'} fdx + gdy = \lim_{\theta \to 0} \int_{C_\varepsilon} fdx + gdy$

Step4.

Let $x = \varepsilon \cos t, \ y = \varepsilon \sin t$ then $dx = -\varepsilon \sin t, \ dy = \varepsilon \cos t$

$\because \lim_{\theta \to 0} \int_{C_\varepsilon} fdx + gdy = \int_0^{2\pi} f(\varepsilon \cos t, \varepsilon \sin t)(-\varepsilon \sin t) + g(\varepsilon \cos t, \varepsilon \sin t)\varepsilon \cos t \, dt$

$\because \lim_{\theta \to 0} \int_{C'} fdx + gdy = \oint_C fdx + gdy$

$\therefore \oint_C fdx + gdy = \int_0^{2\pi} f(\varepsilon \cos t, \varepsilon \sin t)(-\varepsilon \sin t) + g(\varepsilon \cos t, \varepsilon \sin t)\varepsilon \cos t \, dt$

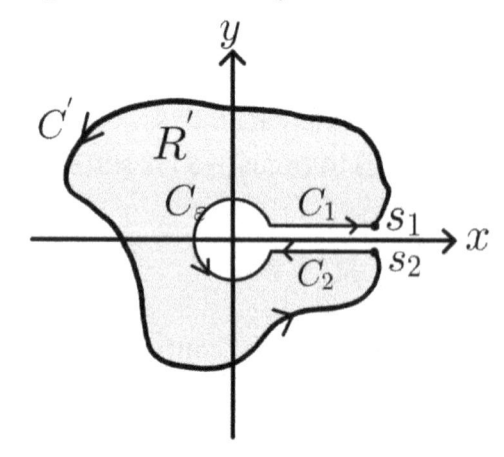

Examples:

Find $\oint_C f(x,y)\,dx + g(x,y)\,dy = ?,$

(I) If $f(x,y) = \dfrac{-y}{x^2+y^2}$ and $g(x,y) = \dfrac{x}{x^2+y^2}$ and C is any closed curve that includes the origin, then the first partial derivatives of f and g do not exist and are not continuous at the origin.

Let $(x,y) \neq (0,0)$ then
$$\frac{\partial g}{\partial x}(x,y) = \frac{x^2+y^2-2x^2}{(x^2+y^2)^2} = \frac{-x^2+y^2}{(x^2+y^2)^2}, \quad \frac{\partial f}{\partial y}(x,y) = \frac{-(x^2+y^2)+2y^2}{(x^2+y^2)^2} = \frac{-x^2+y^2}{(x^2+y^2)^2}$$

$\therefore \dfrac{\partial g}{\partial x}(x,y) - \dfrac{\partial f}{\partial y}(x,y) = 0, \quad \forall (x,y) \neq (0,0)$

(II) If $f(x,y) = \dfrac{x-y}{x^2+y^2}$ and $g(x,y) = \dfrac{x+y}{x^2+y^2}$ and C is any closed curve that includes the origin, then the first partial derivatives of f and g do not exist and are not continuous at the origin.

Let $(x,y) \neq (0,0)$ then
$$\frac{\partial g}{\partial x}(x,y) = \frac{x^2+y^2-2x^2-2xy}{(x^2+y^2)^2} = \frac{-x^2+y^2-2xy}{(x^2+y^2)^2}$$

and $\dfrac{\partial f}{\partial y}(x,y) = \dfrac{-(x^2+y^2)+2y^2-2xy}{(x^2+y^2)^2} = \dfrac{-x^2+y^2-2xy}{(x^2+y^2)^2}$

$\therefore \dfrac{\partial g}{\partial x}(x,y) - \dfrac{\partial f}{\partial y}(x,y) = 0, \quad \forall (x,y) \neq (0,0)$

(III) If $f(x,y) = \dfrac{-y^3}{(x^2+y^2)^2}$ and $g(x,y) = \dfrac{xy^2}{(x^2+y^2)^2}$ and C is any closed curve that includes the origin, then the first partial derivatives of f and g do not exist and are not continuousat the origin.

Let $(x,y) \neq (0,0)$ then
$$\frac{\partial g}{\partial x}(x,y) = \frac{y^2(x^2+y^2)^2 - xy^2 \cdot 2(x^2+y^2)2x}{(x^2+y^2)^4} = \frac{y^2(x^2+y^2) - 4x^2y^2}{(x^2+y^2)^3}$$

and $\dfrac{\partial f}{\partial y}(x,y) = \dfrac{-3y^2(x^2+y^2)^2 + y^3 \cdot 2(x^2+y^2)2y}{(x^2+y^2)^4} = \dfrac{-3y^2(x^2+y^2)+4y^4}{(x^2+y^2)^3}$

$\therefore \dfrac{\partial g}{\partial x}(x,y) - \dfrac{\partial f}{\partial y}(x,y) = 0, \quad \forall (x,y) \neq (0,0)$

Let O_ε be a circle of radius ε that contains the origin, and choose two points on the circle: $(\varepsilon \cos\theta, \varepsilon \sin\theta)(\varepsilon \cos(-\theta), \varepsilon \sin(-\theta))$, where $\theta > 0$.
Assume the horizontal line passing through $(\varepsilon \cos\theta, \varepsilon \sin\theta)$ intersect C at s_1.

9.4.6 Create a Closed Region without Singularities

and the horizontal line passing through $(\varepsilon\cos(-\theta), \varepsilon\sin(-\theta))$ intersect C at s_2

Let C_ε be the counterclockwise arc on O_ε from $(\varepsilon\cos\theta, \varepsilon\sin\theta)$ to $(\varepsilon\cos(-\theta), \varepsilon\sin(-\theta))$.

Let C_1 be the line segment from $(\varepsilon\cos\theta, \varepsilon\sin\theta)$ to s_1.

Let C_2 be the line segment from s_2 to $(\varepsilon\cos(-\theta), \varepsilon\sin(-\theta))$.

Let C' be the counterclockwise arc of the curve from s_1 to s_2.

By Green's Theorem,

$$\int_{C'} fdx + gdy + \int_{C_1} fdx + gdy - \int_{C_\varepsilon} fdx + gdy + \int_{C_2} fdx + gdy$$

$$= \iint_{R'} \frac{\partial g}{\partial x}(x,y) - \frac{\partial f}{\partial y}(x,y)\,dxdy$$

Claim: $\lim_{\theta \to 0} \int_{C'} fdx + gdy = \lim_{\theta \to 0} \int_{C_\varepsilon} fdx + gdy$

$\because \frac{\partial g}{\partial x}(x,y) - \frac{\partial f}{\partial y}(x,y) = 0 \quad \therefore \iint_{R'} \frac{\partial g}{\partial x}(x,y) - \frac{\partial f}{\partial y}(x,y)\,dxdy = 0$

$\Rightarrow \int_{C'} fdx + gdy + \int_{C_1} fdx + gdy - \int_{C_\varepsilon} fdx + gdy + \int_{C_2} fdx + gdy = 0$

$\because \lim_{\theta \to 0} \int_{C_1} fdx + gdy + \int_{C_2} fdx + gdy = 0 \quad \therefore \lim_{\theta \to 0} \int_{C'} fdx + gdy - \int_{C_\varepsilon} fdx + gdy = 0$

$\therefore \lim_{\theta \to 0} \int_{C'} fdx + gdy = \lim_{\theta \to 0} \int_{C_\varepsilon} fdx + gdy$

(I)

Claim: $\oint_C fdx + gdy = 2\pi$

Let $x = \varepsilon\cos t$, $y = \varepsilon\sin t$ then $dx = -\varepsilon\sin t$, $dy = \varepsilon\cos t$

$\because \lim_{\theta \to 0} \int_{C_\varepsilon} fdx + gdy = \int_0^{2\pi} \frac{-\varepsilon\sin t(-\varepsilon\sin t) + \varepsilon^2\cos^2 t}{\varepsilon^2}\,dt = 2\pi$

$\therefore \lim_{\theta \to 0} \int_{C'} fdx + gdy = \lim_{\theta \to 0} \int_{C_\varepsilon} fdx + gdy = 2\pi$

$\because \lim_{\theta \to 0} \int_{C'} fdx + gdy = \oint_C fdx + gdy \quad \therefore \oint_C fdx + gdy = 2\pi$

(II)

Claim: $\oint_C fdx + gdy = 2\pi$

9.4.6 Create a Closed Region without Singularities

Let $x = \varepsilon \cos t$, $y = \varepsilon \sin t$ then $dx = -\varepsilon \sin t$, $dy = \varepsilon \cos t$

$$\because \lim_{\theta \to 0} \int_{C_\varepsilon} f\,dx + g\,dy = \int_0^{2\pi} \frac{(\varepsilon \cos t - \varepsilon \sin t)(-\varepsilon \sin t) + (\varepsilon \cos t + \varepsilon \sin t)\varepsilon \cos t}{\varepsilon^2}\,dt = 2\pi$$

$$\therefore \lim_{\theta \to 0} \int_{C'} f\,dx + g\,dy = \lim_{\theta \to 0} \int_{C_\varepsilon} f\,dx + g\,dy = 2\pi$$

$$\because \lim_{\theta \to 0} \int_{C'} f\,dx + g\,dy = \oint_C f\,dx + g\,dy \quad \therefore \oint_C f\,dx + g\,dy = 2\pi$$

(III)

Claim: $\oint_C f\,dx + g\,dy = \pi$

Let $x = \varepsilon \cos t$, $y = \varepsilon \sin t$ then $dx = -\varepsilon \sin t$, $dy = \varepsilon \cos t$

$$\because \lim_{\theta \to 0} \int_{C_\varepsilon} f\,dx + g\,dy = \int_0^{2\pi} \frac{-(\varepsilon \sin t)^3(-\varepsilon \sin t) + (\varepsilon \cos t \cdot \varepsilon^2 \sin^2 t)\varepsilon \cos t}{\varepsilon^4}\,dt$$

$$= \int_0^{2\pi} \sin^2 t\,dt = \pi$$

$$\therefore \lim_{\theta \to 0} \int_{C'} f\,dx + g\,dy = \lim_{\theta \to 0} \int_{C_\varepsilon} f\,dx + g\,dy = \pi$$

$$\because \lim_{\theta \to 0} \int_{C'} f\,dx + g\,dy = \oint_C f\,dx + g\,dy \quad \therefore \oint_C f\,dx + g\,dy = \pi$$

Example 1.

Let $f(x,y) = \dfrac{-y}{x^2 + y^2}$ and $g(x,y) = \dfrac{x}{x^2 + y^2}$, and C be any closed curve that encloses the origin. Find $\oint_C f(x,y)\,dx + g(x,y)\,dy = ?$

【Solution】

Let O_ε be a circle of radius ε that contains the origin, and choose two points on the circle: $(\varepsilon \cos \theta, \varepsilon \sin \theta)(\varepsilon \cos(-\theta), \varepsilon \sin(-\theta))$, where $\theta > 0$.
Assume the horizontal line passing through $(\varepsilon \cos \theta, \varepsilon \sin \theta)$ intersect C at s_1 and the horizontal line passing through $(\varepsilon \cos(-\theta), \varepsilon \sin(-\theta))$ intersect C at s_2.
Let C_ε be the counterclockwise arc on O_ε from $(\varepsilon \cos \theta, \varepsilon \sin \theta)$ to $(\varepsilon \cos(-\theta), \varepsilon \sin(-\theta))$.
Let C_1 be the line segment from $(\varepsilon \cos \theta, \varepsilon \sin \theta)$ to s_1.
Let C_2 be the line segment from s_2 to $(\varepsilon \cos(-\theta), \varepsilon \sin(-\theta))$.

9.4.6 Create a Closed Region without Singularities

Let C' be the counterclockwise arc of the curve from s_1 to s_2.
Let R' be the closed region enclosed by C_ε、C_1、C_2 and C' that does not contain the origin.
By Green's Theorem,

$$\int_{C'} f\,dx + g\,dy + \int_{C_1} f\,dx + g\,dy - \int_{C_\varepsilon} f\,dx + g\,dy + \int_{C_2} f\,dx + g\,dy$$

$$= \iint_{R'} \frac{\partial g}{\partial x}(x,y) - \frac{\partial f}{\partial y}(x,y)\,dxdy$$

Let $(x,y) \neq (0,0)$ then

$$\frac{\partial g}{\partial x}(x,y) = \frac{x^2+y^2-2x^2}{(x^2+y^2)^2} = \frac{-x^2+y^2}{(x^2+y^2)^2},\quad \frac{\partial f}{\partial y}(x,y) = \frac{-(x^2+y^2)+2y^2}{(x^2+y^2)^2} = \frac{-x^2+y^2}{(x^2+y^2)^2}$$

$$\therefore \frac{\partial g}{\partial x}(x,y) - \frac{\partial f}{\partial y}(x,y) = 0 \Rightarrow \iint_{R'} \frac{\partial g}{\partial x}(x,y) - \frac{\partial f}{\partial y}(x,y)\,dxdy = 0$$

$$\Rightarrow \int_{C'} f\,dx + g\,dy + \int_{C_1} f\,dx + g\,dy - \int_{C_\varepsilon} f\,dx + g\,dy + \int_{C_2} f\,dx + g\,dy = 0$$

$$\because \lim_{\theta \to 0} \int_{C_1} f\,dx + g\,dy + \int_{C_2} f\,dx + g\,dy = 0 \quad \therefore \lim_{\theta \to 0} \int_{C'} f\,dx + g\,dy - \int_{C_\varepsilon} f\,dx + g\,dy = 0$$

$$\therefore \lim_{\theta \to 0} \int_{C'} f\,dx + g\,dy = \lim_{\theta \to 0} \int_{C_\varepsilon} f\,dx + g\,dy$$

Let $x = \varepsilon \cos t, y = \varepsilon \sin t$ then $dx = -\varepsilon \sin t, dy = \varepsilon \cos t$

$$\therefore \lim_{\theta \to 0} \int_{C_\varepsilon} f\,dx + g\,dy = \int_0^{2\pi} \frac{-\varepsilon \sin t(-\varepsilon \sin t) + \varepsilon^2 \cos^2 t}{\varepsilon^2} dt = 2\pi$$

$$\therefore \lim_{\theta \to 0} \int_{C'} f\,dx + g\,dy = \lim_{\theta \to 0} \int_{C_\varepsilon} f\,dx + g\,dy = 2\pi$$

$$\because \lim_{\theta \to 0} \int_{C'} f\,dx + g\,dy = \oint_C f\,dx + g\,dy \quad \therefore \oint_C f\,dx + g\,dy = 2\pi$$

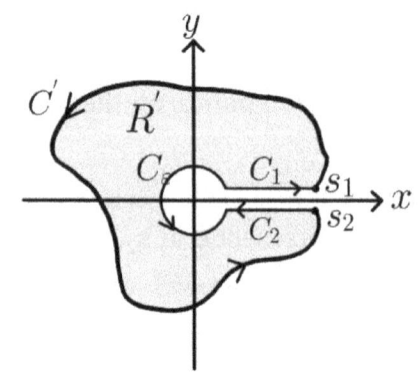

Example 2.

Suppose $f(x,y) = \dfrac{x-y}{x^2+y^2}$ and $g(x,y) = \dfrac{x+y}{x^2+y^2}$, and let C be the counterclockwise rectangular boundary with vertices at $(1,0), (1,1), (-1,1),$ and $(-1,0)$. Find

$$\oint_C f(x,y)\,dx + g(x,y)\,dy = ?$$

【Solution】

Let $s_1 = (-\varepsilon, 0)$ and $s_2 = (\varepsilon, 0) \in C$ where $\varepsilon > 0$.
Let O_ε be a circle centered at the origin with radius ε.
Let C_ε be the counterclockwise semicircular arc of O_ε from $(\varepsilon, 0)$ to $(-\varepsilon, 0)$.
Let C' be the counterclockwise path on curve C from s_2 to s_1.
Let R' be the closed region enclosed by C_ε and C'.

By Green's Theorem, $\displaystyle\int_{C'} f\,dx + g\,dy - \int_{C_\varepsilon} f\,dx + g\,dy = \iint_{R'} \dfrac{\partial g}{\partial x}(x,y) - \dfrac{\partial f}{\partial y}(x,y)\,dxdy$

Let $(x,y) \neq (0,0)$ then

$\dfrac{\partial g}{\partial x}(x,y) = \dfrac{x^2 + y^2 - 2x^2 - 2xy}{(x^2+y^2)^2} = \dfrac{-x^2 + y^2 - 2xy}{(x^2+y^2)^2}$

$\dfrac{\partial f}{\partial y}(x,y) = \dfrac{-(x^2+y^2) + 2y^2 - 2xy}{(x^2+y^2)^2} = \dfrac{-x^2 + y^2 - 2xy}{(x^2+y^2)^2}$

$\therefore \dfrac{\partial g}{\partial x}(x,y) - \dfrac{\partial f}{\partial y}(x,y) = 0 \Rightarrow \displaystyle\iint_{R'} \dfrac{\partial g}{\partial x}(x,y) - \dfrac{\partial f}{\partial y}(x,y)\,dxdy = 0$

$\Rightarrow \displaystyle\int_{C'} f\,dx + g\,dy - \int_{C_\varepsilon} f\,dx + g\,dy = 0 \Rightarrow \int_{C'} f\,dx + g\,dy = \int_{C_\varepsilon} f\,dx + g\,dy$

Let $x = \varepsilon\cos\theta, y = \varepsilon\sin\theta$ then $dx = -\varepsilon\sin\theta, dy = \varepsilon\cos\theta$

$\displaystyle\lim_{\varepsilon \to 0} \int_{C_\varepsilon} f\,dx + g\,dy = \lim_{\varepsilon \to 0} \int_0^\pi \dfrac{\varepsilon(\cos\theta - \sin\theta)(-\varepsilon\sin\theta) + \varepsilon(\cos\theta + \sin\theta)(\varepsilon\cos\theta)}{\varepsilon^2}\,d\theta = \pi$

$\therefore \displaystyle\lim_{\varepsilon \to 0} \int_{C'} f\,dx + g\,dy = \lim_{\varepsilon \to 0} \int_{C_\varepsilon} f\,dx + g\,dy = \pi$

$\because \displaystyle\lim_{\varepsilon \to 0}\int_{C'} f\,dx + g\,dy = \oint_C f\,dx + g\,dy \quad \therefore \oint_C f\,dx + g\,dy = \pi$

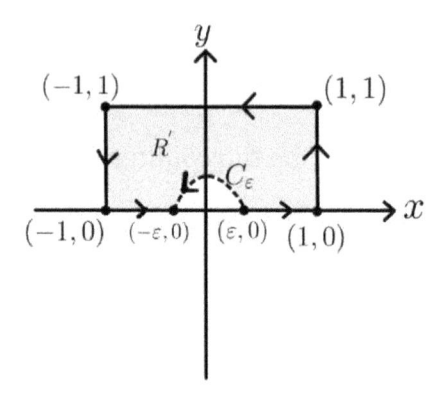

9.4.6 Create a Closed Region without Singularities

Example 3.

Suppose $f(x,y) = \dfrac{x-y}{x^2+y^2}$ and $g(x,y) = \dfrac{x+y}{x^2+y^2}$, and let C be the counterclockwise square boundary with vertices at $(1,1), (-1,1), (-1,-1)$, and $(1,-1)$. Find

$$\oint_C f(x,y)\,dx + g(x,y)\,dy = ?$$

【Solution】

Let O_ε be a circle of radius ε that contains the origin, and choose two points on the circle:
$(\varepsilon\cos\theta, \varepsilon\sin\theta)(\varepsilon\cos(-\theta), \varepsilon\sin(-\theta))$, where $\theta > 0$.
Assume the horizontal line passing through $(\varepsilon\cos\theta, \varepsilon\sin\theta)$ intersect C at s_1
and the horizontal line passing through $(\varepsilon\cos(-\theta), \varepsilon\sin(-\theta))$ intersect C at s_2.
Let C_ε be the counterclockwise arc on O_ε from $(\varepsilon\cos\theta, \varepsilon\sin\theta)$ to $(\varepsilon\cos(-\theta), \varepsilon\sin(-\theta))$.
Let C_1 be the line segment from $(\varepsilon\cos\theta, \varepsilon\sin\theta)$ to s_1.
Let C_2 be the line segment from s_2 to $(\varepsilon\cos(-\theta), \varepsilon\sin(-\theta))$.
Let C' be the counterclockwise arc of the curve from s_1 to s_2.
Let R' be the closed region enclosed by C_ε、C_1、C_2 and C' that does not contain the origin.
By Green's Theorem,

$$\int_{C'} fdx + gdy + \int_{C_1} fdx + gdy - \int_{C_\varepsilon} fdx + gdy + \int_{C_2} fdx + gdy$$

$$= \iint_{R'} \frac{\partial g}{\partial x}(x,y) - \frac{\partial f}{\partial y}(x,y)\,dxdy$$

Let $(x,y) \neq (0,0)$ then

$$\frac{\partial g}{\partial x}(x,y) = \frac{x^2+y^2-2x^2-2xy}{(x^2+y^2)^2} = \frac{-x^2+y^2-2xy}{(x^2+y^2)^2}$$

$$\frac{\partial f}{\partial y}(x,y) = \frac{-(x^2+y^2)+2y^2-2xy}{(x^2+y^2)^2} = \frac{-x^2+y^2-2xy}{(x^2+y^2)^2}$$

$$\therefore \frac{\partial g}{\partial x}(x,y) - \frac{\partial f}{\partial y}(x,y) = 0 \Rightarrow \iint_{R'} \frac{\partial g}{\partial x}(x,y) - \frac{\partial f}{\partial y}(x,y) dx dy = 0$$

$$\Rightarrow \int_{C'} f dx + g dy + \int_{C_1} f dx + g dy - \int_{C_\varepsilon} f dx + g dy + \int_{C_2} f dx + g dy = 0$$

$$\because \lim_{\theta \to 0} \int_{C_1} f dx + g dy + \int_{C_2} f dx + g dy = 0 \quad \therefore \lim_{\theta \to 0} \int_{C'} f dx + g dy - \int_{C_\varepsilon} f dx + g dy = 0$$

$$\therefore \lim_{\theta \to 0} \int_{C'} f dx + g dy = \lim_{\theta \to 0} \int_{C_\varepsilon} f dx + g dy$$

Let $x = \varepsilon \cos t, y = \varepsilon \sin t$ then $dx = -\varepsilon \sin t, dy = \varepsilon \cos t$

$$\therefore \lim_{\theta \to 0} \int_{C_\varepsilon} f dx + g dy = \int_0^{2\pi} \frac{\varepsilon(\cos t - \sin t)(-\varepsilon \sin t) + \varepsilon(\cos t + \sin t)(\varepsilon \cos t)}{\varepsilon^2} dt = 2\pi$$

$$\therefore \lim_{\theta \to 0} \int_{C'} f dx + g dy = \lim_{\theta \to 0} \int_{C_\varepsilon} f dx + g dy = 2\pi$$

$$\because \lim_{\theta \to 0} \int_{C'} f dx + g dy = \oint_C f dx + g dy \quad \therefore \oint_C f dx + g dy = 2\pi$$

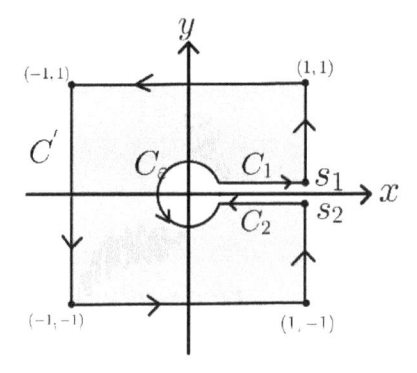

Example 4.

Suppose $f(x,y) = \dfrac{x-y}{x^2+y^2}$ and $g(x,y) = \dfrac{x+y}{x^2+y^2}$, and let C be the counterclockwise elliptical boundary $\dfrac{x^2}{a^2} + \dfrac{y^2}{b^2} = 1$. Find $\oint_C f(x,y) dx + g(x,y) dy = ?$

【Solution】

9.4.6 Create a Closed Region without Singularities

Let O_ε be a circle of radius ε that contains the origin, and choose two points on the circle: $(\varepsilon\cos\theta, \varepsilon\sin\theta)(\varepsilon\cos(-\theta), \varepsilon\sin(-\theta))$, where $\theta > 0$.

Assume the horizontal line passing through $(\varepsilon\cos\theta, \varepsilon\sin\theta)$ intersect C at s_1.
and the horizontal line passing through $(\varepsilon\cos(-\theta), \varepsilon\sin(-\theta))$ intersect C at s_2.
Let C_ε be the counterclockwise arc on O_ε from $(\varepsilon\cos\theta, \varepsilon\sin\theta)$ to $(\varepsilon\cos(-\theta), \varepsilon\sin(-\theta))$.
Let C_1 be the line segment from $(\varepsilon\cos\theta, \varepsilon\sin\theta)$ to s_1.
Let C_2 be the line segment from s_2 to $(\varepsilon\cos(-\theta), \varepsilon\sin(-\theta))$.
Let C' be the counterclockwise arc of the curve from s_1 to s_2.
Let R' be the closed region enclosed by C_ε 、 C_1 、 C_2 and C' that does not contain the origin.

By Green's Theorem,

$$\int_{C'} fdx + gdy + \int_{C_1} fdx + gdy - \int_{C_\varepsilon} fdx + gdy + \int_{C_2} fdx + gdy$$

$$= \iint_{R'} \frac{\partial g}{\partial x}(x,y) - \frac{\partial f}{\partial y}(x,y) dxdy$$

Let $(x, y) \neq (0,0)$ then

$$\frac{\partial g}{\partial x}(x,y) = \frac{x^2 + y^2 - 2x^2 - 2xy}{(x^2 + y^2)^2} = \frac{-x^2 + y^2 - 2xy}{(x^2 + y^2)^2}$$

$$\frac{\partial f}{\partial y}(x,y) = \frac{-(x^2 + y^2) + 2y^2 - 2xy}{(x^2 + y^2)^2} = \frac{-x^2 + y^2 - 2xy}{(x^2 + y^2)^2}$$

$$\therefore \frac{\partial g}{\partial x}(x,y) - \frac{\partial f}{\partial y}(x,y) = 0 \Rightarrow \iint_{R'} \frac{\partial g}{\partial x}(x,y) - \frac{\partial f}{\partial y}(x,y) dxdy = 0$$

$$\Rightarrow \int_{C'} fdx + gdy + \int_{C_1} fdx + gdy - \int_{C_\varepsilon} fdx + gdy + \int_{C_2} fdx + gdy = 0$$

$$\because \lim_{\theta \to 0} \int_{C_1} fdx + gdy + \int_{C_2} fdx + gdy = 0 \quad \therefore \lim_{\theta \to 0} \int_{C'} fdx + gdy - \int_{C_\varepsilon} fdx + gdy = 0$$

$$\therefore \lim_{\theta \to 0} \int_{C'} fdx + gdy = \lim_{\theta \to 0} \int_{C_\varepsilon} fdx + gdy$$

Let $x = \varepsilon\cos t, y = \varepsilon\sin t$ then $dx = -\varepsilon\sin t, dy = \varepsilon\cos t$

$$\therefore \lim_{\theta \to 0} \int_{C_\varepsilon} fdx + gdy = \int_0^{2\pi} \frac{\varepsilon(\cos t - \sin t)(-\varepsilon\sin t) + \varepsilon(\cos t + \sin t)(\varepsilon\cos t)}{\varepsilon^2} dt = 2\pi$$

$$\therefore \lim_{\theta \to 0} \int_{C'} fdx + gdy = \lim_{\theta \to 0} \int_{C_\varepsilon} fdx + gdy = 2\pi$$

$$\because \lim_{\theta \to 0} \int_{C'} fdx + gdy = \oint_C fdx + gdy \quad \therefore \oint_C fdx + gdy = 2\pi$$

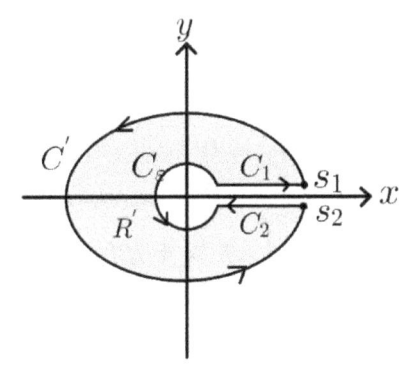

Example 5.

Suppose $f(x,y) = \dfrac{-y}{x^2+y^2}$ and $g(x,y) = \dfrac{x}{x^2+y^2}$, and let C be the counterclockwise circular boundary $4x^2 + y^2 = 4$. Find $\oint_C f(x,y)\,dx + g(x,y)\,dy = ?$

【Solution】

Let O_ε be a circle of radius ε that contains the origin, and choose two points on the circle: $(\varepsilon\cos\theta, \varepsilon\sin\theta)(\varepsilon\cos(-\theta), \varepsilon\sin(-\theta))$, where $\theta > 0$.
Assume the horizontal line passing through $(\varepsilon\cos\theta, \varepsilon\sin\theta)$ intersect C at s_1
and the horizontal line passing through $(\varepsilon\cos(-\theta), \varepsilon\sin(-\theta))$ intersect C at s_2.
Let C_ε be the counterclockwise arc on O_ε from $(\varepsilon\cos\theta, \varepsilon\sin\theta)$ to $(\varepsilon\cos(-\theta), \varepsilon\sin(-\theta))$.
Let C_1 be the line segment from $(\varepsilon\cos\theta, \varepsilon\sin\theta)$ to s_1.
Let C_2 be the line segment from s_2 to $(\varepsilon\cos(-\theta), \varepsilon\sin(-\theta))$.
Let C' be the counterclockwise arc of the curve from s_1 to s_2.
Let R' be the closed region enclosed by C_ε、C_1、C_2 and C' that does not contain the origin.
By Green's Theorem,

$$\int_{C'} f\,dx + g\,dy + \int_{C_1} f\,dx + g\,dy - \int_{C_\varepsilon} f\,dx + g\,dy + \int_{C_2} f\,dx + g\,dy$$

$$= \iint_{R'} \frac{\partial g}{\partial x}(x,y) - \frac{\partial f}{\partial y}(x,y)\,dxdy$$

Let $(x,y) \neq (0,0)$ then

$$\frac{\partial g}{\partial x}(x,y) = \frac{x^2+y^2-2x^2}{(x^2+y^2)^2} = \frac{-x^2+y^2}{(x^2+y^2)^2},\quad \frac{\partial f}{\partial y}(x,y) = \frac{-(x^2+y^2)+2y^2}{(x^2+y^2)^2} = \frac{-x^2+y^2}{(x^2+y^2)^2}$$

9.4.6 Create a Closed Region without Singularities

$$\therefore \frac{\partial g}{\partial x}(x,y) - \frac{\partial f}{\partial y}(x,y) = 0 \Rightarrow \iint_{R'} \frac{\partial g}{\partial x}(x,y) - \frac{\partial f}{\partial y}(x,y) dxdy = 0$$

$$\Rightarrow \int_{C'} fdx + gdy + \int_{C_1} fdx + gdy - \int_{C_\varepsilon} fdx + gdy + \int_{C_2} fdx + gdy = 0$$

$$\because \lim_{\theta \to 0} \int_{C_1} fdx + gdy + \int_{C_2} fdx + gdy = 0 \quad \therefore \lim_{\theta \to 0} \int_{C'} fdx + gdy - \int_{C_\varepsilon} fdx + gdy = 0$$

$$\therefore \lim_{\theta \to 0} \int_{C'} fdx + gdy = \lim_{\theta \to 0} \int_{C_\varepsilon} fdx + gdy$$

Let $x = \varepsilon \cos t, y = \varepsilon \sin t$ then $dx = -\varepsilon \sin t, dy = \varepsilon \cos t$

$$\therefore \lim_{\theta \to 0} \int_{C_\varepsilon} fdx + gdy = \int_0^{2\pi} \frac{-\varepsilon \sin t(-\varepsilon \sin t) + \varepsilon^2 \cos^2 t}{\varepsilon^2} dt = 2\pi$$

$$\therefore \lim_{\theta \to 0} \int_{C'} fdx + gdy = \lim_{\theta \to 0} \int_{C_\varepsilon} fdx + gdy = 2\pi$$

$$\because \lim_{\theta \to 0} \int_{C'} fdx + gdy = \oint_C fdx + gdy \quad \therefore \oint_C fdx + gdy = 2\pi$$

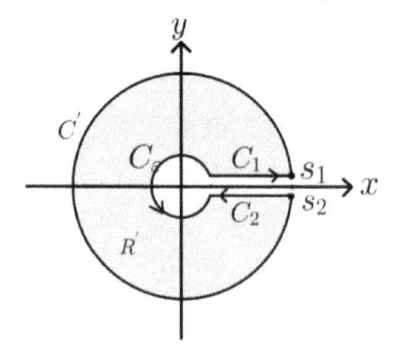

Example 6.

Find $\oint_C \frac{-y}{x^2+y^2} dx + \frac{x}{x^2+y^2} dy$ along the following paths:

(1) C is the counterclockwise closed path along $x^2 + y^2 = 1$.
(2) C is the counterclockwise closed path along $x^2 + y^2 = r$ $(r > 1)$.
(3) C is the counterclockwise closed path along $\left(\frac{x}{4}\right)^2 + \left(\frac{y}{3}\right)^2 = 1$.
(4) C is the counterclockwise closed path along the square with vertices $(1,1), (-1,1), (-1,-1), (1,-1), (1,1)$.

(5) C is the counterclockwise closed path along $(x-5)^2 + (y-5)^2 = 16$.
(6) C is the counterclockwise closed path along $(x-1)^2 + y^2 = 1$.
(7) C is the counterclockwise closed path along $(x-1)^2 + y^2 = a^2, a \in R$.

【Solution】

(1)(2)(3)(4)
Let O_ε be a circle of radius ε that contains the origin, and choose two points on the circle:
$(\varepsilon \cos\theta, \varepsilon \sin\theta)(\varepsilon \cos(-\theta), \varepsilon \sin(-\theta))$, where $\theta > 0$.
Assume the horizontal line passing through $(\varepsilon \cos\theta, \varepsilon \sin\theta)$ intersect C at s_1
and the horizontal line passing through $(\varepsilon \cos(-\theta), \varepsilon \sin(-\theta))$ intersect C at s_2.
Let C_ε be the counterclockwise arc on O_ε from $(\varepsilon \cos\theta, \varepsilon \sin\theta)$ to $(\varepsilon \cos(-\theta), \varepsilon \sin(-\theta))$.
Let C_1 be the line segment from $(\varepsilon \cos\theta, \varepsilon \sin\theta)$ to s_1.
Let C_2 be the line segment from s_2 to $(\varepsilon \cos(-\theta), \varepsilon \sin(-\theta))$.
Let C' be the counterclockwise arc of the curve from s_1 to s_2.
Let R' be the closed region enclosed by C_ε 、 C_1 、 C_2 and C' that does not contain the origin.
By Green's Theorem,

$$\int_{C'} fdx + gdy + \int_{C_1} fdx + gdy - \int_{C_\varepsilon} fdx + gdy + \int_{C_2} fdx + gdy$$

$$= \iint_{R'} \frac{\partial g}{\partial x}(x,y) - \frac{\partial f}{\partial y}(x,y) dxdy$$

Let $(x,y) \neq (0,0)$ then

$$\frac{\partial g}{\partial x}(x,y) = \frac{x^2 + y^2 - 2x^2}{(x^2+y^2)^2} = \frac{-x^2+y^2}{(x^2+y^2)^2}, \quad \frac{\partial f}{\partial y}(x,y) = \frac{-(x^2+y^2)+2y^2}{(x^2+y^2)^2} = \frac{-x^2+y^2}{(x^2+y^2)^2}$$

$$\therefore \frac{\partial g}{\partial x}(x,y) - \frac{\partial f}{\partial y}(x,y) = 0 \Rightarrow \iint_{R'} \frac{\partial g}{\partial x}(x,y) - \frac{\partial f}{\partial y}(x,y) dxdy = 0$$

$$\Rightarrow \int_{C'} fdx + gdy + \int_{C_1} fdx + gdy - \int_{C_\varepsilon} fdx + gdy + \int_{C_2} fdx + gdy = 0$$

$$\because \lim_{\theta \to 0} \int_{C_1} fdx + gdy + \int_{C_2} fdx + gdy = 0 \quad \therefore \lim_{\theta \to 0} \int_{C'} fdx + gdy - \int_{C_\varepsilon} fdx + gdy = 0$$

$$\therefore \lim_{\theta \to 0} \int_{C'} fdx + gdy = \lim_{\theta \to 0} \int_{C_\varepsilon} fdx + gdy$$

Let $x = \varepsilon \cos t, y = \varepsilon \sin t$ then $dx = -\varepsilon \sin t, dy = \varepsilon \cos t$

$$\therefore \lim_{\theta \to 0} \int_{C_\varepsilon} fdx + gdy = \int_0^{2\pi} \frac{-\varepsilon \sin t(-\varepsilon \sin t) + \varepsilon^2 \cos^2 t}{\varepsilon^2} dt = 2\pi$$

9.4.6 Create a Closed Region without Singularities

$$\therefore \lim_{\theta \to 0} \int_{C'} f dx + g dy = \lim_{\theta \to 0} \int_{C_\varepsilon} f dx + g dy = 2\pi$$

$$\because \lim_{\theta \to 0} \int_{C'} f dx + g dy = \oint_C f dx + g dy \quad \therefore \oint_C f dx + g dy = 2\pi$$

(5)

Let R' be the closed region enclosed by C, then then the first partial derivatives f and g exist and are continuous on R'.

By Green's Theorem, $\oint_C f dx + g dy = \iint_{R'} \frac{\partial g}{\partial x}(x,y) - \frac{\partial f}{\partial y}(x,y) dx dy$

$$\because \frac{\partial g}{\partial x}(x,y) = \frac{x^2 + y^2 - 2x^2}{(x^2 + y^2)^2} = \frac{-x^2 + y^2}{(x^2 + y^2)^2}, \frac{\partial f}{\partial y}(x,y) = \frac{-(x^2 + y^2) + 2y^2}{(x^2 + y^2)^2} = \frac{-x^2 + y^2}{(x^2 + y^2)^2}$$

$$\therefore \frac{\partial g}{\partial x}(x,y) - \frac{\partial f}{\partial y}(x,y) = 0 \Rightarrow \iint_{R'} \frac{\partial g}{\partial x}(x,y) - \frac{\partial f}{\partial y}(x,y) dx dy = 0 \quad \therefore \oint_C f dx + g dy = 0$$

(6)

Let O_ε be the circle centered at the origin with radius ε, intersecting C at $s_1 = (\varepsilon \cos \theta_1, \varepsilon \sin \theta_1)$ and $s_2 = (\varepsilon \cos \theta_2, \varepsilon \sin \theta_2)$, where $\theta_2 > \theta_1$.
Let C_ε be the clockwise arc on O_ε from $(\varepsilon \cos \theta_1, \varepsilon \sin \theta_1)$ to $(\varepsilon \cos \theta_2, \varepsilon \sin \theta_2)$.
Let C' be the counterclockwise arc on curve C from s_2 to s_1.
Let R' be the closed region enclosed by C_ε and C'.

By Green's Theorem, $\int_{C'} f dx + g dy + \int_{C_\varepsilon} f dx + g dy = \iint_{R'} \frac{\partial g}{\partial x}(x,y) - \frac{\partial f}{\partial y}(x,y) dx dy$

Let $(x, y) \neq (0,0)$ then

$$\frac{\partial g}{\partial x}(x,y) = \frac{x^2 + y^2 - 2x^2}{(x^2 + y^2)^2} = \frac{-x^2 + y^2}{(x^2 + y^2)^2}, \frac{\partial f}{\partial y}(x,y) = \frac{-(x^2 + y^2) + 2y^2}{(x^2 + y^2)^2} = \frac{-x^2 + y^2}{(x^2 + y^2)^2}$$

$$\therefore \frac{\partial g}{\partial x}(x,y) - \frac{\partial f}{\partial y}(x,y) = 0 \Rightarrow \iint_{R'} \frac{\partial g}{\partial x}(x,y) - \frac{\partial f}{\partial y}(x,y) dx dy = 0$$

$$\Rightarrow \int_{C'} f dx + g dy + \int_{C_\varepsilon} f dx + g dy = 0 \Rightarrow \int_{C'} f dx + g dy = -\int_{C_\varepsilon} f dx + g dy$$

Let $x = \varepsilon \cos \theta, y = \varepsilon \sin \theta$ then $dx = -\varepsilon \sin \theta, dy = \varepsilon \cos \theta$

$$\lim_{\varepsilon \to 0} \int_{C_\varepsilon} f dx + g dy = \lim_{\varepsilon \to 0} \int_{\frac{\pi}{2}}^{-\frac{\pi}{2}} \frac{\varepsilon \sin \theta (\varepsilon \sin \theta) + \varepsilon^2 \cos^2 \theta}{\varepsilon^2} d\theta = -\pi$$

$$\lim_{\varepsilon \to 0} \int_{C'} f dx + g dy = -\lim_{\varepsilon \to 0} \int_{C_\varepsilon} f dx + g dy = \pi$$

$$\because \lim_{\varepsilon \to 0} \int_{C'} f dx + g dy = \oint_C f dx + g dy \quad \therefore \oint_C f dx + g dy = \pi$$
(7)

$$a < 1 \Rightarrow \oint_C f dx + g dy = 0$$

$$a = 1 \Rightarrow \oint_C f dx + g dy = \pi$$

$$a > 1 \Rightarrow \oint_C f dx + g dy = 2\pi$$

9.5 Surface integrals
9.5.1 Representation of Surfaces S in Space and Smooth Surfaces

The types of surface integral problems in exams include finding the surface area of a closed region for a given surface S, calculating the mass for a given scalar function f and surface S, and determining the flux for a given vector field $\vec{F}$ and surface S. A surface in three-dimensional space can be represented using two parameters (as shown in the figure),
$$\vec{r}(s,t) = (x(s,t), y(s,t), z(s,t)).$$

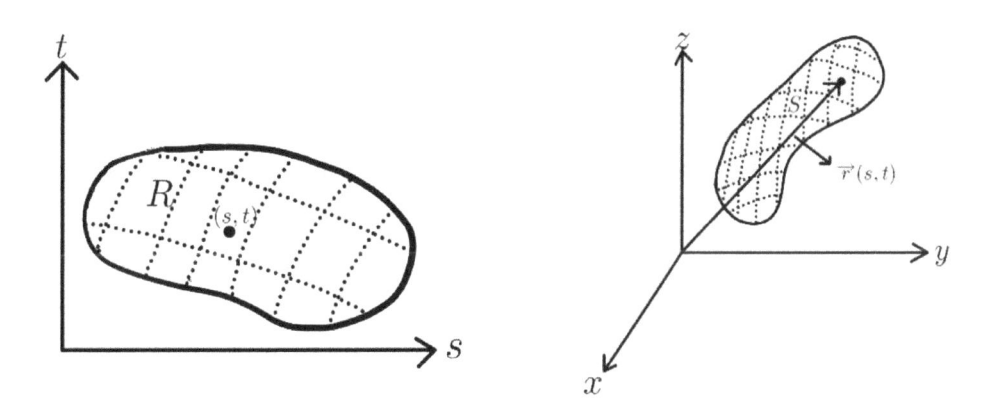

Suppose the function f is a bivariate function, and the equation of the surface S is $z = f(x,y)$. Then the parametric representation of the surface is
$$\vec{r}(x,y) = (x, y, f(x,y)), \forall\, (x,y) \in D_f,$$
where D_f is the domain of f. Similarly, if the equation of the surface S is
$$x = g(y,z),\ \forall\, (y,z) \in D_g \ \text{ or } \ y = h(x,z),\ \forall (x,z) \in D_h,$$

9.5.1 Representation of Surfaces S in Space and Smooth Surfaces

then the parametric representations are
$$\vec{r}(y,z) = (g(y,z), y, z), \ \forall \, (y,z) \in D_g \ \text{ or } \ \vec{r}(x,z) = (x, h(x,z), z), \ \forall (x,z) \in D_h.$$

【Definition】

Suppose the parametric equation of the surface is $\vec{r}(s,t) = (x(s,t), y(s,t), z(s,t))$, then
$$\vec{r}_s = \frac{\partial \vec{r}}{\partial s} := \left(\frac{\partial x}{\partial s}, \frac{\partial y}{\partial s}, \frac{\partial z}{\partial s}\right), \quad \vec{r}_t = \frac{\partial \vec{r}}{\partial t} := \left(\frac{\partial x}{\partial t}, \frac{\partial y}{\partial t}, \frac{\partial z}{\partial t}\right)$$

【Definition】

Suppose the parametric equation of the surface is
$$\vec{r}(s,t) = (x(s,t), y(s,t), z(s,t)).$$
If there exists a point (s,t) in the domain such that $\vec{r}_s \times \vec{r}_t \neq 0$ then $\vec{r}(s,t)$ is called a regular parameterization.

Hereafter, we will only consider regularly parameterized surfaces. The definition of whether a parametric surface is smooth is similar to that of a curve being smooth. If a curve has no sharp corners, then the parametric curve is smooth. If a surface has no sharp corners, then the parametric surface is called smooth.

【Definition】

Suppose the parametric equation of the surface is
$$\vec{r}(s,t) = (x(s,t), y(s,t), z(s,t)).$$
If $\vec{r}_s \times \vec{r}_t \neq 0, \ \forall \, (s,t) \in D$ then the surface is called a smooth surface.

Below are important formulas for calculating surface area, surface integrals of scalar functions, and surface integrals of vector fields. Detailed derivations can be found in the respective sections. Suppose the parametric equation of the surface S is $\vec{r}(s,t)$. The first key calculation is finding the surface area of the closed region on the surface S. The formula is:
$$\iint_S dA = \iint_D |\vec{r}_s \times \vec{r}_t| dA.$$
The second key calculation is finding the surface integral of a scalar function f over the surface S. The formula is
$$\iint_S f(x,y,z) dS = \iint_D f(\vec{r}(s,t)) |\vec{r}_s \times \vec{r}_t| dA.$$

9.5.1 Representation of Surfaces S in Space and Smooth Surfaces

The third key calculation is finding the surface integral of a vector field $\vec{F}$ over the surface S. The formula is

$$\iint_S \vec{F} \cdot d\vec{S} \iint_D \vec{F}(\vec{r}(s,t)) \cdot (\vec{r}_s \times \vec{r}_t) ds dt,$$

where $\vec{r}_s = \left(\dfrac{\partial x}{\partial s}, \dfrac{\partial y}{\partial s}, \dfrac{\partial z}{\partial s}\right)$ and $\vec{r}_t = \left(\dfrac{\partial x}{\partial t}, \dfrac{\partial y}{\partial t}, \dfrac{\partial z}{\partial t}\right).$

From the above, we can observe that given a parameterized surface
$$\vec{r}(s,t) = (x(s,t), y(s,t), z(s,t)),$$
efficiently and accurately computing the cross product $\vec{r}_s \times \vec{r}_t$ and its magnitude $|\vec{r}_s \times \vec{r}_t|$ is crucial.

Example 1.
Assume the parametric representation of the surface is $\vec{r}(s,t) = (3\cos s, t, 3\sin s)$. Find the single equation representation of the surface.

【Solution】

∵ the parametric equations of the surface are $x = 3\cos s, y = t, z = 3\sin s$
∴ $x^2 + z^2 = 9\cos^2 s + 9\sin^2 s = 9.$

Example 2.
Find a vector function that passes through point P_0 with position vector $\vec{r}_0$, and lies in the plane containing two non-parallel vectors $\vec{u}$ and $\vec{v}$.

【Solution】

Let P be an arbitrary point on the plane, then there exist scalars s, t s.t. $\overrightarrow{P_0 P} = s\vec{u} + t\vec{v}$
Let $\vec{r}$ be the position vector of point P then $\vec{r} = \overrightarrow{OP_0} + \overrightarrow{P_0 P} = \vec{r}_0 + s\vec{u} + t\vec{v}$
∴ $\vec{r}(s,t) = \vec{r}_0 + s\vec{u} + t\vec{v}$

Example 3.
Find the parametric representations of the following surface equations:
(1) The equation of a sphere: $x^2 + y^2 + z^2 = r^2$, $r > 0$.
(2) The equation of a cylinder: $x^2 + y^2 = a^2$, where $0 \le z \le 1, a > 0$.
(3) The equation of a cone: $z = 5\sqrt{x^2 + y^2}$.
(4) The equation of an elliptical paraboloid: $z = x^2 + 2y^2$.

9.5.1 Representation of Surfaces S in Space and Smooth Surfaces

(5) The equation of an elliptical paraboloid: $x^2 + y^2 = z^2$ above the plane $z = -3$.

(6) The equation of an ellipsoid: $\dfrac{x^2}{a^2} + \dfrac{y^2}{b^2} + \dfrac{z^2}{c^2} = 1$.

【Solution】

(1)
$\vec{r}(\theta, \varphi) = r\cos\theta \sin\varphi, y = r\sin\theta \sin\varphi, z = r\cos\varphi$

(2)
$\vec{r}(\theta, z) = (a\cos\theta, a\sin\theta, z), \ 0 \le \theta \le 2\pi \text{ and } 0 \le z \le 1$

(3)

The parametric representation is given by $\vec{r}(x, y) = (x, y, 5\sqrt{x^2 + y^2})$
Another parametric representation can be expressed in polar coordinates.

Let $x = r\cos\theta, y = r\sin\theta$ then $z = 5\sqrt{x^2 + y^2} = 5r$

∴ the polar coordinate parametric representation is $\vec{r}(r, \theta) = (r\cos\theta, r\sin\theta, 5r)$

(4)
$\vec{r}(x, y) = (x, y, x^2 + 2y^2)$

(5)
$\vec{r}(s, t) = (s\cos t, s\sin t, s)$ where $-3 \le s < \infty, 0 \le t < 2\pi$

(6)
$\vec{r}(\theta, \varphi) = (a\cos\theta \sin\varphi, b\sin\theta \sin\varphi, c\cos\varphi)$ where $0 \le \theta \le 2\pi, \ 0 \le \varphi \le \pi$

Example 4.

Assume the parametric representation of the surface is $\vec{r}(s, t) = (s\cos t, s\sin t, s^2)$, where $0 \le s < \infty$, and $0 \le t < 2\pi$. Find its single equation representation.

【Solution】

∵ $x^2 + y^2 = (s\cos t)^2 + (s\sin t)^2 = s^2 = z$

∴ the single equation representation is the elliptical paraboloid $x^2 + y^2 = z$

Example 5.

Assume the parametric representation of surface S is
$$\vec{r}(s, t) = ((3 + \cos t)\cos s, (3 + \cos t)\sin s, \sin t),$$
where $0 \le s < 2\pi$, and $0 \le t < 2\pi$. Determine whether the surface is a smooth surface.

【Solution】

$\vec{r}_s = (-(3 + \cos t)\sin s, (3 + \cos t)\cos s, 0)$, and $\vec{r}_t = (-\sin t \cos s, -\sin t \sin s, \cos t)$

9.5.1 Representation of Surfaces S in Space and Smooth Surfaces

$\therefore \vec{r}_s \times \vec{r}_t = ((3 + \cos t) \cos s \cos t, (3 + \cos t) \sin s \cos t, (3 + \cos t)\sin t)$

Let $\vec{r}_s \times \vec{r}_t = \vec{0}$ then $(3 + \cos t)\sin t = 0 \Rightarrow t = 0$ or π

$\because t = 0$ or π and $(3 + \cos t) \sin s \cos t = 0 \quad \therefore s = 0$ or π

$\because (3 + \cos t) \cos s \cos t \neq 0$, for $s = 0$ or π and $t = 0$ or π

$\therefore S$ is a smooth surface

Example 6.

For the following surfaces, find $\dfrac{\partial \vec{r}}{\partial s} \times \dfrac{\partial \vec{r}}{\partial t}$ and $\left|\dfrac{\partial \vec{r}}{\partial s} \times \dfrac{\partial \vec{r}}{\partial t}\right|$, and determine whether the surface is a smooth surface.

(1) Assume the parametric equation of the surface is
$$\vec{r}(s,t) = (kt \cos s, kt \sin s, t).$$

(2) Assume the parametric equation of the surface is
$$\vec{r}(s,t) = (a \cos s \cos t, a \sin s \cos t, a \sin t).$$

【Solution】

(1)

$\because \dfrac{\partial \vec{r}}{\partial s} = (-kt \sin s, kt \cos s, 0), \quad \dfrac{\partial \vec{r}}{\partial t} = (k \cos s, k \sin s, 1)$

$\therefore \dfrac{\partial \vec{r}}{\partial s} \times \dfrac{\partial \vec{r}}{\partial t} = (kt \cos s, kt \sin s, -k^2 t)$ and $\left|\dfrac{\partial \vec{r}}{\partial s} \times \dfrac{\partial \vec{r}}{\partial t}\right| = kt\sqrt{1 + k^2}$

$\because \dfrac{\partial \vec{r}}{\partial s} \times \dfrac{\partial \vec{r}}{\partial t} = (0,0,0)$, as $t = 0 \quad \therefore$ the surface is not smooth

(2)

$\because \dfrac{\partial \vec{r}}{\partial s} \times \dfrac{\partial \vec{r}}{\partial t} = a\cos t \, (a \cos s \cos t, a \sin s \cos t, a \sin t) = a\cos t \, \vec{r}(s,t)$ and

$\left|\dfrac{\partial \vec{r}}{\partial s} \times \dfrac{\partial \vec{r}}{\partial t}\right| = a \cos^2 t$

$\because \dfrac{\partial \vec{r}}{\partial s} \times \dfrac{\partial \vec{r}}{\partial t} = (0,0,0)$, as $t = \dfrac{\pi}{2} \quad \therefore$ the surface is not smooth

Example 7.

The parametric equation of the surface is $\vec{r}(x, \theta) = (x, x^2 \cos \theta, x^2 \sin \theta)$.

(1) Find $\dfrac{\partial \vec{r}}{\partial x} \times \dfrac{\partial \vec{r}}{\partial \theta}$ and $\left|\dfrac{\partial \vec{r}}{\partial x} \times \dfrac{\partial \vec{r}}{\partial \theta}\right|$ (2) Determine whether it is a smooth surface.

9.5.1 Representation of Surfaces S in Space and Smooth Surfaces

【Solution】

(1)

$$\because \frac{\partial \vec{r}}{\partial x} = (1, 2x\cos\theta, 2x\sin\theta), \quad \frac{\partial \vec{r}}{\partial \theta} = (0, -x^2\sin\theta, x^2\cos\theta)$$

$$\therefore \frac{\partial \vec{r}}{\partial x} \times \frac{\partial \vec{r}}{\partial \theta} = (2x^3, -x^2\sin\theta, -x^2\sin\theta) \text{ and } \left|\frac{\partial \vec{r}}{\partial x} \times \frac{\partial \vec{r}}{\partial \theta}\right| = x^2\sqrt{4x^2+1}$$

(2)

$$\because \frac{\partial \vec{r}}{\partial x} \times \frac{\partial \vec{r}}{\partial \theta} = (0,0,0), \text{ as } x = 0 \quad \therefore \text{ the surface is not smooth}$$

Example 8.

For the following surfaces, find $\frac{\partial \vec{r}}{\partial x} \times \frac{\partial \vec{r}}{\partial y}$ and $\left|\frac{\partial \vec{r}}{\partial x} \times \frac{\partial \vec{r}}{\partial y}\right|$, and determine whether the surface is a smooth surface.

(1) Suppose the surface is the plane in space $Ax + By + Cz = D$.

(2) Suppose the surface in space is $z = \frac{2}{3}y^{\frac{3}{2}}$.

(3) Suppose the surface in space is $z = x + ay + b$.

(4) Suppose the surface in space is $z = xy$.

(5) Suppose the surface in space is $z = y^2 + bx + c$.

(6) Suppose the surface in space is $z = x^2 + y^2$.

(7) Suppose the surface in space is $z = 3 - y^2$.

(8) Suppose the surface in space is $z = x^3$.

【Solution】

(1)

Let $\vec{r}(x, y) = \left(x, y, \frac{D - (Ax + By)}{C}\right)$

$$\therefore \frac{\partial \vec{r}}{\partial x} = \left(1, 0, -\frac{A}{C}\right), \frac{\partial \vec{r}}{\partial y} = \left(0, 1, -\frac{B}{C}\right)$$

$$\therefore \frac{\partial \vec{r}}{\partial x} \times \frac{\partial \vec{r}}{\partial y} = \left(\frac{A}{C}, \frac{B}{C}, 1\right) \Rightarrow \left|\frac{\partial \vec{r}}{\partial x} \times \frac{\partial \vec{r}}{\partial y}\right| = \sqrt{1 + \frac{A^2 + B^2}{C^2}} = \sqrt{\frac{A^2 + B^2 + C^2}{C^2}}$$

$\therefore \dfrac{\partial \vec{r}}{\partial x} \times \dfrac{\partial \vec{r}}{\partial y} \neq (0,0,0), \ \forall x, y \in R \quad \therefore$ the surface is smooth

(2)

Let $\vec{r}(x, y) = \left(x, y, \dfrac{2}{3} y^{\frac{3}{2}}\right)$

$\therefore \dfrac{\partial \vec{r}}{\partial x} = (1,0,0), \ \dfrac{\partial \vec{r}}{\partial y} = \left(0, 1, y^{\frac{1}{2}}\right) \ \therefore \dfrac{\partial \vec{r}}{\partial x} \times \dfrac{\partial \vec{r}}{\partial y} = \left(0, -y^{\frac{1}{2}}, 1\right) \Rightarrow \left|\dfrac{\partial \vec{r}}{\partial x} \times \dfrac{\partial \vec{r}}{\partial y}\right| = (y + 1)^{\frac{1}{2}}$

$\therefore \dfrac{\partial \vec{r}}{\partial x} \times \dfrac{\partial \vec{r}}{\partial y} \neq (0,0,0), \ \forall x, y \in R \quad \therefore$ the surface is smooth

(3)

Let $\vec{r}(x, y) = (x, y, x + ay + b)$

$\therefore \dfrac{\partial \vec{r}}{\partial x} = (1,0,1), \ \dfrac{\partial \vec{r}}{\partial y} = (0, 1, a) \ \therefore \dfrac{\partial \vec{r}}{\partial x} \times \dfrac{\partial \vec{r}}{\partial y} = (-1, -a, 1) \Rightarrow \left|\dfrac{\partial \vec{r}}{\partial x} \times \dfrac{\partial \vec{r}}{\partial y}\right| = \sqrt{a^2 + 2}$

$\therefore \dfrac{\partial \vec{r}}{\partial x} \times \dfrac{\partial \vec{r}}{\partial y} \neq (0,0,0), \ \forall x, y \in R \quad \therefore$ the surface is smooth

(4)

Let $\vec{r}(x, y) = (x, y, xy)$

$\therefore \dfrac{\partial \vec{r}}{\partial x} = (1,0,y), \ \dfrac{\partial \vec{r}}{\partial y} = (0, 1, x) \ \therefore \dfrac{\partial \vec{r}}{\partial x} \times \dfrac{\partial \vec{r}}{\partial y} = (-y, -x, 1) \Rightarrow \left|\dfrac{\partial \vec{r}}{\partial x} \times \dfrac{\partial \vec{r}}{\partial y}\right| = (x^2 + y^2 + 1)^{\frac{1}{2}}$

$\therefore \dfrac{\partial \vec{r}}{\partial x} \times \dfrac{\partial \vec{r}}{\partial y} \neq (0,0,0), \ \forall x, y \in R \quad \therefore$ the surface is smooth

(5)

Let $\vec{r}(x, y) = (x, y, y^2 + bx + c)$

$\therefore \dfrac{\partial \vec{r}}{\partial x} = (1,0,b), \ \dfrac{\partial \vec{r}}{\partial y} = (0, 1, 2y) \ \therefore \dfrac{\partial \vec{r}}{\partial x} \times \dfrac{\partial \vec{r}}{\partial y} = (-b, -2y, 1) \Rightarrow \left|\dfrac{\partial \vec{r}}{\partial x} \times \dfrac{\partial \vec{r}}{\partial y}\right| = (b^2 + 1 + 4y^2)^{\frac{1}{2}}$

$\therefore \dfrac{\partial \vec{r}}{\partial x} \times \dfrac{\partial \vec{r}}{\partial y} \neq (0,0,0), \ \forall x, y \in R \quad \therefore$ the surface is smooth

9.5.1 Representation of Surfaces S in Space and Smooth Surfaces 434

(6)

Let $\vec{r}(x,y) = (x, y, x^2 + y^2)$ $\because \dfrac{\partial \vec{r}}{\partial x} = (1,0,2x), \dfrac{\partial \vec{r}}{\partial y} = (0,1,2y)$

$\therefore \dfrac{\partial \vec{r}}{\partial x} \times \dfrac{\partial \vec{r}}{\partial y} = (-2x, -2y, 1) \Rightarrow \left|\dfrac{\partial \vec{r}}{\partial x} \times \dfrac{\partial \vec{r}}{\partial y}\right| = (4x^2 + 4y^2 + 1)^{\frac{1}{2}}$

$\because \dfrac{\partial \vec{r}}{\partial x} \times \dfrac{\partial \vec{r}}{\partial y} \neq (0,0,0), \forall x, y \in R$ $\therefore$ the surface is smooth

(7)

Let $\vec{r}(x,y) = (x, y, 3 - y^2)$

$\because \dfrac{\partial \vec{r}}{\partial x} = (1,0,0), \dfrac{\partial \vec{r}}{\partial y} = (0,1,-2y)$ $\therefore \dfrac{\partial \vec{r}}{\partial x} \times \dfrac{\partial \vec{r}}{\partial y} = (0,2y, 1) \Rightarrow \left|\dfrac{\partial \vec{r}}{\partial x} \times \dfrac{\partial \vec{r}}{\partial y}\right| = \sqrt{1 + 4y^2}$

$\because \dfrac{\partial \vec{r}}{\partial x} \times \dfrac{\partial \vec{r}}{\partial y} \neq (0,0,0), \forall x, y \in R$ $\therefore$ the surface is smooth

(8)

Let $\vec{r}(x,y) = (x, y, x^3)$

$\because \dfrac{\partial \vec{r}}{\partial x} = (1,0, 3x^2), \dfrac{\partial \vec{r}}{\partial y} = (0,1,0)$ $\therefore \dfrac{\partial \vec{r}}{\partial x} \times \dfrac{\partial \vec{r}}{\partial y} = (-3x^2, 0, -1)$

$\left|\dfrac{\partial \vec{r}}{\partial x} \times \dfrac{\partial \vec{r}}{\partial y}\right| = \sqrt{1 + 9x^4}$

$\because \dfrac{\partial \vec{r}}{\partial x} \times \dfrac{\partial \vec{r}}{\partial y} \neq (0,0,0), \forall x, z \in R$ $\therefore$ the surface is smooth

Example 9.

Suppose the sphere $S = \{(x, y, z): x^2 + y^2 + z^2 = r^2, z \geq 0\}$, and let $\vec{r}(\theta, \varphi)$ be the parameterization of the surface. Find $\dfrac{\partial \vec{r}}{\partial \varphi} \times \dfrac{\partial \vec{r}}{\partial \theta}$ and $\left|\dfrac{\partial \vec{r}}{\partial \varphi} \times \dfrac{\partial \vec{r}}{\partial \theta}\right|$.

【Solution】

Let $\vec{r}(\theta, \varphi) = (r \cos \theta \sin \varphi, r \sin \theta \sin \varphi, r\cos \varphi)$

$$\therefore \frac{\partial \vec{r}}{\partial \theta} = (-r\sin\theta\sin\varphi, r\cos\theta\sin\varphi, 0) \text{ and } \frac{\partial \vec{r}}{\partial \varphi} = (r\cos\theta\cos\varphi, r\sin\theta\cos\varphi, -r\sin\varphi)$$

$$\therefore \frac{\partial \vec{r}}{\partial \varphi} \times \frac{\partial \vec{r}}{\partial \theta} = (r^2\cos\theta\sin^2\varphi, r^2\sin\theta\sin^2\varphi, r^2\sin\varphi\cos\varphi) \Rightarrow \left|\frac{\partial \vec{r}}{\partial \varphi} \times \frac{\partial \vec{r}}{\partial \theta}\right| = r^2\sin\varphi$$

Example 10.

For the following surfaces, find $\frac{\partial \vec{r}}{\partial y} \times \frac{\partial \vec{r}}{\partial z}$ and $\left|\frac{\partial \vec{r}}{\partial y} \times \frac{\partial \vec{r}}{\partial z}\right|$, and determine whether the surface is a smooth surface. (1) The surface is given by $x = \sqrt{z^2 - y^2}$ (2) The surface is given by $x - y^2 - z = 2$.

【Solution】

(1)

Let $\vec{r}(x,y) = \left(\sqrt{z^2 - y^2}, y, z\right)$

$$\therefore \frac{\partial \vec{r}}{\partial y} = \left(-y(z^2-y^2)^{-\frac{1}{2}}, 1, 0\right), \quad \frac{\partial \vec{r}}{\partial z} = \left(z(z^2-y^2)^{-\frac{1}{2}}, 0, 1\right),$$

$$\therefore \frac{\partial \vec{r}}{\partial y} \times \frac{\partial \vec{r}}{\partial z} = \left(1, y(z^2-y^2)^{-\frac{1}{2}}, -z(z^2-y^2)^{-\frac{1}{2}}\right) \text{ and } \left|\frac{\partial \vec{r}}{\partial y} \times \frac{\partial \vec{r}}{\partial z}\right| = \sqrt{1 + (z^2+y^2)(z^2-y^2)^{-1}}$$

$$\therefore \frac{\partial \vec{r}}{\partial y} \times \frac{\partial \vec{r}}{\partial z} \neq (0,0,0), \quad \forall x,y \in R \quad \therefore \text{ the surface is smooth}$$

(2)

Let $\vec{r}(y,z) = (y^2 + z + 2, y, z)$

$$\therefore \frac{\partial \vec{r}}{\partial y} = (2y, 1, 0), \quad \frac{\partial \vec{r}}{\partial z} = (1, 0, 1) \quad \therefore \frac{\partial \vec{r}}{\partial y} \times \frac{\partial \vec{r}}{\partial z} = (1, -2y, -1) \Rightarrow \left|\frac{\partial \vec{r}}{\partial y} \times \frac{\partial \vec{r}}{\partial z}\right| = \sqrt{2 + 4y^2}$$

$$\therefore \frac{\partial \vec{r}}{\partial y} \times \frac{\partial \vec{r}}{\partial z} \neq (0,0,0), \quad \forall x,y \in R \quad \therefore \text{ the surface is smooth}$$

Example 11.

9.5.2 Find the Surface Area of a Closed Region on Surface S

Assuming the surface is defined by $y = x^2$, (1) Find $\dfrac{\partial \vec{r}}{\partial x} \times \dfrac{\partial \vec{r}}{\partial z}$ and $\left|\dfrac{\partial \vec{r}}{\partial x} \times \dfrac{\partial \vec{r}}{\partial z}\right|$

(2) Determine whether the surface is a smooth surface.

【Solution】

Let $\vec{r}(x,z) = (x, x^2, z)$

$\because \dfrac{\partial \vec{r}}{\partial x} = (1, 2x, 0)$, $\dfrac{\partial \vec{r}}{\partial z} = (0, 0, 1)$ $\therefore \dfrac{\partial \vec{r}}{\partial x} \times \dfrac{\partial \vec{r}}{\partial z} = (2x, -1, 0)$ and $\left|\dfrac{\partial \vec{r}}{\partial x} \times \dfrac{\partial \vec{r}}{\partial z}\right| = \sqrt{4x^2 + 1}$

(2)

$\because \dfrac{\partial \vec{r}}{\partial x} \times \dfrac{\partial \vec{r}}{\partial z} \neq (0,0,0)$, $\forall x, z \in R$ $\therefore$ the surface is smooth

9.5.2 Find the Surface Area of a Closed Region on Surface S

Given a parametric surface $S: \vec{r}(s,t) = (x(s,t), y(s,t), z(s,t))$, with the domain defined as a rectangular parametric region D, we divide the domain D into several small rectangles D_{ij}. The center point of D_{ij} is denoted as (s_i^*, t_j^*). Suppose the region D_{ij} corresponds to a region S_{ij} on the surface S, and (s_i^*, t_j^*) corresponds to the point E_{ij} on the surface S_{ij}, with the position vector $\vec{r}(s_i^*, t_j^*)$ (as shown in the figure). The vector

$$\vec{r}_s(s_i^*, t_j^*) \times \vec{r}_t(s_i^*, t_j^*)$$

is perpendicular to the surface at the point $\vec{r}(s_i^*, t_j^*)$ and we can parameterize the tangent plane of this region S_{ij} using a linear function.

Let $f(s,t) = s\,\vec{r}_s(s_i^*, t_j^*) + t\vec{r}_t(s_i^*, t_j^*) + \left(\vec{r}(s_i^*, t_j^*) - s_i^*\vec{r}_s(s_i^*, t_j^*) - t_j^*\vec{r}_t(s_i^*, t_j^*)\right).$

Since S_{ij} can be approximated by the parallelogram formed by the two vectors

$$f(s_1 + \Delta s, t_1) - f(s_1, t_1) \text{ and } f(s_1, t_1 + \Delta t) - f(s_1, t_1),$$

we have

$f(s_1 + \Delta s, t_1) - f(s_1, t_1) = \Delta s \vec{r}_s(s_i^*, t_j^*)$ and $f(s_1, t_1 + \Delta t) - f(s_1, t_1) = \Delta t \vec{r}_t(s_i^*, t_j^*).$

Thus, the area of this parallelogram is

$$\left|\Delta s\, \vec{r}_s(s_i^*, t_j^*) \times \Delta t \vec{r}_t(s_i^*, t_j^*)\right| = \left|\vec{r}_s(s_i^*, t_j^*) \times \vec{r}_t(s_i^*, t_j^*)\right| \Delta s \Delta t.$$

So,

$$\Delta S_{ij} \approx \left|\vec{r}_s(s_i^*, t_j^*) \times \vec{r}_t(s_i^*, t_j^*)\right| \Delta s \Delta t.$$

Let $\vec{r}_s^* = \vec{r}_s(s_i^*, t_j^*)$ and $\vec{r}_t^* = \vec{r}_t(s_i^*, t_j^*)$ then

$$\text{Approximate Surface Area of } S \approx \sum_{i=1}^{m}\sum_{j=1}^{n} |\vec{r}_s^* \times \vec{r}_t^*|\Delta s \Delta t.$$

We can treat this double sum as a Riemann sum for the double integral

$$\iint_D |\vec{r}_s \times \vec{r}_t| ds dt$$

which provides the motivation for defining the surface area of the parametric surface.

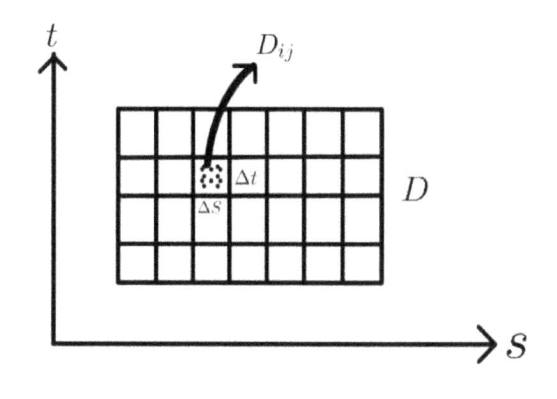

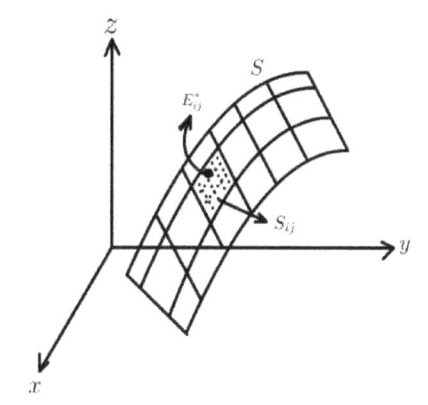

【Definition】

Consider a smooth parametric surface S given by
$$\vec{r}(s,t) = (x(s,t), y(s,t), z(s,t)), \quad \text{where } (s,t) \in D,$$
and suppose that S is covered exactly once over the entire parametric domain D. Then, the surface area of S is given by

$$\text{Surface Area of } S = \iint_S dA = \iint_D |\vec{r}_s \times \vec{r}_t| ds dt,$$

where $\vec{r}_s = \left(\dfrac{\partial x}{\partial s}, \dfrac{\partial y}{\partial s}, \dfrac{\partial z}{\partial s}\right)$ and $\vec{r}_t = \left(\dfrac{\partial x}{\partial t}, \dfrac{\partial y}{\partial t}, \dfrac{\partial z}{\partial t}\right)$

From the above observation, we can see that calculating the surface area is equivalent to evaluating a double integral, where the integrand is the magnitude of the cross product
$$|\vec{r}_s(s,t) \times \vec{r}_t(s,t)|.$$
Therefore, finding the surface area involves first determining the integrand and then computing the double integral. Next, we will provide a preliminary explanation of the different types of surfaces that can be expressed as double integrals.

Consider the surface $S: z = f(x,y)$, where $a \leq x \leq b, \ c \leq y \leq d$. The goal is to calculate

9.5.2 Find the Surface Area of a Closed Region on Surface S

the surface area $\iint_S dA$. Parameterize the surface as $S: \vec{r}(x, y) = (x, y, f(x, y))$, and let $R = \{(x, y): a \leq x \leq b, c \leq y \leq d\}$. Then the surface area of S is given by

$$\iint_S dA = \iint_R \left|\frac{\partial \vec{r}}{\partial x} \times \frac{\partial \vec{r}}{\partial y}\right| dxdy.$$

Since $\frac{\partial \vec{r}}{\partial x} = (1, 0, f_x(x, y))$ and $\frac{\partial \vec{r}}{\partial y} = (0, 1, f_y(x, y))$, their cross product is

$$\frac{\partial \vec{r}}{\partial x} \times \frac{\partial \vec{r}}{\partial y} = (-f_x(x, y), -f_y(x, y), 1).$$

Thus,

$$\left|\frac{\partial \vec{r}}{\partial x} \times \frac{\partial \vec{r}}{\partial y}\right| = \sqrt{1 + (f_x(x, y))^2 + (f_y(x, y))^2}.$$

The surface area formula can be rewritten as

$$\iint_S dA = \iint_R \left|\frac{\partial \vec{r}}{\partial x} \times \frac{\partial \vec{r}}{\partial y}\right| dxdy = \int_a^b \int_c^d \sqrt{1 + (f_x(x, y))^2 + (f_y(x, y))^2}\, dydx.$$

If the surface is given by $z = f(x, y)$ and its projection onto the xy-plane is

$$\{(x, y): x^2 + y^2 \leq a\},$$

we can use polar coordinates to transform the integral. Let

$$x = r\cos\theta, \quad y = r\sin\theta$$

then

$$\iint_S dA = \iint_R \left|\frac{\partial \vec{r}}{\partial x} \times \frac{\partial \vec{r}}{\partial y}\right| dxdy$$

$$= \int_0^{2\pi} \int_0^a \sqrt{1 + (f_x(r\cos\theta, r\sin\theta))^2 + (f_y(r\cos\theta, r\sin\theta))^2}\, rdrd\theta$$

Suppose the surface S is a sphere, given by $S = \{(x, y, z): x^2 + y^2 + z^2 = r^2, r > 0\}$. The goal is to compute the surface area of S, denoted as $\iint_S dA$. The parametric equation for the surface S is

$$\vec{r}(\theta, \varphi) = (r\cos\theta\sin\varphi, r\sin\theta\sin\varphi, r\cos\varphi).$$

Let $R = \{(\theta, \varphi): 0 \leq \theta \leq 2\pi, 0 \leq \varphi \leq \pi\}$ then $\iint_S dA = \iint_R \left|\frac{\partial \vec{r}}{\partial \varphi} \times \frac{\partial \vec{r}}{\partial \theta}\right| d\theta d\varphi.$

Since
$$\frac{\partial \vec{r}}{\partial \theta} = (-r \sin \theta \sin \varphi, r \cos \theta \sin \varphi, 0)$$

and
$$\frac{\partial \vec{r}}{\partial \varphi} = (r \cos \theta \cos \varphi, r \sin \theta \cos \varphi, -r\sin \varphi),$$

the cross product of these two vectors is
$$\frac{\partial \vec{r}}{\partial \varphi} \times \frac{\partial \vec{r}}{\partial \theta} = (r^2 \cos \theta \sin^2 \varphi, r^2 \sin \theta \sin^2 \varphi, r^2 \sin \varphi \cos \varphi).$$

Thus, the magnitude is
$$\left| \frac{\partial \vec{r}}{\partial \varphi} \times \frac{\partial \vec{r}}{\partial \theta} \right| = r^2 \sin \varphi.$$

Therefore,
$$\iint_S dA = \iint_R \left| \frac{\partial \vec{r}}{\partial \varphi} \times \frac{\partial \vec{r}}{\partial \theta} \right| d\theta d\varphi = r^2 \int_0^{2\pi} \int_0^{\pi} \sin \varphi \, d\varphi d\theta = -r^2 \int_0^{2\pi} \cos \varphi |_0^{\pi} \, d\theta = 4r^2 \pi.$$

Question Types:

Type 1.

Given a parameterized surface $S: \vec{r}(s,t)$, $\forall a \leq s \leq b$ and $c \leq t \leq d$, find the surface area $\iint_S dA = ?$

Problem-Solving Process:

Step1.
$$\because \iint_S dA = \iint_D \left| \frac{\partial \vec{r}}{\partial s} \times \frac{\partial \vec{r}}{\partial t} \right| dsdt$$

Step2.

Find $\frac{\partial \vec{r}}{\partial s}, \frac{\partial \vec{r}}{\partial t}, \frac{\partial \vec{r}}{\partial s} \times \frac{\partial \vec{r}}{\partial t},$ and $\left| \frac{\partial \vec{r}}{\partial s} \times \frac{\partial \vec{r}}{\partial t} \right|$

Step3.
$$\because \iint_S dA = \iint_R \left| \frac{\partial \vec{r}}{\partial s} \times \frac{\partial \vec{r}}{\partial t} \right| dsdt = \int_a^b \int_c^d \left| \frac{\partial \vec{r}}{\partial s} \times \frac{\partial \vec{r}}{\partial t} \right| dtds$$

9.5.2 Find the Surface Area of a Closed Region on Surface S

Find $\int_a^b \int_c^d \left| \dfrac{\partial \vec{r}}{\partial s} \times \dfrac{\partial \vec{r}}{\partial t} \right| dtds = ?$

Type 2.
Given a parameterized surface $S: z = f(x, y)$, $\forall a \le x \le b$ and $c \le y \le d$, find the surface area $\iint_S dA = ?$

Problem-Solving Process:

Step1.
Let $\vec{r}(x, y) = (x, y, f(x, y))$ and $R = \{(x, y): a \le x \le b, c \le y \le d\}$ then

$$\iint_S dA = \iint_R \left| \dfrac{\partial \vec{r}}{\partial x} \times \dfrac{\partial \vec{r}}{\partial y} \right| dxdy$$

Step2.
$\because \dfrac{\partial \vec{r}}{\partial x} = (1, 0, f_x(x, y))$, $\dfrac{\partial \vec{r}}{\partial y} = (0, 1, f_y(x, y))$ $\therefore \dfrac{\partial \vec{r}}{\partial x} \times \dfrac{\partial \vec{r}}{\partial y} = (-f_x(x, y), -f_y(x, y), 1)$

$\Rightarrow \left| \dfrac{\partial \vec{r}}{\partial x} \times \dfrac{\partial \vec{r}}{\partial y} \right| = \sqrt{1 + (f_x(x, y))^2 + (f_y(x, y))^2}$

Step3.

$$\iint_S dA = \iint_R \left| \dfrac{\partial \vec{r}}{\partial x} \times \dfrac{\partial \vec{r}}{\partial y} \right| dxdy = \int_a^b \int_c^d \sqrt{1 + (f_x(x, y))^2 + (f_y(x, y))^2} \, dydx$$

Type 3.
Given a parameterized surface $S: z = f(x, y)$, and assuming the surface S is projected onto the xy-plane as $\{(x, y): x^2 + y^2 \le a^2\}$, find the surface area

$$\iint_S dA = ?$$

Problem-Solving Process:

Step1.
Let $\vec{r}(x, y) = (x, y, f(x, y))$ and let R be the region where the surface S is projected onto the xy-plane. Then

$R = \{(x, y): x^2 + y^2 \le a^2\}$ and $\iint_S dA = \iint_R \left| \dfrac{\partial \vec{r}}{\partial x} \times \dfrac{\partial \vec{r}}{\partial y} \right| dxdy$

Step2.

$$\because \frac{\partial \vec{r}}{\partial x} = (1, 0, f_x(x, y)) \text{ and } \frac{\partial \vec{r}}{\partial y} = (0, 1, f_y(x, y)) \quad \therefore \frac{\partial \vec{r}}{\partial x} \times \frac{\partial \vec{r}}{\partial y} = (-f_x(x, y), -f_y(x, y), 1)$$

$$\Rightarrow \left| \frac{\partial \vec{r}}{\partial x} \times \frac{\partial \vec{r}}{\partial y} \right| = \sqrt{1 + (f_x(x, y))^2 + (f_y(x, y))^2}$$

Step3.

$$\therefore \iint_S dA = \iint_R \left| \frac{\partial \vec{r}}{\partial x} \times \frac{\partial \vec{r}}{\partial y} \right| dxdy = \iint_R \sqrt{1 + (f_x(x, y))^2 + (f_y(x, y))^2} \, dxdy$$

Step4.

Let $x = r\cos\theta$, $y = r\sin\theta$ then
$R = \{(x, y): x^2 + y^2 \le a^2\} = \{(r, \theta): 0 \le r \le a, 0 \le \theta \le 2\pi\}$

$$\iint_R \sqrt{1 + (f_x(x, y))^2 + (f_y(x, y))^2} \, dxdy$$

$$= \int_0^{2\pi} \int_0^a \sqrt{1 + (f_x(r\cos\theta, r\sin\theta))^2 + (f_y(r\cos\theta, r\sin\theta))^2} \, rdrd\theta$$

Type 4.

Given the surface $S: \{(x, y, z): x^2 + y^2 + z^2 = r^2, r > 0\}$, find the surface area $\iint_S dA = ?$

Problem-Solving Process:

Step1.

Let $\vec{r}(\theta, \varphi) = (r\cos\theta\sin\varphi, r\sin\theta\sin\varphi, r\cos\varphi)$ and $R = \{(\theta, \varphi): 0 \le \theta \le 2\pi, 0 \le \varphi \le \pi\}$

then $\iint_S dA = \iint_R \left| \frac{\partial \vec{r}}{\partial \varphi} \times \frac{\partial \vec{r}}{\partial \theta} \right| d\theta d\varphi$

Step2.

$$\because \frac{\partial \vec{r}}{\partial \theta} = (-r\sin\theta\sin\varphi, r\cos\theta\sin\varphi, 0) \text{ and } \frac{\partial \vec{r}}{\partial \varphi} = (r\cos\theta\cos\varphi, r\sin\theta\cos\varphi, -r\sin\varphi)$$

$$\therefore \frac{\partial \vec{r}}{\partial \varphi} \times \frac{\partial \vec{r}}{\partial \theta} = (r^2 \cos\theta \sin^2\varphi, r^2 \sin\theta \sin^2\varphi, r^2 \sin\varphi \cos\varphi) \Rightarrow \left| \frac{\partial \vec{r}}{\partial \varphi} \times \frac{\partial \vec{r}}{\partial \theta} \right| = r^2 \sin\varphi$$

$$\therefore \iint_S dA = \iint_R \left| \frac{\partial \vec{r}}{\partial \varphi} \times \frac{\partial \vec{r}}{\partial \theta} \right| d\theta d\varphi = r^2 \int_0^{2\pi} \int_0^{\pi} \sin\varphi \, d\varphi d\theta = -r^2 \int_0^{2\pi} \cos\varphi |_0^{\pi} \, d\theta = 4r^2\pi$$

9.5.2 Find the Surface Area of a Closed Region on Surface S

Type 5.
Given the plane in space $S: Ax + By + Cz = D$
(1) Find the surface area of the region enclosed by $0 \le ax + by \le s$, and $0 \le cx + dy \le t$, $\forall t, s > 0$.
(2) Find the surface area of the region enclosed by $y = x^\alpha$, $y = \beta x^\alpha$, $x = y^\alpha$, and $x = \gamma y^\alpha$, where $\alpha \ne 1, \beta, \gamma > 1$.
(3) Find the surface area of the region enclosed by $ax^2 + bxy + cy^2 = a^2$, where $\alpha > 0$, $b^2 - 4ac < 0$.
(4) Find the surface area of the region enclosed by $y = ax$, $y = bx$, $xy = c$, and $xy = d$, Where $a < b, c < d$.

Problem-Solving Process:

Let $\vec{r}(x, y) = \left(x, y, \dfrac{D - (Ax + By)}{C}\right)$, and let R represent the region projected onto the xy-plane. Then

$$\iint_S dA = \iint_R \left|\dfrac{\partial \vec{r}}{\partial x} \times \dfrac{\partial \vec{r}}{\partial y}\right| dxdy$$

$$\because \dfrac{\partial \vec{r}}{\partial x} = \left(1, 0, -\dfrac{A}{C}\right) \text{ and } \dfrac{\partial \vec{r}}{\partial y} = \left(0, 1, -\dfrac{B}{C}\right) \quad \therefore \dfrac{\partial \vec{r}}{\partial x} \times \dfrac{\partial \vec{r}}{\partial y} = \left(\dfrac{A}{C}, \dfrac{B}{C}, 1\right)$$

$$\Rightarrow \left|\dfrac{\partial \vec{r}}{\partial x} \times \dfrac{\partial \vec{r}}{\partial y}\right| = \sqrt{1 + \dfrac{A^2 + B^2}{C^2}} = \sqrt{\dfrac{A^2 + B^2 + C^2}{C^2}}$$

$$\therefore \iint_S dA = \iint_R \left|\dfrac{\partial \vec{r}}{\partial x} \times \dfrac{\partial \vec{r}}{\partial y}\right| dxdy = \sqrt{\dfrac{A^2 + B^2 + C^2}{C^2}} \iint_R dxdy$$

(1)
Step1.

Let $u = ax + by$, $v = cx + dy$ then $x = \dfrac{du - bv}{ad - bc}$ and $y = \dfrac{av - cu}{ad - bc}$

$$\therefore dxdy = \left\|\begin{matrix}\dfrac{\partial x}{\partial u} & \dfrac{\partial x}{\partial v} \\ \dfrac{\partial y}{\partial u} & \dfrac{\partial y}{\partial v}\end{matrix}\right\| dudv = \dfrac{\left\|\begin{matrix}d & -b \\ -c & a\end{matrix}\right\|}{|ad - bc|^2} dudv = \dfrac{1}{|ad - bc|} dudv$$

$$\therefore \iint_R dxdy = \int_0^t \int_0^s \frac{1}{|ad-bc|} dudv$$

Step2.

Compute $\int_0^t \int_0^s \frac{1}{|ad-bc|} dudv = ?$

$$\therefore \iint_R dxdy = \int_0^t \int_0^s \frac{1}{|ad-bc|} dudv = \frac{ts}{|ad-bc|}$$

$$\therefore \iint_S dA = \sqrt{\frac{A^2+B^2+C^2}{C^2}} \iint_R dxdy = \frac{ts}{|ad-bc|} \sqrt{\frac{A^2+B^2+C^2}{C^2}}$$

(2)

Step1.

Let $u = \frac{y}{x^\alpha}$, $v = \frac{x}{y^\alpha}$ then $x = u^{\frac{\alpha}{1-\alpha^2}} v^{\frac{1}{1-\alpha^2}}$, $y = u^{\frac{1}{1-\alpha^2}} v^{\frac{\alpha}{1-\alpha^2}}$

$$\therefore dxdy = \begin{Vmatrix} \frac{\partial x}{\partial u} & \frac{\partial x}{\partial v} \\ \frac{\partial y}{\partial u} & \frac{\partial y}{\partial v} \end{Vmatrix} dudv = \begin{Vmatrix} \frac{\alpha u^{\left(\frac{\alpha}{1-\alpha^2}-1\right)} v^{\frac{1}{1-\alpha^2}}}{1-\alpha^2} & \frac{u^{\frac{\alpha}{1-\alpha^2}} v^{\left(\frac{1}{1-\alpha^2}-1\right)}}{1-\alpha^2} \\ \frac{u^{\left(\frac{1}{1-\alpha^2}-1\right)} v^{\frac{\alpha}{1-\alpha^2}}}{1-\alpha^2} & \frac{\alpha u^{\frac{1}{1-\alpha^2}} v^{\left(\frac{\alpha}{1-\alpha^2}-1\right)}}{1-\alpha^2} \end{Vmatrix} dudv$$

$$= \frac{\alpha^2 u^{\frac{\alpha-1+\alpha^2+1}{1-\alpha^2}} v^{\frac{1+\alpha-1+\alpha^2}{1-\alpha^2}} - u^{\frac{\alpha-1+\alpha^2+1}{1-\alpha^2}} v^{\frac{1+\alpha-1+\alpha^2}{1-\alpha^2}}}{(1-\alpha^2)^2} dudv$$

$$= \frac{\alpha^2 u^{\frac{\alpha}{1-\alpha}} v^{\frac{\alpha}{1-\alpha}} - u^{\frac{\alpha}{1-\alpha}} v^{\frac{\alpha}{1-\alpha}}}{(1-\alpha^2)^2} dudv = \frac{u^{\frac{\alpha}{1-\alpha}} v^{\frac{\alpha}{1-\alpha}}}{\alpha^2-1} dudv$$

Step2.

Let $R = \left\{(x,y): 1 \le \frac{y}{x^\alpha} \le \beta, 1 \le \frac{x}{y^\alpha} \le \gamma\right\}$ then $R = \{(u,v): 1 \le u \le \beta, 1 \le v \le \gamma\}$

Step3.

Area $= \iint_R dxdy = \int_1^\beta \int_1^\gamma \frac{u^{\frac{\alpha}{1-\alpha}} v^{\frac{\alpha}{1-\alpha}}}{\alpha^2-1} dvdu$

$$\therefore \iint_S dA = \sqrt{\frac{A^2+B^2+C^2}{C^2}} \iint_R dxdy = \sqrt{\frac{A^2+B^2+C^2}{C^2}} \int_1^\beta \int_1^\gamma \frac{u^{\frac{\alpha}{1-\alpha}} v^{\frac{\alpha}{1-\alpha}}}{\alpha^2-1} dvdu$$

Find $\int_1^\beta \int_1^\gamma \frac{u^{\frac{\alpha}{1-\alpha}} v^{\frac{\alpha}{1-\alpha}}}{\alpha^2-1} dvdu$

(3)

9.5.2 Find the Surface Area of a Closed Region on Surface S

Step1.

$$\because ax^2 + bxy + cy^2 = a(x + \frac{by}{2a})^2 + \frac{4ac - b^2}{4a}y^2$$

Let $\sqrt{a}\left(x + \frac{by}{2a}\right) = u, \quad \sqrt{\frac{4ac - b^2}{4a}}y = v$ then

$$dxdy = \begin{Vmatrix} \frac{\partial x}{\partial u} & \frac{\partial x}{\partial v} \\ \frac{\partial y}{\partial u} & \frac{\partial y}{\partial v} \end{Vmatrix} dudv = \begin{Vmatrix} \frac{1}{\sqrt{a}} & 0 \\ \frac{2\sqrt{a}}{b} & \sqrt{\frac{4a}{4ac-b^2}} \end{Vmatrix} dudv = \frac{2}{\sqrt{4ac - b^2}} dudv$$

$$\therefore \iint_R dxdy = \iint_{u^2+v^2 \le \alpha^2} \frac{2}{\sqrt{4ac - b^2}} dudv$$

Step2.

Let $u = r\cos\theta, v = r\sin\theta$ then $\{(u,v): u^2 + v^2 \le \alpha^2\} = \{(r,\theta): 0 \le r \le \alpha, 0 \le \theta \le 2\pi\}$

and $dudv = \begin{Vmatrix} \frac{\partial u}{\partial r} & \frac{\partial u}{\partial \theta} \\ \frac{\partial v}{\partial r} & \frac{\partial v}{\partial \theta} \end{Vmatrix} drd\theta = \begin{Vmatrix} \cos\theta & -r\sin\theta \\ \sin\theta & r\cos\theta \end{Vmatrix} drd\theta = rdrd\theta$

Step3.

$$\therefore \iint_{u^2+v^2 \le \alpha^2} \frac{2}{\sqrt{4ac - b^2}} dudv = \int_0^{2\pi}\int_0^{\alpha} \frac{2}{\sqrt{4ac - b^2}} rdrd\theta = \frac{2}{\sqrt{4ac - b^2}} \int_0^{2\pi} d\theta \int_0^{\alpha} rdr$$

$$= \frac{2\pi\alpha^2}{\sqrt{4ac - b^2}}$$

$$\therefore \iint_S dA = \frac{2\pi\alpha^2}{\sqrt{4ac - b^2}} \sqrt{\frac{A^2 + B^2 + C^2}{C^2}}$$

(4)

Step1.

Let $\frac{y}{x} = u, \ xy = v$, then $x = \sqrt{\frac{v}{u}}, \ y = \sqrt{uv}$

$$\because dxdy = \begin{Vmatrix} \frac{\partial x}{\partial u} & \frac{\partial x}{\partial v} \\ \frac{\partial y}{\partial u} & \frac{\partial y}{\partial v} \end{Vmatrix} dudv = \begin{Vmatrix} \frac{1}{2}\left(\frac{v}{u}\right)^{-\frac{1}{2}}\left(-\frac{v}{u^2}\right) & \frac{1}{2}\left(\frac{v}{u}\right)^{-\frac{1}{2}}\left(\frac{1}{u}\right) \\ \frac{1}{2}(uv)^{-\frac{1}{2}}v & \frac{1}{2}(uv)^{-\frac{1}{2}}u \end{Vmatrix} dudv = \frac{1}{2u}dudv$$

Step2.

Let $R = \{(x,y): a \leq \dfrac{y}{x} \leq b, c \leq xy \leq d\}$ then $R = \{(u,v): a \leq u \leq b, c \leq v \leq d\}$

Step 3.

$$\therefore \iint_R dxdy = \int_a^b \int_c^d \frac{1}{2u} dvdu = \frac{(d-c)\ln\frac{b}{a}}{2}$$

$$\therefore \iint_S dA = \sqrt{\frac{A^2+B^2+C^2}{C^2}} \iint_R dxdy = \sqrt{\frac{A^2+B^2+C^2}{C^2}} \int_a^b \int_c^d \frac{1}{2u} dvdu$$

$$= \frac{(d-c)\ln\frac{b}{a}\sqrt{\dfrac{A^2+B^2+C^2}{C^2}}}{2}$$

Example 1.

Assuming the parameterization of surface S is given by

$\vec{r}(s,t) = (kt\cos s, kt\sin s, t)$, where $0 \leq s \leq 2\pi$, and $0 \leq t \leq h$,

find the surface area of surface S.

【Solution】

$\therefore \dfrac{\partial \vec{r}}{\partial s} = (-kt\sin s, kt\cos s, 0)$ and $\dfrac{\partial \vec{r}}{\partial t} = (k\cos s, k\sin s, 1)$

$\therefore \dfrac{\partial \vec{r}}{\partial s} \times \dfrac{\partial \vec{r}}{\partial t} = (kt\cos s, kt\sin s, -k^2 t)$ and $\left|\dfrac{\partial \vec{r}}{\partial s} \times \dfrac{\partial \vec{r}}{\partial t}\right| = kt\sqrt{1+k^2}$

$\therefore$ surface area $= \iint_S dA = \int_0^h \int_0^{2\pi} kt\sqrt{1+k^2} \, ds \, dt = \pi k h^2 \sqrt{1+k^2}$

Example 2.

Assuming the parameterization of surface S is given by

$\vec{r}(s,t) = (s^2, s^3, 2t), \forall 0 \leq s \leq 1, 0 \leq t \leq 2$,

find the surface area of surface S.

【Solution】

$\therefore \dfrac{\partial \vec{r}}{\partial s} = (2s, 3s^2, 0)$ and $\dfrac{\partial \vec{r}}{\partial t} = (0,0,2)$

$\therefore \dfrac{\partial \vec{r}}{\partial s} \times \dfrac{\partial \vec{r}}{\partial t} = (6s^2, -4s, 0)$ and $\left|\dfrac{\partial \vec{r}}{\partial s} \times \dfrac{\partial \vec{r}}{\partial t}\right| = \sqrt{16s^2 + 36s^4} = 4s\sqrt{1+\dfrac{9s^2}{4}}$

$$\therefore \text{ surface area} = \iint_S dA = \int_0^2 \int_0^1 4s\sqrt{1+\frac{9s^2}{4}}\,dsdt = 8\int_0^1 s\sqrt{1+\frac{9s^2}{4}}\,ds$$

$$= \frac{32}{27}\left(1+\frac{9s^2}{4}\right)^{\frac{3}{2}}\Bigg|_0^1 = \frac{32}{27}\left(\frac{13\sqrt{13}}{8}-1\right)$$

Example 3.

Find the area enclosed by the plane $acx + bcy + abz = abc$ in the first quadrant.

【Solution】

Let $\vec{r}(x,y) = \left(x, y, \dfrac{abc - (acx+bcy)}{ab}\right)$ and let R represent the region projected onto the

xy-plane. Then $R = \{(x,y): acx + bcy \leq abc, x \geq 0, y \geq 0\}$ and

$$\text{surface area} = \iint_S dA = \iint_R \left|\frac{\partial \vec{r}}{\partial x} \times \frac{\partial \vec{r}}{\partial y}\right| dxdy.$$

$$\because \frac{\partial \vec{r}}{\partial x} = \left(1, 0, -\frac{c}{b}\right) \text{ and } \frac{\partial \vec{r}}{\partial y} = \left(0, 1, -\frac{b}{a}\right) \quad \therefore \frac{\partial \vec{r}}{\partial x} \times \frac{\partial \vec{r}}{\partial y} = \left(\frac{c}{b}, \frac{c}{a}, 1\right)$$

$$\Rightarrow \left|\frac{\partial \vec{r}}{\partial x} \times \frac{\partial \vec{r}}{\partial y}\right| = \sqrt{1 + \frac{a^2c^2 + b^2c^2}{a^2b^2}} = \sqrt{\frac{a^2c^2 + b^2c^2 + a^2b^2}{a^2b^2}}$$

$$\therefore \text{ surface area} = \iint_S dA = \iint_R \left|\frac{\partial \vec{r}}{\partial x} \times \frac{\partial \vec{r}}{\partial y}\right| dxdy = \sqrt{\frac{a^2c^2 + b^2c^2 + a^2b^2}{a^2b^2}} \iint_R dxdy$$

$$= \sqrt{\frac{a^2c^2 + b^2c^2 + a^2b^2}{a^2b^2}} \times \frac{ab}{2} = \frac{\sqrt{a^2c^2 + b^2c^2 + a^2b^2}}{2}$$

Example 4.

Assuming the parameterization of surface is given by

$$\vec{r}(s,t) = (a\cos s \cos t, a\sin s \cos t, a\sin t), 0 \leq s \leq 2\pi, -\frac{\pi}{2} \leq t \leq \frac{\pi}{2},$$

find the surface area.

【Solution】

$$\because \frac{\partial \vec{r}}{\partial s} = (-a \sin s \cos t, a \cos s \cos t, 0) \text{ and } \frac{\partial \vec{r}}{\partial t} = (-a \cos s \sin t, -a \sin s \sin t, a \cos t)$$

$$\therefore \frac{\partial \vec{r}}{\partial s} \times \frac{\partial \vec{r}}{\partial t} = a\cos t \,(a\cos s \cos t, a \sin s \cos t, a \sin t) = a\cos t \, \vec{r}(s,t) \text{ and}$$

$$\left| \frac{\partial \vec{r}}{\partial s} \times \frac{\partial \vec{r}}{\partial t} \right| = a\cos^2 t$$

$$\therefore \iint_S dA = \int_0^{2\pi} \int_{-\frac{\pi}{2}}^{\frac{\pi}{2}} \left| \frac{\partial \vec{r}}{\partial s} \times \frac{\partial \vec{r}}{\partial t} \right| dt ds = \int_0^{2\pi} \int_{-\frac{\pi}{2}}^{\frac{\pi}{2}} a\cos^2 t \, dt ds = 2\pi a^2 \int_{-\frac{\pi}{2}}^{\frac{\pi}{2}} a\cos^2 t \, dt = 4\pi a^2$$

Example 5.

Find the area of the surface, $z = a^2 - (x^2 + y^2)$ with $\frac{a^2}{4} \leq x^2 + y^2 \leq a^2$.

【Solution】

Let $S: z = a^2 - (x^2 + y^2)$, $\frac{a^2}{4} \leq x^2 + y^2 \leq a^2$

Let $\vec{r}(x,y) = (x, y, a^2 - (x^2 + y^2))$ and let R represent the region projected onto the xy-plane. Then

$$R = \left\{ (x,y): \frac{a^2}{4} \leq x^2 + y^2 \leq a^2 \right\} \text{ and surface area} = \iint_S dA = \iint_R \left| \frac{\partial \vec{r}}{\partial x} \times \frac{\partial \vec{r}}{\partial y} \right| dxdy.$$

$$\because \frac{\partial \vec{r}}{\partial x} = (1, 0, -2x) \text{ and } \frac{\partial \vec{r}}{\partial y} = (0, 1, -2y) \quad \therefore \frac{\partial \vec{r}}{\partial x} \times \frac{\partial \vec{r}}{\partial y} = (2x, 2y, 1)$$

$$\Rightarrow \left| \frac{\partial \vec{r}}{\partial x} \times \frac{\partial \vec{r}}{\partial y} \right| = (4x^2 + 4y^2 + 1)^{\frac{1}{2}}$$

$$\therefore \iint_S dA = \iint_R \left| \frac{\partial \vec{r}}{\partial x} \times \frac{\partial \vec{r}}{\partial y} \right| dxdy = \iint_{\frac{a^2}{4} \leq x^2 + y^2 \leq a^2} (4x^2 + 4y^2 + 1)^{\frac{1}{2}} dxdy$$

Let $x = r\cos\theta, y = r\sin\theta$ then

9.5.2 Find the Surface Area of a Closed Region on Surface S 448

$$\left\{(x,y): \frac{a^2}{4} \le x^2 + y^2 \le a^2\right\} = \left\{(r,\theta): \frac{a}{2} \le r \le a, 0 \le \theta \le 2\pi\right\}$$

$$\therefore \iint_S dA = \iint_{\frac{a^2}{4} \le x^2+y^2 \le a^2} (4x^2 + 4y^2 + 1)^{\frac{1}{2}} dxdy = \int_0^{2\pi} \int_{\frac{a}{2}}^{a} (4r^2 + 1)^{\frac{1}{2}} r \, dr d\theta$$

$$= \frac{\pi}{6}(4r^2 + 1)^{\frac{3}{2}} \Big|_{\frac{a}{2}}^{a} = \frac{\pi}{6}\left((4a^2 + 1)^{\frac{3}{2}} - (a^2 + 1)^{\frac{3}{2}}\right)$$

Example 6.

Find the area of the surface $z = \frac{2}{3}\left(x^{\frac{3}{2}} + y^{\frac{3}{2}}\right)$ with $0 \le x \le 1, 0 \le y \le 1$.

【Solution】

Let $S: z = \frac{2}{3}\left(x^{\frac{3}{2}} + y^{\frac{3}{2}}\right), \; 0 \le x \le 1, 0 \le y \le 1$

Let $\vec{r}(x,y) = \left(x, y, \frac{2}{3}\left(x^{\frac{3}{2}} + y^{\frac{3}{2}}\right)\right)$ and let R represent the region projected onto the

xy-plane. Then

$R = \{(x,y): 0 \le x \le 1, 0 \le y \le 1\}$ and $\displaystyle\iint_S dA = \iint_R \left|\frac{\partial \vec{r}}{\partial x} \times \frac{\partial \vec{r}}{\partial y}\right| dxdy.$

$\because \dfrac{\partial \vec{r}}{\partial x} = \left(1, 0, x^{\frac{1}{2}}\right)$ and $\dfrac{\partial \vec{r}}{\partial y} = \left(0, 1, y^{\frac{1}{2}}\right) \quad \therefore \dfrac{\partial \vec{r}}{\partial x} \times \dfrac{\partial \vec{r}}{\partial y} = \left(x^{\frac{1}{2}}, y^{\frac{1}{2}}, 1\right)$

$$\Rightarrow \left|\frac{\partial \vec{r}}{\partial x} \times \frac{\partial \vec{r}}{\partial y}\right| = (x+y+1)^{\frac{1}{2}}$$

$$\iint_S dA = \iint_R \left|\frac{\partial \vec{r}}{\partial x} \times \frac{\partial \vec{r}}{\partial y}\right| dxdy = \int_0^1 \int_0^1 (x+y+1)^{\frac{1}{2}} dydx = \frac{2}{3}\int_0^1 (2+y)^{\frac{3}{2}} - (1+y)^{\frac{3}{2}} dx$$

$$= \frac{2}{3}\left(\frac{2}{5}\left((2+y)^{\frac{5}{2}} - (1+y)^{\frac{5}{2}}\right)\right)\Big|_{y=0}^{y=1} = \frac{4}{15}\left(3^{\frac{5}{2}} - 2^{\frac{7}{2}} + 1\right)$$

Example 7.

Find the area of the surface $x^2 + y^2 + z^2 - 4z = 0$ with $0 \le 3(x^2+y^2) \le z^2, z \ge 2$.

【Solution】

Let $S: x^2 + y^2 + z^2 = 4z$, $0 \le 3(x^2+y^2) \le z^2$, $z \ge 2$ then surface area $= \iint_S dA$

Let $\vec{r}(x,y) = \left(x, y, 2+\sqrt{4-(x^2+y^2)}\right)$ and let R represent the region projected onto the xy-plane. Then

$$R = \{(x,y): 0 \le x^2+y^2 \le 3\} \text{ and } \iint_S dA = \iint_R \left|\frac{\partial \vec{r}}{\partial x} \times \frac{\partial \vec{r}}{\partial y}\right| dxdy$$

$$\because \frac{\partial \vec{r}}{\partial x} = \left(1, 0, -x(4-(x^2+y^2))^{-\frac{1}{2}}\right) \text{ and } \frac{\partial \vec{r}}{\partial y} = \left(0, 1, -y(4-(x^2+y^2))^{-\frac{1}{2}}\right),$$

$$\therefore \frac{\partial \vec{r}}{\partial x} \times \frac{\partial \vec{r}}{\partial y} = \left(x(4-(x^2+y^2))^{-\frac{1}{2}}, y(4-(x^2+y^2))^{-\frac{1}{2}}, 1\right)$$

$$\Rightarrow \left|\frac{\partial \vec{r}}{\partial x} \times \frac{\partial \vec{r}}{\partial y}\right| = (x^2(4-(x^2+y^2))^{-1} + y^2(4-(x^2+y^2))^{-1} + 1)^{\frac{1}{2}} = 2(4-(x^2+y^2))^{-\frac{1}{2}}$$

$$\therefore \iint_S dA = \iint_R \left|\frac{\partial \vec{r}}{\partial x} \times \frac{\partial \vec{r}}{\partial y}\right| dxdy = \iint_{x^2+y^2 \le 3} 2(4-(x^2+y^2))^{-\frac{1}{2}} dxdy$$

Let $x = r\cos\theta$, $y = r\sin\theta$ then $\{(x,y): x^2+y^2 \le 3\} = \{(r,\theta): 0 \le r \le \sqrt{3}, 0 \le \theta \le 2\pi\}$

9.5.2 Find the Surface Area of a Closed Region on Surface S 450

$$\iint_{x^2+y^2\le 3} 2(4-(x^2+y^2))^{-\frac{1}{2}}dxdy = \int_0^{2\pi}\int_0^{\sqrt{3}} 2(4-r^2)^{-\frac{1}{2}}rdrd\theta = 4\pi(4-r^2)^{\frac{1}{2}}\Big|_{\sqrt{3}}^0 = 4\pi$$

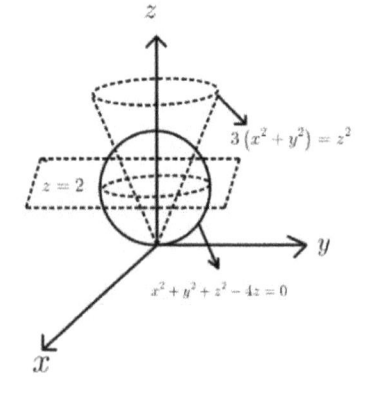

Example 8.

Find the area of the part of the plane $x + 2y + 3z = 1$ that lies within the cylinder $x^2 + y^2 = 3$.

【Solution】

Let $\vec{r}(x,y) = \left(x, y, \dfrac{1-(x+2y)}{3}\right)$ and let R represent the region projected onto the xy-plane. Then

$R = \{(x,y): x^2 + y^2 \le 3\}$ and surface area $= \iint_S dA = \iint_R \left|\dfrac{\partial \vec{r}}{\partial x} \times \dfrac{\partial \vec{r}}{\partial y}\right| dxdy$

$\because \dfrac{\partial \vec{r}}{\partial x} = \left(1, 0, -\dfrac{1}{3}\right)$ and $\dfrac{\partial \vec{r}}{\partial y} = \left(0, 1, -\dfrac{2}{3}\right)$

$\therefore \dfrac{\partial \vec{r}}{\partial x} \times \dfrac{\partial \vec{r}}{\partial y} = \left(\dfrac{1}{3}, \dfrac{2}{3}, 1\right) \Rightarrow \left|\dfrac{\partial \vec{r}}{\partial x} \times \dfrac{\partial \vec{r}}{\partial y}\right| = \sqrt{1 + \dfrac{5}{9}} = \sqrt{\dfrac{14}{9}}$

$\therefore \iint_S dA = \iint_R \left|\dfrac{\partial \vec{r}}{\partial x} \times \dfrac{\partial \vec{r}}{\partial y}\right| dxdy = \sqrt{\dfrac{14}{9}} \iint_{x^2+y^2\le 3} dxdy = \sqrt{\dfrac{14}{9}} \times 3\pi = \sqrt{14}\pi$

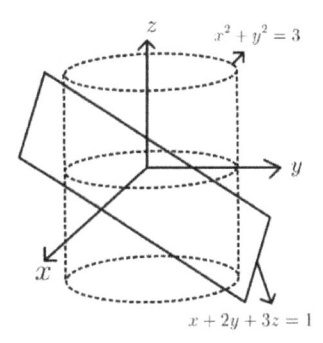

Example 9.

Find the surface area of that part of the parabolic cylinder $z = y^2$ that lies over the triangle with vertices $(0,0), (0,1), (1,1)$ in the xy-plane.

【Solution】

Let $S: z = y^2$, $0 \leq x \leq 1, 0 \leq y \leq 1$ then surface area $= \iint_S dA$

Let $\vec{r}(x,y) = (x, y, y^2)$ and let R represent the region projected onto the xy-plane,

then $R = \{(x,y): 0 \leq x \leq y, 0 \leq y \leq 1\}$ and surface area $= \iint_S dA = \iint_R \left|\frac{\partial \vec{r}}{\partial x} \times \frac{\partial \vec{r}}{\partial y}\right| dxdy$

$\because \frac{\partial \vec{r}}{\partial x} = (1,0,0)$ and $\frac{\partial \vec{r}}{\partial y} = (0,1,2y) \therefore \frac{\partial \vec{r}}{\partial x} \times \frac{\partial \vec{r}}{\partial y} = (0,2y,1) \Rightarrow \left|\frac{\partial \vec{r}}{\partial x} \times \frac{\partial \vec{r}}{\partial y}\right| = (1+4y^2)^{\frac{1}{2}}$

$\iint_R \left|\frac{\partial \vec{r}}{\partial x} \times \frac{\partial \vec{r}}{\partial y}\right| dxdy = \int_0^1 \int_0^y (1+4y^2)^{\frac{1}{2}} dxdy = \int_0^1 y(1+4y^2)^{\frac{1}{2}} dy = \frac{(1+4y^2)^{\frac{3}{2}}}{12} \Bigg|_0^1$

$= \frac{5\sqrt{5} - 1}{12}$

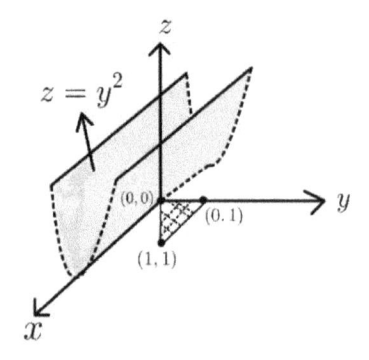

9.5.2 Find the Surface Area of a Closed Region on Surface S

Example 10.

Find the area of the surface with parametric equation $x = s^2$, $y = st$, $z = \dfrac{t^2}{2}$, $0 \le s \le 1, 0 \le t \le 2$.

【Solution】

Let $\vec{r}(s,t) = \left(s^2, st, \dfrac{t^2}{2}\right)$, $\forall\, 0 \le s \le 1, 0 \le t \le 2$ then $\dfrac{\partial \vec{r}}{\partial s} = (2s, t, 0)$ and $\dfrac{\partial \vec{r}}{\partial t} = (0, s, t)$

$\therefore \dfrac{\partial \vec{r}}{\partial s} \times \dfrac{\partial \vec{r}}{\partial t} = (t^2, 2st, 2s^2)$ and $\left|\dfrac{\partial \vec{r}}{\partial s} \times \dfrac{\partial \vec{r}}{\partial t}\right| = \sqrt{t^4 + 4s^2 t^2 + 4s^4} = t^2 + 2s^2$

$\therefore \iint_S dA = \iint_{[0,1]\times[0,2]} \left|\dfrac{\partial \vec{r}}{\partial s} \times \dfrac{\partial \vec{r}}{\partial t}\right| ds\, dt = \int_0^2 \int_0^1 t^2 + 2s^2 ds\, dt = \int_0^2 t^2 + \dfrac{1}{3} dt = 4$

Example 11.

Find the area of the surface $y = 4x + z^2$ that lies between the planes $x = 0, x = 1$, $z = 0,$ and $z = 1$.

【Solution】

Let $S: y = 4x + z^2$, $0 \le x \le 1$, $0 \le z \le 1$ then surface area $= \iint_S dA$

Let $\vec{r}(x, y) = (x, 4x + z^2, z)$ and let R represent the region projected onto the xz-plane,

then $R = \{(x, z): 0 \le x \le 1, 0 \le z \le 1\}$ and $\iint_S dA = \iint_R \left|\dfrac{\partial \vec{r}}{\partial x} \times \dfrac{\partial \vec{r}}{\partial z}\right| dx\, dz$

$$\because \frac{\partial \vec{r}}{\partial x} = (1,4,0) \text{ and } \frac{\partial \vec{r}}{\partial z} = (0,2z,1) \therefore \frac{\partial \vec{r}}{\partial x} \times \frac{\partial \vec{r}}{\partial z} = (4,-1,2z) \Rightarrow \left|\frac{\partial \vec{r}}{\partial x} \times \frac{\partial \vec{r}}{\partial z}\right| = (17+4z^2)^{\frac{1}{2}}$$

$$\therefore \iint_S dA = \iint_R \left|\frac{\partial \vec{r}}{\partial x} \times \frac{\partial \vec{r}}{\partial z}\right| dxdz = \int_0^1 \int_0^1 (17+4z^2)^{\frac{1}{2}} dxdz = \int_0^1 (17+4z^2)^{\frac{1}{2}} dz$$

Let $t = 2z$ then $\int_0^1 (17+4z^2)^{\frac{1}{2}} dz = \frac{1}{2}\int_0^2 (17+t^2)^{\frac{1}{2}} dt$

Let $t = \sqrt{17}\tan\theta$ then $dt = \sqrt{17}\sec^2\theta\, d\theta$

$$\therefore \frac{1}{2}\int_0^2 (17+t^2)^{\frac{1}{2}} dt = \frac{17}{2}\int_0^{\tan^{-1} 2} \sec^3\theta\, d\theta = \frac{1}{2}\left(\frac{t\sqrt{17+t^2}}{2} + \frac{17}{2}\ln\left(\frac{t}{\sqrt{17}} + \frac{\sqrt{17+t^2}}{\sqrt{17}}\right)\right)\bigg|_{t=0}^{t=2}$$

$$= \frac{\sqrt{21}}{2} + \frac{17}{4}\left(\ln(2+\sqrt{21}) - \ln\sqrt{17}\right)$$

Example 12.

Find the surface area of the plane $Ax + By + Cz = D$ inside the elliptical cylinder $\frac{x^2}{a^2} + \frac{y^2}{b^2} = 1$.

【Solution】

Let $\vec{r}(x,y) = \left(x, y, \frac{D-(Ax+By)}{C}\right)$ and let R represent the region projected onto the

xy-plane. Then

$$R = \left\{(x,y): \frac{x^2}{a^2} + \frac{y^2}{b^2} \le 1\right\} \text{ and surface area} = \iint_S dA = \iint_R \left|\frac{\partial \vec{r}}{\partial x} \times \frac{\partial \vec{r}}{\partial y}\right| dxdy$$

$$\because \frac{\partial \vec{r}}{\partial x} = \left(1, 0, -\frac{A}{C}\right) \text{ and } \frac{\partial \vec{r}}{\partial y} = \left(0, 1, -\frac{B}{C}\right) \therefore \frac{\partial \vec{r}}{\partial x} \times \frac{\partial \vec{r}}{\partial y} = \left(\frac{A}{C}, \frac{B}{C}, 1\right)$$

$$\Rightarrow \left|\frac{\partial \vec{r}}{\partial x} \times \frac{\partial \vec{r}}{\partial y}\right| = \sqrt{1 + \frac{A^2+B^2}{C^2}} = \sqrt{\frac{A^2+B^2+C^2}{C^2}}$$

9.5.2 Find the Surface Area of a Closed Region on Surface S

$$\therefore \text{ surface area} = \iint_S dA = \iint_R \left|\frac{\partial \vec{r}}{\partial x} \times \frac{\partial \vec{r}}{\partial y}\right| dxdy = \sqrt{\frac{A^2 + B^2 + C^2}{C^2}} \iint_{\frac{x^2}{a^2}+\frac{y^2}{b^2}\leq 1} dxdy$$

Let $x = r a \cos\theta$, $y = rb \sin\theta$ then

$$\left\{(x,y): \frac{x^2}{a^2} + \frac{y^2}{b^2} \leq 1\right\} = \{(r, \theta): 0 \leq r \leq 1, 0 \leq \theta \leq 2\pi\}$$

$$\therefore \iint_{\frac{x^2}{a^2}+\frac{y^2}{b^2}\leq 1} dxdy = \int_0^{2\pi} \int_0^1 abr\, dr\, d\theta = ab\pi \quad \therefore \text{ surface area} = ab\pi\sqrt{\frac{A^2 + B^2 + C^2}{C^2}}$$

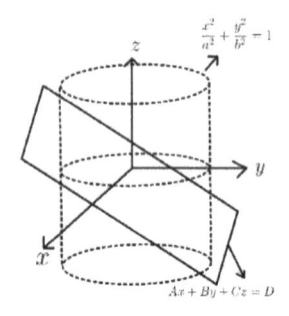

Example 13.

Find the surface area of the plane $Ax + By + Cz = D$ within the region bounded by $y = x^2$, $y = 3x^2$, $x = y^2$, $x = 4y^2$.

【Solution】

Let $\vec{r}(x,y) = \left(x, y, \dfrac{D - (Ax + By)}{C}\right)$ and $R = \{(x,y): x^2 \leq y \leq 3x^2, 4y^2 \leq x \leq y^2\}$,

$$\text{surface area} = \iint_S dA = \iint_R \left|\frac{\partial \vec{r}}{\partial x} \times \frac{\partial \vec{r}}{\partial y}\right| dxdy$$

$$\because \frac{\partial \vec{r}}{\partial x} = \left(1, 0, -\frac{A}{C}\right) \text{ and } \frac{\partial \vec{r}}{\partial y} = \left(0, 1, -\frac{B}{C}\right) \quad \therefore \frac{\partial \vec{r}}{\partial x} \times \frac{\partial \vec{r}}{\partial y} = \left(\frac{A}{C}, \frac{B}{C}, 1\right)$$

$$\Rightarrow \left|\frac{\partial \vec{r}}{\partial x} \times \frac{\partial \vec{r}}{\partial y}\right| = \sqrt{1 + \frac{A^2 + B^2}{C^2}} = \sqrt{\frac{A^2 + B^2 + C^2}{C^2}}$$

$$\therefore \text{ surface area} = \iint_S dA = \iint_R \left|\frac{\partial \vec{r}}{\partial x} \times \frac{\partial \vec{r}}{\partial y}\right| dxdy = \sqrt{\frac{A^2+B^2+C^2}{C^2}} \iint_R dxdy$$

Let $u = \dfrac{y}{x^2}$, $v = \dfrac{x}{y^2}$ then $x = u^{-\frac{2}{3}}v^{-\frac{1}{3}}$, $y = u^{-\frac{1}{3}}v^{-\frac{2}{3}}$

and $dxdy = \begin{Vmatrix} \frac{\partial x}{\partial u} & \frac{\partial x}{\partial v} \\ \frac{\partial y}{\partial u} & \frac{\partial y}{\partial v} \end{Vmatrix} dudv = \begin{Vmatrix} \frac{-2u^{-\frac{5}{3}}v^{-\frac{1}{3}}}{3} & \frac{-u^{-\frac{2}{3}}v^{-\frac{4}{3}}}{3} \\ \frac{-u^{-\frac{4}{3}}v^{-\frac{2}{3}}}{3} & \frac{-2u^{-\frac{1}{3}}v^{-\frac{5}{3}}}{3} \end{Vmatrix} dudv = \frac{u^{-2}v^{-2}}{3} dudv$

$\because R = \left\{(x,y): 1 \leq \dfrac{y}{x^2} \leq 3, 1 \leq \dfrac{x}{y^2} \leq 4\right\}$ $\therefore R = \{(u,v): 1 \leq u \leq 3, 1 \leq v \leq 4\}$

$\therefore \iint_R 1\, dA = \int_1^3 \int_1^4 \dfrac{u^{-2}v^{-2}}{3} dvdu = \dfrac{1}{6}$ $\therefore \iint_S dA = \dfrac{1}{6}\sqrt{\dfrac{A^2+B^2+C^2}{C^2}}$

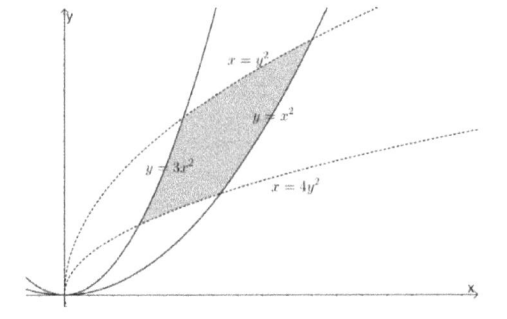

Example 14.

Find the surface area of the plane $Ax + By + Cz = D$ within the region bounded by $x - 2y = -4$, $x - 2y = 1$, $2x - y = 0$, $2x - y = 3$.

【Solution】

Let $\vec{r}(x,y) = \left(x, y, \dfrac{D - (Ax + By)}{C}\right)$ and $R = \{(x,y): -4 \leq x - 2y \leq 1, 0 \leq 2x - y \leq 3\}$,

then surface area $= \iint_S dA = \iint_R \left|\dfrac{\partial \vec{r}}{\partial x} \times \dfrac{\partial \vec{r}}{\partial y}\right| dxdy$

$\therefore \dfrac{\partial \vec{r}}{\partial x} = \left(1, 0, -\dfrac{A}{C}\right)$ and $\dfrac{\partial \vec{r}}{\partial y} = \left(0, 1, -\dfrac{B}{C}\right)$ $\therefore \dfrac{\partial \vec{r}}{\partial x} \times \dfrac{\partial \vec{r}}{\partial y} = \left(\dfrac{A}{C}, \dfrac{B}{C}, 1\right)$

9.5.2 Find the Surface Area of a Closed Region on Surface S

$$\Rightarrow \left|\frac{\partial \vec{r}}{\partial x} \times \frac{\partial \vec{r}}{\partial y}\right| = \sqrt{1 + \frac{A^2 + B^2}{C^2}} = \sqrt{\frac{A^2 + B^2 + C^2}{C^2}}$$

$$\therefore \iint_R \left|\frac{\partial \vec{r}}{\partial x} \times \frac{\partial \vec{r}}{\partial y}\right| dxdy = \sqrt{\frac{A^2 + B^2 + C^2}{C^2}} \iint_R dxdy$$

Let $x - 2y = u$, $2x - y = v$ then $x = -\frac{u}{3} + \frac{2v}{3}$, $y = -\frac{2u}{3} + \frac{v}{3}$

and $dxdy = \left\|\begin{matrix} \frac{\partial x}{\partial u} & \frac{\partial x}{\partial v} \\ \frac{\partial y}{\partial u} & \frac{\partial y}{\partial v} \end{matrix}\right\| dudv = \left\|\begin{matrix} -\frac{1}{3} & \frac{2}{3} \\ -\frac{2}{3} & \frac{1}{3} \end{matrix}\right\| dudv = \frac{1}{3} dudv$

$\because R = \{(x, y): -4 \le x - 2y \le 1, 0 \le 2x - y \le 3\}$ $\therefore R = \{(u, v): -4 \le u \le 1, 0 \le v \le 3\}$

$$\therefore \iint_R 1 dA = \int_{-4}^{1} \int_{0}^{3} \frac{1}{3} dudv = 5 \quad \therefore \text{surface area} = \iint_S dA = 5\sqrt{\frac{A^2 + B^2 + C^2}{C^2}}$$

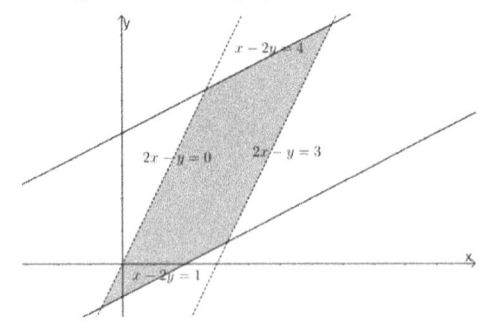

Example 15.

Find the surface area of the plane $Ax + By + Cz = D$ within the region bounded by $y = x$, $y = 2x$, $xy = 1$, $xy = 4$.

【Solution】

Let $\vec{r}(x, y) = \left(x, y, \frac{D - (Ax + By)}{C}\right)$ and let R be the region enclosed by $y = x$,

$y = 2x$, $xy = 1$, $xy = 4$, then surface area $= \iint_S dA = \iint_R \left|\frac{\partial \vec{r}}{\partial x} \times \frac{\partial \vec{r}}{\partial y}\right| dxdy$

$\because \frac{\partial \vec{r}}{\partial x} = \left(1, 0, -\frac{A}{C}\right)$ and $\frac{\partial \vec{r}}{\partial y} = \left(0, 1, -\frac{B}{C}\right)$ $\therefore \frac{\partial \vec{r}}{\partial x} \times \frac{\partial \vec{r}}{\partial y} = \left(\frac{A}{C}, \frac{B}{C}, 1\right)$

$$\Rightarrow \left|\frac{\partial \vec{r}}{\partial x} \times \frac{\partial \vec{r}}{\partial y}\right| = \sqrt{1 + \frac{A^2 + B^2}{C^2}} = \sqrt{\frac{A^2 + B^2 + C^2}{C^2}}$$

$$\therefore \iint_R \left|\frac{\partial \vec{r}}{\partial x} \times \frac{\partial \vec{r}}{\partial y}\right| dxdy = \sqrt{\frac{A^2 + B^2 + C^2}{C^2}} \iint_R dxdy$$

Let $\frac{y}{x} = u$, $xy = v$ then $x = \sqrt{\frac{v}{u}}$, $y = \sqrt{uv}$

and $dxdy = \begin{Vmatrix} \frac{\partial x}{\partial u} & \frac{\partial x}{\partial v} \\ \frac{\partial y}{\partial u} & \frac{\partial y}{\partial v} \end{Vmatrix} dudv = \begin{Vmatrix} \frac{1}{2}\left(\frac{v}{u}\right)^{-\frac{1}{2}}\left(-\frac{v}{u^2}\right) & \frac{1}{2}\left(\frac{v}{u}\right)^{-\frac{1}{2}}\left(\frac{1}{u}\right) \\ \frac{1}{2}(uv)^{-\frac{1}{2}}v & \frac{1}{2}(uv)^{-\frac{1}{2}}u \end{Vmatrix} dudv = \frac{1}{2u} dudv$

$\because R = \left\{(x,y): 1 \leq \frac{y}{x} \leq 2, 1 \leq xy \leq 4\right\}$ $\therefore R = \{(u,v): 1 \leq u \leq 2, 1 \leq v \leq 4\}$

$\therefore \iint_R dxdy = 2\int_1^4 \int_1^2 \frac{1}{2u} dudv = 3\ln 2$ $\therefore$ surface area $= \iint_S dA = 3\ln 2 \sqrt{\frac{A^2 + B^2 + C^2}{C^2}}$

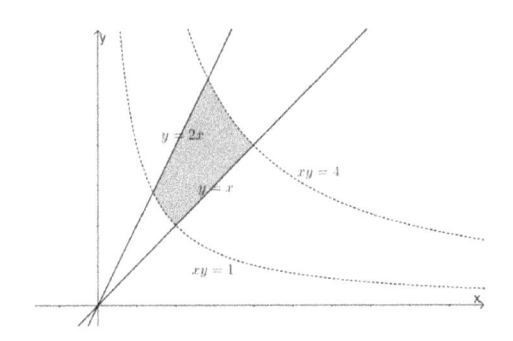

Example 16.

Find the surface area of the plane $Ax + By + Cz = D$ within the region bounded by $ax^2 + bxy + cy^2 = \alpha^2$.

【Solution】

Let $\vec{r}(x,y) = \left(x, y, \frac{D - (Ax + By)}{C}\right)$ and $R = \{(x,y): ax^2 + bxy + cy^2 \leq \alpha^2\}$ then

9.5.2 Find the Surface Area of a Closed Region on Surface S

$$\iint_S dA = \iint_R \left|\frac{\partial \vec{r}}{\partial x} \times \frac{\partial \vec{r}}{\partial y}\right| dxdy.$$

$$\because \frac{\partial \vec{r}}{\partial x} = \left(1, 0, -\frac{A}{C}\right) \text{ and } \frac{\partial \vec{r}}{\partial y} = \left(0, 1, -\frac{B}{C}\right) \quad \therefore \frac{\partial \vec{r}}{\partial x} \times \frac{\partial \vec{r}}{\partial y} = \left(\frac{A}{C}, \frac{B}{C}, 1\right)$$

$$\Rightarrow \left|\frac{\partial \vec{r}}{\partial x} \times \frac{\partial \vec{r}}{\partial y}\right| = \sqrt{1 + \frac{A^2 + B^2}{C^2}} = \sqrt{\frac{A^2 + B^2 + C^2}{C^2}}$$

$$\therefore \iint_R \left|\frac{\partial \vec{r}}{\partial x} \times \frac{\partial \vec{r}}{\partial y}\right| dxdy = \sqrt{\frac{A^2 + B^2 + C^2}{C^2}} \iint_R dxdy$$

$$\because ax^2 + bxy + cy^2 = a\left(x + \frac{by}{2a}\right)^2 + \left(\frac{4ac - b^2}{4a}\right)y^2$$

Let $\sqrt{a}\left(x + \frac{by}{2a}\right) = u$, $\sqrt{\frac{4ac - b^2}{4a}} y = v$ then

$$dxdy = \left\|\begin{matrix} \frac{\partial x}{\partial u} & \frac{\partial x}{\partial v} \\ \frac{\partial y}{\partial u} & \frac{\partial y}{\partial v} \end{matrix}\right\| dudv = \left\|\begin{matrix} \frac{1}{\sqrt{a}} & 0 \\ 0 & \sqrt{\frac{4a}{4ac - b^2}} \end{matrix}\right\| dudv = \frac{2}{\sqrt{4ac - b^2}} dudv$$

$$\therefore \iint_R dA = \iint_{u^2 + v^2 \leq a^2} \frac{2}{\sqrt{4ac - b^2}} dudv$$

Let $u = r\cos\theta, v = r\sin\theta$ then $\{(u, v): u^2 + v^2 \leq \alpha^2\} = \{(r, \theta): 0 \leq r \leq \alpha, 0 \leq \theta \leq 2\pi\}$

and $dudv = \left\|\begin{matrix} \frac{\partial u}{\partial r} & \frac{\partial u}{\partial \theta} \\ \frac{\partial v}{\partial r} & \frac{\partial v}{\partial \theta} \end{matrix}\right\| drd\theta = \left\|\begin{matrix} \cos\theta & -r\sin\theta \\ \sin\theta & r\cos\theta \end{matrix}\right\| drd\theta = rdrd\theta$

$$\therefore \iint_{u^2+v^2\leq \alpha^2} \frac{2}{\sqrt{4ac-b^2}} dudv = \int_0^{2\pi}\int_0^\alpha \frac{2}{\sqrt{4ac-b^2}} rdrd\theta = \frac{2}{\sqrt{4ac-b^2}} \int_0^{2\pi} d\theta \int_0^\alpha rdr$$

$$= \frac{2\pi\alpha^2}{\sqrt{4ac-b^2}}$$

$$\therefore \text{surface area} = \iint_S dA = \iint_R \left|\frac{\partial \vec{r}}{\partial x} \times \frac{\partial \vec{r}}{\partial y}\right| dxdy = \frac{2\pi\alpha^2}{\sqrt{4ac-b^2}} \sqrt{\frac{A^2 + B^2 + C^2}{C^2}}$$

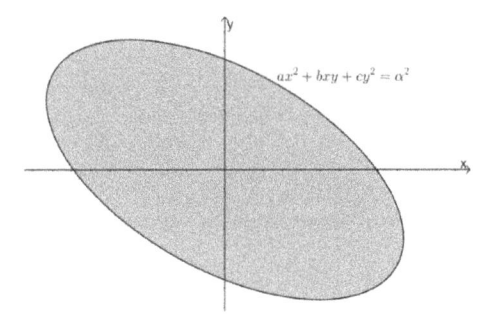

9.5.2 Find the Surface Area of a Closed Region on Surface S

Example 17.

Find the surface area of the surface $z = \frac{2}{3}y^{\frac{3}{2}}$ within the region $0 < x < 2$, and $0 < y < 2$.

【Solution】

Let $S: z = \frac{2}{3}y^{\frac{3}{2}}, \ 0 < x < 2, \ 0 < y < 2$ then surface area $= \iint_S dA$

Let $\vec{r}(x,y) = \left(x, y, \frac{2}{3}y^{\frac{3}{2}}\right)$ and let R represent the region projected onto the xy-plane,

then $R = \{(x,y): 0 < x < 2, 0 < y < 2\}$ and $\iint_S dA = \iint_R \left|\frac{\partial \vec{r}}{\partial x} \times \frac{\partial \vec{r}}{\partial y}\right| dxdy$

$\because \frac{\partial \vec{r}}{\partial x} = (1,0,0)$ and $\frac{\partial \vec{r}}{\partial y} = \left(0, 1, y^{\frac{1}{2}}\right)$ $\therefore \frac{\partial \vec{r}}{\partial x} \times \frac{\partial \vec{r}}{\partial y} = \left(0, -y^{\frac{1}{2}}, 1\right) \Rightarrow \left|\frac{\partial \vec{r}}{\partial x} \times \frac{\partial \vec{r}}{\partial y}\right| = (y+1)^{\frac{1}{2}}$

$\therefore \iint_S dA = \iint_R \left|\frac{\partial \vec{r}}{\partial x} \times \frac{\partial \vec{r}}{\partial y}\right| dxdy = \int_0^2 \int_0^2 (y+1)^{\frac{1}{2}} dxdy = 2 \cdot \frac{2(y+1)^{\frac{3}{2}}}{3}\bigg|_0^2 = \frac{4}{3}(3\sqrt{3} - 1)$

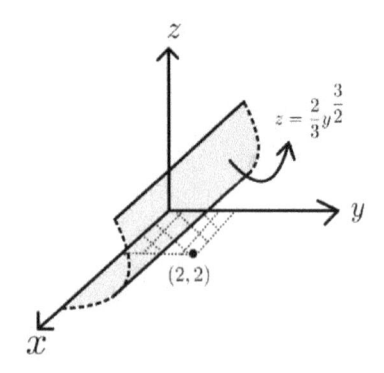

9.5.2 Find the Surface Area of a Closed Region on Surface S

Example 18.

Find the surface area of the surface $z = x + ay + b$ within the region defined by $y > x^2$, $y < 1$, and $x > 0$.

【Solution】

Let $S: z = x + ay + b$, $y > x^2$, $y < 1$, $x > 0$ then area surface $= \iint_S dA$

Let $\vec{r}(x,y) = (x, y, x + ay + b)$ and let R represent the region projected onto the xy-plane.

Then $R = \{(x,y): y > x^2, y < 1, x > 0\}$ and $\iint_S dA = \iint_R \left| \frac{\partial \vec{r}}{\partial x} \times \frac{\partial \vec{r}}{\partial y} \right| dxdy$

$\because \frac{\partial \vec{r}}{\partial x} = (1,0,1)$ and $\frac{\partial \vec{r}}{\partial y} = (0,1,a)$ $\therefore \frac{\partial \vec{r}}{\partial x} \times \frac{\partial \vec{r}}{\partial y} = (-1, -a, 1) \Rightarrow \left| \frac{\partial \vec{r}}{\partial x} \times \frac{\partial \vec{r}}{\partial y} \right| = \sqrt{a^2 + 2}$

$\therefore \iint_S dA = \iint_R \left| \frac{\partial \vec{r}}{\partial x} \times \frac{\partial \vec{r}}{\partial y} \right| dxdy = \int_0^1 \int_{x^2}^1 \sqrt{a^2 + 2}\, dydx = \frac{2\sqrt{a^2 + 2}}{3}$

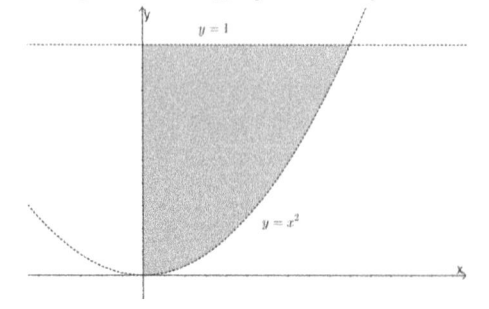

Example 19.

Find the surface area of the surface $z = xy$ within the cylindrical surface $x^2 + y^2 = a^2$, where $x \geq 0$ and $y \geq 0$.

【Solution】

Let $S: z = xy$, $x^2 + y^2 \leq a^2$ then surface area $= \iint_S dA$

Let $\vec{r}(x, y) = (x, y, xy)$ and let R represent the region projected onto the xy-plane.

Then $R = \{(x, y): x^2 + y^2 \leq a^2\}$ and $\iint_S dA = \iint_R \left|\frac{\partial \vec{r}}{\partial x} \times \frac{\partial \vec{r}}{\partial y}\right| dxdy$

$\because \dfrac{\partial \vec{r}}{\partial x} = (1, 0, y)$ and $\dfrac{\partial \vec{r}}{\partial y} = (0, 1, x)$

$\therefore \dfrac{\partial \vec{r}}{\partial x} \times \dfrac{\partial \vec{r}}{\partial y} = (-y, -x, 1) \Rightarrow \left|\dfrac{\partial \vec{r}}{\partial x} \times \dfrac{\partial \vec{r}}{\partial y}\right| = (x^2 + y^2 + 1)^{\frac{1}{2}}$

$\therefore \iint_S dA = \iint_R \left|\dfrac{\partial \vec{r}}{\partial x} \times \dfrac{\partial \vec{r}}{\partial y}\right| dxdy = \iint_{x^2+y^2 \leq a^2} (x^2 + y^2 + 1)^{\frac{1}{2}} dxdy$

Let $x = r \cos \theta$, $y = r \sin \theta$ then

$\{(x, y): x^2 + y^2 \leq a^2, x \geq 0, y \geq 0\} = \{(r, \theta): 0 \leq r \leq a, 0 \leq \theta \leq \dfrac{\pi}{2}\}$

$\therefore \iint_{x^2+y^2 \leq a^2} (x^2 + y^2 + 1)^{\frac{1}{2}} dxdy = \int_0^{\frac{\pi}{2}} \int_0^a (r^2 + 1)^{\frac{1}{2}} r \, dr d\theta = \dfrac{\pi}{6}(r^2 + 1)^{\frac{3}{2}}\Big|_0^a$

$= \dfrac{\pi}{6}((a^2 + 1)^{\frac{3}{2}} - 1)$

Example 20.
Find the area of the region in the xy-plane bounded by the triangle with vertices at $(0,0)$, $(0, a)$, and (a, a) for the surface $z = y^2 + bx + c$.

【Solution】

9.5.2 Find the Surface Area of a Closed Region on Surface S

Let $S: z = y^2 + bx + c$, $0 \le x \le y$, $0 \le y \le a$ then surface area $= \iint_S dA$

Let $\vec{r}(x,y) = (x, y, y^2 + bx + c)$ and let R represent the region projected onto the xy-plane.

Then $R = \{(x, y): 0 \le x \le y, 0 \le y \le a\}$ and $\iint_S dA = \iint_R \left|\frac{\partial \vec{r}}{\partial x} \times \frac{\partial \vec{r}}{\partial y}\right| dxdy$

$\because \frac{\partial \vec{r}}{\partial x} = (1, 0, b), \quad \frac{\partial \vec{r}}{\partial y} = (0, 1, 2y) \quad \therefore \frac{\partial \vec{r}}{\partial x} \times \frac{\partial \vec{r}}{\partial y} = (-b, -2y, 1) \Rightarrow \left|\frac{\partial \vec{r}}{\partial x} \times \frac{\partial \vec{r}}{\partial y}\right| = (b^2 + 1 + 4y^2)^{\frac{1}{2}}$

$\therefore \iint_S dA = \iint_R \left|\frac{\partial \vec{r}}{\partial x} \times \frac{\partial \vec{r}}{\partial y}\right| dxdy = \int_0^a \int_0^y (b^2 + 1 + 4y^2)^{\frac{1}{2}} dxdy = \int_0^a (b^2 + 1 + 4y^2)^{\frac{1}{2}} y\, dy$

$= \frac{(b^2 + 1 + 4y^2)^{\frac{3}{2}}}{12} \Big|_0^a = \frac{1}{12}\left((1 + 4a^2 + b^2)\sqrt{1 + 4a^2 + b^2} - (b^2 + 1)\sqrt{b^2 + 1}\right)$

Example 21.
Find the surface area of the paraboloid $z = x^2 + y^2$ within the region $0 \le z \le 3$.
【Solution】

Let $S: z = x^2 + y^2$, $0 \le z \le 3$ then surface area $= \iint_S dA$

Let $\vec{r}(x, y) = (x, y, x^2 + y^2)$ and let R represent the region projected onto the xy-plane.

Then $R = \{(x, y): x^2 + y^2 \le 3\}$ and $\iint_S dA = \iint_R \left|\frac{\partial \vec{r}}{\partial x} \times \frac{\partial \vec{r}}{\partial y}\right| dxdy$

$\because \frac{\partial \vec{r}}{\partial x} = (1, 0, 2x)$ and $\frac{\partial \vec{r}}{\partial y} = (0, 1, 2y) \quad \therefore \frac{\partial \vec{r}}{\partial x} \times \frac{\partial \vec{r}}{\partial y} = (-2x, -2y, 1)$

$\Rightarrow \left|\frac{\partial \vec{r}}{\partial x} \times \frac{\partial \vec{r}}{\partial y}\right| = (4x^2 + 4y^2 + 1)^{\frac{1}{2}}$

$$\therefore \iint_R \left|\frac{\partial \vec{r}}{\partial x} \times \frac{\partial \vec{r}}{\partial y}\right| dxdy = \iint_{x^2+y^2 \leq 3} (4x^2 + 4y^2 + 1)^{\frac{1}{2}} dxdy$$

Let $x = r\cos\theta$, $y = r\sin\theta$ then $\{(x,y): x^2 + y^2 \leq 3\} = \{(r,\theta): 0 \leq r \leq \sqrt{3}, 0 \leq \theta \leq 2\pi\}$

$$\therefore \iint_{x^2+y^2 \leq 3} (4x^2 + 4y^2 + 1)^{\frac{1}{2}} dxdy = \int_0^{2\pi} \int_0^{\sqrt{3}} (4r^2 + 1)^{\frac{1}{2}} r \, dr \, d\theta = 2\pi \cdot \frac{1}{12} \cdot (4r^2 + 1)^{\frac{3}{2}} \bigg|_0^{\sqrt{3}}$$

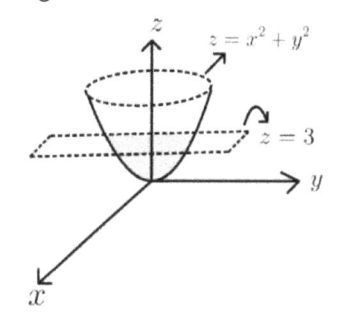

$$= \frac{\pi}{6}(13\sqrt{13} - 1)$$

Example 22.

Find the surface area of the plane $ax + by + cz = 1$ in the first quadrant bounded by $x = 0, y = 0$, and $x^2 + y^2 = r^2$.

【Solution】

Let $S: ax + by + cz = 1$, $x \geq 0$, $y \geq 0$, $x^2 + y^2 \leq r^2$ then surface area $= \frac{1}{4} \iint_S dA$

Let $\vec{r}(x,y) = \left(x, y, \frac{1}{c} - \frac{ax}{c} - \frac{by}{c}\right)$ and let R represent the region projected onto the

xy-plane. Then

$$R = \{(x,y): x \geq 0 \cdot y \geq 0 \cdot x^2 + y^2 \leq r^2\} \text{ and } \iint_S dA = \iint_R \left|\frac{\partial \vec{r}}{\partial x} \times \frac{\partial \vec{r}}{\partial y}\right| dxdy$$

$$\because \frac{\partial \vec{r}}{\partial x} = \left(1, 0, -\frac{a}{c}\right) \text{ and } \frac{\partial \vec{r}}{\partial y} = \left(0, 1, -\frac{b}{c}\right) \therefore \frac{\partial \vec{r}}{\partial x} \times \frac{\partial \vec{r}}{\partial y} = \left(\frac{a}{c}, \frac{b}{c}, 1\right)$$

9.5.2 Find the Surface Area of a Closed Region on Surface S

$$\Rightarrow \left|\frac{\partial \vec{r}}{\partial x} \times \frac{\partial \vec{r}}{\partial y}\right| = \frac{\sqrt{a^2+b^2+c^2}}{c}$$

$$\therefore \frac{1}{4}\iint_S dA = \frac{1}{4}\iint_R \left|\frac{\partial \vec{r}}{\partial x} \times \frac{\partial \vec{r}}{\partial y}\right| dxdy = \frac{1}{4}\iint_{x^2+y^2 \le r^2} \frac{\sqrt{a^2+b^2+c^2}}{c} dxdy$$

$$= \frac{r^2\pi\sqrt{a^2+b^2+c^2}}{4c}$$

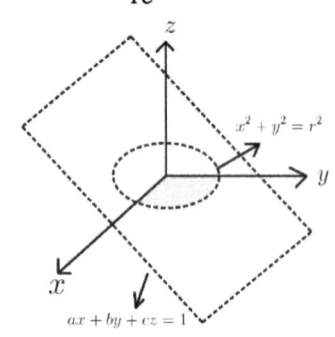

Example 23.

(1) Find the surface area of the sphere $x^2 + y^2 + z^2 = 8$ inside the cone $z = \sqrt{x^2 + y^2}$.

(2) Find the surface area of the sphere $x^2 + y^2 + z^2 = 2a(a > 0)$ inside the cone $z = \sqrt{x^2 + y^2}$.

【Solution】

(1)

Let $S: x^2 + y^2 + z^2 = 8,\ z \le \sqrt{x^2 + y^2}$ then surface area $= \iint_S dA$

Let $\vec{r}(x,y) = \left(x, y, \sqrt{8 - (x^2 + y^2)}\right)$ and let R represent the region projected onto the xy-plane. Then

$R = \{(x, y): x^2 + y^2 \le 4\}$ and $\iint_S dA = \iint_R \left|\frac{\partial \vec{r}}{\partial x} \times \frac{\partial \vec{r}}{\partial y}\right| dxdy$

$$\because \frac{\partial \vec{r}}{\partial x} = \left(1, 0, -x(8-(x^2+y^2))^{-\frac{1}{2}}\right) \text{ and } \frac{\partial \vec{r}}{\partial y} = \left(0, 1, -y(8-(x^2+y^2))^{-\frac{1}{2}}\right)$$

$$\therefore \frac{\partial \vec{r}}{\partial x} \times \frac{\partial \vec{r}}{\partial y} = \left(x(8-(x^2+y^2))^{-\frac{1}{2}}, y(8-(x^2+y^2))^{-\frac{1}{2}}, 1\right)$$

$$\Rightarrow \left|\frac{\partial \vec{r}}{\partial x} \times \frac{\partial \vec{r}}{\partial y}\right| = (x^2(8-(x^2+y^2))^{-1} + y^2(8-(x^2+y^2))^{-1} + 1)^{\frac{1}{2}} = \sqrt{8}(8-(x^2+y^2))^{-\frac{1}{2}}$$

$$\therefore \iint_S dA = \iint_R \left|\frac{\partial \vec{r}}{\partial x} \times \frac{\partial \vec{r}}{\partial y}\right| dxdy = \iint_{x^2+y^2 \le 4} \sqrt{8}(8-(x^2+y^2))^{-\frac{1}{2}} dxdy$$

Let $x = r\cos\theta$, $y = r\sin\theta$ then $\{(x,y): x^2 + y^2 \le 4\} = \{(r,\theta): 0 \le r \le 2, 0 \le \theta \le 2\pi\}$

$$\therefore \iint_{x^2+y^2 \le 4} \sqrt{8}(8-(x^2+y^2))^{-\frac{1}{2}} dxdy = \int_0^{2\pi} \int_0^2 \sqrt{8}(8-r^2)^{-\frac{1}{2}} r\, dr\, d\theta = 4\sqrt{2}\pi (8-r^2)^{\frac{1}{2}} \Big|_2^0$$

$$= \pi(16 - 8\sqrt{2})$$

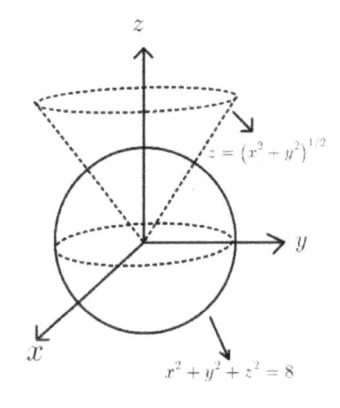

(2)

Let $S: x^2 + y^2 + z^2 = 2a$, $z \le \sqrt{x^2 + y^2}$ then surface area $= \iint_S dA$

Let $\vec{r}(x,y) = \left(x, y, \sqrt{2a - (x^2+y^2)}\right)$ and let R represent the region projected onto the xy-plane. Then

9.5.2 Find the Surface Area of a Closed Region on Surface S 466

$R = \{(x,y): x^2 + y^2 \leq a\}$ and $\iint_S dA = \iint_R \left|\frac{\partial \vec{r}}{\partial x} \times \frac{\partial \vec{r}}{\partial y}\right| dxdy$

$\therefore \frac{\partial \vec{r}}{\partial x} = \left(1, 0, -x(2a - (x^2 + y^2))^{-\frac{1}{2}}\right)$ and $\frac{\partial \vec{r}}{\partial y} = \left(0, 1, -y(2a - (x^2 + y^2))^{-\frac{1}{2}}\right)$

$\therefore \frac{\partial \vec{r}}{\partial x} \times \frac{\partial \vec{r}}{\partial y} = \left(x(2a - (x^2 + y^2))^{-\frac{1}{2}}, y(2a - (x^2 + y^2))^{-\frac{1}{2}}, 1\right)$

$\Rightarrow \left|\frac{\partial \vec{r}}{\partial x} \times \frac{\partial \vec{r}}{\partial y}\right| = (x^2(2a - (x^2 + y^2))^{-1} + y^2(2a - (x^2 + y^2))^{-1} + 1)^{\frac{1}{2}}$

$= \sqrt{2a}(2a - (x^2 + y^2))^{-\frac{1}{2}}$

$\therefore \iint_R \left|\frac{\partial \vec{r}}{\partial x} \times \frac{\partial \vec{r}}{\partial y}\right| dxdy = \iint_{x^2+y^2 \leq a} \sqrt{2a}(2a - (x^2 + y^2))^{-\frac{1}{2}} dxdy$

Let $x = r\cos\theta$, $y = r\sin\theta$ then $\{(x,y): x^2 + y^2 \leq a\} = \{(r,\theta): 0 \leq r \leq \sqrt{a}, 0 \leq \theta \leq 2\pi\}$

$\therefore \iint_{x^2+y^2 \leq a} \sqrt{2a}(2a - (x^2 + y^2))^{-\frac{1}{2}} dxdy = \int_0^{2\pi} \int_0^{\sqrt{a}} \sqrt{2a}(2a - r^2)^{-\frac{1}{2}} r \, dr d\theta$

$= 2\sqrt{2a}\pi(2a - r^2)^{\frac{1}{2}}\Big|_{\sqrt{a}}^{0} = 4a\pi - 2a\sqrt{2}\pi$

Example 24.
(1) Find the surface area of the sphere $x^2 + y^2 + z^2 = 6z$ inside the paraboloid $z = x^2 + y^2$.

(2) Find the surface area of the sphere $x^2 + y^2 + z^2 = 2az$, $\left(a > \frac{1}{2}\right)$ inside the paraboloid $z = x^2 + y^2$.

【Solution】

(1)

Let $S: x^2 + y^2 + z^2 = 6z$, $z \leq x^2 + y^2$ then surface area $= \iint_S dA$

Let $\vec{r}(x,y) = \left(x, y, 3 + \sqrt{9-(x^2+y^2)}\right)$ and let R represent the region projected onto the xy-plane. Then

$$R = \{(x,y): x^2 + y^2 \le 5\} \text{ and } \iint_S dA = \iint_R \left|\frac{\partial \vec{r}}{\partial x} \times \frac{\partial \vec{r}}{\partial y}\right| dxdy$$

$$\because \frac{\partial \vec{r}}{\partial x} = \left(1, 0, -x(9-(x^2+y^2))^{-\frac{1}{2}}\right) \text{ and } \frac{\partial \vec{r}}{\partial y} = \left(0, 1, -y(9-(x^2+y^2))^{-\frac{1}{2}}\right)$$

$$\therefore \frac{\partial \vec{r}}{\partial x} \times \frac{\partial \vec{r}}{\partial y} = \left(x(9-(x^2+y^2))^{-\frac{1}{2}}, y(9-(x^2+y^2))^{-\frac{1}{2}}, 1\right)$$

$$\Rightarrow \left|\frac{\partial \vec{r}}{\partial x} \times \frac{\partial \vec{r}}{\partial y}\right| = (x^2(9-(x^2+y^2))^{-1} + y^2(9-(x^2+y^2))^{-1} + 1)^{\frac{1}{2}} = 3(9-(x^2+y^2))^{-\frac{1}{2}}$$

$$\therefore \iint_R \left|\frac{\partial \vec{r}}{\partial x} \times \frac{\partial \vec{r}}{\partial y}\right| dxdy = \iint_{x^2+y^2 \le 5} 3(9-(x^2+y^2))^{-\frac{1}{2}} dxdy$$

Let $x = r\cos\theta$, $y = r\sin\theta$ then $\{(x,y): x^2+y^2 \le 5\} = \{(r,\theta): 0 \le r \le \sqrt{5}, 0 \le \theta \le 2\pi\}$

$$\therefore \iint_{x^2+y^2 \le 5} 3(9-(x^2+y^2))^{-\frac{1}{2}} dxdy = \int_0^{2\pi} \int_0^{\sqrt{5}} 3(9-r^2)^{-\frac{1}{2}} r\,drd\theta = 6\pi(9-r^2)^{\frac{1}{2}}\Big|_{\sqrt{5}}^{0} = 6\pi$$

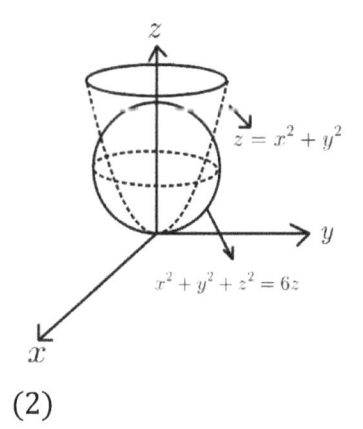

(2)

Let $S: x^2 + y^2 + z^2 = 2az$, $z \le x^2 + y^2$ then surface area $= \iint_S dA$

9.5.2 Find the Surface Area of a Closed Region on Surface S

Let $\vec{r}(x,y) = \left(x, y, a + \sqrt{a^2 - (x^2 + y^2)}\right)$ and let R represent the region projected onto the xy-plane. Then

$$R = \{(x,y): x^2 + y^2 \leq 2a - 1\} \text{ and } \iint_S dA = \iint_R \left|\frac{\partial \vec{r}}{\partial x} \times \frac{\partial \vec{r}}{\partial y}\right| dxdy$$

$$\because \frac{\partial \vec{r}}{\partial x} = \left(1, 0, -x(a^2 - (x^2 + y^2))^{-\frac{1}{2}}\right) \text{ and } \frac{\partial \vec{r}}{\partial y} = \left(0, 1, -y(a^2 - (x^2 + y^2))^{-\frac{1}{2}}\right)$$

$$\therefore \frac{\partial \vec{r}}{\partial x} \times \frac{\partial \vec{r}}{\partial y} = \left(x(a^2 - (x^2 + y^2))^{-\frac{1}{2}}, y(a^2 - (x^2 + y^2))^{-\frac{1}{2}}, 1\right)$$

$$\Rightarrow \left|\frac{\partial \vec{r}}{\partial x} \times \frac{\partial \vec{r}}{\partial y}\right| = (x^2(a^2 - (x^2 + y^2))^{-1} + y^2(a^2 - (x^2 + y^2))^{-1} + 1)^{\frac{1}{2}}$$

$$= a(a^2 - (x^2 + y^2))^{-\frac{1}{2}}$$

$$\therefore \iint_R \left|\frac{\partial \vec{r}}{\partial x} \times \frac{\partial \vec{r}}{\partial y}\right| dxdy = \iint_{x^2+y^2 \leq 2a-1} a(a^2 - (x^2 + y^2))^{-\frac{1}{2}} dxdy$$

Let $x = r\cos\theta, y = r\sin\theta$ then

$$\{(x,y): x^2 + y^2 \leq 2a - 1\} = \{(r,\theta): 0 \leq r \leq \sqrt{2a-1}, 0 \leq \theta \leq 2\pi\}$$

$$\therefore \iint_{x^2+y^2 \leq 2a-1} a(a^2 - (x^2 + y^2))^{-\frac{1}{2}} dxdy = \int_0^{2\pi} \int_0^{\sqrt{2a-1}} a(a^2 - r^2)^{-\frac{1}{2}} r\, drd\theta$$

$$= 2a\pi(a^2 - r^2)^{\frac{1}{2}}\Big|_{\sqrt{2a-1}}^{0} = 2a\pi(a - (a-1)) = 2a\pi$$

Example 25.

(1) Prove that the surface area of a sphere of radius r cut by two parallel planes separated by distance h depends only on h, and not on the position of the cut.

(2) Assume $-r < a < b < r$, and find the surface area of the unit sphere $x^2 + y^2 + z^2 = r^2$ between $z = a$ and $z = b (r > 0)$.

【Solution】
(1)
Let the distance from the center of the sphere to the nearer plane $= b$

Let

$$S: \vec{r}(\theta, \varphi) = (r\cos\theta\sin\varphi, r\sin\theta\sin\varphi, r\cos\varphi), \cos^{-1}\frac{b+h}{r} \leq \varphi \leq \cos^{-1}\frac{b}{r}, 0 \leq \theta \leq 2\pi$$

then surface area $= \iint_S dA$

$\because \dfrac{\partial \vec{r}}{\partial \theta} = (-r\sin\theta\sin\varphi, r\cos\theta\sin\varphi, 0)$ and $\dfrac{\partial \vec{r}}{\partial \varphi} = (r\cos\theta\cos\varphi, r\sin\theta\cos\varphi, -r\sin\varphi)$

$\therefore \dfrac{\partial \vec{r}}{\partial \varphi} \times \dfrac{\partial \vec{r}}{\partial \theta} = (r^2\cos\theta\sin^2\varphi, r^2\sin\theta\sin^2\varphi, r^2\sin\varphi\cos\varphi) \Rightarrow \left|\dfrac{\partial \vec{r}}{\partial \varphi} \times \dfrac{\partial \vec{r}}{\partial \theta}\right| = r^2\sin\varphi$

$\therefore \iint_S dA = \iint_R \left|\dfrac{\partial \vec{r}}{\partial \varphi} \times \dfrac{\partial \vec{r}}{\partial \theta}\right| d\theta d\varphi = \int_0^{2\pi}\int_{\cos^{-1}\frac{b+h}{r}}^{\cos^{-1}\frac{b}{r}} r^2\sin\varphi \, d\varphi d\theta$

$= -r^2\int_0^{2\pi} \cos\varphi\Big|_{\cos^{-1}\frac{b+h}{r}}^{\cos^{-1}\frac{b}{r}} d\theta = 2\pi r h$

∴ The surface area of the spherical cap depends only on h

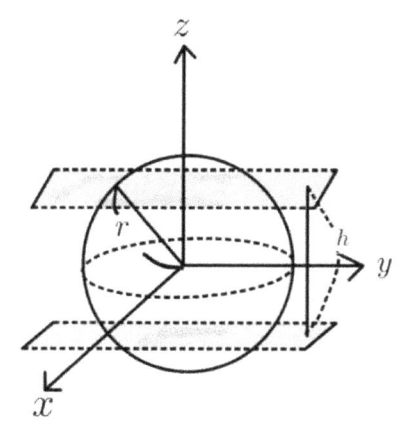

(2)

9.5.2 Find the Surface Area of a Closed Region on Surface S

Let $S: x^2 + y^2 + z^2 = r^2$, $a \leq z \leq b$ then area $= \iint_S dA$

Let $\vec{r}(\theta, \varphi) = (r\cos\theta \sin\varphi, r\sin\theta \sin\varphi, r\cos\varphi)$

Let $R = \left\{(\theta, \varphi): 0 \leq \theta \leq 2\pi, \cos^{-1}\frac{b}{r} \leq \varphi \leq \cos^{-1}\frac{a}{r}\right\}$ then $\iint_S dA = \iint_R \left|\frac{\partial \vec{r}}{\partial \varphi} \times \frac{\partial \vec{r}}{\partial \theta}\right| d\theta d\varphi$

$\because \dfrac{\partial \vec{r}}{\partial \theta} = (-r\sin\theta \sin\varphi, r\cos\theta \sin\varphi, 0)$ and $\dfrac{\partial \vec{r}}{\partial \varphi} = (r\cos\theta \cos\varphi, r\sin\theta \cos\varphi, -r\sin\varphi)$

$\therefore \dfrac{\partial \vec{r}}{\partial \varphi} \times \dfrac{\partial \vec{r}}{\partial \theta} = (r^2 \cos\theta \sin^2\varphi, r^2 \sin\theta \sin^2\varphi, r^2 \sin\varphi \cos\varphi) \Rightarrow \left|\dfrac{\partial \vec{r}}{\partial \varphi} \times \dfrac{\partial \vec{r}}{\partial \theta}\right| = r^2 \sin\varphi$

$\therefore \iint_S dA = \iint_R \left|\dfrac{\partial \vec{r}}{\partial \varphi} \times \dfrac{\partial \vec{r}}{\partial \theta}\right| d\theta d\varphi = r^2 \int_0^{2\pi} \int_{\cos^{-1}\frac{b}{r}}^{\cos^{-1}\frac{a}{r}} \sin\varphi \, d\varphi d\theta = -r^2 \int_0^{2\pi} \cos\varphi \Big|_{\cos^{-1}\frac{b}{r}}^{\cos^{-1}\frac{a}{r}} d\theta$

$= 2\pi r(b - a)$

Example 26.

Let $S_1 = \left\{(x, y, z): x^2 + y^2 + z^2 = 1, z \geq \dfrac{\sqrt{2}}{2}\right\}$, $S_2 = \left\{(x, y, z): x^2 + y^2 \leq 1, z = \dfrac{\sqrt{2}}{2}\right\}$.

Find the surface area of $S_1 \cup S_2$.

【Solution】

Let $S_1 = \left\{(x, y, z): x^2 + y^2 + z^2 = 1, z \geq \dfrac{\sqrt{2}}{2}\right\}$ and $S_2 = \left\{(x, y, z): x^2 + y^2 \leq 1, z = \dfrac{\sqrt{2}}{2}\right\}$

then area $= \iint_{S_1} dA + \iint_{S_2} dA$

Let $\vec{r}(\theta, \varphi) = (\cos\theta \sin\varphi, \sin\theta \sin\varphi, \cos\varphi)$ and $R = \left\{(\theta, \varphi): 0 \leq \theta \leq 2\pi, 0 \leq \varphi \leq \dfrac{\pi}{4}\right\}$

then $\iint_{S_1} dA = \iint_R \left|\dfrac{\partial \vec{r}}{\partial \varphi} \times \dfrac{\partial \vec{r}}{\partial \theta}\right| d\theta d\varphi$

$\therefore \dfrac{\partial \vec{r}}{\partial \theta} = (-\sin\theta \sin\varphi, \cos\theta \sin\varphi, 0)$ and $\dfrac{\partial \vec{r}}{\partial \varphi} = (\cos\theta \cos\varphi, \sin\theta \cos\varphi, -\sin\varphi)$

$\therefore \dfrac{\partial \vec{r}}{\partial \varphi} \times \dfrac{\partial \vec{r}}{\partial \theta} = (\cos\theta \sin^2\varphi, \sin\theta \sin^2\varphi, \sin\varphi \cos\varphi) \Rightarrow \left|\dfrac{\partial \vec{r}}{\partial \varphi} \times \dfrac{\partial \vec{r}}{\partial \theta}\right| = \sin\varphi$

$\therefore \iint_{S_1} dA = \iint_R \left|\dfrac{\partial \vec{r}}{\partial \varphi} \times \dfrac{\partial \vec{r}}{\partial \theta}\right| d\theta d\varphi = \int_0^{2\pi} \int_0^{\frac{\pi}{4}} \sin\varphi \, d\varphi d\theta = 2\pi\left(1 - \dfrac{1}{\sqrt{2}}\right)$

$\therefore S_2 = \left\{(x,y,z): x^2 + y^2 \leq \dfrac{1}{2}, z = \dfrac{\sqrt{2}}{2}\right\} \therefore \iint_{S_2} dA = \dfrac{\pi}{2} \Rightarrow \text{surface area} = \dfrac{\pi}{2} + 2\pi\left(1 - \dfrac{1}{\sqrt{2}}\right)$

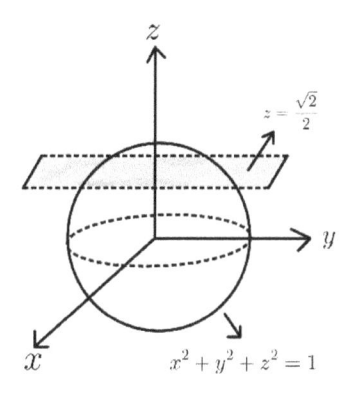

$x^2 + y^2 + z^2 = 1$

Example 27.
Find the surface area of $S = \{(x,y,z): x^2 + y^2 + z^2 = r^2, z \geq 0\}$.

【Solution】

Let $\vec{r}(\theta, \varphi) = (r\cos\theta \sin\varphi, r\sin\theta \sin\varphi, r\cos\varphi)$ and $R = \left\{(\theta, \varphi): 0 \leq \theta \leq 2\pi, 0 \leq \varphi \leq \dfrac{\pi}{2}\right\}$

then $\iint_S dA = \iint_R \left|\dfrac{\partial \vec{r}}{\partial \varphi} \times \dfrac{\partial \vec{r}}{\partial \theta}\right| d\theta d\varphi$

$\therefore \dfrac{\partial \vec{r}}{\partial \theta} = (-r\sin\theta \sin\varphi, r\cos\theta \sin\varphi, 0)$ and $\dfrac{\partial \vec{r}}{\partial \varphi} = (r\cos\theta \cos\varphi, r\sin\theta \cos\varphi, -r\sin\varphi)$

$\therefore \dfrac{\partial \vec{r}}{\partial \varphi} \times \dfrac{\partial \vec{r}}{\partial \theta} = (r^2\cos\theta \sin^2\varphi, r^2\sin\theta \sin^2\varphi, r^2\sin\varphi \cos\varphi) \Rightarrow \left|\dfrac{\partial \vec{r}}{\partial \varphi} \times \dfrac{\partial \vec{r}}{\partial \theta}\right| = r^2\sin\varphi$

9.5.2 Find the Surface Area of a Closed Region on Surface S

$$\therefore \text{ surface area} = \iint_S dA = \iint_R \left|\frac{\partial \vec{r}}{\partial \varphi} \times \frac{\partial \vec{r}}{\partial \theta}\right| d\theta d\varphi = \int_0^{2\pi} \int_0^{\frac{\pi}{2}} r^2 \sin\varphi \, d\varphi d\theta = 2\pi r^2$$

Example 28.

Find the surface area of $S = \left\{(x, y, z): x^2 + y^2 + z^2 = r^2, \frac{r^2}{4} \le x^2 + y^2 \le \frac{3r^2}{4}, z > 0\right\}$.

【Solution】

Let $\vec{r}(\theta, \varphi) = (r\cos\theta\sin\varphi, r\sin\theta\sin\varphi, r\cos\varphi)$

Let $R = \left\{(\theta, \varphi): 0 \le \theta \le 2\pi, \cos^{-1}\frac{\sqrt{3}}{2} \le \varphi \le \cos^{-1}\frac{1}{2}\right\}$ then $\iint_S dA = \iint_R \left|\frac{\partial \vec{r}}{\partial \varphi} \times \frac{\partial \vec{r}}{\partial \theta}\right| d\theta d\varphi$

$\therefore \frac{\partial \vec{r}}{\partial \theta} = (-r\sin\theta\sin\varphi, r\cos\theta\sin\varphi, 0)$ and $\frac{\partial \vec{r}}{\partial \varphi} = (r\cos\theta\cos\varphi, r\sin\theta\cos\varphi, -r\sin\varphi)$

$\therefore \frac{\partial \vec{r}}{\partial \varphi} \times \frac{\partial \vec{r}}{\partial \theta} = (r^2\cos\theta\sin^2\varphi, r^2\sin\theta\sin^2\varphi, r^2\sin\varphi\cos\varphi) \Rightarrow \left|\frac{\partial \vec{r}}{\partial \varphi} \times \frac{\partial \vec{r}}{\partial \theta}\right| = r^2\sin\varphi$

$$\iint_S dA = \iint_R \left|\frac{\partial \vec{r}}{\partial \varphi} \times \frac{\partial \vec{r}}{\partial \theta}\right| d\theta d\varphi = r^2 \int_0^{2\pi} \int_{\cos^{-1}\frac{\sqrt{3}}{2}}^{\cos^{-1}\frac{1}{2}} \sin\varphi \, d\varphi d\theta = -r^2 \int_0^{2\pi} \cos\varphi \Big|_{\cos^{-1}\frac{\sqrt{3}}{2}}^{\cos^{-1}\frac{1}{2}} d\theta$$

$= r^2\pi(\sqrt{3} - 1)$

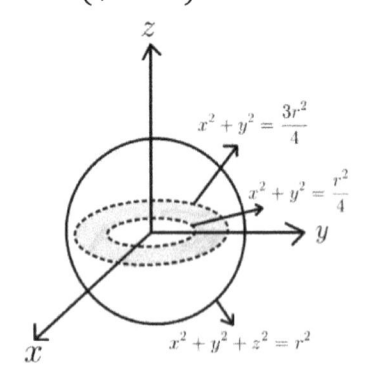

Example 29.

(1) Find the surface area of the sphere $x^2 + y^2 + z^2 = a^2$ inside the cylinder $x^2 + y^2 = b^2$ and above the xy-plane, where $a > b > 0$.

(2) Find the surface area of the sphere $x^2 + y^2 + z^2 = 16$ inside the cylinder

$(x-2)^2 + y^2 = 4$.

(3) Find the surface area of the sphere $x^2 + y^2 + z^2 = 4a^2$ inside the cylinder $x^2 + y^2 = 2ay$.

【Solution】

(1)

Let $S: x^2 + y^2 + z^2 = a^2$, $x^2 + y^2 \le b^2$, $z \ge 0$ then surface area $= \iint_S dA$

Let $\vec{r}(x,y) = \left(x, y, \sqrt{a^2 - (x^2+y^2)}\right)$ and let R represent the region projected onto the xy-plane. Then

$R = \{(x,y): x^2 + y^2 \le b^2\}$ and $\iint_S dA = \iint_R \left|\dfrac{\partial \vec{r}}{\partial x} \times \dfrac{\partial \vec{r}}{\partial y}\right| dxdy$

$\because \dfrac{\partial \vec{r}}{\partial x} = \left(1, 0, -x(a^2 - (x^2+y^2))^{-\frac{1}{2}}\right)$ and $\dfrac{\partial \vec{r}}{\partial y} = \left(0, 1, -y(a^2 - (x^2+y^2))^{-\frac{1}{2}}\right)$

$\therefore \dfrac{\partial \vec{r}}{\partial x} \times \dfrac{\partial \vec{r}}{\partial y} = \left(x(a^2-(x^2+y^2))^{-\frac{1}{2}}, y(a^2-(x^2+y^2))^{-\frac{1}{2}}, 1\right)$

$\Rightarrow \left|\dfrac{\partial \vec{r}}{\partial x} \times \dfrac{\partial \vec{r}}{\partial y}\right| = (x^2(a^2-(x^2+y^2))^{-1} + y^2(a^2-(x^2+y^2))^{-1} + 1)^{\frac{1}{2}}$

$= a(a^2 - (x^2+y^2))^{-\frac{1}{2}}$

$\therefore \iint_S dA = \iint_R \left|\dfrac{\partial \vec{r}}{\partial x} \times \dfrac{\partial \vec{r}}{\partial y}\right| dxdy = \iint_{x^2+y^2 \le b^2} a(a^2 - (x^2+y^2))^{-\frac{1}{2}} dxdy$

Let $x = r\cos\theta$, $y = r\sin\theta$ then $\{(x,y): x^2+y^2 \le b^2\} = \{(r,\theta): 0 \le r \le b, 0 \le \theta \le 2\pi\}$

$\therefore \iint_{x^2+y^2\le b^2} a(a^2-(x^2+y^2))^{-\frac{1}{2}} dxdy = \int_0^{2\pi}\int_0^b a(a^2-r^2)^{-\frac{1}{2}} r\, dr d\theta = 2\pi a(a - \sqrt{a^2-b^2})$

9.5.2 Find the Surface Area of a Closed Region on Surface S 474

[Figure: Sphere $x^2 + y^2 + z^2 = a^2$ intersected by cylinder $x^2 + y^2 = b^2$, with axes x, y, z.]

(2)

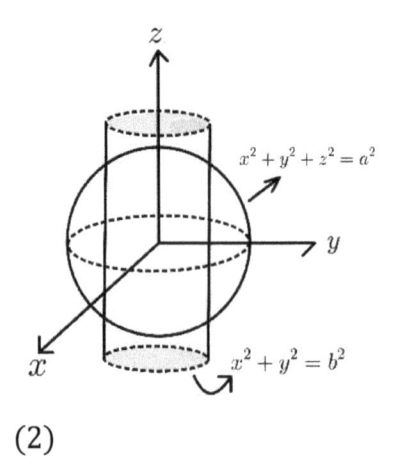

Let $S: x^2 + y^2 + z^2 = 16$, $(x-2)^2 + y^2 \leq 4$, $z \geq 0$ then surface area $= 2\iint_S dA$

Let $\vec{r}(x,y) = \left(x, y, \sqrt{16 - (x^2 + y^2)}\right)$ and let R represent the region projected onto the xy-plane, then

$$R = \{(x,y): (x-2)^2 + y^2 \leq 4\} \text{ and } \iint_S dA = \iint_R \left|\frac{\partial \vec{r}}{\partial x} \times \frac{\partial \vec{r}}{\partial y}\right| dxdy$$

$$\therefore \frac{\partial \vec{r}}{\partial x} = \left(1, 0, -x(16 - (x^2 + y^2))^{-\frac{1}{2}}\right) \text{ and } \frac{\partial \vec{r}}{\partial y} = \left(0, 1, -y(16 - (x^2 + y^2))^{-\frac{1}{2}}\right)$$

$$\therefore \frac{\partial \vec{r}}{\partial x} \times \frac{\partial \vec{r}}{\partial y} = \left(x(16 - (x^2 + y^2))^{-\frac{1}{2}}, y(16 - (x^2 + y^2))^{-\frac{1}{2}}, 1\right)$$

$$\Rightarrow \left|\frac{\partial \vec{r}}{\partial x} \times \frac{\partial \vec{r}}{\partial y}\right| = (x^2(16 - (x^2 + y^2))^{-1} + y^2(16 - (x^2 + y^2))^{-1} + 1)^{\frac{1}{2}}$$

$$= \left(\frac{x^2 + y^2}{16 - (x^2 + y^2)} + 1\right)^{\frac{1}{2}} = \left(\frac{x^2 + y^2 + 16 - (x^2 + y^2)}{16 - (x^2 + y^2)}\right)^{\frac{1}{2}} = 4(16 - (x^2 + y^2))^{-\frac{1}{2}}$$

$$\therefore 2\iint_S dA = 2\iint_R \left|\frac{\partial \vec{r}}{\partial x} \times \frac{\partial \vec{r}}{\partial y}\right| dxdy = 2\iint_R 4(16 - (x^2 + y^2))^{-\frac{1}{2}} dxdy$$

Let $x = r\cos\theta$, $y = r\sin\theta$ then

$\{(x,y): (x-2)^2 + y^2 \leq 4\} = \{(r,\theta): 0 \leq r \leq 4\cos\theta, -\frac{\pi}{2} \leq \theta \leq \frac{\pi}{2}\}$

$\therefore 2 \iint_R 4(16 - (x^2 + y^2))^{-\frac{1}{2}} dxdy = 8 \int_{-\frac{\pi}{2}}^{\frac{\pi}{2}} \int_0^{4\cos\theta} (16 - r^2)^{-\frac{1}{2}} r\, dr\, d\theta$

$= 16 \int_0^{\frac{\pi}{2}} (-1)(16 - r^2)^{\frac{1}{2}} \Big|_0^{4\cos\theta} d\theta = 16 \int_0^{\frac{\pi}{2}} 4 - 4\sin\theta\, d\theta = 32(\pi - 2)$

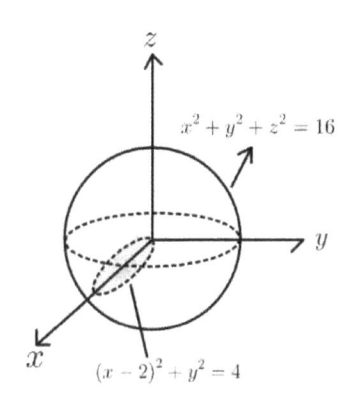

(3)

Let $S: x^2 + y^2 + z^2 = 4a^2$, $x^2 + y^2 \leq 2ay$, $z \geq 0$ then surface area $= 2 \iint_S dA$

Let $\vec{r}(x,y) = \left(x, y, \sqrt{4a^2 - (x^2 + y^2)}\right)$ and let R represent the region projected onto the xy-plane. Then

$R = \{(x,y): x^2 + y^2 \leq 2ay\}$ and $\iint_S dA = \iint_R \left|\frac{\partial \vec{r}}{\partial x} \times \frac{\partial \vec{r}}{\partial y}\right| dxdy$

$\therefore \frac{\partial \vec{r}}{\partial x} = \left(1, 0, -x(4a^2 - (x^2 + y^2))^{-\frac{1}{2}}\right)$ and $\frac{\partial \vec{r}}{\partial y} = \left(0, 1, -y(4a^2 - (x^2 + y^2))^{-\frac{1}{2}}\right)$

$\therefore \frac{\partial \vec{r}}{\partial x} \times \frac{\partial \vec{r}}{\partial y} = \left(x(4a^2 - (x^2 + y^2))^{-\frac{1}{2}}, y(4a^2 - (x^2 + y^2))^{-\frac{1}{2}}, 1\right)$

9.5.2 Find the Surface Area of a Closed Region on Surface S

$$\Rightarrow \left|\frac{\partial \vec{r}}{\partial x} \times \frac{\partial \vec{r}}{\partial y}\right| = (x^2(4a^2-(x^2+y^2))^{-1} + y^2(4a^2-(x^2+y^2))^{-1}+1)^{\frac{1}{2}}$$

$$= 2a(4a^2-(x^2+y^2))^{-\frac{1}{2}}$$

$$\therefore 2\iint_S dA = 2\iint_R \left|\frac{\partial \vec{r}}{\partial x} \times \frac{\partial \vec{r}}{\partial y}\right| dxdy = 2\iint_R 2a(4a^2-(x^2+y^2))^{-\frac{1}{2}}dxdy$$

Let $x = r\cos\theta, y = r\sin\theta$ then

$$\{(x,y): x^2+y^2 \le 2ay\} = \{(r,\theta): 0 \le r \le 2a\sin\theta, 0 \le \theta \le \pi\}$$

$$\therefore 2\iint_R 2a(4a^2-(x^2+y^2))^{-\frac{1}{2}}dxdy = 2\int_0^\pi \int_0^{2a\sin\theta} 2a(4a^2-r^2)^{-\frac{1}{2}} rdrd\theta$$

$$= 8a\int_0^{\frac{\pi}{2}}\int_0^{2a\sin\theta}(4a^2-r^2)^{-\frac{1}{2}}rdrd\theta = -8a\int_0^{\frac{\pi}{2}}(4a^2-r^2)^{\frac{1}{2}}\bigg|_0^{2a\sin\theta} d\theta$$

$$= 16a^2 \int_0^{\frac{\pi}{2}} 1 - \cos\theta \, d\theta = 8a^2(\pi-2)$$

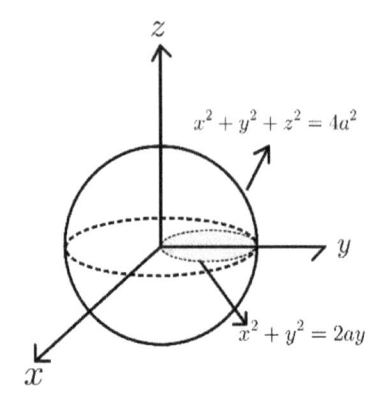

Example 30.

Find the surface area of the surface $z^2 = x^2 + y^2$, $z \ge 0$ inside the region $y^2 + z^2 \le 2$.

【Solution】

9.5.2 Find the Surface Area of a Closed Region on Surface S

Let $S: z^2 = x^2 + y^2,\ z \geq 0,\ y^2 + z^2 \leq 2$ then surface area $= \iint_S dA$

Let $\vec{r}(x,y) = \left(\sqrt{z^2 - y^2}, y, z\right)$ and let R represent the region projected onto the yz-plane.

Then $R = \{(y,z): 0 \leq y \leq 1, y \leq z \leq \sqrt{2-y^2}\}$ and $\iint_S dA = \iint_R \left|\dfrac{\partial \vec{r}}{\partial y} \times \dfrac{\partial \vec{r}}{\partial z}\right| dy\, dz$

$\because \dfrac{\partial \vec{r}}{\partial y} = \left(-y(z^2 - y^2)^{-\frac{1}{2}}, 1, 0\right)$ and $\dfrac{\partial \vec{r}}{\partial z} = \left(z(z^2 - y^2)^{-\frac{1}{2}}, 0, 1\right)$

$\therefore \dfrac{\partial \vec{r}}{\partial y} \times \dfrac{\partial \vec{r}}{\partial z} = \left(1, y(z^2 - y^2)^{-\frac{1}{2}}, -z(z^2 - y^2)^{-\frac{1}{2}}\right)$

$\Rightarrow \left|\dfrac{\partial \vec{r}}{\partial y} \times \dfrac{\partial \vec{r}}{\partial z}\right| = (1 + y^2(z^2 - y^2)^{-1} + z^2(z^2 - y^2)^{-1})^{\frac{1}{2}} = \sqrt{2}z(z^2 - y^2)^{-\frac{1}{2}}$

$\therefore \iint_S dA = \iint_R \left|\dfrac{\partial \vec{r}}{\partial y} \times \dfrac{\partial \vec{r}}{\partial z}\right| dy\, dz = 4 \int_0^1 \int_y^{\sqrt{2-y^2}} \sqrt{2}z(z^2 - y^2)^{-\frac{1}{2}} dz\, dy$

$= 4\sqrt{2} \int_0^1 (z^2 - y^2)^{\frac{1}{2}} \Big|_y^{\sqrt{2-y^2}} dy = 4\sqrt{2} \int_0^1 (2 - 2y^2)^{\frac{1}{2}} dy = 8 \int_0^1 (1 - y^2)^{\frac{1}{2}} dy$

Let $y = \sin \theta$ then $8 \int_0^1 (1 - y^2)^{\frac{1}{2}} dy = 8 \int_0^{\frac{\pi}{2}} \cos^2 \theta\, d\theta = 8 \int_0^{\frac{\pi}{2}} \dfrac{1 + \cos 2\theta}{2} d\theta = 2\pi$

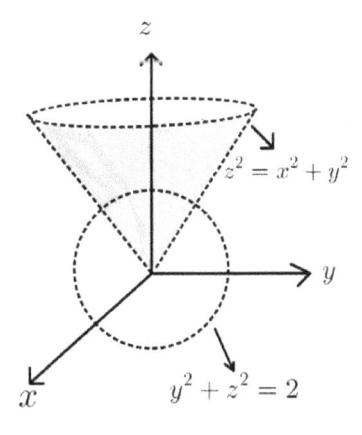

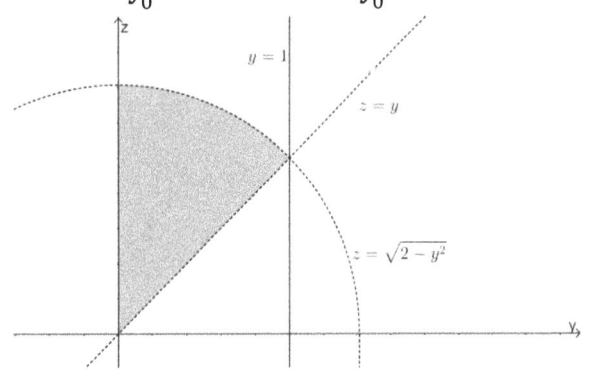

9.5.2 Find the Surface Area of a Closed Region on Surface S

Example 31.
Find the surface area of the surface $z^2 = x^2 + y^2$ inside the cylinder $x^2 + y^2 = 2ay$.
【Solution】

Let $S: z^2 = x^2 + y^2$, $x^2 + y^2 \leq 2ay$, $z \geq 0$ then surface area $= 2\iint_S dA$

Let $\vec{r}(x,y) = \left(x, y, \sqrt{x^2+y^2}\right)$ and let R represent the region projected onto the xy-plane,

then $R = \{(x,y): x^2 + (y-a)^2 \leq a^2\}$ and $\iint_S dA = \iint_R \left|\frac{\partial \vec{r}}{\partial x} \times \frac{\partial \vec{r}}{\partial y}\right| dxdy$

$\because \frac{\partial \vec{r}}{\partial x} = \left(1, 0, x(x^2+y^2)^{-\frac{1}{2}}\right)$ and $\frac{\partial \vec{r}}{\partial y} = \left(0, 1, y(x^2+y^2)^{-\frac{1}{2}}\right)$

$\therefore \frac{\partial \vec{r}}{\partial x} \times \frac{\partial \vec{r}}{\partial y} = \left(x(x^2+y^2)^{-\frac{1}{2}}, y(x^2+y^2)^{-\frac{1}{2}}, 1\right)$

$\Rightarrow \left|\frac{\partial \vec{r}}{\partial x} \times \frac{\partial \vec{r}}{\partial y}\right| = ((x^2+y^2)(x^2+y^2)^{-1} + 1)^{\frac{1}{2}} = \sqrt{2}$

$\therefore 2\iint_S dA = 2\iint_R \left|\frac{\partial \vec{r}}{\partial x} \times \frac{\partial \vec{r}}{\partial y}\right| dxdy = 2\sqrt{2} \iint_{x^2+(y-a)^2 \leq a^2} dxdy = 2\sqrt{2}\pi a^2$

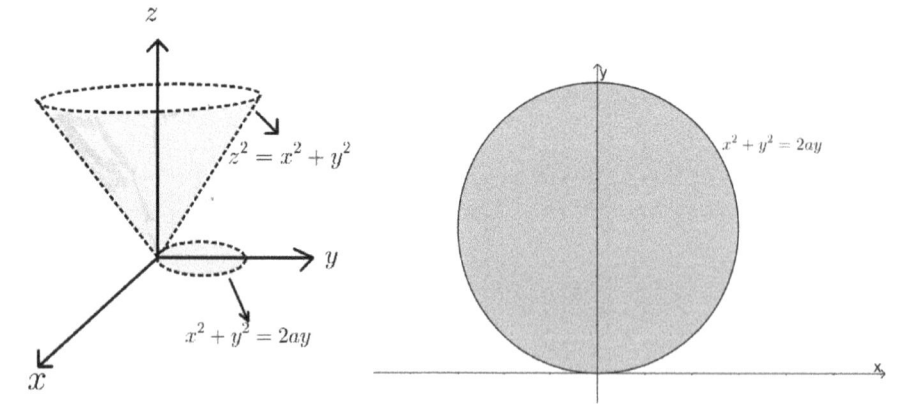

Example 32.
Find the surface area of the surface $z^2 = x^2 + y^2$ inside the elliptical cylinder

$$\frac{x^2}{a^2}+\frac{y^2}{b^2}=1.$$

【Solution】

Let $S: z^2 = x^2 + y^2$, $\frac{x^2}{a^2}+\frac{y^2}{b^2} \le 1$, $z \ge 0$ then surface area $= 2\iint_S dA$

Let $\vec{r}(x,y) = \left(x, y, \sqrt{x^2+y^2}\right)$ and let R represent the region projected onto the xy-plane,

then $R = \left\{(x,y): \frac{x^2}{a^2}+\frac{y^2}{b^2} \le 1\right\}$ and $\iint_S dA = \iint_R \left|\frac{\partial \vec{r}}{\partial x} \times \frac{\partial \vec{r}}{\partial y}\right| dxdy$

$\because \frac{\partial \vec{r}}{\partial x} = \left(1, 0, x(x^2+y^2)^{-\frac{1}{2}}\right)$ and $\frac{\partial \vec{r}}{\partial y} = \left(0, 1, y(x^2+y^2)^{-\frac{1}{2}}\right)$

$\therefore \frac{\partial \vec{r}}{\partial x} \times \frac{\partial \vec{r}}{\partial y} = \left(x(x^2+y^2)^{-\frac{1}{2}}, y(x^2+y^2)^{-\frac{1}{2}}, 1\right)$

$\Rightarrow \left|\frac{\partial \vec{r}}{\partial x} \times \frac{\partial \vec{r}}{\partial y}\right| = ((x^2+y^2)(x^2+y^2)^{-1}+1)^{\frac{1}{2}} = \sqrt{2}$

$\therefore 2\iint_S dA = 2\iint_R \left|\frac{\partial \vec{r}}{\partial x} \times \frac{\partial \vec{r}}{\partial y}\right| dxdy = 2\sqrt{2} \iint_{\frac{x^2}{a^2}+\frac{y^2}{b^2} \le 1} dxdy = 2\sqrt{2}ab\pi$

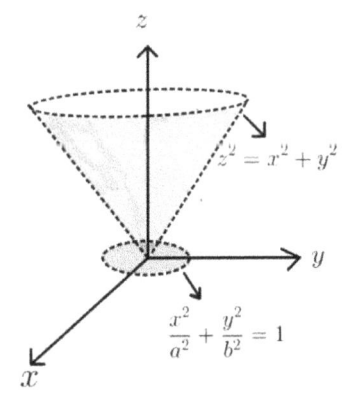

Example 33.
 Find the surface area of the surface $z^2 = x^2 + y^2$ inside the region bounded by $y = x^2$, $y = 3x^2$, $x = y^2$, and $x = 4y^2$.

9.5.2 Find the Surface Area of a Closed Region on Surface S 480

【Solution】

Let $S: z^2 = x^2 + y^2, y \geq x^2, y \leq 3x^2, x \geq y^2, x \leq 4y^2$ then surface area $= 2\iint_S dA$

Let $\vec{r}(x,y) = \left(x, y, \sqrt{x^2+y^2}\right)$ and $R = \left\{(x,y): 1 \leq \dfrac{y}{x^2} \leq 3, 1 \leq \dfrac{x}{y^2} \leq 4\right\}$

then $\iint_S dA = \iint_R \left|\dfrac{\partial \vec{r}}{\partial x} \times \dfrac{\partial \vec{r}}{\partial y}\right| dxdy$

$\because \dfrac{\partial \vec{r}}{\partial x} = \left(1, 0, x(x^2+y^2)^{-\frac{1}{2}}\right)$ and $\dfrac{\partial \vec{r}}{\partial y} = \left(0, 1, y(x^2+y^2)^{-\frac{1}{2}}\right)$

$\therefore \dfrac{\partial \vec{r}}{\partial x} \times \dfrac{\partial \vec{r}}{\partial y} = \left(x(x^2+y^2)^{-\frac{1}{2}}, y(x^2+y^2)^{-\frac{1}{2}}, 1\right)$

$\Rightarrow \left|\dfrac{\partial \vec{r}}{\partial x} \times \dfrac{\partial \vec{r}}{\partial y}\right| = ((x^2+y^2)(x^2+y^2)^{-1} + 1)^{\frac{1}{2}} = \sqrt{2}$

$\therefore 2\iint_S dA = 2\iint_R \left|\dfrac{\partial \vec{r}}{\partial x} \times \dfrac{\partial \vec{r}}{\partial y}\right| dxdy = 2\sqrt{2}\iint_R dxdy$

Let $u = \dfrac{y}{x^2}, v = \dfrac{x}{y^2}$ then $x = u^{-\frac{2}{3}}v^{-\frac{1}{3}}, y = u^{-\frac{1}{3}}v^{-\frac{2}{3}}$

and $dxdy = \begin{Vmatrix} \dfrac{\partial x}{\partial u} & \dfrac{\partial x}{\partial v} \\ \dfrac{\partial y}{\partial u} & \dfrac{\partial y}{\partial v} \end{Vmatrix} dudv = \begin{Vmatrix} \dfrac{-2u^{-\frac{5}{3}}v^{-\frac{1}{3}}}{3} & \dfrac{-u^{-\frac{2}{3}}v^{-\frac{4}{3}}}{3} \\ \dfrac{-u^{-\frac{4}{3}}v^{-\frac{2}{3}}}{3} & \dfrac{-2u^{-\frac{1}{3}}v^{-\frac{5}{3}}}{3} \end{Vmatrix} dudv = \dfrac{u^{-2}v^{-2}}{3} dudv$

$\because R = \left\{(x,y): 1 \leq \dfrac{y}{x^2} \leq 3, 1 \leq \dfrac{x}{y^2} \leq 4\right\}$ $\therefore R = \{(u,v): 1 \leq u \leq 3, 1 \leq v \leq 4\}$

$\therefore \iint_R 1 dA = \int_1^3 \int_1^4 \dfrac{u^{-2}v^{-2}}{3} dvdu = \dfrac{1}{6}$ $\therefore$ surface area $= 2\iint_S dA = 2\sqrt{2}\iint_R dxdy = \dfrac{\sqrt{2}}{3}$

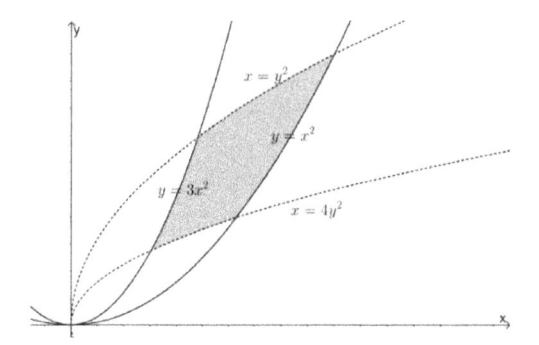

Example 34.

Find the surface area of the surface $z^2 = x^2 + y^2$ inside the region bounded by $y = x$, $y = 2x$, $xy = 1$, and $xy = 4$.

【Solution】

Let $S: z^2 = x^2 + y^2$, $y \geq x, y \leq 2x$, $xy \geq 1$, $xy \leq 4$ then surface area $= 2 \iint_S dA$

Let $\vec{r}(x, y) = \left(x, y, \sqrt{x^2 + y^2}\right)$ and $R = \left\{(x,y): 1 \leq \dfrac{y}{x} \leq 2, 1 \leq xy \leq 4\right\}$ then

$$\iint_S dA = \iint_R \left|\dfrac{\partial \vec{r}}{\partial x} \times \dfrac{\partial \vec{r}}{\partial y}\right| dxdy$$

$\because \dfrac{\partial \vec{r}}{\partial x} = \left(1, 0, x(x^2 + y^2)^{-\frac{1}{2}}\right)$ and $\dfrac{\partial \vec{r}}{\partial y} = \left(0, 1, y(x^2 + y^2)^{-\frac{1}{2}}\right)$

$\therefore \dfrac{\partial \vec{r}}{\partial x} \times \dfrac{\partial \vec{r}}{\partial y} = \left(x(x^2 + y^2)^{-\frac{1}{2}}, y(x^2 + y^2)^{-\frac{1}{2}}, 1\right)$

$\Rightarrow \left|\dfrac{\partial \vec{r}}{\partial x} \times \dfrac{\partial \vec{r}}{\partial y}\right| = \left((x^2 + y^2)(x^2 + y^2)^{-1} + 1\right)^{\frac{1}{2}} = \sqrt{2}$

$\therefore 2 \iint_S dA = 2 \iint_R \left|\dfrac{\partial \vec{r}}{\partial x} \times \dfrac{\partial \vec{r}}{\partial y}\right| dxdy = 2\sqrt{2} \iint_R dxdy$

Let $\dfrac{y}{x} = u$, $xy = v$ then $x = \sqrt{\dfrac{v}{u}}$, $y = \sqrt{uv}$

9.5.2 Find the Surface Area of a Closed Region on Surface S

and $dxdy = \begin{Vmatrix} \dfrac{\partial x}{\partial u} & \dfrac{\partial x}{\partial v} \\ \dfrac{\partial y}{\partial u} & \dfrac{\partial y}{\partial v} \end{Vmatrix} dudv = \begin{Vmatrix} \dfrac{1}{2}\left(\dfrac{v}{u}\right)^{-\frac{1}{2}}\left(-\dfrac{v}{u^2}\right) & \dfrac{1}{2}\left(\dfrac{v}{u}\right)^{-\frac{1}{2}}\left(\dfrac{1}{u}\right) \\ \dfrac{1}{2}(uv)^{-\frac{1}{2}}v & \dfrac{1}{2}(uv)^{-\frac{1}{2}}u \end{Vmatrix} dudv = \dfrac{1}{2u} dudv$

$\because R = \left\{(x,y): 1 \leq \dfrac{y}{x} \leq 2, 1 \leq xy \leq 4\right\} \quad \therefore R = \{(u,v): 1 \leq u \leq 2, 1 \leq v \leq 4\}$

$\therefore \iint_R dxdy = 2\int_1^4 \int_1^2 \dfrac{1}{2u} dudv = 3\ln 2$

$\therefore \text{surface area} = 2\iint_S dA = 2\sqrt{2}\iint_R dxdy = 6\sqrt{2}\ln 2$

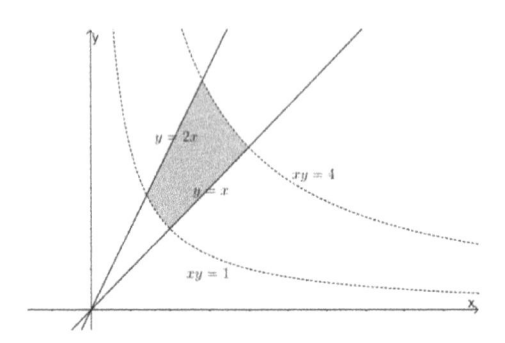

9.5.3 For Scalar Functions, Find the Mass of Surface S

The relationship between surface integrals and surface area is analogous to the relationship between line integrals and arc length. Consider a parametrized surface S,
$$\vec{r}(s,t) = (x(s,t), y(s,t), z(s,t)), \quad \forall (s,t) \in D.$$
Assume the parameter domain D is a rectangle, and divide it into several small rectangles D_{ij}, with length Δs and width Δt. The center of D_{ij} is denoted as (s_i^*, t_j^*). The region on the surface S corresponding to D_{ij} is S_{ij} and the point corresponding to (s_i^*, t_j^*) on the surface S_{ij} is E_{ij}^*, with position vector $\vec{r}(s_i^*, t_j^*)$. Therefore, the surface S is divided into several segments S_{ij}. For each segment, the value of the scalar function f at the point E_{ij}^* is chosen, and multiplying it by the area ΔS_{ij} of the segment producesa Riemann sum:
$$\sum_{i=1}^{m}\sum_{j=1}^{n} f(E_{ij}^*)\Delta S_{ij}.$$
As the number of divisions increases, the surface integral of the scalar function f over the surface S can be defined as the limit of this Riemann sum.

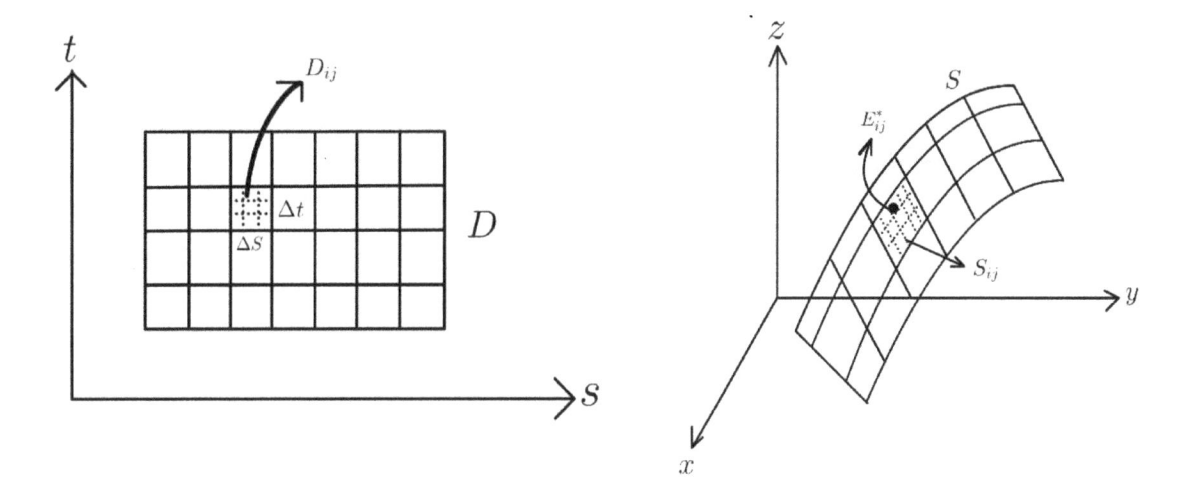

【Definition】

The surface integral of a scalar function f over a smooth surface segment S is defined as follows,

$$\iint_S f(x,y,z)dS = \lim_{m\to\infty, n\to\infty} \sum_{i=1}^{m}\sum_{j=1}^{n} f(E_{ij}^*)\Delta S_{ij},$$

where E_{ij}^* and ΔS_{ij} are defined as above.

It is important to note that since any continuous function defined over a closed interval is Riemann integrable, if the function f defined above is continuous, then the limit of the Riemann sum must exist. To compute the surface integral defined above, consider the area of an approximate parallelogram located in the tangent plane to approximate the region ΔS_{ij}. In previous discussions about surface area, we had an approximate estimate for ΔS_{ij},

$$\Delta S_{ij} \approx |\vec{r}_s(s_i^*, t_j^*) \times \vec{r}_t(s_i^*, t_j^*)|\Delta s \Delta t.$$

Hence,

$$\iint_S f(x,y,z)dS = \lim_{m\to\infty, n\to\infty} \sum_{i=1}^{m}\sum_{j=1}^{n} f(E_{ij}^*)\Delta S_{ij}$$

$$= \lim_{m\to\infty, n\to\infty} \sum_{i=1}^{m}\sum_{j=1}^{n} f(E_{ij}^*)|\vec{r}_s(s_i^*, t_j^*) \times \vec{r}_t(s_i^*, t_j^*)|\Delta s \Delta t = \iint_D f(\vec{r}(s,t))|\vec{r}_s \times \vec{r}_t|ds dt.$$

From the observations above, it can be seen that calculating surface integrals is equivalent to finding double integrals. Notably, the integrand for the double integral is given by

9.5.3 For Scalar Functions, Find the Mass of Surface S

$$f(\vec{r}(s,t))|\vec{r}_s(s,t) \times \vec{r}_t(s,t)|.$$

Thus, computing the surface integral involves first determining the integrand and then calculating the double integral. Below, we will provide a preliminary explanation of the different types of surface integrals transformed into double integrals.

Given the surface $S: z = g(x,y)$, $\forall\, a \le x \le b, c \le y \le d$, the goal is to compute the surface integral,

$$\iint_S f(x,y,z)\, dA.$$

Let $\vec{r}(x,y) = (x,y,g(x,y))$ and $R = \{(x,y): a \le x \le b, c \le y \le d\}$. Then, the surface integral can be expressed as

$$\iint_S f(x,y,z)\, dS = \iint_R f(x,y,g(x,y)) \left|\frac{\partial \vec{r}}{\partial x} \times \frac{\partial \vec{r}}{\partial y}\right| dx\, dy.$$

Since $\dfrac{\partial \vec{r}}{\partial x} = (1, 0, g_x(x,y))$ and $\dfrac{\partial \vec{r}}{\partial y} = (0, 1, g_y(x,y))$, the cross product of the two is

$$\frac{\partial \vec{r}}{\partial x} \times \frac{\partial \vec{r}}{\partial y} = (-g_x(x,y), -g_y(x,y), 1).$$

Thus,

$$\left|\frac{\partial \vec{r}}{\partial x} \times \frac{\partial \vec{r}}{\partial y}\right| = \sqrt{1 + (g_x(x,y))^2 + (g_y(x,y))^2}.$$

Therefore, the surface integral can be rewritten as

$$\iint_S f(x,y,z)\, dA = \int_a^b \int_c^d f(x,y,g(x,y))\sqrt{1 + (g_x(x,y))^2 + (g_y(x,y))^2}\, dy\, dx.$$

Notably, if the surface is given by $z = g(x,y)$ and is projected onto the xy-plane $\{(x,y): x^2 + y^2 \le a\}$, we can use polar coordinate transformation: let $x = r\cos\theta$, $y = r\sin\theta$ then

$$\iint_S f(x,y,z)\, dA$$

$$= \int_0^{2\pi} \int_0^a f(x,y,g(x,y))\sqrt{1 + (g_x(x,y))^2 + (g_y(x,y))^2}\,\bigg|_{x=r\cos\theta,\, y=r\sin\theta}\, r\, dr\, d\theta$$

Assuming the surface S is a sphere defined by the equation $x^2 + y^2 + z^2 = r^2$, where $r > 0$, we can parameterize the surface S as follows,

$$\vec{r}(\theta, \varphi) = (r\cos\theta \sin\varphi, r\sin\theta \sin\varphi, r\cos\varphi),$$

with the domain defined as $R = \{(\theta, \varphi): 0 \leq \theta \leq 2\pi, 0 \leq \varphi \leq \pi\}$. Then the surface integral transforms to

$$\iint_S f(x,y,z)dA = \iint_D f(\vec{r}(\theta,\varphi))|\vec{r}_\theta \times \vec{r}_\varphi|d\theta d\varphi = \iint_R f(\vec{r}(\theta,\varphi))\left|\frac{\partial \vec{r}}{\partial \varphi} \times \frac{\partial \vec{r}}{\partial \theta}\right| d\theta d\varphi.$$

Since

$$\frac{\partial \vec{r}}{\partial \theta} = (-r\sin\theta \sin\varphi, r\cos\theta \sin\varphi, 0) \text{ and } \frac{\partial \vec{r}}{\partial \varphi} = (r\cos\theta \cos\varphi, r\sin\theta \cos\varphi, -r\sin\varphi),$$

we have

$$\frac{\partial \vec{r}}{\partial \varphi} \times \frac{\partial \vec{r}}{\partial \theta} = (r^2 \cos\theta \sin^2\varphi, r^2\sin\theta \sin^2\varphi, r^2\sin\varphi \cos\varphi) \text{ and } \left|\frac{\partial \vec{r}}{\partial \varphi} \times \frac{\partial \vec{r}}{\partial \theta}\right| = r^2\sin\varphi.$$

Therefore, the formula for the surface integral over the sphere can be rewritten as

$$\iint_S f(x,y,z)dS = \iint_R f(r\cos\theta \sin\varphi, r\sin\theta \sin\varphi, r\cos\varphi) r^2\sin\varphi \, d\theta d\varphi.$$

Question Types:

Type 1.

Given the surface $S: \vec{r}(s,t)$, $\forall a \leq s \leq b, c \leq t \leq d$, find the surface intagral

$$\iint_S f(x,y,z)dA = ?$$

Problem-Solving Process:

Step1.

Let $R = \{(x,y): a \leq s \leq b, c \leq t \leq d\}$ then

$$\iint_S f(x,y,z)dA = \iint_R f(\vec{r}(s,t))|\vec{r}_s \times \vec{r}_t|dsdt = \int_a^b \int_c^d f(\vec{r}(s,t))|\vec{r}_s \times \vec{r}_t| \, dtds$$

Step2.

Find $\vec{r}_s$, $\vec{r}_s$, $\vec{r}_s \times \vec{r}_t$, and $|\vec{r}_s \times \vec{r}_t|$

Find $\int_a^b \int_c^d f(\vec{r}(s,t)))|\vec{r}_s \times \vec{r}_t| \, dtds = ?$

Type 2.

Given the surface $S: z = g(x,y)$, $\forall a \leq x \leq b, c \leq y \leq d$, find the surface intagral

$$\iint_S f(x,y,z)dA = ?$$

Problem-Solving Process:

Step1.

9.5.3 For Scalar Functions, Find the Mass of Surface S

Let $\vec{r}(x,y) = (x, y, g(x,y))$ and $R = \{(x,y): a \le x \le b, c \le y \le d\}$ then

$$\iint_S f(x,y,z)dA = \iint_R f(x,y,g(x,y)) \left|\frac{\partial \vec{r}}{\partial x} \times \frac{\partial \vec{r}}{\partial y}\right| dxdy$$

Step2.

$\because \dfrac{\partial \vec{r}}{\partial x} = (1, 0, g_x(x,y))$, $\dfrac{\partial \vec{r}}{\partial y} = (0, 1, g_y(x,y))$ $\therefore \dfrac{\partial \vec{r}}{\partial x} \times \dfrac{\partial \vec{r}}{\partial y} = (-g_x(x,y), -g_y(x,y), 1)$

$\Rightarrow \left|\dfrac{\partial \vec{r}}{\partial x} \times \dfrac{\partial \vec{r}}{\partial y}\right| = \sqrt{1 + (g_x(x,y))^2 + (g_y(x,y))^2}$

Step3.

$$\iint_S f(x,y,z)dA = \int_a^b \int_c^d f(x,y,g(x,y)) \sqrt{1 + (g_x(x,y))^2 + (g_y(x,y))^2} \, dydx$$

Similarly, if the surface S is given by $y = g(x, z)$, $\forall a \le x \le b, c \le z \le d$ then

$$\iint_S f(x,y,z)dA = \int_a^b \int_c^d f(x, g(x,z), z) \sqrt{1 + (g_x(x,z))^2 + (g_z(x,z))^2} \, dzdx$$

If the surface S is given by $x = g(y, z)$, $\forall a \le y \le b, c \le z \le d$ then

$$\iint_S f(x,y,z)dA = \int_a^b \int_c^d f(g(y,z), y, z) \sqrt{1 + (g_y(y,z))^2 + (g_z(y,z))^2} \, dzdy$$

Examples:

Find $\iint_S f(x,y,z)dA = ?$

(I) Assume $f(x,y,z) = x + y + z - 1$, $S:\{(x,y,z): z = x + y + 1, \ 0 \le y \le x, \ 0 \le x \le 2\}$.
Let $\vec{r}(x,y) = (x, y, x + y + z - 1)$

$\because \dfrac{\partial \vec{r}}{\partial x} = (1,0,1)$ and $\dfrac{\partial \vec{r}}{\partial y} = (0,1,1)$ $\therefore \dfrac{\partial \vec{r}}{\partial x} \times \dfrac{\partial \vec{r}}{\partial y} = (-1,-1,1) \Rightarrow \left|\dfrac{\partial \vec{r}}{\partial x} \times \dfrac{\partial \vec{r}}{\partial y}\right| = \sqrt{3}$

$\therefore \iint_S f(x,y,z)\,dA = \int_0^2 \int_0^x (2x + 2y)\sqrt{3}\,dydx = 8\sqrt{3}$

(II) Assume $f(x,y,z) = (1 - x^2)y$, $S:\{(x,y,z): z = 3 - y^2, \ 0 \le x \le 1, \ 0 \le y \le 1\}$.
Let $\vec{r}(x,y) = (x, y, (1 - x^2)y)$

$\because \dfrac{\partial \vec{r}}{\partial x} = (1,0,0)$ and $\dfrac{\partial \vec{r}}{\partial y} = (0,1,-2y)$ $\therefore \dfrac{\partial \vec{r}}{\partial x} \times \dfrac{\partial \vec{r}}{\partial y} = (0, 2y, 1) \Rightarrow \left|\dfrac{\partial \vec{r}}{\partial x} \times \dfrac{\partial \vec{r}}{\partial y}\right| = \sqrt{1 + 4y^2}$

$\therefore \iint_S f(x,y,z)\,dA = \int_0^1 \int_0^1 y(1 - x^2)\sqrt{1 + 4y^2}\,dydx = \dfrac{5\sqrt{5} - 1}{18}$

Type 3.
Given the surface S defined by $z = g(x,y)$, if S is projected onto the xy-plane as $\{(x,y): x^2 + y^2 \leq a^2\}$, find

$$\iint_S f(x,y,z)\,dA = ?$$

Remark:
Examples include: cylinders, elliptical cylinders, cones, and paraboloids.

Problem-Solving Process:

Step1.
Let $\vec{r}(x,y) = (x, y, g(x,y))$ and $R = \{(x,y): x^2 + y^2 \leq a^2\}$ then

$$\iint_S dA = \iint_R \left|\frac{\partial \vec{r}}{\partial x} \times \frac{\partial \vec{r}}{\partial y}\right| dxdy$$

Step2.
$$\because \frac{\partial \vec{r}}{\partial x} = (1, 0, g_x(x,y)) \text{ and } \frac{\partial \vec{r}}{\partial y} = (0, 1, g_y(x,y)) \quad \therefore \frac{\partial \vec{r}}{\partial x} \times \frac{\partial \vec{r}}{\partial y} = (-g_x(x,y), -g_y(x,y), 1)$$

$$\Rightarrow \left|\frac{\partial \vec{r}}{\partial x} \times \frac{\partial \vec{r}}{\partial y}\right| = \sqrt{1 + (g_x(x,y))^2 + (g_y(x,y))^2}$$

Step3.
$$\therefore \iint_S f(x,y,z)\,dA = \iint_R f(x,y,g(x,y)) \left|\frac{\partial \vec{r}}{\partial x} \times \frac{\partial \vec{r}}{\partial y}\right| dxdy$$

$$= \iint_R f(x,y,g(x,y))\sqrt{1 + (g_x(x,y))^2 + (g_y(x,y))^2}\,dxdy$$

Step4.
Let $x = r\cos\theta$, $y = r\sin\theta$ then $\{(x,y): x^2 + y^2 \leq a^2\} = \{(r,\theta): 0 \leq r \leq a, 0 \leq \theta \leq 2\pi\}$

$$\iint_R f(x,y,g(x,y))\sqrt{1 + (g_x(x,y))^2 + (g_y(x,y))^2}\,dxdy$$

$$= \int_0^{2\pi}\int_0^a f(x,y,g(x,y))\sqrt{1 + (g_x(x,y))^2 + (g_y(x,y))^2}\bigg|_{x=r\cos\theta, y=r\sin\theta} r\,dr\,d\theta$$

Type 4.
Given the sphere $S: x^2 + y^2 + z^2 = r^2$, $r > 0$, find $\iint_S f(x,y,z)\,dA = ?$

Problem-Solving Process:

9.5.3 For Scalar Functions, Find the Mass of Surface S

Step 1.
Let $\vec{r}(\theta, \varphi) = (r\cos\theta \sin\varphi, r\sin\theta \sin\varphi, r\cos\varphi)$ and $R = \{(\theta, \varphi): 0 \le \theta \le 2\pi, 0 \le \varphi \le \pi\}$

then $\iint_S f(x,y,z)dA = \iint_R f(r\cos\theta \sin\varphi, r\sin\theta \sin\varphi, r\cos\varphi) \left|\dfrac{\partial \vec{r}}{\partial \varphi} \times \dfrac{\partial \vec{r}}{\partial \theta}\right| d\theta d\varphi$

Step 2.

$\because \dfrac{\partial \vec{r}}{\partial \theta} = (-r\sin\theta \sin\varphi, r\cos\theta \sin\varphi, 0)$ and $\dfrac{\partial \vec{r}}{\partial \varphi} = (r\cos\theta \cos\varphi, r\sin\theta \cos\varphi, -r\sin\varphi)$

$\therefore \dfrac{\partial \vec{r}}{\partial \varphi} \times \dfrac{\partial \vec{r}}{\partial \theta} = (r^2 \cos\theta \sin^2\varphi, r^2\sin\theta \sin^2\varphi, r^2\sin\varphi \cos\varphi) \Rightarrow \left|\dfrac{\partial \vec{r}}{\partial \varphi} \times \dfrac{\partial \vec{r}}{\partial \theta}\right| = r^2 \sin\varphi$

Step 3.

$\therefore \iint_S f(x,y,z)dA = \iint_R f(r\cos\theta \sin\varphi, r\sin\theta \sin\varphi, r\cos\varphi) r^2 \sin\varphi \, d\theta d\varphi$

Find $\iint_R f(r\cos\theta \sin\varphi, r\sin\theta \sin\varphi, r\cos\varphi) r^2 \sin\varphi \, d\theta d\varphi = ?$

Examples:

Find $\iint_S f(x,y,z)dA = ?$, where $S: x^2 + y^2 + z^2 = r^2, \ r > 0$

(I) If $f(x,y,z) = x^2$ then $\iint_S f(x,y,z)dA = \iint_R r^2\cos^2\theta \sin^2\varphi \, r^2\sin\varphi \, d\theta d\varphi = \dfrac{4\pi r^4}{3}$

(II) If $f(x,y,z) = z^2$ then $\iint_S f(x,y,z)dA = \iint_R r^2 \cos^2\varphi \, r^2\sin\varphi \, d\theta d\varphi = \dfrac{4\pi r^4}{3}$

Example 1.

Find $\iint_S 3xy \, dA = ?$, $S: z = xy + \sqrt{7}, \ 0 \le x \le 1, \ 0 \le y \le 1$.

【Solution】

Let $\vec{r}(x,y) = (x, y, xy + \sqrt{7})$ and let R represent the region projected onto the xy-plane

then $R = \{(x,y): 0 \le x \le 1, 0 \le y \le 1\}$ and $\iint_S 3xy \, dA = \iint_R 3xy \left|\dfrac{\partial \vec{r}}{\partial x} \times \dfrac{\partial \vec{r}}{\partial y}\right| dxdy$

$$\because \frac{\partial \vec{r}}{\partial x} = (1,0,y) \text{ and } \frac{\partial \vec{r}}{\partial y} = (0,1,x) \quad \therefore \frac{\partial \vec{r}}{\partial x} \times \frac{\partial \vec{r}}{\partial y} = (-y,-x,1) \Rightarrow \left|\frac{\partial \vec{r}}{\partial x} \times \frac{\partial \vec{r}}{\partial y}\right| = (x^2+y^2+1)^{\frac{1}{2}}$$

$$\therefore 3\iint_S xy \, dA = \iint_R 3xy \left|\frac{\partial \vec{r}}{\partial x} \times \frac{\partial \vec{r}}{\partial y}\right| dxdy = 3\iint_R xy(x^2+y^2+1)^{\frac{1}{2}} dxdy$$

Let $u = x^2$, $v = y^2$ then

$R = \{(x,y): 0 \le x \le 1, 0 \le y \le 1\} = \{(u,v): 0 \le u \le 1, 0 \le v \le 1\}$

$$\therefore 3\iint_R xy(x^2+y^2+1)^{\frac{1}{2}} dxdy = \frac{3}{4}\int_0^1 \int_0^1 (u+v+1)^{\frac{1}{2}} dudv = \frac{9\sqrt{3}-8\sqrt{2}+1}{5}$$

Example 2.
Let $f(x,y,z) = x+y+z-1$, $S: z = x+y+1$, $0 \le y \le x$, $0 \le x \le 2$.

Find $\iint_S f(x,y,z) \, dA = ?$

【Solution】

Let $\vec{r}(x,y) = (x,y,x+y+1)$ and let R represent the region projected onto the xy-plane

then $R = \{(x,y): 0 \le y \le x, 0 \le x \le 2\}$

and $\iint_S f(x,y,z) \, dA = \iint_S f(x,y,x+y+1) \, dA = \iint_R f(x,y,x+y+1) \left|\frac{\partial \vec{r}}{\partial x} \times \frac{\partial \vec{r}}{\partial y}\right| dxdy$

$$\because \frac{\partial \vec{r}}{\partial x} = (1,0,1) \text{ and } \frac{\partial \vec{r}}{\partial y} = (0,1,1) \quad \therefore \frac{\partial \vec{r}}{\partial x} \times \frac{\partial \vec{r}}{\partial y} = (-1,-1,1) \Rightarrow \left|\frac{\partial \vec{r}}{\partial x} \times \frac{\partial \vec{r}}{\partial y}\right| = \sqrt{3}$$

$$\therefore \iint_S f(x,y,z) \, dA = \iint_R f(x,y,x+y+1) \left|\frac{\partial \vec{r}}{\partial x} \times \frac{\partial \vec{r}}{\partial y}\right| dxdy = \int_0^2 \int_0^x (2x+2y)\sqrt{3} \, dydx$$

$$= \sqrt{3} \int_0^2 2xy+y^2 |_0^x \, dx = \sqrt{3} \int_0^2 3x^2 \, dx = 8\sqrt{3}$$

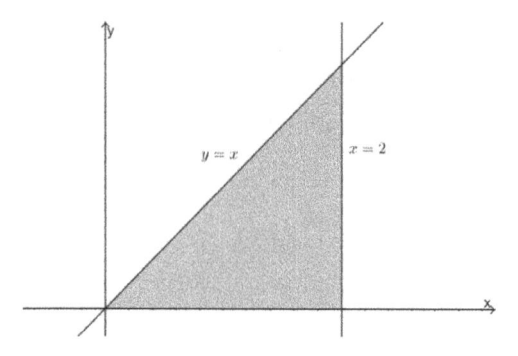

9.5.3 For Scalar Functions, Find the Mass of Surface S

Example 3.

Evaluate $\iint_S \sqrt{x^2 + y^2}\, dA = ?$, over the surface $S: z = xy$, $0 \leq x^2 + y^2 \leq 1$.

【Solution】

Let $\vec{r}(x,y) = (x, y, xy)$ and let R represent the region projected onto the xy-plane

then $R = \{(x,y): 0 \leq x^2 + y^2 \leq 1\}$ and $\iint_S \sqrt{x^2 + y^2}\, dA = \iint_R \sqrt{x^2 + y^2}\, \left|\dfrac{\partial \vec{r}}{\partial x} \times \dfrac{\partial \vec{r}}{\partial y}\right| dxdy$

$\because \dfrac{\partial \vec{r}}{\partial x} = (1, 0, y)$ and $\dfrac{\partial \vec{r}}{\partial y} = (0, 1, x)$ $\therefore \dfrac{\partial \vec{r}}{\partial x} \times \dfrac{\partial \vec{r}}{\partial y} = (-y, -x, 1) \Rightarrow \left|\dfrac{\partial \vec{r}}{\partial x} \times \dfrac{\partial \vec{r}}{\partial y}\right| = \sqrt{1 + x^2 + y^2}$

$\therefore \iint_R \sqrt{x^2 + y^2}\, \left|\dfrac{\partial \vec{r}}{\partial x} \times \dfrac{\partial \vec{r}}{\partial y}\right| dxdy = \iint_{0 \leq \sqrt{x^2+y^2} \leq 1} \sqrt{x^2 + y^2}\sqrt{1 + x^2 + y^2}\, dxdy$

Let $x = r\cos\theta$, $y = r\sin\theta$ and $R = \{(r, \theta): 0 \leq r \leq 1, 0 \leq \theta \leq 2\pi\}$

$\iint_{0 \leq \sqrt{x^2+y^2} \leq 1} \sqrt{x^2 + y^2}\sqrt{1 + x^2 + y^2}\, dxdy = \int_0^{2\pi} \int_0^1 r^2 \sqrt{1 + r^2}\, dr d\theta = 2\pi \int_0^1 r^2 \sqrt{1 + r^2}\, dr$

Let $r = \tan\alpha$ then

$2\pi \int_0^1 r^2 \sqrt{1 + r^2}\, dr = 2\pi \int_0^{\frac{\pi}{4}} \tan^2\alpha \sec^3\alpha\, d\alpha = 2\pi \int_0^{\frac{\pi}{4}} \sec^5\alpha - \sec^3\alpha\, d\alpha$

$= \dfrac{\pi}{4}(3\sqrt{2} - \ln(1 + \sqrt{2}))$

Example 4.

Evaluate $\iint_S xy\, dA = ?$, over the surface $\vec{r}(s,t) = s\vec{a} + t\vec{b}$, where $0 \le s \le 1$, $0 \le t \le 1$, $\vec{a} = (a_1, a_2, a_3)$, $\vec{b} = (b_1, b_2, b_3)$.

【Solution】

Let $\vec{r}(s,t) = s\vec{a} + t\vec{b} = (sa_1 + tb_1, sa_2 + tb_2, sa_3 + tb_3)$ and
$R = \{(s,t): 0 \le s \le 1, 0 \le t \le 1\}$

$\because \dfrac{\partial \vec{r}}{\partial s} = (a_1, a_2, a_3)$ and $\dfrac{\partial \vec{r}}{\partial t} = (b_1, b_2, b_3)$ $\therefore \dfrac{\partial \vec{r}}{\partial s} \times \dfrac{\partial \vec{r}}{\partial t} = \vec{a} \times \vec{b}$ and $\left|\dfrac{\partial \vec{r}}{\partial s} \times \dfrac{\partial \vec{r}}{\partial t}\right| = |\vec{a} \times \vec{b}|$

$$\iint_S xy\, dA = \iint_R (sa_1 + tb_1)(sa_2 + tb_2) \left|\dfrac{\partial \vec{r}}{\partial s} \times \dfrac{\partial \vec{r}}{\partial t}\right| ds\, dt$$

$$= |\vec{a} \times \vec{b}| \int_0^1 \int_0^1 (sa_1 + tb_1)(sa_2 + tb_2)\, ds\, dt$$

$$= |\vec{a} \times \vec{b}| \int_0^1 \int_0^1 s^2 a_1 a_2 + st(a_1 b_2 + a_2 b_1) + t^2 b_1 b_2\, ds\, dt$$

$$= |\vec{a} \times \vec{b}| \left(\dfrac{a_1 a_2}{3} + \dfrac{a_1 b_2 + a_2 b_1}{4} + \dfrac{b_1 b_2}{3}\right)$$

Example 5.

Find $\iint_S x^2\, dA = ?$, $S: x^2 + y^2 + z^2 = r^2$.

【Solution】

Let $\vec{r}(\theta, \varphi) = (r\cos\theta \sin\varphi, r\sin\theta \sin\varphi, r\cos\varphi)$ and $R = \{(\theta, \varphi): 0 \le \theta \le 2\pi, 0 \le \varphi \le \pi\}$

then $\iint_S x^2\, dA = \iint_R r^2 \cos^2\theta \sin^2\varphi \left|\dfrac{\partial \vec{r}}{\partial \varphi} \times \dfrac{\partial \vec{r}}{\partial \theta}\right| d\theta\, d\varphi$

$\because \dfrac{\partial \vec{r}}{\partial \theta} = (-r\sin\theta \sin\varphi, r\cos\theta \sin\varphi, 0)$ and $\dfrac{\partial \vec{r}}{\partial \varphi} = (r\cos\theta \cos\varphi, r\sin\theta \cos\varphi, -r\sin\varphi)$

9.5.3 For Scalar Functions, Find the Mass of Surface S

$$\therefore \frac{\partial \vec{r}}{\partial \varphi} \times \frac{\partial \vec{r}}{\partial \theta} = (r^2 \cos \theta \sin^2 \varphi, r^2 \sin \theta \sin^2 \varphi, r^2 \sin \varphi \cos \varphi) \Rightarrow \left| \frac{\partial \vec{r}}{\partial \varphi} \times \frac{\partial \vec{r}}{\partial \theta} \right| = r^2 \sin \varphi$$

$$\therefore \iint_S x^2 \, dA = \iint_R r^2 \cos^2 \theta \sin^2 \varphi \left| \frac{\partial \vec{r}}{\partial \varphi} \times \frac{\partial \vec{r}}{\partial \theta} \right| d\theta d\varphi = r^4 \int_0^{2\pi} \int_0^{\pi} \cos^2 \theta \sin^2 \varphi \sin \varphi \, d\varphi d\theta$$

$$= r^4 \int_0^{2\pi} \cos^2 \theta \, d\theta \int_0^{\pi} \sin^3 \varphi \, d\varphi = r^4 \cdot \pi \cdot \frac{4}{3} = \frac{4\pi r^4}{3}$$

Example 6.

Find $\iint_S y \, dA = ?$, $S: x - y^2 - z = 2$, $0 \le y \le 1$, $0 \le z \le 1$.

【Solution】

Let $\vec{r}(x,y) = (y^2 + z + 2, y, z)$ and let R represent the region projected onto the yz-plane

then $R = \{(y,z): 0 \le y \le 1, 0 \le z \le 1\}$ and $\iint_S y \, dA = \iint_R y \left| \frac{\partial \vec{r}}{\partial y} \times \frac{\partial \vec{r}}{\partial z} \right| dydz$

$\therefore \frac{\partial \vec{r}}{\partial y} = (2y, 1, 0)$ and $\frac{\partial \vec{r}}{\partial z} = (1, 0, 1) \quad \therefore \frac{\partial \vec{r}}{\partial y} \times \frac{\partial \vec{r}}{\partial z} = (1, -2y, -1) \Rightarrow \left| \frac{\partial \vec{r}}{\partial y} \times \frac{\partial \vec{r}}{\partial z} \right| = \sqrt{2 + 4y^2}$

$\therefore \iint_S y \, dA = \iint_R y \left| \frac{\partial \vec{r}}{\partial y} \times \frac{\partial \vec{r}}{\partial z} \right| dxdy = \int_0^1 \int_0^1 y\sqrt{2 + 4y^2} \, dydz = \frac{3\sqrt{6} - \sqrt{2}}{6}$

Example 7.

Evaluate $\iint_S \sqrt{x^2 + y^2} \, dA = ?$, over the surface $\vec{r}(s,t) = (s \cos wt, s \sin wt, bt)$,

where $0 \le s \le k, 0 \le t \le \frac{2\pi}{w}$.

【Solution】

$\therefore \frac{\partial \vec{r}}{\partial s} = (\cos wt, \sin wt, 0)$ and $\frac{\partial \vec{r}}{\partial t} = (-ws \sin wt, ws \cos wt, b)$

$\therefore \frac{\partial \vec{r}}{\partial s} \times \frac{\partial \vec{r}}{\partial t} = (b \sin wt, -b \cos wt, ws)$ and $\left| \frac{\partial \vec{r}}{\partial s} \times \frac{\partial \vec{r}}{\partial t} \right| = \sqrt{b^2 + w^2 s^2}$

Let $R = \{(s,t): 0 \le s \le k, 0 \le t \le \frac{2\pi}{w}\}$ then

$$\therefore \iint_S \sqrt{x^2+y^2}\, dA = \iint_R \sqrt{(s\cos wt)^2+(s\sin wt)^2}\left|\frac{\partial \vec{r}}{\partial s} \times \frac{\partial \vec{r}}{\partial t}\right| dsdt$$

$$= \iint_R \sqrt{(s\cos wt)^2+(s\sin wt)^2}\sqrt{b^2+w^2s^2}\, dA = \iint_R s\sqrt{b^2+w^2s^2}\, dA$$

$$= \int_0^{\frac{2\pi}{w}} \int_0^k s\sqrt{b^2+w^2s^2}\, dsdt = \frac{2\pi}{w}\int_0^k s\sqrt{b^2+w^2s^2}\, ds = \frac{2\pi}{3w^3}\left((b^2+w^2k^2)^{\frac{3}{2}} - b^3\right)$$

Example 8.

Evaluate $\iint_S 3y\, dA =?$, over the surface S: $z = \frac{y^2}{2}$, $0 \le x \le 1$, $0 \le y \le 1$.

【Solution】

Let $\vec{r}(x,y) = \left(x, y, \frac{y^2}{2}\right)$ and let R represent the region projected onto the xy-plane

then $R = \{(x,y): 0 \le x \le 1, 0 \le y \le 1\}$ and $\iint_S 3y\, dA = \iint_R 3y \left|\frac{\partial \vec{r}}{\partial x} \times \frac{\partial \vec{r}}{\partial y}\right| dxdy$

$\therefore \frac{\partial \vec{r}}{\partial x} = (1,0,0)$ and $\frac{\partial \vec{r}}{\partial y} = (0,1,y)$ $\therefore \frac{\partial \vec{r}}{\partial x} \times \frac{\partial \vec{r}}{\partial y} = (0,-y,1) \Rightarrow \left|\frac{\partial \vec{r}}{\partial x} \times \frac{\partial \vec{r}}{\partial y}\right| = (y^2+1)^{\frac{1}{2}}$

$\iint_R 3y\left|\frac{\partial \vec{r}}{\partial x} \times \frac{\partial \vec{r}}{\partial y}\right| dxdy = 3\iint_R y(y^2+1)^{\frac{1}{2}} dxdy = (y^2+1)^{\frac{3}{2}}\Big|_0^1 = 2\sqrt{2} - 1$

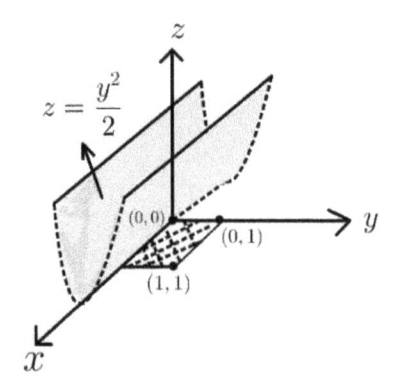

9.5.3 For Scalar Functions, Find the Mass of Surface S

Example 9.

Let $f(x, y, z) = (1 - x^2)y$, $S: z = 3 - y^2$, $0 \leq x \leq 1$, $0 \leq y \leq 1$. Find

$$\iint_S f(x, y, z)\, dA = ?$$

【Solution】

Let $\vec{r}(x, y) = (x, y, 3 - y^2)$ and let R represent the region projected onto the xy-plane then $R = \{(x, y): 0 \leq x \leq 1, 0 \leq y \leq 1\}$ and

$$\iint_S f(x, y, z)\, dA = \iint_R f(x, y, z) \left|\frac{\partial \vec{r}}{\partial x} \times \frac{\partial \vec{r}}{\partial y}\right| dxdy$$

$\because \dfrac{\partial \vec{r}}{\partial x} = (1, 0, 0)$ and $\dfrac{\partial \vec{r}}{\partial y} = (0, 1, -2y)$ $\therefore \dfrac{\partial \vec{r}}{\partial x} \times \dfrac{\partial \vec{r}}{\partial y} = (0, 2y, 1) \Rightarrow \left|\dfrac{\partial \vec{r}}{\partial x} \times \dfrac{\partial \vec{r}}{\partial y}\right| = \sqrt{1 + 4y^2}$

$\therefore \iint_S f(x, y, z)\, dA = \iint_R f(x, y, z) \left|\dfrac{\partial \vec{r}}{\partial x} \times \dfrac{\partial \vec{r}}{\partial y}\right| dxdy = \int_0^1 \int_0^1 y(1 - x^2)\sqrt{1 + 4y^2}\, dydx$

$= \int_0^1 y\sqrt{1 + 4y^2}\, dy \int_0^1 (1 - x^2)\, dx$

$\because \int_0^1 y\sqrt{1 + 4y^2}\, dy = \dfrac{1}{12}(1 + 4y^2)^{\frac{3}{2}}\bigg|_0^1 = \dfrac{5\sqrt{5} - 1}{12}$

Let $x = \sin\theta$ then $dx = \cos\theta\, d\theta$

$\therefore \int_0^1 (1 - x^2)\, dx = \int_0^{\frac{\pi}{2}} \cos^3\theta\, dx = \int_0^{\frac{\pi}{2}} (1 - \sin^2\theta)\cos\theta\, dx = \left(\sin\theta - \dfrac{\sin^3\theta}{3}\right)\bigg|_0^{\frac{\pi}{2}} = \dfrac{2}{3}$

$$\therefore \iint_S f(x,y,z)\, dA = \int_0^1 y\sqrt{1+4y^2}\, dy \int_0^1 (1-x^2)\, dx = \frac{5\sqrt{5}-1}{12} \cdot \frac{2}{3} = \frac{5\sqrt{5}-1}{18}$$

Example 10.

Find $\iint_S y^2 + 2yz\, dA = ?$, where S is the region bounded by the plane

$2x + y + 2z = 4$ in the first quadrant.

【Solution】

Let $\vec{r}(x,y) = \left(x, y, 2 - x - \frac{y}{2}\right)$ and let R represent the region projected onto the xy-plane

then $R = \{(x,y): 2x + y \leq 4, x \geq 0, y \geq 0\}$

and $\iint_S y^2 + 2yz\, dA = \iint_R \left(y^2 + 2y\left(2 - x - \frac{y}{2}\right)\right) \left|\frac{\partial \vec{r}}{\partial x} \times \frac{\partial \vec{r}}{\partial y}\right| dxdy$

$\therefore \dfrac{\partial \vec{r}}{\partial x} = (1, 0, -1)$ and $\dfrac{\partial \vec{r}}{\partial y} = \left(0, 1, -\dfrac{1}{2}\right)$ $\therefore \dfrac{\partial \vec{r}}{\partial x} \times \dfrac{\partial \vec{r}}{\partial y} = \left(1, \dfrac{1}{2}, 1\right) \Rightarrow \left|\dfrac{\partial \vec{r}}{\partial x} \times \dfrac{\partial \vec{r}}{\partial y}\right| = \dfrac{3}{2}$

$\therefore \iint_S y^2 + 2yz\, dA = \iint_R \left(y^2 + 2y\left(2 - x - \frac{y}{2}\right)\right) \left|\frac{\partial \vec{r}}{\partial x} \times \frac{\partial \vec{r}}{\partial y}\right| dxdy$

$= \dfrac{3}{2} \int_0^2 \int_0^{4-2x} y^2 + 2y \cdot \left(2 - x - \dfrac{y}{2}\right) dydx$

$\therefore \int_0^{4-2x} y^2 + 2y \cdot \left(2 - x - \dfrac{y}{2}\right) dy = \int_0^{4-2x} 4y - 2xy\, dy = 2y^2 - xy^2\big|_0^{4-2x} = 4(2-x)^3$

$\therefore \int_0^2 \int_0^{4-2x} y^2 + 2y \cdot \dfrac{1}{2} \cdot (4 - 2x - y)\, dydx = \int_0^2 4(2-x)^3\, dx = -(2-x)^4\big|_0^2 = 16$

$\therefore \iint_S y^2 + 2yz\, dA = 24$

9.5.3 For Scalar Functions, Find the Mass of Surface S

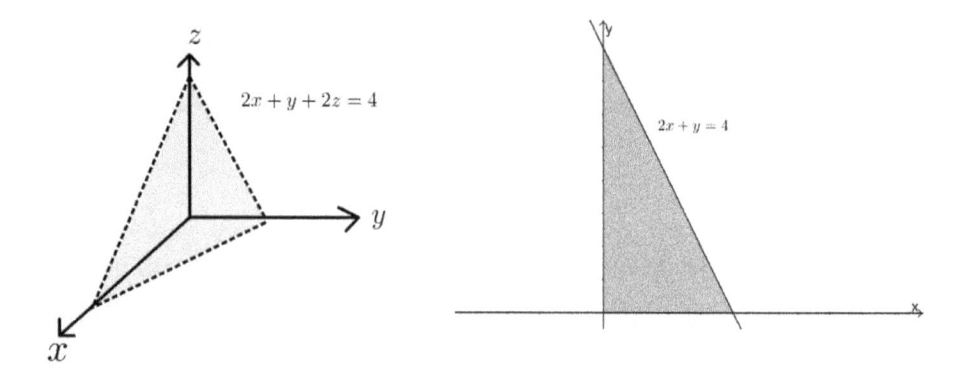

Example 11.

$$\text{Find } \iint_S z^2 \, dA = ?, \quad S: z = \sqrt{r^2 - x^2 - y^2}.$$

【Solution】

Let $\vec{r}(\theta, \varphi) = (r \cos \theta \sin \varphi, r \sin \theta \sin \varphi, r \cos \varphi)$ and $R = \left\{ (\theta, \varphi): 0 \leq \theta \leq 2\pi, 0 \leq \varphi \leq \dfrac{\pi}{2} \right\}$

then $\iint_S z^2 \, dA = \iint_R r^2 \cos^2 \varphi \left| \dfrac{\partial \vec{r}}{\partial \varphi} \times \dfrac{\partial \vec{r}}{\partial \theta} \right| d\theta d\varphi$

$\because \dfrac{\partial \vec{r}}{\partial \theta} = (-r \sin \theta \sin \varphi, r \cos \theta \sin \varphi, 0)$ and $\dfrac{\partial \vec{r}}{\partial \varphi} = (r \cos \theta \cos \varphi, r \sin \theta \cos \varphi, -r \sin \varphi)$

$\therefore \dfrac{\partial \vec{r}}{\partial \varphi} \times \dfrac{\partial \vec{r}}{\partial \theta} = (r^2 \cos \theta \sin^2 \varphi, r^2 \sin \theta \sin^2 \varphi, r^2 \sin \varphi \cos \varphi) \Rightarrow \left| \dfrac{\partial \vec{r}}{\partial \varphi} \times \dfrac{\partial \vec{r}}{\partial \theta} \right| = r^2 \sin \varphi$

$\therefore \iint_S z^2 \, dA = \iint_R r^2 \cos^2 \varphi \left| \dfrac{\partial \vec{r}}{\partial \varphi} \times \dfrac{\partial \vec{r}}{\partial \theta} \right| d\theta d\varphi = r^4 \int_0^{2\pi} \int_0^{\frac{\pi}{2}} \cos^2 \varphi \sin \varphi \, d\varphi d\theta = \dfrac{2\pi r^4}{3}$

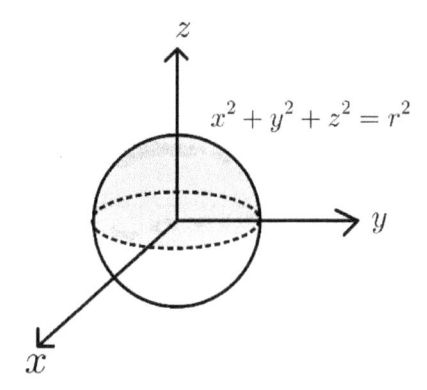

Example 12.

Evaluate $\iint_S \sqrt{2z} \, dA = ?$, over the surface S: $z = \dfrac{y^2}{2}$, $0 \le x \le 1$, $0 \le y \le 1$.

【Solution】

Let $\vec{r}(x, y) = \left(x, y, \dfrac{y^2}{2}\right)$ and let R represent the region projected onto the xy-plane

then $R = \{(x, y): 0 \le x \le 1, 0 \le y \le 1\}$ and $\iint_S \sqrt{2z} \, dA = \iint_R y \left|\dfrac{\partial \vec{r}}{\partial x} \times \dfrac{\partial \vec{r}}{\partial y}\right| dxdy$

$\because \dfrac{\partial \vec{r}}{\partial x} = (1, 0, 0)$ and $\dfrac{\partial \vec{r}}{\partial y} = (0, 1, y)$ $\therefore \dfrac{\partial \vec{r}}{\partial x} \times \dfrac{\partial \vec{r}}{\partial y} = (0, -y, 1) \Rightarrow \left|\dfrac{\partial \vec{r}}{\partial x} \times \dfrac{\partial \vec{r}}{\partial y}\right| = (y^2 + 1)^{\frac{1}{2}}$

$\iint_R y \left|\dfrac{\partial \vec{r}}{\partial x} \times \dfrac{\partial \vec{r}}{\partial y}\right| dxdy = \iint_R y(y^2 + 1)^{\frac{1}{2}} dxdy = \dfrac{(y^2 + 1)^{\frac{3}{2}}}{3} \bigg|_0^1 = \dfrac{2\sqrt{2} - 1}{3}$

Example 13.

Find $\iint_S xy \, dA = ?$, where S is the region bounded by the plane $x + 2y + 3z = 6$ in the first quadrant.

【Solution】

9.5.3 For Scalar Functions, Find the Mass of Surface S 498

Let $\vec{r}(x,y) = \left(x, y, 2 - \dfrac{x+2y}{3}\right)$ and let R represent the region projected onto the xy-plane

then $R = \{(x,y): x + 2y \le 6, x \ge 0, y \ge 0\}$ and $\iint_S xy\, dA = \iint_R xy \left|\dfrac{\partial \vec{r}}{\partial x} \times \dfrac{\partial \vec{r}}{\partial y}\right| dxdy$

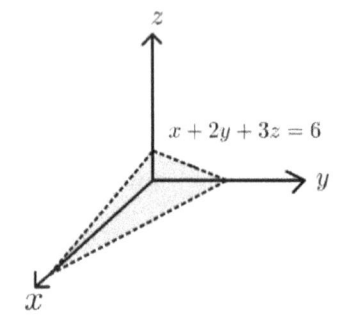

$\because \dfrac{\partial \vec{r}}{\partial x} = \left(1, 0, -\dfrac{1}{3}\right)$ and $\dfrac{\partial \vec{r}}{\partial y} = \left(0, 1, -\dfrac{2}{3}\right) \quad \therefore \dfrac{\partial \vec{r}}{\partial x} \times \dfrac{\partial \vec{r}}{\partial y} = \left(\dfrac{1}{3}, \dfrac{2}{3}, 1\right) \Rightarrow \left|\dfrac{\partial \vec{r}}{\partial x} \times \dfrac{\partial \vec{r}}{\partial y}\right| = \dfrac{\sqrt{14}}{3}$

$\therefore \iint_S xy\, dA = \iint_R xy \left|\dfrac{\partial \vec{r}}{\partial x} \times \dfrac{\partial \vec{r}}{\partial y}\right| dxdy = \dfrac{\sqrt{14}}{3}\int_0^3 \int_0^{6-2y} xy\, dx\, dy = \dfrac{\sqrt{14}}{3}\int_0^3 y \cdot \dfrac{x^2}{2}\bigg|_0^{6-2y} dxdy$

$= \dfrac{\sqrt{14}}{3}\int_0^3 18y - 12y^2 + 2y^3\, dy = \dfrac{\sqrt{14}}{3}\left(9y^2 - 4y^3 + \dfrac{y^4}{2}\right)\bigg|_0^3 = \dfrac{\sqrt{14}}{3} \cdot \dfrac{27}{2} = \dfrac{9\sqrt{14}}{2}$

$x + 2y + 3z = 6$

Example 14.

Find $\iint_S x^2 + y^2\, dA = ?,\quad S: z = \sqrt{r^2 - x^2 - y^2}.$

【Solution】

Let $\vec{r}(\theta, \varphi) = (r\cos\theta \sin\varphi, r\sin\theta \sin\varphi, r\cos\varphi)$ and $R = \left\{(\theta, \varphi): 0 \le \theta \le 2\pi, 0 \le \varphi \le \dfrac{\pi}{2}\right\}$

then $\iint_S x^2 + y^2\, dA = \iint_R r^2 \sin^2\varphi \left|\dfrac{\partial \vec{r}}{\partial \varphi} \times \dfrac{\partial \vec{r}}{\partial \theta}\right| d\theta d\varphi$

$\because \dfrac{\partial \vec{r}}{\partial \theta} = (-r\sin\theta \sin\varphi, r\cos\theta \sin\varphi, 0)$ and $\dfrac{\partial \vec{r}}{\partial \varphi} = (r\cos\theta \cos\varphi, r\sin\theta \cos\varphi, -r\sin\varphi)$

$$\therefore \frac{\partial \vec{r}}{\partial \varphi} \times \frac{\partial \vec{r}}{\partial \theta} = (r^2 \cos\theta \sin^2\varphi, r^2 \sin\theta \sin^2\varphi, r^2 \sin\varphi \cos\varphi) \Rightarrow \left|\frac{\partial \vec{r}}{\partial \varphi} \times \frac{\partial \vec{r}}{\partial \theta}\right| = r^2 \sin\varphi$$

$$\therefore \iint_S x^2 + y^2 \, dA = \iint_R r^2 \sin^2\varphi \left|\frac{\partial \vec{r}}{\partial \varphi} \times \frac{\partial \vec{r}}{\partial \theta}\right| d\theta d\varphi = r^4 \int_0^{2\pi} \int_0^{\frac{\pi}{2}} \sin^3\varphi \, d\varphi d\theta$$

$$= r^4 \, 2\pi \left(-\cos\varphi + \frac{\cos^3\varphi}{3}\right)\Big|_0^{\frac{\pi}{2}} = \frac{4\pi r^4}{3}$$

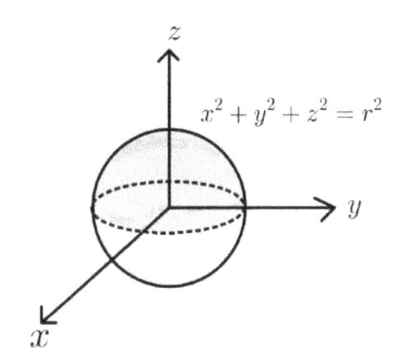

Example 15.

Evaluate $\iint_S z \, dA = ?$, over the conical surface $z = \sqrt{x^2 + y^2}$ between $z = 0$ and $z = 1$.

【Solution】

Let $\vec{r}(x, y) = \left(x, y, \sqrt{x^2 + y^2}\right)$ and let R represent the region projected onto the xy-plane

then $R = \{(x, y) : 0 \leq x^2 + y^2 \leq 1\}$ and $\iint_S z \, dA = \iint_R \sqrt{x^2 + y^2} \left|\frac{\partial \vec{r}}{\partial x} \times \frac{\partial \vec{r}}{\partial y}\right| dxdy$

$$\therefore \frac{\partial \vec{r}}{\partial x} = \left(1, 0, \frac{x}{\sqrt{x^2 + y^2}}\right) \text{ and } \frac{\partial \vec{r}}{\partial y} = \left(0, 1, \frac{y}{\sqrt{x^2 + y^2}}\right)$$

$$\therefore \frac{\partial \vec{r}}{\partial x} \times \frac{\partial \vec{r}}{\partial y} = \left(\frac{x}{\sqrt{x^2 + y^2}}, -\frac{y}{\sqrt{x^2 + y^2}}, 1\right) \Rightarrow \left|\frac{\partial \vec{r}}{\partial x} \times \frac{\partial \vec{r}}{\partial y}\right| = \sqrt{1 + \frac{x^2}{x^2 + y^2} + \frac{y^2}{x^2 + y^2}} = \sqrt{2}$$

9.5.3 For Scalar Functions, Find the Mass of Surface S

$$\therefore \iint_S z\, dA = \iint_{\sqrt{x^2+y^2}\le 1} \sqrt{x^2+y^2}\left|\frac{\partial \vec{r}}{\partial x} \times \frac{\partial \vec{r}}{\partial y}\right| dxdy = \iint_{\sqrt{x^2+y^2}\le 1} \sqrt{x^2+y^2}\sqrt{2}\,dxdy$$

Let $x = r\cos\theta$, $y = r\sin\theta$ and $R = \{(r,\theta): 0 \le r \le 1, 0 \le \theta \le 2\pi\}$

$$\therefore \iint_{\sqrt{x^2+y^2}\le 1} \sqrt{x^2+y^2}\sqrt{2}\,dxdy = \sqrt{2}\int_0^{2\pi}\int_0^1 r\sqrt{2}\,r\,drd\theta = \frac{2\sqrt{2}\pi}{3}$$

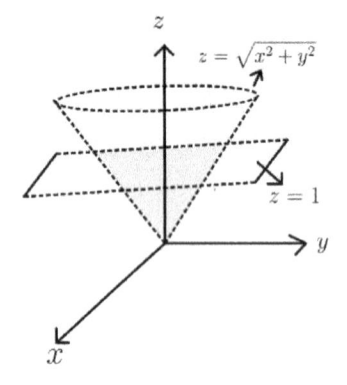

Example 16.

Evaluate $\iint_S z\, dA = ?$, where S is the hyperbolic $z^2 = 1 + x^2 + y^2$ between $z = 0$ and $z = \sqrt{5}$.

【Solution】

Let $\vec{r}(x,y) = \left(x, y, \sqrt{1+x^2+y^2}\right)$ and let R represent the region projected onto the xy-plane. Then

$R = \{(x,y): 0 \le x^2 + y^2 \le 4\}$ and $\iint_S z\, dA = \iint_R \sqrt{x^2+y^2}\left|\frac{\partial \vec{r}}{\partial x} \times \frac{\partial \vec{r}}{\partial y}\right| dxdy$

$\therefore \dfrac{\partial \vec{r}}{\partial x} = \left(1, 0, \dfrac{x}{\sqrt{1+x^2+y^2}}\right)$ and $\dfrac{\partial \vec{r}}{\partial y} = \left(0, 1, \dfrac{y}{\sqrt{1+x^2+y^2}}\right)$

$$\therefore \frac{\partial \vec{r}}{\partial x} \times \frac{\partial \vec{r}}{\partial y} = \left(\frac{x}{\sqrt{1+x^2+y^2}}, \frac{y}{\sqrt{1+x^2+y^2}}, 1 \right) \Rightarrow \left| \frac{\partial \vec{r}}{\partial x} \times \frac{\partial \vec{r}}{\partial y} \right| = \sqrt{1 + \frac{x^2+y^2}{1+x^2+y^2}}$$

$$\therefore \iint_S z \, dA = \iint_{\sqrt{x^2+y^2} \le 2} \sqrt{1+x^2+y^2} \left| \frac{\partial \vec{r}}{\partial x} \times \frac{\partial \vec{r}}{\partial y} \right| dxdy$$

$$= \iint_{\sqrt{x^2+y^2} \le 2} \sqrt{1+x^2+y^2} \sqrt{1 + \frac{x^2+y^2}{1+x^2+y^2}} \, dxdy$$

Let $x = r\cos\theta$, $y = r\sin\theta$ and $R = \{(r,\theta) : 0 \le r \le 2, 0 \le \theta \le 2\pi\}$

$$\iint_{\sqrt{x^2+y^2} \le 2} \sqrt{1+x^2+y^2} \sqrt{1 + \frac{x^2+y^2}{1+x^2+y^2}} \, dxdy = \int_0^{2\pi} \int_0^2 r\sqrt{1+2r^2} \, drd\theta$$

$$= 2\pi \times \left. \frac{(1+2r^2)^{\frac{3}{2}}}{6} \right|_0^2 = \frac{26\pi}{3}$$

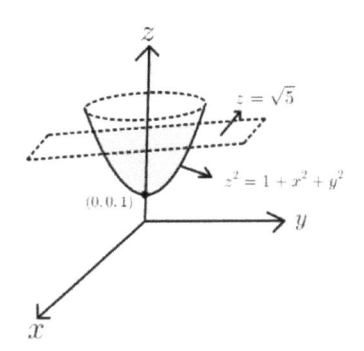

Example 17.

Find $\iint_S x^2 + y^2 \, dA = ?$, $S: z = 1 - (x^2+y^2)$, $z > 0$.

【Solution】

Let $\vec{r}(x,y) = (x, y, 1 - (x^2+y^2))$ and let R represent the region projected onto the

9.5.3 For Scalar Functions, Find the Mass of Surface S

xy-plane. Then

$$R = \{(x,y): 0 \leq x^2 + y^2 \leq 1\} \text{ and } \iint_S x^2 + y^2 \, dA = \iint_R (x^2 + y^2) \left| \frac{\partial \vec{r}}{\partial x} \times \frac{\partial \vec{r}}{\partial y} \right| dxdy$$

$$\because \frac{\partial \vec{r}}{\partial x} = (1,0,-2x) \text{ and } \frac{\partial \vec{r}}{\partial y} = (0,1,-2y) \therefore \frac{\partial \vec{r}}{\partial x} \times \frac{\partial \vec{r}}{\partial y} = (2x, 2y, 1)$$

$$\Rightarrow \left| \frac{\partial \vec{r}}{\partial x} \times \frac{\partial \vec{r}}{\partial y} \right| = \sqrt{1 + 4x^2 + 4y^2}$$

Let $x = r\cos\theta, y = r\sin\theta$ and $R = \{(r,\theta): 0 \leq r \leq 1, 0 \leq \theta \leq 2\pi\}$

$$\therefore \iint_S x^2 + y^2 \, dA = \iint_R (x^2 + y^2) \left| \frac{\partial \vec{r}}{\partial x} \times \frac{\partial \vec{r}}{\partial y} \right| dxdy$$

$$= \iint_{0 \leq x^2 + y^2 \leq 1} (x^2 + y^2)\sqrt{1 + 4x^2 + 4y^2} \, dxdy = \int_0^{2\pi} \int_0^1 r^2 \sqrt{1 + 4r^2} \, rdrd\theta$$

$$= 2\pi \int_0^1 r^3 \sqrt{1 + 4r^2} \, dr$$

Let $t = \sqrt{1 + 4r^2}$ then $\int_0^1 r^3 \sqrt{1 + 4r^2} \, dr = \int_0^{\sqrt{5}} \frac{t^2 - 1}{4} \cdot t \cdot \frac{t}{4} dt = \frac{1}{16}\left(\frac{10\sqrt{5}}{3} + \frac{2}{15}\right)$

$$\therefore \iint_S x^2 + y^2 \, dA = \frac{\pi}{8}\left(\frac{10\sqrt{5}}{3} + \frac{2}{15}\right)$$

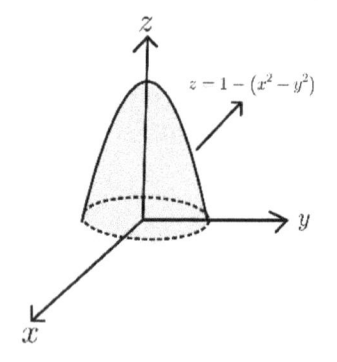

$z = 1 - (x^2 - y^2)$

Example 18.

Find $\iint_S x^3 \sin y \, dA = ?$, $S: z = x^3, 0 \leq x \leq 1, 0 \leq y \leq \frac{\pi}{2}$.

【Solution】

Let $\vec{r}(x,y) = (x, y, x^3)$ and let R represent the region projected onto the xy-plane. Then

$$R = \{(x,y): 0 \le x \le 1, 0 \le y \le \frac{\pi}{2}\} \text{ and } \iint_S x^3 \sin y \, dA = \iint_R x^3 \sin y \left|\frac{\partial \vec{r}}{\partial x} \times \frac{\partial \vec{r}}{\partial y}\right| dxdy$$

$$\because \frac{\partial \vec{r}}{\partial x} = (1, 0, 3x^2) \text{ and } \frac{\partial \vec{r}}{\partial y} = (0, 1, 0) \therefore \frac{\partial \vec{r}}{\partial x} \times \frac{\partial \vec{r}}{\partial y} = (-3x^2, 0, 1) \Rightarrow \left|\frac{\partial \vec{r}}{\partial x} \times \frac{\partial \vec{r}}{\partial y}\right| = \sqrt{1 + 9x^4}$$

$$\iint_S x^3 \sin y \, dA = \iint_R x^3 \sin y \left|\frac{\partial \vec{r}}{\partial x} \times \frac{\partial \vec{r}}{\partial y}\right| dxdy = \int_0^{\frac{\pi}{2}} \int_0^1 x^3 \sin y \sqrt{1 + 9x^4} \, dxdy$$

$$= \int_0^{\frac{\pi}{2}} \frac{1}{54}(1 + 9x^4)^{\frac{3}{2}}\bigg|_0^1 \sin y \, dy = \frac{1}{54}(10\sqrt{10} - 1)$$

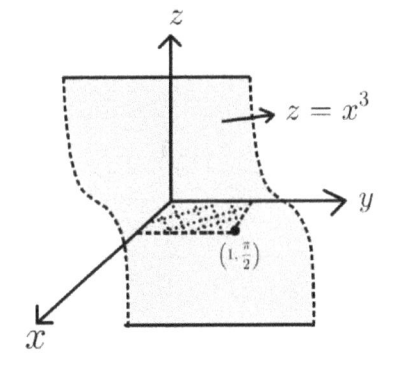

Example 19.

Find $\iint_S \frac{xy}{z} dA = ?$, $S: z = x^2 + y^2$, $4 \le x^2 + y^2 \le 16$, $x, y \ge 0$.

【Solution】

Let $\vec{r}(x, y) = (x, y, x^2 + y^2)$ and $R = \{(x,y): 4 \le x^2 + y^2 \le 16, x, y \ge 0\}$. Then

$$\iint_S \frac{xy}{z} dA = \iint_R \frac{xy}{x^2 + y^2} \left|\frac{\partial \vec{r}}{\partial x} \times \frac{\partial \vec{r}}{\partial y}\right| dxdy$$

$$\because \frac{\partial \vec{r}}{\partial x} = (1, 0, 2x) \text{ and } \frac{\partial \vec{r}}{\partial y} = (0, 1, 2y) \therefore \frac{\partial \vec{r}}{\partial x} \times \frac{\partial \vec{r}}{\partial y} = (-2x, -2y, 1)$$

$$\therefore \left|\frac{\partial \vec{r}}{\partial x} \times \frac{\partial \vec{r}}{\partial y}\right| = \sqrt{1 + 4x^2 + 4y^2}$$

$$\therefore \iint_S \frac{xy}{z} dA = \iint_R \frac{xy}{x^2 + y^2} \left|\frac{\partial \vec{r}}{\partial x} \times \frac{\partial \vec{r}}{\partial y}\right| dxdy = \iint_{4 \le x^2+y^2 \le 16} \frac{xy\sqrt{1 + 4x^2 + 4y^2}}{x^2 + y^2} dxdy$$

Let $x = r\cos\theta$, $y = r\sin\theta$ and $R = \left\{(r,\theta): 2 \le r \le 4, 0 \le \theta \le \frac{\pi}{2}\right\}$ then

$$\iint_{4 \le x^2+y^2 \le 16} \frac{xy\sqrt{1 + 4x^2 + 4y^2}}{x^2 + y^2} dxdy = \int_0^{\frac{\pi}{2}} \int_2^4 \frac{r\cos\theta \, r\sin\theta \sqrt{1 + 4r^2}}{r^2} r \, drd\theta$$

$$= \int_0^{\frac{\pi}{2}} \int_2^4 r\cos\theta \sin\theta \sqrt{1 + 4r^2} \, drd\theta = \int_2^4 r\sqrt{1 + 4r^2} \, dr \int_0^{\frac{\pi}{2}} \cos\theta \sin\theta \, d\theta$$

$$= \frac{(1 + 4r^2)^{\frac{3}{2}}}{12} \Bigg|_2^4 \cdot \left(-\frac{\cos 2\theta}{4}\right) \Bigg|_0^{\frac{\pi}{2}} = \frac{65\sqrt{65} - 17\sqrt{17}}{24}$$

9.5.4 For Vector Functions, Find the Flux across Surface S

When calculating the flux through a surface, we typically consider orientable surfaces. Below is an introduction to the definition of an orientable surface.

【Definition】

If at every point on the surface S, there exists a continuous unit normal vector $\vec{n}$, then surface S is called an orientable surface.

Assume that S is an orientable surface, described by the parametric equation:
$$\vec{r}(s,t) = (x(s,t), y(s,t), z(s,t)), \quad \forall (s,t) \in D,$$
where $\vec{n}$ is a unit normal vector. Suppose the parameter domain D is a rectangle,

subdivided into several small rectangles D_{ij}, with length Δs and width Δt. The center of D_{ij} is (s_i^*, t_j^*), and the region corresponding to surface S is S_{ij}. The point (s_i^*, t_j^*) on the surface S_{ij} corresponds to point E_{ij}^*, with position vector $\vec{r}(s_i^*, t_j^*)$. Thus, the surface S is divided into several segments S_{ij}, as shown in the figure below.

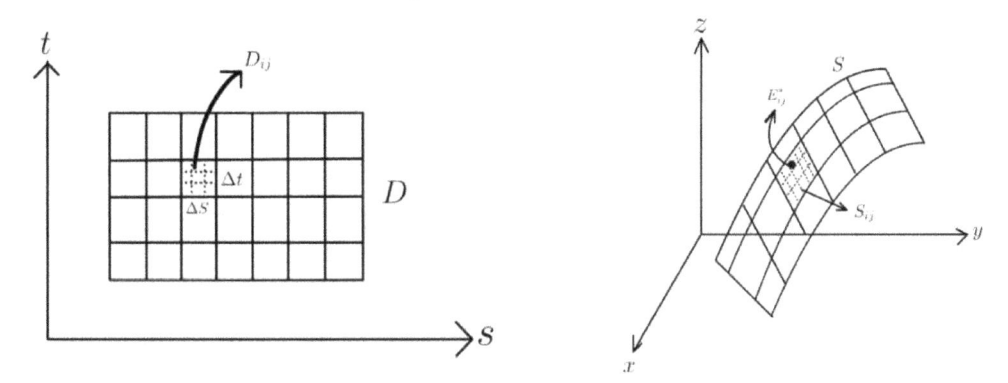

Consider a fluid with density $\rho(x, y, z)$ and velocity field $v(x, y, z)$ passing through surface S. The flux per unit area is represented by ρv. If we divide D into countless small rectangles, the surface S is similarly divided into countless small surface patches S_{ij}, and each S_{ij} is approximately a plane. Then, the mass of fluid passing through S_{ij} per unit time in the direction of the normal vector $\vec{n}$ can be approximated as:

$$\rho v \cdot \vec{n} \Delta S_{ij}$$

where ρ, v and $\vec{n}$ are evaluated at some point on S_{ij}. By summing the mass of fluid passing through all the surface patches S_{ij}, we get an approximate estimate for the mass of fluid passing through surface S,

$$\sum_{i=1}^{m} \sum_{j=1}^{n} \rho v \cdot \vec{n} \Delta S_{ij}.$$

As the size of each S_{ij} becomes arbitrarily small, this sum approaches the total mass of fluid passing through surface S. In the limit, we obtain the surface integral of the function $\rho v \cdot \vec{n}$ over S, i.e.,

$$\iint_S \rho v \cdot \vec{n} dS = \iint_S \rho(x, y, z) v(x, y, z) \cdot \vec{n}(x, y, z) dS = \lim_{m \to \infty, n \to \infty} \sum_{i=1}^{m} \sum_{j=1}^{n} \rho v \cdot \vec{n} \Delta S_{ij}.$$

Physically, this represents the flow rate across surface S. Defining $\vec{F} = \rho v$, where $\vec{F}$ is a vector field in R^3, we have:

$$\iint_S \rho(x, y, z) v(x, y, z) \cdot \vec{n}(x, y, z) dS = \iint_S \vec{F} \cdot \vec{n} dS.$$

Thus, using the above Riemann sum definition, the surface integral of the vector function $\vec{F}$

over surface S can be expressed.

【Definition】

Assume that a continuous vector field $\vec{F}$ is defined on an orientable surface S and has a unit normal vector $\vec{n}$. Then, the surface integral of $\vec{F}$ over S is given by

$$\iint_S \vec{F} \cdot d\vec{S} = \iint_S \vec{F} \cdot \vec{n} dS.$$

This integral is called the flux of $\vec{F}$ across surface S.

The next key point is how to find the unit normal vector given an orientable parametric surface $\vec{r}(s,t)$. Suppose the parametric surface is

$$\vec{r}(s,t) = (x(s,t), y(s,t), z(s,t)), \quad \forall (s,t) \in D.$$

We can find two continuous and opposite unit normal vector fields, which are

$$\frac{\vec{r}_s(s,t) \times \vec{r}_t(s,t)}{|\vec{r}_s(s,t) \times \vec{r}_t(s,t)|} \quad \text{and} \quad -\frac{\vec{r}_s(s,t) \times \vec{r}_t(s,t)}{|\vec{r}_s(s,t) \times \vec{r}_t(s,t)|}.$$

【Definition】

Suppose S is an orientable surface.

(i) If the z-coordinate of $\dfrac{\vec{r}_s(s,t) \times \vec{r}_t(s,t)}{|\vec{r}_s(s,t) \times \vec{r}_t(s,t)|}$ is positive, $\forall (s,t) \in D$

then S is oriented upward.

(ii) If the z-coordinate of $\dfrac{\vec{r}_s(s,t) \times \vec{r}_t(s,t)}{|\vec{r}_s(s,t) \times \vec{r}_t(s,t)|}$ is negative, $\forall (s,t) \in D$

then S is oriented downward.

In general, the problem will tell us whether S is oriented upward or downward. Similarly, if the y-coordinate of $\dfrac{\vec{r}_s(s,t) \times \vec{r}_t(s,t)}{|\vec{r}_s(s,t) \times \vec{r}_t(s,t)|}$ is positive, then it is oriented in the positive y-axis direction, and so on.

【Definition】

Suppose S is a closed surface.

(i) If the unit normal vector points outward, toward the non-enclosed region, then S is oriented outward or positively oriented.

(ii) If the unit normal vector points inward, toward the enclosed region, then S is oriented inward or negatively oriented.

In general, if we are calculating the flux through a closed surface S, and there is no specific indication whether S is oriented outward or inward, it is conventionally assumed to be oriented outward.

To compute the surface integral defined above, consider the area of a parallelogram lying within the tangent plane to approximate the region ΔS_{ij}. As discussed previously in the context of surface area, there is an approximate estimate for ΔS_{ij},

$$\Delta S_{ij} \approx |\vec{r}_s(s_i^*, t_j^*) \times \vec{r}_t(s_i^*, t_j^*)|\Delta s \Delta t.$$

Thus,

$$\iint_S \vec{F} \cdot \vec{n} dS = \lim_{m \to \infty, n \to \infty} \sum_{i=1}^{m} \sum_{j=1}^{n} \vec{F} \cdot \vec{n} \Delta S_{ij} = \lim_{m \to \infty, n \to \infty} \sum_{i=1}^{m} \sum_{j=1}^{n} \vec{F} \cdot \vec{n} |\vec{r}_s(s_i^*, t_j^*) \times \vec{r}_t(s_i^*, t_j^*)|\Delta s \Delta t$$

$$= \lim_{m \to \infty, n \to \infty} \sum_{i=1}^{m} \sum_{j=1}^{n} \vec{F} \cdot \frac{\vec{r}_s(s_i^*, t_j^*) \times \vec{r}_t(s_i^*, t_j^*)}{|\vec{r}_s(s_i^*, t_j^*) \times \vec{r}_t(s_i^*, t_j^*)|} \cdot |\vec{r}_s(s_i^*, t_j^*) \times \vec{r}_t(s_i^*, t_j^*)|\Delta s \Delta t$$

$$= \lim_{m \to \infty, n \to \infty} \sum_{i=1}^{m} \sum_{j=1}^{n} \vec{F}(s_i^*, t_j^*) \cdot \left(\vec{r}_s(s_i^*, t_j^*) \times \vec{r}_t(s_i^*, t_j^*)\right) \Delta s \Delta t = \iint_D \vec{F}(\vec{r}(s,t)) \cdot (\vec{r}_s \times \vec{r}_t) ds dt$$

Based on the above observation, calculating the surface integral of a vector function is equivalent to calculating a double integral. Specifically, the integrand for the double integral is

$$\vec{F}(\vec{r}(s,t)) \cdot (\vec{r}_s \times \vec{r}_t).$$

Thus, finding the surface integral of a vector function is equivalent to first determining the integrand and then evaluating the double integral. Below, we provide a preliminary explanation of the different types of surface integrals that can be transformed into double integrals. Consider the surface $S: z = g(x, y)$, then then the normal vector is

$$\vec{n} = \frac{\vec{r}_x(x, y) \times \vec{r}_y(x, y)}{|\vec{r}_x(x, y) \times \vec{r}_y(x, y)|}.$$

Since

$$\frac{\partial \vec{r}}{\partial x} = (1, 0, g_x(x, y)), \quad \frac{\partial \vec{r}}{\partial y} = (0, 1, g_y(x, y)) \text{ and } \frac{\partial \vec{r}}{\partial x} \times \frac{\partial \vec{r}}{\partial y} = (-g_x(x, y), -g_y(x, y), 1),$$

we have

$$\vec{n} = \frac{\vec{r}_x(x, y) \times \vec{r}_y(x, y)}{|\vec{r}_x(x, y) \times \vec{r}_y(x, y)|} = \frac{1}{\sqrt{1 + (g_x(x, y))^2 + (g_1(x, y))^2}} (-g_x(x, y), -g_y(x, y), 1).$$

$\because$ the z-coordinate of $\dfrac{\vec{r}_x(x,y) \times \vec{r}_y(x,y)}{|\vec{r}_x(x,y) \times \vec{r}_y(x,y)|}$ is positive, it follows that S is oriented upward.

9.5.4 For Vector Functions, Find the Flux across Surface S

Surfaces that are oriented upward come in various forms. Below are some common examples of surfaces that are oriented upward.

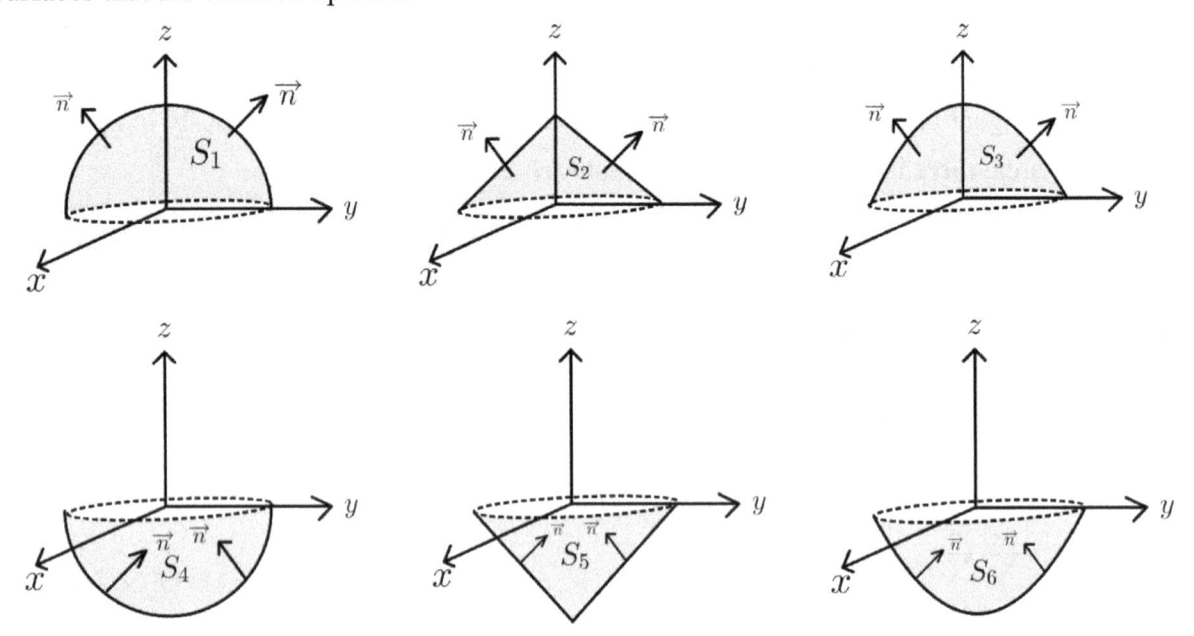

Surfaces that are oriented downward come in various forms. Below are some common examples of surfaces that are oriented downward.

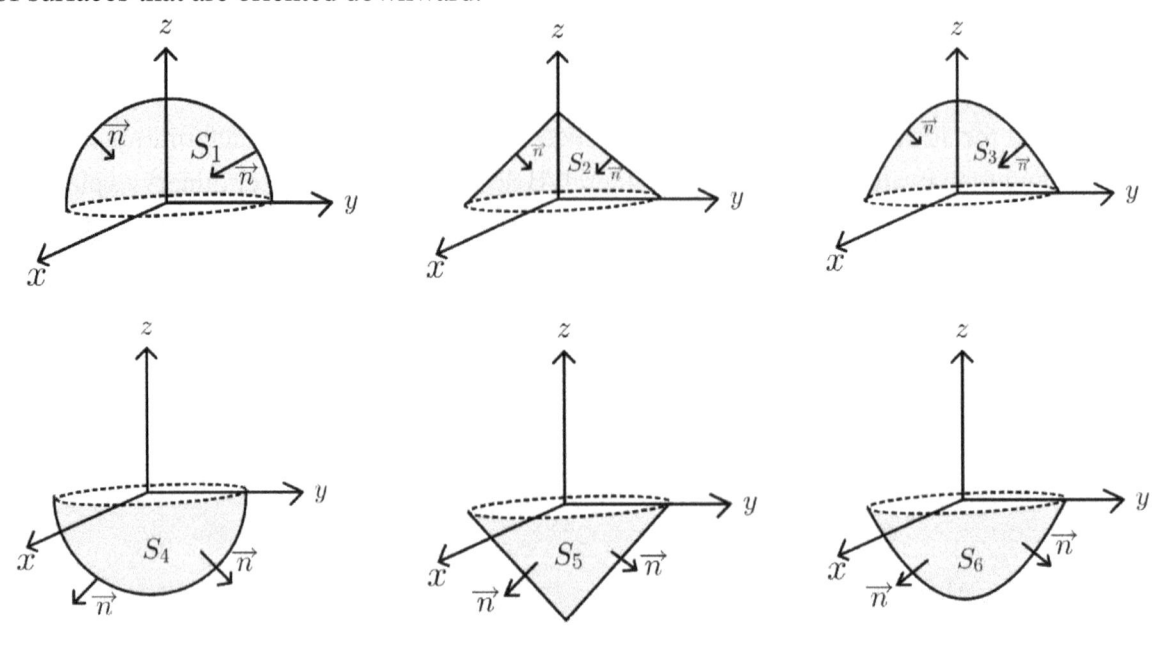

Suppose the surface $S: z = g(x, y)$ is oriented upward, with the domain $a \leq x \leq b$ and $c \leq y \leq d$. Let $\vec{F}(x, y, z)$ be a vector-valued function in three-dimensional space. Let $\vec{r}(x, y) = (x, y, g(x, y))$, and R be the region where the surface S is projected onto

the xy-plane, i. e.,
$$R = \{(x,y): a \leq x \leq b, c \leq y \leq d\}.$$
Then the surface integral of $\vec{F} \cdot \vec{n}$ over the surface S is given by
$$\iint_S \vec{F} \cdot \vec{n} \, dA = \iint_R \vec{F}(\vec{r}(x,y)) \cdot \frac{\partial \vec{r}}{\partial x} \times \frac{\partial \vec{r}}{\partial y} dxdy = \iint_R \vec{F}(x,y,g(x,y)) \cdot \frac{\partial \vec{r}}{\partial x} \times \frac{\partial \vec{r}}{\partial y} dxdy.$$

Since $\dfrac{\partial \vec{r}}{\partial x} = \left(1, 0, g_x(x,y)\right)$, $\dfrac{\partial \vec{r}}{\partial y} = \left(0, 1, g_y(x,y)\right)$ and $\dfrac{\partial \vec{r}}{\partial x} \times \dfrac{\partial \vec{r}}{\partial y} = \left(-g_x(x,y), -g_y(x,y), 1\right)$,
the formula can be rewritten as
$$\iint_S \vec{F} \cdot \vec{n} \, dA = \iint_R \vec{F}(x,y,g(x,y)) \cdot \left(-g_x(x,y), -g_y(x,y), 1\right) dxdy.$$
Therefore, if S is oriented downward, then
$$\iint_S \vec{F} \cdot \vec{n} \, dA = \iint_R \vec{F}(x,y,g(x,y)) \cdot \left(g_x(x,y), g_y(x,y), -1\right) dxdy.$$

In particular, if the surface S is given by $z = g(x,y)$, and its projection onto the xy-plane is $\{(x,y): x^2 + y^2 \leq a\}$, using polar coordinates, we have
$$x = r\cos\theta, \; y = r\sin\theta.$$
If S is oriented upward, then
$$\iint_S \vec{F} \cdot \vec{n} \, dA = \int_0^{2\pi} \int_0^a \vec{F}(x,y,g(x,y)) \cdot \left(-g_x(x,y), -g_y(x,y), 1\right)\Big|_{x=r\cos\theta, y=r\sin\theta} drd\theta.$$
If S is oriented downward, then
$$\iint_S \vec{F} \cdot \vec{n} \, dA = \int_0^{2\pi} \int_0^a \vec{F}(x,y,g(x,y)) \cdot \left(g_x(x,y), g_y(x,y), -1\right)\Big|_{x=r\cos\theta, y=r\sin\theta} drd\theta.$$

Consider the surface S to be a sphere, $x^2 + y^2 + z^2 = r^2$, and a given three-dimensional vector function $\vec{F}(x,y,z)$. The goal is to calculate the flux
$$\iint_S \vec{F} \cdot \vec{n} dA.$$
Let $\vec{r}(\theta, \varphi) = (r\cos\theta \sin\varphi, r\sin\theta \sin\varphi, r\cos\varphi)$. Then
$$\iint_S \vec{F} \cdot \vec{n} \, dA = \iint_R \vec{F}(\vec{r}(\theta, \varphi)) \cdot \frac{\partial \vec{r}}{\partial \varphi} \times \frac{\partial \vec{r}}{\partial \theta} d\theta d\varphi.$$
The cross product is
$$\frac{\partial \vec{r}}{\partial \varphi} \times \frac{\partial \vec{r}}{\partial \theta} = (r^2 \cos\theta \sin^2\varphi, r^2 \sin\theta \sin^2\varphi, r^2 \sin\varphi \cos\varphi).$$
Substituting this cross product into the integral expression, we get

9.5.4 For Vector Functions, Find the Flux across Surface S

$$\iint_S \vec{F} \cdot \vec{n}\, dA = \iint_R \vec{F}(\vec{r}(\theta, \varphi)) \cdot \frac{\partial \vec{r}}{\partial \varphi} \times \frac{\partial \vec{r}}{\partial \theta}\, d\theta d\varphi$$

$$= \iint_R \vec{F}(r\cos\theta \sin\varphi, r\sin\theta \sin\varphi, r\cos\varphi) \cdot$$

$$(r^2 \cos\theta \sin^2\varphi, r^2\sin\theta \sin^2\varphi, r^2\sin\varphi \cos\varphi)\, d\theta d\varphi,$$

where $R = \{(\theta, \varphi): 0 \leq \theta \leq 2\pi, 0 \leq \varphi \leq \pi\}$.

The formulas above present how to calculate the surface integral of a vector function over a surface in different contexts. Two important theorems are closely related to the calculation of surface flux: Stokes' Theorem and Gauss's Theorem. When the surface is an open surface, Stokes' Theorem serves as a crucial bridge for computing the flux through open surfaces and for calculating the line integral along closed space curves. When the surface is a closed surface, Gauss's Theorem becomes the key tool for computing the flux through closed surfaces and for evaluating triple integrals of scalar functions. Therefore, whether the surface is open or closed, calculating the flux across the surface is an important step, and readers should become well-versed in various types of problems.

Question Types:

Type 1.

Given the surface $S: \vec{r}(s, t)$, $\forall\, a \leq s \leq b, c \leq t \leq d$, where S is oriented upward, and $\vec{F}(x, y, z)$ is a vector function in three-dimensional space, find the flux through surface S,

$$\iint_S \vec{F} \cdot \vec{n}\, dA = ?$$

Problem-Solving Process:

Step 1.

Let $R = \{(s, t): a \leq s \leq b, c \leq t \leq d\}$ then $\iint_S \vec{F} \cdot \vec{n}\, dA = \iint_R \vec{F} \cdot \frac{\partial \vec{r}}{\partial s} \times \frac{\partial \vec{r}}{\partial t}\, dsdt$

Step 2.

Find $\frac{\partial \vec{r}}{\partial s} \times \frac{\partial \vec{r}}{\partial t} = ?$, $\iint_R \vec{F} \cdot \frac{\partial \vec{r}}{\partial s} \times \frac{\partial \vec{r}}{\partial t}\, dsdt = ?$

If S is oriented downward, then calculate:

$$-\iint_R \vec{F} \cdot \frac{\partial \vec{r}}{\partial s} \times \frac{\partial \vec{r}}{\partial t}\, dsdt = ?$$

Type 2.

Given the surface $S: z = g(x,y), \forall\, a \le x \le b, c \le y \le d$, S is upward oriented and $\vec{F}(x,y,z)$ is a vector function in three-dimensional space. Find the flux through the surface S,

$$\iint_S \vec{F}\cdot\vec{n}\,dA = ?$$

Problem-Solving Process:

Step1.

Let $\vec{r}(x,y) = (x,y,g(x,y))$ and $R = \{(x,y): a \le x \le b, c \le y \le d\}$ then

$$\iint_S \vec{F}\cdot\vec{n}\,dA = \iint_R \vec{F}\cdot\frac{\partial \vec{r}}{\partial x}\times\frac{\partial \vec{r}}{\partial y}\,dxdy = \iint_R \vec{F}(x,y,g(x,y))\cdot\frac{\partial \vec{r}}{\partial x}\times\frac{\partial \vec{r}}{\partial y}\,dxdy$$

Step2.

$$\because \frac{\partial \vec{r}}{\partial x} = (1,0,g_x(x,y)) \text{ and } \frac{\partial \vec{r}}{\partial y} = (0,1,g_y(x,y)) \therefore \frac{\partial \vec{r}}{\partial x}\times\frac{\partial \vec{r}}{\partial y} = (-g_x(x,y),-g_y(x,y),1)$$

Step3.

$$\therefore \iint_S \vec{F}\cdot\vec{n}\,dA = \iint_R \vec{F}(x,y,g(x,y))\cdot(-g_x(x,y),-g_y(x,y),1)\,dxdy$$

Step4.

By Fubini's Theorem,

$$\iint_R \vec{F}(x,y,g(x,y))\cdot(-g_x(x,y),-g_y(x,y),1)\,dxdy$$
$$= \int_a^b\int_c^d \vec{F}(x,y,g(x,y))\cdot(-g_x(x,y),-g_y(x,y),1)\,dxdy$$

If S is oriented downward, then

$$\iint_S \vec{F}\cdot\vec{n}\,dA = -\int_a^b\int_c^d \vec{F}(x,y,g(x,y))\cdot(-g_x(x,y),-g_y(x,y),1)\,dxdy$$

Examples:

Find $\iint_S \vec{F}\cdot\vec{n}\,dA = ?$

(I) If $\vec{F}(x,y,z) = (6y^2, 0, 2x^2 + \sqrt{xy})$, the surface S is the upward-oriented region of $x+y+z=1$ in the first octant.

Let

$$\vec{r}(x,y,z) = (x,y,1-x-y),$$

and let R be the closed region obtainedby projecting surface S onto the xy-plane.

Then $R = \{(x,y): x+y \leq 1, x \geq 0, y \geq 0\}$ and $\iint_S \vec{F} \cdot \vec{n}\, dA = \iint_R \vec{F} \cdot \dfrac{\partial \vec{r}}{\partial x} \times \dfrac{\partial \vec{r}}{\partial y}\, dxdy$

$\because \dfrac{\partial \vec{r}}{\partial x} = (1,0,-1)$ and $\dfrac{\partial \vec{r}}{\partial y} = (0,1,-1)$ $\quad \therefore \dfrac{\partial \vec{r}}{\partial x} \times \dfrac{\partial \vec{r}}{\partial y} = (1,1,1)$

$\because \vec{F} \cdot \dfrac{\partial \vec{r}}{\partial x} \times \dfrac{\partial \vec{r}}{\partial y} = (6y^2, 0, 2x^2) \cdot \dfrac{\partial \vec{r}}{\partial x} \times \dfrac{\partial \vec{r}}{\partial y} = 2(x^2 + 3y^2), \; \forall (x,y,z) \in S$

$\therefore \iint_S \vec{F} \cdot \vec{n}\, dA = \iint_R \vec{F} \cdot \dfrac{\partial \vec{r}}{\partial x} \times \dfrac{\partial \vec{r}}{\partial y}\, dxdy = 2\iint_R (x^2 + 3y^2)\, dxdy$

$= 2\int_0^1 \int_0^{1-y} x^2 + 3y^2\, dxdy = 2\int_0^1 \left(\dfrac{x^3}{3} + 3y^2 x\right)\Big|_{x=0}^{x=1-y} dy$

$= 2\int_0^1 \left(\dfrac{(1-y)^3}{3} + 3y^2(1-y)\right) dy = \dfrac{2}{3}$

(II) If $\vec{F}(x,y,z) = (e^y, e^x + 6y, 12y),$ the surface S is the upward-oriented region of $x + y + z = 6$ in the first octant.

Let
$$\vec{r}(x,y,z) = (x, y, 6-x-y)$$
and let R be the closed region obtained by projecting surface S onto the xy-plane.

Then $R = \{(x,y): x+y \leq 6, x \geq 0, y \geq 0\}$ and $\iint_S \vec{F} \cdot \vec{n}\, dA = \iint_R \vec{F} \cdot \dfrac{\partial \vec{r}}{\partial x} \times \dfrac{\partial \vec{r}}{\partial y}\, dxdy$

$\because \dfrac{\partial \vec{r}}{\partial x} = (1,0,-1)$ and $\dfrac{\partial \vec{r}}{\partial y} = (0,1,-1)$ $\quad \therefore \dfrac{\partial \vec{r}}{\partial x} \times \dfrac{\partial \vec{r}}{\partial y} = (1,1,1)$

$\because \vec{F} \cdot \dfrac{\partial \vec{r}}{\partial x} \times \dfrac{\partial \vec{r}}{\partial y} = e^y + e^x + 18y, \; \forall (x,y,z) \in S$

$\therefore \iint_S \vec{F} \cdot \vec{n}\, dA = \iint_R \vec{F} \cdot \dfrac{\partial \vec{r}}{\partial x} \times \dfrac{\partial \vec{r}}{\partial y}\, dxdy = \iint_R e^y + e^x + 18y\, dxdy$

$= \int_0^6 \int_0^{6-x} e^y + e^x + 18y\, dydx = 634 + 2e^6$

-

Type 3.

Given the surface $S: z = g(x,y)$, where S is upward oriented and its projection onto the xy-plane is defined by $\{(x,y): x^2 + y^2 \leq a^2\}$. Let the vector field be
$$\vec{F} = (f_1(x,y,z), f_2(x,y,z), f_3(x,y,z)).$$
Find the flux through the surface S,
$$\iint_S \vec{F} \cdot \vec{n}\, dA = ?$$

Remark

The surface S is typically an ellipsoid, sphere, elliptic cone, cone, or elliptic paraboloid.

Problem-Solving Process:

Step 1.

Let $\vec{r}(x,y) = (x, y, g(x,y))$ and $R = \{(x,y): x^2 + y^2 \leq a^2\}$ then
$$\iint_S \vec{F} \cdot \vec{n}\, dA = \iint_R \vec{F} \cdot \frac{\partial \vec{r}}{\partial x} \times \frac{\partial \vec{r}}{\partial y}\, dxdy$$

Step 2.

$$\because \frac{\partial \vec{r}}{\partial x} = (1, 0, g_x(x,y)), \quad \frac{\partial \vec{r}}{\partial y} = (0, 1, g_y(x,y)) \quad \therefore \frac{\partial \vec{r}}{\partial x} \times \frac{\partial \vec{r}}{\partial y} = (-g_x(x,y), -g_y(x,y), 1)$$

$$\therefore \iint_S \vec{F} \cdot \vec{n}\, dA = \iint_R \vec{F} \cdot \frac{\partial \vec{r}}{\partial x} \times \frac{\partial \vec{r}}{\partial y}\, dxdy = \iint_R \vec{F} \cdot (-g_x(x,y), -g_y(x,y), 1)\, dxdy$$

$$= \iint_R (-f_1 \cdot g_x(x,y) - f_2 \cdot g_y(x,y) + f_3)\big|_{z=g(x,y)}\, dxdy$$

Step 3.

Let $x = r\cos\theta, y = r\sin\theta$ then $\{(x,y): x^2 + y^2 \leq a^2\} = \{(r,\theta): 0 \leq r \leq a, 0 \leq \theta \leq 2\pi\}$

and $dxdy = \begin{Vmatrix} \frac{\partial x}{\partial r} & \frac{\partial x}{\partial \theta} \\ \frac{\partial y}{\partial r} & \frac{\partial y}{\partial \theta} \end{Vmatrix} drd\theta = \begin{Vmatrix} \cos\theta & -r\sin\theta \\ \sin\theta & r\cos\theta \end{Vmatrix} drd\theta = rdrd\theta$

Step 4.

$$\iint_R (-f_1 \cdot g_x(x,y) - f_2 \cdot g_y(x,y) + f_3)\big|_{z=g(x,y)}\, dxdy$$

$$= \int_0^{2\pi} \int_0^a -f_1(r\cos\theta, r\sin\theta, g(r\cos\theta, r\sin\theta)) \cdot g_x(r\cos\theta, r\sin\theta) \cdot r$$
$$\quad - f_2(r\cos\theta, r\sin\theta, g(r\cos\theta, r\sin\theta)) \cdot g_y(r\cos\theta, r\sin\theta) \cdot r$$
$$\quad + f_3(r\cos\theta, r\sin\theta, g(r\cos\theta, r\sin\theta)) \cdot rdrd\theta$$

If S is oriented downward, then

9.5.4 For Vector Functions, Find the Flux across Surface S

$$\iint_S \vec{F} \cdot \vec{n}\, dA$$

$$= -\int_0^{2\pi}\int_0^a -f_1(r\cos\theta, r\sin\theta, g(r\cos\theta, r\sin\theta)) \cdot g_x(r\cos\theta, r\sin\theta) \cdot r$$
$$\quad -f_2(r\cos\theta, r\sin\theta, g(r\cos\theta, r\sin\theta)) \cdot g_y(r\cos\theta, r\sin\theta) \cdot r$$
$$\quad +f_3(r\cos\theta, r\sin\theta, g(r\cos\theta, r\sin\theta)) \cdot r\, dr d\theta$$

Examples:

Find $\iint_S \vec{F} \cdot \vec{n}\, dA =?$

(I) If $\vec{F}(x, y, z) = (yz, -1, 1)$ and the surface S is upward oriented:

$$z = \sqrt{x^2 + y^2}, \ \forall\ x^2 + y^2 \leq a^2.$$

Let $R = \{(x, y): x^2 + y^2 \leq a^2\}$ and $x = r\cos\theta, y = r\sin\theta$ then

$$\iint_S \vec{F} \cdot \vec{n}\, dA = \iint_R \vec{F} \cdot \frac{\partial \vec{r}}{\partial x} \times \frac{\partial \vec{r}}{\partial y} dxdy = \iint_{x^2+y^2 \leq a^2} -xy + y(x^2+y^2)^{-\frac{1}{2}} + 1\, dxdy$$

$$= \int_0^{2\pi}\int_0^a (-r^2\cos\theta\sin\theta + \sin\theta + 1)\, rdrd\theta = \pi a^2$$

Type 4.

Given the spherical surface $S: x^2 + y^2 + z^2 = r^2$, let $\vec{F}(x, y, z)$ be a vector function in three-dimensional space. Find the flux through the surface S,

$$\iint_S \vec{F} \cdot \vec{n} dA =?$$

Problem-Solving Process:

Step1.

Let $\vec{r}(\theta, \varphi) = (r\cos\theta\sin\varphi, r\sin\theta\sin\varphi, r\cos\varphi)$ and $R = \{(\theta, \varphi): 0 \leq \theta \leq 2\pi, 0 \leq \varphi \leq \pi\}$

then $\iint_S \vec{F} \cdot \vec{n} dA = \iint_R \vec{F}(r\cos\theta\sin\varphi, r\sin\theta\sin\varphi, r\cos\varphi) \cdot \frac{\partial \vec{r}}{\partial \varphi} \times \frac{\partial \vec{r}}{\partial \theta} d\theta d\varphi$

Step2.

$\because \dfrac{\partial \vec{r}}{\partial \theta} = (-r\sin\theta\sin\varphi, r\cos\theta\sin\varphi, 0)$ and $\dfrac{\partial \vec{r}}{\partial \varphi} = (r\cos\theta\cos\varphi, r\sin\theta\cos\varphi, -r\sin\varphi)$

$\therefore \dfrac{\partial \vec{r}}{\partial \varphi} \times \dfrac{\partial \vec{r}}{\partial \theta} = (r^2\cos\theta\sin^2\varphi, r^2\sin\theta\sin^2\varphi, r^2\sin\varphi\cos\varphi)$

Step3.

Substituting $\dfrac{\partial \vec{r}}{\partial \varphi} \times \dfrac{\partial \vec{r}}{\partial \theta} = (r^2 \cos\theta \sin^2\varphi, r^2 \sin\theta \sin^2\varphi, r^2 \sin\varphi \cos\varphi)$

into $\displaystyle\iint_R \vec{F}(r\cos\theta \sin\varphi, r\sin\theta \sin\varphi, r\cos\varphi) \cdot \dfrac{\partial \vec{r}}{\partial \varphi} \times \dfrac{\partial \vec{r}}{\partial \theta} d\theta d\varphi$

$\therefore \displaystyle\iint_S \vec{F} \cdot \vec{n}\, dA = \iint_R \vec{F}(\vec{r}(\theta, \varphi)) \cdot \dfrac{\partial \vec{r}}{\partial \varphi} \times \dfrac{\partial \vec{r}}{\partial \theta} d\theta d\varphi$

$= \displaystyle\iint_R \vec{F}(r\cos\theta \sin\varphi, r\sin\theta \sin\varphi, r\cos\varphi)$

$\cdot (r^2 \cos\theta \sin^2\varphi, r^2 \sin\theta \sin^2\varphi, r^2 \sin\varphi \cos\varphi) d\theta d\varphi$

Examples:

Find $\displaystyle\iint_S \vec{F} \cdot \vec{n}\, dA = ?$, where $S: x^2 + y^2 + z^2 = r^2$.

(I) Assume $\vec{F}(x, y, z) = (z, y, x)$. Let $R = \{(\theta, \varphi): 0 \leq \theta \leq 2\pi, 0 \leq \varphi \leq \pi\}$ then

$\displaystyle\iint_S \vec{F} \cdot \vec{n}\, dA = \iint_R (r\cos\varphi, r\sin\theta \sin\varphi, r\cos\theta \sin\varphi)$

$\cdot (r^2 \cos\theta \sin^2\varphi, r^2 \sin\theta \sin^2\varphi, r^2 \sin\varphi \cos\varphi) d\theta d\varphi = \dfrac{4\pi r^3}{3}$

(II) Assume $\vec{F}(x, y, z) = (x, y, z)$. Let $R = \{(\theta, \varphi): 0 \leq \theta \leq 2\pi, 0 \leq \varphi \leq \pi\}$ then

$\displaystyle\iint_S \vec{F} \cdot \vec{n}\, dA = \iint_R (r\cos\theta \sin\varphi, r\sin\theta \sin\varphi, r\cos\varphi)$

$\cdot (r^2 \cos\theta \sin^2\varphi, r^2 \sin\theta \sin^2\varphi, r^2 \sin\varphi \cos\varphi) d\theta d\varphi = 4\pi r^3$

Example 1.

Assume $\vec{F} = (y + 2, 2 + 4x, xyz)$. Find the flux through the upward-oriented surface $S: y = x^2,\ 0 \leq x \leq 2, 0 \leq z \leq 2$, $\displaystyle\iint_S \vec{F} \cdot \vec{n}\, dA = ?$

【Solution】

$\because S$ is oriented upward

Let $\vec{r}(x, y, z) = (x, x^2, z)$ and let R represent the region projected onto the xz-plane

then $R = \{(x, z): 0 \leq x \leq 2, 0 \leq z \leq 2\}$ and $\displaystyle\iint_S \vec{F} \cdot \vec{n}\, dA = \iint_R \vec{F} \cdot \dfrac{\partial \vec{r}}{\partial x} \times \dfrac{\partial \vec{r}}{\partial z} dx dz$

$\because \dfrac{\partial \vec{r}}{\partial x} = (1, 2x, 0)$ and $\dfrac{\partial \vec{r}}{\partial z} = (0,0,1)$ $\therefore \dfrac{\partial \vec{r}}{\partial x} \times \dfrac{\partial \vec{r}}{\partial z} = (2x, -1, 0)$

$\Rightarrow \vec{F} \cdot \dfrac{\partial \vec{r}}{\partial x} \times \dfrac{\partial \vec{r}}{\partial z} = (y+2, 2+4x, xyz) \cdot (2x, -1, 0) = 2xy + 4x - 2 - 4x = 2x^3 - 2,$

$\forall (x, y, z) \in S$

$\therefore \iint_S \vec{F} \cdot \vec{n}\, dA = \iint_R \vec{F} \cdot \dfrac{\partial \vec{r}}{\partial x} \times \dfrac{\partial \vec{r}}{\partial z}\, dxdz = \int_0^2 \int_0^2 2x^3 - 2\, dxdz = 8$

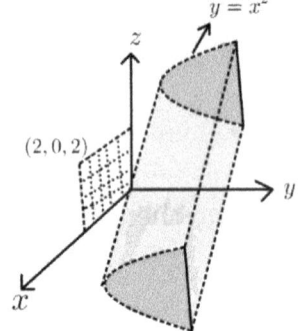

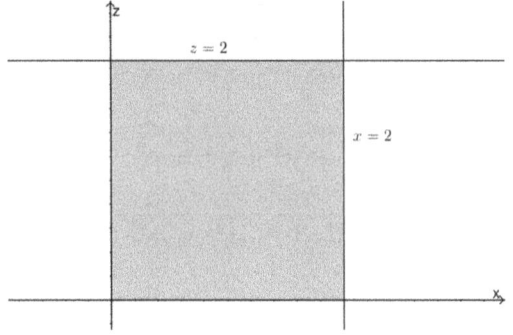

Example 2.

Assume $\vec{F} = (yz, -1, 1)$. Find the flux through the downward-oriented surface $S: z = \sqrt{x^2 + y^2},\ x^2 + y^2 \leq a^2,\ \iint_S \vec{F} \cdot \vec{n}\, dA = ?$

【Solution】

$\because$ S is oriented downward

Let $\vec{r}(x, y, z) = \left(x, y, (x^2 + y^2)^{\frac{1}{2}}\right)$ and let R represent the region projected onto the xy-plane. Then

$R = \{(x, y): x^2 + y^2 \leq a^2\}$ and $\iint_S \vec{F} \cdot \vec{n}\, dA = -\iint_R \vec{F} \cdot \dfrac{\partial \vec{r}}{\partial x} \times \dfrac{\partial \vec{r}}{\partial y}\, dxdy$

$\because \dfrac{\partial \vec{r}}{\partial x} = \left(1, 0, x(x^2 + y^2)^{-\frac{1}{2}}\right)$ and $\dfrac{\partial \vec{r}}{\partial y} = \left(0, 1, y(x^2 + y^2)^{-\frac{1}{2}}\right)$

$$\therefore \frac{\partial \vec{r}}{\partial x} \times \frac{\partial \vec{r}}{\partial y} = \left(-x(x^2+y^2)^{-\frac{1}{2}}, -y(x^2+y^2)^{-\frac{1}{2}}, 1\right)$$

$$\Rightarrow \vec{F} \cdot \frac{\partial \vec{r}}{\partial x} \times \frac{\partial \vec{r}}{\partial y} = (yz, -1, 1) \cdot \left(-x(x^2+y^2)^{-\frac{1}{2}}, -y(x^2+y^2)^{-\frac{1}{2}}, 1\right)$$

$$= -xy + y(x^2+y^2)^{-\frac{1}{2}} + 1, \ \forall (x,y,z) \in S$$

$$\therefore \iint_S \vec{F} \cdot \vec{n} \, dA = -\iint_R \vec{F} \cdot \frac{\partial \vec{r}}{\partial x} \times \frac{\partial \vec{r}}{\partial y} \, dxdy = \iint_{x^2+y^2 \leq a^2} xy - y(x^2+y^2)^{-\frac{1}{2}} - 1 \, dxdy$$

Let $x = r\cos\theta$, $y = r\sin\theta$ then $\{(x,y): x^2+y^2 \leq a^2\} = \{(r,\theta): 0 \leq r \leq a, 0 \leq \theta \leq 2\pi\}$

$$\therefore \iint_{x^2+y^2 \leq a^2} xy - y(x^2+y^2)^{-\frac{1}{2}} - 1 \, dxdy = \int_0^{2\pi} \int_0^a (r^2 \cos\theta \sin\theta - \sin\theta - 1) \, rdrd\theta$$

$$= -\pi a^2$$

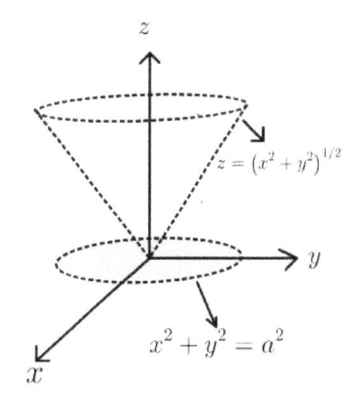

Example 3.

Assume $\vec{F} = \left(6y^2 - \sqrt{xy}, 0, 2x^2 + \sqrt{xy}\right)$. Let S be the region defined by $x + y + z = 1$ in the first octant. Find the flux through the upward-oriented plane S, $\iint_S \vec{F} \cdot \vec{n} \, dA = ?$

【Solution】

∵ S is oriented upward

Let $\vec{r}(x,y,z) = (x, y, 1-x-y)$ and let R represent the region projected onto the xy-plane. Then

$R = \{(x, y): x + y \leq 1, x \geq 0, y \geq 0\}$ and $\iint_S \vec{F} \cdot \vec{n} \, dA = \iint_R \vec{F} \cdot \dfrac{\partial \vec{r}}{\partial x} \times \dfrac{\partial \vec{r}}{\partial y} dxdy$

$\because \dfrac{\partial \vec{r}}{\partial x} = (1,0,-1)$ and $\dfrac{\partial \vec{r}}{\partial y} = (0,1,-1)$ $\therefore \dfrac{\partial \vec{r}}{\partial x} \times \dfrac{\partial \vec{r}}{\partial y} = (1,1,1)$

$\because \vec{F} \cdot \dfrac{\partial \vec{r}}{\partial x} \times \dfrac{\partial \vec{r}}{\partial y} = (6y^2 - \sqrt{xy}, 0, 2x^2 + \sqrt{xy}) \cdot \dfrac{\partial \vec{r}}{\partial x} \times \dfrac{\partial \vec{r}}{\partial y} = 2(x^2 + 3y^2), \ \forall (x,y,z) \in S$

$\therefore \iint_S \vec{F} \cdot \vec{n} \, dA = \iint_R \vec{F} \cdot \dfrac{\partial \vec{r}}{\partial x} \times \dfrac{\partial \vec{r}}{\partial y} dxdy = 2 \iint_R x^2 + 3y^2 \, dxdy = 2 \int_0^1 \int_0^{1-y} x^2 + 3y^2 \, dxdy$

$= 2 \int_0^1 \left(\dfrac{x^3}{3} + 3y^2 x \right) \Big|_{x=0}^{x=1-y} dy = 2 \int_0^1 \dfrac{(1-y)^3}{3} + 3y^2(1-y) \, dy = \dfrac{2}{3}$

Example 4.

Assume $\vec{F} = (xy, yz, xz)$ and S is the part of the paraboloid $z = 4 - x^2 - y^2$ that lies above the square $0 \leq x \leq 1, 0 \leq y \leq 1$ with upward orientation. Find $\iint_S \vec{F} \cdot \vec{n} \, dA$.

【Solution】

$\because S$ is oriented upward

Let $\vec{r}(x, y, z) = (x, y, 4 - x^2 - y^2)$ and $R = \{(x, y): 0 \leq x \leq 1, 0 \leq y \leq 1\}$ then

$\iint_S \vec{F} \cdot \vec{n} \, dA = \iint_R \vec{F} \cdot \dfrac{\partial \vec{r}}{\partial x} \times \dfrac{\partial \vec{r}}{\partial y} dxdy$

$\because \dfrac{\partial \vec{r}}{\partial x} = (1,0,-2x)$ and $\dfrac{\partial \vec{r}}{\partial y} = (0,1,-2y)$ $\therefore \dfrac{\partial \vec{r}}{\partial x} \times \dfrac{\partial \vec{r}}{\partial y} = (2x, 2y, 1)$

$$\because \vec{F} \cdot \frac{\partial \vec{r}}{\partial x} \times \frac{\partial \vec{r}}{\partial y} = (xy, yz, xz) \cdot (2x, 2y, 1) = 2x^2y + 2y^2z + xz = z(2y^2 + x) + 2x^2y$$

$$= (4 - x^2 - y^2)(2y^2 + x) + +2x^2y = 8y^2 + 4x - 2x^2y^2 - x^3 - 2y^4 - xy^2 + 2x^2y,$$

$$\forall (x, y, z) \in S$$

$$\therefore \iint_S \vec{F} \cdot \vec{n}\, dA = \iint_R \vec{F} \cdot \frac{\partial \vec{r}}{\partial x} \times \frac{\partial \vec{r}}{\partial y} dxdy$$

$$= \int_0^1 \int_0^1 8y^2 + 4x - 2x^2y^2 - x^3 - 2y^4 - xy^2 + 2x^2y\, dydx = \int_0^1 \frac{34}{15} + \frac{11x}{3} + \frac{x^2}{3} - x^3 dx$$

$$= \frac{34}{15} + \frac{11}{6} + \frac{1}{9} - \frac{1}{4} = \frac{713}{180}$$

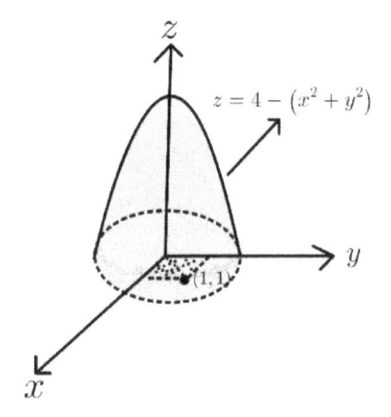

Example 5.

Assume $\vec{F} = (x, -z, y)$. Find the flux through the surface $S: x^2 + y^2 + z^2 = r^2$ in the first octant oriented inward, $\iint_S \vec{F} \cdot \vec{n}\, dA = ?$

【Solution】

$\because$ S is oriented inward

Let $\vec{r}(\theta, \varphi) = (r\cos\theta \sin\varphi, r\sin\theta \sin\varphi, r\cos\varphi)$ and $R = \left\{(\theta, \varphi): 0 \le \theta \le \frac{\pi}{2}, 0 \le \varphi \le \frac{\pi}{2}\right\}$

then $\oiint_S \vec{F} \cdot \vec{n}dA = -\iint_R \vec{F} \cdot \frac{\partial \vec{r}}{\partial \varphi} \times \frac{\partial \vec{r}}{\partial \theta} d\theta d\varphi$

$\because \frac{\partial \vec{r}}{\partial \theta} = (-r\sin\theta \sin\varphi, r\cos\theta \sin\varphi, 0)$ and $\frac{\partial \vec{r}}{\partial \varphi} = (r\cos\theta \cos\varphi, r\sin\theta \cos\varphi, -r\sin\varphi)$

$$\therefore \frac{\partial \vec{r}}{\partial \varphi} \times \frac{\partial \vec{r}}{\partial \theta} = (r^2 \cos \theta \sin^2 \varphi, r^2 \sin \theta \sin^2 \varphi, r^2 \sin \varphi \cos \varphi)$$

$$\therefore \vec{F} \cdot \frac{\partial \vec{r}}{\partial \varphi} \times \frac{\partial \vec{r}}{\partial \theta}$$

$$= (r\cos \theta \sin \varphi, -r\cos \varphi, r\sin \theta \sin \varphi) \cdot (r^2 \cos \theta \sin^2 \varphi, r^2 \sin \theta \sin^2 \varphi, r^2 \sin \varphi \cos \varphi)$$

$$= r^3(\cos^2 \theta \sin^3 \varphi - \sin \theta \sin^2 \varphi \cos \varphi + \sin \theta \sin^2 \varphi \cos \varphi) = r^3 \cos^2 \theta \sin^3 \varphi$$

$$\therefore \oiint_S \vec{F} \cdot \vec{n} dA = -r^3 \int_0^{\frac{\pi}{2}} \int_0^{\frac{\pi}{2}} \cos^2 \theta \sin^3 \varphi \, d\varphi d\theta = -r^3 \int_0^{\frac{\pi}{2}} \cos^2 \theta \, d\theta \int_0^{\frac{\pi}{2}} \sin^3 \varphi \, d\varphi$$

$$= -r^3 \int_0^{\frac{\pi}{2}} \frac{1 + \cos 2\theta}{2} d\theta \int_0^{\frac{\pi}{2}} \sin^3 \varphi \, d\varphi = -r^3 \cdot \frac{\pi}{4} \cdot \left(-\cos \varphi + \frac{\cos^3 \varphi}{3} \right) \Big|_0^{\frac{\pi}{2}} = \frac{\pi r^3}{6}$$

Example 6.

Let $\vec{F} = (0, y, -z)$. Assume S consists of the paraboloid S_1: $y = x^2 + z^2, 0 \le y \le 1$ and the disk S_2: $x^2 + z^2 \le 1, y = 1$ with outward orientation. Find $\oiint_S \vec{F} \cdot \vec{n} dA =?$

【Solution】

Let S_1: $y = x^2 + z^2, 0 \le y \le 1$, S_2: $x^2 + z^2 \le 1, y = 1$

$\because S_1$ is oriented in the negative direction of the y-axis

Let $\vec{r}(x, y, z) = (x, x^2 + z^2, z)$ and let R be the closed region of the projection of the surface S_1 onto the xz-plane. Then

$$R = \{(x, y): x^2 + z^2 \le 1\} \text{ and } \iint_{S_1} \vec{F} \cdot \vec{n} \, dA = -\iint_R \vec{F} \cdot \frac{\partial \vec{r}}{\partial x} \times \frac{\partial \vec{r}}{\partial z} dxdz$$

$\because \frac{\partial \vec{r}}{\partial x} = (1, 2x, 0)$ and $\frac{\partial \vec{r}}{\partial z} = (0, 2z, 1)$ $\therefore \frac{\partial \vec{r}}{\partial x} \times \frac{\partial \vec{r}}{\partial z} = (2x, -1, 2z)$

$\therefore \vec{F} \cdot \frac{\partial \vec{r}}{\partial x} \times \frac{\partial \vec{r}}{\partial z} = (0, y, -z) \cdot (2x, -1, 2z) = -y - 2z^2 = -x^2 - 3z^2, \ \forall (x, y, z) \in S_1$

$\therefore \iint_{S_1} \vec{F} \cdot \vec{n} \, dA = -\iint_R \vec{F} \cdot \frac{\partial \vec{r}}{\partial x} \times \frac{\partial \vec{r}}{\partial z} dxdz = \iint_{x^2+z^2 \le 1} x^2 + 3z^2 dxdz$

Let $x = r \cos \theta, z = r \sin \theta$ then $\{(x, y): x^2 + z^2 \le 1\} = \{(r, \theta): 0 \le r \le 1, 0 \le \theta \le 2\pi\}$

and $dxdz = \begin{Vmatrix} \frac{\partial x}{\partial r} & \frac{\partial x}{\partial \theta} \\ \frac{\partial z}{\partial r} & \frac{\partial z}{\partial \theta} \end{Vmatrix} drd\theta = \begin{Vmatrix} \cos\theta & -r\sin\theta \\ \sin\theta & r\cos\theta \end{Vmatrix} drd\theta = rdrd\theta$

$\therefore \iint_{x^2+z^2\leq 1} x^2 + 3z^2 dxdz = \int_0^{2\pi}\int_0^1 r^3\cos^2\theta + 3r^3\sin^2\theta \, drd\theta$

$= \int_0^{2\pi} \frac{1}{4}\left(\frac{1+\cos\theta}{2}\right) + \frac{3}{4}\left(\frac{1-\cos\theta}{2}\right) d\theta = \pi$

Let $\vec{s}(x,y,z) = (x,1,z)$

$\therefore \frac{\partial \vec{s}}{\partial x} = (1,0,0)$ and $\frac{\partial \vec{s}}{\partial z} = (0,0,1) \quad \therefore \frac{\partial \vec{r}}{\partial x} \times \frac{\partial \vec{r}}{\partial z} = (0,-1,0)$

$\therefore \iint_{S_2} \vec{F}\cdot\vec{n}dA = \iint_{S_2} dA = \iint_{x^2+z^2\leq 1} (0,y,-z)\cdot(0,-1,0)dxdz = -\pi$

$\therefore \oiint_S \vec{F}\cdot\vec{n}dA = \iint_{S_1}\vec{F}\cdot\vec{n}dA + \iint_{S_2}\vec{F}\cdot\vec{n}dA = 0$

$y = x^2 + z^2$

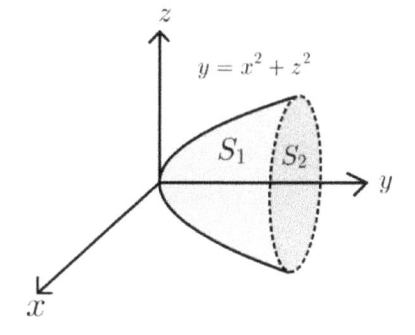

$S_1 \quad S_2$

Example 7.

Assume $\vec{F} = (e^y, e^x + 6y, 12y)$. Let S be the region defined by $x + y + z = 6$ in the first octant. Find the flux through the upward-oriented plane S, $\iint_S \vec{F}\cdot\vec{n}\,dA = ?$

【Solution】

$\because$ S is oriented upward

Let $\vec{r}(x,y,z) = (x,y,6-x-y)$ and let R be the closed region of the projection of S onto the xy-plane. Then

$R = \{(x,y): x+y \leq 6, x \geq 0, y \geq 0\}$ and $\iint_S \vec{F} \cdot \vec{n} \, dA = \iint_R \vec{F} \cdot \dfrac{\partial \vec{r}}{\partial x} \times \dfrac{\partial \vec{r}}{\partial y} \, dxdy$

$\because \dfrac{\partial \vec{r}}{\partial x} = (1,0,-1)$ and $\dfrac{\partial \vec{r}}{\partial y} = (0,1,-1)$ $\therefore \dfrac{\partial \vec{r}}{\partial x} \times \dfrac{\partial \vec{r}}{\partial y} = (1,1,1)$

$\because \vec{F} \cdot \dfrac{\partial \vec{r}}{\partial x} \times \dfrac{\partial \vec{r}}{\partial y} = (e^y, e^x + 6y, 12y) \cdot (1,1,1) = e^y + e^x + 18y, \quad \forall (x,y,z) \in S$

$\therefore \iint_S \vec{F} \cdot \vec{n} \, dA = \iint_R \vec{F} \cdot \dfrac{\partial \vec{r}}{\partial x} \times \dfrac{\partial \vec{r}}{\partial y} \, dxdy = \iint_R e^y + e^x + 18y \, dxdy$

$= \displaystyle\int_0^6 \int_0^{6-x} e^y + e^x + 18y \, dydx = 634 + 2e^6$

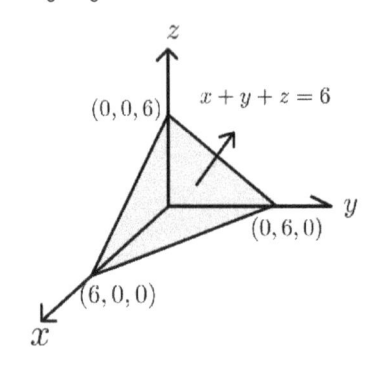

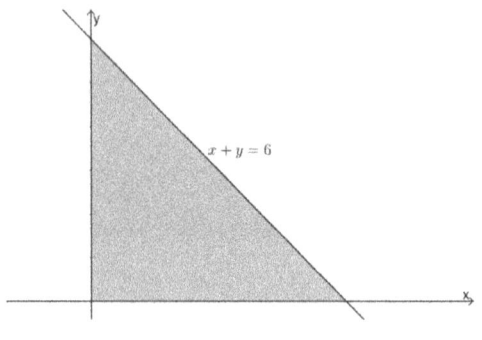

Example 8.

Assume $\vec{F} = (y, x, z)$. The surface S is the closed surface formed by $z = 4 - x^2 - y^2$, and $z = 0$. Find the flux through the surface S, $\oiint_S \vec{F} \cdot \vec{n} \, dA = ?$

【Solution】

Let $S_1: z = 4 - x^2 - y^2, z \geq 0$

$\because S_1$ is oriented upward

Let $\vec{r}(x,y,z) = (x, y, 4 - x^2 - y^2)$ and let R be the closed region of the projection of S onto the xy-plane. Then

$R = \{(x,y): x^2 + y^2 \leq 4\}$ and $\iint_{S_1} \vec{F} \cdot \vec{n} \, dA = \iint_R \vec{F} \cdot \dfrac{\partial \vec{r}}{\partial x} \times \dfrac{\partial \vec{r}}{\partial y} \, dxdy$

$$\because \frac{\partial \vec{r}}{\partial x} = (1,0,-2x) \text{ and } \frac{\partial \vec{r}}{\partial y} = (0,1,-2y) \quad \therefore \frac{\partial \vec{r}}{\partial x} \times \frac{\partial \vec{r}}{\partial y} = (2x, 2y, 1)$$

$$\because \vec{F} \cdot \frac{\partial \vec{r}}{\partial x} \times \frac{\partial \vec{r}}{\partial y} = (y, x, z) \cdot (2x, 2y, 1) = 4xy + z = 4xy + 4 - x^2 - y^2, \quad \forall (x, y, z) \in S_1$$

$$\therefore \iint_{S_1} \vec{F} \cdot \vec{n} \, dA = \iint_{x^2+y^2 \leq 4} 4xy + 4 - x^2 - y^2 \, dxdy$$

Let $x = r \cos \theta$, $y = r \sin \theta$ then $\{(x, y): x^2 + y^2 \leq 4\} = \{(r, \theta): 0 \leq r \leq 2, 0 \leq \theta \leq 2\pi\}$

$$\therefore \iint_{x^2+y^2 \leq 4} 4xy + 4 - x^2 - y^2 \, dxdy = \int_0^{2\pi} \int_0^2 (4r^2 \cos \theta \sin \theta + 4 - r^2) r \, dr d\theta$$

$$= 2\pi \left(2r^2 - \frac{r^4}{4} \right) \Big|_{r=0}^{r=2} = 8\pi$$

Let $S_2: z = 0, x^2 + y^2 \leq 4$, $\because S_2$ is oriented downward

Let $\vec{s}(x, y, z) = (x, y, 0)$ then $\iint_{S_2} \vec{F} \cdot \vec{n} \, dA = -\iint_{S_2} \vec{F} \cdot \frac{\partial \vec{s}}{\partial x} \times \frac{\partial \vec{s}}{\partial y} \, dxdy$

$$\because \frac{\partial \vec{s}}{\partial x} = (1,0,0) \text{ and } \frac{\partial \vec{s}}{\partial y} = (0,1,0) \quad \therefore \frac{\partial \vec{s}}{\partial x} \times \frac{\partial \vec{s}}{\partial y} = (0,0,1)$$

$$\therefore \iint_{S_2} \vec{F} \cdot \vec{n} \, dA = \iint_{S_2} 0 \, dA = 0 \quad \therefore \oiint_S \vec{F} \cdot \vec{n} dA = \iint_{S_1} \vec{F} \cdot \vec{n} \, dA + \iint_{S_2} \vec{F} \cdot \vec{n} \, dA = 8\pi$$

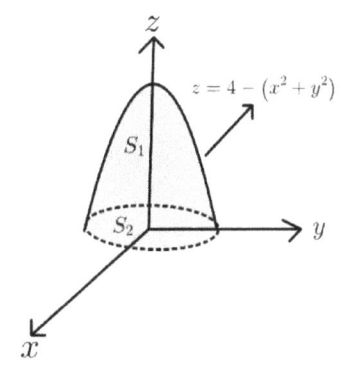

Example 9.

Let $\vec{F} = (x, y, z)$. Determine the flux across the triangle $S: (a, 0,0), (0, a, 0), (0,0, a)$,

$a > 0$, with upward orientation.

【Solution】

∵ S is oriented upward

Let $\vec{r}(x, y, z) = (x, y, a - x - y)$ and $R = \{(x, y): 0 \leq x \leq a, 0 \leq y \leq a - x\}$ then

$$\iint_S \vec{F} \cdot \vec{n}\, dA = \iint_R \vec{F} \cdot \frac{\partial \vec{r}}{\partial x} \times \frac{\partial \vec{r}}{\partial y} dxdy$$

∵ $\dfrac{\partial \vec{r}}{\partial x} = (1, 0, -1)$ and $\dfrac{\partial \vec{r}}{\partial y} = (0, 1, -1)$ ∴ $\dfrac{\partial \vec{r}}{\partial x} \times \dfrac{\partial \vec{r}}{\partial y} = (1, 1, 1)$

∴ $\vec{F} \cdot \dfrac{\partial \vec{r}}{\partial x} \times \dfrac{\partial \vec{r}}{\partial y} = (x, y, z) \cdot (1, 1, 1) = x + y + z = x + y + a - x - y = a$, $\forall (x, y, z) \in S$

$$\iint_R \vec{F} \cdot \frac{\partial \vec{r}}{\partial x} \times \frac{\partial \vec{r}}{\partial y} dxdy = \int_0^a \int_0^{a-x} a\, dydx = a \int_0^a a - x\, dx = a\left(ax - \frac{x^2}{2}\right)\bigg|_0^a = \frac{a^3}{2}$$

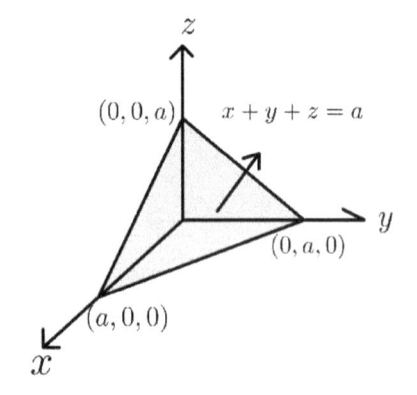

Example 10.

Let $\vec{F} = (x^2, -y^2, 0)$. Determine the flux across the triangle S: $(a, 0, 0), (0, a, 0), (0, 0, a)$, $a > 0$, with upward orientation.

【Solution】

∵ S is oriented upward

Let $\vec{r}(x, y, z) = (x, y, a - x - y)$ and $R = \{(x, y): 0 \leq x \leq a, 0 \leq y \leq a - x\}$ then

$$\iint_S \vec{F} \cdot \vec{n}\, dA = \iint_R \vec{F} \cdot \frac{\partial \vec{r}}{\partial x} \times \frac{\partial \vec{r}}{\partial y} dxdy$$

$$\because \frac{\partial \vec{r}}{\partial x} = (1,0,-1) \text{ and } \frac{\partial \vec{r}}{\partial y} = (0,1,-1) \quad \therefore \frac{\partial \vec{r}}{\partial x} \times \frac{\partial \vec{r}}{\partial y} = (1,1,1)$$

$$\therefore \vec{F} \cdot \frac{\partial \vec{r}}{\partial x} \times \frac{\partial \vec{r}}{\partial y} = (x^2, -y^2, 0) \cdot (1,1,1) = x^2 - y^2, \quad \forall (x,y,z) \in S$$

$$\therefore \iint_R \vec{F} \cdot \frac{\partial \vec{r}}{\partial x} \times \frac{\partial \vec{r}}{\partial y} dxdy = \int_0^a \int_0^{a-x} x^2 - y^2 \, dydx = \int_0^a x^2(a-x) - \frac{(a-x)^3}{3} dx$$

$$= \int_0^a -\frac{a^3}{3} + a^2 x - \frac{2x^3}{3} dx = a^4 \left(-\frac{1}{3} + \frac{1}{2} - \frac{1}{6} \right) = 0$$

Example 11.

Assume $\vec{F} = (z, y, x)$. Find the flux through the outward-oriented surface

$$S: x^2 + y^2 + z^2 = r^2, \quad \oiint_S \vec{F} \cdot \vec{n} dA = ?$$

【Solution】

$\because$ S is oriented outward

Let $\vec{r}(\theta, \varphi) = (r\cos\theta \sin\varphi, r\sin\theta \sin\varphi, r\cos\varphi)$ and $R = \{(\theta, \varphi): 0 \leq \theta \leq 2\pi, 0 \leq \varphi \leq \pi\}$

then $\oiint_S \vec{F} \cdot \vec{n} dA = \iint_R \vec{F} \cdot \frac{\partial \vec{r}}{\partial \varphi} \times \frac{\partial \vec{r}}{\partial \theta} d\theta d\varphi$

$$\because \frac{\partial \vec{r}}{\partial \theta} = (-r\sin\theta \sin\varphi, r\cos\theta \sin\varphi, 0) \text{ and } \frac{\partial \vec{r}}{\partial \varphi} = (r\cos\theta \cos\varphi, r\sin\theta \cos\varphi, -r\sin\varphi)$$

$$\therefore \frac{\partial \vec{r}}{\partial \varphi} \times \frac{\partial \vec{r}}{\partial \theta} = (r^2 \cos\theta \sin^2\varphi, r^2 \sin\theta \sin^2\varphi, r^2 \sin\varphi \cos\varphi)$$

$$\therefore \vec{F} \cdot \frac{\partial \vec{r}}{\partial \varphi} \times \frac{\partial \vec{r}}{\partial \theta}$$

$= (r\cos\varphi, r\sin\theta \sin\varphi, r\cos\theta \sin\varphi) \cdot (r^2 \cos\theta \sin^2\varphi, r^2 \sin\theta \sin^2\varphi, r^2 \sin\varphi \cos\varphi)$

$= r^3 (2\cos\theta \sin^2\varphi \cos\varphi + \sin^3\varphi \sin^2\theta)$

$$\therefore \oiint_S \vec{F}\cdot\vec{n}dA = \iint_R \vec{F}\cdot\frac{\partial \vec{r}}{\partial \varphi}\times\frac{\partial \vec{r}}{\partial \theta}d\theta d\varphi$$

$$= r^3 \int_0^{2\pi}\int_0^{\pi} 2\cos\theta\sin^2\varphi\cos\varphi + \sin^3\varphi\sin^2\theta\, d\varphi d\theta$$

$$\because \int_0^{\pi}\sin^2\varphi\cos\varphi\, d\varphi = \left.\frac{\sin^3\varphi}{3}\right|_0^{\pi} = 0 \quad \therefore \int_0^{2\pi}\int_0^{\pi} 2\cos\theta\sin^2\varphi\cos\varphi\, d\varphi d\theta = 0$$

$$\therefore \int_0^{2\pi}\int_0^{\pi}(2\cos\theta\sin^2\varphi\cos\varphi + \sin^3\varphi\sin^2\theta)d\varphi d\theta = \int_0^{2\pi}\int_0^{\pi}\sin^3\varphi\sin^2\theta\, d\varphi d\theta$$

$$= \int_0^{2\pi}\sin^2\theta\, d\theta \int_0^{\pi}\sin^3\varphi\, d\varphi = \int_0^{2\pi}\frac{1-\cos 2\theta}{2}d\theta \int_0^{\pi}\sin\varphi(1-\cos^2\varphi)d\varphi$$

$$= \pi \int_0^{\pi}\sin\varphi(1-\cos^2\varphi)d\varphi$$

$$\because \int_0^{\pi}\sin\varphi(1-\cos^2\varphi)d\varphi = \left.-\cos\varphi\right|_0^{\pi} + \left.\frac{\cos^3\varphi}{3}\right|_0^{\pi} = \frac{4}{3}$$

$$\therefore \oiint_S \vec{F}\cdot\vec{n}dA = r^3\int_0^{2\pi}\int_0^{\pi}(2\cos\theta\sin^2\varphi\cos\varphi + \sin^3\varphi\sin^2\theta)d\varphi d\theta = \frac{4\pi r^3}{3}$$

Example 12.

Assume $\vec{F} = (-xze^y, xze^y, 2z)$. Find the flux through the upward-oriented surface $S: x+y+z=1$ in the first octant, $\iint_S \vec{F}\cdot\vec{n}\, dA =?$

【Solution】

∵ S is oriented upward

Let $\vec{r}(x,y,z) = (x, y, 1-x-y)$ and $R = \{(x,y): x+y \leq 1, x \geq 0, y \geq 0\}$ then

$$\iint_S \vec{F}\cdot\vec{n}\, dA = \iint_R \vec{F}\cdot\frac{\partial \vec{r}}{\partial x}\times\frac{\partial \vec{r}}{\partial y}dxdy$$

$$\because \frac{\partial \vec{r}}{\partial x} = (1,0,-1) \text{ and } \frac{\partial \vec{r}}{\partial y} = (0,1,-1) \quad \therefore \frac{\partial \vec{r}}{\partial x}\times\frac{\partial \vec{r}}{\partial y} = (1,1,1)$$

$$\because \vec{F} \cdot \frac{\partial \vec{r}}{\partial x} \times \frac{\partial \vec{r}}{\partial y} = (-xze^y, xze^y, 2z) \cdot (1,1,1) = 2z = 2(1-x-y), \quad \forall (x,y,z) \in S$$

$$\therefore \iint_S \vec{F} \cdot \vec{n} \, dA = \iint_R \vec{F} \cdot \frac{\partial \vec{r}}{\partial x} \times \frac{\partial \vec{r}}{\partial y} dxdy = 2 \iint_R 1 - x - y \, dxdy$$

$$= 2 \int_0^1 \int_0^{1-x} 1 - x - y \, dydx = 2 \int_0^1 y - xy - \frac{y^2}{2} \Big|_{y=0}^{y=1-x} dx$$

$$= 2 \int_0^1 (1-x) - x(1-x) - \frac{(1-x)^2}{2} dx = 2 \int_0^1 \left(\frac{x^2}{2} - x + \frac{1}{2} \right) dx = \frac{1}{3}$$

Example 13.

Let $\vec{F} = \left(x, -y, \frac{3z}{2} \right)$. Determine the flux across $S: z = \frac{2}{3}\left(x^{\frac{3}{2}} + y^{\frac{3}{2}} \right)$ with $0 \leq x \leq 1$, $0 \leq y \leq 1 - x$, with upward orientation.

【Solution】

$\because$ S is oriented upward

Let $\vec{r}(x,y,z) = \left(x, y, \frac{2}{3}\left(x^{\frac{3}{2}} + y^{\frac{3}{2}} \right) \right)$ and $R = \{(x,y): 0 \leq x \leq 1, 0 \leq y \leq 1 - x\}$ then

$$\iint_S \vec{F} \cdot \vec{n} \, dA = \iint_R \vec{F} \cdot \frac{\partial \vec{r}}{\partial x} \times \frac{\partial \vec{r}}{\partial y} dxdy$$

$\because \frac{\partial \vec{r}}{\partial x} = \left(1, 0, x^{\frac{1}{2}} \right)$ and $\frac{\partial \vec{r}}{\partial y} = \left(0, 1, y^{\frac{1}{2}} \right) \quad \therefore \frac{\partial \vec{r}}{\partial x} \times \frac{\partial \vec{r}}{\partial y} = \left(-x^{\frac{1}{2}}, -y^{\frac{1}{2}}, 1 \right)$

$$\therefore \vec{F} \cdot \frac{\partial \vec{r}}{\partial x} \times \frac{\partial \vec{r}}{\partial y} = \left(x, -y, \frac{3z}{2}\right) \cdot \left(-x^{\frac{1}{2}}, -y^{\frac{1}{2}}, 1\right) = \left(x, -y, x^{\frac{3}{2}} + y^{\frac{3}{2}}\right) \cdot \left(-x^{\frac{1}{2}}, -y^{\frac{1}{2}}, 1\right) = 2y^{\frac{3}{2}},$$

$\forall (x, y, z) \in S$

$$\iint_R \vec{F} \cdot \frac{\partial \vec{r}}{\partial x} \times \frac{\partial \vec{r}}{\partial y} dxdy = \int_0^1 \int_0^{1-x} 2y^{\frac{3}{2}} dydx = \frac{4}{5} \int_0^1 (1-x)^{\frac{5}{2}} dx$$

Let $1 - x = u$ then $\iint_S \vec{F} \cdot \vec{n} \, dA = \frac{4}{5} \int_0^1 (1-x)^{\frac{5}{2}} dx = -\frac{4}{5} \int_1^0 u^{\frac{5}{2}} du = \frac{8}{35}$

Example 14.

Let $\vec{F} = (0, y^2, 0)$. Determine the flux across $S: z = \frac{2}{3}\left(x^{\frac{3}{2}} + y^{\frac{3}{2}}\right)$ with $0 \leq x \leq 1$, $0 \leq y \leq 1 - x$, with upward orientation

【Solution】

$\because S$ is oriented upward

Let $\vec{r}(x, y, z) = \left(x, y, \frac{2}{3}\left(x^{\frac{3}{2}} + y^{\frac{3}{2}}\right)\right)$ and $R = \{(x, y): 0 \leq x \leq 1, 0 \leq y \leq 1 - x\}$ then

$$\iint_S \vec{F} \cdot \vec{n} \, dA = \iint_R \vec{F} \cdot \frac{\partial \vec{r}}{\partial x} \times \frac{\partial \vec{r}}{\partial y} dxdy$$

$\because \frac{\partial \vec{r}}{\partial x} = \left(1, 0, x^{\frac{1}{2}}\right)$ and $\frac{\partial \vec{r}}{\partial y} = \left(0, 1, y^{\frac{1}{2}}\right)$ $\therefore \frac{\partial \vec{r}}{\partial x} \times \frac{\partial \vec{r}}{\partial y} = \left(-x^{\frac{1}{2}}, -y^{\frac{1}{2}}, 1\right)$

$\therefore \vec{F} \cdot \frac{\partial \vec{r}}{\partial x} \times \frac{\partial \vec{r}}{\partial y} = (0, y^2, 0) \cdot \left(-x^{\frac{1}{2}}, -y^{\frac{1}{2}}, 1\right) = -y^{\frac{5}{2}}, \forall (x, y, z) \in S$

$$\iint_R \vec{F} \cdot \frac{\partial \vec{r}}{\partial x} \times \frac{\partial \vec{r}}{\partial y} dxdy = \int_0^1 \int_0^{1-x} -y^{\frac{5}{2}} dydx = -\frac{2}{7} \int_0^1 y^{\frac{7}{2}} \Big|_0^{1-x} dx = -\frac{2}{7} \int_0^1 (1-x)^{\frac{7}{2}} dx$$

Let $1 - x = u$ then $\iint_S \vec{F} \cdot \vec{n} \, dA = -\frac{2}{7} \int_0^1 (1-x)^{\frac{7}{2}} dx = \frac{2}{7} \int_1^0 u^{\frac{7}{2}} du = -\frac{4}{63}$

Example 15.

Assume $\vec{F} = (x, y, z)$. Find the flux through the outward-oriented surface

$S: x^2 + y^2 + z^2 = r^2$, $\oiint_S \vec{F} \cdot \vec{n} dA = ?$

【Solution】

∵ S is oriented outward

Let $\vec{r}(r, \theta, \varphi) = (r \cos \theta \sin \varphi, r \sin \theta \sin \varphi, r \cos \varphi)$ and let R be the closed region of the projection of S onto the $\theta\varphi$-plane. Then

$R = \{(\theta, \varphi): 0 \leq \theta \leq 2\pi, 0 \leq \varphi \leq \pi\}$ and $\oiint_S \vec{F} \cdot \vec{n} dA = \iint_R \vec{F} \cdot \dfrac{\partial \vec{r}}{\partial \varphi} \times \dfrac{\partial \vec{r}}{\partial \theta} d\theta d\varphi$

∵ $\dfrac{\partial \vec{r}}{\partial \theta} = (-r \sin \theta \sin \varphi, r \cos \theta \sin \varphi, 0)$ and $\dfrac{\partial \vec{r}}{\partial \varphi} = (r \cos \theta \cos \varphi, r \sin \theta \cos \varphi, -r \sin \varphi)$

∴ $\dfrac{\partial \vec{r}}{\partial \varphi} \times \dfrac{\partial \vec{r}}{\partial \theta} = (r^2 \cos \theta \sin^2 \varphi, r^2 \sin \theta \sin^2 \varphi, r^2 \sin \varphi \cos \varphi)$

∴ $\vec{F} \cdot \dfrac{\partial \vec{r}}{\partial \varphi} \times \dfrac{\partial \vec{r}}{\partial \theta}$

$= (r \cos \theta \sin \varphi, r \sin \theta \sin \varphi, r \cos \varphi) \cdot (r^2 \cos \theta \sin^2 \varphi, r^2 \sin \theta \sin^2 \varphi, r^2 \sin \varphi \cos \varphi)$

$= r^3(\cos^2 \theta \sin^3 \varphi + \sin^3 \varphi \sin^2 \theta + \sin \varphi \cos^2 \varphi) = r^3(\sin^3 \varphi + \sin \varphi \cos^2 \varphi) = r^3 \sin \varphi$

∴ $\oiint_S \vec{F} \cdot \vec{n} dA = \iint_R \vec{F} \cdot \dfrac{\partial \vec{r}}{\partial \varphi} \times \dfrac{\partial \vec{r}}{\partial \theta} d\theta d\varphi = r^3 \int_0^{2\pi} \int_0^{\pi} \sin \varphi \, d\varphi d\theta = r^3 \cdot 2\pi \cdot 2 = 4\pi r^3$

Example 16.

Find $\oiint_S \vec{F} \cdot \vec{n} dA = ?$, $\vec{F} = (0, 4yz, 0)$, $S: x^2 + y^2 + z^2 = r^2$.

【Solution】

∵ S is oriented outward

Let $\vec{r}(\theta, \varphi) = (r\cos \theta \sin \varphi, r\sin \theta \sin \varphi, r\cos \varphi)$ and $R = \{(\theta, \varphi): 0 \leq \theta \leq 2\pi, 0 \leq \varphi \leq \pi\}$

then $\oiint_S \vec{F} \cdot \vec{n} dA = \iint_R \vec{F} \cdot \dfrac{\partial \vec{r}}{\partial \varphi} \times \dfrac{\partial \vec{r}}{\partial \theta} d\theta d\varphi$

9.5.4 For Vector Functions, Find the Flux across Surface S 530

$$\because \frac{\partial \vec{r}}{\partial \theta} = (-r\sin\theta \sin\varphi, r\cos\theta \sin\varphi, 0) \text{ and } \frac{\partial \vec{r}}{\partial \varphi} = (r\cos\theta \cos\varphi, r\sin\theta \cos\varphi, -r\sin\varphi)$$

$$\therefore \frac{\partial \vec{r}}{\partial \varphi} \times \frac{\partial \vec{r}}{\partial \theta} = (r^2 \cos\theta \sin^2\varphi, r^2\sin\theta \sin^2\varphi, r^2\sin\varphi \cos\varphi)$$

$$\therefore \vec{F} \cdot \frac{\partial \vec{r}}{\partial \varphi} \times \frac{\partial \vec{r}}{\partial \theta} = (0, 4r^2\sin\theta \sin\varphi \cos\varphi, 0) \cdot (r^2\cos\theta \sin^2\varphi, r^2\sin\theta \sin^2\varphi, r^2\sin\varphi \cos\varphi)$$

$$= 4r^4 \sin^3\varphi \sin^2\theta \cos\varphi$$

$$\therefore \oiint_S \vec{F} \cdot \vec{n} dA = \int_0^{2\pi} \int_0^{\pi} 4r^4 \sin^3\varphi \sin^2\theta \cos\varphi \, d\varphi dr d\theta$$

$$\because \int_0^{\pi} \sin^3\varphi \cos\varphi \, d\varphi = \left.\frac{\sin^4\varphi}{4}\right|_0^{\pi} = 0 \quad \therefore \oiint_S \vec{F} \cdot \vec{n} dA = 0$$

Example 17.

Find $\oiint_S \vec{F} \cdot \vec{n} dA = ?$, where $\vec{F} = (x - y + z, 2x, 1)$, $S = S_1 \cup S_1$ is the outward-oriented surface, $S_1: z = x^2 + y^2, z \leq 1$, and $S_2: x^2 + y^2 \leq 1, z = 1$.

【Solution】

∵ S_1 is oriented downward

Let $\vec{r}(x, y, z) = (x, y, x^2 + y^2)$ and let R be the closed region of the projection of S_1 onto the xy-plane. Then

$$R = \{(x, y): x^2 + y^2 \leq 1\} \text{ and } \iint_{S_1} \vec{F} \cdot \vec{n} dA = -\iint_R \vec{F} \cdot \frac{\partial \vec{r}}{\partial x} \times \frac{\partial \vec{r}}{\partial y} dxdy$$

$$\because \frac{\partial \vec{r}}{\partial x} = (1, 0, 2x) \text{ and } \frac{\partial \vec{r}}{\partial y} = (0, 1, 2y) \quad \therefore \frac{\partial \vec{r}}{\partial x} \times \frac{\partial \vec{r}}{\partial y} = (-2x, -2y, 1)$$

$$\therefore \vec{F} \cdot \frac{\partial \vec{r}}{\partial x} \times \frac{\partial \vec{r}}{\partial y} = (x - y + z, 2x, 1) \cdot (-2x, -2y, 1) = -2x(x - y + x^2 + y^2) - 4xy + 1,$$

$\forall (x, y, z) \in S_1$

$$\therefore \iint_{S_1} \vec{F} \cdot \vec{n} \, dA = -\iint_R \vec{F} \cdot \frac{\partial \vec{r}}{\partial x} \times \frac{\partial \vec{r}}{\partial y} dxdy = \iint_{x^2+y^2 \leq 1} 2x^2 + 2xy + 2x(x^2 + y^2) - 1 dxdy$$

Let $x = r\cos\theta, y = r\sin\theta$ then $\{(x, y): x^2 + y^2 \leq 1\} = \{(r, \theta): 0 \leq r \leq 1, 0 \leq \theta \leq 2\pi\}$

and $dxdy = \begin{Vmatrix} \dfrac{\partial x}{\partial r} & \dfrac{\partial x}{\partial \theta} \\ \dfrac{\partial y}{\partial r} & \dfrac{\partial y}{\partial \theta} \end{Vmatrix} drd\theta = \begin{Vmatrix} \cos\theta & -r\sin\theta \\ \sin\theta & r\cos\theta \end{Vmatrix} drd\theta = rdrd\theta$

$\therefore \iint_{x^2+y^2 \le 1} 2x^2 + 2xy + 2x(x^2+y^2) - 1 \, dxdy$

$= \int_0^{2\pi} \int_0^1 (2r^2 \cos^2\theta + 2r^2 \cos\theta \sin\theta + 2r^3 \cos\theta - 1) r \, drd\theta = -\dfrac{\pi}{2}$

$\because \iint_{S_2} \vec{F} \cdot \vec{n} dA = \iint_{S_2} dA = \pi \quad \therefore \iint_{S_1} \vec{F} \cdot \vec{n} dA + \iint_{S_2} \vec{F} \cdot \vec{n} dA = \dfrac{\pi}{2}$

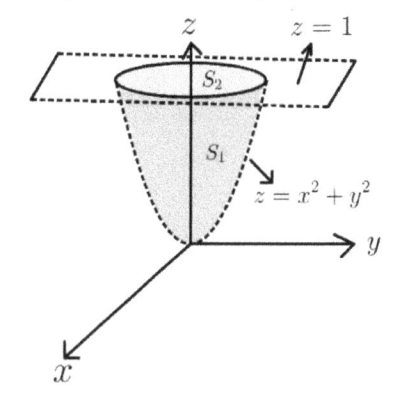

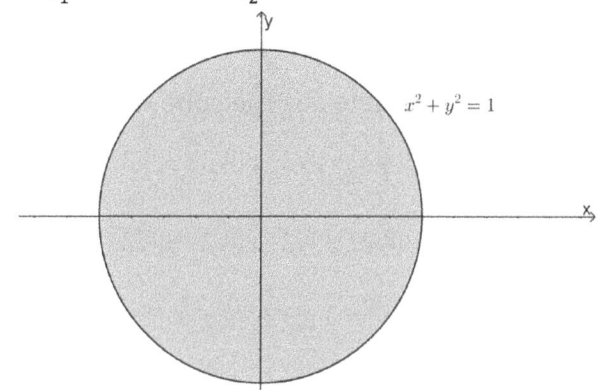

Example 18.

Assume the surface $S: \vec{r}(s,t) = (s+t, s-t, 1+2s+t)$, $0 \le s \le 2, 0 \le t \le 1$ and $\vec{F} = (ze^{xy}, -3ze^{xy}, xy)$. Find the flux through the upward-oriented surface S,

$$\iint_S \vec{F} \cdot \vec{n} \, dA = ?$$

【Solution】

∵ S is oriented upward

Let $\vec{r}(s,t) = (s+t, s-t, 1+2s+t)$ and $R = \{(s,t): 0 \le s \le 2, 0 \le t \le 1\}$ then

$$\iint_S \vec{F} \cdot \vec{n} \, dA = \iint_R \vec{F} \cdot \dfrac{\partial \vec{r}}{\partial s} \times \dfrac{\partial \vec{r}}{\partial t} \, dsdt$$

9.5.4 For Vector Functions, Find the Flux across Surface S

$$\because \frac{\partial \vec{r}}{\partial s} = (1,1,2) \text{ and } \frac{\partial \vec{r}}{\partial t} = (1,-1,1) \quad \therefore \frac{\partial \vec{r}}{\partial s} \times \frac{\partial \vec{r}}{\partial t} = (3,1,-2)$$

$$\because \vec{F} \cdot \frac{\partial \vec{r}}{\partial s} \times \frac{\partial \vec{r}}{\partial t} = (ze^{xy}, -3ze^{xy}, xy) \cdot (3,1,-2) = -2xy = -2(s^2 - t^2), \ \forall (x,y,z) \in S$$

$$\therefore \iint_S \vec{F} \cdot \vec{n} \, dA = \iint_R \vec{F} \cdot \frac{\partial \vec{r}}{\partial s} \times \frac{\partial \vec{r}}{\partial t} \, dsdt = -2 \iint_R s^2 - t^2 \, dsdt = -2 \int_0^2 \int_0^1 s^2 - t^2 \, dtds$$

$$= -2 \int_0^2 s^2 - \frac{1}{3} \, ds = -2 \left(\frac{s^3 - s}{3} \right) \Big|_0^2 = -4$$

Example 19.

Let $\vec{F} = (x, 2y, 3z)$. Assume S is the cube with the vertices $(\pm 1, \pm 1, \pm 1)$ with outward orientation. Find $\oiint_S \vec{F} \cdot \vec{n} dA = ?$

【Solution】

Let $S_1: x = 1, \ S_2: x = -1, \ S_3: y = 1, \ S_4: y = -1, \ S_5: z = 1, \ S_6: z = -1$, then
$S = S_1 \cup S_2 \cup S_3 \cup S_4 \cup S_5 \cup S_6$

As $x = 1, \vec{n} = (1,0,0), \quad \iint_{S_1} \vec{F} \cdot \vec{n} \, dA = \int_{-1}^{1} \int_{-1}^{1} x \, dydz = 4x = 4$

As $x = -1, \vec{n} = (-1,0,0), \quad \iint_{S_2} \vec{F} \cdot \vec{n} \, dA = \int_{-1}^{1} \int_{-1}^{1} -x \, dydz = -4x = 4$

As $y = 1, \vec{n} = (0,1,0), \quad \iint_{S_3} \vec{F} \cdot \vec{n} \, dA = \int_{-1}^{1} \int_{-1}^{1} 2y \, dxdz = 8y = 8$

As $y = -1, \vec{n} = (0,-1,0), \quad \iint_{S_4} \vec{F} \cdot \vec{n} \, dA = \int_{-1}^{1} \int_{-1}^{1} -2y \, dxdz = -8y = 8$

As $z = 1, \vec{n} = (0,0,1), \quad \iint_{S_5} \vec{F} \cdot \vec{n} \, dA = \int_{-1}^{1} \int_{-1}^{1} 3z \, dxdy = 12z = 12$

As $z = -1, \vec{n} = (0,0,-1), \quad \iint_{S_6} \vec{F} \cdot \vec{n} \, dA = \int_{-1}^{1} \int_{-1}^{1} -3z \, dxdy = -12z = 12$

$$\therefore \oiint_S \vec{F} \cdot \vec{n} dA = 48$$

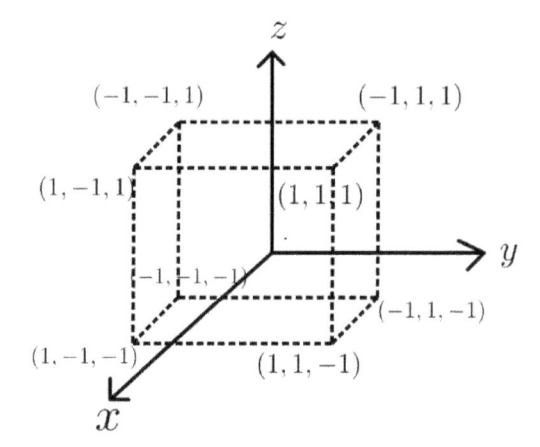

[Figure: A cube with vertices labeled $(-1,-1,1)$, $(-1,1,1)$, $(1,-1,1)$, $(1,1,1)$, $(-1,-1,-1)$, $(-1,1,-1)$, $(1,-1,-1)$, $(1,1,-1)$, shown in a 3D coordinate system with axes x, y, z.]

Example 20.

Let $\vec{F} = (x^2, y^2, z^2)$. Assume S is the boundary of the solid half-cylinder

$0 \leq z \leq \sqrt{1-y^2}, 0 \leq x \leq 2$ with outward orientation. Find $\iint_S \vec{F} \cdot \vec{n} \, dA = ?$

【Solution】

Let $S_1: z = \sqrt{1-y^2}, 0 \leq x \leq 2, 0 \leq y \leq 1$, $S_2: y^2 + z^2 \leq 1, x = 0$, $S_3: y^2 + z^2 \leq 1, x = 2$
then $S = S_1 \cup S_2 \cup S_3$
$\because S_1$ is oriented upward

Let $\vec{r}(x, y, z) = \left(x, y, \sqrt{1-y^2}\right)$ and $R = \{(x, y): 0 \leq x \leq 2, 0 \leq y \leq 1\}$ then

$$\iint_{S_1} \vec{F} \cdot \vec{n} \, dA = \iint_R \vec{F} \cdot \frac{\partial \vec{r}}{\partial x} \times \frac{\partial \vec{r}}{\partial y} \, dxdy$$

$\because \frac{\partial \vec{r}}{\partial x} = (1,0,0)$ and $\frac{\partial \vec{r}}{\partial y} = \left(0, 1, -y(1-y^2)^{-\frac{1}{2}}\right)$ $\therefore \frac{\partial \vec{r}}{\partial x} \times \frac{\partial \vec{r}}{\partial y} = \left(0, -y(1-y^2)^{-\frac{1}{2}}, 1\right)$

$\because \vec{F} \cdot \frac{\partial \vec{r}}{\partial x} \times \frac{\partial \vec{r}}{\partial y} = (x^2, y^2, 1-y^2) \cdot \left(0, -y(1-y^2)^{-\frac{1}{2}}, 1\right) = -y^3(1-y^2)^{-\frac{1}{2}} + 1 - y^2$,

$\forall (x, y, z) \in S_1$
Let $y = \sin \theta$ then $dy = \cos \theta \, d\theta$

$$\iint_R \vec{F} \cdot \frac{\partial \vec{r}}{\partial x} \times \frac{\partial \vec{r}}{\partial y} \, dxdy = \int_0^2 \int_0^1 -y^3(1-y^2)^{-\frac{1}{2}} + 1 - y^2 \, dydx$$

$$= 2\int_0^1 -y^3(1-y^2)^{-\frac{1}{2}} + 1 - y^2\, dy = 2\int_0^{\frac{\pi}{2}} -\sin^3\theta + \cos^3\theta\, d\theta$$

$$= 2\cdot\left(\left(-\cos\theta + \frac{\cos^3\theta}{3}\right)\Big|_0^{\frac{\pi}{2}} + \left(\sin\theta + \frac{\sin^3\theta}{3}\right)\Big|_0^{\frac{\pi}{2}}\right) = \frac{8}{3}$$

$$\because \iint_{S_2} \vec{F}\cdot\vec{n}dA = \iint_{S_2} dA = 0 \text{ and } \iint_{S_3} \vec{F}\cdot\vec{n}dA = \iint_{S_3} dA = 2\pi$$

$$\therefore \iint_S \vec{F}\cdot\vec{n}\, dA = \frac{8}{3} + 2\pi$$

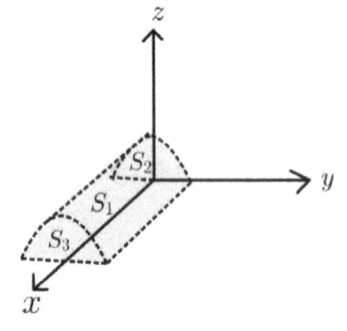

Example 21.

Let $\vec{F} = (0, xz, -xy)$. Determine the flux across $S: z = xy$ with $0 \le x \le 1, 0 \le y \le 2$, with upward orientation.

【Solution】

∵ S is oriented upward

Let $\vec{r}(x,y,z) = (x,y,xy)$ and $R = \{(x,y): 0 \le x \le 1, 0 \le y \le 2\}$ then

$$\iint_S \vec{F}\cdot\vec{n}\, dA = \iint_R \vec{F}\cdot\frac{\partial\vec{r}}{\partial x}\times\frac{\partial\vec{r}}{\partial y}dxdy$$

$$\because \frac{\partial\vec{r}}{\partial x} = (1,0,y) \text{ and } \frac{\partial\vec{r}}{\partial y} = (0,1,x) \quad \therefore \frac{\partial\vec{r}}{\partial x}\times\frac{\partial\vec{r}}{\partial y} = (-y,-x,1)$$

$$\because \vec{F}\cdot\frac{\partial\vec{r}}{\partial x}\times\frac{\partial\vec{r}}{\partial y} = (0,xz,-xy)\cdot(-y,-x,1) = -x^2z - xy = -x^3y - xy, \quad \forall(x,y,z) \in S$$

$$\iint_R \vec{F}\cdot\frac{\partial\vec{r}}{\partial x}\times\frac{\partial\vec{r}}{\partial y}dxdy = \int_0^1\int_0^2 -x^3y - xy\, dydx = \int_0^a -2x^3 - 2xdx = \left(-\frac{x^4}{2} - x^2\right)\Big|_0^1 = -\frac{3}{2}$$

Example 22.

Calculate the flux of $\vec{F} = (x, y, z)$ out of the cylindrical surface with parameterization
$S: \vec{r}(s, t) = (a \cos s, a \sin s, t), 0 \leq s \leq 2\pi, 0 \leq t \leq k$.

【Solution】

Let $\vec{r}(s, t) = (a\cos s, a \sin s, t)$ and $R = \{(s, t): 0 \leq s \leq 2\pi, 0 \leq t \leq k\}$ then

$$\iint_S \vec{F} \cdot \vec{n} \, dA = \iint_R \vec{F} \cdot \frac{\partial \vec{r}}{\partial s} \times \frac{\partial \vec{r}}{\partial t} \, ds dt$$

$\because \dfrac{\partial \vec{r}}{\partial s} = (-a\sin s, a \cos s, 0)$ and $\dfrac{\partial \vec{r}}{\partial t} = (0,0,1)$ $\therefore \dfrac{\partial \vec{r}}{\partial s} \times \dfrac{\partial \vec{r}}{\partial t} = (a \cos s, a\sin s, 0)$

$\because \vec{F} \cdot \dfrac{\partial \vec{r}}{\partial s} \times \dfrac{\partial \vec{r}}{\partial t} = (a\cos s, a\sin s, t) \cdot (a\cos s, a\sin s, 0) = a^2, \ \forall (x,y,z) \in S$

$$\iint_R \vec{F} \cdot \frac{\partial \vec{r}}{\partial s} \times \frac{\partial \vec{r}}{\partial t} \, ds dt = \int_0^{2\pi} \int_0^k a^2 \, dt ds = 2\pi a^2 k$$

Example 23.

Assume $\vec{F} = (y, z, x)$. The surface S is the upward-oriented plane formed by the points $(a, 0,0), (0, a, 0), (0,0, a)$ in the first octant. Find

$$\iint_S \vec{F} \cdot \vec{n} dA = ?$$

【Solution】

$\because$ S is oriented upward

Let $\vec{r}(x, y, z) = (x, y, a - x - y)$ and $R = \{(x, y): 0 \leq x \leq a, 0 \leq y \leq a - x\}$ then

$$\iint_S \vec{F} \cdot \vec{n} dA = \iint_R \vec{F} \cdot \frac{\partial \vec{r}}{\partial x} \times \frac{\partial \vec{r}}{\partial y} \, dx dy$$

$\because \dfrac{\partial \vec{r}}{\partial x} = (1,0, -1)$ and $\dfrac{\partial \vec{r}}{\partial y} = (0,1, -1)$ $\therefore \dfrac{\partial \vec{r}}{\partial x} \times \dfrac{\partial \vec{r}}{\partial y} = (1,1,1)$

$$\because \vec{F}\cdot\frac{\partial \vec{r}}{\partial x}\times\frac{\partial \vec{r}}{\partial y} = (y,z,x)\cdot(1,1,1) = x+y+z = x+y+a-x-y = a, \quad \forall (x,y,z)\in S$$

$$\iint_R \vec{F}\cdot\frac{\partial \vec{r}}{\partial x}\times\frac{\partial \vec{r}}{\partial y}\,dxdy = \int_0^a\int_0^{a-x} a\,dydx = a\int_0^a a-x\,dx = a\left(ax-\frac{x^2}{2}\right)\Big|_0^a = \frac{a^3}{2}$$

9.6 Stokes' Theorem

As previously mentioned, the curl form of Green's Theorem is expressed as

$$\oint_C \vec{F}\cdot d\vec{r} = \iint_R (\nabla\times\vec{F})\cdot \mathbf{k}\,dA.$$

If extended to three-dimensional space, the intuitive inference is

$$\oint_C \vec{F}\cdot d\vec{r} = \iint_S (\nabla\times\vec{F})\cdot \vec{n}\,dA,$$

where S is an open surface in space, and C is the boundary of S (a closed curve in space).

Stokes' Theorem is not only a higher-dimensional extension of the Fundamental Theorem of Calculus, but also a generalization of Green's Theorem to higher dimensions. Green's Theorem transforms the double integral over a region in the plane into a line integral over a closed curve in two-dimensional space. Similarly, Stokes' Theorem transforms the line integral over a closed curve in three-dimensional space into a surface integral over a surface S. Assuming that the boundary of the open surface S is a closed curve C in space, Stokes' Theorem states that the line integral of $\vec{F}$ along the closed curve C can be converted into the surface integral of the curl of $\vec{F}$ over the open surface S. Conversely, one can also compute the surface integral of the curl of $\vec{F}$ over S by evaluating the line integral of $\vec{F}$ along the closed curve C, as shown in the figure below.

An important condition for the validity of Stokes' Theorem is that the surface S must be open, and its boundary must be a closed curve in space. Furthermore, the orientation of the surface S is closely related to whether the boundary curve C is positively oriented (counterclockwise). If the orientation of the surface S is given, then the direction of the normal vector is determined. By applying the right-hand rule, with the thumb pointing in the direction of the normal vector, the direction in which the other four fingers curl indicates the orientation of the curve C. Conversely, if the direction of the closed curve C is given, the right-hand rule can be used to determine the direction of the corresponding normal vector, allowing one to judge whether the surface is oriented upwards or downwards, as illustrated in the figure.

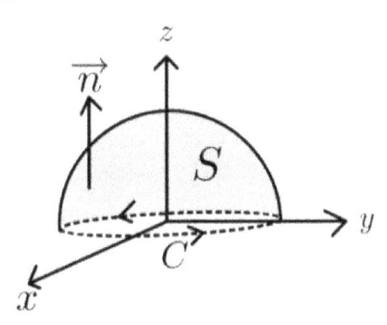

Originally, to calculate flux, we project the surface onto the xy-plane and then compute a double integral. However, **Stokes' Theorem** transforms the problem into evaluating the line integral over a closed curve in three-dimensional space. Conversely, evaluating the line integral of a closed curve in three-dimensional space can be converted, using **Stokes' Theorem**, into a flux calculation over an open surface in space (as shown in the figure).

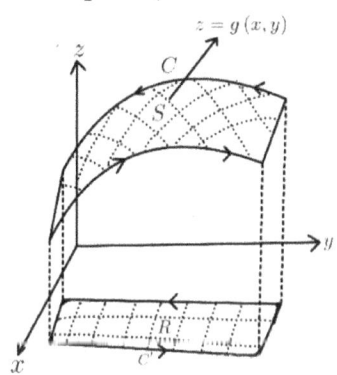

To illustrate Stokes' Theorem, let's first review the formula for surface integrals. For a surface S oriented upwards, given by $z = g(x, y), \forall\ a \leq x \leq b, c \leq y \leq d,$ assume that $\vec{F}(x, y, z)$ is a vector field in three-dimensional space. Then,

$$\iint_S \vec{F} \cdot \vec{n}\, dA = \int_a^b \int_c^d \vec{F}(x, y, g(x, y))\, (-g_x(x, y), -g_y(x, y), 1)\, dxdy.$$

If $\vec{F} = (F_1, F_2, F_3)$ then $\iint_S \vec{F} \cdot \vec{n}\, dA = \iint_D -F_1 \dfrac{\partial g}{\partial x} - F_2 \dfrac{\partial g}{\partial y} + F_3\, dA.$

$$\therefore \text{curl } \vec{F} = \nabla \times \vec{F} = \begin{Vmatrix} i & j & k \\ \frac{\partial}{\partial x} & \frac{\partial}{\partial y} & \frac{\partial}{\partial z} \\ F_1 & F_2 & F_3 \end{Vmatrix} = \left(\frac{\partial F_3}{\partial y} - \frac{\partial F_2}{\partial z}\right)i + \left(\frac{\partial F_1}{\partial z} - \frac{\partial F_3}{\partial x}\right)j + \left(\frac{\partial F_2}{\partial x} - \frac{\partial F_1}{\partial y}\right)k$$

Now, if we replace $\vec{F}$ with curl $\vec{F}$, then

$$\iint_S (\nabla \times \vec{F}) \cdot \vec{n} \, dA = \iint_D -\left(\frac{\partial F_3}{\partial y} - \frac{\partial F_2}{\partial z}\right)\frac{\partial z}{\partial x} - \left(\frac{\partial F_1}{\partial z} - \frac{\partial F_3}{\partial x}\right)\frac{\partial z}{\partial y} + \left(\frac{\partial F_2}{\partial x} - \frac{\partial F_1}{\partial y}\right) dA.$$

【Theorem】 Stokes' Theorem

Assume
1. The surface S is a smooth, non-closed surface oriented upwards, and the boundary curve C is a simple, closed, and piecewise smooth positively oriented curve.
2. $\vec{F}$ is a vector field whose components have continuous partial derivatives on an open set containing S. Then:

$$\oint_C \vec{F} \cdot d\vec{r} = \iint_S (\nabla \times \vec{F}) \cdot \vec{n} \, dA.$$

Proof:

Assume the surface S is defined by $z = g(x, y)$, where $(x, y) \in D$, and R is the projection of the surface S onto the xy-plane, with its closed boundary curve denoted as C'. By the surface integral formula for vector fields, we have

$$\iint_S \vec{F} \cdot \vec{n} \, dA = \iint_R \vec{F}(x, y, g(x, y)) \cdot (-g_x(x, y), -g_y(x, y), 1) \, dxdy.$$

Replacing $\vec{F}$ with $\nabla \times \vec{F}$, and using the fact that:

$$\therefore \text{curl } \vec{F} = \nabla \times \vec{F} = \begin{Vmatrix} i & j & k \\ \frac{\partial}{\partial x} & \frac{\partial}{\partial y} & \frac{\partial}{\partial z} \\ F_1 & F_2 & F_3 \end{Vmatrix} = \left(\frac{\partial F_3}{\partial y} - \frac{\partial F_2}{\partial z}\right)i + \left(\frac{\partial F_1}{\partial z} - \frac{\partial F_3}{\partial x}\right)j + \left(\frac{\partial F_2}{\partial x} - \frac{\partial F_1}{\partial y}\right)k$$

$$\therefore \iint_S (\nabla \times \vec{F}) \cdot \vec{n} \, dA = \iint_R (\nabla \times \vec{F})(x, y, g(x, y)) \cdot (-g_x(x, y), -g_y(x, y), 1) \, dxdy$$

$$= \iint_R \left(\frac{\partial F_3}{\partial y} - \frac{\partial F_2}{\partial z}, \frac{\partial F_1}{\partial z} - \frac{\partial F_3}{\partial x}, \frac{\partial F_2}{\partial x} - \frac{\partial F_1}{\partial y}\right) \cdot (-g_x(x, y), -g_y(x, y), 1) \, dxdy$$

$$= \iint_R -\left(\frac{\partial F_3}{\partial y} - \frac{\partial F_2}{\partial z}\right)\frac{\partial z}{\partial x} - \left(\frac{\partial F_1}{\partial z} - \frac{\partial F_3}{\partial x}\right)\frac{\partial z}{\partial y} + \left(\frac{\partial F_2}{\partial x} - \frac{\partial F_1}{\partial y}\right) dA,$$

where F_1, F_2, F_3 are evaluated at $(x, y, g(x, y))$. Let the parametric equations of the curve C' be: $x = x(t), y = y(t), \ a \leq t \leq b$. Then the parametric equation of the curve is given by

$$C: x = x(t), y = y(t), z = g(x(t), y(t)), \quad a \leq t \leq b.$$

By Chain Rule,

$$\oint_C \vec{F} \cdot d\vec{r} = \int_a^b F_1 \frac{dx}{dt} + F_2 \frac{dy}{dt} + F_3 \frac{dz}{dt} \, dt = \int_a^b F_1 \frac{dx}{dt} + F_2 \frac{dy}{dt} + F_3 \left(\frac{\partial z}{\partial x} \frac{dx}{dt} + \frac{\partial z}{\partial y} \frac{dy}{dt} \right) dt$$

$$= \int_a^b \frac{dx}{dt} \left(F_1 + F_3 \frac{\partial z}{\partial x} \right) + \frac{dy}{dt} \left(F_2 + F_3 \frac{\partial z}{\partial x} \right) dt = \int_{C'} \left(F_1 + F_3 \frac{\partial z}{\partial x} \right) dx + \left(F_2 + F_3 \frac{\partial z}{\partial x} \right) dy.$$

By Green's Theorem,

$$= \int_{C'} \left(F_1 + F_3 \frac{\partial z}{\partial x} \right) dx + \left(F_2 + F_3 \frac{\partial z}{\partial x} \right) dy = \iint_R \frac{\partial}{\partial x} \left(F_2 + F_3 \frac{\partial z}{\partial x} \right) - \frac{\partial}{\partial y} \left(F_1 + F_3 \frac{\partial z}{\partial x} \right) dA.$$

∵ F_1, F_2, F_3 are functions of $x, y,$ and z and z is a function of x and y

Applying the Chain Rule to right-hand side integral, we have

$$\iint_R \frac{\partial}{\partial x} \left(F_2 + F_3 \frac{\partial z}{\partial x} \right) - \frac{\partial}{\partial y} \left(F_1 + F_3 \frac{\partial z}{\partial x} \right) dA$$

$$= \iint_R \left(\frac{\partial F_2}{\partial x} + \frac{\partial F_2}{\partial z} \frac{\partial z}{\partial x} + \frac{\partial F_3}{\partial x} \frac{\partial z}{\partial y} + \frac{\partial F_3}{\partial z} \frac{\partial z}{\partial x} \frac{\partial z}{\partial y} + F_3 \frac{\partial^2 z}{\partial x \partial y} \right)$$

$$- \left(\frac{\partial F_1}{\partial y} + \frac{\partial F_1}{\partial z} \frac{\partial z}{\partial y} + \frac{\partial F_3}{\partial y} \frac{\partial z}{\partial x} + \frac{\partial F_3}{\partial z} \frac{\partial z}{\partial y} \frac{\partial z}{\partial x} + F_3 \frac{\partial^2 z}{\partial y \partial x} \right) dA$$

$$= \iint_R -\left(\frac{\partial F_3}{\partial y} - \frac{\partial F_2}{\partial z} \right) \frac{\partial z}{\partial x} - \left(\frac{\partial F_1}{\partial z} - \frac{\partial F_3}{\partial x} \right) \frac{\partial z}{\partial y} + \left(\frac{\partial F_2}{\partial x} - \frac{\partial F_1}{\partial y} \right) dA = \iint_S (\nabla \times \vec{F}) \cdot \vec{n} \, dA$$

$$\therefore \oint_C \vec{F} \cdot d\vec{r} = \iint_S (\nabla \times \vec{F}) \cdot \vec{n} \, dA$$

There is a noteworthy correspondence in **Stokes' Theorem** concerning the relationship between the surface's normal vector and the orientation of the closed curve. If the surface S is oriented upwards, then the closed curve C will have a counterclockwise (positive) orientation. By applying the right-hand rule, if the thumb points in the direction of the surface's upward normal vector, the other four fingers will curl in the direction of the curve's traversal. Therefore, if the surface is oriented upwards, the closed curve C (in the direction of the curled fingers) will have a counterclockwise (positive) orientation, and

$$\oint_C \vec{F} \cdot d\vec{r} = \iint_S (\nabla \times \vec{F}) \cdot \vec{n} \, dA.$$

If the surface is oriented downwards, then the closed curve C' (in the direction of the curled fingers) will have a negative orientation, and

9.6 Stokes' Theorem

$$\oint_C \vec{F} \cdot d\vec{r} = -\iint_S (\nabla \times \vec{F}) \cdot \vec{n} dA = -\oint_{C'} \vec{F} \cdot d\vec{r}.$$

It is crucial to clearly understand this correspondence, otherwise, an incorrect sign in the calculation could lead to the wrong answer. Additionally, another important aspect worth discussing is that different surfaces can share the same boundary. According to **Stokes' Theorem**, these different surfaces with the same boundary will have the same flux. To help the reader grasp these subtle rules and the applications of **Stokes' Theorem**, we will illustrate with diagrams. Assume S_1, S_2, S_3 represent the upper hemisphere, the upper cone, and the upper paraboloid surface, respectively, all sharing the same boundary C. By the right-hand rule, the orientation of C is counterclockwise (positive orientation), and by **Stokes' Theorem**, we have

$$\oint_C \vec{F} \cdot d\vec{r} = \iint_{S_1} (\nabla \times \vec{F}) \cdot \vec{n} dA = \iint_{S_2} (\nabla \times \vec{F}) \cdot \vec{n} dA = \iint_{S_3} (\nabla \times \vec{F}) \cdot \vec{n} dA.$$

As shown in the figure below, let S_4, S_5, and S_6 represent the lower hemisphere, lower cone, and lower paraboloid surface, respectively, all sharing the same boundary C'. According to the right-hand rule, the orientation of C' is clockwise (negative orientation), and C and C' are the same curve but with opposite directions. By **Stokes' Theorem**, we have

$$\oint_{C'} \vec{F} \cdot d\vec{r} = \iint_{S_4} (\nabla \times \vec{F}) \cdot \vec{n} dA = \iint_{S_5} (\nabla \times \vec{F}) \cdot \vec{n} dA = \iint_{S_6} (\nabla \times \vec{F}) \cdot \vec{n} dA = -\oint_C \vec{F} \cdot d\vec{r}.$$

From these observations, we can draw an intuitive conclusion about **Stokes' Theorem**: a closed space curve can lead to various non-closed surfaces, but the flux values of these surfaces will either be the same or differ by a negative sign.

Types of problems involving **Stokes' Theorem** on exams include:
- Using **Stokes' Theorem** to show that the line integral is independent of the path,
- Using **Stokes' Theorem** to convert a closed line integral into a surface integral over a non-closed surface,
- Using **Stokes' Theorem** to convert a surface integral over a non-closed surface into a closed line integral.

9.6.1 When $\nabla \times \vec{F} = 0$, the Line Integral Is Path-Independent

Given the vector-valued function $\vec{F}$, prove that the line integral $\int_C \vec{F} \cdot d\vec{r}$ is path-independent.

Question Types:

Type 1.

Given (x_0, y_0, z_0) and (x_1, y_1, z_1), prove that the line integral $\int_{(x_0,y_0,z_0)}^{(x_1,y_1,z_1)} \vec{F} \cdot d\vec{r}$ is path-independent and compute this integral.

Problem-Solving Process:

Step1.

Let C_1 and C_2 be any two curves from (x_0, y_0, z_0) to (x_1, y_1, z_1).
Let C be the counterclockwise closed curve formed by C_1 and C_2, and let S be the enclosed surface.

Step2.

Claim: $\nabla \times \vec{F} = 0$

By Stokes' Theorem, $\oint_C \vec{F} \cdot d\vec{r} = \iint_S (\nabla \times \vec{F}) \cdot \vec{n} dA = 0$

$\therefore \int_{C_1} \vec{F} \cdot d\vec{r} - \int_{C_2} \vec{F} \cdot d\vec{r} = 0 \Rightarrow \int_{C_1} \vec{F} \cdot d\vec{r} = \int_{C_2} \vec{F} \cdot d\vec{r} \quad \therefore \int_C \vec{F} \cdot d\vec{r}$ is path-independent

Step3.

$\because \int_C \vec{F} \cdot d\vec{r}$ is path-independent $\quad \therefore \exists G(x, y, z)$ s. t. $\nabla G(x, y, z) = \vec{F}(x, y, z)$

9.6.1 $\nabla \times = 0$, the Line Integral Is Path-Independent 542

and $\int_{(x_0,y_0,z_0)}^{(x_1,y_1,z_1)} \vec{F} \cdot d\vec{r} = \int_{(x_0,y_0,z_0)}^{(x_1,y_1,z_1)} \nabla G \cdot d\vec{r} = \int_{(x_0,y_0,z_0)}^{(x_1,y_1,z_1)} dG = G(x_1,y_1,z_1) - G(x_0,y_0,z_0)$

Step4.

∵ $\nabla G(x,y,z) = \vec{F}(x,y,z) = (f_1, f_2, f_3)$

∵ $\begin{cases} \frac{\partial G}{\partial x} = f_1(x,y,z) \\ \frac{\partial G}{\partial y} = f_2(x,y,z) \\ \frac{\partial G}{\partial z} = f_3(x,y,z) \end{cases}$ ∴ $\begin{cases} G(x,y,z) = \int f_1(x,y,z)dx + c \\ G(x,y,z) = \int f_2(x,y,z)dy + c \\ G(x,y,z) = \int f_3(x,y,z)dz + c \end{cases}$

Find a specific $G(x,y,z)$ that simultaneously satisfies the above three equations, and then calculate $G(x_1,y_1,z_1) - G(x_0,y_0,z_0)$.

Examples:

Given (x_0, y_0, z_0) and (x_1, y_1, z_1), prove that the line integral $\int_{(x_0,y_0,z_0)}^{(x_1,y_1,z_1)} \vec{F} \cdot d\vec{r}$ is path-independent.

Let C_1 and C_2 be any two curves from (x_0, y_0, z_0) to (x_1, y_1, z_1).

Let C be the counterclockwise closed curve formed by C_1 and C_2 and let S be the enclosed surface.

(I) If $\vec{F} = (2xy - y^4 + 3, x^2 - 4xy^3, 0)$ then $\nabla \times \vec{F} = (0,0,0)$

(II) If $\vec{F} = (4x^3y, x^4, 0)$ then $\nabla \times \vec{F} = (0,0,0)$

(III) If $\vec{F} = (10x^4 - 2xy^3, -3x^2y^2, 0)$ then $\nabla \times \vec{F} = (0,0,0)$

(IV) If $\vec{F} = \left(yz^2, xz^2 + ze^{yz}, 2xyz + ye^{yz} + \frac{1}{1+z}\right)$ then $\nabla \times \vec{F} = (0,0,0)$

(V) If $\vec{F} = (2xyz, x^2z, x^2y)$ then $\nabla \times \vec{F} = (0,0,0)$

By Stokes' Theorem, $\oint_C \vec{F} \cdot d\vec{r} = \iint_S (\nabla \times \vec{F}) \cdot \vec{n} dA = 0$

∴ $\int_{C_1} \vec{F} \cdot d\vec{r} - \int_{C_2} \vec{F} \cdot d\vec{r} = 0 \Rightarrow \int_{C_1} \vec{F} \cdot d\vec{r} = \int_{C_2} \vec{F} \cdot d\vec{r}$ ∴ $\int_C \vec{F} \cdot d\vec{r}$ is path-independent

Example 1.

Find $\oint_C (x^2y \cos x + 2xy \sin x - y^2 e^x)dx + (x^2 \sin x - 2ye^x)dy = ?$, $C: x^{\frac{2}{3}} + y^{\frac{2}{3}} = a^{\frac{2}{3}}$, where C is a counterclockwise-oriented closed curve.

【Solution】

Let $\vec{F} = (x^2 y \cos x + 2xy \sin x - y^2 e^x, x^2 \sin x - 2y e^x, 0)$

Claim: $\nabla \times \vec{F} = 0$

$$\nabla \times \vec{F} = \begin{vmatrix} \vec{i} & \vec{j} & \vec{k} \\ \dfrac{\partial}{\partial x} & \dfrac{\partial}{\partial y} & \dfrac{\partial}{\partial z} \\ x^2 y \cos x + 2xy \sin x - y^2 e^x & x^2 \sin x - 2y e^x & 0 \end{vmatrix}$$

$= (0, 0, 2x \sin x + x^2 \cos x - 2y e^x - (x^2 \cos x + 2x \sin x - 2y e^x)) = (0,0,0)$

∵ the first-order partial derivatives of the components of $\vec{F}$ exist and are continuous

By Stokes' Theorem, $\oint_C \vec{F} \cdot d\vec{r} = \iint_S (\nabla \times \vec{F}) \cdot \vec{n} \, dA = 0$

Example 2.

Prove that $\displaystyle\int_{(0,0)}^{(3,1)} (2xy - y^4 + 3) dx + (x^2 - 4xy^3) dy$ is independent of the path from $(0,0)$ to $(3,1)$ and find the value of the integral.

【Solution】

Let C_1 and C_2 be any two curves from $(0,0)$ to $(3,1)$. Let C be the counterclockwise closed curve formed by C_1 and C_2, and let S be the enclosed region.

Let $\vec{F} = (2xy - y^4 + 3, x^2 - 4xy^3, 0)$

Claim: $\nabla \times \vec{F} = 0$

$$\nabla \times \vec{F} = \begin{vmatrix} \vec{i} & \vec{j} & \vec{k} \\ \dfrac{\partial}{\partial x} & \dfrac{\partial}{\partial y} & \dfrac{\partial}{\partial z} \\ 2xy - y^4 + 3 & x^2 - 4xy^3 & 0 \end{vmatrix} = \vec{i} \cdot 0 + \vec{j} \cdot 0 + \vec{k} \cdot 0 = (0,0,0)$$

∵ the first-order partial derivatives of the components of $\vec{F}$ exist and are continuous

By Stokes' Theorem, $\oint_C \vec{F} \cdot d\vec{r} = \iint_S (\nabla \times \vec{F}) \cdot \vec{n} \, dA = 0$

∴ $\displaystyle\int_{C_1} \vec{F} \cdot d\vec{r} - \int_{C_2} \vec{F} \cdot d\vec{r} = 0 \Rightarrow \int_{C_1} \vec{F} \cdot d\vec{r} = \int_{C_2} \vec{F} \cdot d\vec{r}$

∴ $\displaystyle\int_{(0,0)}^{(3,1)} (2xy - y^4 + 3) dx + (x^2 - 4xy^3) dy$ is path-independent

9.6.1 ∇×=0, the Line Integral Is Path-Independent 544

∴ ∃$G(x, y, z)$ s.t. $\nabla G(x, y, z) = \vec{F}$ and

$$\int_{(0,0)}^{(3,1)} \vec{F} \cdot d\vec{r} = \int_{(0,0)}^{(3,1)} \nabla G \cdot d\vec{r} = \int_{(0,0)}^{(3,1)} dG = G(3,1) - G(0,0)$$

∵ $\begin{cases} \dfrac{\partial G}{\partial x} = 2xy - y^4 + 3 \\ \dfrac{\partial G}{\partial y} = x^2 - 4xy^3 \\ \dfrac{\partial G}{\partial z} = 0 \end{cases}$ ∴ $G(x, y, z) = x^2 y - xy^4 + 3x + c$

∴ $\int_{(0,0)}^{(3,1)} (2xy - y^4 + 3)dx + (x^2 - 4xy^3)dy = x^2 y - xy^4 + 3x \big|_{(0,0)}^{(3,1)} = 15$

Example 3.

Prove that $\int_{(1,2)}^{(3,4)} (6xy^2 - y^3)dx + (6x^2 y - 3xy^2)dy$ is independent of the path and find the value of the integral.

【Solution】

Let C_1 and C_2 be any two curves from $(3,4)$ to $(1,2)$. Let C be the counterclockwise closed curve formed by C_1 and C_2, and let S be the enclosed region.

Let $\vec{F} = (6xy^2 - y^3, 6x^2 y - 3xy^2, 0)$

Claim: $\nabla \times \vec{F} = 0$

$$\nabla \times \vec{F} = \begin{vmatrix} \vec{i} & \vec{j} & \vec{k} \\ \dfrac{\partial}{\partial x} & \dfrac{\partial}{\partial y} & \dfrac{\partial}{\partial z} \\ 6xy^2 - y^3 & 6x^2 y - 3xy^2 & 0 \end{vmatrix} = (0, 0, 12x - 3y^2 - (12xy - 3y^2)) = (0,0,0)$$

∵ the first-order partial derivatives of the components of $\vec{F}$ exist and are continuous

By Stokes' Theorem, $\oint_C \vec{F} \cdot d\vec{r} = \iint_S (\nabla \times \vec{F}) \cdot \vec{n} dA = 0$

∴ $\int_{C_1} \vec{F} \cdot d\vec{r} - \int_{C_2} \vec{F} \cdot d\vec{r} = 0 \Rightarrow \int_{C_1} \vec{F} \cdot d\vec{r} = \int_{C_2} \vec{F} \cdot d\vec{r}$

∴ $\int_{(1,2)}^{(3,4)} (6xy^2 - y^3)dx + (6x^2 y - 3xy^2)dy$ is path-independent

∴ ∃$G(x, y, z)$ s.t. $\nabla G(x, y, z) = \vec{F}$ and

$$\int_{(1,2)}^{(3,4)} \vec{F} \cdot d\vec{r} = \int_{(1,2)}^{(3,4)} \nabla G \cdot d\vec{r} = \int_{(1,2)}^{(3,4)} dG = G(3,4) - G(1,2)$$

$$\because \begin{cases} \dfrac{\partial G}{\partial x} = 6xy^2 - y^3 \\ \dfrac{\partial G}{\partial y} = 6x^2y - 3xy^2 \\ \dfrac{\partial G}{\partial z} = 0 \end{cases} \therefore G(x,y,z) = 3x^2y^2 - xy^3 + c$$

$$\therefore \int_C \vec{F} \cdot d\vec{r} = 3x^2y^2 - xy^3 \big|_{(1,2)}^{(3,4)} = 236$$

Example 4.

Prove that $\displaystyle\int_{(0,0)}^{(a,b)} 4x^3y\,dx + x^4\,dy$ is independent of the path from $(0,0)$ to (a,b)

and find the value of the integral.

【Solution】

Let C_1 and C_2 be any two curves from $(0,0)$ to (a,b). Let C be the counterclockwise closed curve formed by C_1 and C_2, and let S be the enclosed region.

Let $\vec{F} = (4x^3y, x^4, 0)$

Claim: $\nabla \times \vec{F} = 0$

$$\nabla \times \vec{F} = \begin{vmatrix} \vec{i} & \vec{j} & \vec{k} \\ \dfrac{\partial}{\partial x} & \dfrac{\partial}{\partial y} & \dfrac{\partial}{\partial z} \\ 4x^3y & x^4 & 0 \end{vmatrix} = (0, 0, 4x^3 - 4x^3) = (0,0,0)$$

∵ the first-order partial derivatives of the components of $\vec{F}$ exist and are continuous

By Stokes' Theorem, $\displaystyle\oint_C \vec{F} \cdot d\vec{r} = \iint_S (\nabla \times \vec{F}) \cdot \vec{n}\,dA = 0$

$$\therefore \int_{C_1} \vec{F} \cdot d\vec{r} - \int_{C_2} \vec{F} \cdot d\vec{r} = 0 \Rightarrow \int_{C_1} \vec{F} \cdot d\vec{r} = \int_{C_2} \vec{F} \cdot d\vec{r}$$

$\therefore \displaystyle\int_{(0,0)}^{(a,b)} 4x^3y\,dx + x^4\,dy$ is path-independent

$\therefore \exists G(x,y,z)$ s.t. $\nabla G(x,y,z) = \vec{F}$ and

$$\int_{(0,0)}^{(a,b)} \vec{F} \cdot d\vec{r} = \int_{(0,0)}^{(a,b)} \nabla G \cdot d\vec{r} = \int_{(0,0)}^{(a,b)} dG = G(a,b) - G(0,0)$$

9.6.1 ∇×=0, the Line Integral Is Path-Independent

$\because \begin{cases} \dfrac{\partial G}{\partial x} = 4x^3 y \\ \dfrac{\partial G}{\partial y} = x^4 \\ \dfrac{\partial G}{\partial z} = 0 \end{cases}$ $\therefore G(x,y,z) = x^4 y + c$

$\therefore \int_{(0,0)}^{(a,b)} 4x^3 y\, dx + x^4 dy = x^4 y + c|_{(0,0)}^{(a,b)} = a^4 b$

Example 5.

Find $\int_{(0,0)}^{(-2,-1)} (10x^4 - 2xy^3)dx - 3x^2 y^2 dy = ?$, where the integral path is given by $x^4 - 6xy^3 = 4y^2$.

【Solution】

Let C_1 and C_2 be any two curves from $(0,0)$ to $(-2,-1)$. Let C be the counterclockwise closed curve formed by C_1 and C_2, and let S be the enclosed region.

Let $\vec{F} = (10x^4 - 2xy^3, -3x^2 y^2, 0)$

Claim: $\nabla \times \vec{F} = 0$

$\nabla \times \vec{F} = \begin{vmatrix} \vec{i} & \vec{j} & \vec{k} \\ \dfrac{\partial}{\partial x} & \dfrac{\partial}{\partial y} & \dfrac{\partial}{\partial z} \\ 10x^4 - 2xy^3 & -3x^2 y^2 & 0 \end{vmatrix} = (0, 0, -6xy^2 + 6xy^2) = (0,0,0)$

$\because$ the first-order partial derivatives of the components of $\vec{F}$ exist and are continuous

By Stokes' Theorem, $\oint_C \vec{F} \cdot d\vec{r} = \iint_S (\nabla \times \vec{F}) \cdot \vec{n}\, dA = 0$

$\therefore \int_{C_1} \vec{F} \cdot d\vec{r} - \int_{C_2} \vec{F} \cdot d\vec{r} = 0 \Rightarrow \int_{C_1} \vec{F} \cdot d\vec{r} = \int_{C_2} \vec{F} \cdot d\vec{r}$

$\therefore \int_{(0,0)}^{(-2,-1)} (10x^4 - 2xy^3)dx - 3x^2 y^2 dy$ is path-independent

$\therefore \exists G(x,y,z)$ s.t. $\nabla G(x,y,z) = \vec{F}$ and

$\int_{(0,0)}^{(-2,-1)} \vec{F} \cdot d\vec{r} = \int_{(0,0)}^{(-2,-1)} \nabla G \cdot d\vec{r} = \int_{(0,0)}^{(-2,-1)} dG = G(-2,-1) - G(0,0)$

$$\because \begin{cases} \dfrac{\partial G}{\partial x} = 10x^4 - 2xy^3 \\ \dfrac{\partial G}{\partial y} = -3x^2 y^2 \\ \dfrac{\partial G}{\partial z} = 0 \end{cases} \quad \therefore G(x,y,z) = 2x^5 - x^2 y^3 + c$$

$$\therefore \int_{(0,0)}^{(-2,-1)} (10x^4 - 2xy^3)dx - 3x^2 y^2 dy = 2x^5 - x^2 y^3 \big|_{(0,0)}^{(-2,-1)} = -60$$

Example 6.

Find $\int_{(0,0)}^{(\frac{\pi}{2},1)} (x^2 + 6xy - 2y^2)dx + (3x^2 - 4xy + 2y)dy = ?$, assuming the integral path is given by $y = \sin x$.

【Solution】

Let C_1 and C_2 be any two curves from $(0,0)$ to $\left(\dfrac{\pi}{2}, 1\right)$. Let C be the counterclockwise closed curve formed by C_1 and C_2, and let S be the enclosed region.

Let $\vec{F} = (x^2 + 6xy - 2y^2, 3x^2 - 4xy + 2y, 0)$

Claim: $\nabla \times \vec{F} = 0$

$$\nabla \times \vec{F} = \begin{vmatrix} \vec{i} & \vec{j} & \vec{k} \\ \dfrac{\partial}{\partial x} & \dfrac{\partial}{\partial y} & \dfrac{\partial}{\partial z} \\ x^2 + 6xy - 2y^2 & 3x^2 - 4xy + 2y & 0 \end{vmatrix} = (0, 0, 6x - 4y - (6x - 4y)) = (0,0,0)$$

∵ the first-order partial derivatives of the components of $\vec{F}$ exist and are continuous

By Stokes' Theorem, $\oint_C \vec{F} \cdot d\vec{r} = \iint_S (\nabla \times \vec{F}) \cdot \vec{n} dA = 0$

$$\therefore \int_{C_1} \vec{F} \cdot d\vec{r} - \int_{C_2} \vec{F} \cdot d\vec{r} = 0 \Rightarrow \int_{C_1} \vec{F} \cdot d\vec{r} = \int_{C_2} \vec{F} \cdot d\vec{r}$$

$\therefore \int_{(0,0)}^{(\frac{\pi}{2},1)} (x^2 + 6xy - 2y^2)dx + (3x^2 - 4xy + 2y)dy$ is path-independent

$\therefore \exists G(x,y,z)$ s.t. $\nabla G(x,y,z) = \vec{F}$ and

$$\int_{(0,0)}^{(\frac{\pi}{2},1)} \vec{F} \cdot d\vec{r} = \int_{(0,0)}^{(\frac{\pi}{2},1)} \nabla G \cdot d\vec{r} = \int_{(0,0)}^{(\frac{\pi}{2},1)} dG = G\left(\dfrac{\pi}{2}, 1\right) - G(0,0)$$

9.6.1 ∇×=0, the Line Integral Is Path-Independent

$$\therefore \begin{cases} \dfrac{\partial G}{\partial x} = x^2 + 6xy - 2y^2 \\ \dfrac{\partial G}{\partial y} = 3x^2 - 4xy + 2y \\ \dfrac{\partial G}{\partial z} = 0 \end{cases} \quad \therefore G(x,y,z) = 3x^2y - 2xy^2 + y^2 + \dfrac{x^3}{3} + c$$

$$\therefore \int_{(0,0)}^{(\frac{\pi}{2},1)} (x^2 + 6xy - 2y^2)dx + (3x^2 - 4xy + 2y)dy = 3x^2y - 2xy^2 + y^2 + \dfrac{x^3}{3}\Big|_{(0,0)}^{(\frac{\pi}{2},1)}$$

$$= 3\left(\dfrac{\pi}{2}\right)^2 - 2\left(\dfrac{\pi}{2}\right) + 1 + \dfrac{\left(\dfrac{\pi}{2}\right)^3}{3}$$

Example 7.

Find $\int_C \vec{F} \cdot d\vec{r} =?$, $\vec{F} = (2xy + z^3, x^2, 3xz^2)$, where C is any curve from $(1,-2,1)$ to $(3,1,4)$.

【Solution】

Let C_1 and C_2 be any two curves from $(1,-2,1)$ to $(3,1,4)$. Let C be the counterclockwise closed curve formed by C_1 and C_2, and let S be the enclosed region.

Claim: $\nabla \times \vec{F} = 0$

$$\nabla \times \vec{F} = \begin{vmatrix} \vec{i} & \vec{j} & \vec{k} \\ \dfrac{\partial}{\partial x} & \dfrac{\partial}{\partial y} & \dfrac{\partial}{\partial z} \\ 2xy+z^3 & x^2 & 3xz^2 \end{vmatrix} = (0, 3z^2 - 3z^2, 2x - 2x) = (0,0,0)$$

∵ the first-order partial derivatives of the components of $\vec{F}$ exist and are continuous

By Stokes' Theorem, $\oint_C \vec{F} \cdot d\vec{r} = \iint_S (\nabla \times \vec{F}) \cdot \vec{n} dA = 0$

$$\therefore \int_{C_1} \vec{F} \cdot d\vec{r} - \int_{C_2} \vec{F} \cdot d\vec{r} = 0 \Rightarrow \int_{C_1} \vec{F} \cdot d\vec{r} = \int_{C_2} \vec{F} \cdot d\vec{r} \quad \therefore \int_C \vec{F} \cdot d\vec{r} \text{ is path-independent}$$

∴ ∃$G(x,y,z)$ s. t. $\nabla G(x,y,z) = \vec{F}$ and

$$\int_{(1,-2,1)}^{(3,1,4)} \vec{F} \cdot d\vec{r} = \int_{(1,-2,1)}^{(3,1,4)} \nabla G \cdot d\vec{r} = \int_{(1,-2,1)}^{(3,1,4)} dG = G(3,1,4) - G(1,-2,1)$$

$$\therefore \begin{cases} \dfrac{\partial G}{\partial x} = 2xy + z^3 \\ \dfrac{\partial G}{\partial y} = x^2 \\ \dfrac{\partial G}{\partial z} = 3xz^2 \end{cases} \quad \therefore G(x,y,z) = x^2y + xz^3 + c$$

$$\therefore \int_C \vec{F} \cdot d\vec{r} = x^2y + xz^3 \Big|_{(1,-2,1)}^{(3,1,4)} = 202$$

Example 8.

Find $\displaystyle\int_C yz^2 dx + (xz^2 + ze^{yz})dy + \left(2xyz + ye^{yz} + \dfrac{1}{1+z}\right)dz$, where the curve $C = \{(t, t^2, t^3): 0 \le t \le 1\}$.

【Solution】

Let C_1 and C_2 be any two curves from $(0,0,0)$ to $(1,1,1)$. Let C be the counterclockwise closed curve formed by C_1 and C_2, and let S be the enclosed region.

Let $\vec{F} = \left(yz^2, xz^2 + ze^{yz}, 2xyz + ye^{yz} + \dfrac{1}{1+z}\right)$

Claim: $\nabla \times \vec{F} = 0$

$$\nabla \times \vec{F} = \begin{vmatrix} \vec{i} & \vec{j} & \vec{k} \\ \dfrac{\partial}{\partial x} & \dfrac{\partial}{\partial y} & \dfrac{\partial}{\partial z} \\ yz^2 & xz^2 + ze^{yz} & 2xyz + ye^{yz} + \dfrac{1}{1+z} \end{vmatrix}$$

$= (2xz + e^{yz} + zye^{yz} - (2xz + e^{yz} + yze^{yz}), 2yz - 2yz, z^2 - z^2) = (0,0,0)$

∵ the first-order partial derivatives of the components of $\vec{F}$ exist and are continuous

By Stokes' Theorem, $\displaystyle\oint_C \vec{F} \cdot d\vec{r} = \iint_S (\nabla \times \vec{F}) \cdot \vec{n} dA = 0$

$$\therefore \int_{C_1} \vec{F} \cdot d\vec{r} - \int_{C_2} \vec{F} \cdot d\vec{r} = 0 \Rightarrow \int_{C_1} \vec{F} \cdot d\vec{r} = \int_{C_2} \vec{F} \cdot d\vec{r}$$

$\therefore \displaystyle\int_C yz^2 dx + (xz^2 + ze^{yz})dy + \left(2xyz + ye^{yz} + \dfrac{1}{1+z}\right) dz$ is path-independent

$\therefore \exists G(x,y,z)$ s.t. $\nabla G(x,y,z) = \vec{F}$ and

$$\int_{(0,0,0)}^{(1,1,1)} \vec{F} \cdot d\vec{r} = \int_{(0,0,0)}^{(1,1,1)} \nabla G \cdot d\vec{r} = \int_{(0,0,0)}^{(1,1,1)} dG = G(1,1,1) - G(0,0,0)$$

9.6.1 ∇×=0, the Line Integral Is Path-Independent

$$\because \begin{cases} \dfrac{\partial G}{\partial x} = yz^2 \\ \dfrac{\partial G}{\partial y} = xz^2 + ze^{yz} \\ \dfrac{\partial G}{\partial z} = 2xyz + ye^{yz} + \dfrac{1}{1+z} \end{cases} \quad \therefore G(x,y,z) = xyz^2 + e^{yz} + \ln(1+z) + c$$

$$\therefore \int_C yz^2 dx + (xz^2 + ze^{yz})dy + \left(2xyz + ye^{yz} + \dfrac{1}{1+z}\right)dz$$

$$= xyz^2 + e^{yz} + \ln(1+z)\big|_{(0,0,0)}^{(1,1,1)} = e + \ln 2$$

Example 9.

Find $\int_C \vec{F} \cdot d\vec{r} = ?$, $\vec{F} = (2xyz, x^2z, x^2y)$, where C is any curve from $(0,0,0)$ to $(1,1,1)$.

【Solution】

Let C_1 and C_2 be any two curves from $(0,0,0)$ to $(1,1,1)$. Let C be the counterclockwise closed curve formed by C_1 and C_2, and let S be the enclosed region.

Let $\vec{F} = (2xyz, x^2z, x^2y)$

Claim: $\nabla \times \vec{F} = 0$

$$\nabla \times \vec{F} = \begin{vmatrix} \vec{i} & \vec{j} & \vec{k} \\ \dfrac{\partial}{\partial x} & \dfrac{\partial}{\partial y} & \dfrac{\partial}{\partial z} \\ 2xyz & x^2z & x^2y \end{vmatrix} = (x^2 - x^2, 2xy - 2xy, 2xz - 2xz) = (0,0,0)$$

∵ the first-order partial derivatives of the components of $\vec{F}$ exist and are continuous

By Stokes' Theorem, $\oint_C \vec{F} \cdot d\vec{r} = \iint_S (\nabla \times \vec{F}) \cdot \vec{n} dA = 0$

$$\therefore \int_{C_1} \vec{F} \cdot d\vec{r} - \int_{C_2} \vec{F} \cdot d\vec{r} = 0 \Rightarrow \int_{C_1} \vec{F} \cdot d\vec{r} = \int_{C_2} \vec{F} \cdot d\vec{r} \therefore \int_C \vec{F} \cdot d\vec{r} \text{ is path-independent}$$

$\therefore \exists G(x,y,z) \text{ s.t. } \nabla G(x,y,z) = \vec{F}$ and

$$\int_{(0,0,0)}^{(1,1,1)} \vec{F} \cdot d\vec{r} = \int_{(0,0,0)}^{(1,1,1)} \nabla G \cdot d\vec{r} = \int_{(0,0,0)}^{(1,1,1)} dG = G(1,1,1) - G(0,0,0)$$

$$\because \begin{cases} \dfrac{\partial G}{\partial x} = 2xyz \\ \dfrac{\partial G}{\partial y} = x^2 z \\ \dfrac{\partial G}{\partial z} = x^2 y \end{cases} \therefore G(x,y,z) = x^2 yz + c \quad \therefore \int_C \vec{F} \cdot d\vec{r} = x^2 yz \big|_{(0,0,0)}^{(1,1,1)} = 1$$

9.6.2 Convert Closed Line Integrals into Surface Integrals

Application Context: If the integrand of the line integral is relatively complex, using Stokes' Theorem to convert it to a surface integral may result in a cleaner integrand.

Question Types:

Type 1.

Let C be a counterclockwise closed curve in space. Prove that $\oint_C \vec{F} \cdot d\vec{r}$ only depends on the area enclosed by C and not on the position or shape of C.

Problem-Solving Process:

Claim: $(\nabla \times \vec{F}) \cdot \vec{n} = $ constant

Assume $\vec{F} = (f_1, f_2, f_3)$, and let the closed region enclosed by C be R.

By Stokes' Theorem,

$$\oint_C \vec{F} \cdot d\vec{r} = \iint_R (\nabla \times \vec{F}) \cdot \vec{n} \, dA,$$

where $\vec{n}$ is the unit normal vector and $\nabla \times \vec{F} = \begin{vmatrix} \vec{i} & \vec{j} & \vec{k} \\ \dfrac{\partial}{\partial x} & \dfrac{\partial}{\partial y} & \dfrac{\partial}{\partial z} \\ f_1 & f_2 & f_3 \end{vmatrix}$.

If $(\nabla \times \vec{F}) \cdot \vec{n} = c$ (c is constant) then $\oint_C \vec{F} \cdot d\vec{r} = \iint_R (\nabla \times \vec{F}) \cdot \vec{n} \, dA = \iint_R c \, dA$.

$\therefore \oint_C \vec{F} \cdot d\vec{r}$ only depends on the area enclosed by C and is independent of the shape and position of C.

Examples:

Let C be a counterclockwise closed curve in space. Prove that $\oint_C \vec{F} \cdot d\vec{r}$ only depends on

9.6.2 Convert Closed Line Integrals into Surface Integrals

the area enclosed by C and not on the position or shape of C. Let the closed region enclosed by C be R. By Stokes' Theorem,

$$\oint_C \vec{F} \cdot d\vec{r} = \iint_R (\nabla \times \vec{F}) \cdot \vec{n} \, dA,$$

where $\vec{n}$ is the unit normal vector and $\nabla \times \vec{F} = \begin{vmatrix} \vec{i} & \vec{j} & \vec{k} \\ \frac{\partial}{\partial x} & \frac{\partial}{\partial y} & \frac{\partial}{\partial z} \\ f_1 & f_2 & f_3 \end{vmatrix}$.

(I) If $\vec{F} = (3y, -5z, x)$ and C is any counterclockwise closed curve on the plane defined by $x + 2y + 2z = 5$. Then $(\nabla \times \vec{F}) \cdot \vec{n} = (5, -1, -3) \cdot \left(\frac{1}{3}, \frac{2}{3}, \frac{2}{3}\right) = -1$

By Stokes' Theorem, $\oint_C \vec{F} \cdot d\vec{r} = \iint_R (\nabla \times \vec{F}) \cdot \vec{n} \, dA = \iint_R -1 \, dA$

$\therefore \oint_C \vec{F} \cdot d\vec{r}$ only depends on the area enclosed by C and is independent of the shape and position of C.

(II) If $\vec{F} = (z, -2x, 4y)$ and C is any counterclockwise closed curve on the plane defined by $x + y + z = 2$. Then $(\nabla \times \vec{F}) \cdot \vec{n} = (4, 1, -2) \cdot \left(\frac{1}{\sqrt{3}}, \frac{1}{\sqrt{3}}, \frac{1}{\sqrt{3}}\right) = \sqrt{3}$

By Stokes' Theorem, $\oint_C \vec{F} \cdot d\vec{r} = \iint_R (\nabla \times \vec{F}) \cdot \vec{n} \, dA = \iint_R \sqrt{3} \, dA$

$\therefore \oint_C \vec{F} \cdot d\vec{r}$ only depends on the area enclosed by C and is independent of the shape and position of C.

Type 2.
Given a non-closed, upward-oriented surface S in space: $z = g(x, y), \forall\, a \le x \le b, c \le y \le d$, let C be the boundry of S(a counterclockwise closed curve). If the first-order partial derivatives of the components of $\vec{F}$ exist and are continuous, find $\oint_C \vec{F} \cdot d\vec{r} = ?$

Problem-Solving Process:
Step1.
By Stokes' Theorem, $\oint_C \vec{F} \cdot d\vec{r} = \iint_S (\nabla \times \vec{F}) \cdot \vec{n} \, dA$
Step2.

∵ S is upward-oriented

Let $\vec{r}(x,y) = (x, y, g(x,y))$ and let R be the closed region of the projection of surface S onto the xy-plane. Then

$R = \{(x,y): a \le x \le b, c \le y \le d\}$ and $\iint_S (\nabla \times \vec{F}) \cdot \vec{n}\, dA = \iint_R (\nabla \times \vec{F}) \cdot \dfrac{\partial \vec{r}}{\partial x} \times \dfrac{\partial \vec{r}}{\partial y}\, dxdy$

Step3.

∵ $\dfrac{\partial \vec{r}}{\partial x} = (1,0, g_x(x,y))$, $\dfrac{\partial \vec{r}}{\partial y} = (0,1, g_y(x,y))$ ∴ $\dfrac{\partial \vec{r}}{\partial x} \times \dfrac{\partial \vec{r}}{\partial y} = (-g_x(x,y), -g_y(x,y), 1)$

∴ $\iint_R (\nabla \times \vec{F}) \cdot \dfrac{\partial \vec{r}}{\partial x} \times \dfrac{\partial \vec{r}}{\partial y}\, dxdy = \iint_R (\nabla \times \vec{F}) \cdot (-g_x(x,y), -g_y(x,y), 1)\, dxdy$

Step4.

By Fubini's Theorem

$\iint_R (\nabla \times \vec{F}) \cdot (-g_x(x,y), -g_y(x,y), 1)\, dxdy$

$= \displaystyle\int_c^d \int_a^b (\nabla \times \vec{F}) \cdot (-g_x(x,y), -g_y(x,y), 1)\, dxdy$

$= \displaystyle\int_c^d \int_a^b -f_1 \cdot g_x(x,y) - f_2 \cdot g_y(x,y) + f_3\, dxdy$

where $\nabla \times \vec{F} = (f_1(x,y,g(x,y)), f_2(x,y,g(x,y)), f_3(x,y,g(x,y)))$

Step5.

Find $\displaystyle\int_c^d \int_a^b -f_1 \cdot g_x(x,y) - f_2 \cdot g_y(x,y) + f_3\, dxdy = ?$

If S is downward-oriented then

$\displaystyle\oint_C \vec{F} \cdot d\vec{r} = \int_c^d \int_a^b f_1 \cdot g_x(x,y) + f_2 \cdot g_y(x,y) - f_3\, dxdy$

Examples:

Find $\displaystyle\oint_C \vec{F} \cdot d\vec{r} = ?$

(I) If $\vec{F} = (z^2, y^2, x)$, and C is the counterclockwise closed boundary enclosed by the points $(a,0,0), (0,a,0)$ and $(0,0,a)$.

Let $\vec{r}(x,y) = (x, y, a - x - y)$ then $\dfrac{\partial \vec{r}}{\partial x} = (1,0,-1)$, $\dfrac{\partial \vec{r}}{\partial y} = (0,1,-1)$ ∴ $\dfrac{\partial \vec{r}}{\partial x} \times \dfrac{\partial \vec{r}}{\partial y} = (1,1,1)$

9.6.2 Convert Closed Line Integrals into Surface Integrals

$$\because \nabla \times \vec{F} = \begin{vmatrix} \vec{i} & \vec{j} & \vec{k} \\ \dfrac{\partial}{\partial x} & \dfrac{\partial}{\partial y} & \dfrac{\partial}{\partial z} \\ z^2 & y^2 & x \end{vmatrix} = (0, 2z - 1, 0)$$

By Stokes' Theorem,

$$\oint_C \vec{F} \cdot d\vec{r} = \iint_S (\nabla \times \vec{F}) \cdot \vec{n} dA = \iint_{0 \le x+y \le a} (0, 2z-1, 0) \cdot (1,1,1) dxdy$$

$$= \iint_{0 \le x+y \le a} 2z - 1 \, dxdy = \iint_{0 \le x+y \le a} 2(a - x - y) - 1 \, dxdy = \dfrac{a^3}{3} - \dfrac{a^2}{2}$$

Type 3.
Given a non-closed, upward-oriented surface $S: z = g(x, y)$ in space, with the projection of S onto the xy-plane given by $\{(x, y): x^2 + y^2 \le a^2\}$. Let C be the boundary of S (a counterclockwise closed curve). If the first-order partial derivatives of the components of $\vec{F}$ exist and are continuous, find $\oint_C \vec{F} \cdot d\vec{r} = ?$

Problem-Solving Process:

Step1.

By Stokes' Theorem, $\oint_C \vec{F} \cdot d\vec{r} = \iint_S (\nabla \times \vec{F}) \cdot \vec{n} dA$

Step2.
$\because$ S is upward-oriented
Let $\vec{r}(x, y) = (x, y, g(x, y))$ and let R be the closed region of the projection of surface S onto the xy-plane. Then

$R = \{(x, y): x^2 + y^2 \le a^2\}$ and $\iint_S (\nabla \times \vec{F}) \cdot \vec{n} \, dA = \iint_R (\nabla \times \vec{F}) \cdot \dfrac{\partial \vec{r}}{\partial x} \times \dfrac{\partial \vec{r}}{\partial y} dxdy$

Step3.

$\because \dfrac{\partial \vec{r}}{\partial x} = (1, 0, g_x(x, y)), \quad \dfrac{\partial \vec{r}}{\partial y} = (0, 1, g_y(x, y)) \quad \therefore \dfrac{\partial \vec{r}}{\partial x} \times \dfrac{\partial \vec{r}}{\partial y} = (-g_x(x, y), -g_y(x, y), 1)$

$\therefore \iint_R (\nabla \times \vec{F}) \cdot \dfrac{\partial \vec{r}}{\partial x} \times \dfrac{\partial \vec{r}}{\partial y} dxdy = \iint_R (\nabla \times \vec{F}) \cdot (-g_x(x, y), -g_y(x, y), 1) dxdy$

$= \iint_R -f_1 \cdot g_x(x, y) - f_2 \cdot g_y(x, y) + f_3 dxdy$

where $\nabla \times \vec{F} = (f_1(x, y, g(x, y)), f_2(x, y, g(x, y)), f_3(x, y, g(x, y)))$

Step4.

Let $x = r\cos\theta, y = r\sin\theta$ then $\{(x,y): x^2 + y^2 \leq a^2\} = \{(r,\theta): 0 \leq r \leq a, 0 \leq \theta \leq 2\pi\}$

and $dxdy = \begin{Vmatrix} \frac{\partial x}{\partial r} & \frac{\partial x}{\partial \theta} \\ \frac{\partial y}{\partial r} & \frac{\partial y}{\partial \theta} \end{Vmatrix} drd\theta = \begin{Vmatrix} \cos\theta & -r\sin\theta \\ \sin\theta & r\cos\theta \end{Vmatrix} drd\theta = rdrd\theta$

Step5.

$$\iint_R -f_1 \cdot g_x(x,y) - f_2 \cdot g_y(x,y) + f_3 dxdy$$

$$= \int_0^{2\pi} \int_0^a (-f_1 \cdot g_x(r\cos\theta, r\sin\theta) - f_2 \cdot g_y(r\cos\theta, r\sin\theta) + f_3) \cdot r \, drd\theta$$

If S is downward-oriented,

then $\oint_C \vec{F} \cdot d\vec{r} = \int_0^{2\pi} \int_0^a (f_1 \cdot g_x(r\cos\theta, r\sin\theta) + f_2 \cdot g_y(r\cos\theta, r\sin\theta) - f_3) \cdot r \, drd\theta$

Examples:

(I) Let $\vec{F} = (y^3, -x^3, 0)$ and let C be the counterclockwise closed curve defined by

$x^2 + y^2 = 4, \ z = 0$. Find $\oint_C \vec{F} \cdot d\vec{r} = ?$

Let $S = \{(x,y,z): x^2 + y^2 \leq 4, z = 0\}$

∵ the first-order partial derivatives of the components of $\vec{F}$ exist and are continuous

By Stokes' Theorem, $\oint_C \vec{F} \cdot d\vec{r} = \iint_S (\nabla \times \vec{F}) \cdot \vec{n} dA$

$\iint_S (\nabla \times \vec{F}) \cdot \vec{n} dA = \iint_S (0,0,-3x^2 - 3y^2) \cdot (0,0,1) dxdy = -3 \iint_{x^2+y^2 \leq 4} x^2 + y^2 dxdy$

$= -24\pi$

(II) Let $\vec{F} = (e^x y - y^3, e^x + x^3, 0)$ and let C be the counterclockwise closed curve

defined by $x^2 + y^2 = 6, \ z = 0$. Find $\oint_C \vec{F} \cdot d\vec{r} = ?$

Let $S = \{(x,y,z): x^2 + y^2 \leq 6, z = 0\}$

∵ the first-order partial derivatives of the components of $\vec{F}$ exist and are continuous

By Stokes' Theorem, $\oint_C \vec{F} \cdot d\vec{r} = \iint_S (\nabla \times \vec{F}) \cdot \vec{n} dA$

$\iint_S (\nabla \times \vec{F}) \cdot \vec{n} dA = \iint_S (0,0,3x^2 + 3y^2) \cdot (0,0,1) dxdy = 3 \iint_{x^2+y^2 \leq 36} x^2 + y^2 dxdy$

$$= 1944\pi$$

Example 1.

Using Stokes' Theorem, find $\oint_C \vec{F} \cdot d\vec{r} = ?$, $C: x^2 + y^2 = 4$, $z = 1$, $\vec{F} = (x, 2z - x, y^2)$, where C is a counterclockwise-oriented closed curve.

【Solution】

Let $S = \{(x, y, z): x^2 + y^2 \le 4, z = 1\}$ be oriented upwards
∵ the first-order partial derivatives of the components of $\vec{F}$ exist and are continuous

By Stokes' Theorem, $\oint_C \vec{F} \cdot d\vec{r} = \iint_S (\nabla \times \vec{F}) \cdot \vec{n} dA$

∵ S is upward-oriented

Let $\vec{r}(x, y) = (x, y, 1)$, $R = \{(x, y): x^2 + y^2 \le 4\}$ then

$$\iint_S (\nabla \times \vec{F}) \cdot \vec{n} dA = \iint_R (\nabla \times \vec{F}) \cdot \frac{\partial \vec{r}}{\partial x} \times \frac{\partial \vec{r}}{\partial y} dx dy$$

∵ $\frac{\partial \vec{r}}{\partial x} = (1,0,0)$, $\frac{\partial \vec{r}}{\partial y} = (0,1,0)$ ∴ $\frac{\partial \vec{r}}{\partial x} \times \frac{\partial \vec{r}}{\partial y} = (0,0,1)$

∵ $\nabla \times \vec{F} = \begin{vmatrix} \vec{i} & \vec{j} & \vec{k} \\ \frac{\partial}{\partial x} & \frac{\partial}{\partial y} & \frac{\partial}{\partial z} \\ x & 2z - x & y^2 \end{vmatrix} = (2y + 2, 0, -1)$

∴ $\iint_R (\nabla \times \vec{F}) \cdot \frac{\partial \vec{r}}{\partial x} \times \frac{\partial \vec{r}}{\partial y} dx dy = \iint_R (2y + 2, 0, -1) \cdot (0,0,1) dx dy = -\iint_{x^2+y^2 \le 4} dx dy$

$= -4\pi$

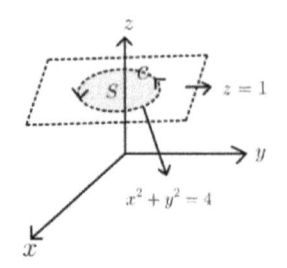

Example 2.

Assume C is any positively oriented closed curve on the plane $x + 2y + 2z = 5$.

Prove that $\oint_C 3y\,dx - 5z\,dy + x\,dz$ depends only on the area enclosed by C, regardless of the shape and position of C.

【Solution】

Let the closed curve C enclose the region R on the plane $x + 2y + 2z = 5$

Let $\vec{F} = (3y, -5z, x)$

∵ the first-order partial derivatives of the components of $\vec{F}$ exist and are continuous

By Stokes' Theorem, $\oint_C 3y\,dx - 5z\,dy + x\,dz = \oint_C \vec{F} \cdot d\vec{r} = \iint_R (\nabla \times \vec{F}) \cdot \vec{n}\,dA$

Let $\vec{n}$ be the unit normal vector to the plane $x + 2y + 2z = 5$ then $\vec{n} = \left(\frac{1}{3}, \frac{2}{3}, \frac{2}{3}\right)$

∵ $\nabla \times \vec{F} = \begin{vmatrix} \vec{i} & \vec{j} & \vec{k} \\ \dfrac{\partial}{\partial x} & \dfrac{\partial}{\partial y} & \dfrac{\partial}{\partial z} \\ 3y & -5z & x \end{vmatrix} = (5, -1, -3)$

∴ $\iint_R (\nabla \times \vec{F}) \cdot \vec{n}\,dA = \iint_R (5, -1, -3) \cdot \left(\frac{1}{3}, \frac{2}{3}, \frac{2}{3}\right) dA = -\iint_R dA$

∴ $\oint_C 3y\,dx - 5z\,dy + x\,dz = -\iint_R dA$

∴ The line integral $\oint_C 3y\,dx - 5z\,dy + x\,dz$ depends only on the area enclosed by C, and is independent of the shape and position of C

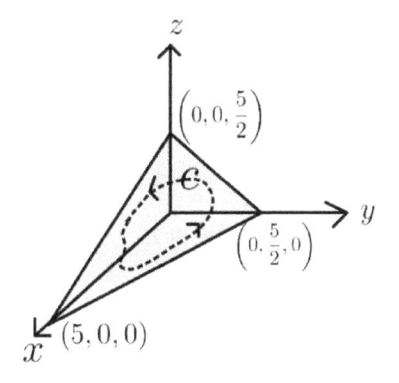

Example 3.

Assume $\vec{F} = (-y^2, x, z^2)$, and let C be the counterclockwise-oriented closed curve

9.6.2 Convert Closed Line Integrals into Surface Integrals

that is the intersection of $y + z = 2$ and $x^2 + y^2 = 1$. Use Stokes' Theorem to find

$$\oint_C \vec{F} \cdot d\vec{r} = ?$$

【Solution】

Let $S = \{(x, y, z): y + z = 2, x^2 + y^2 \leq 1\}$ be oriented upwards

∵ the first-order partial derivatives of the components of $\vec{F}$ exist and are continuous

By Stokes' Theorem, $\oint_C \vec{F} \cdot d\vec{r} = \iint_S (\nabla \times \vec{F}) \cdot \vec{n} dA$

∵ S is upward-oriented

Let $\vec{r}(x, y) = (x, y, 2 - y)$ and $R = \{(x, y): x^2 + y^2 \leq 1\}$ then

$$\iint_S (\nabla \times \vec{F}) \cdot \vec{n} dA = \iint_R (\nabla \times \vec{F}) \cdot \frac{\partial \vec{r}}{\partial x} \times \frac{\partial \vec{r}}{\partial y} dxdy$$

∵ $\frac{\partial \vec{r}}{\partial x} = (1,0,0)$, $\frac{\partial \vec{r}}{\partial y} = (0,1,-1)$ ∴ $\frac{\partial \vec{r}}{\partial x} \times \frac{\partial \vec{r}}{\partial y} = (0,1,1)$

∵ $\nabla \times \vec{F} = \begin{vmatrix} \vec{i} & \vec{j} & \vec{k} \\ \frac{\partial}{\partial x} & \frac{\partial}{\partial y} & \frac{\partial}{\partial z} \\ -y^2 & x & z^2 \end{vmatrix} = (0,0,1 - 2y)$

∴ $\iint_R (\nabla \times \vec{F}) \cdot \frac{\partial \vec{r}}{\partial x} \times \frac{\partial \vec{r}}{\partial y} dxdy = \iint_R (0,0,1 - 2y) \cdot (0,1,1) dxdy = \iint_{x^2+y^2 \leq 1} 1 - 2y \, dxdy$

Let $x = r \cos \theta$, $y = r \sin \theta$ then $\{(x, y): x^2 + y^2 \leq 1\} = \{(r, \theta): 0 \leq r \leq 1, 0 \leq \theta \leq 2\pi\}$

and $dxdy = \begin{Vmatrix} \frac{\partial x}{\partial r} & \frac{\partial x}{\partial \theta} \\ \frac{\partial y}{\partial r} & \frac{\partial y}{\partial \theta} \end{Vmatrix} drd\theta = \begin{Vmatrix} \cos \theta & -r \sin \theta \\ \sin \theta & r \cos \theta \end{Vmatrix} drd\theta = rdrd\theta$

∴ $\iint_{x^2+y^2 \leq 1} 1 - 2ydxdy = \int_0^{2\pi} \int_0^1 (1 - 2r \sin \theta) r \, drd\theta = \int_0^{2\pi} \frac{1}{2} - \frac{2\sin \theta}{3} d\theta = \pi$

∴ $\oint_C \vec{F} \cdot d\vec{r} = \pi$

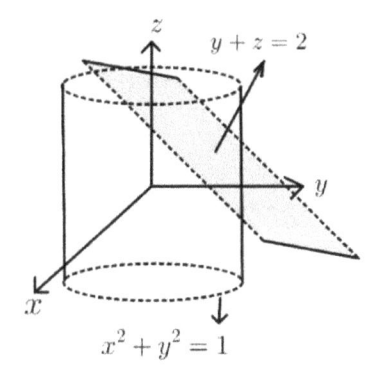

9.6.2 Convert Closed Line Integrals into Surface Integrals

Example 4.

Given $\vec{F} = \left(-z, x, \dfrac{y^2 z}{2}\right)$, assume the curve C is the counterclockwise-oriented closed intersection line between $z = a$ and $z = \sqrt{x^2 + y^2}$. Find $\oint_C \vec{F} \cdot d\vec{r} = ?$:

(1) Direct calculation (2) Using Stoke's Theorem

【Solution】

(1)

Let $C = \{(x, y, z): \sqrt{x^2 + y^2} = a, z = a\}$

Let $\vec{r}(\theta) = (a\cos\theta, a\sin\theta, a)$, $0 \le \theta \le 2\pi$ then $\vec{r}'(\theta) = (-a\sin\theta, a\cos\theta, 0)$

$$\oint_C \vec{F} \cdot d\vec{r} = \int_0^{2\pi} \vec{F}(\vec{r}(\theta)) \cdot \vec{r}'(\theta)\, d\theta = \int_0^{2\pi} \left(-a, a\cos\theta, \dfrac{a^3 \sin^2\theta}{2}\right) \cdot (-a\sin\theta, a\cos\theta, 0)\, d\theta$$

$$= \int_0^{2\pi} a^2 \sin\theta + a^2 \cos^2\theta\, d\theta = \int_0^{2\pi} a^2 \cos^2\theta\, d\theta = a^2 \int_0^{2\pi} \dfrac{1 + \cos\theta}{2}\, d\theta = \pi a^2$$

(2)

$$\nabla \times \vec{F} = \begin{Vmatrix} i & j & k \\ \dfrac{\partial}{\partial x} & \dfrac{\partial}{\partial y} & \dfrac{\partial}{\partial z} \\ F_1 & F_2 & F_3 \end{Vmatrix} = \begin{Vmatrix} i & j & k \\ \dfrac{\partial}{\partial x} & \dfrac{\partial}{\partial y} & \dfrac{\partial}{\partial z} \\ -z & x & \dfrac{y^2 z}{2} \end{Vmatrix} = (yz, -1, 1)$$

Let $S: z = \sqrt{x^2 + y^2}$, $z \le a$ be oriented upwards

By Stokes' Theorem, $\oint_C \vec{F} \cdot d\vec{r} = \iint_S (\nabla \times \vec{F}) \cdot \vec{n}\, dA$

∵ S is upward-oriented

9.6.2 Convert Closed Line Integrals into Surface Integrals 560

Let $\vec{r}(x, y, z) = \left(x, y, (x^2 + y^2)^{\frac{1}{2}}\right)$ and let R be the closed region of the projection of S onto the xy-plane. Then

$R = \{(x, y): x^2 + y^2 \leq a^2\}$ and $\iint_S (\nabla \times \vec{F}) \cdot \vec{n} \, dA = \iint_R (\nabla \times \vec{F}) \cdot \frac{\partial \vec{r}}{\partial x} \times \frac{\partial \vec{r}}{\partial y} dxdy$

$\because \frac{\partial \vec{r}}{\partial x} = \left(1, 0, x(x^2 + y^2)^{-\frac{1}{2}}\right)$ and $\frac{\partial \vec{r}}{\partial y} = \left(0, 1, y(x^2 + y^2)^{-\frac{1}{2}}\right)$

$\therefore \frac{\partial \vec{r}}{\partial x} \times \frac{\partial \vec{r}}{\partial y} = \left(-x(x^2 + y^2)^{-\frac{1}{2}}, -y(x^2 + y^2)^{-\frac{1}{2}}, 1\right)$

$\Rightarrow (\nabla \times \vec{F}) \cdot \frac{\partial \vec{r}}{\partial x} \times \frac{\partial \vec{r}}{\partial y} = (yz, -1, 1) \cdot \left(-x(x^2 + y^2)^{-\frac{1}{2}}, -y(x^2 + y^2)^{-\frac{1}{2}}, 1\right)$

$= -xy + y(x^2 + y^2)^{-\frac{1}{2}} + 1, \quad \forall (x, y, z) \in S$

$\therefore \iint_S (\nabla \times \vec{F}) \cdot \vec{n} \, dA = \iint_R (\nabla \times \vec{F}) \cdot \frac{\partial \vec{r}}{\partial x} \times \frac{\partial \vec{r}}{\partial y} dxdy$

$= \iint_{x^2+y^2 \leq a^2} -xy + y(x^2 + y^2)^{-\frac{1}{2}} + 1 \, dxdy$

Let $x = r\cos\theta$, $y = r\sin\theta$ then $\{(x, y): x^2 + y^2 \leq a^2\} = \{(r, \theta): 0 \leq r \leq a, 0 \leq \theta \leq 2\pi\}$

$\therefore \iint_{x^2+y^2 \leq a^2} -xy + y(x^2 + y^2)^{-\frac{1}{2}} + 1 \, dxdy = \int_0^{2\pi} \int_0^a (-r^2 \cos\theta \sin\theta + \sin\theta + 1) \, r dr d\theta$

$= \pi a^2$

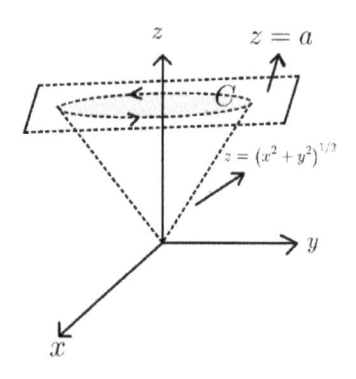

Example 5.

Let $\vec{F} = \left(0, \frac{2x^3}{3}, 2y^3\right)$. The curve C is the counterclockwise-oriented closed curve bounded by the points (1,0,0), (0,1,0), (0,0,1). Find $\oint_C \vec{F} \cdot d\vec{r}$: (1)Direct calculation (2)Using Stoke's Theorem.

【Solution】

(1)

Let $C_1: (-t + 1, t, 0)$, $C_2: (0, -t + 1, t)$, $C_3: (t, 0, -t + 1)$ then $C = C_1 \cup C_2 \cup C_3$

$$\therefore \oint_C \vec{F} \cdot d\vec{r} = \int_{C_1} \vec{F} \cdot d\vec{r} + \int_{C_2} \vec{F} \cdot d\vec{r} + \int_{C_3} \vec{F} \cdot d\vec{r}$$

$$\therefore \int_{C_1} \vec{F} \cdot d\vec{r} = \int_0^1 \frac{2(1-t)^3}{3} dt = -\int_1^0 \frac{2u^3}{3} du = \frac{1}{6}$$

$$\int_{C_2} \vec{F} \cdot d\vec{r} = \int_0^1 2(1-t)^3 dt = -\int_1^0 2u^3 du = \frac{1}{2} \quad \text{and} \quad \int_{C_3} \vec{F} \cdot d\vec{r} = 0$$

$$\therefore \oint_C \vec{F} \cdot d\vec{r} = \frac{2}{3}$$

(2)

$$\nabla \times \vec{F} = \begin{vmatrix} i & j & k \\ \frac{\partial}{\partial x} & \frac{\partial}{\partial y} & \frac{\partial}{\partial z} \\ F_1 & F_2 & F_3 \end{vmatrix} = \begin{vmatrix} i & j & k \\ \frac{\partial}{\partial x} & \frac{\partial}{\partial y} & \frac{\partial}{\partial z} \\ 0 & \frac{2x^3}{3} & 2y^3 \end{vmatrix} = (6y^2, 0, 2x^2)$$

Let S be the upward-oriented plane enclosed by (1,0,0), (0,1,0), (0,0,1) in the first quadrant

By Stokes' Theorem, $\oint_C \vec{F} \cdot d\vec{r} = \iint_S (\nabla \times \vec{F}) \cdot \vec{n} dA$

$\because$ S is upward-oriented

Let $\vec{r}(x, y, z) = (x, y, 1 - x - y)$ and let R be the closed region of the projection of S onto the xy-plane, then

$$R = \{(x,y): x + y \le 1, x \ge 0, y \ge 0\} \text{ and } \iint_S (\nabla \times \vec{F}) \cdot \vec{n} \, dA = \iint_R (\nabla \times \vec{F}) \cdot \frac{\partial \vec{r}}{\partial x} \times \frac{\partial \vec{r}}{\partial y} dxdy$$

$\because \frac{\partial \vec{r}}{\partial x} = (1, 0, -1)$ and $\frac{\partial \vec{r}}{\partial y} = (0, 1, -1)$ $\therefore \frac{\partial \vec{r}}{\partial x} \times \frac{\partial \vec{r}}{\partial y} = (1,1,1)$

$\therefore (\nabla \times \vec{F}) \cdot \frac{\partial \vec{r}}{\partial x} \times \frac{\partial \vec{r}}{\partial y} = (6y^2, 0, 2x^2) \cdot \frac{\partial \vec{r}}{\partial x} \times \frac{\partial \vec{r}}{\partial y} = 2(x^2 + 3y^2), \ \forall (x, y, z) \in S$

$\therefore \iint_S (\nabla \times \vec{F}) \cdot \vec{n} \, dA = \iint_R (\nabla \times \vec{F}) \cdot \frac{\partial \vec{r}}{\partial x} \times \frac{\partial \vec{r}}{\partial y} dxdy = 2\iint_R x^2 + 3y^2 \, dxdy$

9.6.2 Convert Closed Line Integrals into Surface Integrals

$$= 2\int_0^1 \int_0^{1-y} x^2 + 3y^2 \, dxdy = 2\int_0^1 \left(\frac{x^3}{3} + 3y^2 x\right)\Big|_{x=0}^{x=1-y} dy = 2\int_0^1 \frac{(1-y)^3}{3} + 3y^2(1-y) \, dy$$

$$= \frac{2}{3}$$

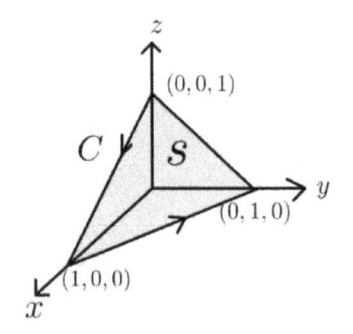

Example 6.

Let $\vec{F} = (yz, 0, 0)$. The curve C is the counterclockwise-oriented closed circle given by $x^2 + z^2 = 1, y = 1$. Find $\oint_C \vec{F} \cdot d\vec{r} = ?$: (1) Direct calculation (2) Using Stoke's Theorem.

【Solution】

(1)

Let $C = \{(x, y, z): x^2 + z^2 = 1, y = 1\}$ and let $\vec{r}(\theta) = (\cos\theta, 1, \sin\theta)$ then $\vec{r}'(\theta) = (-\sin\theta, 0, \cos\theta)$

$$\oint_C \vec{F} \cdot d\vec{r} = \int_0^{2\pi} \vec{F}(\vec{r}(\theta)) \cdot \vec{r}'(\theta) \, d\theta = \int_0^{2\pi} (\sin\theta, 0, 0) \cdot (-\sin\theta, 0, \cos\theta) \, d\theta$$

$$= -\int_0^{2\pi} \sin^2\theta \, d\theta = \int_0^{2\pi} \frac{1 - \cos\theta}{2} \, d\theta = -\pi$$

(2)

$$\nabla \times \vec{F} = \begin{Vmatrix} i & j & k \\ \frac{\partial}{\partial x} & \frac{\partial}{\partial y} & \frac{\partial}{\partial z} \\ F_1 & F_2 & F_3 \end{Vmatrix} = \begin{Vmatrix} i & j & k \\ \frac{\partial}{\partial x} & \frac{\partial}{\partial y} & \frac{\partial}{\partial z} \\ yz & 0 & 0 \end{Vmatrix} = (0, y, -z)$$

Let $S = \{(x, y, z): y = x^2 + z^2, 0 \le y \le 1\}$ be oriented in the positive direction of the y-axis

By Stokes' Theorem, $\oint_C \vec{F} \cdot d\vec{r} = \iint_S (\nabla \times \vec{F}) \cdot \vec{n} dA$

Let $\vec{r}(x, y, z) = (x, x^2 + z^2, z)$ and let R be the closed region of the projection of S

onto the xz-plane. Then

$$R = \{(x,y): x^2 + z^2 \leq 1\} \text{ and } \iint_S (\nabla \times \vec{F}) \cdot \vec{n} dA = \iint_R (\nabla \times \vec{F}) \cdot \frac{\partial \vec{r}}{\partial x} \times \frac{\partial \vec{r}}{\partial z} dxdz$$

$$\because \frac{\partial \vec{r}}{\partial x} = (1, 2x, 0) \text{ and } \frac{\partial \vec{r}}{\partial z} = (0, 2z, 1) \quad \therefore \frac{\partial \vec{r}}{\partial x} \times \frac{\partial \vec{r}}{\partial z} = (2x, -1, 2z)$$

$$\therefore (\nabla \times \vec{F}) \cdot \frac{\partial \vec{r}}{\partial x} \times \frac{\partial \vec{r}}{\partial z} = (0, y, -z) \cdot (2x, -1, 2z) = -y - 2z^2 = -x^2 - 3z^2, \ \forall (x,y,z) \in S$$

$$\therefore \iint_S (\nabla \times \vec{F}) \cdot \vec{n} dA = \iint_R (\nabla \times \vec{F}) \cdot \frac{\partial \vec{r}}{\partial x} \times \frac{\partial \vec{r}}{\partial z} dxdz = \iint_{x^2 + z^2 \leq 1} -x^2 - 3z^2 dxdz$$

Let $x = r\cos\theta, z = r\sin\theta$ then $\{(x,y): x^2 + z^2 \leq 1\} = \{(r, \theta): 0 \leq r \leq 1, 0 \leq \theta \leq 2\pi\}$

and $dxdz = \begin{Vmatrix} \dfrac{\partial x}{\partial r} & \dfrac{\partial x}{\partial \theta} \\ \dfrac{\partial z}{\partial r} & \dfrac{\partial z}{\partial \theta} \end{Vmatrix} drd\theta = \begin{Vmatrix} \cos\theta & -r\sin\theta \\ \sin\theta & r\cos\theta \end{Vmatrix} drd\theta = rdrd\theta$

$$\therefore \iint_{x^2+z^2 \leq 1} -x^2 - 3z^2 dxdz = \int_0^{2\pi} \int_0^1 -r^3 \cos^2\theta - 3r^3 \sin^2\theta \, drd\theta$$

$$= \int_0^{2\pi} -\frac{1}{4}\left(\frac{1+\cos\theta}{2}\right) - \frac{3}{4}\left(\frac{1-\cos\theta}{2}\right) d\theta = -\pi$$

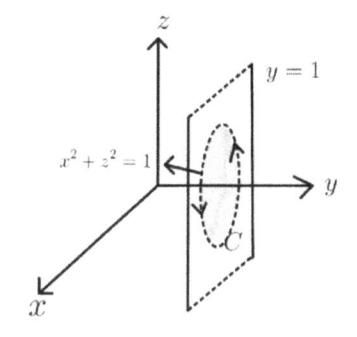

Example 7.
Assume C is any positively oriented closed curve on the plane $x + y + z = 2$. Prove that

$$\oint_C zdx - 2xdy + 4ydz$$ depends only on the area enclosed by C, regardless of the shape and position of C.

【Solution】

Let the closed curve C enclose the region R on the plane $x + y + z = 2$
Let $\vec{F} = (z, -2x, 4y)$,

9.6.2 Convert Closed Line Integrals into Surface Integrals

∵ the first-order partial derivatives of the components of $\vec{F}$ exist and are continuous

By Stokes' Theorem, $\oint_C z\,dx - 2x\,dy + 4y\,dz = \oint_C \vec{F} \cdot d\vec{r} = \iint_R (\nabla \times \vec{F}) \cdot \vec{n}\,dA$

Let $\vec{n}$ be the unit normal vector to the plane $x + y + z = 2$ then $\vec{n} = \left(\dfrac{1}{\sqrt{3}}, \dfrac{1}{\sqrt{3}}, \dfrac{1}{\sqrt{3}}\right)$

$\because \nabla \times \vec{F} = \begin{vmatrix} \vec{i} & \vec{j} & \vec{k} \\ \dfrac{\partial}{\partial x} & \dfrac{\partial}{\partial y} & \dfrac{\partial}{\partial z} \\ z & -2x & 4y \end{vmatrix} = (4, 1, -2)$

$\therefore \iint_R (\nabla \times \vec{F}) \cdot \vec{n}\,dA = \iint_R (4, 1, -2) \cdot \left(\dfrac{1}{\sqrt{3}}, \dfrac{1}{\sqrt{3}}, \dfrac{1}{\sqrt{3}}\right) dA = \dfrac{3}{\sqrt{3}} \iint_R dA$

$\therefore \oint_C z\,dx - 2x\,dy + 4y\,dz = \sqrt{3} \iint_R dA$

∴ The line integral $\oint_C z\,dx - 2x\,dy + 4y\,dz$ depends only on the area enclosed by C, and is independent of the shape and position of C

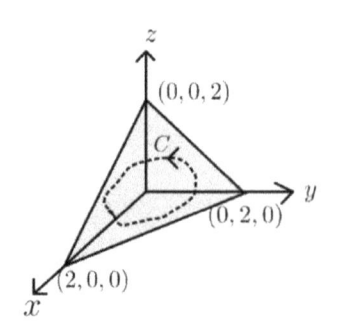

Example 8.

Using Stokes' Theorem, find $\oint_C \vec{F} \cdot d\vec{r} = ?$, $\vec{F} = (z^2, y^2, x)$, where C is the counterclockwise-oriented closed curve formed by the points $(a, 0,0), (0, a, 0), (0,0, a)$.

【Solution】

Let $S = \{(x, y, z): x, y, z \geq 0, x + y + z = a\}$ be the upward-oriented plane

∵ the first-order partial derivatives of the components of $\vec{F}$ exist and are continuous

By Stokes' Theorem, $\oint_C z^2 dx + y^2 dy + x dz = \oint_C \vec{F} \cdot d\vec{r} = \iint_S (\nabla \times \vec{F}) \cdot \vec{n} dA$

$\because$ S is upward-oriented

Let $\vec{r}(x,y) = (x, y, a-x-y)$ and R = $\{(x,y): 0 \leq x+y \leq a\}$ then

$$\iint_S (\nabla \times \vec{F}) \cdot \vec{n} dA = \iint_R (\nabla \times \vec{F}) \cdot \frac{\partial \vec{r}}{\partial x} \times \frac{\partial \vec{r}}{\partial y} dx dy$$

$\because \frac{\partial \vec{r}}{\partial x} = (1, 0, -1), \quad \frac{\partial \vec{r}}{\partial y} = (0, 1, -1) \quad \therefore \frac{\partial \vec{r}}{\partial x} \times \frac{\partial \vec{r}}{\partial y} = (1, 1, 1)$

$\because \nabla \times \vec{F} = \begin{vmatrix} \vec{i} & \vec{j} & \vec{k} \\ \frac{\partial}{\partial x} & \frac{\partial}{\partial y} & \frac{\partial}{\partial z} \\ z^2 & y^2 & x \end{vmatrix} = (0, 2z-1, 0) = (0, 2(a-x-y)-1, 0), \quad \forall (x,y,z) \in S$

$\therefore \iint_R (\nabla \times \vec{F}) \cdot \frac{\partial \vec{r}}{\partial x} \times \frac{\partial \vec{r}}{\partial y} dx dy = \iint_{0 \leq x+y \leq a} 2(a-x-y) - 1 dx dy$

By Fubini's Theorem,

$$\iint_{0 \leq x+y \leq a} 2(a-x-y) - 1 dx dy = \int_0^a \int_0^{a-y} (2a-1) - 2x - 2y \, dx dy$$

$$= \int_0^a a(a-1) - y(2a-1) + y^2 dy = \frac{a^3}{3} - \frac{a^2}{2}$$

Example 9.

Let $\vec{F} = \left(x^2 z, 0, -\frac{y^3}{3}\right)$. The curve C is the counterclockwise-oriented closed curve formed by the points $(a, 0, 0), (0, a, 0), (0, 0, a)$ (where $a > 0$). Find $\oint_C \vec{F} \cdot d\vec{r} =$?:

(1) Direct calculation (2) Using Stoke's Theorem.

【Solution】

(1)

Let $C_1: (-at + a, at, 0), \quad C_2: (0, -at + a, at), \quad C_3: (at, 0, -at + a)$ then $C = C_1 \cup C_2 \cup C_3$

$\therefore \oint_C \vec{F} \cdot d\vec{r} = \int_{C_1} \vec{F} \cdot d\vec{r} + \int_{C_2} \vec{F} \cdot d\vec{r} + \int_{C_3} \vec{F} \cdot d\vec{r}$

$\because \int_{C_1} \vec{F} \cdot d\vec{r} = -\int_0^1 \frac{(at)^3}{3} dt = 0$

9.6.2 Convert Closed Line Integrals into Surface Integrals

$$\int_{C_2} \vec{F} \cdot d\vec{r} = -a^3 \int_0^1 \frac{(1-t)^3}{3} dt = -\frac{a^3}{12} \quad \text{and} \quad \int_{C_3} \vec{F} \cdot d\vec{r} = a^3 \int_0^1 t^2(1-t)dt = \frac{a^3}{12}$$

$$\therefore \oint_C \vec{F} \cdot d\vec{r} = \int_{C_1} \vec{F} \cdot d\vec{r} + \int_{C_2} \vec{F} \cdot d\vec{r} + \int_{C_3} \vec{F} \cdot d\vec{r} = 0$$

(2)

$$\nabla \times \vec{F} = \begin{vmatrix} i & j & k \\ \frac{\partial}{\partial x} & \frac{\partial}{\partial y} & \frac{\partial}{\partial z} \\ F_1 & F_2 & F_3 \end{vmatrix} = \begin{vmatrix} i & j & k \\ \frac{\partial}{\partial x} & \frac{\partial}{\partial y} & \frac{\partial}{\partial z} \\ x^2 z & 0 & -\frac{y^3}{3} \end{vmatrix} = (-y^2, x^2, 0)$$

Let S be the upward-oriented plane enclosed by $(a,0,0)$, $(0,a,0)$, and $(0,0,a)$ in the first quadrant

By Stokes' Theorem, $\oint_C \vec{F} \cdot d\vec{r} = \iint_S (\nabla \times \vec{F}) \cdot \vec{n} dA$

∵ S is upward-oriented

Let $\vec{r}(x,y,z) = (x, y, a-x-y)$ and $R = \{(x,y): 0 \le x \le a, 0 \le y \le a-x\}$ then

$$\iint_S (\nabla \times \vec{F}) \cdot \vec{n} \, dA = \iint_R (\nabla \times \vec{F}) \cdot \frac{\partial \vec{r}}{\partial x} \times \frac{\partial \vec{r}}{\partial y} dxdy$$

∵ $\frac{\partial \vec{r}}{\partial x} = (1, 0, -1)$ and $\frac{\partial \vec{r}}{\partial y} = (0, 1, -1)$ ∴ $\frac{\partial \vec{r}}{\partial x} \times \frac{\partial \vec{r}}{\partial y} = (1,1,1)$

∵ $(\nabla \times \vec{F}) \cdot \frac{\partial \vec{r}}{\partial x} \times \frac{\partial \vec{r}}{\partial y} = (-y^2, x^2, 0) \cdot (1,1,1) = x^2 - y^2, \; \forall (x,y,z) \in S$

$$\therefore \iint_R (\nabla \times \vec{F}) \cdot \frac{\partial \vec{r}}{\partial x} \times \frac{\partial \vec{r}}{\partial y} dxdy = \int_0^a \int_0^{a-x} x^2 - y^2 \, dydx = \int_0^a x^2(a-x) - \frac{(a-x)^3}{3} dx$$

$$= \int_0^a -\frac{a^3}{3} + a^2 x - \frac{2x^3}{3} dx = a^4 \left(-\frac{1}{3} + \frac{1}{2} - \frac{1}{6}\right) = 0$$

Example 10.

Let $\vec{F} = \left(0, -\frac{x^2 y}{2}, -\frac{x^2 z}{2}\right)$. Given that $S: z = xy, 0 \le x \le 1, 0 \le y \le 2$ and C is the positively oriented boundary curve of surface S, use Stoke's Theorem to find

$$\oint_C \vec{F} \cdot d\vec{r} = ?$$

【Solution】

$$\nabla \times \vec{F} = \begin{Vmatrix} i & j & k \\ \frac{\partial}{\partial x} & \frac{\partial}{\partial y} & \frac{\partial}{\partial z} \\ F_1 & F_2 & F_3 \end{Vmatrix} = \begin{Vmatrix} i & j & k \\ \frac{\partial}{\partial x} & \frac{\partial}{\partial y} & \frac{\partial}{\partial z} \\ 0 & -\frac{x^2 y}{2} & -\frac{x^2 z}{2} \end{Vmatrix} = (0, xz, -xy)$$

By Stokes' Theorem, $\oint_C \vec{F} \cdot d\vec{r} = \iint_S (\nabla \times \vec{F}) \cdot \vec{n} dA$

Let $\vec{r}(x, y, z) = (x, y, xy)$ and $R = \{(x, y): 0 \le x \le 1, 0 \le y \le 2\}$ then

$$\iint_S (\nabla \times \vec{F}) \cdot \vec{n} \, dA = \iint_R (\nabla \times \vec{F}) \cdot \frac{\partial \vec{r}}{\partial x} \times \frac{\partial \vec{r}}{\partial y} dx dy$$

$\because \frac{\partial \vec{r}}{\partial x} = (1, 0, y)$ and $\frac{\partial \vec{r}}{\partial y} = (0, 1, x)$ $\therefore \frac{\partial \vec{r}}{\partial x} \times \frac{\partial \vec{r}}{\partial y} = (-y, -x, 1)$

$\therefore (\nabla \times \vec{F}) \cdot \frac{\partial \vec{r}}{\partial x} \times \frac{\partial \vec{r}}{\partial y} = (0, xz, -xy) \cdot (-y, -x, 1) = -x^2 z - xy = -x^3 y - xy, \ \forall (x, y, z) \in S$

$$\iint_R (\nabla \times \vec{F}) \cdot \frac{\partial \vec{r}}{\partial x} \times \frac{\partial \vec{r}}{\partial y} dx dy = \int_0^1 \int_0^2 -x^3 y - xy \, dy dx = \int_0^a -2x^3 - 2x \, dx = \left(-\frac{x^4}{2} - x^2 \right) \Big|_0^1$$

$= -\frac{3}{2}$

Example 11.

Use Stokes' Theorem to find $\oint_C (e^x y - y^3) dx + (e^x + x^3) dy = ?$, where C is a counterclockwise-oriented closed circle with radius 6: $x = 6 \cos \theta$, $y = 6 \sin \theta$, $0 \le \theta \le 2\pi$

【Solution】

Let $S = \{(x, y, z): x^2 + y^2 \le 36, z = 0\}$ be oriented upwards
Let $\vec{F} = (e^x y - y^3, e^x + x^3, 0)$,
∵ the first-order partial derivatives of the components of $\vec{F}$ exist and are continuous

9.6.2 Convert Closed Line Integrals into Surface Integrals

By Stokes' Theorem, $\oint_C \vec{F} \cdot d\vec{r} = \iint_S (\nabla \times \vec{F}) \cdot \vec{n} dA$

∵ S is upward-oriented

Let $\vec{r}(x,y) = (x, y, 0)$ and $R = \{(x,y): x^2 + y^2 \leq 36\}$ then

$$\iint_S (\nabla \times \vec{F}) \cdot \vec{n} dA = \iint_R (\nabla \times \vec{F}) \cdot \frac{\partial \vec{r}}{\partial x} \times \frac{\partial \vec{r}}{\partial y} dxdy$$

∵ $\frac{\partial \vec{r}}{\partial x} = (1,0,0)$, $\frac{\partial \vec{r}}{\partial y} = (0,1,0)$ ∴ $\frac{\partial \vec{r}}{\partial x} \times \frac{\partial \vec{r}}{\partial y} = (0,0,1)$

∵ $\nabla \times \vec{F} = \begin{vmatrix} \vec{i} & \vec{j} & \vec{k} \\ \frac{\partial}{\partial x} & \frac{\partial}{\partial y} & \frac{\partial}{\partial z} \\ e^x y - y^3 & e^x + x^3 & 0 \end{vmatrix} = (0, 0, 3x^2 + 3y^2)$

∴ $\iint_R (\nabla \times \vec{F}) \cdot \frac{\partial \vec{r}}{\partial x} \times \frac{\partial \vec{r}}{\partial y} dxdy = \iint_R (0,0,3x^2+3y^2) \cdot (0,0,1) dxdy$

$= 3 \iint_{x^2+y^2 \leq 36} x^2 + y^2 dxdy$

Let $x = r\cos\theta$, $y = r\sin\theta$ then $\{(x,y): x^2+y^2 \leq 36\} = \{(r,\theta): 0 \leq r \leq 6, 0 \leq \theta \leq 2\pi\}$

and $dxdy = \begin{Vmatrix} \frac{\partial x}{\partial r} & \frac{\partial x}{\partial \theta} \\ \frac{\partial y}{\partial r} & \frac{\partial y}{\partial \theta} \end{Vmatrix} drd\theta = \begin{Vmatrix} \cos\theta & -r\sin\theta \\ \sin\theta & r\cos\theta \end{Vmatrix} drd\theta = rdrd\theta$

∴ $3 \iint_{x^2+y^2 \leq 36} x^2 + y^2 dxdy = 3 \int_0^{2\pi} \int_0^6 r^3 drd\theta = 1944\pi$

Example 12.

Use Stokes' Theorem to find $\oint_C y^3 dx - x^3 dy = ?$, where C is a counterclockwise-oriented closed circle with radius 2: $x^2 + y^2 = 4$.

【Solution】

Let $S = \{(x,y,z): x^2 + y^2 \leq 4, z = 0\}$ be oriented upwards

Let $\vec{F} = (y^3, -x^3, 0)$,

∵ the first-order partial derivatives of the components of $\vec{F}$ exist and are continuous

By Stokes' Theorem, $\oint_C \vec{F} \cdot d\vec{r} = \iint_S (\nabla \times \vec{F}) \cdot \vec{n} dA$

∵ S is upward-oriented

Let $\vec{r}(x, y) = (x, y, 0)$ and $R = \{(x, y): x^2 + y^2 \leq 4\}$ then

$$\iint_S (\nabla \times \vec{F}) \cdot \vec{n} dA = \iint_R (\nabla \times \vec{F}) \cdot \frac{\partial \vec{r}}{\partial x} \times \frac{\partial \vec{r}}{\partial y} dxdy$$

∵ $\frac{\partial \vec{r}}{\partial x} = (1,0,0)$, $\frac{\partial \vec{r}}{\partial y} = (0,1,0)$ ∴ $\frac{\partial \vec{r}}{\partial x} \times \frac{\partial \vec{r}}{\partial y} = (0,0,1)$

∵ $\nabla \times \vec{F} = \begin{vmatrix} \vec{i} & \vec{j} & \vec{k} \\ \frac{\partial}{\partial x} & \frac{\partial}{\partial y} & \frac{\partial}{\partial z} \\ y^3 & -x^3 & 0 \end{vmatrix} = (0,0, -3x^2 - 3y^2)$

∴ $\iint_R (\nabla \times \vec{F}) \cdot \frac{\partial \vec{r}}{\partial x} \times \frac{\partial \vec{r}}{\partial y} dxdy = \iint_R (0,0, -3x^2 - 3y^2) \cdot (0,0,1) dxdy$

$= -3 \iint_{x^2+y^2 \leq 4} x^2 + y^2 dxdy$

Let $x = r\cos\theta$, $y = r\sin\theta$ then $\{(x,y): x^2 + y^2 \leq 4\} = \{(r, \theta): 0 \leq r \leq 2, 0 \leq \theta \leq 2\pi\}$

and $dxdy = \begin{Vmatrix} \frac{\partial x}{\partial r} & \frac{\partial x}{\partial \theta} \\ \frac{\partial y}{\partial r} & \frac{\partial y}{\partial \theta} \end{Vmatrix} drd\theta = \begin{Vmatrix} \cos\theta & -r\sin\theta \\ \sin\theta & r\cos\theta \end{Vmatrix} drd\theta = r drd\theta$

∴ $-3 \iint_{x^2+y^2 \leq 4} x^2 + y^2 dxdy = -3 \int_0^{2\pi} \int_0^2 r^3 drd\theta = -24\pi$

9.6.3 Convert Surface Integrals into Closed Line Integrals

Similarly, Stokes' Theorem can be used to convert the problem of a non-closed surface integral into a closed curve integral.

Question Types:

Type 1.

Use Stokes' Theorem to solve the following

9.6.3 Convert Surface Integrals into Closed Line Integrals

$$\iint_S (\nabla \times \vec{F}) \cdot \vec{n} dA = ?,$$

where S is an upward-oriented, non-closed smooth surface with boundary:
$$\partial S = \{(x, y, z): x^2 + y^2 = a^2, z = c\}.$$

The first partial derivatives of each component of $\vec{F}$ exist and are continuous.

Problem-Solving Process:

Step1.
Let $C = \{(x, y, z): x^2 + y^2 = a^2, z = c\}$

Step2.
Let $x = a\cos\theta$, $y = a\sin\theta$ then $\vec{r}(\theta) = (a\cos\theta, a\sin\theta, c)$
$\therefore \vec{r}'(\theta) = (-a\sin\theta, a\cos\theta, 0)$

Step3.
By Stokes' Theorem

$$\iint_S (\nabla \times \vec{F}) \cdot \vec{n} dA = \oint_C \vec{F} \cdot d\vec{r} = \int_0^{2\pi} \vec{F}(\vec{r}(\theta)) \cdot \vec{r}'(\theta) d\theta$$

$$= \int_0^{2\pi} (f_1(a\cos\theta, a\sin\theta, c), f_2(a\cos\theta, a\sin\theta, c)) \cdot (-a\sin\theta, a\cos\theta) d\theta$$

where $\vec{F} = (f_1(x, y, z), f_2(x, y, z), f_3(x, y, z))$

Step4.

Find $\int_0^{2\pi} (f_1(a\cos\theta, a\sin\theta, c), f_2(a\cos\theta, a\sin\theta, c)) \cdot (-a\sin\theta, a\cos\theta) d\theta = ?$

If S is an downward-oriented, non-closed smooth surface, then calculate

$$-\int_0^{2\pi} (f_1(a\cos\theta, a\sin\theta, c), f_2(a\cos\theta, a\sin\theta, c)) \cdot (-a\sin\theta, a\cos\theta) d\theta$$

Examples:

(1) $\vec{F} = (-y^3 \cos z, x^3 e^z, -e^z)$ and S is the upward-oriented hemisphere defined by the equation $x^2 + y^2 + (z-a)^2 = 2a^2, z > 0$.

$$\text{Find } \iint_S (\nabla \times \vec{F}) \cdot \vec{n} dA = ?$$

Let $C = \{(x, y, z): x^2 + y^2 = a^2, z = 0\}$
Let $x = a\cos\theta$, $y = a\sin\theta$ then $\vec{r}(\theta) = (a\cos\theta, a\sin\theta, 0)$
$\therefore \vec{r}'(\theta) = (-a\sin\theta, a\cos\theta, 0)$

By Stokes' Theorem,

$$\iint_S (\nabla \times \vec{F}) \cdot \vec{n} dA = \oint_C \vec{F} \cdot d\vec{r} = \int_0^{2\pi} \vec{F}(\vec{r}(\theta)) \cdot \vec{r}'(\theta) d\theta$$

$$= \int_0^{2\pi} (-(a\sin\theta)^3, (a\cos\theta)^3, 1) \cdot (-a\sin\theta, a\cos\theta, 0)d\theta = \frac{3a^4\pi}{2}$$

(II) $\vec{F} = (2x - y, 2y + z, xyz)$ and S is the upward-oriented hemisphere defined by the equation $\frac{x^2}{a^2} + \frac{y^2}{b^2} + \frac{z^2}{c^2} = 1, \ z > 0$.

Find $\iint_S (\nabla \times \vec{F}) \cdot \vec{n} dA = ?$

Let $C = \{(x,y,z): \frac{x^2}{a^2} + \frac{y^2}{b^2} = 1, z = 0\}$

Let $x = a\cos\theta, \ y = b\sin\theta$ then $\vec{r}(\theta) = (a\cos\theta, b\sin\theta, 0)$

$\therefore \vec{r}'(\theta) = (-a\sin\theta, b\cos\theta, 0)$

By Stokes' Theorem,

$$\iint_S (\nabla \times \vec{F}) \cdot \vec{n} dA = \oint_C \vec{F} \cdot d\vec{r} = \int_0^{2\pi} \vec{F}(\vec{r}(\theta)) \cdot \vec{r}'(\theta) d\theta$$

$$= \int_0^{2\pi} (2a\cos\theta - b\sin\theta, 2b\sin\theta, 0) \cdot (-a\sin\theta, b\cos\theta, 0) d\theta = ab\pi$$

Type 2.

Use Stokes' Theorem to find

$$\iint_S (\nabla \times \vec{F}) \cdot \vec{n} dA = ?,$$

where S is an upward-oriented, non-closed smooth surface, and the closed boundary C of S is a piecewise smooth curve, that is, $C = \bigcup_{i=1}^{n} C_i$ where $C_i : \vec{r}_i(t)$. The first partial derivatives of each component of $\vec{F}$ exist and are continuous.

Problem-Solving Process:

Step 1.

By Stokes' Theorem, $\iint_S (\nabla \times \vec{F}) \cdot \vec{n} dA = \oint_C \vec{F} \cdot d\vec{r} = \sum_{i=1}^{n} \int_{C_i} \vec{F}(\vec{r}_i(t)) \cdot \vec{r}_i'(t) dt$

Step 2.

Find $\sum_{i=1}^{n} \int_{C_i} \vec{F}(\vec{r}_i(t)) \cdot \vec{r}_i'(t) dt$

9.6.3 Convert Surface Integrals into Closed Line Integrals

Example 1.

Find $\iint_S (\nabla \times \vec{F}) \cdot \vec{n} dA = ?$, where $S = \{(x,y,z): x^2 + y^2 \leq 1, z = 0\}$ with upward orientation, and $\vec{F} = (y^2, x^2, 0)$: (1)Direct calculation (2)Using Stoke's Theorem.

【Solution】

(1)

$$\therefore \nabla \times \vec{F} = \begin{vmatrix} \vec{i} & \vec{j} & \vec{k} \\ \frac{\partial}{\partial x} & \frac{\partial}{\partial y} & \frac{\partial}{\partial z} \\ y^2 & x^2 & 0 \end{vmatrix} = (0, 0, 2x - 2y)$$

∵ S is upward-oriented

Let $\vec{r}(x, y) = (x, y, 0)$, and let R be the closed region of the projection of S onto the xy-plane. Then

$$R = \{(x,y): x^2 + y^2 \leq 1\} \text{ and } \iint_S (\nabla \times \vec{F}) \cdot \vec{n} dA = \iint_R (\nabla \times \vec{F}) \cdot \frac{\partial \vec{r}}{\partial x} \times \frac{\partial \vec{r}}{\partial y} dxdy$$

$\because \frac{\partial \vec{r}}{\partial x} = (1,0,0), \frac{\partial \vec{r}}{\partial y} = (0,1,0) \ \therefore \frac{\partial \vec{r}}{\partial x} \times \frac{\partial \vec{r}}{\partial y} = (0,0,1)$

$\therefore \iint_S (\nabla \times \vec{F}) \cdot \vec{n} dA = 2 \iint_{x^2+y^2 \leq 1} x - y \, dxdy$

Let $x = r\cos\theta, y = r\sin\theta$ then $2 \iint_{x^2+y^2 \leq 1} x - y \, dxdy = 2 \int_0^{2\pi} \int_0^1 (r\cos\theta - r\sin\theta) r \, dr d\theta$

$\because \int_0^{2\pi} \cos\theta - \sin\theta \, d\theta = 0 \ \therefore \iint_S (\nabla \times \vec{F}) \cdot \vec{n} dA = 2 \int_0^{2\pi} \int_0^1 (r\cos\theta - r\sin\theta) r \, dr d\theta = 0$

(2)

Let C be the counterclockwise closed curve formed by the intersection of the surface S and $z = 0$ then $C = \{(x, y, z): x^2 + y^2 = 1, z = 0\}$

Let $\vec{r}(\theta) = (\cos\theta, \sin\theta, 0), \ 0 \leq \theta \leq 2\pi \ \therefore \vec{r}'(\theta) = (-\sin\theta, \cos\theta, 0)$

∵ the first-order partial derivatives of the components of $\vec{F}$ exist and are continuous

By Stokes' Theorem,

$$\iint_S (\nabla \times \vec{F}) \cdot \vec{n} dA = \oint_C \vec{F} \cdot d\vec{r} = \int_0^{2\pi} \vec{F}(\vec{r}(\theta)) \cdot \vec{r}'(\theta) \, d\theta$$

$$= \int_0^{2\pi} (\sin^2, \cos^2, 0) \cdot (-\sin\theta, \cos\theta, 0) \, d\theta = \int_0^{2\pi} -\sin^3\theta + \cos^3\theta \, d\theta$$

$\because -\sin^3\theta + \cos^3\theta = (\cos\theta - \sin\theta)(\cos^2\theta + \sin\theta\cos\theta + \sin^2\theta)$
$= (\cos\theta - \sin\theta)(1 + \sin\theta\cos\theta) = (\cos\theta - \sin\theta) - \sin^2\theta\cos\theta + \sin\theta\cos^2\theta$

$$\therefore \int_0^{2\pi} -\sin^3\theta + \cos^3\theta \, d\theta = \int_0^{2\pi} (\cos\theta - \sin\theta) - \sin^2\theta\cos\theta + \sin\theta\cos^2\theta \, d\theta$$

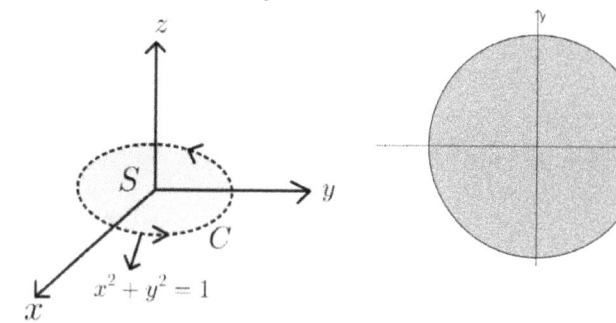

$$= -\frac{\sin^3\theta}{3} + \frac{\cos^3\theta}{3} \Big|_0^{2\pi} = 0$$

Example 2.

Let $\vec{F} = (0, -xz, -xy)$. The surface S is a non-closed parabolic surface oriented downward along the negative y-axis defined by $y = x^2 + z^2$, with $0 \leq y \leq 1$.

Find $\iint_S (\nabla \times \vec{F}) \cdot \vec{n} \, dA = ?$: (1) Direct calculation (2) Using Stokes' Theorem

【Solution】

(1)

$$\because \operatorname{curl} \vec{F} = \nabla \times \vec{F} = \begin{vmatrix} i & j & k \\ \dfrac{\partial}{\partial x} & \dfrac{\partial}{\partial y} & \dfrac{\partial}{\partial z} \\ 0 & -xz & -xy \end{vmatrix} = (0, y, -z)$$

$\because S$ is oriented in the negative direction of the y-axis

Let $\vec{r}(x, y, z) = (x, x^2 + z^2, z)$ and let R be the closed region of the projection of S onto the xz-plane. Then

$$R = \{(x, y): x^2 + z^2 \leq 1\} \text{ and } \iint_S (\nabla \times \vec{F}) \cdot \vec{n} \, dA = -\iint_R (\nabla \times \vec{F}) \cdot \frac{\partial \vec{r}}{\partial x} \times \frac{\partial \vec{r}}{\partial z} \, dx \, dz$$

9.6.3 Convert Surface Integrals into Closed Line Integrals 574

$\because \dfrac{\partial \vec{r}}{\partial x} = (1, 2x, 0)$ and $\dfrac{\partial \vec{r}}{\partial y} = (0, 2z, 1) \quad \therefore \dfrac{\partial \vec{r}}{\partial x} \times \dfrac{\partial \vec{r}}{\partial y} = (2x, -1, 2z)$

$\therefore (\nabla \times \vec{F}) \cdot \dfrac{\partial \vec{r}}{\partial x} \times \dfrac{\partial \vec{r}}{\partial z} = (0, y, -z) \cdot (2x, -1, 2z) = -y - 2z^2 = -x^2 - 3z^2, \quad \forall (x, y, z) \in S$

$\therefore \iint_S (\nabla \times \vec{F}) \cdot \vec{n}\, dA = -\iint_R (\nabla \times \vec{F}) \cdot \dfrac{\partial \vec{r}}{\partial x} \times \dfrac{\partial \vec{r}}{\partial z}\, dxdz = \iint_{x^2+z^2 \le 1} x^2 + 3z^2\, dxdz$

Let $x = r\cos\theta, z = r\sin\theta$ then $\{(x,y): x^2 + z^2 \le 1\} = \{(r,\theta): 0 \le r \le 1, 0 \le \theta \le 2\pi\}$

and $dxdz = \begin{Vmatrix} \dfrac{\partial x}{\partial r} & \dfrac{\partial x}{\partial \theta} \\ \dfrac{\partial z}{\partial r} & \dfrac{\partial z}{\partial \theta} \end{Vmatrix} drd\theta = \begin{Vmatrix} \cos\theta & -r\sin\theta \\ \sin\theta & r\cos\theta \end{Vmatrix} drd\theta = r\, drd\theta$

$\therefore \iint_{x^2+z^2 \le 1} x^2 + 3z^2\, dxdz = \int_0^{2\pi}\int_0^1 r^3\cos^2\theta + 3r^3\sin^2\theta\, drd\theta$

$= \int_0^{2\pi} \dfrac{1}{4}\left(\dfrac{1+\cos\theta}{2}\right) + \dfrac{3}{4}\left(\dfrac{1-\cos\theta}{2}\right) d\theta = \pi$

(2)
Let $C: x^2 + z^2 = 1, \ y = 1,$ be the counterclockwise closed curve
$\because$ S is oriented in the negative direction of the y-axis and the first-order partial derivatives of the components of $\vec{F}$ exist and are continuous

By Stokes' Theorem, $\iint_S (\nabla \times \vec{F}) \cdot \vec{n}\, dA = -\oint_C \vec{F} \cdot d\vec{r}$

Let $\vec{r}(t) = (\cos t, 1, \sin t)$ then $\vec{r}'(t) = (-\sin t, 0, \cos t)$

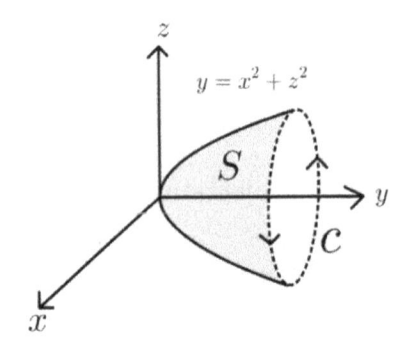

$\therefore -\oint_C \vec{F} \cdot d\vec{r} = -\int_0^{2\pi} \vec{F}(\vec{r}(t)) \cdot \vec{r}'(t)\, dt$

$= -\int_0^{2\pi} (0, -\cos t \sin t, -\cos t) \cdot (-\sin t, 0, \cos t)\, dt = \int_0^{2\pi} \cos^2 t\, dt = \pi$

Example 3.

Use Stokes' Theorem to find $\iint_S (\nabla \times \vec{F}) \cdot \vec{n} dA$, where $\vec{F} = (-y^3 \cos z, x^3 e^z, -e^z)$, $S: x^2 + y^2 + (z-a)^2 = 2a^2$, $z > 0$ is the upper hemisphere oriented upward.

【Solution】

Let C be the counterclockwise closed curve formed by the intersection of the surface S and $z = 0$ then $C = \{(x, y, z): x^2 + y^2 = a^2, z = 0\}$
Let $x = a\cos\theta$, $y = a\sin\theta$ then $\vec{r}(\theta) = (a\cos\theta, a\sin\theta, 0)$
∴ $\vec{r}'(\theta) = (-a\sin\theta, a\cos\theta, 0)$
∵ S is upward-oriented
∵ the first-order partial derivatives of the components of $\vec{F}$ exist and are continuous
By Stokes' Theorem,

$$\iint_S (\nabla \times \vec{F}) \cdot \vec{n} dA = \oint_C \vec{F} \cdot d\vec{r} = \int_0^{2\pi} \vec{F}(\vec{r}(\theta)) \cdot \vec{r}'(\theta) \, d\theta$$

$$= \int_0^{2\pi} (-(a\sin\theta)^3, (a\cos\theta)^3, 1) \cdot (-a\sin\theta, a\cos\theta, 0) \, d\theta = a^4 \int_0^{2\pi} \sin^4\theta + \cos^4\theta \, d\theta$$

∵ $\sin^4\theta + \cos^4\theta = (\sin^2\theta + \cos^2\theta)^2 - 2\sin^2\theta\cos^2\theta = 1 - 2(\sin\theta\cos\theta)^2$

$$= 1 - 2\left(\frac{\sin 2\theta}{2}\right)^2 = 1 - \frac{\sin^2 2\theta}{2} = 1 - \frac{1 - \cos 4\theta}{4} = \frac{3 - \cos 4\theta}{4}$$

and $\int_0^{2\pi} \sin^4\theta + \cos^4\theta \, d\theta = \int_0^{2\pi} \frac{3 - \cos 4\theta}{4} d\theta = \frac{3\pi}{2}$

∴ $\iint_S (\nabla \times \vec{F}) \cdot \vec{n} dA = \frac{3a^4\pi}{2}$

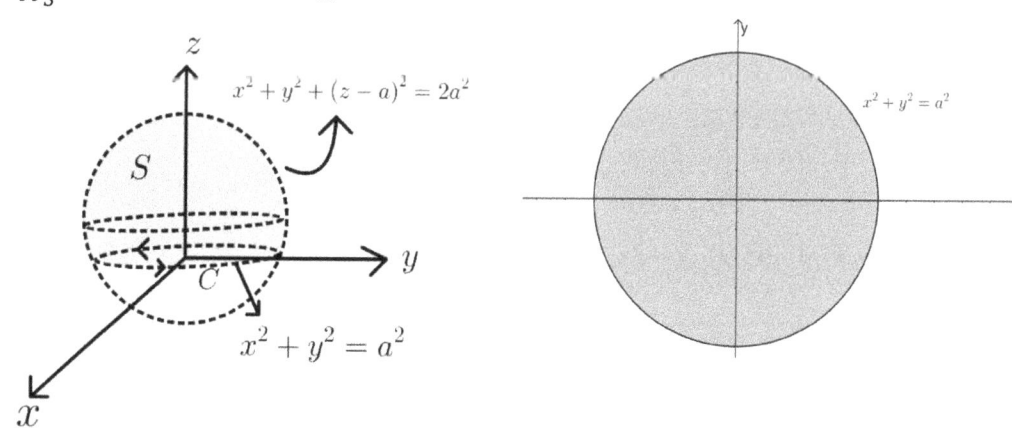

Example 4.

9.6.3 Convert Surface Integrals into Closed Line Integrals

Find $\iint_S (\nabla \times \vec{F}) \cdot \vec{n} dA =?$, $S = \{(x,y,z): x^2 + y^2 \leq 4, z = 1\}$ is an upward-oriented plane, and $\vec{F} = (x, 2z - x, y^2)$: (1) Direct calculation (2) Using Stoke's Theorem.

【Solution】

(1)

$$\because \nabla \times \vec{F} = \begin{vmatrix} \vec{i} & \vec{j} & \vec{k} \\ \dfrac{\partial}{\partial x} & \dfrac{\partial}{\partial y} & \dfrac{\partial}{\partial z} \\ x & 2z-x & y^2 \end{vmatrix} = (2y - 2, 0, -1)$$

$\because$ S is upward-oriented

Let $\vec{r}(x,y) = (x, y, 1)$ and let R be the closed region of the projection of S onto the xy-plane. Then

$$R = \{(x,y): x^2 + y^2 \leq 4\} \text{ and } \iint_S (\nabla \times \vec{F}) \cdot \vec{n} dA = \iint_R (\nabla \times \vec{F}) \cdot \frac{\partial \vec{r}}{\partial x} \times \frac{\partial \vec{r}}{\partial y} dx dy$$

$\because \dfrac{\partial \vec{r}}{\partial x} = (1,0,0), \quad \dfrac{\partial \vec{r}}{\partial y} = (0,1,0) \quad \therefore \dfrac{\partial \vec{r}}{\partial x} \times \dfrac{\partial \vec{r}}{\partial y} = (0,0,1)$

$$\iint_S (\nabla \times \vec{F}) \cdot \vec{n} dA = \iint_R (2y-2, 0, -1) \cdot (0,0,1) dx dy = \iint_{x^2+y^2 \leq 4} -1 dA = -4\pi$$

(2)

Let $C = \{(x,y,z): x^2 + y^2 = 4, z = 1\}$ be the counterclockwise closed curve

Let $x = 2\cos\theta$, $y = 2\sin\theta$ then $\vec{r}(\theta) = (2\cos\theta, 2\sin\theta, 0)$ $\therefore \vec{r}'(\theta) = (-2\sin\theta, 2\cos\theta, 0)$

$\because$ the first-order partial derivatives of the components of $\vec{F}$ exist and are continuous

By Stokes' Theorem,

$$\iint_S (\nabla \times \vec{F}) \cdot \vec{n} dA = \oint_C \vec{F} \cdot d\vec{r} = \int_0^{2\pi} \vec{F}(\vec{r}(\theta)) \cdot \vec{r}'(\theta) d\theta$$

$$= \int_0^{2\pi} (2\cos\theta, 2 - 2\cos\theta, 4\sin^2\theta) \cdot (-2\sin\theta, 2\cos\theta, 0) d\theta$$

$$= \int_0^{2\pi} -4\sin\theta \cos\theta + 4\cos\theta - 4\cos^2\theta \, d\theta = \int_0^{2\pi} -4\cos^2\theta \, d\theta = -4 \int_0^{2\pi} \frac{1 + \cos 2\theta}{2} d\theta$$

$$= -4\pi$$

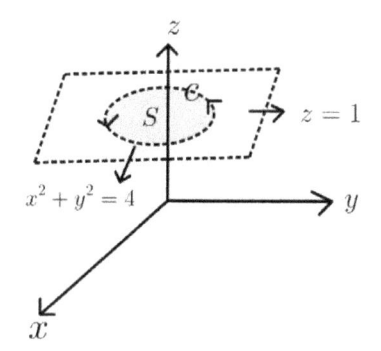

9.6.3 Convert Surface Integrals into Closed Line Integrals

Example 5.

Let $\vec{F} = \left(0, \dfrac{2x^3}{3}, 2y^3\right)$. The surface S is the plane defined by $x + y + z = 1$ in the first octant oriented upward. Find $\iint_S (\nabla \times \vec{F}) \cdot \vec{n}\, dA = ?$: (1) Direct calculation (2) Using Stokes' Theorem.

【Solution】

(1)

$$\because \operatorname{curl} \vec{F} = \nabla \times \vec{F} = \begin{Vmatrix} i & j & k \\ \dfrac{\partial}{\partial x} & \dfrac{\partial}{\partial y} & \dfrac{\partial}{\partial z} \\ 0 & \dfrac{2x^3}{3} & 2y^3 \end{Vmatrix} = (6y^2, 0, 2x^2)$$

$\because$ S is upward-oriented

Let $\vec{r}(x, y, z) = (x, y, 1 - x - y)$ and let R be the closed region of the projection of S onto the xy-plane. Then

$$R = \{(x, y): x + y \leq 1, x \geq 0, y \geq 0\} \text{ and } \iint_S (\nabla \times \vec{F}) \cdot \vec{n}\, dA = \iint_R (\nabla \times \vec{F}) \cdot \dfrac{\partial \vec{r}}{\partial x} \times \dfrac{\partial \vec{r}}{\partial y}\, dxdy$$

$\because \dfrac{\partial \vec{r}}{\partial x} = (1, 0, -1)$ and $\dfrac{\partial \vec{r}}{\partial y} = (0, 1, -1)$ $\therefore \dfrac{\partial \vec{r}}{\partial x} \times \dfrac{\partial \vec{r}}{\partial y} = (1,1,1)$

$\therefore (\nabla \times \vec{F}) \cdot \dfrac{\partial \vec{r}}{\partial x} \times \dfrac{\partial \vec{r}}{\partial y} = (6y^2, 0, 2x^2) \cdot \dfrac{\partial \vec{r}}{\partial x} \times \dfrac{\partial \vec{r}}{\partial y} = 2(x^2 + 3y^2), \ \forall (x, y, z) \in S$

9.6.3 Convert Surface Integrals into Closed Line Integrals

$$\therefore \iint_S (\nabla \times \vec{F}) \cdot \vec{n}\, dA = \iint_R (\nabla \times \vec{F}) \cdot \frac{\partial \vec{r}}{\partial x} \times \frac{\partial \vec{r}}{\partial y}\, dxdy = 2\iint_R x^2 + 3y^2\, dxdy$$

$$= 2\int_0^1 \int_0^{1-y} x^2 + 3y^2\, dxdy = 2\int_0^1 \left(\frac{x^3}{3} + 3y^2 x\right)\Big|_{x=0}^{x=1-y} dy = 2\int_0^1 \frac{(1-y)^3}{3} + 3y^2(1-y)\, dy$$

$$= \frac{2}{3}$$

(2)

Let $C_1: \vec{r}_1(t) = (1-t, t, 0)$, $C_2: \vec{r}_2(t) = (0, 1-t, t)$, $C_3: \vec{r}_3(t) = (t, 0, 1-t)$, $0 \leq t \leq 1$

then $\vec{r}_1'(t) = (-1,1,0)$, $\vec{r}_2'(t) = (0,-1,1)$, $\vec{r}_3'(t) = (1,0,-1)$ and $C = C_1 \cup C_2 \cup C_3$

By Stokes' Theorem, $\iint_S (\nabla \times \vec{F}) \cdot \vec{n}\, dA = \oint_C \vec{F} \cdot d\vec{r}$

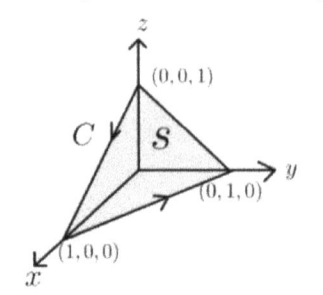

$$\therefore \oint_C \vec{F} \cdot d\vec{r} = \int_{C_1} \vec{F}(\vec{r}_1(t)) \cdot \vec{r}_1'(t)\, dt + \int_{C_2} \vec{F}(\vec{r}_2(t)) \cdot \vec{r}_2'(t)\, dt + \int_{C_3} \vec{F}(\vec{r}_3(t)) \cdot \vec{r}_3'(t)\, dt$$

$$= \frac{2}{3}\int_0^1 (1-t)^3\, dt + 2\int_0^1 (1-t)^3\, dt = \frac{2}{3}$$

Example 6.

Let $\vec{F} = \left(\frac{z^2}{2}, \frac{x^2}{2}, \frac{y^2}{2}\right)$. The surface S is the plane in the first octant oriented upward, bounded by the points $(a, 0, 0)$, $(0, a, 0)$, $(0, 0, a)$. Find $\iint_S (\nabla \times \vec{F}) \cdot \vec{n}\, dA = ?$:

(1) Direct calculation (2) Using Stokes' Theorem.

【Solution】

(1)

$$\text{curl } \vec{F} = \nabla \times \vec{F} = \begin{Vmatrix} i & j & k \\ \dfrac{\partial}{\partial x} & \dfrac{\partial}{\partial y} & \dfrac{\partial}{\partial z} \\ \dfrac{z^2}{2} & \dfrac{x^2}{2} & \dfrac{y^2}{2} \end{Vmatrix} = (y, z, x)$$

∵ S is upward-oriented

Let $\vec{r}(x, y, z) = (x, y, a - x - y)$ and $R = \{(x, y): 0 \le x \le a, 0 \le y \le a - x\}$ then

$$\iint_S (\nabla \times \vec{F}) \cdot \vec{n} \, dA = \iint_R (\nabla \times \vec{F}) \cdot \dfrac{\partial \vec{r}}{\partial x} \times \dfrac{\partial \vec{r}}{\partial y} dx dy$$

∵ $\dfrac{\partial \vec{r}}{\partial x} = (1, 0, -1)$ and $\dfrac{\partial \vec{r}}{\partial y} = (0, 1, -1)$ ∴ $\dfrac{\partial \vec{r}}{\partial x} \times \dfrac{\partial \vec{r}}{\partial y} = (1, 1, 1)$

∴ $(\nabla \times \vec{F}) \cdot \dfrac{\partial \vec{r}}{\partial x} \times \dfrac{\partial \vec{r}}{\partial y} = (y, z, x) \cdot (1, 1, 1) = x + y + z = x + y + a - x - y = a, \quad \forall (x, y, z) \in S$

$$\iint_R (\nabla \times \vec{F}) \cdot \dfrac{\partial \vec{r}}{\partial x} \times \dfrac{\partial \vec{r}}{\partial y} dx dy = \int_0^a \int_0^{a-x} a \, dy dx = a \int_0^a a - x \, dx = a \left(ax - \dfrac{x^2}{2} \right) \Big|_0^a = \dfrac{a^3}{2}$$

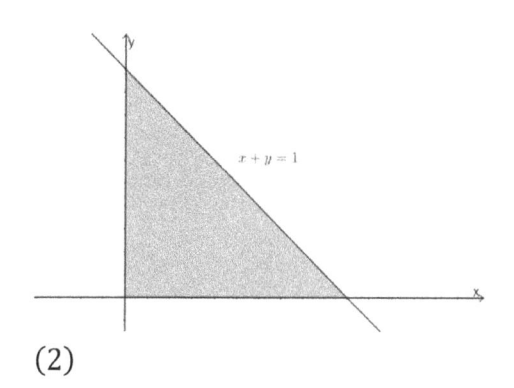

(2)

Let $C_1: \vec{r}_1(t) = (a - at, at, 0), C_2: \vec{r}_2(t) = (0, a - at, at), C_3: \vec{r}_3(t) = (at, 0, a - at), 0 \le t \le 1$

then $\vec{r}_1'(t) = (-a, a, 0), \vec{r}_2'(t) = (0, -a, a), \vec{r}_3'(t) = (a, 0, -a)$ and $C = C_1 \cup C_2 \cup C_3$

By Stokes' Theorem, $\iint_S (\nabla \times \vec{F}) \cdot \vec{n} dA = \oint_C \vec{F} \cdot d\vec{r}$

∴ $\oint_C \vec{F} \cdot d\vec{r} = \int_{C_1} \vec{F}(\vec{r}_1(t)) \cdot \vec{r}_1'(t) dt + \int_{C_2} \vec{F}(\vec{r}_2(t)) \cdot \vec{r}_2'(t) dt + \int_{C_3} \vec{F}(\vec{r}_3(t)) \cdot \vec{r}_3'(t) dt$

$$= \frac{a}{2}\int_0^1 (a-at)^3\, dt \cdot 3 = \frac{a^3}{2}$$

Example 7.

Using Stokes' Theorem, find $\iint_S (\nabla \times \vec{F}) \cdot \vec{n}\, dA = ?$, where $\vec{F} = (2x - y, 2y + z, xyz)$,

$S: \dfrac{x^2}{a^2} + \dfrac{y^2}{b^2} + \dfrac{z^2}{c^2} = 1$, $z > 0$ is the upper half-ellipsoid oriented upward.

【Solution】

Let C be the counterclockwise closed curve formed by the intersection of the surface S and $z = 0$ then $C = \{(x, y, z): \dfrac{x^2}{a^2} + \dfrac{y^2}{b^2} = 1, z = 0\}$

Let $x = a\cos\theta$, $y = b\sin\theta$ then $\vec{r}(\theta) = (a\cos\theta, b\sin\theta, 0)$

$\therefore \vec{r}'(\theta) = (-a\sin\theta, b\cos\theta, 0)$

∵ S is upward-oriented

∵ the first-order partial derivatives of the components of $\vec{F}$ exist and are continuous

By Stokes' Theorem,

$$\iint_S (\nabla \times \vec{F}) \cdot \vec{n}\, dA = \oint_C \vec{F} \cdot d\vec{r} = \int_0^{2\pi} \vec{F}(\vec{r}(\theta)) \cdot \vec{r}'(\theta)\, d\theta$$

$$= \int_0^{2\pi} (2a\cos\theta - b\sin\theta, 2b\sin\theta, 0) \cdot (-a\sin\theta, b\cos\theta, 0)\, d\theta$$

$$= \int_0^{2\pi} -2a^2 \sin\theta \cos\theta + ab\sin^2\theta + 2b^2 \sin\theta \cos\theta\, d\theta$$

$$= \int_0^{2\pi} 2(b^2 - a^2)\sin\theta\cos\theta + ab\sin^2\theta\, d\theta = \int_0^{2\pi} (b^2 - a^2)\sin 2\theta + ab\left(\dfrac{1 - \cos 2\theta}{2}\right)d\theta$$

$$= ab\pi$$

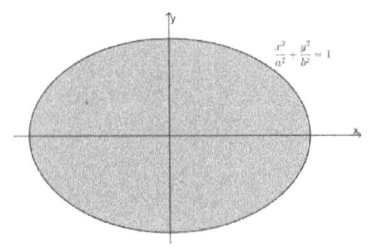

Example 8.

Using Stokes' Theorem, find $\iint_S (\nabla \times \vec{F}) \cdot d\vec{S} = ?$, where $\vec{F} = (xz, yz, xy)$, and S is the upward-oriented portion of the sphere defined by $x^2 + y^2 + z^2 = 4$ inside the cylinder $x^2 + y^2 = 1$ with $z > 0$.

【Solution】

Let C be the counterclockwise closed curve formed by the intersection of the surface S and $x^2 + y^2 = 1$ then $C = \{(x, y, z) : x^2 + y^2 = 1, z = \sqrt{3}\}$

Let $x = \cos\theta, y = \sin\theta$ then $\vec{r}(\theta) = (\cos\theta, \sin\theta, \sqrt{3}) \quad \therefore \vec{r}'(\theta) = (-\sin\theta, \cos\theta, 0)$

∵ S is upward-oriented

∵ the first-order partial derivatives of the components of $\vec{F}$ exist and are continuous

By Stokes' Theorem,

$$\iint_S (\nabla \times \vec{F}) \cdot \vec{n} dA = \oint_C \vec{F} \cdot d\vec{r} = \int_0^{2\pi} \vec{F}(\vec{r}(\theta)) \cdot \vec{r}'(\theta) d\theta$$

$$= \int_0^{2\pi} (\sqrt{3}\cos\theta, \sqrt{3}\sin\theta, \sin\theta\cos\theta) \cdot (-\sin\theta, \cos\theta, 0) d\theta = \sqrt{3} \int_0^{2\pi} 0 d\theta = 0$$

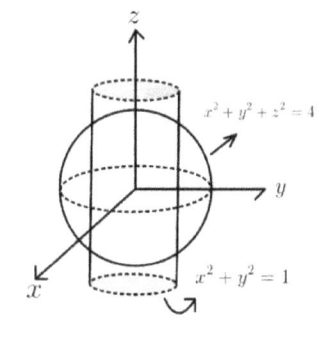

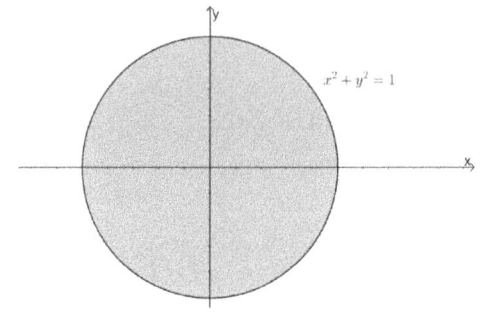

Example 9.

Let $\vec{F} = (0, e^x + 16xy, e^y + 6y^2)$. The surface S is the plane defined by $x + y + z = 6$ oriented upward in the first octant. Find $\iint_S (\nabla \times \vec{F}) \cdot \vec{n} dA = ?$: (1) Direct calculation (2) Using Stokes' Theorem.

【Solution】

(1)

9.6.3 Convert Surface Integrals into Closed Line Integrals

$$\text{curl } \vec{F} = \nabla \times \vec{F} = \begin{vmatrix} i & j & k \\ \dfrac{\partial}{\partial x} & \dfrac{\partial}{\partial y} & \dfrac{\partial}{\partial z} \\ 0 & e^x + 16xy & e^y + 6y^2 \end{vmatrix} = (e^y + 12y, 0, e^x + 6y)$$

∵ S is upward-oriented

Let $\vec{r}(x, y, z) = (x, y, 6 - x - y)$ and let R be the closed region of the projection of S onto the xy-plane. Then

$$R = \{(x, y): x + y \leq 6, x \geq 0, y \geq 0\} \text{ and } \iint_S (\nabla \times \vec{F}) \cdot \vec{n} \, dA = \iint_R (\nabla \times \vec{F}) \cdot \dfrac{\partial \vec{r}}{\partial x} \times \dfrac{\partial \vec{r}}{\partial y} dxdy$$

∵ $\dfrac{\partial \vec{r}}{\partial x} = (1, 0, -1)$ and $\dfrac{\partial \vec{r}}{\partial y} = (0, 1, -1)$ ∴ $\dfrac{\partial \vec{r}}{\partial x} \times \dfrac{\partial \vec{r}}{\partial y} = (1, 1, 1)$

∵ $(\nabla \times \vec{F}) \cdot \dfrac{\partial \vec{r}}{\partial x} \times \dfrac{\partial \vec{r}}{\partial y} = (e^y + 12y, 0, e^x + 6y) \cdot (1, 1, 1) = e^y + e^x + 18y, \ \forall (x, y, z) \in S$

∴ $\iint_S (\nabla \times \vec{F}) \cdot \vec{n} \, dA = \iint_R (\nabla \times \vec{F}) \cdot \dfrac{\partial \vec{r}}{\partial x} \times \dfrac{\partial \vec{r}}{\partial y} dxdy = \iint_R e^y + e^x + 18y \, dxdy$

$$= \int_0^6 \int_0^{6-x} e^y + e^x + 18y \, dydx = 634 + 2e^6$$

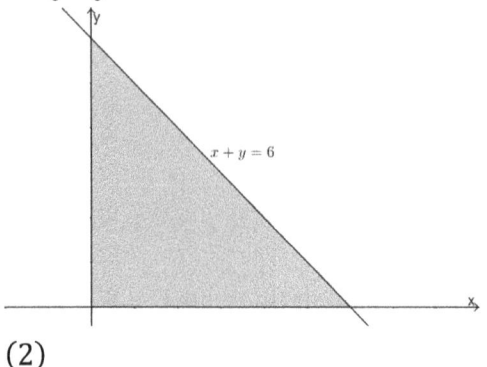

(2)

Let $C_1: \vec{r}_1(t) = (6 - 6t, 6t, 0), C_2: \vec{r}_2(t) = (0, 6 - 6t, 6t), C_3: \vec{r}_3(t) = (6t, 0, 6 - 6t)$ then

$\vec{r}_1'(t) = (-6, 6, 0), \ \vec{r}_2'(t) = (0, -6, 6), \ \vec{r}_3'(t) = (6, 0, -6)$ and $C = C_1 \cup C_2 \cup C_3$

By Stokes' Theorem, $\iint_S (\nabla \times \vec{F}) \cdot \vec{n} dA = \oint_C \vec{F} \cdot d\vec{r}$

$$\therefore \oint_C \vec{F} \cdot d\vec{r} = \int_{C_1} \vec{F}(\vec{r}_1(t)) \cdot \vec{r}_1'(t)dt + \int_{C_2} \vec{F}(\vec{r}_2(t)) \cdot \vec{r}_2'(t)dt + \int_{C_3} \vec{F}(\vec{r}_3(t)) \cdot \vec{r}_3'(t)dt$$

$$\because \int_{C_1} \vec{F}(\vec{r}_1(t)) \cdot \vec{r}_1'(t)dt = 6\int_0^1 e^{6-6t} + 96t(6-6t)\, dt = e^6 - 1 + 576$$

$$\int_{C_2} \vec{F}(\vec{r}_2(t)) \cdot \vec{r}_2'(t)dt = -6\int_0^1 dt + 6\int_0^1 e^{6-6t} + 6(6-6t)^2\, dt = e^6 + 65$$

$$\int_{C_3} \vec{F}(\vec{r}_3(t)) \cdot \vec{r}_3'(t)dt = -6\int_0^1 dt = -6$$

$$\therefore \oint_C \vec{F} \cdot d\vec{r} = 634 + 2e^6$$

Example 10.

Find the surface integral of the curl of $\vec{F} = (2x - y, -yz^2, -zy^2)$ over the upward normal component of the upper surface of the sphere $S: x^2 + y^2 + z^2 = 1$:
(1) Direct calculation (2) Using Stoke's Theorem

【Solution】

(1)

$$\because \nabla \times \vec{F} = \begin{vmatrix} \vec{i} & \vec{j} & \vec{k} \\ \dfrac{\partial}{\partial x} & \dfrac{\partial}{\partial y} & \dfrac{\partial}{\partial z} \\ 2x-y & -yz^2 & -zy^2 \end{vmatrix} = (-2zy - (-2yz), 0, 1) = (0,0,1)$$

$\because$ S is upward-oriented

Let $\vec{r}(x,y) = \left(x, y, \sqrt{1-(x^2+y^2)}\right)$

Let R be the closed region of the projection of S onto the xy-plane then

$R = \{(x,y): x^2 + y^2 \leq 1\}$ and $\iint_S (\nabla \times \vec{F}) \cdot \vec{n}\, dA = \iint_R (\nabla \times \vec{F}) \cdot \dfrac{\partial \vec{r}}{\partial x} \times \dfrac{\partial \vec{r}}{\partial y}\, dxdy$

9.6.3 Convert Surface Integrals into Closed Line Integrals 584

$$\because \frac{\partial \vec{r}}{\partial x} = \left(1, 0, -x(1-(x^2+y^2))^{-\frac{1}{2}}\right), \frac{\partial \vec{r}}{\partial y} = \left(0, 1, -y(1-(x^2+y^2))^{-\frac{1}{2}}\right),$$

$$\therefore \frac{\partial \vec{r}}{\partial x} \times \frac{\partial \vec{r}}{\partial y} = \left(x(1-(x^2+y^2))^{-\frac{1}{2}}, y(1-(x^2+y^2))^{-\frac{1}{2}}, 1\right)$$

$$\therefore (\nabla \times \vec{F}) \cdot \frac{\partial \vec{r}}{\partial x} \times \frac{\partial \vec{r}}{\partial y} = (0,0,1) \cdot \left(x(1-(x^2+y^2))^{-\frac{1}{2}}, y(1-(x^2+y^2))^{-\frac{1}{2}}, 1\right) = 1$$

$$\therefore \iint_S (\nabla \times \vec{F}) \cdot \vec{n} \, dA = \iint_R (\nabla \times \vec{F}) \cdot \frac{\partial \vec{r}}{\partial x} \times \frac{\partial \vec{r}}{\partial y} \, dxdy = \iint_R dxdy = \pi$$

(2)
Let C be the counterclockwise closed curve formed by the intersection of the surface S and $z = 0$ then $C = \{(x, y, z): x^2 + y^2 = 1, z = 0\}$
Let $x = \cos\theta$, $y = \sin\theta$ then $\vec{r}(\theta) = (\cos\theta, \sin\theta, 0)$ $\therefore \vec{r}'(\theta) = (-\sin\theta, \cos\theta, 0)$
∵ S is upward-oriented
∵ the first-order partial derivatives of the components of $\vec{F}$ exist and are continuous
By Stokes' Theorem,

$$\iint_S (\nabla \times \vec{F}) \cdot \vec{n} \, dA = \oint_C \vec{F} \cdot d\vec{r} = \int_0^{2\pi} \vec{F}(\vec{r}(\theta)) \cdot \vec{r}'(\theta) \, d\theta$$

$$= \int_0^{2\pi} (2\cos\theta - \sin\theta, 0, 0) \cdot (-\sin\theta, \cos\theta, 0) d\theta = \int_0^{2\pi} -2\sin\theta\cos\theta + \sin^2\theta \, d\theta$$

$$= \int_0^{2\pi} -2\sin\theta\cos\theta + \frac{1-\cos 2\theta}{2} \, d\theta = \pi$$

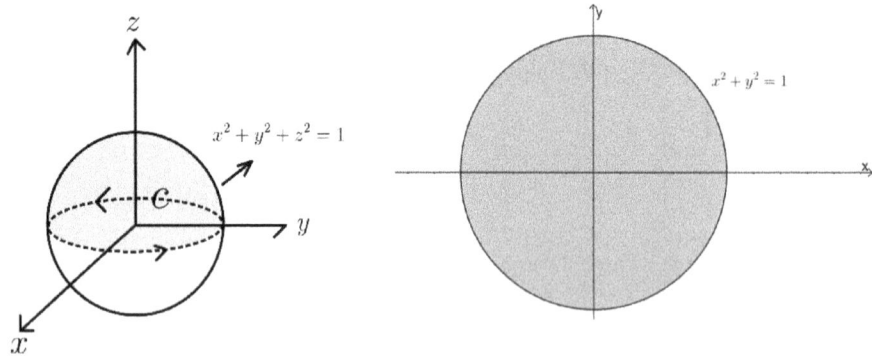

Example 11.
Find the surface integral of the curl of $\vec{F} = (2x - y, -yz^2, -zy^2)$ over the downward normal component of the lower surface of the sphere $S: x^2 + y^2 + z^2 = 1$:

(1)Direct calculation (2)Using Stoke's Theorem.

【Solution】

(1)

$$\because \nabla \times \vec{F} = \begin{vmatrix} \vec{i} & \vec{j} & \vec{k} \\ \dfrac{\partial}{\partial x} & \dfrac{\partial}{\partial y} & \dfrac{\partial}{\partial z} \\ 2x - y & -yz^2 & -zy^2 \end{vmatrix} = (-2zy - (-2yz), 0, 1) = (0,0,1)$$

∵ S is downward-oriented

Let $\vec{r}(x, y) = \left(x, y, -\sqrt{1 - (x^2 + y^2)}\right)$.

Let R be the closed region of the projection of S onto the xy-plane then

$$R = \{(x, y): x^2 + y^2 \leq 1\} \text{ and } \iint_S (\nabla \times \vec{F}) \cdot \vec{n} dA = -\iint_R (\nabla \times \vec{F}) \cdot \dfrac{\partial \vec{r}}{\partial x} \times \dfrac{\partial \vec{r}}{\partial y} dxdy$$

$$\because \dfrac{\partial \vec{r}}{\partial x} = \left(1, 0, x(1 - (x^2 + y^2))^{-\frac{1}{2}}\right), \dfrac{\partial \vec{r}}{\partial y} = \left(0, 1, y(1 - (x^2 + y^2))^{-\frac{1}{2}}\right)$$

$$\therefore \dfrac{\partial \vec{r}}{\partial x} \times \dfrac{\partial \vec{r}}{\partial y} = \left(-x(1 - (x^2 + y^2))^{-\frac{1}{2}}, -y(1 - (x^2 + y^2))^{-\frac{1}{2}}, 1\right)$$

$$\therefore -(\nabla \times \vec{F}) \cdot \dfrac{\partial \vec{r}}{\partial x} \times \dfrac{\partial \vec{r}}{\partial y} = -(0,0,1) \cdot \left(x(1 - (x^2 + y^2))^{-\frac{1}{2}}, y(1 - (x^2 + y^2))^{-\frac{1}{2}}, 1\right) = -1$$

$$\therefore \iint_S (\nabla \times \vec{F}) \cdot \vec{n} dA = -\iint_R (\nabla \times \vec{F}) \cdot \dfrac{\partial \vec{r}}{\partial x} \times \dfrac{\partial \vec{r}}{\partial y} dxdy = -\iint_R dxdy = -\pi$$

(2)

Let C be the counterclockwise closed curve formed by the intersection of the surface S and $z = 0$ then $C = \{(x, y, z): x^2 + y^2 = 1, z = 0\}$

Let $x = \cos \theta, y = \sin \theta$ then $\vec{r}(\theta) = (\cos \theta, \sin \theta, 0)$ ∴ $\vec{r}'(\theta) = (-\sin \theta, \cos \theta, 0)$

∵ S is oriented downward

∵ the first-order partial derivatives of the components of $\vec{F}$ exist and are continuous

By Stokes' Theorem,

$$\iint_S (\nabla \times \vec{F}) \cdot \vec{n} dA = -\oint_C \vec{F} \cdot d\vec{r} = -\int_0^{2\pi} \vec{F}(\vec{r}(\theta)) \cdot \vec{r}'(\theta) d\theta$$

$$= -\int_0^{2\pi} (2 \cos \theta - \sin \theta, 0, 0) \cdot (-\sin \theta, \cos \theta, 0) d\theta = -\int_0^{2\pi} -2 \sin \theta \cos \theta + \sin^2 \theta \, d\theta$$

9.6.3 Convert Surface Integrals into Closed Line Integrals

$$= -\int_0^{2\pi} -2\sin\theta\cos\theta + \frac{1-\cos 2\theta}{2} d\theta = -\pi$$

Example 12.

Find $\iint_S (\nabla \times \vec{F}) \cdot \vec{n} dA = ?$, where $S = \{(x, y, z): 4z = x^2 + y^2, z \leq 1\}$ is a downward-oriented surface, and $\vec{F} = (x, 2z - x, y^2)$: (1)Direct calculation (2)Using Stoke's Theorem

【Solution】

(1)

$$\because \nabla \times \vec{F} = \begin{vmatrix} \vec{i} & \vec{j} & \vec{k} \\ \dfrac{\partial}{\partial x} & \dfrac{\partial}{\partial y} & \dfrac{\partial}{\partial z} \\ x & 2z-x & y^2 \end{vmatrix} = (2y - 2, 0, -1)$$

∵ S is oriented downward

Let $\vec{r}(x, y) = \left(x, y, \dfrac{x^2 + y^2}{4}\right)$

Let R be the closed region of the projection of S onto the xy-plane then

$R = \{(x, y): x^2 + y^2 \leq 4\}$ and $\iint_S (\nabla \times \vec{F}) \cdot \vec{n} dA = -\iint_R (\nabla \times \vec{F}) \cdot \dfrac{\partial \vec{r}}{\partial x} \times \dfrac{\partial \vec{r}}{\partial y} dxdy$

$\because \dfrac{\partial \vec{r}}{\partial x} = \left(1, 0, \dfrac{x}{2}\right)$ and $\dfrac{\partial \vec{r}}{\partial y} = \left(0, 1, \dfrac{y}{2}\right)$ ∴ $\dfrac{\partial \vec{r}}{\partial x} \times \dfrac{\partial \vec{r}}{\partial y} = \left(-\dfrac{x}{2}, -\dfrac{y}{2}, 1\right)$

$$\iint_S (\nabla \times \vec{F}) \cdot \vec{n} dA = -\iint_R (\nabla \times \vec{F}) \cdot \frac{\partial \vec{r}}{\partial x} \times \frac{\partial \vec{r}}{\partial y} dxdy$$

$$= -\iint_R (2y-2, 0, -1) \cdot \left(-\frac{x}{2}, -\frac{y}{2}, 1\right) dxdy = \iint_{x^2+y^2 \leq 4} xy - x + 1 dxdy$$

Let $x = r\cos\theta, y = r\sin\theta$ then

$$\iint_{x^2+y^2 \leq 4} xy - x + 1 dxdy = \int_0^{2\pi} \int_0^2 (r^2 \cos\theta \sin\theta - r\cos\theta + 1) r dr d\theta = 4\pi$$

(2)
Let $C = \{(x,y,z): x^2 + y^2 = 4, z = 1\}$ be the counterclockwise closed curve
Let $x = 2\cos\theta, y = 2\sin\theta$ then $\vec{r}(\theta) = (2\cos\theta, 2\sin\theta, 1)$ ∴ $\vec{r}'(\theta) = (-2\sin\theta, 2\cos\theta, 0)$
∵ S is oriented downward
∵ the first-order partial derivatives of the components of $\vec{F}$ exist and are continuous
By Stokes' Theorem,

$$\iint_S (\nabla \times \vec{F}) \cdot \vec{n} dA = -\oint_C \vec{F} \cdot d\vec{r} = -\int_0^{2\pi} \vec{F}(\vec{r}(\theta)) \cdot \vec{r}'(\theta) d\theta$$

$$= -\int_0^{2\pi} (2\cos\theta, 2 - 2\cos\theta, 4\sin^2\theta) \cdot (-2\sin\theta, 2\cos\theta, 0) d\theta$$

$$= \int_0^{2\pi} 4\sin\theta\cos\theta - 4\cos\theta + 4\cos^2\theta \, d\theta = \int_0^{2\pi} 4\cos^2\theta \, d\theta = 4\int_0^{2\pi} \frac{1 + \cos 2\theta}{2} d\theta = 4\pi$$

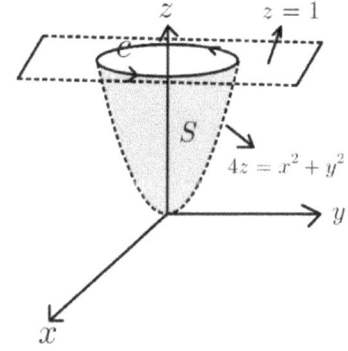

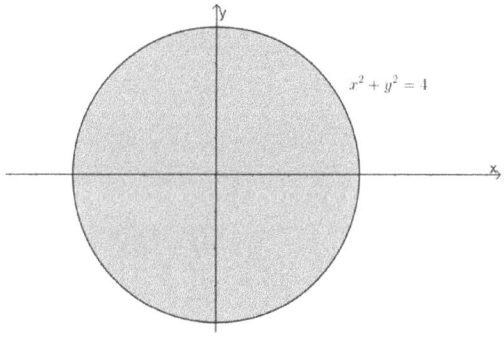

Example 13.

Let $\vec{F} = \left(-z, x, \frac{y^2 z}{2}\right)$. The surface S is defined by $z = \sqrt{x^2 + y^2}$, $z \leq a$ with downward orientation. Find $\iint_S (\nabla \times \vec{F}) \cdot \vec{n} dA = ?$: (1)Direct calculation

9.6.3 Convert Surface Integrals into Closed Line Integrals

(2) Using Stokes' Theorem

【Solution】

(1)

$$\because \text{curl } \vec{F} = \nabla \times \vec{F} = \begin{vmatrix} i & j & k \\ \dfrac{\partial}{\partial x} & \dfrac{\partial}{\partial y} & \dfrac{\partial}{\partial z} \\ -z & x & \dfrac{y^2 z}{2} \end{vmatrix} = (yz, -1, 1)$$

∵ S is oriented downward

Let $\vec{r}(x, y, z) = \left(x, y, (x^2 + y^2)^{\frac{1}{2}}\right)$

Let R be the closed region of the projection of S onto the xy-plane then

$$R = \{(x, y): x^2 + y^2 \leq a^2\} \text{ and } \iint_S (\nabla \times \vec{F}) \cdot \vec{n} \, dA = -\iint_R (\nabla \times \vec{F}) \cdot \dfrac{\partial \vec{r}}{\partial x} \times \dfrac{\partial \vec{r}}{\partial y} dxdy$$

$$\because \dfrac{\partial \vec{r}}{\partial x} = \left(1, 0, x(x^2 + y^2)^{-\frac{1}{2}}\right) \text{ and } \dfrac{\partial \vec{r}}{\partial y} = \left(0, 1, y(x^2 + y^2)^{-\frac{1}{2}}\right)$$

$$\therefore \dfrac{\partial \vec{r}}{\partial x} \times \dfrac{\partial \vec{r}}{\partial y} = \left(-x(x^2 + y^2)^{-\frac{1}{2}}, -y(x^2 + y^2)^{-\frac{1}{2}}, 1\right)$$

$$\Rightarrow (\nabla \times \vec{F}) \cdot \dfrac{\partial \vec{r}}{\partial x} \times \dfrac{\partial \vec{r}}{\partial y} = (yz, -1, 1) \cdot \left(-x(x^2 + y^2)^{-\frac{1}{2}}, -y(x^2 + y^2)^{-\frac{1}{2}}, 1\right)$$

$$= -xy + y(x^2 + y^2)^{-\frac{1}{2}} + 1, \ \forall (x, y, z) \in S$$

$$\therefore \iint_S (\nabla \times \vec{F}) \cdot \vec{n} \, dA = -\iint_R (\nabla \times \vec{F}) \cdot \dfrac{\partial \vec{r}}{\partial x} \times \dfrac{\partial \vec{r}}{\partial y} dxdy = \iint_R xy - y(x^2 + y^2)^{-\frac{1}{2}} - 1 \, dxdy$$

Let $x = r \cos \theta, y = r \sin \theta$ then $\{(x, y): x^2 + y^2 \leq a^2\} = \{(r, \theta): 0 \leq r \leq a, 0 \leq \theta \leq 2\pi\}$

$$\therefore \iint_R xy - y(x^2 + y^2)^{-\frac{1}{2}} - 1 \, dxdy = \int_0^{2\pi} \int_0^a (r^2 \cos \theta \sin \theta - \sin \theta - 1) r \, dr d\theta = -\pi a^2$$

(2)

Let $C: z = a, \ x^2 + y^2 = a^2$ be the counterclockwise closed curve

∵ S is oriented downward

∵ the first-order partial derivatives of the components of $\vec{F}$ exist and are continuous

By Stokes' Theorem, $\iint_S (\nabla \times \vec{F}) \cdot \vec{n} dA = -\oint_C \vec{F} \cdot d\vec{r}$

Let $\vec{r}(t) = (a\cos t, a\sin t, a)$, $0 \leq t \leq 2\pi$ then $\vec{r}'(t) = (-a\sin t, a\cos t, 0)$

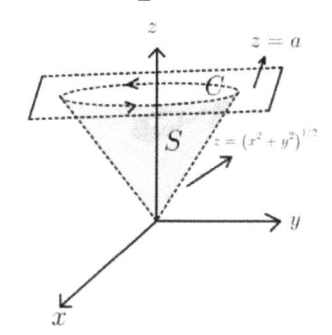

$$\therefore \oint_C \vec{F} \cdot d\vec{r} = -\int_0^{2\pi} \vec{F}(\vec{r}(t)) \cdot \vec{r}'(t) \, dt$$

$$= -\int_0^{2\pi} \left(-a, a\cos t, \frac{a^3 \sin^2 t}{2}\right) \cdot (-a\sin t, a\cos t, 0) \, dt = -\int_0^{2\pi} a^2 \sin t + a^2 \cos^2 t \, dt$$

$$= -2\pi \cdot \frac{a^2}{2} = -\pi a^2$$

9.7 Gauss's Theorem

Earlier, when explaining Green's Theorem, it was mentioned that the divergence form of Green's Theorem can be rewritten as

$$\oint_C \vec{F} \cdot \vec{n} ds = \iint_R \nabla \cdot \vec{F}(x, y) dA,$$

where C is the positively oriented (counterclockwise) boundary curve of the plane region R. An intuitive extension of this to three-dimensional space is

$$\oiint_S \vec{F} \cdot \vec{n} dA = \iiint_V \nabla \cdot \vec{F}(x, y, z) \, dV,$$

where S is the boundary surface of the closed volume V.

If the first-order partial derivatives of $\vec{F}$ exist and are continuous, then the above equality holds and is referred to as the Divergence Theorem (or Gauss's Theorem). It is worth noting that the Divergence Theorem, Stokes' Theorem, and Green's Theorem are all extensions of the Fundamental Theorem of Calculus. Furthermore, the Divergence Theorem allows us to convert the flux of the vector field $\vec{F}$ through the closed surface S (surface integral) into the triple integral of the divergence of $\vec{F}$ over the closed volume V. For example, if a closed surface is composed of S_1, S_2 and S_3 (as shown in the figure below), the original method for calculating flux involves finding the individual fluxes and summing them. However, by using the Divergence Theorem, the problem can be transformed into a volume integral of a scalar function. Conversely, the divergence of $\vec{F}$ in the closed volume V can also be determined by calculating the flux of $\vec{F}$ through the closed surface S (surface integral).

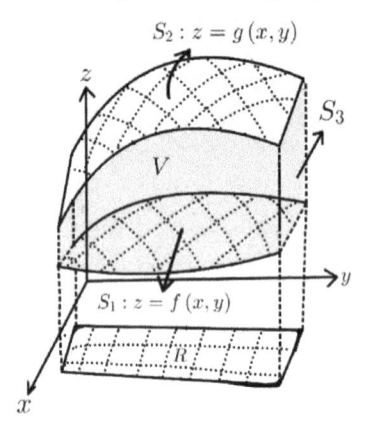

The application scenario of the Divergence Theorem involves closed surfaces, while Stokes' Theorem applies to non-closed surfaces, and Green's Theorem deals with two-dimensional closed curves. Below is a demonstration of common closed surfaces.

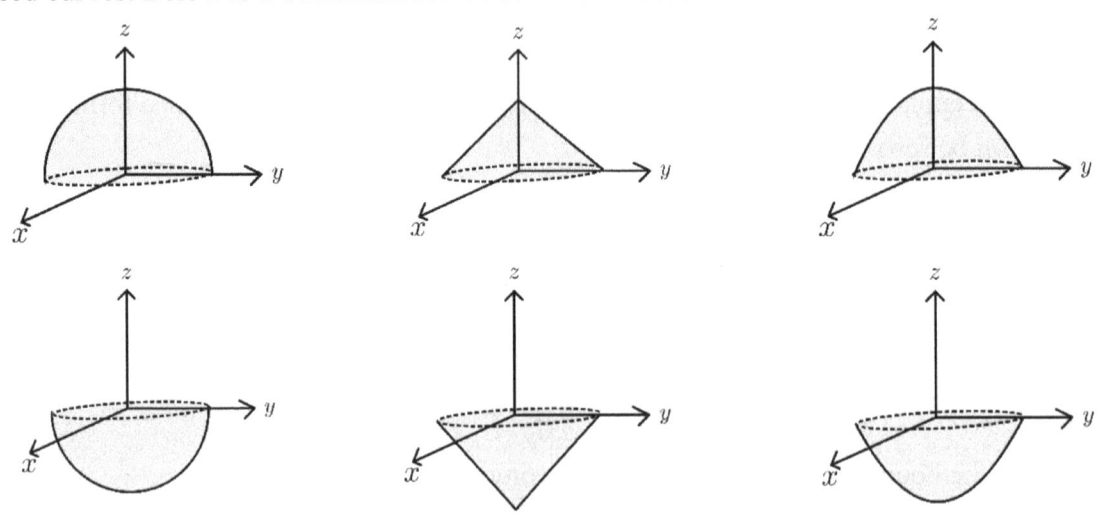

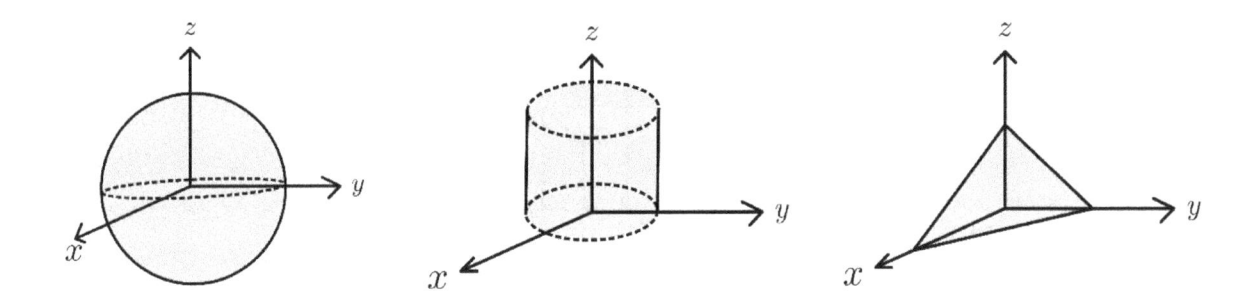

From an intuitive perspective, when the flux of $\vec{F}$ through a closed surface is converted into a triple integral, the integrand should naturally take the form of the divergence of $\vec{F}$. Suppose the volume V is divided into an infinite number of small rectangular prisms. Choosing a small rectangular prism with a vertex at (x,y,z) and edge lengths Δx, Δy, Δz as an example, the flux through this small rectangular prism is the sum of the fluxes through its six faces. Consider the edges parallel to the yz-plane: the flux through the bottom face is approximately
$$F_1(x,y,z)\Delta y \Delta z$$
and the flux through the top face is approximately
$$F_1(x+\Delta x,y,z)\Delta y \Delta z.$$
The difference in flux between the top and bottom faces is
$$\big(F_1(x+\Delta x,y,z) - F_1(x,y,z)\big)\Delta y \Delta z.$$
Using a linear approximation,
$$F_1(x+\Delta x,y,z) \approx F_1(x,y,z) + \frac{\partial F_1}{\partial x}\Delta x.$$
The approximate difference in flux between the top and bottom faces is
$$\big(F_1(x+\Delta x,y,z) - F_1(x,y,z)\big)\Delta y \Delta z = \left(F_1(x,y,z) + \frac{\partial F_1}{\partial x}\Delta x - F_1(x,y,z)\right)\Delta y \Delta z$$
$$= \frac{\partial F_1}{\partial x}\Delta x \Delta y \Delta z.$$
Similarly, the approximate differences for the other two pairs of faces are
$$\frac{\partial F_2}{\partial y}\Delta x \Delta y \Delta z \quad \text{and} \quad \frac{\partial F_3}{\partial z}\Delta x \Delta y \Delta z.$$
Summing these up, we get
$$\left(\frac{\partial F_1}{\partial x} + \frac{\partial F_2}{\partial y} + \frac{\partial F_3}{\partial z}\right)\Delta x \Delta y \Delta z.$$
Thus, we obtain the form of the divergence of $\vec{F}$.

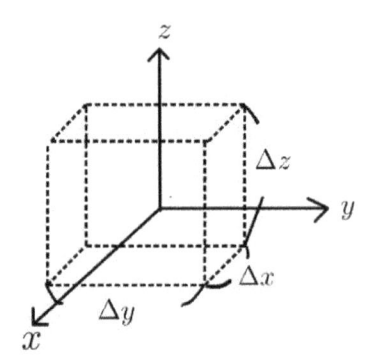

The types of problems involving the Divergence Theorem include converting a closed surface integral into a directly computable volume integral, converting a closed surface integral into a volume integral with coordinate transformations, and converting a volume integral back into a closed surface integral.

【Definition】 Divergence of $\vec{F}$

Suppose $\vec{F}(x, y, z) = F_1\mathbf{i} + F_2\mathbf{j} + F_3\mathbf{k}$ is a continuously defined vector field on V, then the divergence of $\vec{F}$ is defined as

$$\nabla \cdot \vec{F} = \frac{\partial F_1}{\partial x} + \frac{\partial F_2}{\partial y} + \frac{\partial F_3}{\partial z}.$$

【Definition】 Simple Solid Regions

A solid region V is said to be of
 (i) type I if ∃ continuous functions $f(x, y), g(x, y)$ such that
$V = \{(x, y, z): (x, y) \in R, f(x, y) \leq z \leq g(x, y)\}$, R is the projection of V onto the xy-plane.
 (ii) type II if ∃ continuous functions $f(y, z), g(y, z)$ such that
$V = \{(x, y, z): (y, z) \in R, f(y, z) \leq x \leq g(y, z)\}$, R is the projection of V onto the yz-plane.
 (ii) type III if ∃ continuous functions $f(x, z), g(x, z)$ such that
$V = \{(x, y, z): (x, z) \in R, f(x, z) \leq y \leq g(x, z)\}$, R is the projection of V onto the xz-plane.
 (iv) simple solid region if it is concurrently of types I, II, and III.

Using a simple solid region as an example, the figure is shown below. Notably, the plane S_3, which is perpendicular to the xy-plane, may sometimes be absent.

9.7 Gauss's Theorem

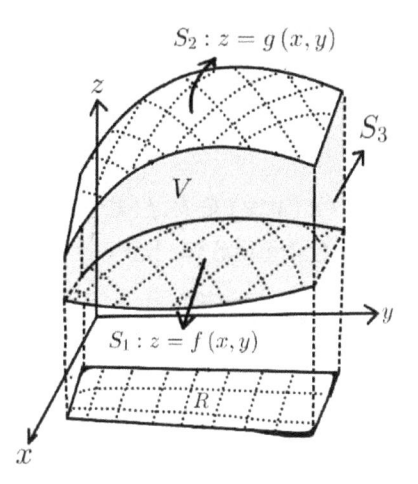

【Theorem】The Divergence Theorem

Suppose
1. S is a closed surface with an outward orientation, enclosing a region V and V is a simple solid region.
2. The vectoe fielf $F(x, y, z) = F_1\mathbf{i} + F_2\mathbf{j} + F_3\mathbf{k}$ has continuously differentiable component functions in an open set containing V.

$$\oiint_S \vec{F} \cdot \vec{n} dA = \iiint_V \nabla \cdot \vec{F}\, dV.$$

Proof:

$$\because \oiint_S \vec{F} \cdot \vec{n} dA = \iint_S F_1\mathbf{i} \cdot \vec{n} dA + \iint_S F_2\mathbf{j} \cdot \vec{n} dA + \iint_S F_3\mathbf{k} \cdot \vec{n} dA$$

and

$$\iiint_V \nabla \cdot \vec{F}\, dV = \iiint_V \frac{\partial F_1}{\partial x} dV + \iiint_V \frac{\partial F_2}{\partial y} dV + \iiint_V \frac{\partial F_3}{\partial z} dV$$

Hence,

If $\begin{cases} \iint_S F_1\mathbf{i} \cdot \vec{n} dA = \iiint_V \frac{\partial F_1}{\partial x} dV \\ \iint_S F_2\mathbf{j} \cdot \vec{n} dA = \iiint_V \frac{\partial F_2}{\partial y} dV \\ \iint_S F_3\mathbf{k} \cdot \vec{n} dA = \iiint_V \frac{\partial F_3}{\partial z} dV \end{cases}$ then $\oiint_S \vec{F} \cdot \vec{n} dA = \iiint_V \nabla \cdot \vec{F}\, dV$

Claim: $\iint_S F_3\mathbf{k} \cdot \vec{n} dA = \iiint_V \frac{\partial F_3}{\partial z} dV$

Let V be a solid region of type I that includes a plane perpendicular to the xy-plane. Let R be

9.7 Gauss's Theorem

the closed region in the xy-plane that is the projection of the closed surface S, where the surface S consists of three parts: S_1, S_2 and S_3. Here, S_1 is the lower surface: $z = f(x,y)$, S_2 is the upper surface: $z = g(x,y)$, and S_3 is the surface perpendicular to the xy-plane. The volume V is given by

$$V = \{(x,y,z): (x,y) \in R, f(x,y) \leq z \leq g(x,y)\}.$$

Using the Fundamental Theorem of Calculus, we have:

$$\iiint_V \frac{\partial F_3}{\partial z} dV = \iint_R \int_{f(x,y)}^{g(x,y)} \frac{\partial F_3}{\partial z} dz\, dx\, dy = \iint_R F_3(x,y,g) - F_3(x,y,f) dx\, dy.$$

On the other hand,

$$\iint_S F_3 \mathbf{k} \cdot \vec{n}\, dA = \iint_{S_1} F_3 \mathbf{k} \cdot \vec{n}\, dA + \iint_{S_2} F_3 \mathbf{k} \cdot \vec{n}\, dA + \iint_{S_3} F_3 \mathbf{k} \cdot \vec{n}\, dA.$$

∵ $\iint_{S_2} F_3 \mathbf{k} \cdot \vec{n}\, dA = \iint_R F_3(x,y,g) dx\, dy$ and $\iint_{S_1} F_3 \mathbf{k} \cdot \vec{n}\, dA = -\iint_R F_3(x,y,f) dx\, dy$

∵ $\mathbf{k}$ is the direction along the z-axis and $\vec{n}$ on S_3 is horizontal

∴ $\iint_{S_3} F_3 \mathbf{k} \cdot \vec{n}\, dA = 0$

∴ $\iint_S F_3 \mathbf{k} \cdot \vec{n}\, dA = \iint_R F_3(x,y,g) dx\, dy - \iint_R F_3(x,y,f) dx\, dy$

$$= \iint_R F_3(x,y,g) - F_3(x,y,f) dx\, dy = \iiint_V \frac{\partial F_3}{\partial z} dV$$

∴ $\iint_S F_3 \mathbf{k} \cdot \vec{n}\, dA = \iiint_V \frac{\partial F_3}{\partial z} dV.$

Thus, if V is a simple solid region of type I and does not include any plane perpendicular to the xy-plane and is only composed of S_1 and S_2, the above equation still holds.

Claim: $\iint_S F_1 \mathbf{i} \cdot \vec{n}\, dA = \iiint_V \frac{\partial F_1}{\partial x} dV$

Let V be solid region of type II. Similarly, $\iint_S F_1 \mathbf{i} \cdot \vec{n}\, dA = \iiint_V \frac{\partial F_1}{\partial x} dV$

Claim: $\iint_S F_2 \mathbf{j} \cdot \vec{n}\, dA = \iiint_V \frac{\partial F_2}{\partial y} dV$

Let V be solid region of type III. Similarly, $\iint_S F_2 \mathbf{j} \cdot \vec{n}\, dA = \iiint_V \frac{\partial F_2}{\partial y} dV$

$\Rightarrow \oiint_S \vec{F} \cdot \vec{n}\, dA = \iiint_V \nabla \cdot \vec{F}\, dV$

9.7.1 Convert Closed Surface Integrals into Volume Integrals($\nabla \cdot \vec{F} = c$)

Use Gauss's Theorem to convert the surface integral over a closed surface into a volume integral that can be directly computed.

Question Types:

Type 1.

Assume the first-order partial derivatives of each component of $\vec{F}$ exist and are continuous, and $\nabla \cdot \vec{F} = c$ (constant), $S: x^2 + y^2 + z^2 = r^2$. Then,

$$\text{the flux flowing out of } S \text{ is } \oiint_S \vec{F} \cdot \vec{n} dA = \frac{4c\pi r^3}{3}.$$

Problem-Solving Process:

Step1.

$\because$ the first-order partial derivatives of each component of $\vec{F}$ exist and are continuous

By Gauss's Theorem,

$$\oiint_S \vec{F} \cdot \vec{n} dA = \iiint_V \nabla \cdot \vec{F} \, dV$$

Step2.

$\because \nabla \cdot \vec{F} = c \quad \therefore \iiint_V \nabla \cdot \vec{F} \, dV = \iiint_{x^2+y^2+z^2 \leq r^2} c \, dxdydz = c \cdot \dfrac{4\pi}{3} \cdot r^3$

Examples:

(I) Assume $\vec{F} = (x^2 y, 2y + z^2, 2z - 2xyz)$, and $S: x^2 + y^2 + z^2 = r^2$ is outwardly oriented. Find the surface integral $\oiint_S \vec{F} \cdot \vec{n} dA = ?$

$\because$ the first-order partial derivatives of each component of $\vec{F}$ exist and are continuous

By Gauss's Theorem,

$\oiint_S \vec{F} \cdot \vec{n} dA = \iiint_V \nabla \cdot \vec{F} \, dV = \iiint_{x^2+y^2+z^2 \leq r^2} 4 \, dxdydz = 4 \cdot \dfrac{4\pi}{3} \cdot r^3 = \dfrac{16\pi r^3}{3}$

(II) Assume $\vec{F} = \left(x + y + \sin z^2, y + e^{x^2}, z + \ln(x^2 + y^2 + 1)\right)$, and $S: x^2 + y^2 + z^2 = r^2$ is outwardly oriented. Find the surface integral $\oiint_S \vec{F} \cdot \vec{n} dA = ?$

9.7.1 Convert Closed Surface Integrals into Volume Integrals ($\nabla \cdot F = c$)

∵ the first-order partial derivatives of each component of $\vec{F}$ exist and are continuous

By Gauss's Theorem,

$$\oiint_S \vec{F} \cdot \vec{n} dA = \iiint_V \nabla \cdot \vec{F}\, dV = \iiint_{x^2+y^2+z^2 \le r^2} 3\, dxdydz = 3 \cdot \frac{4\pi}{3} \cdot r^3 = 4\pi r^3$$

Type 2.

Suppose the first-order partial derivatives of each component of $\vec{F}$ exist and are continuous, and $\nabla \cdot \vec{F} = c$, where $c \in R$. Let $S: x^2 + y^2 = a^2, 0 \le z \le b$ be a closed surface oriented outward, with $a, b > 0$. Then

$$\oiint_S \vec{F} \cdot \vec{n} dA = \pi a^2 bc.$$

Problem-Solving Process:

Step1.

∵ the first-order partial derivatives of each component of $\vec{F}$ exist and are continuous

By Gauss's Theorem, $\oiint_S \vec{F} \cdot \vec{n} dA = \iiint_V \nabla \cdot \vec{F}\, dV$

Step2.

∵ $\nabla \cdot \vec{F} = c$ ∴ $\iiint_V \nabla \cdot \vec{F}\, dV = \iiint_V c\, dxdydz = c \cdot \pi a^2 b$

Examples:

(I) Assume $\vec{F} = (x, e^x + y, 2z)$, $S: x^2 + y^2 = a^2$, $0 \le z \le b$. Find $\oiint_S \vec{F} \cdot \vec{n} dA =?$

By Gauss's Theorem,

$$\oiint_S \vec{F} \cdot \vec{n} dA = \iiint_V \nabla \cdot \vec{F}\, dV = \iiint_V 4\, dxdydz = 4\pi a^2 b$$

Example 1.

Find $\oiint_{S_1} \vec{F} \cdot \vec{n} dA - \oiint_{S_2} \vec{F} \cdot \vec{n} dA =?$, where $\vec{F} = (x, 2y + z, z + x^2)$,

$S_1: x^2 + y^2 + z^2 = 4, S_2: x^2 + y^2 + z^2 = 1$, both surfaces are outward-oriented.

【Solution】

9.7.1 Convert Closed Surface Integrals into Volume Integrals ($\nabla \cdot F = c$)

Let $V = \{(x, y, z): 1 \leq x^2 + y^2 + z^2 \leq 4\}$

∵ the first-order partial derivatives of the components of $\vec{F}$ exist and are continuous

By Gauss's Theorem, $\oiint_{S_1} \vec{F} \cdot \vec{n} dA - \oiint_{S_2} \vec{F} \cdot \vec{n} dA = \iiint_V \nabla \cdot \vec{F} \, dV$

∵ $\nabla \cdot \vec{F} = 1 + 2 + 1 = 4$

∴ $\iiint_V \nabla \cdot \vec{F} \, dV = \iiint_{1 \leq x^2+y^2+z^2 \leq 4} 4 \, dxdydz = 4 \left(\frac{4\pi}{3} \cdot 2^3 - \frac{4\pi}{3} \cdot 1^3 \right) = \frac{112\pi}{3}$

Example 2.

Find the flux of the vector field $\vec{F} = (x^2 y, 2y + z^2, 2z - 2xyz)$ across the surface $S = ?$, $S: x^2 + y^2 + z^2 = r^2$.

【Solution】

∵ the first-order partial derivatives of the components of $\vec{F}$ exist and are continuous

By Gauss's Theorem, $\oiint_S \vec{F} \cdot \vec{n} dA = \iiint_V \nabla \cdot \vec{F} \, dV, \quad V: x^2 + y^2 + z^2 \leq r^2$

∵ $\nabla \cdot \vec{F} = 2xy + 2 + 2 - 2xy = 4$

∴ $\iiint_V \nabla \cdot \vec{F} \, dV = \iiint_{x^2+y^2+z^2 \leq r^2} 4 \, dxdydz = 4 \cdot \frac{4\pi}{3} \cdot r^3 = \frac{16\pi r^3}{3}$

Example 3.

Find $\oiint_S \vec{F} \cdot \vec{n} dA = ?$, where $\vec{F} = (x, y, z)$, $S: x^2 + y^2 + z^2 = 1$ is outward-oriented.

【Solution】

∵ the first-order partial derivatives of the components of $\vec{F}$ exist and are continuous

By Gauss's Theorem, $\oiint_S \vec{F} \cdot \vec{n} dA = \iiint_V \nabla \cdot \vec{F} \, dV, \quad V: x^2 + y^2 + z^2 \leq 1$

∵ $\nabla \cdot \vec{F} = 1 + 1 + 1 = 3$

∴ $\iiint_V \nabla \cdot \vec{F} \, dV = \iiint_{x^2+y^2+z^2 \leq 1} 3 \, dxdydz = 3 \left(\frac{4\pi}{3} \cdot 1^3 \right) = 4\pi$

Example 4.

Find the flux of the vector field $\vec{F} = \left(x + y + \sin z^2, y + e^{x^2}, z + \ln(x^2 + y^2 + 1) \right)$

9.7.1 Convert Closed Surface Integrals into Volume Integrals ($\nabla \cdot \mathbf{F} = c$)

across the surface S, where $S: x^2 + y^2 + z^2 = r^2$.

【Solution】

∵ the first-order partial derivatives of the components of $\vec{F}$ exist and are continuous

By Gauss's Theorem, $\oiint_S \vec{F} \cdot \vec{n} dA = \iiint_V \nabla \cdot \vec{F} \, dV, \quad V: x^2 + y^2 + z^2 \leq r^2$

∵ $\nabla \cdot \vec{F} = 1 + 1 + 1 = 3$

∴ $\iiint_V \nabla \cdot \vec{F} \, dV = \iiint_{x^2+y^2+z^2 \leq r^2} 3 \, dxdydz = 3 \cdot \frac{4\pi}{3} \cdot r^3 = 4\pi r^3$

Example 5.

Find $\oiint_S \vec{F} \cdot \vec{n} dA = ?$, where $\vec{F} = (2x + ye^z, e^x - ye^z, e^z + 3z - xy)$,

$S: x^2 + y^2 + z^2 = 9$ is outward-oriented.

【Solution】

∵ the first-order partial derivatives of the components of $\vec{F}$ exist and are continuous

By Gauss's Theorem, $\oiint_S \vec{F} \cdot \vec{n} dA = \iiint_V \nabla \cdot \vec{F} \, dV, \quad V: x^2 + y^2 + z^2 \leq 9$

∵ $\nabla \cdot \vec{F} = 2 - e^z + e^z + 3 = 5$

∴ $\iiint_V \nabla \cdot \vec{F} \, dV = \iiint_{x^2+y^2+z^2 \leq 9} 5 \, dxdydz = 5 \left(\frac{4\pi}{3} \cdot 3^3 \right) = 180\pi$

Example 6.

Find $\oiint_S \vec{F} \cdot \vec{n} dA = ?$, where $\vec{F} = (x, e^x + y, 2z)$, $S: x^2 + y^2 = a^2$, $0 \leq z \leq b$

and the surface is outward-oriented.

【Solution】

Let $V = \{(x, y, z): x^2 + y^2 \leq a^2, 0 \leq z \leq b\}$

∵ the first-order partial derivatives of the components of $\vec{F}$ exist and are continuous

By Gauss's Theorem, $\oiint_S \vec{F} \cdot \vec{n} dA = \iiint_V \nabla \cdot \vec{F} \, dV$

∵ $\nabla \cdot \vec{F} = 1 + 1 + 2 = 4$ ∴ $\iiint_V \nabla \cdot \vec{F} \, dV = \iiint_V 4 \, dxdydz = 4\pi a^2 b$

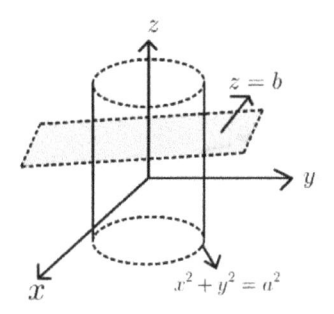

Example 7.

Find $\oiint_S (2x + 2y + z^2)dA = ?$, where $S: x^2 + y^2 + z^2 = r^2$ is outward-oriented.

【Solution】

Let $\vec{F} = (2, 2, z)$

∵ the first-order partial derivatives of the components of $\vec{F}$ exist and are continuous

By Gauss's Theorem, $\oiint_S \vec{F} \cdot \vec{n} dA = \iiint_V \nabla \cdot \vec{F} \, dV$, $V: x^2 + y^2 + z^2 \leq r^2$

∵ $\nabla \cdot \vec{F} = 0 + 0 + 1 = 1$ ∴ $\iiint_V \nabla \cdot \vec{F} \, dV = \iiint_V 1 \, dxdydz = \frac{4\pi}{3} \cdot r^3 = \frac{4\pi r^3}{3}$

Example 8.

Let $\vec{F} = (x, 2y, 3z)$. Assume S is the cube with the vertices $(\pm 1, \pm 1, \pm 1)$ with outward orientation. Find $\oiint_S \vec{F} \cdot \vec{n} dA$ by using Gauss's Theorem.

【Solution】

Let $V = \{(x, y, z): -1 \leq x \leq 1, -1 \leq y \leq 1, -1 \leq z \leq 1\}$

∵ the first-order partial derivatives of the components of $\vec{F}$ exist and are continuous

By Gauss's Theorem, $\oiint_S \vec{F} \cdot \vec{n} dA = \iiint_V \nabla \cdot \vec{F} \, dV$

∵ $\nabla \cdot \vec{F} = 1 + 2 + 3 = 6$ ∴ $\iiint_V \nabla \cdot \vec{F} \, dV = \iiint_V 6 \, dxdydz = 6 \cdot 2 \cdot 2 \cdot 2 = 48$

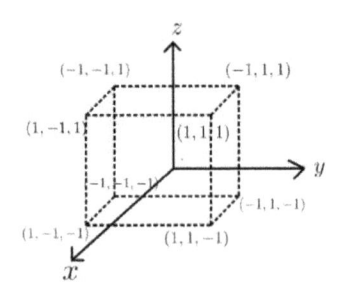

9.7.2 Convert Closed Surface Integrals into Volume Integrals($\nabla \cdot \vec{F} \neq c$)

Application Scenario: The integrand vector-valued function $\vec{F}$ in the surface integral over the closed surface is relatively complex, but its divergence $\nabla \cdot \vec{F}$ is simpler.

Question Types:

Type 1.

Suppose the first-order partial derivatives of each component of $\vec{F}$ exist and are continuous, and $S: x^2 + y^2 + z^2 = r^2$, where $r > 0$, and S is an outward-oriented surface. Find

$$\oiint_S \vec{F} \cdot \vec{n} dA = ?$$

Problem-Solving Process:

Step1.

∵ the first-order partial derivatives of each component of $\vec{F}$ exist and are continuous

By Gauss's Theorem, $\oiint_S \vec{F} \cdot \vec{n} dA = \iiint_V \nabla \cdot \vec{F} \, dV$

Step2.

$$\iiint_V \nabla \cdot \vec{F} \, dV = \iiint_{x^2+y^2+z^2 \leq r^2} f(x,y,z) \, dxdydz, \quad \text{where } \nabla \cdot \vec{F} = f(x,y,z)$$

Step3.

Let $x = \rho \sin \varphi \cos \theta$, $y = \rho \sin \varphi \sin \theta$, $z = \rho \cos \varphi$ then

$$dxdydz = \begin{Vmatrix} \dfrac{\partial x}{\partial \rho} & \dfrac{\partial x}{\partial \varphi} & \dfrac{\partial x}{\partial \theta} \\ \dfrac{\partial y}{\partial \rho} & \dfrac{\partial y}{\partial \varphi} & \dfrac{\partial y}{\partial \theta} \\ \dfrac{\partial z}{\partial \rho} & \dfrac{\partial z}{\partial \varphi} & \dfrac{\partial z}{\partial \theta} \end{Vmatrix} d\rho d\varphi d\theta$$

9.7.2 Convert Closed Surface Integrals into Volume Integrals($\nabla \cdot \mathbf{F} \neq c$)

$$= \begin{Vmatrix} \sin\varphi\cos\theta & \rho\cos\varphi\cos\theta & -\rho\sin\varphi\sin\theta \\ \sin\varphi\sin\theta & \rho\cos\varphi\sin\theta & -\rho\sin\varphi\cos\theta \\ \cos\varphi & -\rho\sin\varphi & 0 \end{Vmatrix} d\rho d\varphi d\theta = \rho^2 \sin\varphi \, d\rho d\varphi d\theta$$

Step4.

$$\therefore \iiint_{x^2+y^2+z^2 \le r^2} f(x,y,z) \, dxdydz$$

$$= \int_0^\pi \int_0^{2\pi} \int_0^r f(\rho\sin\varphi\cos\theta, \rho\sin\varphi\sin\theta, \rho\cos\varphi) \cdot \rho^2 \sin\varphi \, d\rho d\theta d\varphi$$

Find $\int_0^\pi \int_0^{2\pi} \int_0^r f(\rho\sin\varphi\cos\theta, \rho\sin\varphi\sin\theta, \rho\cos\varphi) \cdot \rho^2 \sin\varphi \, d\rho d\theta d\varphi = ?$

Examples:

(I) Assume $\vec{F} = (2x + 3xy^2 + z, xy^2 + 3x^2y + e^z, z^3 - 2xyz)$, and S is outward-oriented: $x^2 + y^2 + z^2 = r^2$.

$$\text{Find } \oiint_S \vec{F} \cdot \vec{n} dA = ?$$

∵ the first-order partial derivatives of each component of $\vec{F}$ exist and are continuous

By Gauss's Theorem,

$$\oiint_S \vec{F} \cdot \vec{n} dA = \iiint_V \nabla \cdot \vec{F} \, dV = \iiint_{x^2+y^2+z^2 \le r^2} 2 + 3x^2 + 3y^2 + 3z^2 \, dxdydz$$

Let $x = \rho\sin\varphi\cos\theta$, $y = \rho\sin\varphi\sin\theta$, $z = \rho\cos\varphi$ then $dxdydz = \rho^2 \sin\varphi \, d\rho d\varphi d\theta$

$$\therefore \iiint_{x^2+y^2+z^2 \le r^2} 2 + 3x^2 + 3y^2 + 3z^2 \, dxdydz = \frac{8\pi r^3}{3} + \int_0^\pi \int_0^{2\pi} \int_0^r 3\rho^2 \cdot \rho^2 \sin\varphi \, d\rho d\theta d\varphi$$

$$= \frac{8\pi r^3}{3} + \frac{12\pi r^5}{5}$$

(II) Assume $\vec{F} = (x^3 + z, y^3 + e^{-z}, z^3)$, and S is outward-oriented: $x^2 + y^2 + z^2 = a^2$.

$$\text{Find } \oiint_S \vec{F} \cdot \vec{n} dA = ?$$

∵ the first-order partial derivatives of each component of $\vec{F}$ exist and are continuous

By Gauss's Theorem,

$$\oiint_S \vec{F} \cdot \vec{n} dA = \iiint_V \nabla \cdot \vec{F} \, dV = \iiint_{x^2+y^2+z^2 \le a^2} 3x^2 + 3y^2 + 3z^2 \, dxdydz$$

Let $x = \rho\sin\varphi\cos\theta$, $y = \rho\sin\varphi\sin\theta$, $z = \rho\cos\varphi$ then $dxdydz = \rho^2 \sin\varphi \, d\rho d\varphi d\theta$

9.7.2 Convert Closed Surface Integrals into Volume Integrals($\nabla \cdot \mathbf{F} \neq c$) 602

$$\therefore \iiint_{x^2+y^2+z^2 \leq a^2} 3x^2 + 3y^2 + 3z^2 \, dxdydz = 3 \int_0^\pi \int_0^{2\pi} \int_0^a \rho^2 \cdot \rho^2 \sin\varphi \, d\rho d\theta d\varphi = \frac{12\pi a^5}{5}$$

Type 2.

Suppose the first-order partial derivatives of each component of the vector field $\vec{F}$ exist and are continuous, and S is the closed surface defined by $x^2 + y^2 = a^2$, $0 \leq z \leq b$, oriented outward.

$$\text{Find } \oiint_S \vec{F} \cdot \vec{n} dA = ?$$

Problem-Solving Process:

Step1.

$\because$ the first-order partial derivatives of each component of $\vec{F}$ exist and are continuous

By Gauss's Theorem, $\oiint_S \vec{F} \cdot \vec{n} dA = \iiint_V \nabla \cdot \vec{F} \, dV$

Step2.

$$\iiint_V \nabla \cdot \vec{F} \, dV = \iiint_V f(x,y) \, dxdydz = \iint_R \int_0^b f(x,y) dz dxdy = b \iint_R f(x,y) \, dxdy$$

where $\nabla \cdot \vec{F} = f(x,y)$, $R = \{(x,y): x^2 + y^2 \leq a^2\}$

Step3.

Let $x = r\cos\theta$, $y = r\sin\theta$ then $\{(x,y): x^2 + y^2 \leq a^2\} = \{(r,\theta): 0 \leq r \leq a, 0 \leq \theta \leq 2\pi\}$

and $dxdy = \begin{Vmatrix} \frac{\partial x}{\partial r} & \frac{\partial x}{\partial \theta} \\ \frac{\partial y}{\partial r} & \frac{\partial y}{\partial \theta} \end{Vmatrix} drd\theta = \begin{Vmatrix} \cos\theta & -r\sin\theta \\ \sin\theta & r\cos\theta \end{Vmatrix} drd\theta = rdrd\theta$

$$\therefore b \iint_R f(x,y) \, dxdy = b \int_0^{2\pi} \int_0^a f(r\cos\theta, r\sin\theta) r dr d\theta$$

Step4.

Find $\int_0^{2\pi} \int_0^a f(r\cos\theta, r\sin\theta) r dr d\theta = ?$

Examples:

(1) Let $\vec{F} = (2x^3, x^2y, x^2z)$, $S: x^2 + y^2 = a^2$, $0 \leq z \leq b$. Find $\oiint_S \vec{F} \cdot \vec{n} dA = ?$

Let $R = \{(x,y): x^2 + y^2 \leq a^2\}$ and $x = r\cos\theta$, $y = r\sin\theta$ then
$\{(x,y): x^2 + y^2 \leq a^2\} = \{(r,\theta): 0 \leq r \leq a, 0 \leq \theta \leq 2\pi\}$.

By Gauss's Theorem,
$$\oiint_S \vec{F} \cdot \vec{n} dA = \iiint_V \nabla \cdot \vec{F}\, dV = \iint_R \int_0^b 8x^2 dz dx dy = b \iint_{x^2+y^2 \le a^2} 8x^2 dx dy$$
$$= b \int_0^{2\pi} \int_0^a 8r^2 \cos^2\theta\, r dr d\theta = 2\pi a^4 b$$

Type 3.
Suppose the first-order partial derivatives of each component of the vector field $\vec{F}$ exist and are continuous. Let S be the closed surface oriented outward, bounded by the surface $z = g(x, y)$ and $z = 0$, where the projection of $z = g(x, y)$ onto the xy-plane is $\{(x, y): x^2 + y^2 \le a^2\}$.

$$\text{Find } \oiint_S \vec{F} \cdot \vec{n} dA =?$$

Problem-Solving Process:
Step1.
Let $R = \{(x, y): x^2 + y^2 \le a^2\}$ and $x = r \cos\theta, y = r \sin\theta$ then
$\{(x, y): x^2 + y^2 \le a^2\} = \{(r, \theta): 0 \le r \le a, 0 \le \theta \le 2\pi\}$
Step2.
∵ the first-order partial derivatives of each component of $\vec{F}$ exist and are continuous
By Gauss's Theorem,
$$\oiint_S \vec{F} \cdot \vec{n} dA = \iiint_V \nabla \cdot \vec{F}\, dV = \iint_R \int_0^{g(x,y)} f(x, y, z) dz dx dy, \text{ where } \nabla \cdot \vec{F} = f(x, y, z)$$
Step3.
$$\iint_R \int_0^{g(x,y)} f(x, y, z) dz dx dy = \iint_R h(x, y)\, dx dy = \int_0^{2\pi} \int_0^a h(r \cos\theta, r \sin\theta) r dr d\theta$$
where $h(x, y) = \int_0^{g(x,y)} f(x, y, z) dz$

Find $\int_0^{2\pi} \int_0^a h(r \cos\theta, r \sin\theta) r dr d\theta$

Examples:
(I)Given $\vec{F} = (x^2, xy, z)$, and S is the closed surface bounded by $z = 4 - x^2 - y^2$ and the xy-plane, find $\oiint_S \vec{F} \cdot \vec{n} dA =?$

Let $R = \{(x, y): x^2 + y^2 \le 4\}$ and $x = r \cos\theta, y = r \sin\theta$ then
$\{(x, y): x^2 + y^2 \le 4\} = \{(r, \theta): 0 \le r \le 2, 0 \le \theta \le 2\pi\}$

9.7.2 Convert Closed Surface Integrals into Volume Integrals($\nabla \cdot \mathbf{F} \neq c$) 604

By Gauss's Theorem,

$$\oiint_S \vec{F} \cdot \vec{n} dA = \iiint_V \nabla \cdot \vec{F}\, dV = \iint_R \int_0^{4-x^2-y^2} 3x + 1\, dzdxdy$$

$$= \iint_R (3x+1)(4-x^2-y^2)\, dxdy = \int_0^{2\pi} \int_0^2 (3\cos\theta + 1)(4-r^2)r\, drd\theta = 8\pi$$

Type 4.

Suppose the first-order partial derivatives of each component of the vector field $\vec{F}$ exist and are continuous. Let S be a closed surface oriented outward, enclosing a volume V.

$$\text{Find } \oiint_S \vec{F} \cdot \vec{n} dA =?$$

where $V = \{(x,y,z): x_0 \le x \le x_1, y_0 \le y \le y_1, z_0 \le z \le z_1\}$.

Problem-Solving Process:

By Gauss's Theorem, $\oiint_S \vec{F} \cdot \vec{n} dA = \iiint_V \nabla \cdot \vec{F}\, dV = \int_{x_0}^{x_1} \int_{y_0}^{y_1} \int_{z_0}^{z_1} f(x,y,z)\, dzdydx$

where $\nabla \cdot \vec{F} = f(x,y,z)$

Examples:

Find $\oiint_S \vec{F} \cdot \vec{n} dA =?$

(I) $\vec{F} = (xy + 3, y^2 + e^{zx}, \cos xy)$ and S is the closed surface bounded by $z = 1 - x^2$, $y = 0, z = 0, \ y + z = 2$.

By Gauss's Theorem, $\oiint_S \vec{F} \cdot \vec{n} dA = \iiint_V \nabla \cdot \vec{F}\, dV = 3\int_{-1}^1 \int_0^{1-x} \int_0^{2-z} y\, dydzdx = \dfrac{184}{35}$

(II) $\vec{F} = (e^x \sin y, e^x \cos y, yz^2 + xy)$ and S is the closed surface bounded by $0 \le x \le 1$, $0 \le y \le 1, 0 \le z \le 2$.

By Gauss's Theorem, $\oiint_S \vec{F} \cdot \vec{n} dA = \iiint_V \nabla \cdot \vec{F}\, dV = \int_0^1 \int_0^1 \int_0^2 2yz\, dzdydx = 2$

Example 1.

Let $\vec{F} = (2x + x^3 + ye^z, e^x - ye^z + y^3, e^z + 3z - xy + z^3)$, $S: x^2 + y^2 + z^2 = 4$.

Use Gauss's Theorem to find $\oiint_S \vec{F} \cdot \vec{n} dA =?$

【Solution】

Let $V: x^2 + y^2 + z^2 \leq 4$

∵ the first-order partial derivatives of the components of $\vec{F}$ exist and are continuous

By Gauss's Theorem, $\oiint_S \vec{F} \cdot \vec{n} dA = \iiint_V \nabla \cdot \vec{F} \, dV$

∵ $\nabla \cdot \vec{F} = 5 + 3x^2 + 3y^2 + 3z^2$ ∴ $\iiint_V \nabla \cdot \vec{F} \, dV = \iiint_{x^2+y^2+z^2 \leq 4} 5 + 3x^2 + 3y^2 + 3z^2 \, dxdydz$

Let $x = \rho \sin\varphi \cos\theta$, $y = \rho \sin\varphi \sin\theta$, $z = \rho \cos\varphi$ then

$$dxdydz = \begin{Vmatrix} \dfrac{\partial x}{\partial \rho} & \dfrac{\partial x}{\partial \varphi} & \dfrac{\partial x}{\partial \theta} \\ \dfrac{\partial y}{\partial \rho} & \dfrac{\partial y}{\partial \varphi} & \dfrac{\partial y}{\partial \theta} \\ \dfrac{\partial z}{\partial \rho} & \dfrac{\partial z}{\partial \varphi} & \dfrac{\partial z}{\partial \theta} \end{Vmatrix} d\rho d\varphi d\theta$$

$$= \begin{Vmatrix} \sin\varphi \cos\theta & \rho\cos\varphi \cos\theta & -\rho\sin\varphi \sin\theta \\ \sin\varphi \sin\theta & \rho\cos\varphi \sin\theta & -\rho\sin\varphi \cos\theta \\ \cos\varphi & -\rho\sin\varphi & 0 \end{Vmatrix} d\rho d\varphi d\theta = \rho^2 \sin\varphi \, d\rho d\varphi d\theta$$

where $0 \leq \rho \leq 2$, $0 \leq \varphi \leq \pi$, $0 \leq \theta \leq 2\pi$

∴ $\iiint_{x^2+y^2+z^2 \leq 4} 5 + 3x^2 + 3y^2 + 3z^2 \, dxdydz$

$= 5 \cdot \dfrac{4\pi}{3} \cdot 2^3 + \int_0^\pi \int_0^{2\pi} \int_0^3 3\rho^2 \cdot \rho^2 \sin\varphi \, d\rho d\theta d\varphi = \dfrac{160\pi}{3} + \dfrac{1152\pi}{5} = \dfrac{1952\pi}{15}$

Example 2.

Use Gauss's Theorem to find the flux of the vector field
$$\vec{F} = (2x + 3xy^2 + z, xy^2 + 3x^2y + e^z, z^3 - 2xyz)$$
across the surface S, where $S: x^2 + y^2 + z^2 = r^2$.

【Solution】

Let $V: x^2 + y^2 + z^2 \leq r^2$

∵ the first-order partial derivatives of the components of $\vec{F}$ exist and are continuous

By Gauss Theorem, $\oiint_S \vec{F} \cdot \vec{n} dA = \iiint_V \nabla \cdot \vec{F} \, dV$

∵ $\nabla \cdot \vec{F} = 2 + 3y^2 + 2xy + 3x^2 + 3z^2 - 2xy = 2 + 3x^2 + 3y^2 + 3z^2$

9.7.2 Convert Closed Surface Integrals into Volume Integrals($\nabla \cdot \vec{F} \neq c$)

$$\therefore \iiint_V \nabla \cdot \vec{F}\, dV = \iiint_{x^2+y^2+z^2 \leq r^2} 2 + 3x^2 + 3y^2 + 3z^2\, dxdydz$$

Let $x = \rho \sin\varphi \cos\theta$, $y = \rho \sin\varphi \sin\theta$, $z = \rho \cos\varphi$ then

$$dxdydz = \begin{Vmatrix} \dfrac{\partial x}{\partial \rho} & \dfrac{\partial x}{\partial \varphi} & \dfrac{\partial x}{\partial \theta} \\ \dfrac{\partial y}{\partial \rho} & \dfrac{\partial y}{\partial \varphi} & \dfrac{\partial y}{\partial \theta} \\ \dfrac{\partial z}{\partial \rho} & \dfrac{\partial z}{\partial \varphi} & \dfrac{\partial z}{\partial \theta} \end{Vmatrix} d\rho d\varphi d\theta$$

$$= \begin{Vmatrix} \sin\varphi \cos\theta & \rho\cos\varphi \cos\theta & -\rho\sin\varphi \sin\theta \\ \sin\varphi \sin\theta & \rho\cos\varphi \sin\theta & -\rho\sin\varphi \cos\theta \\ \cos\varphi & -\rho\sin\varphi & 0 \end{Vmatrix} d\rho d\varphi d\theta = \rho^2 \sin\varphi\, d\rho d\varphi d\theta$$

where $0 \leq \rho \leq r$, $0 \leq \varphi \leq \pi$, $0 \leq \theta \leq 2\pi$

$$\therefore \iiint_{x^2+y^2+z^2 \leq r^2} 2 + 3x^2 + 3y^2 + 3z^2\, dxdydz$$

$$= 2 \cdot \frac{4\pi \cdot r^3}{3} + \int_0^\pi \int_0^{2\pi} \int_0^r 3\rho^2 \cdot \rho^2 \sin\varphi\, d\rho d\theta d\varphi = \frac{8\pi r^3}{3} + \frac{12\pi r^5}{5}$$

Example 3.

Find $\oiint_S \vec{F} \cdot \vec{n}\, dA =?$, where $\vec{F} = (x^2 y + x^3, 2y + z^2 + y^3, 4z - 2xyz + z^3)$,

$S: x^2 + y^2 + z^2 = 1$ is outward-oriented.

【Solution】

Let $V: x^2 + y^2 + z^2 \leq 1$

$\because$ the first-order partial derivatives of the components of $\vec{F}$ exist and are continuous

By Gauss Theorem, $\oiint_S \vec{F} \cdot \vec{n}\, dA = \iiint_V \nabla \cdot \vec{F}\, dV$

$\because \nabla \cdot \vec{F} = 6 + 3x^2 + 3y^2 + 3z^2$ $\therefore \iiint_V \nabla \cdot \vec{F}\, dV = \iiint_{x^2+y^2+z^2 \leq 1} 6 + 3x^2 + 3y^2 + 3z^2\, dxdydz$

Let $x = \rho \sin\varphi \cos\theta$, $y = \rho \sin\varphi \sin\theta$, $z = \rho \cos\varphi$ then

$$dxdydz = \begin{Vmatrix} \dfrac{\partial x}{\partial \rho} & \dfrac{\partial x}{\partial \varphi} & \dfrac{\partial x}{\partial \theta} \\ \dfrac{\partial y}{\partial \rho} & \dfrac{\partial y}{\partial \varphi} & \dfrac{\partial y}{\partial \theta} \\ \dfrac{\partial z}{\partial \rho} & \dfrac{\partial z}{\partial \varphi} & \dfrac{\partial z}{\partial \theta} \end{Vmatrix} d\rho d\varphi d\theta$$

$$= \begin{Vmatrix} \sin\varphi\cos\theta & \rho\cos\varphi\cos\theta & -\rho\sin\varphi\sin\theta \\ \sin\varphi\sin\theta & \rho\cos\varphi\sin\theta & -\rho\sin\varphi\cos\theta \\ \cos\varphi & -\rho\sin\varphi & 0 \end{Vmatrix} d\rho d\varphi d\theta = \rho^2\sin\varphi\, d\rho d\varphi d\theta$$

where $0 \leq \rho \leq 1$, $0 \leq \varphi \leq \pi$, $0 \leq \theta \leq 2\pi$

$$\therefore \iiint_{x^2+y^2+z^2 \leq 1} 6 + 3x^2 + 3y^2 + 3z^2 \, dxdydz = 6\cdot\frac{4\pi}{3} + \int_0^\pi \int_0^{2\pi} \int_0^1 3\rho^2 \cdot \rho^2 \sin\varphi\, d\rho d\theta d\varphi$$

$$= 8\pi + \frac{12\pi}{5} = \frac{52\pi}{15}$$

Example 4.

Use Gauss's Theorem to find the flux of the vector field $\vec{F} = (0, 4yz, 0)$ across the surface $S: x^2 + y^2 + z^2 = a^2$, $\oiint_S \vec{F} \cdot \vec{n} dA =?$

【Solution】

Let $V: x^2 + y^2 + z^2 \leq a^2$

∵ the first-order partial derivatives of the components of $\vec{F}$ exist and are continuous

By Gauss's Theorem, $\oiint_S \vec{F} \cdot \vec{n} dA = \iiint_V \nabla \cdot \vec{F}\, dV$

∵ $\nabla \cdot \vec{F} = 4z$ ∴ $\iiint_V \nabla \cdot \vec{F}\, dV = \iiint_{x^2+y^2+z^2 \leq a^2} 4z\, dxdydz$

Let $x = \rho\sin\varphi\cos\theta$, $y = \rho\sin\varphi\sin\theta$, $z = \rho\cos\varphi$ then

$$dxdydz = \begin{Vmatrix} \dfrac{\partial x}{\partial \rho} & \dfrac{\partial x}{\partial \varphi} & \dfrac{\partial x}{\partial \theta} \\ \dfrac{\partial y}{\partial \rho} & \dfrac{\partial y}{\partial \varphi} & \dfrac{\partial y}{\partial \theta} \\ \dfrac{\partial z}{\partial \rho} & \dfrac{\partial z}{\partial \varphi} & \dfrac{\partial z}{\partial \theta} \end{Vmatrix} d\rho d\varphi d\theta$$

9.7.2 Convert Closed Surface Integrals into Volume Integrals($\nabla \cdot \mathbf{F} \neq c$) 608

$$= \begin{Vmatrix} \sin\varphi\cos\theta & \rho\cos\varphi\cos\theta & -\rho\sin\varphi\sin\theta \\ \sin\varphi\sin\theta & \rho\cos\varphi\sin\theta & -\rho\sin\varphi\cos\theta \\ \cos\varphi & -\rho\sin\varphi & 0 \end{Vmatrix} d\rho d\varphi d\theta = \rho^2 \sin\varphi \, d\rho d\varphi d\theta$$

where $0 \le \rho \le a$, $0 \le \varphi \le \pi$, $0 \le \theta \le 2\pi$

$$\therefore \iiint_{x^2+y^2+z^2 \le a^2} 4z \, dxdydz = \int_0^\pi \int_0^{2\pi} \int_0^a 4\rho\cos\varphi \cdot \rho^2 \sin\varphi \, d\rho d\theta d\varphi = 0$$

Example 5.

Use Gauss's Theorem to find $\oiint_S \vec{F} \cdot \vec{n} dA =?$, where $\vec{F} = (x - y + z, 2x, 1)$, S is the outward-oriented closed surface bounded by $z = x^2 + y^2, z = 1$.

【Solution】

Let $S_1: z = x^2 + y^2, z < 1$, and $S_2: x^2 + y^2 \le 1, z = 1$ then $S = S_1 \cup S_2$
Let $V = \{(x, y, z): x^2 + y^2 \le z, 0 \le z \le 1\}$
∵ the first-order partial derivatives of the components of $\vec{F}$ exist and are continuous

By Gauss's Theorem, $\oiint_S \vec{F} \cdot \vec{n} dA = \iiint_V \nabla \cdot \vec{F} \, dV$

∵ $\nabla \cdot \vec{F} = 1$ ∴ $\iiint_V \nabla \cdot \vec{F} \, dV = \iiint_V 1 \, dxdydz$

Let $R = \{(x, y): x^2 + y^2 \le 1\}$ then

$$\iiint_V 1 \, dxdydz = \iint_R \int_0^{1-x^2-y^2} dz dxdy = \iint_R 1 - x^2 - y^2 \, dxdy$$

Let $x = r\cos\theta, y = r\sin\theta$ then $\{(x,y): x^2+y^2 \le 1\} = \{(r,\theta): 0 \le r \le 1, 0 \le \theta \le 2\pi\}$

where $dxdy = \begin{Vmatrix} \frac{\partial x}{\partial r} & \frac{\partial x}{\partial \theta} \\ \frac{\partial y}{\partial r} & \frac{\partial y}{\partial \theta} \end{Vmatrix} drd\theta = \begin{Vmatrix} \cos\theta & -r\sin\theta \\ \sin\theta & r\cos\theta \end{Vmatrix} drd\theta = rdrd\theta$

$$\therefore \iint_R 1 - x^2 - y^2 \, dxdy = \int_0^{2\pi} \int_0^1 (1 - r^2) r dr d\theta = 2\pi \cdot \left(\frac{r^2}{2} - \frac{r^4}{4} \right) \Big|_0^1 = \frac{\pi}{2}$$

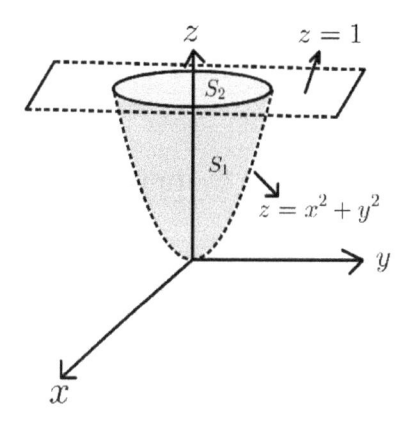

Example 6.

Let $\vec{F} = (x, 2y, 3z)$. Assume V is the volume of cube with the vertices $(\pm 1, \pm 1, \pm 1)$ and S is the boundary of V. Find $\oiint_S \vec{F} \cdot \vec{n} dA$ by using Gauss's Theorem.

【Solution】

Let $V = \{(x, y, z): -1 \le x \le 1, -1 \le y \le 1, -1 \le z \le 1\}$

∵ the first-order partial derivatives of the components of $\vec{F}$ exist and are continuous

By Gauss's Theorem, $\oiint_S \vec{F} \cdot \vec{n} dA = \iiint_V \nabla \cdot \vec{F} \, dV$

∵ $\nabla \cdot \vec{F} = 6$ ∴ $\iiint_V \nabla \cdot \vec{F} \, dV = \iiint_V 6 \, dx dy dz = 6 \cdot 2 \cdot 2 \cdot 2 = 48$

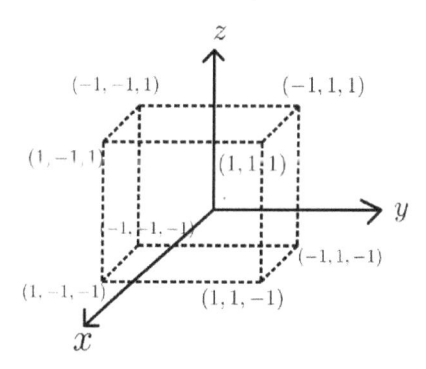

Example 7.

Use Gauss's Theorem to find $\oiint_S \vec{F} \cdot \vec{n} dA =?$, where $\vec{F} = (x^3 + z, y^3 + e^{-z}, z^3)$,

9.7.2 Convert Closed Surface Integrals into Volume Integrals ($\nabla \cdot \mathbf{F} \neq c$)

$S: x^2 + y^2 + z^2 = a^2$

【Solution】

Let $V: x^2 + y^2 + z^2 \leq a^2$

∵ the first-order partial derivatives of the components of $\vec{F}$ exist and are continuous

By Gauss's Theorem, $\oiint_S \vec{F} \cdot \vec{n} dA = \iiint_V \nabla \cdot \vec{F} \, dV$

∵ $\nabla \cdot \vec{F} = 3x^2 + 3y^2 + 3z^2$ ∴ $\iiint_V \nabla \cdot \vec{F} \, dV = \iiint_{x^2+y^2+z^2 \leq a^2} 3x^2 + 3y^2 + 3z^2 \, dxdydz$

Let $x = \rho \sin\varphi \cos\theta$, $y = \rho \sin\varphi \sin\theta$, $z = \rho \cos\varphi$ then

$$dxdydz = \begin{Vmatrix} \dfrac{\partial x}{\partial \rho} & \dfrac{\partial x}{\partial \varphi} & \dfrac{\partial x}{\partial \theta} \\ \dfrac{\partial y}{\partial \rho} & \dfrac{\partial y}{\partial \varphi} & \dfrac{\partial y}{\partial \theta} \\ \dfrac{\partial z}{\partial \rho} & \dfrac{\partial z}{\partial \varphi} & \dfrac{\partial z}{\partial \theta} \end{Vmatrix} d\rho d\varphi d\theta$$

$$= \begin{Vmatrix} \sin\varphi \cos\theta & \rho\cos\varphi \cos\theta & -\rho\sin\varphi \sin\theta \\ \sin\varphi \sin\theta & \rho\cos\varphi \sin\theta & -\rho\sin\varphi \cos\theta \\ \cos\varphi & -\rho\sin\varphi & 0 \end{Vmatrix} d\rho d\varphi d\theta = \rho^2 \sin\varphi \, d\rho d\varphi d\theta$$

where $0 \leq \rho \leq a$, $0 \leq \varphi \leq \pi$, $0 \leq \theta \leq 2\pi$

∴ $\iiint_{x^2+y^2+z^2 \leq a^2} 3x^2 + 3y^2 + 3z^2 \, dxdydz = 3 \int_0^\pi \int_0^{2\pi} \int_0^a \rho^2 \cdot \rho^2 \sin\varphi \, d\rho d\theta d\varphi = \dfrac{12\pi a^5}{5}$

Example 8.

Use Gauss's Theorem to find $\oiint_S \vec{F} \cdot \vec{n} dA = ?$, where $\vec{F} = (xy + 3, y^2 + e^{zx}, \cos xy)$, S is the outward-oriented closed surface bounded by $z = 1 - x^2$, $y = 0$, $z = 0$, and $y + z = 2$.

【Solution】

Let $V = \{(x, y, z): -1 \leq x \leq 1, 0 \leq z \leq 1 - x, 0 \leq y \leq 2 - z\}$

∵ the first-order partial derivatives of the components of $\vec{F}$ exist and are continuous

By Gauss's Theorem, $\oiint_S \vec{F} \cdot \vec{n} dA = \iiint_V \nabla \cdot \vec{F} \, dV$

$\because \nabla \cdot \vec{F} = y + 2y + 0 = 3y \quad \therefore \iiint_V \nabla \cdot \vec{F} \, dV = 3 \int_{-1}^{1} \int_{0}^{1-x} \int_{0}^{2-z} y \, dy \, dz \, dx = \dfrac{184}{35}$

Example 9.

Use Gauss's Theorem to find $\oiint_S \vec{F} \cdot \vec{n} dA = ?$, where $\vec{F} = (2x^3, x^2 y, x^2 z)$, S is the outward-oriented closed surface bounded by $x^2 + y^2 = a^2, z = 0, z = b$.

【Solution】

Let $S_1: x^2 + y^2 = a^2$, $0 \le z \le b$, $S_2: x^2 + y^2 \le a^2$, $z = 0$, $S_3: x^2 + y^2 \le a^2$, $z = b$
then $S = S_1 \cup S_2 \cup S_3$
Let V be the volume enclosed by S, then $V = \{(x, y, z): x^2 + y^2 \le a^2, 0 \le z \le b\}$
$\because$ the first-order partial derivatives of the components of $\vec{F}$ exist and are continuous

By Gauss's Theorem, $\oiint_S \vec{F} \cdot \vec{n} dA = \iiint_V \nabla \cdot \vec{F} \, dV$

$\because \nabla \cdot \vec{F} = 6x^2 + x^2 + x^2 = 8x^2 \quad \therefore \iiint_V \nabla \cdot \vec{F} \, dV = \iiint_V 8x^2 \, dx dy dz$

Let $R = \{(x, y): x^2 + y^2 \le a^2\}$ then

$\iiint_V 8x^2 \, dx dy dz = 8 \iint_R \int_0^b x^2 \, dz dx dy = 8b \iint_{x^2+y^2 \le a^2} x^2 \, dx dy$

Let $x = r \cos\theta$, $y = r \sin\theta$ then $\{(x,y): x^2 + y^2 \le a^2\} = \{(r, \theta): 0 \le r \le a, 0 \le \theta \le 2\pi\}$

and $dxdy = \begin{Vmatrix} \dfrac{\partial x}{\partial r} & \dfrac{\partial x}{\partial \theta} \\ \dfrac{\partial y}{\partial r} & \dfrac{\partial y}{\partial \theta} \end{Vmatrix} drd\theta = \begin{Vmatrix} \cos\theta & -r\sin\theta \\ \sin\theta & r\cos\theta \end{Vmatrix} drd\theta = rdrd\theta$

$\therefore 8b \iint_{x^2+y^2 \le a^2} x^2 \, dx dy = 8b \int_0^{2\pi} \int_0^a r^2 \cos^2\theta \, r dr d\theta = 8b \int_0^{2\pi} \cos^2\theta \, d\theta \int_0^a r^3 dr$

$\because \int_0^{2\pi} \cos^2\theta \, d\theta = \int_0^{2\pi} \dfrac{1 + \cos 2\theta}{2} d\theta = \pi$ and $\int_0^a r^3 dr = \dfrac{a^4}{4} \quad \therefore \oiint_S \vec{F} \cdot \vec{n} dA = 2\pi a^4 b$

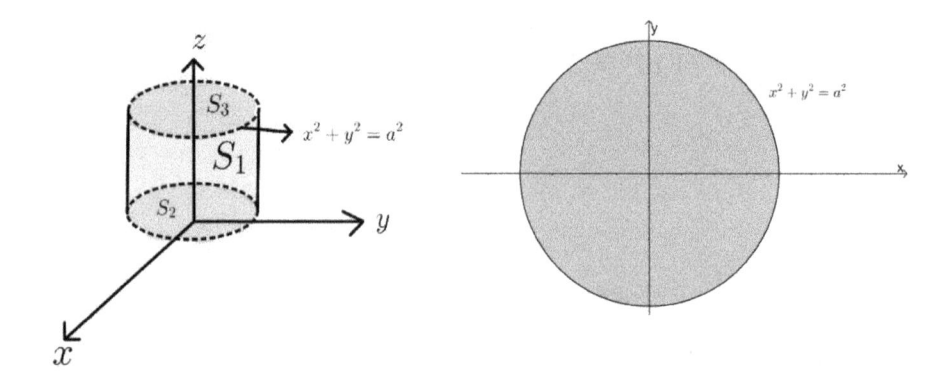

9.7.2 Convert Closed Surface Integrals into Volume Integrals($\nabla \cdot \mathbf{F} \neq c$) 612

Example 10.

Use Gauss's Theorem to find $\oiint_S \vec{F} \cdot \vec{n} dA =?$, where $\vec{F} = (xye^z, xy^2z^3, -ye^z)$,

S is the outward-oriented closed surface bounded by the coordinate planes and $x = 3, y = 2, z = 1$.

【Solution】

Let V be the volume enclosed by S, then $V = \{(x,y,z): 0 \leq x \leq 3, 0 \leq y \leq 2, 0 \leq z \leq 1\}$

∵ the first-order partial derivatives of the components of $\vec{F}$ exist and are continuous

By Gauss's Theorem, $\oiint_S \vec{F} \cdot \vec{n} dA = \iiint_V \nabla \cdot \vec{F} \, dV$

$\iiint_V \nabla \cdot \vec{F} \, dV = \iiint_V ye^z + 2xyz^3 - ye^z \, dV = 2\iiint_V xyz^3 \, dV = 2\int_0^3 \int_0^2 \int_0^1 xyz^3 \, dzdydx$

$= 2 \cdot \dfrac{9}{2} \cdot \dfrac{4}{2} \cdot \dfrac{1}{4} = \dfrac{9}{2}$

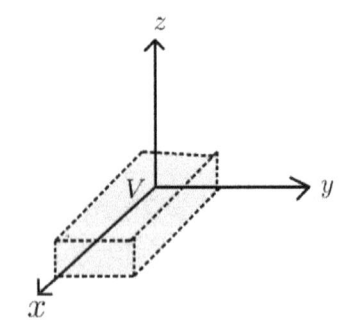

Example 11.

Use Gauss's Theorem to find $\oiint_S \vec{F}\cdot\vec{n}dA =?$, where $\vec{F} = (3xy^2, xe^z, z^3)$, S is the outward-oriented closed surface bounded by $y^2 + z^2 = 1$, $x = -1$, and $x = 2$.

【Solution】

Let V be the volume enclosed by S, then $V = \{(x,y,z): -1 \le x \le 2,\ y^2 + z^2 \le 1\}$

∵ the first-order partial derivatives of the components of $\vec{F}$ exist and are continuous

By Gauss's Theorem, $\oiint_S \vec{F}\cdot\vec{n}dA = \iiint_V \nabla\cdot\vec{F}\,dV$

$\iiint_V \nabla\cdot\vec{F}\,dV = \iiint_V 3y^2 + 3z^2\,dV = \iint_R \int_{-1}^{2} 3y^2 + 3z^2\,dxdydz = 9\iint_R y^2 + z^2\,dydx$

Let $y = r\cos\theta$, $z = r\sin\theta$, $R = \{(y,z): y^2 + z^2 \le 1\} = \{(r,\theta): 0 \le r \le 1, 0 \le \theta \le 2\pi\}$

$\therefore 9\iint_R y^2 + z^2\,dydx = 9\int_0^{2\pi}\int_0^1 r^3\,drd\theta = 9\cdot 2\pi \cdot \frac{1}{4} = \frac{9\pi}{2}$

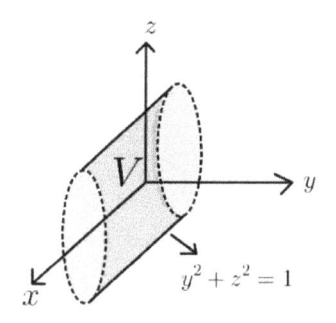

Example 12.

Use Gauss's Theorem to find $\oiint_S x^4 + y^4 + z^4 dA =?$, $S: x^2 + y^2 + z^2 = a^2$.

【Solution】

∵ the unit normal vector of the sphere is $\vec{n} = (x,y,z)$

Let $\vec{F} = (x^3, y^3, z^3)$ then $\vec{F}\cdot\vec{n} = x^4 + y^4 + z^4$

Let $V: x^2 + y^2 + z^2 \le a^2$

∵ the first-order partial derivatives of the components of $\vec{F}$ exist and are continuous

9.7.2 Convert Closed Surface Integrals into Volume Integrals($\nabla \cdot \mathbf{F} \neq c$) 614

By Gauss's Theorem, $\oiint_S \vec{F} \cdot \vec{n} dA = \iiint_V \nabla \cdot \vec{F} \, dV$

$\because \nabla \cdot \vec{F} = 3x^2 + 3y^2 + 3z^2 \quad \therefore \iiint_V \nabla \cdot \vec{F} \, dV = \iiint_{x^2+y^2+z^2 \leq a^2} 3x^2 + 3y^2 + 3z^2 \, dxdydz$

Let $x = \rho \sin \varphi \cos \theta, y = \rho \sin \varphi \sin \theta, z = \rho \cos \varphi$ then

$$dxdydz = \begin{Vmatrix} \dfrac{\partial x}{\partial \rho} & \dfrac{\partial x}{\partial \varphi} & \dfrac{\partial x}{\partial \theta} \\ \dfrac{\partial y}{\partial \rho} & \dfrac{\partial y}{\partial \varphi} & \dfrac{\partial y}{\partial \theta} \\ \dfrac{\partial z}{\partial \rho} & \dfrac{\partial z}{\partial \varphi} & \dfrac{\partial z}{\partial \theta} \end{Vmatrix} d\rho d\varphi d\theta$$

$$= \begin{Vmatrix} \sin \varphi \cos \theta & \rho \cos \varphi \cos \theta & -\rho \sin \varphi \sin \theta \\ \sin \varphi \sin \theta & \rho \cos \varphi \sin \theta & -\rho \sin \varphi \cos \theta \\ \cos \varphi & -\rho \sin \varphi & 0 \end{Vmatrix} d\rho d\varphi d\theta = \rho^2 \sin \varphi \, d\rho d\varphi d\theta$$

where $0 \leq \rho \leq a, 0 \leq \varphi \leq \pi, 0 \leq \theta \leq 2\pi$

$\therefore \iiint_{x^2+y^2+z^2 \leq a^2} 3x^2 + 3y^2 + 3z^2 \, dxdydz = 3 \int_0^\pi \int_0^{2\pi} \int_0^a \rho^2 \cdot \rho^2 \sin \varphi \, d\rho d\theta d\varphi = \dfrac{12\pi a^5}{5}$

Example 13.

Use Gauss's Theorem to find $\oiint_S \vec{F} \cdot \vec{n} dA$, $\vec{F} = (xy^2, yz^2, zx^2)$, $S: x^2 + y^2 + z^2 = r^2$.

【Solution】

Let $V: x^2 + y^2 + z^2 \leq r^2$

$\because$ the first-order partial derivatives of the components of $\vec{F}$ exist and are continuous

By Gauss's Theorem, $\oiint_S \vec{F} \cdot \vec{n} dA = \iiint_V \nabla \cdot \vec{F} \, dV$

$\because \nabla \cdot \vec{F} = x^2 + y^2 + z^2 \quad \therefore \iiint_V \nabla \cdot \vec{F} \, dV = \iiint_{x^2+y^2+z^2 \leq r^2} x^2 + y^2 + z^2 \, dxdydz$

Let $x = \rho \sin \varphi \cos \theta, y = \rho \sin \varphi \sin \theta, z = \rho \cos \varphi$ then

$$dxdydz = \begin{Vmatrix} \dfrac{\partial x}{\partial \rho} & \dfrac{\partial x}{\partial \varphi} & \dfrac{\partial x}{\partial \theta} \\ \dfrac{\partial y}{\partial \rho} & \dfrac{\partial y}{\partial \varphi} & \dfrac{\partial y}{\partial \theta} \\ \dfrac{\partial z}{\partial \rho} & \dfrac{\partial z}{\partial \varphi} & \dfrac{\partial z}{\partial \theta} \end{Vmatrix} d\rho d\varphi d\theta$$

$$= \begin{Vmatrix} \sin\varphi\cos\theta & \rho\cos\varphi\cos\theta & -\rho\sin\varphi\sin\theta \\ \sin\varphi\sin\theta & \rho\cos\varphi\sin\theta & -\rho\sin\varphi\cos\theta \\ \cos\varphi & -\rho\sin\varphi & 0 \end{Vmatrix} d\rho d\varphi d\theta = \rho^2 \sin\varphi\, d\rho d\varphi d\theta$$

where $0 \le \rho \le r, 0 \le \varphi \le \pi, 0 \le \theta \le 2\pi$

$$\therefore \iiint_{x^2+y^2+z^2 \le r^2} x^2 + y^2 + z^2\, dxdydz = \int_0^\pi \int_0^{2\pi} \int_0^r \rho^2 \cdot \rho^2 \sin\varphi\, d\rho d\theta d\varphi = \frac{4\pi r^5}{5}$$

Example 14.

Use Gauss's Theorem to find $\oiint_S \vec{F} \cdot \vec{n} dA = ?$, $\vec{F} = (e^x \sin y, e^x \cos y, yz^2 + xy)$,

$S: 0 \le x \le 1,\ 0 \le y \le 1,\ 0 \le z \le 2$.

【Solution】

Let $V = \{(x, y, z): 0 \le x \le 1,\ 0 \le y \le 1,\ 0 \le z \le 2\}$

∵ the first-order partial derivatives of the components of $\vec{F}$ exist and are continuous

By Gauss's Theorem, $\oiint_S \vec{F} \cdot \vec{n} dA = \iiint_V \nabla \cdot \vec{F}\, dV$

∵ $\nabla \cdot \vec{F} = e^x \sin y - e^x \sin y + 2yz = 2yz$ ∴ $\iiint_V \nabla \cdot \vec{F}\, dV = \int_0^1 \int_0^1 \int_0^2 2yz\, dzdydx = 2$

Example 15.

Use Gauss's Theorem to find $\oiint_S \vec{F} \cdot \vec{n} dA = ?$, where $\vec{F} = (x^2 z^3, 2xyz^3, xz^4)$,

S is the surface of the rectangular box formed by the vertices $(\pm 1, \pm 2, \pm 3)$.

【Solution】

Let V be the volume enclosed by S then

$V = \{(x, y, z): -1 \le x \le 1,\ -2 \le y \le 3,\ -3 \le z \le 3\}$

∵ the first-order partial derivatives of the components of $\vec{F}$ exist and are continuous

9.7.2 Convert Closed Surface Integrals into Volume Integrals($\nabla \cdot \mathbf{F} \neq c$)

By Gauss's Theorem, $\oiint_S \vec{F} \cdot \vec{n} dA = \iiint_V \nabla \cdot \vec{F}\, dV$

$\iiint_V \nabla \cdot \vec{F}\, dV = \iiint_V 2xz^3 + 2xz^3 + 4xz^3\, dV = 8\iiint_V xz^3\, dV = 8\int_{-1}^{1}\int_{-1}^{2}\int_{-3}^{3} xz^3\, dzdydx$
$= 0$

Example 16.

Use Gauss's Theorem to find $\oiint_S \vec{F} \cdot \vec{n} dA = ?$, where $\vec{F} = (x^4, -x^3z^2, 4xy^2z)$, S is the outward-oriented closed surface bounded by $x^2 + y^2 = 1$, $z = x + 2$ and $z = 0$.

【Solution】

Let V be the volume enclosed by S then $V = \{(x, y, z): x^2 + y^2 \leq 1, 0 \leq z \leq x + 2\}$

∵ the first-order partial derivatives of the components of $\vec{F}$ exist and are continuous

By Gauss's Theorem, $\oiint_S \vec{F} \cdot \vec{n} dA = \iiint_V \nabla \cdot \vec{F}\, dV$

∴ $\iiint_V \nabla \cdot \vec{F}\, dV = \iiint_V 4x^3 + 4xy^2\, dV = \iint_R \int_0^{x+2} 4x^3 + 4xy^2\, dzdxdy$

$= 4\iint_R x(x^2 + y^2)(x + 2)\, dydx$

Let $x = r\cos\theta, y = r\sin\theta$ then

$R = \{(x, y): x^2 + y^2 \leq 1\} = \{(r, \theta): 0 \leq r \leq 1, 0 \leq \theta \leq 2\pi\}$

∴ $4\iint_R x(x^2 + y^2)(x + 2)\, dydx = 4\int_0^{2\pi}\int_0^1 r^5 \cos^2\theta\, drd\theta = 4 \cdot 2\pi \cdot \dfrac{1}{6} \cdot \dfrac{1}{2} = \dfrac{2\pi}{3}$

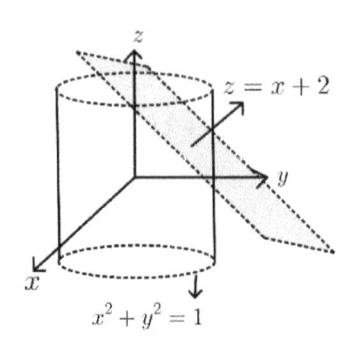

Example 17.

Use Gauss's Theorem to find the flux of the vector field $\vec{F} = (2x, xz, z^2)$ across the surface S, where S is the region bounded by the paraboloid $z = 9 - x^2 - y^2$ and the xy-plane.

【Solution】

Let V be the volume enclosed by S, then $V = \{(x, y, z): x^2 + y^2 \leq 9, 0 \leq z \leq 9 - x^2 - y^2\}$

∵ the first-order partial derivatives of the components of $\vec{F}$ exist and are continuous

By Gauss's Theorem, $\oiint_S \vec{F} \cdot \vec{n} dA = \iiint_V \nabla \cdot \vec{F} \, dV$

∵ $\nabla \cdot \vec{F} = 2 + 0 + 2z = 2 + 2z$ ∴ $\iiint_V \nabla \cdot \vec{F} \, dV = \iiint_V 2 + 2z \, dxdydz$

Let $R = \{(x, y): x^2 + y^2 \leq 9\}$ then

$$\iiint_V 2 + 2z \, dxdydz = \iint_R \int_0^{9-x^2-y^2} 2 + 2z \, dz dxdy = \iint_R 2z + z^2 \big|_0^{9-x^2-y^2} dxdy$$

$$= \iint_R 2(9 - x^2 - y^2) + (9 - x^2 - y^2)^2 \, dxdy$$

Let $x = r\cos\theta, y = r\sin\theta$ then $\{(x, y): x^2 + y^2 \leq 9\} = \{(r, \theta): 0 \leq r \leq 3, 0 \leq \theta \leq 2\pi\}$

and $dxdy = \begin{Vmatrix} \frac{\partial x}{\partial r} & \frac{\partial x}{\partial \theta} \\ \frac{\partial y}{\partial r} & \frac{\partial y}{\partial \theta} \end{Vmatrix} drd\theta = \begin{Vmatrix} \cos\theta & -r\sin\theta \\ \sin\theta & r\cos\theta \end{Vmatrix} drd\theta = rdrd\theta$

∴ $\iint_R 2(9 - x^2 - y^2) + (9 - x^2 - y^2)^2 \, dxdy = \int_0^{2\pi} d\theta \int_0^3 r^5 - 20r^3 + 99r dr = 2\pi \cdot 162$

$= 324\pi$

∴ $\oiint_S \vec{F} \cdot \vec{n} dA = 324\pi$

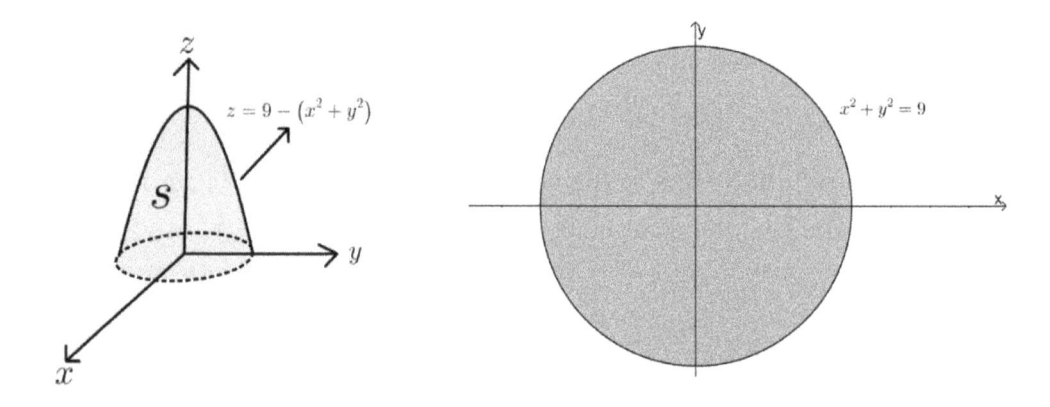

9.7.2 Convert Closed Surface Integrals into Volume Integrals($\nabla \cdot \mathbf{F} \neq c$) 618

Example 18.

Use Gauss's Theorem to find the flux of the vector field $\vec{F} = (x^2, xy, z)$ across the surface S, where S is the region bounded by the paraboloid $z = 4 - x^2 - y^2$ and the xy-plane.

【Solution】

Let V be the volume enclosed by S, then $V = \{(x, y, z): x^2 + y^2 \leq 4, 0 \leq z \leq 4 - x^2 - y^2\}$

By Gauss's Theorem, $\oiint_S \vec{F} \cdot \vec{n} dA = \iiint_V \nabla \cdot \vec{F} \, dV$

$\because \nabla \cdot \vec{F} = 2x + x + 1 = 3x + 1 \quad \therefore \iiint_V \nabla \cdot \vec{F} \, dV = \iiint_V 3x + 1 \, dxdydz$

Let $R = \{(x, y): x^2 + y^2 \leq 4\}$ then

$\iiint_V 3x + 1 \, dxdydz = \iint_R \int_0^{4-x^2-y^2} 3x + 1 \, dzdxdy = \iint_R (3x + 1)(4 - x^2 - y^2) dxdy$

Let $x = r \cos \theta, y = r \sin \theta$ then
$\{(x, y): x^2 + y^2 \leq 4\} = \{(r, \theta): 0 \leq r \leq 2, 0 \leq \theta \leq 2\pi\}$

and $dxdy = \begin{Vmatrix} \dfrac{\partial x}{\partial r} & \dfrac{\partial x}{\partial \theta} \\ \dfrac{\partial y}{\partial r} & \dfrac{\partial y}{\partial \theta} \end{Vmatrix} drd\theta = \begin{Vmatrix} \cos \theta & -r \sin \theta \\ \sin \theta & r \cos \theta \end{Vmatrix} drd\theta = rdrd\theta$

$\therefore \iint_R (3x + 1)(4 - x^2 - y^2) \, dxdy = \int_0^{2\pi} \int_0^2 (3 \cos \theta + 1)(4 - r^2) r dr d\theta$

$\because \int_0^{2\pi} \int_0^2 3 \cos \theta \, (4 - r^2) r dr d\theta = 0$

$$\therefore \int_0^{2\pi}\int_0^2 (3\cos\theta + 1)(4-r^2) r\, dr\, d\theta = \int_0^{2\pi}\int_0^2 (4-r^2) r\, dr\, d\theta = 8\pi$$

Example 19.

Use Gauss's Theorem to find the flux of the vector field $\vec{F} = (y, x, z)$ across the closed surface S bounded by $z = 4 - x^2 - y^2$, $z = 0$. Find $\oiint_S \vec{F}\cdot\vec{n}\, dA = ?$

【Solution】

Let V be the volume enclosed by S then $V = \{(x,y,z): x^2 + y^2 \le 4, 0 \le z \le 4 - x^2 - y^2\}$

∵ the first-order partial derivatives of the components of $\vec{F}$ exist and are continuous

By Gauss's Theorem, $\oiint_S \vec{F}\cdot\vec{n}\, dA = \iiint_V \nabla\cdot\vec{F}\, dV$

∵ $\nabla\cdot\vec{F} = 1$ ∴ $\iiint_V \nabla\cdot\vec{F}\, dV = \iiint_V 1\, dxdydz$

Let $R = \{(x,y): x^2 + y^2 \le 4\}$ then

$$\iiint_V 1\, dxdydz = \iint_R \int_0^{4-x^2-y^2} dz\, dxdy = \iint_R 4 - x^2 - y^2\, dxdy$$

Let $x = r\cos\theta, y = r\sin\theta$ then $\{(x,y): x^2 + y^2 \le 4\} = \{(r,\theta): 0 \le r \le 2, 0 \le \theta \le 2\pi\}$

and $dxdy = \left\| \begin{matrix} \frac{\partial x}{\partial r} & \frac{\partial x}{\partial \theta} \\ \frac{\partial y}{\partial r} & \frac{\partial y}{\partial \theta} \end{matrix} \right\| drd\theta = \left\| \begin{matrix} \cos\theta & -r\sin\theta \\ \sin\theta & r\cos\theta \end{matrix} \right\| drd\theta = rdrd\theta$

$$\therefore \iint_R 4 - x^2 - y^2\, dxdy = \int_0^{2\pi}\int_0^2 (4-r^2) r\, drd\theta = 2\pi\cdot 4 = 8\pi$$

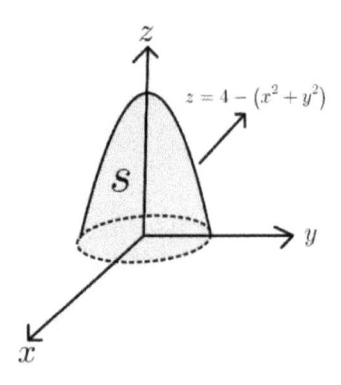

9.7.2 Convert Closed Surface Integrals into Volume Integrals($\nabla \cdot \mathbf{F} \neq c$)

Example 20.

Use Gauss's Theorem to find $\oiint_S \vec{F} \cdot \vec{n} dA = ?$, where $\vec{F} = (x + x^3, y + y^3, z + z^3)$, $S: x^2 + y^2 + z^2 = r^2$.

【Solution】

Let $V: x^2 + y^2 + z^2 \leq r^2$

∵ the first-order partial derivatives of the components of $\vec{F}$ exist and are continuous

By Gauss's Theorem, $\oiint_S \vec{F} \cdot \vec{n} dA = \iiint_V \nabla \cdot \vec{F} \, dV$

∵ $\nabla \cdot \vec{F} = 3 + 3x^2 + 3y^2 + 3z^2$

∴ $\iiint_V \nabla \cdot \vec{F} \, dV = \iiint_{x^2+y^2+z^2 \leq r^2} 3 + 3x^2 + 3y^2 + 3z^2 \, dxdydz$

Let $x = \rho \sin\varphi \cos\theta, y = \rho \sin\varphi \sin\theta, z = \rho \cos\varphi$ then

$$dxdydz = \begin{Vmatrix} \frac{\partial x}{\partial \rho} & \frac{\partial x}{\partial \varphi} & \frac{\partial x}{\partial \theta} \\ \frac{\partial y}{\partial \rho} & \frac{\partial y}{\partial \varphi} & \frac{\partial y}{\partial \theta} \\ \frac{\partial z}{\partial \rho} & \frac{\partial z}{\partial \varphi} & \frac{\partial z}{\partial \theta} \end{Vmatrix} d\rho d\varphi d\theta$$

$$= \begin{Vmatrix} \sin\varphi \cos\theta & \rho\cos\varphi \cos\theta & -\rho\sin\varphi \sin\theta \\ \sin\varphi \sin\theta & \rho\cos\varphi \sin\theta & -\rho\sin\varphi \cos\theta \\ \cos\varphi & -\rho\sin\varphi & 0 \end{Vmatrix} d\rho d\varphi d\theta = \rho^2 \sin\varphi \, d\rho d\varphi d\theta$$

where $0 \leq \rho \leq r, 0 \leq \varphi \leq \pi, 0 \leq \theta \leq 2\pi$

$$\therefore \iiint_{x^2+y^2+z^2 \leq r^2} 3 + 3x^2 + 3y^2 + 3z^2 \, dxdydz$$

$$= 3 \cdot \frac{4\pi \cdot r^3}{3} + \int_0^\pi \int_0^{2\pi} \int_0^r 3\rho^2 \cdot \rho^2 \sin\varphi \, d\rho d\theta d\varphi = 4\pi r^3 + \frac{12\pi r^5}{5}$$

Example 21.

Let $S: x^2 + y^2 + z^2 = 1$ and $\vec{F} = (x + x^3 + y + \sin z^2, y + y^3 + e^{x^2}, z + z^3 + \ln x^2 y^2)$.

Use Gauss's Theorem to find $\oiint_S \vec{F} \cdot \vec{n} dA = ?$

【Solution】

Let $V: x^2 + y^2 + z^2 \leq 1$

$\because$ the first-order partial derivatives of the components of $\vec{F}$ exist and are continuous

By Gauss's Theorem, $\oiint_S \vec{F} \cdot \vec{n} dA = \iiint_V \nabla \cdot \vec{F} \, dV$

$\because \nabla \cdot \vec{F} = 3 + 3x^2 + 3y^2 + 3z^2 \therefore \iiint_V \nabla \cdot \vec{F} \, dV = \iiint_{x^2+y^2+z^2 \leq 1} 3 + 3x^2 + 3y^2 + 3z^2 \, dxdydz$

Let $x = \rho \sin\varphi \cos\theta, y = \rho \sin\varphi \sin\theta, z = \rho \cos\varphi$ then

$$dxdydz = \begin{Vmatrix} \dfrac{\partial x}{\partial \rho} & \dfrac{\partial x}{\partial \varphi} & \dfrac{\partial x}{\partial \theta} \\ \dfrac{\partial y}{\partial \rho} & \dfrac{\partial y}{\partial \varphi} & \dfrac{\partial y}{\partial \theta} \\ \dfrac{\partial z}{\partial \rho} & \dfrac{\partial z}{\partial \varphi} & \dfrac{\partial z}{\partial \theta} \end{Vmatrix} d\rho d\varphi d\theta$$

$$= \begin{Vmatrix} \sin\varphi \cos\theta & \rho\cos\varphi \cos\theta & -\rho\sin\varphi \sin\theta \\ \sin\varphi \sin\theta & \rho\cos\varphi \sin\theta & -\rho\sin\varphi \cos\theta \\ \cos\varphi & -\rho\sin\varphi & 0 \end{Vmatrix} d\rho d\varphi d\theta = \rho^2 \sin\varphi \, d\rho d\varphi d\theta$$

where $0 \leq \rho \leq 1, 0 \leq \varphi \leq \pi, 0 \leq \theta \leq 2\pi$

$$\iiint_{x^2+y^2+z^2 \leq 1} 3 + 3x^2 + 3y^2 + 3z^2 \, dxdydz = 3 \cdot \frac{4\pi}{3} + \int_0^\pi \int_0^{2\pi} \int_0^1 3\rho^2 \cdot \rho^2 \sin\varphi \, d\rho d\theta d\varphi$$

$$= 4\pi + \frac{12\pi}{5} = \frac{32\pi}{5}$$

9.7.2 Convert Closed Surface Integrals into Volume Integrals($\nabla \cdot \mathbf{F} \neq c$)

Example 22.

Let $\vec{F} = (x^2 \sin y, x \cos y, -xz \sin y)$, $S: x^2 + y^2 + z^2 = 8$. Use Gauss's Theorem to find $\oiint_S \vec{F} \cdot \vec{n} dA =?$

【Solution】

Let V be the volume enclosed by S then $V: x^2 + y^2 + z^2 \leq 8$

∵ the first-order partial derivatives of the components of $\vec{F}$ exist and are continuous

By Gauss's Theorem, $\oiint_S \vec{F} \cdot \vec{n} dA = \iiint_V \nabla \cdot \vec{F} dV$

∴ $\iiint_V \nabla \cdot \vec{F} dV = \iiint_V 2x \sin y - x \sin y - x \sin y \, dV = 0$

Example 23.

Use Gauss's Theorem to find $\oiint_S \vec{F} \cdot \vec{n} dA =?$, where $\vec{F} = (x^3 - 3y, 2yz + 1, xyz)$, S is the outward-oriented closed surface bounded by the planes $x = \pm 1, y = \pm 1$, and $z = \pm 1$.

【Solution】

Let V be the volume enclosed by S then
$V = \{(x, y, z): -1 \leq x \leq 1, -1 \leq y \leq 1, -1 \leq z \leq 1\}$

∵ the first-order partial derivatives of the components of $\vec{F}$ exist and are continuous

By Gauss's Theorem, $\oiint_S \vec{F} \cdot \vec{n} dA = \iiint_V \nabla \cdot \vec{F} dV$

$\iiint_V \nabla \cdot \vec{F} dV = \iiint_V 3x^2 + 2z + xy \, dV$

$= \int_{-1}^{1} \int_{-1}^{1} \int_{-1}^{1} 3x^2 \, dxdydz + \int_{-1}^{1} \int_{-1}^{1} \int_{-1}^{1} 2z \, dzdydx + \int_{-1}^{1} \int_{-1}^{1} \int_{-1}^{1} xy \, dxdydz$

∵ $\int_{-1}^{1} \int_{-1}^{1} \int_{-1}^{1} 3x^2 \, dxdydz = 8$, $\int_{-1}^{1} \int_{-1}^{1} \int_{-1}^{1} 2z \, dzdydx = 0$, $\int_{-1}^{1} \int_{-1}^{1} \int_{-1}^{1} xy \, dxdydz = 0$

∴ $\oiint_S \vec{F} \cdot \vec{n} dA = \iiint_V \nabla \cdot \vec{F} dV = \iiint_V 3x^2 + 2z + xy \, dV = 8$

Example 24.

Use Gauss's Theorem to find $\oiint_S \vec{F}\cdot\vec{n}dA =?$, where $\vec{F} = (x^4, -x^3z^2, 4xy^2z)$, S is the outward-oriented closed surface bounded by $x^2 + y^2 = 1$, $z = x + 2$, $z = 0$.

【Solution】

Let V be the volume enclosed by S, then $V = \{(x,y,z): x^2 + y^2 \leq 1, 0 \leq z \leq x + 2\}$

∵ the first-order partial derivatives of the components of $\vec{F}$ exist and are continuous

By Gauss's Theorem, $\oiint_S \vec{F}\cdot\vec{n}dA = \iiint_V \nabla\cdot\vec{F}\, dV$

∵ $\nabla\cdot\vec{F} = 4x^3 + 0 + 4xy^2 = 4x(x^2 + y^2)$ ∴ $\iiint_V \nabla\cdot\vec{F}\, dV = \iiint_V 4x(x^2+y^2)\, dxdydz$

Let $R = \{(x,y): x^2 + y^2 \leq 1\}$ then

$$\iiint_V 4x(x^2+y^2)\, dxdydz = \iint_R \int_0^{x+2} 4x(x^2+y^2)\, dzdxdy$$

$$= \iint_R 4x(x^2+y^2)(x+2)\, dxdy$$

Let $x = r\cos\theta, y = r\sin\theta$ then $\{(x,y): x^2 + y^2 \leq 1\} = \{(r,\theta): 0 \leq r \leq 1, 0 \leq \theta \leq 2\pi\}$

and $dxdy = \begin{Vmatrix}\dfrac{\partial x}{\partial r} & \dfrac{\partial x}{\partial \theta}\\ \dfrac{\partial y}{\partial r} & \dfrac{\partial y}{\partial \theta}\end{Vmatrix} drd\theta = \begin{Vmatrix}\cos\theta & -r\sin\theta\\ \sin\theta & r\cos\theta\end{Vmatrix} drd\theta = rdrd\theta$

∴ $\iint_R 4x(x^2+y^2)(x+2))dxdy = \int_0^{2\pi}\int_0^1 4r^3\cos\theta\,(r\cos\theta + 2)rdrd\theta$

∵ $\int_0^{2\pi}\int_0^1 4r^3\cos\theta\cdot 2r\, drd\theta = 0$

∴ $\int_0^{2\pi}\int_0^1 4r^3\cos\theta\,(r\cos\theta + 2)rdrd\theta = \int_0^{2\pi}\int_0^1 4r^3\cos\theta\,(r\cos\theta)rdrd\theta$

$= 4\int_0^{2\pi}\int_0^1 r^5\dfrac{1+\cos 2\theta}{2}drd\theta = 2\int_0^{2\pi}\int_0^1 r^5 drd\theta = \dfrac{2\pi}{3}$

∴ $\oiint_S \vec{F}\cdot\vec{n}dA = \dfrac{2\pi}{3}$

9.7.2 Convert Closed Surface Integrals into Volume Integrals($\nabla \cdot \mathbf{F} \neq c$) 624

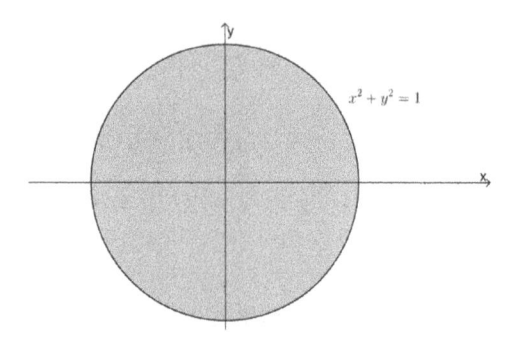

Example 25.

Let $\vec{F} = (xz^2 + \sin z, x^2y - z^3 + \cos x, 2xy + y^2z)$, $S: x^2 + y^2 + z^2 = a^2, z \geq 0$.

Use Gauss's Theorem to find $\oiint_S \vec{F} \cdot \vec{n} dA =?$

【Solution】

∵ the unit normal vector of the sphere is $\vec{n} = (x, y, z)$

Let $\vec{F} = (xz^2 + \sin z, x^2y - z^3 + \cos x, 2xy + y^2z)$

Let $V: x^2 + y^2 + z^2 \leq a^2, z \geq 0$

∵ the first-order partial derivatives of the components of $\vec{F}$ exist and are continuous

By Gauss's Theorem, $\oiint_S \vec{F} \cdot \vec{n} dA = \iiint_V \nabla \cdot \vec{F} \, dV$

∵ $\nabla \cdot \vec{F} = x^2 + y^2 + z^2$ ∴ $\iiint_V \nabla \cdot \vec{F} \, dV = \iiint_{x^2+y^2+z^2 \leq a^2} x^2 + y^2 + z^2 \, dxdydz$

Let $x = \rho \sin \varphi \cos \theta$, $y = \rho \sin \varphi \sin \theta$, $z = \rho \cos \varphi$ then

$$dxdydz = \begin{Vmatrix} \frac{\partial x}{\partial \rho} & \frac{\partial x}{\partial \varphi} & \frac{\partial x}{\partial \theta} \\ \frac{\partial y}{\partial \rho} & \frac{\partial y}{\partial \varphi} & \frac{\partial y}{\partial \theta} \\ \frac{\partial z}{\partial \rho} & \frac{\partial z}{\partial \varphi} & \frac{\partial z}{\partial \theta} \end{Vmatrix} d\rho d\varphi d\theta$$

$$= \begin{Vmatrix} \sin \varphi \cos \theta & \rho \cos \varphi \cos \theta & -\rho \sin \varphi \sin \theta \\ \sin \varphi \sin \theta & \rho \cos \varphi \sin \theta & -\rho \sin \varphi \cos \theta \\ \cos \varphi & -\rho \sin \varphi & 0 \end{Vmatrix} d\rho d\varphi d\theta = \rho^2 \sin \varphi \, d\rho d\varphi d\theta$$

where $0 \leq \rho \leq a, 0 \leq \varphi \leq \frac{\pi}{2}, 0 \leq \theta \leq 2\pi$

$$\therefore \iiint_{x^2+y^2+z^2\le a^2} x^2+y^2+z^2 \, dxdydz = \int_0^{\frac{\pi}{2}} \int_0^{2\pi} \int_0^a \rho^2 \cdot \rho^2 \sin\varphi \, d\rho d\theta d\varphi = \frac{2\pi a^5}{5}$$

9.7.3 Convert Volume Integrals into Closed Surface Integrals

Question Types:

Type 1.

Suppose the first-order partial derivatives of each component of the vector field $\vec{F}$ exist and are continuous. Let S be the surface defined by $x^2 + y^2 + z^2 = r^2$, and let V be the volume enclosed by S. Use Gauss's Theorem to find $\iiint_V \nabla \cdot \vec{F} \, dV = ?$

Problem-Solving Process:

Step1.

By Gauss's Theorem, $\iiint_V \nabla \cdot \vec{F} \, dV = \oiint_S \vec{F} \cdot \vec{n} dA$

Step2.

Let $\vec{r}(\theta, \varphi) = (r\cos\theta \sin\varphi, r\sin\theta \sin\varphi, r\cos\varphi)$ and $R = \{(\theta, \varphi): 0 \le \theta \le 2\pi, 0 \le \varphi \le \pi\}$

then $\oiint_S \vec{F} \cdot \vec{n} dA = \iint_R \vec{F} \cdot \frac{\partial \vec{r}}{\partial \varphi} \times \frac{\partial \vec{r}}{\partial \theta} d\theta d\varphi$

Step3.

$\because \frac{\partial \vec{r}}{\partial \theta} = (-r\sin\theta \sin\varphi, r\cos\theta \sin\varphi, 0)$ and $\frac{\partial \vec{r}}{\partial \varphi} = (r\cos\theta \cos\varphi, r\sin\theta \cos\varphi, -r\sin\varphi)$

$\therefore \frac{\partial \vec{r}}{\partial \varphi} \times \frac{\partial \vec{r}}{\partial \theta} = (r^2 \cos\theta \sin^2\varphi, r^2 \sin\theta \sin^2\varphi, r^2 \sin\varphi \cos\varphi)$

Step4.

$\oiint_S \vec{F} \cdot \vec{n} dA = \int_0^{2\pi} \int_0^{\pi} (f_1(x,y,z) r^2 \cos\theta \sin^2\varphi + f_2(x,y,z) r^2 \sin\theta \sin^2\varphi$
$+ f_3(x,y,z) r^2 \sin\varphi \cos\varphi)|_{x=r\cos\theta \sin\varphi, y=r\sin\theta \sin\varphi, z=r\cos\varphi} \, d\varphi d\theta$

where $\vec{F} = (f_1(x,y,z), f_2(x,y,z), f_3(x,y,z))$

Examples:

(1) Find $\iiint_V \nabla \cdot \vec{F} \, dV = ?$, where $\vec{F} = (0, 4yz, 0)$, $S: x^2 + y^2 + z^2 = r^2$, V is the volume enclosed by S.

$\because$ the first-order partial derivatives of each component of $\vec{F}$ exist and are continuous

9.7.3 Convert Volume Integrals into Closed Surface Integrals 626

By Gauss's Theorem, $\iiint_V \nabla \cdot \vec{F}\, dV = \oiint_S \vec{F} \cdot \vec{n}\, dA$

Let $\vec{r}(\theta, \varphi) = (r\cos\theta \sin\varphi, r\sin\theta \sin\varphi, r\cos\varphi)$ and $R = \{(\theta, \varphi): 0 \leq \theta \leq 2\pi, 0 \leq \varphi \leq \pi\}$

then $\oiint_S \vec{F} \cdot \vec{n}\, dA = \iint_R \vec{F} \cdot \dfrac{\partial \vec{r}}{\partial \varphi} \times \dfrac{\partial \vec{r}}{\partial \theta}\, d\theta d\varphi$

$\because \dfrac{\partial \vec{r}}{\partial \theta} = (-r\sin\theta \sin\varphi, r\cos\theta \sin\varphi, 0)$ and $\dfrac{\partial \vec{r}}{\partial \varphi} = (r\cos\theta \cos\varphi, r\sin\theta \cos\varphi, -r\sin\varphi)$

$\therefore \dfrac{\partial \vec{r}}{\partial \varphi} \times \dfrac{\partial \vec{r}}{\partial \theta} = (r^2 \cos\theta \sin^2\varphi, r^2\sin\theta \sin^2\varphi, r^2\sin\varphi \cos\varphi)$

$\vec{F} \cdot \dfrac{\partial \vec{r}}{\partial \varphi} \times \dfrac{\partial \vec{r}}{\partial \theta} = (0, 4r^2\sin\theta \sin\varphi \cos\varphi, 0) \cdot (r^2\cos\theta \sin^2\varphi, r^2\sin\theta \sin^2\varphi, r^2\sin\varphi \cos\varphi)$

$= 4r^4 \sin^3\varphi \sin^2\theta \cos\varphi$

$\therefore \oiint_S \vec{F} \cdot \vec{n}\, dA = \int_0^{2\pi} \int_0^\pi 4r^4 \sin^3\varphi \sin^2\theta \cos\varphi\, d\varphi d\theta = 0$

(II) Find $\iiint_V \nabla \cdot \vec{F}\, dV = ?$, where $\vec{F} = (x, y, z)$, $S: x^2 + y^2 + z^2 = r^2$, V is the volume enclosed by S.

∵ the first-order partial derivatives of each component of $\vec{F}$ exist and are continuous

By Gauss's Theorem, $\iiint_V \nabla \cdot \vec{F}\, dV = \oiint_S \vec{F} \cdot \vec{n}\, dA$

Let $\vec{r}(\theta, \varphi) = (r\cos\theta \sin\varphi, r\sin\theta \sin\varphi, r\cos\varphi)$ and $R = \{(\theta, \varphi): 0 \leq \theta \leq 2\pi, 0 \leq \varphi \leq \pi\}$

then $\oiint_S \vec{F} \cdot \vec{n}\, dA = \iint_R \vec{F} \cdot \dfrac{\partial \vec{r}}{\partial \varphi} \times \dfrac{\partial \vec{r}}{\partial \theta}\, d\theta d\varphi$

$\because \dfrac{\partial \vec{r}}{\partial \theta} = (-r\sin\theta \sin\varphi, r\cos\theta \sin\varphi, 0)$ and $\dfrac{\partial \vec{r}}{\partial \varphi} = (r\cos\theta \cos\varphi, r\sin\theta \cos\varphi, -r\sin\varphi)$

$\therefore \dfrac{\partial \vec{r}}{\partial \varphi} \times \dfrac{\partial \vec{r}}{\partial \theta} = (r^2 \cos\theta \sin^2\varphi, r^2\sin\theta \sin^2\varphi, r^2\sin\varphi \cos\varphi)$

$$\vec{F} \cdot \frac{\partial \vec{r}}{\partial \varphi} \times \frac{\partial \vec{r}}{\partial \theta}$$

$$= (r\cos\theta \sin\varphi, r\sin\theta \sin\varphi, r\cos\varphi) \cdot (r^2 \cos\theta \sin^2\varphi, r^2\sin\theta \sin^2\varphi, r^2\sin\varphi \cos\varphi)$$

$$= r^3(\cos^2\theta \sin^3\varphi + \sin^3\varphi \sin^2\theta + \sin\varphi \cos^2\varphi) = r^3 \sin\varphi$$

$$\therefore \oiint_S \vec{F} \cdot \vec{n} dA = r^3 \int_0^{2\pi} \int_0^{\pi} \sin\varphi \, d\varphi d\theta = 4\pi r^3$$

Type 2.

Assume that the first-order partial derivatives of the components of the vector field $\vec{F}$ exist and are continuous. Let S be a closed outward-oriented surface defined by $z = g(x, y)$ and $z = 0$, enclosing a volume V. The projection of S onto the xy-plane is the region
$$\{(x, y): x^2 + y^2 \leq a^2\}.$$
We want to evaluate the integral
$$\iiint_V \nabla \cdot \vec{F} \, dV.$$

Remark:

The surface S is typically an ellipsoid, sphere, elliptical cone, cone, or elliptical paraboloid.

Problem-Solving Process:

Step1.

Let $S_1: z = g(x, y)$, $S_2: z = 0$ and $x^2 + y^2 \leq a^2$

By Gauss's Theorem, $\iiint_V \nabla \cdot \vec{F} \, dV = \oiint_S \vec{F} \cdot \vec{n} dA = \iint_{S_1} \vec{F} \cdot \vec{n} \, dA + \iint_{S_2} \vec{F} \cdot \vec{n} \, dA$

Step2.

Assume that S_1 is upward-oriented.

Let $\vec{r}(x, y) = (x, y, g(x, y))$ and let R be the closed region in the xy plane projected from S_1 given by $\{(x, y): x^2 + y^2 \leq a^2\}$.

Then $\iint_{S_1} \vec{F} \cdot \vec{n} \, dA = \iint_R \vec{F} \cdot \frac{\partial \vec{r}}{\partial x} \times \frac{\partial \vec{r}}{\partial y} dxdy$

Step3.

$\because \frac{\partial \vec{r}}{\partial x} = (1, 0, g_x(x, y))$ and $\frac{\partial \vec{r}}{\partial y} = (0, 1, g_y(x, y))$ $\therefore \frac{\partial \vec{r}}{\partial x} \times \frac{\partial \vec{r}}{\partial y} = (-g_x(x, y), -g_y(x, y), 1)$

9.7.3 Convert Volume Integrals into Closed Surface Integrals

$$\iint_{S_1} \vec{F} \cdot \vec{n}\, dA = \iint_R \vec{F} \cdot \frac{\partial \vec{r}}{\partial x} \times \frac{\partial \vec{r}}{\partial y} dxdy = \iint_R \vec{F} \cdot (-g_x(x,y), -g_y(x,y), 1) dxdy$$

$$= \iint_R (-f_1 \cdot g_x(x,y) - f_2 \cdot g_y(x,y) + f_3)\big|_{z=g(x,y)} dxdy$$

where $\vec{F} = (f_1, f_2, f_3)$

Step4.

Let $x = r\cos\theta, y = r\sin\theta$ then $\{(x,y): x^2 + y^2 \le a^2\} = \{(r,\theta): 0 \le r \le a, 0 \le \theta \le 2\pi\}$

and $dxdy = \begin{Vmatrix} \frac{\partial x}{\partial r} & \frac{\partial x}{\partial \theta} \\ \frac{\partial y}{\partial r} & \frac{\partial y}{\partial \theta} \end{Vmatrix} drd\theta = \begin{Vmatrix} \cos\theta & -r\sin\theta \\ \sin\theta & r\cos\theta \end{Vmatrix} drd\theta = rdrd\theta$

Step5.

$$\iint_R (-f_1 \cdot g_x(x,y) - f_2 \cdot g_y(x,y) + f_3)\big|_{z=g(x,y)} dxdy$$

$$= \int_0^{2\pi} \int_0^a -f_1(r\cos\theta, r\sin\theta, g(r\cos\theta, r\sin\theta)) \cdot g_x(r\cos\theta, r\sin\theta) \cdot r$$
$$-f_2(r\cos\theta, r\sin\theta, g(r\cos\theta, r\sin\theta)) \cdot g_y(r\cos\theta, r\sin\theta) \cdot r$$
$$+f_3(r\cos\theta, r\sin\theta, g(r\cos\theta, r\sin\theta)) \cdot rdrd\theta$$

Step6.

Find $\iint_{S_2} \vec{F} \cdot \vec{n}\, dA = ?$

Examples:

(I) Assume $\vec{F} = (y, x, z)$, and V is the closed volume enclosed by $z = 4 - x^2 - y^2$ and $z = 0$. Use Gauss's Theorem to find $\iiint_V \nabla \cdot \vec{F}\, dV = ?$

Let $S_1: z = 4 - x^2 - y^2$, $S_2: z = 0$ and $x^2 + y^2 \le 4$

By Gauss's Theorem, $\iiint_V \nabla \cdot \vec{F}\, dV = \iint_{S_1} \vec{F} \cdot \vec{n}\, dA + \iint_{S_2} \vec{F} \cdot \vec{n}\, dA$

$$\iint_{S_1} \vec{F} \cdot \vec{n}\, dA = \iint_{x^2+y^2 \le 4} 4xy + z\, dxdy = \iint_{x^2+y^2 \le 4} 4xy + 4 - x^2 - y^2\, dxdy$$

Let $x = r\cos\theta, y = r\sin\theta$

$$\iint_{x^2+y^2\leq 4} 4xy + 4 - x^2 - y^2 \, dxdy = \int_0^{2\pi} \int_0^2 (4r^2 \cos\theta \sin\theta + 4 - r^2) \, rdrd\theta = 8\pi$$

$$\because \frac{\partial \vec{s}}{\partial x} = (1,0,0) \text{ and } \frac{\partial \vec{s}}{\partial y} = (0,1,0) \quad \therefore \frac{\partial \vec{s}}{\partial x} \times \frac{\partial \vec{s}}{\partial y} = (0,0,1) \quad \therefore \iint_{S_2} \vec{F} \cdot \vec{n} \, dA = \iint_{S_2} 0 \, dA = 0$$

$$\therefore \iint_{S_1} \vec{F} \cdot \vec{n} \, dA + \iint_{S_2} \vec{F} \cdot \vec{n} \, dA = 8\pi$$

Type 3.

Assume that the first-order partial derivatives of the components of $\vec{F}$ exist and are continuous. Let S be a closed surface composed of several surfaces, $S = \bigcup_{j=1}^n S_j$, where $n \geq 2$. Let V be the closed volume enclosed by S. Using Gauss's Theorem, we want to find

$$\iiint_V \nabla \cdot \vec{F} \, dV = ?$$

Remark:

For example, the volume V could be a rectangular prism.

Problem-Solving Process:

Step1.

By Gauss's Theorem, $\iiint_V \nabla \cdot \vec{F} \, dV = \oiint_S \vec{F} \cdot \vec{n} dA = \sum_{j=1}^n \iint_{S_j} \vec{F} \cdot \vec{n} \, dA$

Step2.

If each S_j is a plane parallel to the coordinate planes, then we can directly compute

$$\iint_{S_j} \vec{F} \cdot \vec{n} \, dA.$$

Furthermore, if S_j is an upward-oriented surface given by $z = g(x,y)$, with the domain defined as $a \leq x \leq b$ and $c \leq y \leq d$ then let $\vec{r}(x,y) = (x, y, g(x,y))$ and R be the closed region in the xy-plane projected from the surface S, given by

$$R = \{(x,y): a \leq x \leq b, c \leq y \leq d\}.$$

Then the formula can be rewritten as

9.7.3 Convert Volume Integrals into Closed Surface Integrals

$$\iint_{S_j} \vec{F} \cdot \vec{n}\, dA = \iint_R \vec{F}(x, y, g(x, y)) \cdot (-g_x(x, y), -g_y(x, y), 1)\, dxdy.$$

Example 1.

Use Gauss's Theorem to find $\iiint_V \nabla \cdot \vec{F}\, dV = ?$, where $\vec{F} = (0, 4yz, 0)$, and $V: x^2 + y^2 + z^2 \leq r^2$.

【Solution】

Let $S: x^2 + y^2 + z^2 = r^2$

∵ the first-order partial derivatives of the components of $\vec{F}$ exist and are continuous

By Gauss's Theorem, $\iiint_V \nabla \cdot \vec{F}\, dV = \oiint_S \vec{F} \cdot \vec{n}\, dA$

Let $\vec{r}(\theta, \varphi) = (r\cos\theta \sin\varphi, r\sin\theta \sin\varphi, r\cos\varphi)$

Let $R = \{(\theta, \varphi): 0 \leq \theta \leq 2\pi, 0 \leq \varphi \leq \pi\}$ then $\oiint_S \vec{F} \cdot \vec{n}\, dA = \iint_R \vec{F} \cdot \dfrac{\partial \vec{r}}{\partial \varphi} \times \dfrac{\partial \vec{r}}{\partial \theta}\, d\theta d\varphi$

∵ $\dfrac{\partial \vec{r}}{\partial \theta} = (-r\sin\theta \sin\varphi, r\cos\theta \sin\varphi, 0)$ and $\dfrac{\partial \vec{r}}{\partial \varphi} = (r\cos\theta \cos\varphi, r\sin\theta \cos\varphi, -r\sin\varphi)$

∴ $\dfrac{\partial \vec{r}}{\partial \varphi} \times \dfrac{\partial \vec{r}}{\partial \theta} = (r^2 \cos\theta \sin^2\varphi, r^2\sin\theta \sin^2\varphi, r^2\sin\varphi \cos\varphi)$

∴ $\vec{F} \cdot \dfrac{\partial \vec{r}}{\partial \varphi} \times \dfrac{\partial \vec{r}}{\partial \theta} = (0, 4r^2\sin\theta \sin\varphi \cos\varphi, 0) \cdot (r^2\cos\theta \sin^2\varphi, r^2\sin\theta \sin^2\varphi, r^2\sin\varphi \cos\varphi)$

$= 4r^4 \sin^3\varphi \sin^2\theta \cos\varphi$

∴ $\oiint_S \vec{F} \cdot \vec{n}\, dA = \int_0^{2\pi} \int_0^{\pi} 4r^4 \sin^3\varphi \sin^2\theta \cos\varphi\, d\varphi d\theta$

∵ $\int_0^{\pi} \sin^3\varphi \cos\varphi\, d\varphi = \left.\dfrac{\sin^4\varphi}{4}\right|_0^{\pi} = 0$ ∴ $\oiint_S \vec{F} \cdot \vec{n}\, dA = 0$

Example 2.

Let $\vec{F} = (x, 2y, 3z)$. Assume V is the volume of cube with the vertices $(\pm 1, \pm 1, \pm 1)$.

Find $\iiint_V \nabla \cdot \vec{F}\, dV$ by using Gauss's Theorem.

【Solution】

Let $S_1: x = 1,\ S_2: x = -1,\ S_3: y = 1,\ S_4: y = -1,\ S_5: z = 1,\ S_6: z = -1$ and $S = \bigcup_{j=1}^{6} S_j$

∵ the first-order partial derivatives of the components of $\vec{F}$ exist and are continuous

By Gauss's Theorem, $\iiint_V \nabla \cdot \vec{F}\, dV = \oiint_S \vec{F} \cdot \vec{n}\, dA$

As $x = 1, \vec{n} = (1,0,0),\ \iint_{S_1} \vec{F} \cdot \vec{n}\, dA = \int_{-1}^{1}\int_{-1}^{1} x\, dydz = 4x = 4$

As $x = -1, \vec{n} = (-1,0,0),\ \iint_{S_2} \vec{F} \cdot \vec{n}\, dA = \int_{-1}^{1}\int_{-1}^{1} -x\, dydz = -4x = 4$

As $y = 1, \vec{n} = (0,1,0),\ \iint_{S_3} \vec{F} \cdot \vec{n}\, dA = \int_{-1}^{1}\int_{-1}^{1} 2y\, dxdz = 8y = 8$

As $y = -1, \vec{n} = (0,-1,0),\ \iint_{S_4} \vec{F} \cdot \vec{n}\, dA = \int_{-1}^{1}\int_{-1}^{1} -2y\, dxdz = -8y = 8$

As $z = 1, \vec{n} = (0,0,1),\ \iint_{S_5} \vec{F} \cdot \vec{n}\, dA = \int_{-1}^{1}\int_{-1}^{1} 3z\, dxdy = 12z = 12$

As $z = -1, \vec{n} = (0,0,-1),\ \iint_{S_6} \vec{F} \cdot \vec{n}\, dA = \int_{-1}^{1}\int_{-1}^{1} -3z\, dxdy = -12z = 12$

∴ $\oiint_S \vec{F} \cdot \vec{n}\, dA = 48$

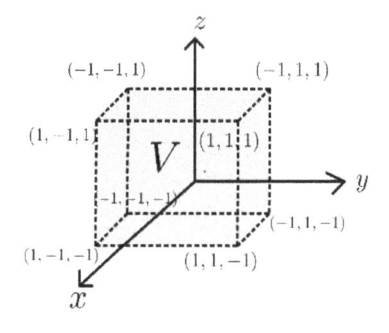

Example 3.

9.7.3 Convert Volume Integrals into Closed Surface Integrals

Find $\iiint_V \nabla \cdot \vec{F}\, dV = ?$, $\vec{F} = (x, y, z)$, $V: x^2 + y^2 + z^2 \leq r^2$: (1) Using Gauss's Theorem (2) Direct calculation.

【Solution】

(1)

Let $S: x^2 + y^2 + z^2 = r^2$

∵ the first-order partial derivatives of the components of $\vec{F}$ exist and are continuous

By Gauss's Theorem, $\iiint_V \nabla \cdot \vec{F}\, dV = \oiint_S \vec{F} \cdot \vec{n}\, dA$

Let $\vec{r}(r, \theta, \varphi) = (r\cos\theta \sin\varphi, r\sin\theta \sin\varphi, r\cos\varphi)$

Let $R = \{(\theta, \varphi): 0 \leq \theta \leq 2\pi, 0 \leq \varphi \leq \pi\}$ then $\oiint_S \vec{F} \cdot \vec{n}\, dA = \iint_R \vec{F} \cdot \frac{\partial \vec{r}}{\partial \varphi} \times \frac{\partial \vec{r}}{\partial \theta}\, d\theta d\varphi$

∵ $\frac{\partial \vec{r}}{\partial \theta} = (-r\sin\theta \sin\varphi, r\cos\theta \sin\varphi, 0)$ and $\frac{\partial \vec{r}}{\partial \varphi} = (r\cos\theta \cos\varphi, r\sin\theta \cos\varphi, -r\sin\varphi)$

∴ $\frac{\partial \vec{r}}{\partial \varphi} \times \frac{\partial \vec{r}}{\partial \theta} = (r^2 \cos\theta \sin^2\varphi, r^2 \sin\theta \sin^2\varphi, r^2 \sin\varphi \cos\varphi)$

∴ $\vec{F} \cdot \frac{\partial \vec{r}}{\partial \varphi} \times \frac{\partial \vec{r}}{\partial \theta}$

$= (r\cos\theta \sin\varphi, r\sin\theta \sin\varphi, r\cos\varphi) \cdot (r^2 \cos\theta \sin^2\varphi, r^2 \sin\theta \sin^2\varphi, r^2 \sin\varphi \cos\varphi)$

$= r^3(\cos^2\theta \sin^3\varphi + \sin^3\varphi \sin^2\theta + \sin\varphi \cos^2\varphi) = r^3 \sin\varphi$

∴ $\oiint_S \vec{F} \cdot \vec{n}\, dA = \iint_R \vec{F} \cdot \frac{\partial \vec{r}}{\partial \varphi} \times \frac{\partial \vec{r}}{\partial \theta}\, d\theta d\varphi = r^3 \int_0^{2\pi} \int_0^{\pi} \sin\varphi\, d\varphi d\theta = 4\pi r^3$

∴ $\iiint_V \nabla \cdot \vec{F}\, dV = \oiint_S \vec{F} \cdot \vec{n}\, dA = 4\pi r^3$

(2)

$\iiint_V \nabla \cdot \vec{F}\, dV = \iiint_{x^2+y^2+z^2 \leq r^2} 3\, dxdydz = 4\pi r^3$

9.7.3 Convert Volume Integrals into Closed Surface Integrals

Example 4.

Let $\vec{F} = (yz, -1, 1)$, and V is the volume bounded by $z = \sqrt{x^2 + y^2}$ and $z = a$.
Use Gauss's Theorem to find

$$\iiint_V \nabla \cdot \vec{F}\, dV = ?$$

【Solution】

Let $S_1: z = \sqrt{x^2 + y^2}, \ z \leq a, \ S_2: z = a, \ x^2 + y^2 \leq a^2$

∵ the first-order partial derivatives of the components of $\vec{F}$ exist and are continuous

By Gauss's Theorem, $\iiint_V \nabla \cdot \vec{F}\, dV = \iint_{S_1} \vec{F} \cdot \vec{n}\, dA + \iint_{S_2} \vec{F} \cdot \vec{n}\, dA$

∵ S_1 is oriented downward

Let $\vec{r}(x, y, z) = \left(x, y, (x^2 + y^2)^{\frac{1}{2}}\right)$ and let R be the closed region of the projection of

S onto the xy-plane. Then

$R = \{(x, y): x^2 + y^2 \leq a^2\}$ and $\iint_{S_1} \vec{F} \cdot \vec{n}\, dA = -\iint_R \vec{F} \cdot \dfrac{\partial \vec{r}}{\partial x} \times \dfrac{\partial \vec{r}}{\partial y}\, dx dy$

∵ $\dfrac{\partial \vec{r}}{\partial x} = \left(1, 0, x(x^2 + y^2)^{-\frac{1}{2}}\right)$ and $\dfrac{\partial \vec{r}}{\partial y} = \left(0, 1, y(x^2 + y^2)^{-\frac{1}{2}}\right)$

∴ $\dfrac{\partial \vec{r}}{\partial x} \times \dfrac{\partial \vec{r}}{\partial y} = \left(-x(x^2 + y^2)^{-\frac{1}{2}}, -y(x^2 + y^2)^{-\frac{1}{2}}, 1\right)$

$\Rightarrow \vec{F} \cdot \dfrac{\partial \vec{r}}{\partial x} \times \dfrac{\partial \vec{r}}{\partial y} = (yz, -1, 1) \cdot \left(-x(x^2 + y^2)^{-\frac{1}{2}}, -y(x^2 + y^2)^{-\frac{1}{2}}, 1\right)$

$= -xy + y(x^2 + y^2)^{-\frac{1}{2}} + 1, \ \forall (x, y, z) \in S_1$

∴ $\iint_{S_1} \vec{F} \cdot \vec{n}\, dA = -\iint_R \vec{F} \cdot \dfrac{\partial \vec{r}}{\partial x} \times \dfrac{\partial \vec{r}}{\partial y}\, dx dy = \iint_{x^2 + y^2 \leq a^2} xy - y(x^2 + y^2)^{-\frac{1}{2}} - 1\, dx dy$

Let $x = r \cos \theta, y = r \sin \theta$ then $\{(x, y): x^2 + y^2 \leq a^2\} = \{(r, \theta): 0 \leq r \leq a, 0 \leq \theta \leq 2\pi\}$

9.7.3 Convert Volume Integrals into Closed Surface Integrals 634

$$\therefore \iint_{x^2+y^2 \leq a^2} xy - y(x^2+y^2)^{-\frac{1}{2}} - 1 \, dxdy = \int_0^{2\pi} \int_0^a (r^2 \cos\theta \sin\theta - \sin\theta - 1) r \, dr d\theta$$

$$= -\pi a^2$$

Let $S_2: z = a, x^2 + y^2 \leq a^2$

Let $\vec{s}(x,y,z) = (x,y,a)$ then $\iint_{S_2} \vec{F} \cdot \vec{n} \, dA = \iint_{S_2} \vec{F} \cdot \frac{\partial \vec{s}}{\partial x} \times \frac{\partial \vec{s}}{\partial y} \, dxdy$

$\therefore \dfrac{\partial \vec{s}}{\partial x} = (1,0,0)$ and $\dfrac{\partial \vec{s}}{\partial y} = (0,1,0)$ $\quad \therefore \dfrac{\partial \vec{s}}{\partial x} \times \dfrac{\partial \vec{s}}{\partial y} = (0,0,1)$

$\therefore \iint_{S_2} \vec{F} \cdot \vec{n} \, dA = \iint_{S_2} 1 \, dA = \pi a^2$ $\quad \therefore \iint_S \vec{F} \cdot \vec{n} \, dA = 0$

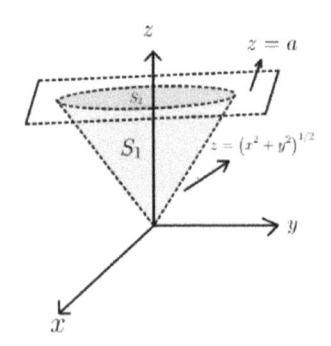

Example 5.

Use Gauss's Theorem to find $\iiint_V \nabla \cdot \vec{F} \, dV = ?$, where $\vec{F} = (x - y + z, 2x, 1)$,

V is the volume bounded by the paraboloid $z = x^2 + y^2$ and $z = 1$.

【Solution】

Let $S_1: z = x^2 + y^2, \ z \leq 1, \ S_2: z = 1, \ x^2 + y^2 \leq 1$

∵ the first-order partial derivatives of the components of $\vec{F}$ exist and are continuous

By Gauss's Theorem, $\iiint_V \nabla \cdot \vec{F} \, dV = \iint_{S_1} \vec{F} \cdot \vec{n} \, dA + \iint_{S_2} \vec{F} \cdot \vec{n} \, dA$

∵ S_1 is oriented downward

Let $\vec{r}(x,y,z) = (x, y, x^2 + y^2)$ and let R be the closed region of the projection of S_1 onto the xy-plane. Then

$$R = \{(x,y): x^2 + y^2 \leq 1\} \text{ and } \iint_{S_1} \vec{F} \cdot \vec{n}\, dA = -\iint_R \vec{F} \cdot \frac{\partial \vec{r}}{\partial x} \times \frac{\partial \vec{r}}{\partial y} dxdy$$

$$\because \frac{\partial \vec{r}}{\partial x} = (1, 0, 2x) \text{ and } \frac{\partial \vec{r}}{\partial y} = (0, 1, 2y) \quad \therefore \frac{\partial \vec{r}}{\partial x} \times \frac{\partial \vec{r}}{\partial y} = (-2x, -2y, 1)$$

$$\therefore -\vec{F} \cdot \frac{\partial \vec{r}}{\partial x} \times \frac{\partial \vec{r}}{\partial y} = (x - y + z, 2x, 1) \cdot (-2x, -2y, 1) = 2x^2 + 2xy + 2x(x^2 + y^2) - 1,$$

$\forall (x, y, z) \in S_1$

$$\therefore \iint_{S_1} \vec{F} \cdot \vec{n}\, dA = -\iint_R \vec{F} \cdot \frac{\partial \vec{r}}{\partial x} \times \frac{\partial \vec{r}}{\partial y} dxdy = \iint_{x^2+y^2 \leq 1} 2x^2 + 2xy + 2x(x^2 + y^2) - 1 dxdy$$

Let $x = r\cos\theta, y = r\sin\theta$ then $\{(x,y): x^2 + y^2 \leq 1\} = \{(r, \theta): 0 \leq r \leq 1, 0 \leq \theta \leq 2\pi\}$

and $dxdy = \begin{Vmatrix} \frac{\partial x}{\partial r} & \frac{\partial x}{\partial \theta} \\ \frac{\partial y}{\partial r} & \frac{\partial y}{\partial \theta} \end{Vmatrix} drd\theta = \begin{Vmatrix} \cos\theta & -r\sin\theta \\ \sin\theta & r\cos\theta \end{Vmatrix} drd\theta = rdrd\theta$

$$\therefore \iint_{x^2+y^2 \leq 1} 2x^2 + 2xy + 2x(x^2 + y^2) - 1 dxdy$$

$$= \int_0^{2\pi} \int_0^1 (2r^2 \cos^2\theta + 2r^2 \cos\theta \sin\theta + 2r^3 \cos\theta - 1) rdrd\theta = -\frac{\pi}{2}$$

$$\because \iint_{S_2} \vec{F} \cdot \vec{n}\, dA = \iint_{S_2} dA = \pi \quad \therefore \iiint_V \nabla \cdot \vec{F}\, dV = \iint_{S_1} \vec{F} \cdot \vec{n}\, dA + \iint_{S_2} \vec{F} \cdot \vec{n}\, dA = \frac{\pi}{2}$$

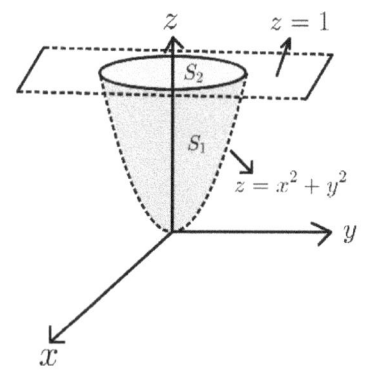

9.7.3 Convert Volume Integrals into Closed Surface Integrals

Example 6.

Find $\iiint_V \nabla \cdot \vec{F}\, dV = ?$, where $\vec{F} = (y, x, z)$, V is the closed region bounded by $z = 4 - x^2 - y^2$ and $z \geq 0$: (1)Using Gauss's Theorem (2)Direct computation.

【Solution】

(1)

Let $S_1: z = 4 - x^2 - y^2, z \geq 0$, $S_2: x^2 + y^2 \leq 4$, $z = 0$

∵ the first-order partial derivatives of the components of $\vec{F}$ exist and are continuous

By Gauss's Theorem, $\iiint_V \nabla \cdot \vec{F}\, dV = \iint_{S_1} \vec{F} \cdot \vec{n}\, dA + \iint_{S_2} \vec{F} \cdot \vec{n}\, dA$

Let $S_1: z = 4 - x^2 - y^2$, Let $\vec{r}(x, y, z) = (x, y, 4 - x^2 - y^2)$

∵ S_1 is oriented upward

Let R be the closed region of the projection of S_1 onto the xy-plane. Then

$R = \{(x, y): x^2 + y^2 \leq 4\}$ and $\iint_{S_1} \vec{F} \cdot \vec{n}\, dA = \iint_R \vec{F} \cdot \dfrac{\partial \vec{r}}{\partial x} \times \dfrac{\partial \vec{r}}{\partial y}\, dxdy$

∵ $\dfrac{\partial \vec{r}}{\partial x} = (1, 0, -2x)$ and $\dfrac{\partial \vec{r}}{\partial y} = (0, 1, -2y)$ ∴ $\dfrac{\partial \vec{r}}{\partial x} \times \dfrac{\partial \vec{r}}{\partial y} = (2x, 2y, 1)$

∴ $\iint_{S_1} \vec{F} \cdot \vec{n}\, dA = \iint_{x^2+y^2 \leq 4} (4xy + z)\, dxdy = \iint_{x^2+y^2 \leq 4} 4xy + 4 - x^2 - y^2\, dxdy$

Let $x = r\cos\theta, y = r\sin\theta$ then $\{(x, y): x^2 + y^2 \leq 4\} = \{(r, \theta): 0 \leq r \leq 2, 0 \leq \theta \leq 2\pi\}$

∴ $\iint_{x^2+y^2 \leq 4} 4xy + 4 - x^2 - y^2\, dxdy$

$= \int_0^{2\pi} \int_0^2 (4r^2 \cos\theta \sin\theta + 4 - r^2)\, rdrd\theta = 2\pi \left(2r^2 - \dfrac{r^4}{4}\right)\Big|_{r=0}^{r=2} = 8\pi$

Let $S_2: z = 0, x^2 + y^2 \leq 4$, ∵ S_2 is oriented downward

Let $\vec{s}(x,y,z) = (x,y,0)$ then $\iint_{S_2} \vec{F} \cdot \vec{n} \, dA = -\iint_{S_2} \vec{F} \cdot \dfrac{\partial \vec{s}}{\partial x} \times \dfrac{\partial \vec{s}}{\partial y} dxdy$

$\because \dfrac{\partial \vec{s}}{\partial x} = (1,0,0)$ and $\dfrac{\partial \vec{s}}{\partial y} = (0,1,0)$ $\therefore \dfrac{\partial \vec{s}}{\partial x} \times \dfrac{\partial \vec{s}}{\partial y} = (0,0,1)$

$\therefore -\iint_{S_2} \vec{F} \cdot \dfrac{\partial \vec{s}}{\partial x} \times \dfrac{\partial \vec{s}}{\partial y} dxdy = \iint_{S_2} 0 \, dA = 0$

$\therefore \iiint_V \nabla \cdot \vec{F} \, dV = \iint_{S_1} \vec{F} \cdot \vec{n} \, dA + \iint_{S_2} \vec{F} \cdot \vec{n} \, dA = 8\pi$

(2)

Let $R = \{(x,y): x^2 + y^2 \leq 4\}$ then

$\iiint_V \nabla \cdot \vec{F} \, dV = \iint_R \int_0^{4-(x^2+y^2)} dz\,dx\,dy = \iint_R 4 - (x^2+y^2) \, dxdy$

Let $x = r\cos\theta$, $y = r\sin\theta$ then $\{(x,y): x^2+y^2 \leq 4\} = \{(r,\theta): 0 \leq r \leq 2, 0 \leq \theta \leq 2\pi\}$

$\therefore \iint_R 4 - (x^2+y^2) \, dxdy = \int_0^{2\pi} \int_0^2 (4-r^2) r \, dr\, d\theta = 2\pi \cdot \dfrac{-(4-r^2)^2}{4} \bigg|_0^2 = 8\pi$

$\therefore \iiint_V \nabla \cdot \vec{F} \, dV = 8\pi$

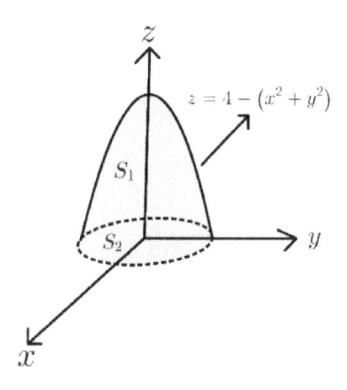

9.7.3 Convert Volume Integrals into Closed Surface Integrals

Example 7.

Let $\vec{F} = (x^3 - 3y, 2yz + 1, xyz)$. Assume V is the cube bounded by planes $x = \pm 1$, $y = \pm 1, z = \pm 1$. Find $\iiint_V \nabla \cdot \vec{F} \, dV$ by using Gauss's Theorem.

【Solution】

Let $S_1: x = 1$, $S_2: x = -1$, $S_3: y = 1$, $S_4: y = -1$, $S_5: z = 1$, $S_6: z = -1$ and $S = \bigcup_{i=1}^{6} S_i$

∵ the first-order partial derivatives of the components of $\vec{F}$ exist and are continuous

By Gauss's Theorem, $\iiint_V \nabla \cdot \vec{F} \, dV = \oiint_S \vec{F} \cdot \vec{n} \, dA$

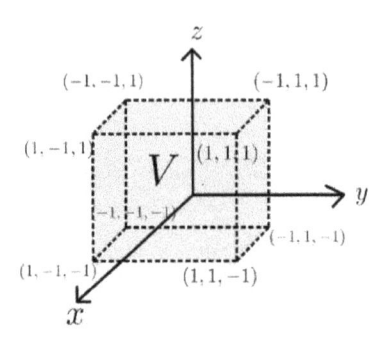

As $x = 1, \vec{n} = (1,0,0)$, $\iint_{S_1} \vec{F} \cdot \vec{n} \, dA = \int_{-1}^{1}\int_{-1}^{1} 1 - 3y \, dydz = 4$

As $x = -1, \vec{n} = (-1,0,0)$, $\iint_{S_2} \vec{F} \cdot \vec{n} \, dA = -\int_{-1}^{1}\int_{-1}^{1}(-1 - 3y) \, dydz = 4$

As $y = 1, \vec{n} = (0,1,0)$, $\iint_{S_3} \vec{F} \cdot \vec{n} \, dA = \int_{-1}^{1}\int_{-1}^{1} 2z + 1 \, dxdz = 2$

As $y = -1, \vec{n} = (0,-1,0)$, $\iint_{S_4} \vec{F} \cdot \vec{n} \, dA = -\int_{-1}^{1}\int_{-1}^{1} -2z + 1 \, dxdz = -2$

As $z = 1, \vec{n} = (0,0,1)$, $\iint_{S_5} \vec{F} \cdot \vec{n} \, dA = \int_{-1}^{1}\int_{-1}^{1} xy \, dxdy = 0$

As $z = -1, \vec{n} = (0,0,-1)$, $\iint_{S_6} \vec{F} \cdot \vec{n} \, dA = -\int_{-1}^{1}\int_{-1}^{1} xy \, dxdy = 0$

∴ $\oiint_S \vec{F} \cdot \vec{n} \, dA = 8$

Example 8.

Find $\iiint_V \nabla \cdot \vec{F}\, dV =?$, $\vec{F} = (z, y, x)$, $V: x^2 + y^2 + z^2 \leq r^2$: (1) Using Gauss's Theorem

(2) Direct calculation.

【Solution】

(1)

Let $S: x^2 + y^2 + z^2 = r^2$

∵ the first-order partial derivatives of the components of $\vec{F}$ exist and are continuous

By Gauss's Theorem, $\iiint_V \nabla \cdot \vec{F}\, dV = \oiint_S \vec{F} \cdot \vec{n}\, dA$

Let $\vec{r}(r, \theta, \varphi) = (r\cos\theta \sin\varphi, r\sin\theta \sin\varphi, r\cos\varphi)$

and $R = \{(\theta, \varphi): 0 \leq \theta \leq 2\pi, 0 \leq \varphi \leq \pi\}$. Then

$$\oiint_S \vec{F} \cdot \vec{n}\, dA = \iint_R \vec{F} \cdot \frac{\partial \vec{r}}{\partial \varphi} \times \frac{\partial \vec{r}}{\partial \theta} d\theta d\varphi$$

∵ $\dfrac{\partial \vec{r}}{\partial \theta} = (-r\sin\theta \sin\varphi, r\cos\theta \sin\varphi, 0)$ and $\dfrac{\partial \vec{r}}{\partial \varphi} = (r\cos\theta \cos\varphi, r\sin\theta \cos\varphi, -r\sin\varphi)$

∴ $\dfrac{\partial \vec{r}}{\partial \varphi} \times \dfrac{\partial \vec{r}}{\partial \theta} = (r^2 \cos\theta \sin^2\varphi, r^2 \sin\theta \sin^2\varphi, r^2 \sin\varphi \cos\varphi)$

∴ $\vec{F} \cdot \dfrac{\partial \vec{r}}{\partial \varphi} \times \dfrac{\partial \vec{r}}{\partial \theta}$

$= (r\cos\varphi, r\sin\theta \sin\varphi, r\cos\theta \sin\varphi) \cdot (r^2\cos\theta \sin^2\varphi, r^2\sin\theta \sin^2\varphi, r^2\sin\varphi \cos\varphi)$

$= r^3(2\cos\theta \sin^2\varphi \cos\varphi + \sin^3\varphi \sin^2\theta)$

∴ $\iint_R \vec{F} \cdot \dfrac{\partial \vec{r}}{\partial \varphi} \times \dfrac{\partial \vec{r}}{\partial \theta} d\theta d\varphi = r^3 \int_0^{2\pi}\int_0^\pi 2\cos\theta \sin^2\varphi \cos\varphi + \sin^3\varphi \sin^2\theta\, d\varphi d\theta$

∵ $\int_0^\pi \sin^2\varphi \cos\varphi\, d\varphi = \left.\dfrac{\sin^3\varphi}{3}\right|_0^\pi = 0$ ∴ $\int_0^{2\pi}\int_0^\pi 2\cos\theta \sin^2\varphi \cos\varphi\, d\varphi d\theta = 0$

9.7.3 Convert Volume Integrals into Closed Surface Integrals 640

$$\therefore \int_0^{2\pi}\int_0^{\pi}(2\cos\theta\sin^2\varphi\cos\varphi+\sin^3\varphi\sin^2\theta)d\varphi d\theta = \int_0^{2\pi}\int_0^{\pi}\sin^3\varphi\sin^2\theta \, d\varphi d\theta$$

$$= \int_0^{2\pi}\sin^2\theta \, d\theta \int_0^{\pi}\sin^3\varphi \, d\varphi = \int_0^{2\pi}\left(\frac{1-\cos 2\theta}{2}\right)d\theta \int_0^{\pi}\sin\varphi(1-\cos^2\varphi)d\varphi$$

$$= \pi \int_0^{\pi}\sin\varphi(1-\cos^2\varphi)d\varphi$$

$$\because \int_0^{\pi}\sin\varphi(1-\cos^2\varphi)d\varphi = -\cos\varphi|_0^{\pi} + \frac{\cos^3\varphi}{3}\bigg|_0^{\pi} = \frac{4}{3}$$

$$\therefore \oiint_S \vec{F}\cdot\vec{n}dA = r^3\int_0^{2\pi}\int_0^{\pi}(2\cos\theta\sin^2\varphi\cos\varphi+\sin^3\varphi\sin^2\theta)d\varphi d\theta = \frac{4\pi r^3}{3}$$

$$\therefore \iiint_V \nabla\cdot\vec{F}\,dV = \frac{4\pi r^3}{3}$$

(2)

$$\iiint_V \nabla\cdot\vec{F}\,dV = \iiint_{x^2+y^2+z^2\le r^2} 1\,dxdydz = \frac{4\pi r^3}{3}$$

Example 9.

Given $\vec{F} = (0, y, -z)$ and V is the volume bounded by the paraboloid $y = x^2 + z^2$ and $y = 1$. Find $\iiint_V \nabla\cdot\vec{F}\,dV$: (1)Using Gauss's Theorem (2) Direct computation.

【Solution】

(1)
Let $S_1: y = x^2 + z^2, 0 \le y \le 1$, $S_2: x^2 + z^2 \le 1, y = 1$
∵ the first-order partial derivatives of the components of $\vec{F}$ exist and are continuous

By Gauss's Theorem, $\iiint_V \nabla\cdot\vec{F}\,dV = \iint_{S_1}\vec{F}\cdot\vec{n}\,dA + \iint_{S_2}\vec{F}\cdot\vec{n}\,dA$

∵ S_1 is oriented in the negative direction of the y-axis

Let $\vec{r}(x, y, z) = (x, x^2 + z^2, z)$ and let R be the closed region of the projection of S_1 onto the xz-plane. Then

$R = \{(x,y): x^2 + z^2 \le 1\}$ and $\iint_{S_1}\vec{F}\cdot\vec{n}\,dA = -\iint_R \vec{F}\cdot\frac{\partial\vec{r}}{\partial x}\times\frac{\partial\vec{r}}{\partial z}dxdz$

$$\because \frac{\partial \vec{r}}{\partial x} = (1, 2x, 0) \text{ and } \frac{\partial \vec{r}}{\partial z} = (0, 2z, 1) \quad \therefore \frac{\partial \vec{r}}{\partial x} \times \frac{\partial \vec{r}}{\partial z} = (2x, -1, 2z)$$

$$\therefore \vec{F} \cdot \frac{\partial \vec{r}}{\partial x} \times \frac{\partial \vec{r}}{\partial z} = (0, y, -z) \cdot (2x, -1, 2z) = -y - 2z^2 = -x^2 - 3z^2, \quad \forall (x, y, z) \in S_1$$

$$\therefore \iint_{S_1} \vec{F} \cdot \vec{n} \, dA = -\iint_R \vec{F} \cdot \frac{\partial \vec{r}}{\partial x} \times \frac{\partial \vec{r}}{\partial z} dxdz = \iint_{x^2+z^2 \leq 1} x^2 + 3z^2 dxdz$$

Let $x = r\cos\theta, z = r\sin\theta$ then $\{(x,y): x^2 + z^2 \leq 1\} = \{(r,\theta): 0 \leq r \leq 1, 0 \leq \theta \leq 2\pi\}$

and $dxdz = \left\| \begin{matrix} \frac{\partial x}{\partial r} & \frac{\partial x}{\partial \theta} \\ \frac{\partial z}{\partial r} & \frac{\partial z}{\partial \theta} \end{matrix} \right\| drd\theta = \left\| \begin{matrix} \cos\theta & -r\sin\theta \\ \sin\theta & r\cos\theta \end{matrix} \right\| drd\theta = rdrd\theta$

$$\therefore \iint_{x^2+z^2 \leq 1} x^2 + 3z^2 dxdz = \int_0^{2\pi} \int_0^1 r^3 \cos^2\theta + 3r^3 \sin^2\theta \, drd\theta$$

$$= \int_0^{2\pi} \frac{1}{4}\left(\frac{1+\cos\theta}{2}\right) + \frac{3}{4}\left(\frac{1-\cos\theta}{2}\right) d\theta = \pi$$

Let $\vec{s}(x, y, z) = (x, 1, z)$

$$\because \frac{\partial \vec{s}}{\partial x} = (1, 0, 0) \text{ and } \frac{\partial \vec{s}}{\partial z} = (0, 0, 1) \quad \therefore \frac{\partial \vec{r}}{\partial x} \times \frac{\partial \vec{r}}{\partial z} = (0, -1, 0)$$

$$\therefore \iint_{S_2} \vec{F} \cdot \vec{n} dA = \iint_{S_2} dA = \iint_{x^2+z^2 \leq 1} (0, y, -z) \cdot (0, -1, 0) dxdz = -\pi$$

$$\therefore \iint_{S_1} \vec{F} \cdot \vec{n} dA + \iint_{S_2} \vec{F} \cdot \vec{n} dA = 0$$

(2)

$$\oiint_S \vec{F} \cdot \vec{n} dA = \iiint_V \nabla \cdot \vec{F} \, dV = \iiint_V 0 \, dV = 0$$

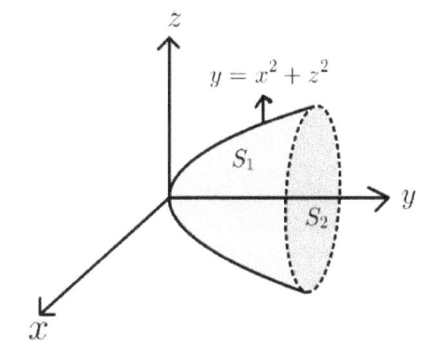

Example 10.

Let $\vec{F} = (x^2, y^2, z^2)$. Assume V is bounded by the solid half-cylinder $0 \leq z \leq \sqrt{1-y^2}$,

9.7.3 Convert Volume Integrals into Closed Surface Integrals

and the two disk planes: $x = 0, x = 2$. Find $\iiint_V \nabla \cdot \vec{F} \, dV$ by using Gauss's Theorem.

【Solution】

Let $S_1: 0 \leq z \leq \sqrt{1 - y^2}, 0 \leq x \leq 2$, $S_2: y^2 + z^2 \leq 1, x = 0$, $S_3: y^2 + z^2 \leq 1, x = 2$

and $S = \bigcup_{i=1}^{3} S_i$

∵ the first-order partial derivatives of the components of $\vec{F}$ exist and are continuous

By Gauss's Theorem, $\iiint_V \nabla \cdot \vec{F} \, dV = \sum_{j=1}^{3} \iint_{S_j} \vec{F} \cdot \vec{n} \, dA$

∵ S_1 is oriented upward

Let $\vec{r}(x, y, z) = \left(x, y, \sqrt{1 - y^2}\right)$ and $R = \{(x, y): 0 \leq x \leq 2, 0 \leq y \leq 1\}$ then

$$\iint_{S_1} \vec{F} \cdot \vec{n} \, dA = \iint_R \vec{F} \cdot \frac{\partial \vec{r}}{\partial x} \times \frac{\partial \vec{r}}{\partial y} \, dxdy$$

∵ $\frac{\partial \vec{r}}{\partial x} = (1, 0, 0)$ and $\frac{\partial \vec{r}}{\partial y} = \left(0, 1, -y(1 - y^2)^{-\frac{1}{2}}\right)$ ∴ $\frac{\partial \vec{r}}{\partial x} \times \frac{\partial \vec{r}}{\partial y} = \left(0, -y(1 - y^2)^{-\frac{1}{2}}, 1\right)$

∵ $\vec{F} \cdot \frac{\partial \vec{r}}{\partial x} \times \frac{\partial \vec{r}}{\partial y} = (x^2, y^2, 1 - y^2) \cdot \left(0, -y(1 - y^2)^{-\frac{1}{2}}, 1\right) = -y^3(1 - y^2)^{-\frac{1}{2}} + 1 - y^2$,

$\forall (x, y, z) \in S_1$
Let $y = \sin \theta$ then $dy = \cos \theta \, d\theta$

$$\iint_R \vec{F} \cdot \frac{\partial \vec{r}}{\partial x} \times \frac{\partial \vec{r}}{\partial y} \, dxdy = \int_0^2 \int_0^1 -y^3(1 - y^2)^{-\frac{1}{2}} + 1 - y^2 \, dydx$$

$$= 2 \int_0^1 -y^3(1 - y^2)^{-\frac{1}{2}} + 1 - y^2 \, dy = 2 \int_0^{\frac{\pi}{2}} -\sin^3 \theta + \cos^3 \theta \, d\theta$$

$$= 2 \cdot \left(\left(-\cos \theta + \frac{\cos^3 \theta}{3}\right)\Big|_0^{\frac{\pi}{2}} + \left(\sin \theta + \frac{\sin^3 \theta}{3}\right)\Big|_0^{\frac{\pi}{2}}\right) = \frac{8}{3}$$

$\iint_{S_2} \vec{F} \cdot \vec{n} dA = \iint_{S_2} dA = 0$ and $\iint_{S_3} \vec{F} \cdot \vec{n} dA = \iint_{S_3} dA = 2\pi$

$$\iiint_V \nabla \cdot \vec{F}\, dV = \frac{8}{3} + 2\pi$$

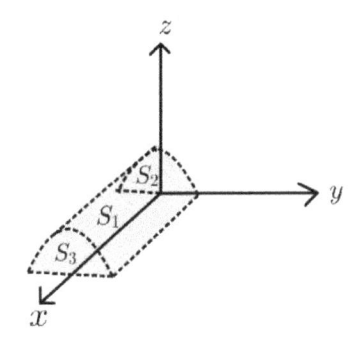

Example 11.

Assume $\vec{F} = (6y^2, 0, 2x^2)$, and V is the volume in the first octant bounded by $x + y + z = 1$. Find $\iiint_V \nabla \cdot \vec{F}\, dV$: (1)Direct computation (2)Using Gauss's Theorem.

【Solution】

(1)

$\because \nabla \cdot \vec{F} = 0 \quad \therefore \iiint_V \nabla \cdot \vec{F}\, dV = 0$

(2)

Let $S_1: x + y + z = 1$, $S_2: x = 0$, $S_3: y = 0$, $S_4: z = 0$ and $S = \bigcup_{j=1}^{4} S_j$

∵ the first-order partial derivatives of the components of $\vec{F}$ exist and are continuous
By Gauss's Theorem,

$$\iiint_V \nabla \cdot \vec{F}\, dV = \sum_{j=1}^{4} \iint_{S_j} \vec{F} \cdot \vec{n}\, dA$$

∵ S_1 is oriented upward

Let $\vec{r}(x, y, z) = (x, y, 1 - x - y)$ and let R be the closed region of the projection of S_1 onto the xy-plane. Then

9.7.3 Convert Volume Integrals into Closed Surface Integrals 644

$R = \{(x,y): x+y \leq 1, x \geq 0, y \geq 0\}$ and $\iint_{S_1} \vec{F} \cdot \vec{n} \, dA = \iint_R \vec{F} \cdot \dfrac{\partial \vec{r}}{\partial x} \times \dfrac{\partial \vec{r}}{\partial y} dxdy$

$\because \dfrac{\partial \vec{r}}{\partial x} = (1,0,-1)$ and $\dfrac{\partial \vec{r}}{\partial y} = (0,1,-1)$ $\quad \therefore \dfrac{\partial \vec{r}}{\partial x} \times \dfrac{\partial \vec{r}}{\partial y} = (1,1,1)$

$\because \vec{F} \cdot \dfrac{\partial \vec{r}}{\partial x} \times \dfrac{\partial \vec{r}}{\partial y} = (6y^2, 0, 2x^2) \cdot \dfrac{\partial \vec{r}}{\partial x} \times \dfrac{\partial \vec{r}}{\partial y} = 2(x^2 + 3y^2), \ \forall (x,y,z) \in S_1$

$\therefore \iint_{S_1} \vec{F} \cdot \vec{n} \, dA = \iint_R \vec{F} \cdot \dfrac{\partial \vec{r}}{\partial x} \times \dfrac{\partial \vec{r}}{\partial y} dxdy = 2\iint_R x^2 + 3y^2 \, dxdy = 2\int_0^1 \int_0^{1-y} x^2 + 3y^2 \, dxdy$

$= 2\int_0^1 \left(\dfrac{x^3}{3} + 3y^2 x\right)\Big|_{x=0}^{x=1-y} dy = 2\int_0^1 \dfrac{(1-y)^3}{3} + 3y^2(1-y) \, dy = \dfrac{2}{3}$

As $x = 0, n = (-1,0,0)$,

$\iint_{S_2} \vec{F} \cdot \vec{n} \, dA = -\int_0^1 \int_0^{1-z} 6y^2 \, dydz = -2\int_0^1 (1-z)^3 \, dz = -\dfrac{1}{2}$

As $y = 0, n = (0,1,0)$, $\iint_{S_3} \vec{F} \cdot \vec{n} \, dA = 0$

As $z = 0, n = (0,0,-1)$,

$\iint_{S_4} \vec{F} \cdot \vec{n} \, dA = -\int_0^1 \int_0^{1-x} 2x^2 \, dydx = -\dfrac{1}{6}$

$\therefore \sum_{j=1}^{4} \iint_{S_j} \vec{F} \cdot \vec{n} \, dA = 0$

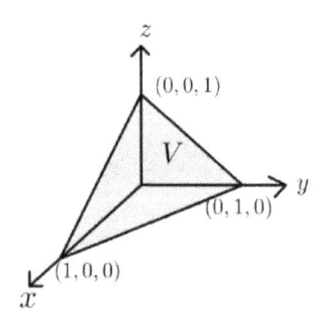

Example 12.

Given $\vec{F} = (x, y, z)$, and V is the volume bounded by $(a, 0, 0), (0, a, 0), (0, 0, a)$, and the coordinate planes. Find $\iiint_V \nabla \cdot \vec{F}\,dV$: (1) Direct computation (2) Using Gauss's Theorem.

【Solution】

(1)

$$\iiint_V \nabla \cdot \vec{F}\,dV = \iiint_V 3\,dV = \frac{a^3}{2}$$

(2)

Let $S_1: x + y + z = a, S_2: x = 0, S_3: y = 0, S_4: z = 0$ and $S = \bigcup_{j=1}^{4} S_j$

∵ the first-order partial derivatives of the components of $\vec{F}$ exist and are continuous

By Gauss's Theorem, $\iiint_V \nabla \cdot \vec{F}\,dV = \sum_{j=1}^{4} \iint_{S_j} \vec{F} \cdot \vec{n}\,dA$

∵ S_1 is oriented upward

Let $\vec{r}(x,y,z) = (x, y, a-x-y)$ and $R = \{(x,y): 0 \le x \le a, 0 \le y \le a-x\}$ then

$$\iint_{S_1} \vec{F} \cdot \vec{n}\,dA = \iint_R \vec{F} \cdot \frac{\partial \vec{r}}{\partial x} \times \frac{\partial \vec{r}}{\partial y}\,dxdy$$

∵ $\dfrac{\partial \vec{r}}{\partial x} = (1, 0, -1)$ and $\dfrac{\partial \vec{r}}{\partial y} = (0, 1, -1)$ ∴ $\dfrac{\partial \vec{r}}{\partial x} \times \dfrac{\partial \vec{r}}{\partial y} = (1, 1, 1)$

∵ $\vec{F} \cdot \dfrac{\partial \vec{r}}{\partial x} \times \dfrac{\partial \vec{r}}{\partial y} = (x, y, z) \cdot (1, 1, 1) = x + y + z = x + y + a - x - y = a$, $\forall (x,y,z) \in S_1$

$$\iint_R \vec{F} \cdot \frac{\partial \vec{r}}{\partial x} \times \frac{\partial \vec{r}}{\partial y}\,dxdy = \int_0^a \int_0^{a-x} a\,dydx = a\int_0^a a - x\,dx = a\left(ax - \frac{x^2}{2}\right)\bigg|_0^a = \frac{a^3}{2}$$

As $x = 0, n = (-1, 0, 0)$, $\iint_{S_2} \vec{F} \cdot \vec{n}\,dA = -\int_0^1 \int_0^{1-z} x\,dydz = 0$

As $y = 0, n = (0, 1, 0)$, $\iint_{S_3} \vec{F} \cdot \vec{n}\,dA = -\int_0^1 \int_0^{1-z} y\,dxdz = 0$

As $z = 0, n = (0, 0, -1)$, $\iint_{S_4} \vec{F} \cdot \vec{n}\,dA = -\int_0^1 \int_0^{1-x} z\,dxdy = 0$

9.7.3 Convert Volume Integrals into Closed Surface Integrals

$$\therefore \sum_{j=1}^{4} \iint_{S_j} \vec{F} \cdot \vec{n} \, dA = \frac{a^3}{2}$$

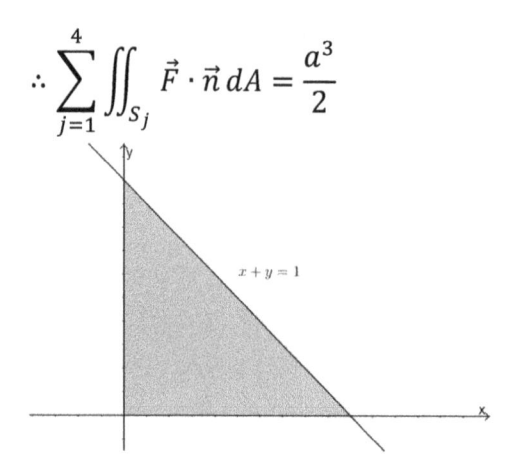

Example 13.

Given $\vec{F} = (e^y, e^x + 6y, 12y)$, and V is the volume in the first octant bounded by $x + y + z = 6$. Find $\iiint_V \nabla \cdot \vec{F} \, dV$: (1) Direct computation (2) Using Gauss's Theorem.

【Solution】

(1)

$$\iiint_V \nabla \cdot \vec{F} \, dV = \iiint_V 6 \, dV = 6 \cdot \frac{1}{3} \cdot \frac{1}{2} \cdot 6 \cdot 6 \cdot 6 = 216$$

(2)

Let $S_1: x + y + z = 6, S_2: x = 0, S_3: y = 0, S_4: z = 0$ and $S = \bigcup_{j=1}^{4} S_j$

∵ the first-order partial derivatives of the components of $\vec{F}$ exist and are continuous

By Gauss's Theorem, $\iiint_V \nabla \cdot \vec{F} \, dV = \sum_{j=1}^{4} \iint_{S_j} \vec{F} \cdot \vec{n} \, dA$

∵ S_1 is oriented upward

Let $\vec{r}(x, y, z) = (x, y, 6 - x - y)$ and let R be the closed region of the projection of S_1 onto the xy-plane. Then

$R = \{(x,y): x + y \leq 6, x \geq 0, y \geq 0\}$ and $\iint_{S_1} \vec{F} \cdot \vec{n} \, dA = \iint_R \vec{F} \cdot \dfrac{\partial \vec{r}}{\partial x} \times \dfrac{\partial \vec{r}}{\partial y} dxdy$

$\because \dfrac{\partial \vec{r}}{\partial x} = (1, 0, -1)$ and $\dfrac{\partial \vec{r}}{\partial y} = (0, 1, -1)$ $\therefore \dfrac{\partial \vec{r}}{\partial x} \times \dfrac{\partial \vec{r}}{\partial y} = (1,1,1)$

$\because \vec{F} \cdot \dfrac{\partial \vec{r}}{\partial x} \times \dfrac{\partial \vec{r}}{\partial y} = (e^y, e^x + 6y, 12y) \cdot (1,1,1) = e^y + e^x + 18y, \ \forall (x,y,z) \in S_1$

$\therefore \iint_{S_1} \vec{F} \cdot \vec{n} \, dA = \iint_R \vec{F} \cdot \dfrac{\partial \vec{r}}{\partial x} \times \dfrac{\partial \vec{r}}{\partial y} dxdy = \iint_R e^y + e^x + 18y \, dxdy$

$= \int_0^6 \int_0^{6-x} e^y + e^x + 18y \, dydx = 634 + 2e^6$

As $x = 0, n = (-1, 0, 0)$,

$\iint_{S_2} \vec{F} \cdot \vec{n} \, dA = -\int_0^6 \int_0^{6-z} e^y \, dydz = -\int_0^6 e^{6-z} - 1 \, dz = 7 - e^6$

As $y = 0, n = (0, -1, 0)$,

$\iint_{S_3} \vec{F} \cdot \vec{n} \, dA = -\int_0^6 \int_0^{6-z} e^x \, dxdz = -\int_0^6 e^{6-z} - 1 \, dz = 7 - e^6$

As $z = 0, n = (0, 0, -1)$,

$\iint_{S_4} \vec{F} \cdot \vec{n} \, dA = -\int_0^6 \int_0^{6-x} 12y \, dydx = -432$

$\therefore \sum_{j=1}^{4} \iint_{S_j} \vec{F} \cdot \vec{n} \, dA = 216$

Example 14.

9.7.3 Convert Volume Integrals into Closed Surface Integrals

Given $\vec{F} = (0, xz, -xy)$ and V is the volume bounded by $z = xy, 0 \leq x \leq 1, 0 \leq y \leq 2$.

Find $\iiint_V \nabla \cdot \vec{F} \, dV$: (1) Direct computation (2) Using Gauss's Theorem

【Solution】

(1)

$\because \nabla \cdot \vec{F} = 0 \quad \therefore \iiint_V \nabla \cdot \vec{F} \, dV = 0$

(2)

Let $S_1: z = xy, S_2: x = 1, S_3: y = 2, S_4: z = 0$ and $S = \bigcup_{j=1}^{4} S_j$

$\because$ the first-order partial derivatives of the components of $\vec{F}$ exist and are continuous

By Gauss's Theorem, $\iiint_V \nabla \cdot \vec{F} \, dV = \sum_{j=1}^{4} \iint_{S_j} \vec{F} \cdot \vec{n} \, dA$

$\because S_1$ is oriented upward

Let $\vec{r}(x, y, z) = (x, y, xy)$ and $R = \{(x, y): 0 \leq x \leq 1, 0 \leq y \leq 2\}$ then

$\iint_{S_1} \vec{F} \cdot \vec{n} \, dA = \iint_R \vec{F} \cdot \frac{\partial \vec{r}}{\partial x} \times \frac{\partial \vec{r}}{\partial y} \, dxdy$

$\because \frac{\partial \vec{r}}{\partial x} = (1, 0, y)$ and $\frac{\partial \vec{r}}{\partial y} = (0, 1, x) \quad \therefore \frac{\partial \vec{r}}{\partial x} \times \frac{\partial \vec{r}}{\partial y} = (-y, -x, 1)$

$\therefore \vec{F} \cdot \frac{\partial \vec{r}}{\partial x} \times \frac{\partial \vec{r}}{\partial y} = (0, xz, -xy) \cdot (-y, -x, 1) = -x^2 z - xy = -x^3 y - xy, \quad \forall (x, y, z) \in S_1$

$\iint_R \vec{F} \cdot \frac{\partial \vec{r}}{\partial x} \times \frac{\partial \vec{r}}{\partial y} \, dxdy = \int_0^1 \int_0^2 -x^3 y - xy \, dy dx = \int_0^a -2x^3 - 2x \, dx = \left(-\frac{x^4}{2} - x^2 \right) \Big|_0^1 = -\frac{3}{2}$

As $x = 1, n = (1, 0, 0), \quad \iint_{S_2} \vec{F} \cdot \vec{n} \, dA = 0$

As $y = 2, n = (0, 1, 0), \quad \iint_{S_3} \vec{F} \cdot \vec{n} \, dA = \int_0^6 \int_0^{2x} xz \, dz dx = \frac{1}{2}$

As $z = 0, n = (0,0,-1)$, $\iint_{S_4} \vec{F} \cdot \vec{n}\, dA = \int_0^2 \int_0^1 xy\, dx dy = 1$

$\therefore \iiint_V \nabla \cdot \vec{F}\, dV = \sum_{j=1}^{4} \iint_{S_j} \vec{F} \cdot \vec{n}\, dA = 0$

Milton Keynes UK
Ingram Content Group UK Ltd.
UKHW032254111124
451073UK00009B/311

9 786260 134570